NATURAL
HISTORY

![DK] SMITHSONIAN ✺

NATURAL

THE ULTIMATE VISUAL GUIDE TO EVERYTHING ON EARTH

HISTORY

LONDON, NEW YORK, MELBOURNE,
MUNICH, AND DELHI

DK PUBLISHING

SENIOR PROJECT EDITOR Kathryn Hennessy
PROJECT EDITOR Victoria Wiggins

SENIOR ART EDITORS Gadi Farfour,
Helen Spencer

EDITORS Becky Alexander, Ann Baggaley,
Kim Dennis-Bryan, Ferdie McDonald,
Elizabeth Munsey, Peter Preston,
Cressida Tuson, Anne Yelland
US EDITORS Jill Hamilton, Christine Heilman,
Jane Perlmutter

DESIGNERS Paul Drislane, Nicola Erdpresser,
Phil Fitzgerald, Anna Hall, Richard Horsford,
Stephen Knowlden, Dean Morris, Amy
Orsborne, Steve Woosnam-Savage

SPECIAL PHOTOGRAPHY Gary Ombler
PICTURE RESEARCH Neil Fletcher,
Peter Cross, Julia Harris-Voss, Sarah Hopper,
Liz Moore, Rebecca Sodergren, Jo Walton,
Debra Weatherley, and Suzanne Williams
DK PICTURE LIBRARY Claire Bowers
DATABASE Peter Cook, David Roberts

PRODUCTION EDITOR Tony Phipps
SENIOR PRODUCTION CONTROLLER
Inderjit Bhullar

MANAGING EDITOR Camilla Hallinan
MANAGING ART EDITOR Karen Self
ART DIRECTOR Phil Ormerod
ASSOCIATE PUBLISHER Liz Wheeler
REFERENCE PUBLISHER Jonathan Metcalf

DK INDIA

MANAGING EDITOR Rohan Sinha
ART DIRECTOR Shefali Upadhyay
PROJECT MANAGER Malavika Talukder
PROJECT EDITOR Kingshuk Ghoshal
PROJECT ART EDITOR Mitun Banerjee
EDITORS Alka Ranjan, Samira Sood,
Garima Sharma
ART EDITORS Ivy Roy, Mahua Mandal,
Neerja Rawat
PRODUCTION MANAGER Pankaj Sharma
DTP COORDINATOR Sunil Sharma
SENIOR DTP DESIGNERS Dheeraj Arora,
Jagtar Singh, Pushpak Tyagi

First American Edition, 2010
First published in the United States by
DK Publishing
375 Hudson Street
New York, New York 10014

10 11 12 13 14 10 9 8 7 6 5 4 3 2 1
LD096—09/10

Published in Great Britain by
Dorling Kindersley Limited

A catalog record for this book is available from
the Library of Congress.

ISBN 978-07566-6752-8

DK books are available at special discounts when
purchased in bulk for sales promotions, premiums,
fund-raising, or educational use. For details, contact:
DK Publishing Special Markets, 375 Hudson Street,
New York, New York, 10014 or SpecialSales@dk.com

Printed and bound in China by
Leo Paper Products Ltd

Discover more at
www.dk.com

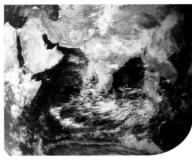

CONTENTS

SMITHSONIAN INSTITUTION

Established in 1846, the Smithsonian Institution—the world's
largest museum and research complex—includes 19 museums
and galleries and the National Zoological Park. The total number
of objects, works of art, and specimens in the Smithsonian's
collections is estimated at 137 million, the bulk of which is
contained in the National Museum of Natural History, which holds
more than 126 million specimens and objects. The Smithsonian
is a renowned research center, dedicated to public education,
national service, and scholarship in the arts, sciences, and history.

SMITHSONIAN CONSULTANTS

Dr. Stephen Cairns, Dr. Allen Collins, Dana M. De Roche,
Dr. Carla Dove, Leslie Hale, Dr. M. G. (Jerry) Harasewych, Gary
Hevel, Dr. Rafael Lemaitre, Dr. Chris Meyer, Dr. Jon Norenburg,
Dr. David L. Pawson, Paul Pohwat, Dr. Jeffrey E. Post, Dr. Klaus
Rutzler, Dr. Hans-Dieter Sues, Dr. Michael Vecchione, Dr. Warren
Wagner, Dr. Jeffrey T. Williams, Dr. Don E. Wilson, Dr. George Zug

ADDITIONAL CONSULTANTS

Dr. Matthew D. Kane, Dr. James D. Lawrey, Dr. Diana Lipscomb,
Dr. Robert Lücking, Dr. Thorsten Lumbsch, Andrew M. Minnis,
Dr. Ashleigh Smythe, Dr. William B. Whitman

CONSULTANT EDITOR

David Burnie is a former
winner of the Aventis Prize for
Science Books, and the editor
of DK's highly successful *Animal*.
He has written or contributed
to more than 100 books and is a
fellow of the Zoological Society
of London.

CONTRIBUTORS

Richard Beatty, Amy-Jane Beer,
Dr. Charles Deeming, Dr. Kim
Dennis-Bryan, Dr. Frances Dipper,
Dr. Chris Gibson, Derek Harvey,
Professor Tim Halliday, Geoffrey
Kibby, Joel Levy, Felicity Maxwell,
Dr. George C. McGavin, Dr. Pat
Morris, Dr. Douglas Palmer,
Dr. Katie Parsons, Chris Pellant,
Helen Pellant, Michael Scott,
Carol Usher

PLANTS

FUNGI

ANIMALS

FOREWORD

We share this planet with millions of species of plants, animals and microorganisms, and our lives are intimately tied to them. Just take a moment to look around you and you will see that we are interacting with them every day, from the food we eat and the clothes we wear to the microbes that live inside our bodies, the air we breathe, and the water we drink. We are one small twig in a large and complex tree of life, a tree where most of the branches have been lost over time.

This book provides a window into the remarkable diversity and natural history of the world around us. It has been a long journey, going back 4.6 billion years to the formation of planet Earth itself. Although astronomers have discovered several hundred planets in other solar systems over the past decade, our home is unique given its location in the solar system, its geological history, and the evolution of life. Any one of these things could have changed and we would not be here today.

The study of these species, and the interactions among them and with their surrounding environment, is our own natural history. More than 1.9 million living species have been described to date, and more than 20,000 new species are discovered and described every year. Each one of them has a unique story, and is the result of millions of years of evolution through natural selection and adaptation to their environment. Their lives are intertwined into a gigantic web of life, with multiple, constantly changing connections between them. We are just one species, albeit one that is having an ever-increasing impact on this planet and beyond.

Fossils give us a rare window into the past, a form of time travel if you wish. We know that most species that have lived in the past 530 million years have become extinct, and that there have been mass extinctions when as many as 90 percent of all species disappeared. For example, fossil leaves from Wyoming show evidence of rapidly changing environments, from temperate grasslands to tropical forests, in the same place over time. Some of the fossil leaves even show the bite marks left by feeding insects 50 million years ago. By comparing these fossil communities in space and time we can see that species and their natural history have continuously responded to environmental change. Some of them made it; most did not. The study of these changes may give us insights into the past, present, and future of life on Earth.

The publication of this book coincides with the centennial anniversary of the Smithsonian's National Museum of Natural History. Our collections are pages of a gigantic encyclopedia of life, and the story is told through the insights of our scientists and educators. I trust you will enjoy this magnificent volume and accept it as an invitation to explore the Smithsonian and to explore the natural history of the world around you.

CRISTIÁN SAMPER

DIRECTOR, NATIONAL MUSEUM OF NATURAL HISTORY,
SMITHSONIAN INSTITUTION

ABOUT THIS BOOK

Natural History begins with a general introduction to life on Earth: the geological foundations of life, the evolution of life forms, and how organisms are classified. The next five chapters form an extensive and accessible catalog of species and specimens—from mineral to mammals—interspersed with fact-filled introductions to each group and in-depth feature profiles.

for easy reference, visual contents panels list the subgroups within each section, and the page number where each subgroup can be found

SECTION INTRODUCTION >
Each chapter is divided into sections representing major taxonomic groupings. The section introduction highlights the characteristics and behaviors that define the group, and discusses their evolution over time.

on each introduction, classification boxes display the current taxonomic hierarchy: the level of the group under discussion is highlighted

PHYLUM	CHORDATA
CLASS	REPTILIA
ORDERS	4
FAMILIES	60
SPECIES	About 7,700

debate boxes tackle scientific controversies and taxonomic discussions arising from new discoveries

a male of the species ♂ *a female of the species ♀*

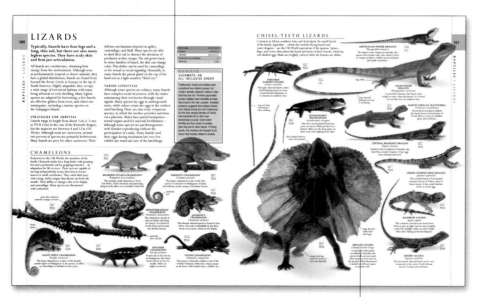

∧ GROUP INTRODUCTION
Within each section—for example, reptiles—lower-ranking taxonomic groups, such as lizards, are explored. Key features are described, including their distribution, habitat, physical characteristics, life cycle, behavior, and reproductive habits.

species-specific information accompanies each image

SPECIES CATALOG >
Pictorial galleries profile around 5,000 species, showing the distinctive visual features of each one. Closely related species are placed together for useful comparison, while essential data highlights unique and interesting aspects of each organism.

data sets display key details at a glance, such as size, habitat, distribution, and diet

SIZE 4½–9¼ ft (1.4–2.8 m)
HABITAT Forests, swamps, scrub thickets, savanna, and rocky landscapes
DISTRIBUTION India to China, Siberia, the Malay peninsula, and Sumatra
DIET Mainly hoofed animals, such as deer and pigs; may also catch smaller mammals and birds

∨ FEATURE PROFILE

Zooming in on single specimens, feature profiles use close-up photographs to provide in-depth portraits of some of the world's most spectacular species.

each feature includes a side profile of the animal or plant

MEASUREMENTS

The approximate sizes of the organisms in this book are given in data sets and size boxes. Below is a list of the dimensions used:

MICROSCOPIC LIFE
Length

PLANTS
Maximum height above ground, except for:
Height above water rushes
Spread aquatic plants

FUNGI
Width (of widest part), except for:
Height stinkhorn, dog stinkhorn

INVERTEBRATES
Adult body length, except for:
Height sponges, feather stars, common hydra, feather hydroid, pink-hearted hydroid, fire "coral" hydroid, tall anemones, tall corals
Diameter snailfur, blue button, eight-ribbed hydromedusa, cup hydromedusa, freshwater jellyfish
Diameter, excluding spines echinoderms
Diameter of medusa jellyfish
Wingspan butterflies and moths
Length of colony bryozoans
Length of shell mollusks, shelled gastropods
Spread of tentacles octopuses

FISH, AMPHIBIANS, AND REPTILES
Adult body length from head to tail

BIRDS
Adult body length from bill to tail

MAMMALS
Adult body length, excluding tail, except for:
Height to shoulder elephants, apes, even-toed ungulates, odd-toed ungulates

PLANT ICONS

The basic shape of all trees, shrubs, and woody plants is described using one of the following symbols. Herbaceous perennials which die back each winter are not given symbols.

TREES
- Broadly columnar
- Broadly conical
- Large weeping
- Small weeping
- Multistemmed tree
- Narrowly columnar, flame-shaped
- Narrowly columnar
- Narrowly conical
- Rounded, broadly columnar
- Rounded, broadly spreading
- Single-stemmed palm
- Multistemmed palm, cycad or similar

SHRUBS
- Bushy, mound-forming
- Bushy, suckering
- Compact, bushy
- Erect, treelike
- Loose, open
- Open, spreading
- Rounded, bushy
- Spreading, prostrate
- Upright
- Upright, arching
- Upright, vigorous, bushy
- Sprawling, climbing

ABBREVIATIONS

SP: species (used where species name is unknown)
MYA: million years ago
H: hardness of a mineral, measured on the Mohs scale
SG: specific gravity—a mineral's density is measured by comparing its weight to that of an equal volume of water

species' common names are highlighted in bold, scientific names are in italics; in some cases the family (F) name is given below

ST VINCENT PARROT
Amazona guildingii
F: Psittacidae

12 in
30 cm

size boxes give the most appropriate measurement for the organism they accompany (see panel, right)

LIVING EARTH

Our blue planet, spinning in the vastness of space, is the only proven home of living things. Over nearly four billion years, life has evolved from the simplest of beginnings. Many species have become extinct, but life itself has flourished and endlessly diversified. The result is an extraordinary variety of living things, which scientists continue to study as they piece together the story of life on Earth.

A LIVING PLANET

Earth is uniquely equipped to support a wide diversity of life, both on land and in the seas. Without heat and light from the Sun, plentiful supplies of water, the protection provided by the atmosphere, and the rocks and minerals that make up the basis of Earth's ecosystems, life would perish.

DYNAMIC EARTH

Within our Solar System, Earth seems to be uniquely placed to support abundant life. The third planet from the Sun, Earth is neither too close to nor too far from the Sun's heat. It therefore retains an outer atmosphere of oxygen and other gases, and a hydrosphere of plentiful surface water. Together, these form a protective, insulating layer that enables life to flourish. In contrast, the other planets in the Solar System are either too hot or too cold, and devoid of the levels of water and oxygen required to support detectable life.

Earth has a layered structure, with an extremely hot, solid metallic core at its center, surrounded by an outer molten layer. This, in turn, is surrounded by a thick and hot silicate mantle, which rises to a thin, cool, and brittle outer crust. The mantle is constantly churned by heat rising from the core, and this puts pressure on the crust, which is broken into crustal "plates." Over geological time, the drift of these plates, both toward and away from one another, has changed Earth's geography and living environments. Oceans, mountains, and landscapes are constantly formed and destroyed, and life has had to adapt to these changes.

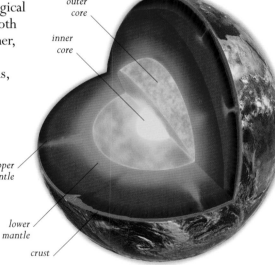

outer
core

inner
core

upper
mantle

lower
mantle

crust

EARTH'S STRUCTURE >
The liquid mantle is constantly stirred by heat rising from the core. This moves the plates of the outer crust, causing earthquakes and volcanic eruptions on the surface.

SUN AND MOON

The Sun and the Moon both have a direct impact on life on Earth. Without the Sun's energy, in the form of heat and light, there would be no life. Solar energy heats the atmosphere, the oceans, and the land, producing our varied climate. Because Earth rotates at an angle while it orbits the Sun, the Sun's radiant energy is unevenly distributed over Earth's surface. This results in daily, seasonal, and annual variations in light, heat, and living conditions for plants and animals. Even at the equator, there are marked temperature changes between day and night. The orbit of Earth's satellite moon and its gravitational pull raise tides in Earth's oceans and seas. Tidal cycles are especially influential on coastal life, which has to adapt to changing conditions.

∧ SOLAR FLARES
The Sun's energy is dramatically released from the surface in periodic explosions, which heat its atmosphere to form solar flares of hot ionized gas.

WATER AND LIFE
Life depends upon water, which forms over 50 percent of all living tissues. Most rainfall is driven by evaporation from the oceans, which hold 97 percent of Earth's surface water, and a vital network of rivers flow in all but the world's hottest, coldest, and driest places.

FRAGILE ATMOSPHERE

Earth's atmosphere is 75 miles (120 km) thick. It is made up of several layers, each with its own temperature and gas composition. Its density decreases with height, until it becomes the sparse, outermost layer, called the ionosphere. The ozone layer, in the lower atmosphere, plays a vital role in protecting life because it absorbs harmful radiation such as ultra-violet light, which damages living cells. Before the ozone layer formed, life was confined to the seas, whose waters offered some protection against ultraviolet light.

The majority of water vapor and weather activity is restricted to the lowest 10 miles (16 km) of atmosphere, known as the troposphere. Earth's surface water and gaseous atmosphere interact to recycle water from the surface into the atmosphere and—through clouds, rain, and snow—redistribute it over the land and sea. From the land, water flows back into the sea, although large quantities are held back in lakes, ice, and under the ground.

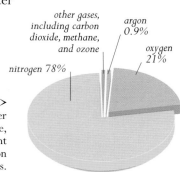

other gases, including carbon dioxide, methane, and ozone

argon 0.9%

oxygen 21%

nitrogen 78%

ATMOSPHERIC GASES >
Nitrogen and oxygen make up over 99 percent of Earth's atmosphere, along with small but important volumes of water vapor, carbon dioxide, and several other gases.

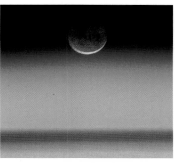

∧ BLUE PLANET
Around two-thirds of Earth's surface is covered with water, which supports the abundance and diversity of life.

∧ LAYERS OF ATMOSPHERE
Earth is surrounded by a thin, layered atmosphere, composed of water vapor and various gases, which trap solar energy and heat the surface.

VARIED ROCKS

There are around 500 different kinds of rocks on Earth, made up of varying combinations of thousands of naturally occurring minerals. All rocks have a specific composition and properties, and can be divided into three main categories: igneous rocks were originally molten; sedimentary rocks are deposited at Earth's surface; and metamorphic rocks result from the alteration of existing rocks within Earth's crust. These different types of rock are exposed on the surface by a mixture of uplift, driven by Earth's moving crust, and surface processes such as weathering and erosion. Erosion also modifies the rocks to produce multiple kinds of landforms, soils, and sediments. These are the inorganic elements on which life depends.

IGNEOUS ROCKS
The cooling and solidification of molten rock produces crystalline igneous rocks. Their composition and texture vary—rapid cooling produces fine grained rocks, and slow cooling produces coarse grained rocks.

BASALT

METAMORPHIC ROCKS
The application of heat and pressure to existing rocks deep within Earth's crust can change their form and mineral composition, resulting in metamorphic rocks such as slate, schist, and marble.

GARNET SCHIST

SEDIMENTARY ROCKS
Layers of sand and dead animal bones constantly settle on sea and river beds. After millennia buried under the weight of subsequent layers, and that of the water above them, these sediments compact and harden into rocks.

SANDSTONE

ACTIVE EARTH

Earth's surface is constantly changing, thanks to dynamic geological processes driven by its internal heat energy. The plates of Earth's brittle outer crust are always in motion, altering the shape of oceans and continents as they do so.

PLATE TECTONICS

Over geological time, Earth's surface—and the distribution and size of the continents and oceans—has constantly changed, driven by the process of plate tectonics. The cool and brittle rocks of Earth's outermost crust are broken into a number of semirigid slabs known as tectonic plates. There are seven major continent-sized plates and about a dozen smaller ones. Over time, these crustal plates have been jostled against one another by the motion of the underlying mantle.

As plates are dragged apart, molten magma from the lower mantle forms new crust. This occurs at divergent boundaries, which are mainly beneath the oceans. And, since Earth itself cannot expand, the creation of new oceanic crust requires that the crust is shortened elsewhere by the same amount. This reduction occurs at convergent boundaries where either one plate overrides the other—a process known as subduction—or the plate margins are compressed and buckled to form mountains.

∧ SAN ANDREAS FAULT
Stretching some 810 miles (1,300 km) through California, this dramatic fault is the product of a transform boundary between the Pacific and North American plates, which slide against one another.

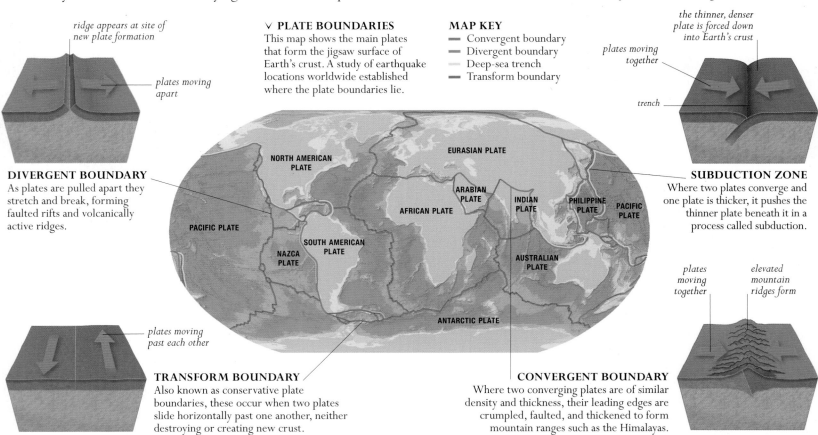

ridge appears at site of new plate formation

plates moving apart

∨ PLATE BOUNDARIES
This map shows the main plates that form the jigsaw surface of Earth's crust. A study of earthquake locations worldwide established where the plate boundaries lie.

MAP KEY
— Convergent boundary
— Divergent boundary
— Deep-sea trench
— Transform boundary

the thinner, denser plate is forced down into Earth's crust

plates moving together

trench

DIVERGENT BOUNDARY
As plates are pulled apart they stretch and break, forming faulted rifts and volcanically active ridges.

NORTH AMERICAN PLATE

EURASIAN PLATE

ARABIAN PLATE

AFRICAN PLATE

INDIAN PLATE

PHILIPPINE PLATE

PACIFIC PLATE

PACIFIC PLATE

SOUTH AMERICAN PLATE

NAZCA PLATE

AUSTRALIAN PLATE

ANTARCTIC PLATE

SUBDUCTION ZONE
Where two plates converge and one plate is thicker, it pushes the thinner plate beneath it in a process called subduction.

plates moving together

elevated mountain ridges form

plates moving past each other

TRANSFORM BOUNDARY
Also known as conservative plate boundaries, these occur when two plates slide horizontally past one another, neither destroying or creating new crust.

CONVERGENT BOUNDARY
Where two converging plates are of similar density and thickness, their leading edges are crumpled, faulted, and thickened to form mountain ranges such as the Himalayas.

< FOLD MOUNTAINS
The intense pressure caused when plate margins converge can create an incredible folding and faulting of the crust as the rocks are pushed upward to form mountains.

ACTIVE VOLCANOES >
Most volcanoes form at plate margins. Deep down, rocks melt to produce hot magma which rises and erupts at the surface. Even dormant volcanoes may one day erupt as plates shift beneath them.

MOUNTAINS AND VOLCANOES

One of the major factors controlling the movement and distribution of life on Earth is its varied topography—its surface features—including the towering mountains and volcanoes on land and under the sea. On land, mountains not only hinder the movement of wildlife but also alter weather, climate, and local plant life, which in turn impacts animal life. Active volcanoes also affect their surroundings when they erupt—initially by destroying life, but also, in the longer term, when the weathering and erosion of erupted lava and ash provide new mineral nutrients that fertilize the area. Mountains under the sea affect the movement of marine life, and submarine volcanic eruptions affect the fertility of ocean waters.

WORN AWAY
Weathering by wind and water can result in the significant erosion of rocks and the sculpting of land surfaces, as seen here in the dramatic landscape of Bryce Canyon, Utah.

WEATHERING AND EROSION

Many rocks are formed below Earth's surface, and when they are exposed by pressures in Earth's crust or by retreating seas or rivers, they react in many ways with the atmosphere, water, and living organisms. The physical and chemical processes resulting from the interaction of rocks and minerals with the atmosphere are known as weathering. The processes by which rock material is loosened, dissolved, and transported away is known as erosion. The combination of weathering and erosion wears down Earth's rocky surfaces, layer by layer. Exposed rocks on mountain tops and on the exterior of buildings, for example, are subject to chemical weathering by acid rain and to physical weathering by temperature change and the splintering freeze-thaw effect of ice formation. Bare rock surfaces may also be physically eroded by sand particles carried by the wind. The combined effect of weathering and erosion dissolves some rocks, and reduces others to fragments. As rock debris is broken down and transported by wind, water, and ice, the sediment created becomes increasingly available to life forms, providing important mineral nutrients and a new surface for anchorage and growth.

< LANDSLIDE, RIO DE JANEIRO
Even with a well-developed plant cover, the impact of high rainfall on an area of steep slopes has the potential to cause landscape-altering and even life-threatening occurrences such as avalanches and landslides.

SOIL FORMATION

The production of a soil requires the initial weathering and erosion of the parent rock, which is broken down into small, mineral-bearing particles known as regolith. The addition of humus—organic matter formed from the remains of plants and animals—forms the basis for a soil. The soil in turn becomes the bed in which more life flourishes.

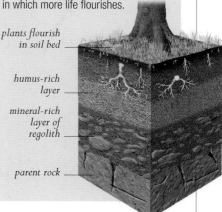

plants flourish in soil bed

humus-rich layer

mineral-rich layer of regolith

parent rock

CHANGING CLIMATES

The features of our seasons—for example, hot, dry summers and cold, icy winters—make up the climate of each region. Earth's climate has always changed from place to place and over time, and this variation in conditions has considerable and continuing impact upon the evolution of life.

ARCTIC FOX
IN SUMMER

ARCTIC FOX
IN WINTER

WHAT IS CLIMATE?

Climate is a region's average weather over a long period of time, produced by all the atmospheric conditions such as temperature, rainfall, wind strength, and pressure. The climate of any given place is also partly controlled by a number of other factors such as its height above sea level, local topography, proximity to seas and oceans—with their prevailing winds and water currents—and, most importantly, its latitudinal position between the equator and pole. Latitude controls the amount of solar radiation received in different regions of the world. For instance there is a major difference between climates of polar regions, which receive the least light and heat, and of tropical regions around the equator, which receive the most.

< CHANGING TREE LINE
As the air temperature declines with increasing altitude, plant life changes. Broadleaved trees are replaced by conifers, then shrubs.

CHANGING CONDITIONS

Global climates are generally classified according to the average temperature and the amount of rainfall each area receives, and their combined effect on plant growth. For instance, equatorial regions at present are hot and wet because oceans dominate there, whereas deserts are dry and polar regions are cold. However, this has not always been the case. Factors controlling weather conditions over geological time have affected the climate of the planet, from ice ages through to global warming.

∧ SEASONAL ADAPTATION
Annual climate change can bring acutely different living conditions from one season to the next. Animals and plants have various means of adapting to these changes; for example, the Arctic fox grows a thick coat in winter, which it sheds in summer.

LIFE IN THE DESERT
Plants such as cacti have adapted to life in semiarid environments by slowing growth, reducing the leaves to spines (to stop water loss from transpiration), and evolving special tissues that retain water.

< **EVIDENCE OF CLIMATE CHANGE**
Studies of polar ice core samples have revealed details of past climate change. The chemical analysis of trapped gas bubbles helps provide an approximate measure of the temperature of the atmosphere at the time the ice was formed.

< **ICE CORE SAMPLE**
This is a close-up of an ice core sample recovered from Lake Bonney, Antarctica, which has a permanent ice cover. It shows trapped bubbles of air and sediment particles from the lake bed.

CLIMATE CYCLES

There is clear evidence from rocks and fossils that Earth's climate has changed significantly over time and that this has affected the evolution and distribution of life, and led to the extinction of many species. There are a number of causes of this natural climate change, including volcanic activity, which pollutes the atmosphere with ejected gas and dust, and changes in ocean currents, which move heat around the globe. Change is also driven by the orbital and rotational cycles of the planet, which affect the amount of solar radiation that reaches Earth's surface. This influences the planet's temperature and climate, and triggers Earth's cycles of ice ages and greenhouse periods.

CHANGING GEOGRAPHY
Over time, continents have been displaced as oceans have expanded and contracted due to plate tectonics. As continents moved from one hemisphere to another they passed through different climatic zones and, at times, formed supercontinents. The size of these huge landmasses affected regional climates. In addition, the changing shape of ocean basins has altered water circulation, which in turn changes the temperature and humidity of the atmosphere above, and so affects climate.

GREENHOUSE AND ICE AGES
Long-term changes in climate are divided into cold periods—ice ages, when there were long-lasting ice sheets at the poles—and warmer, greenhouse periods with largely ice-free poles. Warm phases are linked to the release of greenhouse gases, such as carbon dioxide, by plants into the atmosphere. These trapped heat in Earth's atmosphere and produced huge, shallow seas, arid zones, and lush forests that, in the time of the dinosaurs, provided them with plentiful food. Ice ages, lasting millions of years, can be traced from the impact that glaciation has left on the landscape. Fossils show how rapid climate change associated with ice ages had a dramatic impact upon life on a global scale.

SCIENCE
STUDYING STOMATA

Plant growth depends upon an exchange of gases between the atmosphere and the plant tissue, through special gaps between the leaves' cells called stomata. The stomata open and close to take in carbon dioxide for photosynthesis and allow waste water and oxygen to escape. In general, plants adapt to high levels of atmospheric carbon dioxide, associated with warm "greenhouse" climates, by evolving high densities of stomata on the leaf surface. Fossil evidence of changes in the stomatal density of certain plants tracks changes in atmospheric carbon dioxide over time.

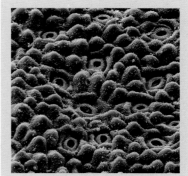

EUCALYPTUS STOMATA

∧ **DEVONIAN CORAL REEF**
Earth's changing climate is illustrated by this limestone outcrop in The Kimberleys, Western Australia. In the Devonian Period (400 million years ago), the area was under water and the cliff was a barrier reef.

∨ **CARBON DIOXIDE AND TEMPERATURE**
Gas bubbles trapped in polar ice cores indicate the fluctuating temperature of the planet—the higher the amount of carbon dioxide detected in the ice core, the higher the atmospheric temperature would have been at that time.

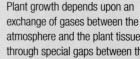

HABITATS FOR LIFE

Earth's uniquely varied habitats enable it to support an abundantly rich diversity of animal and plant life—from the extreme depths of the ocean floors to the highest mountains, and from arid deserts and grasslands to the warm, wet tropics.

Each life form has its preferred habitat—the one to which it has adapted over thousands or even millions of years. However, Earth's varied environments allow many different kinds of animals and plants to live in the same habitat—a phenomenon known as biodiversity. As life evolved and diversified over geological time, it was able to extend beyond the seas and colonize more and more of Earth's different land habitats. The presence of these pioneering organisms in turn produced changes in the environments they colonized, by forming soils for instance, and these modifications encouraged further colonization by new life forms.

Differences in habitats are produced by many factors, ranging from an area's height above sea level, its distance from the equator, and its topography (or physical shape). Some areas of the globe are biodiversity "hotspots," rich in animal and plant life—most notably tropical reefs and rain forests—while other areas with more extreme conditions support only a few, though often heavily populated, species.

SCIENCE
LEVELS OF LIFE

Organisms rarely exist entirely on their own, even in the remotest places on Earth. The interactions between wildlife create different levels—from the individual organism to the overall ecosystem that it inhabits and shares with others.

INDIVIDUAL
A single, usually independent and habitat-restricted member of a population.

POPULATION
A group of individuals of the same species that occupy the same area and interbreed.

COMMUNITY
A naturally occurring collection of plant and animal populations living within the same area.

ECOSYSTEM
A biological community and its physical surroundings, which support one another.

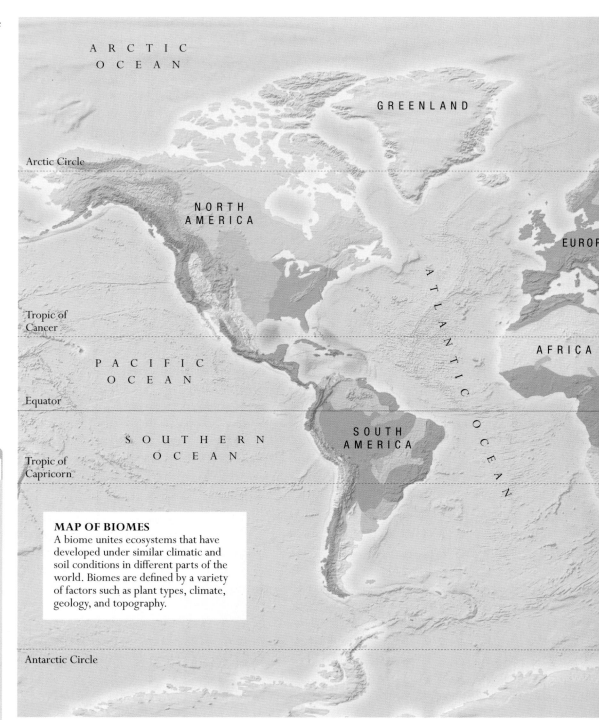

MAP OF BIOMES
A biome unites ecosystems that have developed under similar climatic and soil conditions in different parts of the world. Biomes are defined by a variety of factors such as plant types, climate, geology, and topography.

GRASSLAND
The evolution of grass plants some 20 million years ago, and the colonization by grazing mammals, transformed Earth's landscapes. Temperate grasslands are generally treeless and have extremely fertile soils. Savanna grasslands, as shown here, are more like open woodland, featuring scatterings of trees and scrubs.

BISON

DESERT

Extreme lack of rain and soils for sustained plant growth creates deserts, which at present account for about a third of Earth's landscapes, although this proportion is increasing. The largest desert is the African Sahara.

RATTLESNAKE

TROPICAL FOREST

The richest wildlife habitats on land are found in the forests of the tropics —Earth's hottest areas, which lie by the equator. Their numerous ecosystems are important yet increasingly vulnerable biodiversity hotspots.

STRAWBERRY POISON DART FROG

TEMPERATE FOREST

Temperate environments lie between the tropics and the polar regions. The influence of both tropical and polar air masses promotes vast forests with considerable biodiversity. However, clearing by humans has greatly reduced their extent.

RED DEER

CONIFEROUS FOREST

Coniferous forest trees, such as redwoods, spruce, and fir, belong to an ancient plant group and are the world's toughest trees. Evergreen with small leaves, they thrive where few other trees can, in cold areas and in mountain ranges.

BROWN BEAR

MOUNTAINS

Reaching as high as 5½ miles (9 km) above sea level, Earth's mountains are home to many different environments. A single mountain can rise from temperate woodland up to arctic conditions as the climate changes with altitude.

PEREGRINE FALCON

ARCTIC OCEAN

ASIA

PACIFIC OCEAN

INDIAN OCEAN

AUSTRALIA

SOUTHERN OCEAN

ANTARCTICA

RIVERS AND WETLAND

A wide range of animal and plant life thrives in Earth's rivers and lakes. Landscapes saturated with water, either permanently or seasonally, form distinctive wetlands, where areas of open water mix with dense vegetation.

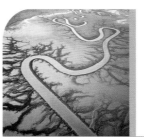

DRAGONFLY

CORAL REEF

Coral reefs develop from the skeletons of marine organisms in shallow, sunlit tropical waters. Supporting an immense variety of life, they are the rain forests of the aquatic world.

YELLOW TANG FISH

POLAR REGIONS

The Arctic and Antarctic experience extreme seasons, with 24-hour daylight in summer and perpetual darkness in winter. They are dominated by large amounts of snow and ice, but also contain vast, dry areas of polar desert, which receive little annual rainfall.

ROCKHOPPER PENGUIN

OCEANS

Life in oceans is found at all levels, from the sunlit surface to the deepest depths. Covering two-thirds of the planet, oceans form the largest continuous habitat on Earth and support very varied life forms— from microscopic plankton to the largest living mammal, the blue whale.

LOBSTER

∧ SCARRED LANDSCAPE
The growth of industry has required
the exploitation of raw materials. Their
extraction, such as at this copper mine,
has changed global landscapes forever.

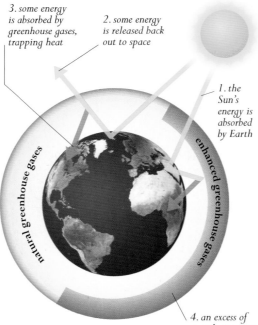

∧ ATMOSPHERIC POLLUTION
Slash-and-burn forest clearing for
agriculture not only releases atmospheric
pollutants but also diminishes the capture
of carbon dioxide by the plant life.

*3. some energy
is absorbed by
greenhouse gases,
trapping heat*

*2. some energy
is released back
out to space*

*1. the
Sun's
energy is
absorbed
by Earth*

natural greenhouse gases

enhanced greenhouse gases

*4. an excess of
greenhouse gases
traps too much of
the Sun's energy,
leading to a steep
rise in Earth's
temperature*

∧ GREENHOUSE EFFECT
An excess of greenhouse gases in the
atmosphere creates a shield which
prevents some of the Sun's energy
from radiating back into space.

HUMAN IMPACT

**The rapid growth of the human population has had an
enormous impact upon Earth's natural environments,
affecting the climate and countless species of plant
and animal life. Some of these changes are irreversible.**

ENVIRONMENTAL CHANGE

Earth has a long history of climate change, which has ranged from
glacial "ice ages" to warm "greenhouse" climates with widespread
forestation and no ice in polar regions. Global warming is known to
be linked to high levels of greenhouse gases in the atmosphere, such
as carbon dioxide and methane, which trap incoming solar energy
and raise the temperature of the oceans, land, and air. In the past,
natural increases in atmospheric carbon dioxide have been balanced
by the development of forests on land and lime-rich sediments in the
seas, which eventually became coal and limestone and stored excess
carbon dioxide effectively. Since the industrial revolution of the
nineteenth century, human activity has released huge amounts of
carbon dioxide and other greenhouse gases back into the atmosphere
through the mining and burning of fossil fuels, the
clearing of forests, and the rearing of cattle.

THE OCEANS

The health of Earth's oceans is vital to
all life. Marine life depends upon the
circulation of oceanic waters that contain
enough oxygen and nutrients to support
the food chain from plankton and
shellfish to all the other animals that
depend upon them for food. Fossil
records show that when the oceanic
environment deteriorated in the past, it
led to extinctions of life. Today, human
activities such as over-fishing
and pollution are affecting
the condition of the seas.

∧ OIL SLICK VICTIM
Oil spills, which float on water,
cause havoc when they reach
land. They saturate coastlines
and devastate coastal wildlife.

THE ATMOSPHERE

For thousands of years human activity has affected the
atmosphere. Initially, it was restricted to the release of
pollutants from domestic fires and forest clearance. In
Roman times metal production released the first industrial
pollutants into the atmosphere, leaving traces that can be
seen in polar ice cores. Over the last 200 years, pollution
by gases and particulates has risen steeply. This has
produced acid rain and smog, and greenhouse gases that
are linked to global warming and depletion of the ozone
layer, which helps screen out harmful ultraviolet radiation.

THE LAND

Since settlements and agriculture became widespread
8,000 years ago, humans have had a growing impact upon
Earth's landscapes. With worldwide population growth,
many regions are now densely settled, with few or no
untouched landscapes in between. Greater awareness of
the impact of human activity on the environment is now
leading to efforts to conserve natural habitats.

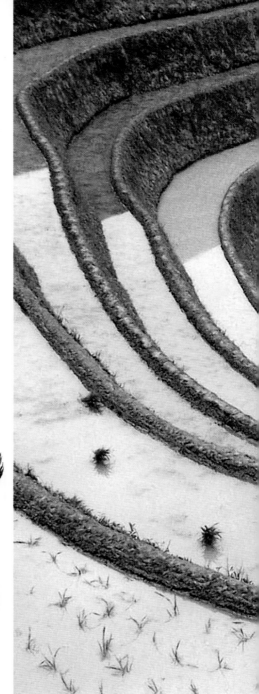

∧ POLAR ICE SHELF COLLAPSE
Rising temperatures are leading to the
collapse of the polar ice shelves. The release
of such large volumes of water is raising sea
levels, which in turn threatens coastlines.

EXTINCTION

The inability of many organisms to adapt to environmental change has led to a huge turnover of species throughout geological time. In fact, the vast majority of nature's species are now extinct. Only the fittest organisms survive, usually through gradual change, but sometimes by the sudden elimination of competitors. For example, when a huge asteroid hit Earth 65 million years ago, it set in motion a chain of events that led to the demise of many life forms, including the dinosaurs on land and the ammonites in the seas. But it also saw the survival of mammal species that led to the human race. More recently, the arrival of modern humans in different regions around the world has contributed to the extinction of particular species, such as the woolly mammoth in Europe and Asia (see far right). Today, as the human population expands, our actions are placing a growing number of species, such as the tiger, in imminent danger of extinction.

∧ DECLINE OF THE CRESTED IBIS
Once widespread in Asia, hunting and habitat loss has restricted the crested ibis to a small wild population in China. Captive breeding has allowed it to be reintroduced to Japan.

< PERE DAVID'S DEER
Extinct in the wild, this east Asian deer has survived only in captive herds bred in England since 1900. Some captive offspring were reintroduced to nature reserves in China in the 1980s.

SCIENCE
MAMMOTHS NO MORE

Woolly mammoths were elephants that were adapted to the cold. They migrated across ice-age Europe and Asia in vast herds. Archaeological evidence such as cave paintings shows that they were actively hunted by humans around 30,000 years ago, which may have contributed to the extinction of most woolly mammoths by 11,000 years ago.

CAVE PAINTING, PECH-MERLE, FRANCE

ORIGINS OF LIFE

The fossil record shows that life first appeared on Earth at least 3.8 billion years ago, and that all complex life has evolved from these first simple forms. Today, diverse life forms range from single-celled organisms to mammals with complex anatomies, such as giant whales.

WHAT IS LIFE?

There are several features that define life and distinguish an active, organic organism from inanimate, inorganic matter. These include an ability to take in and expend energy, to grow and change, to reproduce, to adapt to its environment, and—in more complex living organisms—to communicate.

The cell is the fundamental unit of life, capable of replicating itself and carrying out all living processes. Even the smallest independent organisms are made up of at least one cell, and almost every cell of every living organism has its own set of molecular instructions. Within each cell, the threadlike chromosomes carry hereditary information in the form of genes that are responsible for

˄ VITAL ENERGY
To sustain itself, life must obtain energy from the environment. Energy flows through a food chain from photosynthesis in plants to the animals that feed upon them and are, in turn, eaten by other animals.

SCIENCE

VIRUSES

Viruses are the most abundant biological entity on Earth, and they lie at the boundary between living and nonliving. While they have features in common with living organisms—they are made of genetic material and are protected by a protein coat—viruses are parasitic and can only reproduce within the living cells of other organisms. They are packets of chemicals that copy themselves, without truly being alive.

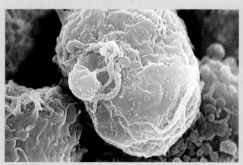

HIV-1 VIRUS

˄ PHOTOSYNTHESIS
Plants use the pigment chlorophyll to capture light energy and convert water and carbon dioxide into sugars and oxygen. This benefits other life forms that feed on the plants and breathe in oxygen.

the particular characteristics of an organism. The set of instructions in a gene are mainly recorded in the form of a molecule called chromosomal deoxyribonucleic acid (DNA).

The DNA of an organism carries information from one generation to another, allowing certain characteristics to be passed on from parent to offspring.

DIVISIONS OF LIFE

The huge array of life on Earth is divided into three domains or superkingdoms—Archaea, Bacteria, and Eukaryota —which encompass all life forms from plants and fungi to animals. The first two domains are formed of prokaryotes—primitive organisms that were probably the earliest form of life on Earth. The more advanced eukaryotes are distinguished from the prokaryotes by having a cell nucleus, which contains the cell's genetic material, DNA. Eukaryotes vary enormously in shape and size, ranging from single-celled organisms to complex, multicelled plants and animals.

< GROWTH
A capacity to grow and repair is a one of the key defining features of life. All organisms, from simple fungi to mammals, grow by means of increase in cell size and cell division.

EARLY LIFE

The very earliest forms of life evolved in the seas, and evidence of this comes from two main sources: living primitive organisms and the fossil record. The most primitive life forms today are single-celled prokaryotes, which can survive even in extreme temperatures and acidic conditions. Such microorganisms may be similar to those that first evolved in the extreme environments of early Earth.

Fossil evidence for early life on Earth consists of the remains of organims dating from around 3.8 billion years ago. They may have originated as microscopic prokaryotes that lived in the first oceans. Some of the most ancient records of life are layered, mound-shaped stromatolites (see right).

FLOURISHING LIFE
Ever since life first appeared on Earth, it has thrived in the sea, evolving into the biodiversity hotspots of modern, sunlit reefs, second only to rain forests for the variety and density of life they sustain.

< STROMATOLITES

For billions of years these layered structures have built up in shallow tropical seas by alternating layers of sediment and sheets of microorganisms, including cyanobacteria (blue-green algae).

By about 750–550 million years ago, the first sponges with colonies of cells had evolved, growing to 4 in (10 cm) high with spiny skeletons for support and protection. By the beginning of the Cambrian period, 545 million years ago, numerous multicelled marine organisms had evolved, including burrowing worms and a variety of small, shelled mollusks, whose bodies had muscle tissues and organs such as gills for respiration. Around 510 million years ago, the first vertebrates appeared, with an internal skeletal support for the body. By late Devonian times, around 380 million years ago, vertebrates had begun to emerge from the seas onto land.

From the first appearance of simple life, it took another 2.5 billion years before complex life forms appeared. The fossil of a microscopic and multicelled red algae called *Bangiomorpha* provides the first evidence of the existence of specialized cells. These cells evolved for sexual reproduction, and also for the development of a holdfast to attach the algae to the sea floor.

∧ BURGESS SHALE

The fossils of the Burgess Shale in Canada show that marine life diversified rapidly in Cambrian times, from sponges and arthropods to vertebrates.

WIWAXIA >

Found at the Burgess Shale, *Wiwaxia* was a 2-in-long (5cm), mollusklike seabed crawler with spines and scales, and a soft undersurface.

EVOLUTION AND DIVERSITY

Until the nineteenth century, when a number of theories were proposed, it was a matter of speculation as to how such remarkably diverse life forms had developed on Earth. Today, the theory of evolution and diversification, alongside geological evidence for changes in the distribution of continents, give a fascinating insight into the ever-changing life on our planet.

CHANGE OVER TIME

All living things have the capacity to change and adapt to their surroundings. Tiny, subtle changes that are passed down from generation to generation are hard to see, but over time—sometimes thousands or even millions of years—they can alter the way a certain species looks or behaves. This process is known as evolution.

The study of fossils to unravel the history of life was in its early stages in Charles Darwin's day (see p.25), and since then a vast amount of information supporting the theory of evolution has emerged. We now know that life evolved in the oceans some 3.8 billion years ago, and that it was from these early simple life forms that all current life on Earth evolved—including plants, fungi, and animals.

As life forms became more complex and moved from sea to land, the first forests and land-living invertebrates evolved. The Mesozoic era, around 250 millions years ago, with its successions of evolving plants and animals, produced the dominant dinosaur reptiles and their bird descendents. These reptiles were largely replaced by mammals both in the seas and on land in Cenozoic times—from 65 million years ago to the present—when flowering plants and their pollinating insects also became abundant and diverse.

< GIANT SALAMANDER
This extremely rare fossil skeleton of an *Andrias* (giant salamander) was mistaken for a human victim of the biblical Flood until French anatomist Georges Cuvier identified it as an amphibian in 1812.

EVIDENCE OF EVOLUTION
Comparison of the anatomy of vertebrate limb bones from different species show that, despite different appearances and functions, they derive from the same basic developmental plan and the same genes.

FROG
The frog's leg, arm, and finger bones are modified for swimming. Large muscles enable it to jump powerfully—essential for catching prey and escaping from predators.

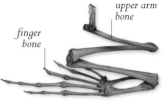
upper arm bone / *finger bone*

OWL
The wing of a bird is powered by flight muscles attached to the upper arm and bones of the wrist, with greatly modified and extended fingers.

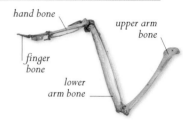
hand bone / *upper arm bone* / *finger bone* / *lower arm bone*

CHIMPANZEE
The arm of a chimpanzee is anatomically very similar to our own, but has slightly different proportions, with elongated fingers and a short thumb.

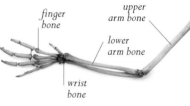
finger bone / *upper arm bone* / *lower arm bone* / *wrist bone*

DOLPHIN
The arm bones of whales and dolphins form a flipper—with shortened, flattened, and strengthened arm bones and greatly lengthened second and third fingers.

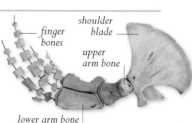
finger bones / *shoulder blade* / *upper arm bone* / *lower arm bone*

LAMARCK LEADS THE WAY

The eighteenth-century French biologist Jean-Baptiste Lamarck developed the first comprehensive evolutionary theory that higher forms of life had "evolved" from simpler organisms. Based on his extensive research of invertebrates in particular, Lamarck argued that necessary characteristics could be acquired during an organism's lifetime—through the desire to gain food, shelter, and mates—and those not needed could be lost, with the resulting transformation passed on to its offspring. Although modern genetics disproves this theory of "soft inheritance," Lamarck's concepts were an important starting point, and were developed further by the Scottish anatomist Robert Grant, who tutored Charles Darwin in Edinburgh. Darwin himself did not entirely rule out a Lamarckian mechanism, which he thought might be supplementary to natural selection.

< ^ INNER STRIVING
Lamarck believed that evolution worked through a process of "inner striving." Thus the giraffe developed a long neck to reach the leaves of trees, and the heron grew long legs in order to wade.

∧ GALAPAGOS FINCHES
On his voyages, Darwin collected many different specimens of Galapagos finches which he believed may have descended from a common ancestor.

butterfly wing

specimen box

DARWIN AND WALLACE

In the mid nineteenth century, British naturalists Charles Darwin and Alfred Russel Wallace independently came up with the theory of evolution by natural selection. They both had experience of fieldwork in the tropics—an environment boasting high biodiversity, competition for resources, and marked differences in the organisms living in separate areas. They both wondered how and why such natural phenomena had arisen. On his travels Wallace collected specimens for study and sale, and it was in the Malay Archipelago that he formulated

his explanations for the geographical distribution of organisms —biogeography—and realized the role of natural selection in evolution. Meanwhile, Darwin's five-year voyage on *HMS Beagle* around the Southern Hemisphere as a gentleman naturalist provided him with much material to formulate his own theory of evolution. In 1858, Wallace and Darwin produced a joint publication on natural selection, and the following year Darwin expanded the theory to produce his famous and influential book, *On the Origin of Species*.

∧ COLLECTION BOX
Both Darwin and Wallace were fascinated by insect diversity, especially as found in the tropics, and were keen collectors.

< BIOGEOGRAPHY
The distribution of certain reptile and plant fossils across the southern continents shows that they were once joined as a supercontinent, called Gondwana.

Cynognathus
reptile fossils from the Triassic Period

AFRICA

INDIA

Lystrosaurus
reptile fossils from the Triassic

Mesosaurus
reptile fossils from the Permian Period

SOUTH AMERICA

AUSTRALIA

Glossopteris
plant fossils from the Permian Period

ANTARCTICA

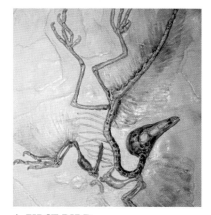

∧ FIRST BIRD
The discovery of the fossil *Archaeopteryx* in 1861 revealed characteristics that provided an evolutionary link between two major groups: reptiles and birds.

EVOLUTION IN PROGRESS

Darwin and Wallace proposed the theory of natural selection, but it was the discovery of the gene that gave scientists the mechanism by which selection takes place. Understanding the gene has since become the key to understanding evolution.

NATURAL SELECTION

A key evolutionary mechanism, natural selection favours the survival of the fittest. In other words, individuals that possess characteristics best adapted to their current environment have a better chance of surviving to reproduce and pass on those favorable traits to another generation. Natural genetic variation within populations produces differences such as size, shape, and color, and some of these may help promote survival. For instance, a particular body coloration may provide a camouflage that better hides the animal from predators than other colors would. If this leads to the survival of that animal and to its reproduction, that same improved coloration will be passed on to some of its offspring. If the environment were to change over time,

∧ INDIVIDUAL VARIATION
Litters of domestic cats commonly include individuals with varied coloration, especially where the parents' colors differ.

a different coloration might be more beneficial, and so natural selection will ensure another change. A split into two populations may even occur after a geographic rift, with each new population becoming adapted to slightly different conditions. Eventually, this may lead to one species becoming two, in a process known as speciation.

< ∨ SEXUAL DIMORPHISM
In many species, there are marked differences between males and females. Male frigatebirds use their inflatable throat pouch to attract the females.

CREATIONISM: BELIEF VERSUS SCIENCE

Most of the world's religions provide a theory of creation, which gives an explanation for the formation of Earth and life. Many of these creation stories originated long before scientific data and theories were available to offer alternative explanations and understanding. Some believers in the Western Judeo-Christian tradition maintain that the complexity of "design" in many organisms implies that there must have been a "designer"—God—behind their creation, and for this reason they dispute the theory of evolution.

GENES AND INHERITANCE

Particular traits pass from parents to offspring through the transmission of genetic material. Genes preserve, encoded in their DNA, all the information necessary for the replication of a cell's structure and its maintenance. Genes are therefore the basic units of heredity. Individual chromosomes—the thread-like part of a cell—hold thousands of genes on long strands of DNA. During sexual reproduction, the fusion of sperm and egg cells produces two complete sets of gene-bearing chromosomes: one copy from the father and one from the mother.

a forest fire wipes out large populations of butterflies

∨ GENES AND CHANCE
Sometimes individuals are eliminated at random, with the result that their genes are not passed on to the next generation.

by chance, the survivors are mostly yellow butterflies

only the genes of the survivors are passed on

next generation sees few purple butterflies

a chance event leads to total loss of purple butterflies

ISLAND EVOLUTION

Isolated islands provide natural laboratories for unusually fast evolution, with intense competition for limited resources leading to rapid speciation. In 1835, Darwin's visit to the Galapagos Islands allowed him to collect many bird specimens, particularly finches. He noted slight variations between the specimens from island to island. He also heard about the differences between the giant tortoises on separate islands, and subsequent visits to other Pacific islands made him wonder about the possibility of new species evolving from a common ancestor. The ornithologist John Gould was able to identify Darwin's finches as a new group of 12 separate species, rather than just varieties of the same species.

This persuaded Darwin that species could change under certain conditions such as island isolation. Island wildlife continues to be an important research area for modern evolutionary biologists.

< FLIGHTLESS BIRDS
Many flightless birds, such as the kiwi, evolved on the islands of New Zealand, which lacked serious predators until humans arrived.

ARTIFICIAL SELECTION

Over millennia, humans have domesticated many different kinds of animals and plants, from dogs and cattle to fruit trees and cereal crops. Before the discovery of the gene, this was achieved simply by selectively interbreeding organisms bearing the desired characteristics—such as the ability to run fast or produce more succulent fruit—over many generations, until the selected traits became dominant. Today, biotechnology achieves the same result much faster by directly manipulating genes, both to enhance beneficial traits and to remove problematic ones.

∧ GENETIC MODIFICATION
Alteration to the genetic makeup of organisms can remove undesired characteristics and add more useful ones, such as resistance to disease.

∧ CLONING
Genetically identical individuals can be created by the transfer of a nucleus—with its genetic information—from an adult cell into a host egg cell.

CLASSIFICATION

Global diversity is estimated to range from two to 100 million species. Only 1.4 million have been described, but many new species are added each year. All are named and classified using a system devised over 250 years ago.

For centuries people have studied the natural world. Initially they were limited to what they could find locally and to reports from travelers because it was impossible to preserve and send specimens any distance. Later, as travel became easier, explorers were paid to collect plants and animals and ship's artists would draw them. By the early 1600s natural history collections in Europe were substantial, and many specimens had been described, but there was no formal arrangement that made these specimens or accounts of them easily accessible.

The aim of the early taxonomists, or scientists who describe and classify species, was simple—to organize living things so that they reflected God's plan of creation. Between 1660 and 1713 John Ray published works on plants, insects, birds, fish, and mammals, forming his groups on the basis of morphological (structural) similarities. The

science of morphology, along with other criteria such as behavior and modern genetics, forms the basis of classification today. In 1758 the tenth edition of *Systema Naturae* was published. It was written by the Swedish botanist Carolus Linnaeus. He and his friend Peter Artedi had decided to divide the natural world between them and classify everything in it, fitting the 7,300 described species into the same hierarchical framework. Although Artedi died before his book was finished, Linnaeus completed the work and published it along with his own.

LATIN NAMES

All living things now have a unique Latin name—such as *Panthera leo* for the lion— which is made up of the genus name, starting with a capital letter, and a descriptive species name. Linnaeus devised this binomial

method for identifying different organisms to replace the arbitrary descriptions that existed previously. The new method put an end to confusion caused by the same name being given to several species, or to single species being know by several names.

Within a species, distinct subspecies can sometimes be recognized in different locations. In the 1800s Elliot Coues and Walter Rothschild adopted a trinomial Latin system to accommodate them. This convention for naming species and subspecies is still used today.

"DOG ROSE" >
Known as the dog rose, wild briar, witches' briar, dogberry, eglantine gall, and the hip tree, this plant has only one Latin name, *Rosa canina*, which identifies it for everyone everywhere.

< "PUMA"
The puma is also known as the cougar or mountain lion. Its Latin name, *Puma concolor*, alludes to its uniform color.

TRADITIONAL CLASSIFICATION

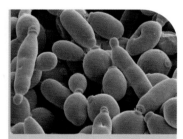

DOMAIN
Eukaryota
The latest taxonomic unit to be created is the domain. It is based on whether organisms possess cells with a nucleus (eukaryotes: protoctists, plants, fungi, and animals) or not (prokaryotes: Archaea and Bacteria).

KINGDOM
Animalia
In recent years, the traditional kingdoms of plants and animals have been further subdivided. Kingdom Animalia now includes only multicellular organisms that must eat other species in order to survive.

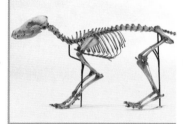

PHYLUM
Chordata
A phylum (known as a division in plants) is a major subdivision of a kingdom and is made up of one or more classes that share certain features. Members of the phylum Chordata have a notochord—the precursor of the backbone.

CLASS
Mammalia
Introduced by Carl Linnaeus, a class contains one or more orders. Class Mammalia includes only animals that are warm-blooded, have fur, a single jaw bone, and that suckle their young on milk.

> CONTRIBUTING TO CLASSIFICATION
Over the years many scientists have attempted to organize the natural world, combining earlier ideas with new research, culminating in the traditional system of classification shown above and the use of bi- and trinomial Latin names. Some scientists were particularly influential, making significant contributions to taxonomy.

ANIMALS AND PLANTS

Aristotle was the first person to classify living things, and he introduced the term *genos* (meaning race, stock, kin)—genus in Latin. He separated animals into those with blood and those without, not realizing blood need not be red. This division is very close to the modern classification of vertebrates and invertebrates.

ARISTOTLE, 384 BCE–322 BCE

ORDER FROM CHAOS

John Ray classified organisms based on their overall morphology rather than just a part of it. In doing so he could establish relationships between species more easily and organize them into groups more effectively. He also divided the flowering plants into two major groups of orders— monocotyledons and dicotyledons.

JOHN RAY, 1627–1705

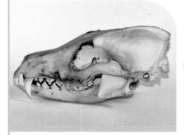 →  →  →

ORDER
Carnivora
Orders are the next level down in Linnaeus's hierarchy, and contain one or more families. The Carnivora have modified cheek teeth (carnassials) and large canine teeth specialized for biting and shearing.

FAMILY
Canidae
A subdivision of an order, a family is made up of genera and the species within them. The family Canidae has 35 living species, all with nonretractile claws and two fused wrist bones. All but one species have long, bushy tails.

GENUS
Vulpes
A term first used by Aristotle in Ancient Greece, a genus identifies subdivisions of a family. *Vulpes* is a genus within the family Canidae. All foxes have large, triangular ears and a long, narrow, pointed snout.

SPECIES
Vulpes vulpes
The basic unit of taxonomy, species are populations of similar animals that breed only with one another. *Vulpes vulpes*, known for its bright red fur, breeds only with other European red foxes.

ANIMAL, VEGETABLE, OR MINERAL

Linnaeus divided the natural world into three kingdoms—animals, plants, and minerals. He then devised a hierarchical system of classification based on class, order, family, genus, and species, and established the convention of giving species binomial Latin names.

CARL LINNAEUS, 1707–1778

A NEW KINGDOM

Historically organisms were classed as being either animals or plants, but in 1866 Ernst Haeckel argued that microscopic organisms formed a separate group, which he called Protista (now Protoctista). There were now three kingdoms of life: Animalia, Plantae, and Protista.

ERNST HAECKEL, 1834–1919

INTRODUCING THE ARCHAEA

Recognized by Carl Woese and George Fox in 1977, the Archaea are microscopic organisms that live in very extreme environments. Initially grouped with the Bacteria, their DNA turned out to be so unique that a new three-domain taxonomic system was introduced.

CARL WOESE, BORN 1928

ANIMAL GENEALOGY

In the 1950s a revolutionary new way of classifying organisms was proposed. Called phylogenetics, it allowed taxonomists to investigate evolutionary relationships between species by placing them in hierarchical groups called clades.

Phylogenetics, also known as cladistics, is based on the work of entomologist Willi Hennig (1913–1976). He assumed that organisms with the same morphological characters must be more closely related to each other than to those that lacked them. Therefore, these organisms must also share the same evolutionary history and have a more recent ancestor in common. Like the traditional taxonomy of Linnaeus, this method of classifying organisms is hierarchical, but due to the volume of data involved, computers are used to generate the family trees, known as cladograms.

For a morphological feature to be useful in cladistic analysis, it must have altered in some way from the so-called "primitive" ancestral condition to a "derived" one. For example, the legs and paws of most carnivores are considered primitive in the cladogram on the opposite page, compared to the derived condition of flippers in the seals, fur seals and sea lions, and walruses. This derived character, called a synapomorphy, is useful because it is shared between at least two taxonomic groups, suggesting

they are more closely related to each other than to groups without flippers. Characters that are unique to one group are useful for recognition purposes but say nothing about relationships. So cladistic analysis is based entirely on the identification of synapomorphic characters.

UNDERSTANDING ANCESTRY

The more derived characters organisms have in common, the closer their relationship is assumed to be. For instance, brothers and sisters look more alike than other children—they may have the same eyes, the same chin, and so on. This is because in terms of common ancestry they have the same parents but are only distantly related to other people.

Cladistics is now most commonly based on genetics—except in the study of fossils—and has exposed some unexpected shared ancestry. For example, a genetic cladogram surprisingly revealed that the whale is most closely linked to the land-dwelling hippopotamus—a relationship that Linnaeus would certainly not have expected.

∨ CLOSE RELATIONS
The giraffes and kudus are both even-toed ungulates, so are more closely related to each other than to the odd-toed zebras. As fur-bearing mammals, all three are more closely related to each other than to the feathered birds flying around them.

LOOKING AT CLADOGRAMS

To do a cladistic analyisis, the different groups of organisms are scored on a set of characters that are either primitive or derived. Their distribution is not always as straightforward as is shown in the diagrams below. Often the resulting cladogram can be constructed in a number of different ways and taxonomists have to choose between them. To do this they adopt the principle of parsimony—they choose the cladogram involving the least number of steps or character transformations to explain the observed relationships between the groups.

< MILKY DIET
All mammals possess mammary glands. Unique to the class Mammalia, this feature is synapomorphic at this taxonomic level. To find further familial relationships within the class, characters that are synapomorphic at familial level are used.

CHARACTER	CANID	BEAR	SEAL	FUR SEAL & SEA LION	WALRUS
Feeding young on milk	1	1	1	1	1
Short tail	0	1	1	1	1
Forelimbs modified into flippers	0	0	1	1	1
Very flexible spine	0	0	1	1	1
Hind limbs turn forwards under body	0	0	0	1	1
Presence of tusks	0	0	0	0	1

< CHARACTER SET
Most modern cladograms are based on genetics using DNA codes. The codes that were used to generate the cladogram below have been replaced with more familiar morphological descriptions in this character set, left. One character, feeding young on milk, is shared by all the groups shown; some characters are found in only some of the groups; and one character, the presence of tusks, is unique to the walrus.

∨ CLADOGRAM
In this cladogram the canids are considered the most primitive group, known as the outgroup, and the walrus is the most derived. All characters on the cladogram are shared by the groups to the right of each number—for example, a short tail is shared by the bears, seals, fur seals and sealions, and the walrus.

KEY

0	primitive character
1	derived character

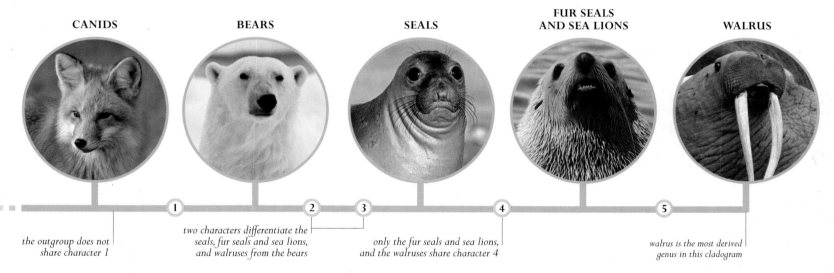

CANIDS   BEARS   SEALS   FUR SEALS AND SEA LIONS   WALRUS

the outgroup does not share character 1

two characters differentiate the seals, fur seals and sea lions, and walruses from the bears

only the fur seals and sea lions, and the walruses share character 4

walrus is the most derived genus in this cladogram

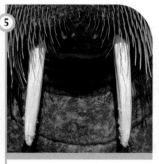

SHORT TAIL
Bears, seals, fur seals and sea lions, and the walrus all have short tails (character 1). Canids, however, display the primitive condition and have long, bushy tails. Character 1 is therefore a synapomorphic character shared by all the carnivore families shown except the canids.

FLIPPERS
Among carnivores, the modified limbs of seals, fur seals and sea lions, and the walrus are unique. Character 2 (forelimbs modified into flippers) is therefore synapomorphic at this level, suggesting these three groups are more closely related to one another than to the bears.

FLEXIBLE SPINE
Character 3 (flexible spine) operates at the same level as character 2, further supporting the relationship indicated by possession of flippers. The more synapomorphic characters that appear at a particular level, the more convincing the suggested relationship becomes.

SUPPORTING LIMBS
In both the fur seals and sea lions and the walrus, the pelvic girdle can be rotated to allow the hind limbs to aid locomotion on land. This suggests they share a more recent common ancestor with each other than they do with the seals, which are far less mobile when out of water.

TUSKS
This is a unique character (autapomorphy) of the walrus that reveals nothing about its relationship to other mammal families. The same is true of the character "feed young on milk"—present in all the groups, it provides no clues as to how they interrelate, so it is not plotted on the cladogram.

TREE OF LIFE

Using a branching tree as a way of showing the diversity of life was first suggested by the German naturalist Peter Pallas in 1766. Since then many such trees have been constructed. Initially treelike, complete with bark and leaves, they later became more diagrammatic, and took account of evolutionary theories. Modern, computer-generated trees of life present many different ideas of how living organisms are related.

DARWIN'S FIRST TREE

The first person to produce a tree of life that reflected the concept of evolution was Charles Darwin. He sketched the first of his ten evolutionary trees of life in 1837. It was a simple branching diagram, which he developed further before publishing it in the *Origin of Species* in 1859. The lettered branches show how he thought his theory might work—the more branch points separating an organism from its ancestor (numbered 1), the more different the organism will be. In 1879, Ernst Haeckel took the idea further, with a tree that showed animals evolving from single-celled organisms. Today DNA and protein analyses as well as morphology are used to construct evolutionary trees and establish the genetic relationships between organisms. Vast data sets require computers to generate the trees, which are continually refined as new species and information are discovered.

Trees of life inevitably place most emphasis on vertebrate groups within chordates because their relationships are well known. The many microscopic prokaryotes (archaea and bacteria) and protists (those eukaryotes not classified as plants, animals or fungi) are often under-represented because their relationships are more problematic. As more is learned about microscopic life, the trees change.

MASS EXTINCTIONS

Mapping all life forms that have ever existed on a tree is difficult because, over time, more than 95 percent of all species have become extinct. A mass extinction occurs when a large number of species dies off at the same time. This has happened five times in the past. The best-known extinction, which wiped out the dinosaurs, occurred at the end of the Cretaceous Period; it is thought to have been caused by a meteor impact combined with volcanic activity. Because habitats are rapidly destroyed by human activity, it is likely that there will be another extinction event in the future.

EXTINCTION TIMELINE

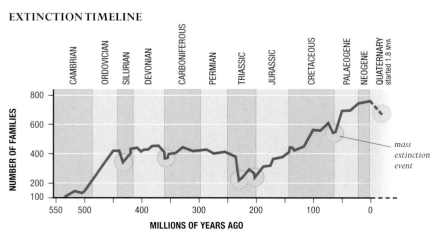

CAMBRIAN ORDOVICIAN SILURIAN DEVONIAN CARBONIFEROUS PERMIAN TRIASSIC JURASSIC CRETACEOUS PALAEOGENE NEOGENE QUATERNARY started 1.8 MYA

NUMBER OF FAMILIES

800
600
400
200
100

550 500 400 300 200 100 0

MILLIONS OF YEARS AGO

mass extinction event

READING THE TREE

This diagram shows how life evolved from simple organisms, such as the Archaea—which appeared about 3.4 billion years ago—to complex life forms, such as animals, which appeared 540 million years ago. It also shows the diversity of the vertebrates (see pp. 34–5), which have a disproportionate representation here. The circles indicate points where two or more groups of organisms have branched from a common ancestor at about the same time. Only extant species are shown.

LIFE BEGINS

ARCHAEA

PROKARYOTES

BACTERIA

STRUCTURE OF LIFE

All forms of life are either prokaryotic or eukaryotic. Prokaryotes lack a cell nucleus and are usually unicellular. Eukaryotic organisms tend to be multicellular; each cell contains a nucleus, within which DNA is stored. This table shows which of these two groups the six kingdoms belong to. Despite appearances, most organisms are prokaryotic. The Archaea and Bacteria are the largest groups—although only about 10,000 species have been described, estimates exceed 10 million species. Among eukaryotes, the phyla that make up the protists and invertebrates are far more numerous in terms of species than vertebrate groups.

PROKARYOTES	EUKARYOTES
ARCHAEA	PROTISTS
BACTERIA	PLANTS
	LIVERWORTS
	MOSSES
	FERNS AND RELATIVES
	CYCADS, GINKGOS,
	GNETOPHYTES
	FLOWERING PLANTS
	FUNGI
	MUSHROOMS
	SAC FUNGI
	LICHENS
	ANIMALS
	INVERTEBRATES
	CHORDATES

CYANOBACTERIA

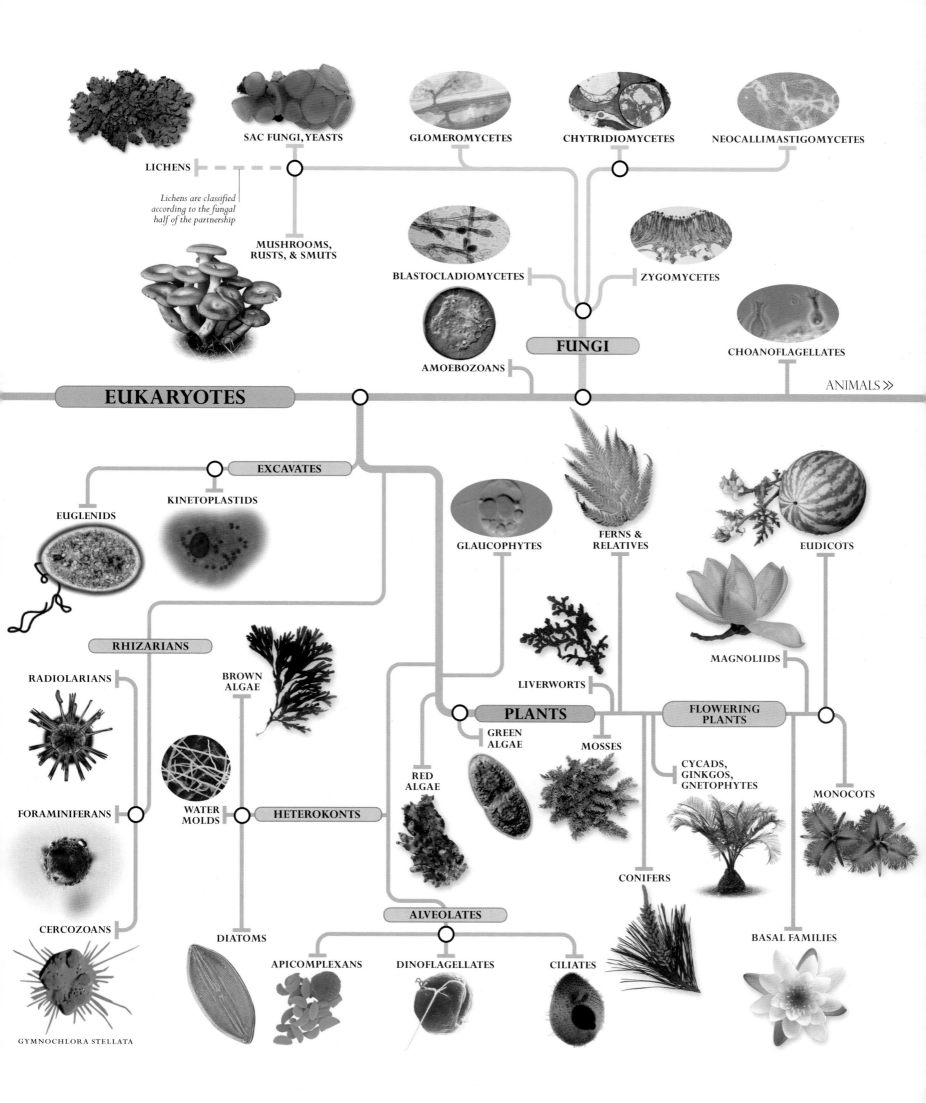

LICHENS

Lichens are classified
according to the fungal
half of the partnership

SAC FUNGI, YEASTS

GLOMEROMYCETES

CHYTRIDIOMYCETES

NEOCALLIMASTIGOMYCETES

MUSHROOMS,
RUSTS, & SMUTS

BLASTOCLADIOMYCETES

ZYGOMYCETES

FUNGI

AMOEBOZOANS

CHOANOFLAGELLATES

EUKARYOTES

ANIMALS »

EXCAVATES

EUGLENIDS

KINETOPLASTIDS

GLAUCOPHYTES

FERNS &
RELATIVES

EUDICOTS

MAGNOLIIDS

RHIZARIANS

RADIOLARIANS

BROWN
ALGAE

LIVERWORTS

PLANTS

FLOWERING
PLANTS

GREEN
ALGAE

RED
ALGAE

MOSSES

FORAMINIFERANS

WATER
MOLDS

HETEROKONTS

CYCADS,
GINKGOS,
GNETOPHYTES

MONOCOTS

CERCOZOANS

DIATOMS

ALVEOLATES

CONIFERS

APICOMPLEXANS

DINOFLAGELLATES

CILIATES

BASAL FAMILIES

GYMNOCHLORA STELLATA

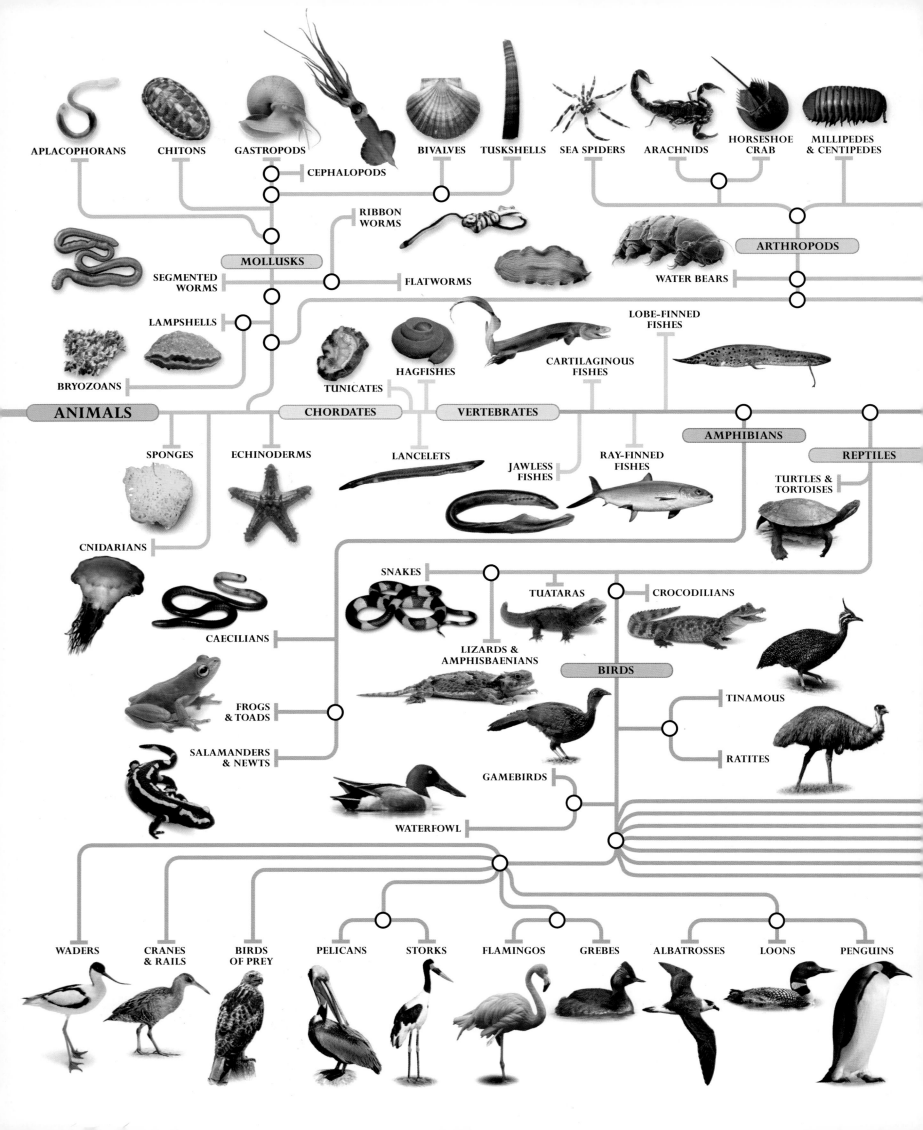

APLACOPHORANS

CHITONS

GASTROPODS

CEPHALOPODS

BIVALVES

TUSKSHELLS

SEA SPIDERS

ARACHNIDS

HORSESHOE CRAB

MILLIPEDES & CENTIPEDES

RIBBON WORMS

MOLLUSKS

ARTHROPODS

SEGMENTED WORMS

FLATWORMS

WATER BEARS

LAMPSHELLS

LOBE-FINNED FISHES

BRYOZOANS

TUNICATES

HAGFISHES

CARTILAGINOUS FISHES

ANIMALS

CHORDATES

VERTEBRATES

AMPHIBIANS

SPONGES

ECHINODERMS

LANCELETS

JAWLESS FISHES

RAY-FINNED FISHES

REPTILES

TURTLES & TORTOISES

CNIDARIANS

SNAKES

TUATARAS

CROCODILIANS

CAECILIANS

LIZARDS & AMPHISBAENIANS

BIRDS

FROGS & TOADS

TINAMOUS

SALAMANDERS & NEWTS

GAMEBIRDS

RATITES

WATERFOWL

WADERS

CRANES & RAILS

BIRDS OF PREY

PELICANS

STORKS

FLAMINGOS

GREBES

ALBATROSSES

LOONS

PENGUINS

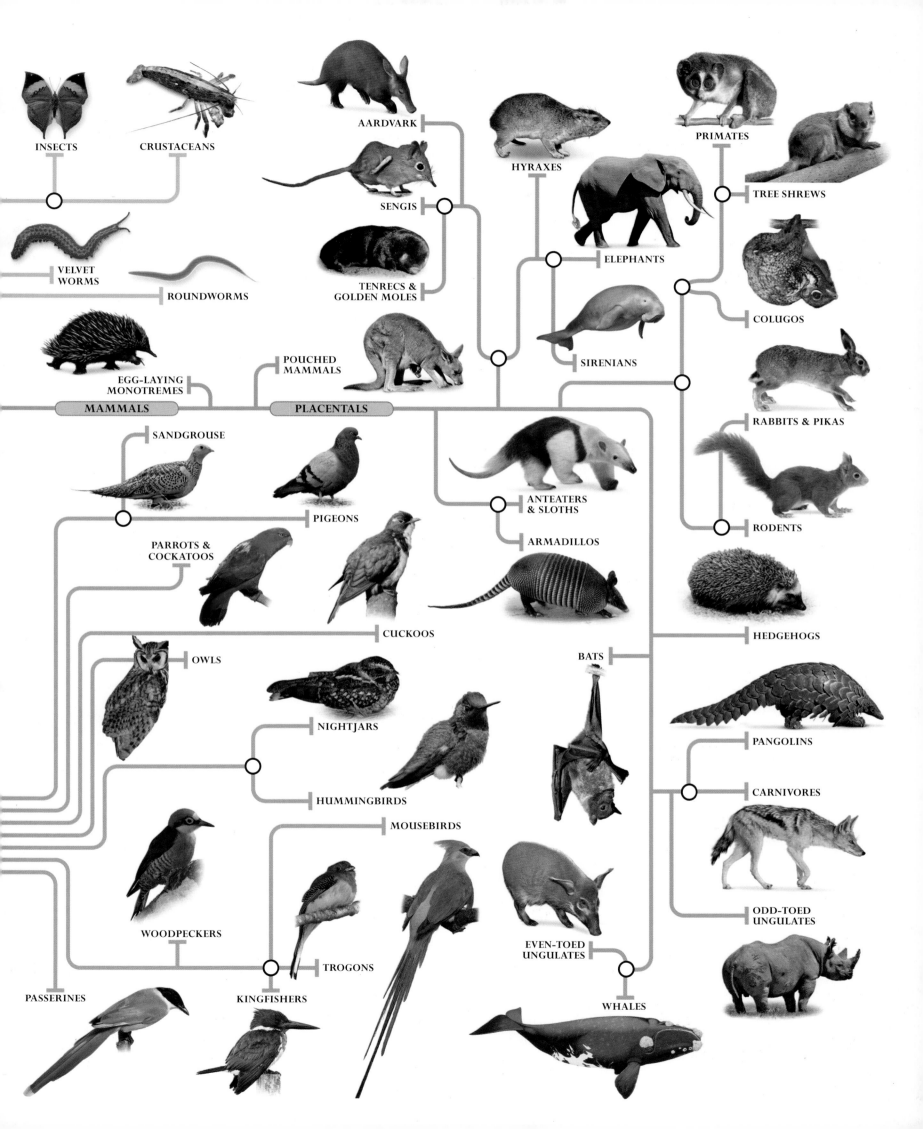

INSECTS

CRUSTACEANS

VELVET
WORMS

ROUNDWORMS

EGG-LAYING
MONOTREMES

MAMMALS

POUCHED
MAMMALS

PLACENTALS

AARDVARK

SENGIS

TENRECS &
GOLDEN MOLES

HYRAXES

ELEPHANTS

SIRENIANS

PRIMATES

TREE SHREWS

COLUGOS

RABBITS & PIKAS

RODENTS

HEDGEHOGS

PANGOLINS

CARNIVORES

ODD-TOED
UNGULATES

SANDGROUSE

PIGEONS

PARROTS &
COCKATOOS

CUCKOOS

OWLS

NIGHTJARS

HUMMINGBIRDS

MOUSEBIRDS

WOODPECKERS

TROGONS

PASSERINES

KINGFISHERS

ANTEATERS
& SLOTHS

ARMADILLOS

BATS

EVEN-TOED
UNGULATES

WHALES

MINERA
ROCKS,
FOSSILS

LS, &

Life on Earth is shaped by the rocks that lie beneath our feet. Made of different combinations of minerals, they have a far-reaching influence on the landscape, on vegetation, and on the soil. Preserved within these rocks, fossils form a highly detailed record of life in the distant past, showing the path that evolution has followed over hundreds of millions of years.

» 38
MINERALS

The building blocks of rocks, minerals typically have a crystalline structure. Several thousand exist in Earth's crust, but fewer than 50 minerals are common and widespread.

» 62
ROCKS

Classified according to how they formed, rocks are constantly broken down and reformed. The oldest rocks date back 3.8 billion years, when Earth's crust first solidified.

» 74
FOSSILS

Most fossils preserve hard remains, such as teeth and bones. They can also include traces such as footprints, and carbon impressions in rock, which reveal the past existence of living things.

MINERALS

Minerals are the materials from which rocks are made. More than 4,000 different kinds exist, each one a single chemical substance found naturally on Earth. Most minerals are hard and crystalline, and, while some are extremely abundant, others—including diamond—are very rare, and highly prized.

Copper often appears in a dendritic form, branching like a tree. It is a very important economic mineral.

Malachite can be botryoidal, with small, rounded shapes, or massive, with no definite shape.

Crocoite, which is lead chromate, frequently occurs as slender, elongated prismatic crystals.

Minerals are of great economic significance. They provide us with countless useful materials, from metals to industrial catalysts, and they also include objects of great intrinsic beauty, particularly when cut and polished as gems. But on an even broader scale, minerals are essential for life itself. In soil and in water, soluble types release a steady stream of chemical nutrients that plants and other organisms need to grow. Without them, the world's ecosystems would no longer operate.

Minerals are classified according to their chemistry. A few, such as gold, silver, and sulfur, can exist in a native state, meaning that they contain a pure chemical element and nothing else. All other minerals are chemical compounds. Quartz, for example, contains two elements—silicon and oxygen—which are bound together very tightly, giving quartz its exceptional hardness and strength. Quartz belongs to the silicate group of minerals, which makes up about 75 percent of the Earth's crust. Other common mineral groups include halides, such as halite, or common salt, and phosphates and carbonates. These last two groups are particularly important to animals, which use phosphate and carbonate minerals—for example, calcite—to construct hard body parts, such as teeth, bones, and shells.

IDENTIFYING MINERALS

With experience, many minerals can be identified by their appearance and texture alone. Important clues include their color, their luster—the way light reflects from their surface—and above all, their habit, or crystal form. Crystals are grouped into six systems according to their symmetry (see panel, below). In addition, the crystals themselves can be arranged in different ways: many are parallel, but they can be dendritic (branching), or even botryoidal (like a bunch of grapes).

Minerals also differ in density, or specific gravity (SG)—measured by comparing the weight of a mineral to that of an equal volume of water—and in their hardness (H). On the 10-point Mohs scale, which measures hardness, talc is rated 1, whereas diamond, the hardest mineral, is rated 10. A fingernail (hardness 2), copper coin (3.5), and steel knife blade (6) all make useful benchmarks for testing hardness. Surprisingly, size is a less useful clue. Gypsum crystals, for example, are usually less than ⅜ in (1 cm) long, but the largest specimens ever discovered are the size of a two-story house.

VOLCANIC MINERALS >
The ground at Dallol in Ethiopia's Danakil Desert is pockmarked by volcanic vents and covered in sulfur, a native element.

CRYSTAL SYSTEMS

Cubic systems are common and easily recognized. Such crystals have three axes at right angles; shapes include the cube and the eight-faced octahedron.

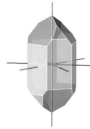

Hexagonal and trigonal systems are very similar to each other, with four axes of symmetry. Their crystals are often six-sided prisms with pyramidal tops (left).

Tetragonal systems have three axes of symmetry, all at right angles and two of equal length. In a tall prism (left), the vertical axis is longer. Squat prisms are also common.

Monoclinic systems have three unequal axes of symmetry, only two at right angles. Tabular (flattened, left) and prismatic habits are common.

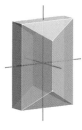

Orthorhombic systems are similar to the monoclinic system, but all three axes are at right angles; the usual habits are tabular and prismatic (left).

Triclinic systems have a low degree of symmetry, because the three axes are unequal in length and none of them are at right angles. Prismatic habits are common.

NATIVE ELEMENTS

Of the 88 natural elements, only about 20 are found in the native state; that is, uncombined with other elements. They are divided into three groups. The metals rarely form as distinct crystals, tend to have a high specific gravity, and are soft. Semimetals, such as antimony and arsenic, commonly occur in rounded masses. Nonmetals, including sulfur and carbon, usually form as crystals.

ANTIMONY
Hexagonal/trigonal
H 3–3½ • SG 6.6–6.7
This rare semimetal occurs in hydrothermal veins, often with arsenic and silver minerals. The silvery gray masses are coated white when oxidized.

GRAPHITE
Hexagonal • H 1–2 • SG 2.1–2.3
A form of pure carbon common in metamorphic rocks, graphite is dark, soft, and greasy, and makes ideal pencil leads.

NATIVE COPPER

COPPER
Cubic • H 2½–3 • SG 8.9
Native copper occurs mainly as irregular masses or branching or wirelike forms. It is most notably associated with basaltic lava. It is a good conductor and is widely used in the electrical industry.

COPPER ON GOETHITE GROUNDMASS

dendritic habit

NICKEL-IRON
Cubic
H 4–5 • SG 7.3–8.2
Iron is invariably alloyed to nickel. The mineral kamacite can have up to 7.5 percent nickel, and taenite up to 50 percent.

distinct, isolated diamond crystal

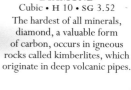

DIAMOND
Cubic • H 10 • SG 3.52
The hardest of all minerals, diamond, a valuable form of carbon, occurs in igneous rocks called kimberlites, which originate in deep volcanic pipes.

rock groundmass

resinous luster

uneven surface

ARSENIC
Hexagonal/trigonal • H 3½ • SG 5.7
Highly poisonous arsenic usually forms as pale gray, rounded masses in hydrothermal veins. If heated, it smells of garlic.

PLATINUM NUGGET

PLATINUM
Cubic • H 4–4½ • SG 21.4
A rare metal, native platinum forms as scales, grains, and nuggets in igneous rocks and in alluvial sands. Its high melting point means it is useful in industry, for example in aircraft spark plugs.

SULFUR
Orthorhombic • H 1½–2½ • SG 2.0–2.1
Native sulfur forms striking yellow crystals and powdery crusts around volcanic vents. It is mined for use in sulfuric acid, dyes, insecticides, and fertilizers.

PLATINUM

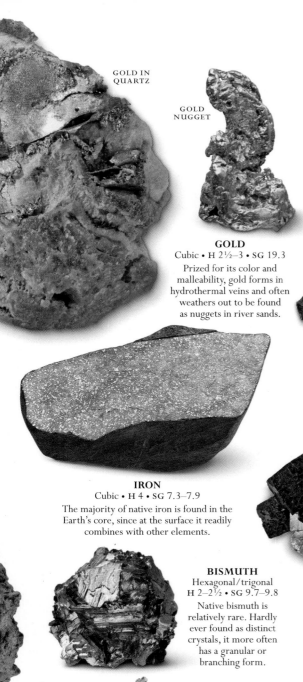

GOLD IN QUARTZ

GOLD NUGGET

GOLD
Cubic • H 2½–3 • SG 19.3

Prized for its color and malleability, gold forms in hydrothermal veins and often weathers out to be found as nuggets in river sands.

IRON
Cubic • H 4 • SG 7.3–7.9

The majority of native iron is found in the Earth's core, since at the surface it readily combines with other elements.

BISMUTH
Hexagonal/trigonal
H 2–2½ • SG 9.7–9.8

Native bismuth is relatively rare. Hardly ever found as distinct crystals, it more often has a granular or branching form.

mercury globules in rock cavities

MERCURY
Hexagonal/trigonal
H Liquid • SG 13.6–14.4

This is the only metal that is liquid at normal temperatures. In liquid form, mercury appears as silvery globules.

SILVER
Cubic • H 2½–3 • SG 10.5

Widely distributed but rarely found in abundance, native silver occurs mainly as twisted wires, scales, and branching masses.

SULFIDES

The sulfides are a large group of minerals, in which sulfur is combined with one or more metals. Many sulfides have a high specific gravity and a metallic luster. They often form as excellent crystals. Sulfides occur in many geological situations, but frequently in hydrothermal veins. The group includes the majority of the economically important metal ore minerals.

COBALTITE
Orthorhombic • H 5½ • SG 6.3

Cobaltite is an uncommon sulfide of arsenic and cobalt. It is an important cobalt ore in, for example, Sweden and Norway.

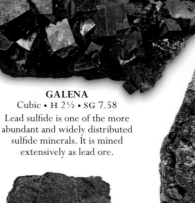

GALENA
Cubic • H 2½ • SG 7.58

Lead sulfide is one of the more abundant and widely distributed sulfide minerals. It is mined extensively as lead ore.

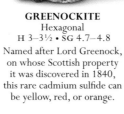

GREENOCKITE
Hexagonal
H 3–3½ • SG 4.7–4.8

Named after Lord Greenock, on whose Scottish property it was discovered in 1840, this rare cadmium sulfide can be yellow, red, or orange.

USUAL, CRYSTALLINE FORM OF SPHALERITE

SPHALERITE
Cubic • H 3½–4 • SG 3.9–4.1

A sulfide of zinc with variable iron content, sphalerite is the most heavily mined zinc-containing ore.

MASSIVE SPHALERITE

CINNABAR
Hexagonal/trigonal
H 2–2½ • SG 8.0–8.2

Red-colored mercury sulfide has been the main source of mercury through the centuries. It occurs around hot springs and volcanic vents.

long, curved crystals resemble blades or swords

STIBNITE
Orthorhombic • H 2 • SG 4.63–4.66

A sulfide of antimony, this dark gray mineral is the main ore of antimony. Large deposits occur in the western US, China, and Japan.

BORNITE CRYSTALS

indistinguishable crystals form large masses

BORNITE
Tetragonal • H 3 • SG 5.0–5.1

This copper iron sulfide is a coppery red color, tarnishing to shimmering purple and blue. It is an important copper ore.

MASSIVE BORNITE

COMMON, MASSIVE FORM OF CHALCOPYRITE

CHALCOPYRITE
Tetragonal
H 3½–4 • SG 4.3–4.4

A sulfide of copper and iron, chalcopyrite has a deep brassy yellow color. It has significant value as a copper ore.

CHALCOPYRITE CRYSTALS

ACANTHITE
Monoclinic
H 2–2½ • SG 7.22

A sulfide occurring as dark, metallic, sometimes spiky crystals, acanthite is the main ore of silver.

»

» SULFIDES

ORPIMENT
Monoclinic
H 1½–2 • SG 3.4–3.5
Named from the Latin for "golden paint," this arsenic sulfide occurs as foliated, columnar masses around hot springs.

REALGAR
Monoclinic • H 1½–2 • SG 3.56
A bright orange-red sulfide of arsenic, realgar was historically used as a pigment.

GLAUCODOT
Orthorhombic • H 5 • SG 5.9–6.1
This sulfide of cobalt, iron, and arsenic occurs as silver-white brittle masses, which have no external crystal form.

MOLYBDENITE
Hexagonal/trigonal
H 1–1½ • SG 4.62–5.06
Molybdenum sulfide is lead-gray in color. It has an oily feel, which is due to weak bonds within a layered crystal structure.

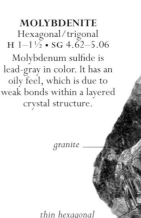

granite

thin hexagonal crystals in layers

MARCASITE
Orthorhombic
H 6–6½ • SG 4.92
A sulfide of iron that is lighter and more brittle than pyrite, marcasite frequently occurs as cockscombs and spear-shaped "twins."

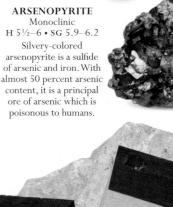

indigo-blue covellite

COVELLITE
Hexagonal • H 1½–2 • SG 4.6–4.8
Covellite is a not particularly common copper sulfide. Its shining indigo-blue color makes it attractive to mineral collectors.

slender prismatic crystals

ARSENOPYRITE
Monoclinic
H 5½–6 • SG 5.9–6.2
Silvery-colored arsenopyrite is a sulfide of arsenic and iron. With almost 50 percent arsenic content, it is a principal ore of arsenic which is poisonous to humans.

STANNITE
Tetragonal • H 4 • SG 4.4
Stannite is a sulfide of tin, copper, and iron mined for the tin. Its name comes from the Latin for tin, *stannum*.

HAUERITE
Cubic • H 4 • SG 3.46
Hauerite is a very rare sulfide of manganese. The brown octahedral crystals can form when certain minerals are altered in the caps of salt domes.

PENTLANDITE
Cubic
H 3½–4 • SG 4.6–5.0
This nickel and iron sulfide is found in basic igneous rocks. It is an important source of nickel.

calcite groundmass

PYRITE
Cubic
H 6–6½ • SG 5
Nicknamed "fools gold" because of its light goldish color, this iron sulfide is the most common of all sulfide minerals.

PYRRHOTITE
Monoclinic • H 3½–4½ • SG 4.53–4.77
An iron sulfide with variable iron content, pyrrhotite has a magnetism that increases as the iron content decreases.

MILLERITE
Hexagonal/trigonal
H 3–3½ • SG 5.3–5.6
This sulfide of nickel occurs in limestones and ultramafic rocks. It is sought after as a nickel ore.

CHALCOCITE
Monoclinic • H 2½–3 • SG 5.5–5.8
Dark gray to black, copper sulfide has been mined for centuries. It is one of the most profitable copper ores.

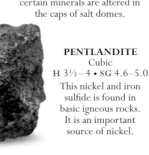

BISMUTHINITE
Orthorhombic • H 2 • SG 6.8
This sulfide of bismuth is an important
ore. Much of the bismuth recovered is
used in medicines and cosmetics.

SULFOSALTS

Sulfosalts are a group of about 200 mainly
rare minerals, structurally related to standard
sulfides and with many of the same properties.
In these compounds sulfur is combined with
a metallic element—commonly silver, copper,
lead, or iron—and a semimetal, often antimony
or arsenic. Sulfosalts frequently occur in
hydrothermal veins, usually in small amounts.

PYRARGYRITE
Hexagonal/trigonal
H 2½ • SG 5.8–5.9
Also called ruby silver,
this sulfide of silver and
antimony is red-black, but
thin splinters appear
deep ruby-red.

POLYBASITE
Monoclinic
H 2–3 • SG 6.0–6.3
Somewhat uncommon,
polybasite is a sulfide of silver,
copper, antimony, and arsenic.
It can yield worthwhile
quantities of silver locally.

BOULANGERITE
Monoclinic
H 2½–3 • SG 5.8–6.2
A bluish gray sulfide of lead
and antimony, boulangerite
is one of few sulfide
minerals to form fine,
hairlike crystals.

*striated
prismatic
crystals*

STEPHANITE
Orthorhombic
H 2–2½ • SG 6.25
Stephanite is an opaque
black sulfide of silver and
antimony. It can be an
important silver ore—for
example, in Nevada.

JAMESONITE
Monoclinic • H 2½ • SG 5.63
A sulfide of lead, iron, and antimony,
jamesonite's dark gray crystals can be fine
and hairlike or larger and prismatic.

PROUSTITE
Hexagonal/trigonal • H 2–2½ • SG 5.55–5.64
This sulfide of silver and arsenic is also called
light ruby silver. The transparent crystals
are bright scarlet-red.

ZINKENITE
Hexagonal • H 3–3½ • SG 5.3
Zinkenite is a sulfide of
lead and antimony. It occurs
as steel-gray hair- or
needlelike crystals.

TETRAHEDRITE
Cubic
H 3–4½ • SG 4.6–5.1
This sulfide of copper, iron,
and antimony is named
for its tetrahedron-
shaped crystals
(crystals with four
triangular faces).

*bright,
metallic
luster*

*radiating,
needlelike
crystals*

TENNANTITE
Cubic • H 3–4½ • SG 4.59–4.75
Tennantite, a sulfide of copper,
iron, and arsenic, is dark gray or
black. It can look very similar
to tetrahedrite.

ENARGITE
Orthorhombic • H 3 • SG 4.4–4.5
A steel-gray colored sulfide of copper and
arsenic, enargite has a metallic luster. Crystals
are usually small, tabular, or prismatic.

BOURNONITE
Orthorhombic • H 2½–3 • SG 5.7–5.9
Black or steel-gray bournonite is a
sulfide of lead, copper, and antimony.
Crystals are tabular to prismatic.

OXIDES

Oxides are compounds of oxygen and other elements. Some oxides are very hard and have a high specific gravity, and many are brightly colored. The group includes the chief ores of iron, manganese, aluminum, tin, and chromium. Some oxide minerals are sought-after gems. Oxides can occur in hydrothermal veins, igneous and metamorphic rocks, and also, because they can be resistant to weathering and transportation, in layers of sediment.

CUPRITE
Cubic • H 3½–4 • SG 6.14
Found in various shades of red, this copper oxide forms near the Earth's surface by oxidation of copper minerals.

PEROVSKITE
Orthorhombic
H 5½ • SG 4.01
Discovered in Russia in 1839, this dark-colored oxide of calcium and titanium forms in igneous and metamorphic rocks.

octahedral franklinite crystal

FRANKLINITE
Cubic
H 5½–6½ • SG 5.07–5.22
This black or brown zinc manganese iron oxide is found in metamorphosed limestones, notably those in Franklin, New Jersey.

ILMENITE
Hexagonal/trigonal
H 5–6 • SG 4.72
Iron titanium oxide is the principal ore of titanium, a high strength, low density metal used in aircraft and rocket construction.

URANINITE
Cubic • H 5–6 • SG 6.5–10.0
Highly radioactive, black- or brown-colored uranium oxide is the main ore of uranium, which is used in nuclear reactors to produce electricity, and in the construction of nuclear weapons.

CASSITERITE
Tetragonal • H 6–7 • SG 7
Almost the sole source of the world's tin, this tin oxide is mainly found as small grains among river gravels.

striated crystal face

vitreous luster

SAMARSKITE
Orthorhombic
H 5–6 • SG 5.15–5.69
Minerals known as samarskite—a radioactive oxide of various metals including yttrium, iron, tantalum, and niobium—occur in igneous rocks and alluvial sands.

GAHNITE
Cubic • H 7½–8 • SG 4.6
A rare oxide of aluminum and zinc found mainly in metamorphic rocks, gahnite can form dark green or blue to black crystals.

CORUNDUM
Hexagonal/trigonal
H 9 • SG 4.0–4.1
Corundum is an aluminum oxide, second only in hardness to diamond. Ruby-red and sapphire-blue varieties are used as gemstones.

CHROMITE
Cubic • H 5½ • SG 4.5–4.8
This iron chromium oxide is the only important source of chromium, an element used in making chrome- and stainless-steel.

bright, metallic luster

HEMATITE
Hexagonal • H 5–6 • SG 5.26
Widespread and abundant,
iron oxide is mined extensively
for iron. Various forms range
in color from metallic gray
to earthy red.

HYDROXIDES

Hydroxide minerals are compounds of a
metallic element and the hydroxyl radical
(OH). They are common minerals and
often form through a chemical reaction
between an existing oxide and fluids rich
in water, seeping through the Earth's
crust. Many hydroxide minerals are
quite soft. Hydroxides tend to
occur in the altered parts of
hydrothermal veins and in
metamorphic rocks.

GIBBSITE
Monoclinic
H 2½–3½ • SG 2.4
Gibbsite is one of three
essential aluminum
hydroxides in the
aluminum ore bauxite.
It also occurs in
hydrothermal veins.

STIBICONITE
Cubic • H 4–5½ • SG 3.3–5.5
An uncommon hydroxide of antimony,
stibiconite is white or yellowish brown,
and forms by the alteration of other
antimony minerals, especially stibnite.

FERGUSONITE
Tetragonal
H 5½–6½ • SG 4.2–5.7
Fergusonite is the group
name for an oxide of
many metals, including
yttrium, lanthanum
niobium, and cerium.

prismatic crystal

PYROLUSITE
Tetragonal
H 2–6½ • SG 5.06
Pyrolusite is a common
manganese oxide. It is
the primary ore of
manganese, an essential
element in
steel production.

LEPIDOCROCITE
Orthorhombic • H 5 • SG 3.9
This relatively rare iron hydroxide
may occur with goethite. It is
reddish brown and can form
irregular and fibrous shapes.

DIASPORE
Orthorhombic • H 6½–7 • SG 3.3–3.5
Diaspore and its variety böhmite are essential
aluminum hydroxides in bauxite. Diaspore also
occurs in marble and altered igneous rocks.

RUTILE
Tetragonal
H 6–6½ • SG 4.23
A source of titanium, this oxide
of titanium often forms impressive
displays of thin, translucent
needles in quartz crystals.

MASSIVE
ROMANÈCHITE

MASSIVE
GOETHITE

ROMANÈCHITE
Orthorhombic • H 5–6 • SG 4.7
Dark and opaque, this oxide of
manganese with barium is usually
found in aggregates or in massive
forms. Crystals are rare.

BOTRYOIDAL
GOETHITE

GOETHITE
Orthorhombic • H 5–5½ • SG 3.3–4.3
A common iron hydroxide, goethite gives a
yellow-brown color to soil and rocks that
have been exposed to the atmosphere.

CHRYSOBERYL
Orthorhombic • H 8½ • SG 3.7–3.8
Chrysoberyl is beryllium aluminum oxide. It
is a prized gemstone, known for its exceptional
hardness and green- or yellow-brown color.

ZINCITE
Trigonal • H 4–4½ • SG 5.68
Zincite is a rare oxide of zinc and
manganese. Its only significant deposit
in the US has been exhausted.

BRUCITE
Hexagonal/trigonal • H 2½ • SG 2.38–2.40
Brucite is a magnesium hydroxide, and its color is white,
gray, blue, or green. It occurs in metamorphic rocks.

BAUXITE
Amorphous mixture
H 1–3 • SG 2.3–2.7
The primary aluminum ore,
bauxite is not a single mineral,
but rather an assemblage of
aluminum hydroxides and
iron oxides.

HALIDES

When metallic elements combine with halogen elements, halides are formed. The halogens are iodine, fluorine, chlorine, and bromine. Halide minerals are commonly very soft and of low specific gravity, often having crystals classified in the cubic system. Many of these minerals, such as halite and sylvite, form in evaporite sequences by the drying out of saline waters. Other halides—for example, fluorite—occur in hydrothermal veins.

YELLOW FLUORITE

PURPLE FLUORITE

cubic crystal

FLUORITE
Cubic • H 4 • SG 3.18
Calcium fluoride often forms transparent to translucent crystals of various colors. Large quantities are used for making hydrofluoric acid.

GREEN FLUORITE

SYLVITE
Cubic • H 2 • SG 1.99
Sylvite is potassium chloride, a salt similar to halite. It occurs with halite in evaporite deposits. Sylvite is used to make potash fertilizers.

granular carnallite

CARNALLITE
Orthorhombic • H 2 • SG 1.6
Carnallite is a hydrated chloride of magnesium and potassium, formed by the evaporation of saline water. It is important in fertilizer manufacture.

vitreous luster

transparent cubic crystal

ORANGE HALITE

DIABOLEITE
Tetragonal • H 2½ • SG 5.42
A lead copper chloride hydroxide with a light to dark blue color, diaboleite forms by the alteration of other minerals.

BOLEITE
Cubic • H 3–3½ • SG 5.0–5.1
Deep blue boleite is a rare hydroxide of lead, silver, copper, and chlorine. It occurs where lead and copper deposits have been altered.

HALITE
Cubic • H 2 • SG 2.1–2.2
Common salt, or sodium chloride, occurs in extensive beds formed by the evaporation of sea water. It can be colored or colorless.

HALITE CRYSTALS

JARLITE
Monoclinic
H 4–4½ • SG 3.87
Usually white, jarlite is found in igneous rocks. It is a rare sodium strontium magnesium aluminum fluoride hydroxide.

chlorargyrite crust

CHLORARGYRITE
Cubic • H 2½ • SG 5.55
This silver chloride is typically scaly, platelike, or, in masses, resembling wax. It occurs where silver deposits are altered.

dark green, tabulate crystals of atacamite

ATACAMITE
Orthorhombic • H 3–3½ • SG 3.76
Green atacamite is copper chloride hydroxide, occurring where copper deposits have been oxidized. It is a minor copper ore.

CALOMEL
Tetragonal
H 1–2 • SG 6.5
Rarely found, this mercury chloride is a white to gray or brown mineral. Its color deepens on exposure to light.

CRYOLITE
Monoclinic
H 2½ • SG 2.97
A rare aluminum sodium fluoride, cryolite often has an icelike appearance. It is found in granitic pegmatites and granites.

CARBONATES

Carbonate minerals are compounds of metallic or semimetallic elements combined with the carbonate radical (CO_3). Over 70 carbonate minerals are known, but calcite, dolomite, and siderite account for most of the carbonate in the Earth's crust. Carbonates usually form as "good" crystals with regular shapes and no foreign substances enclosed within them. Many carbonates are pale colored, but some, such as rhodochrosite, smithsonite, and malachite, are brightly colored.

vitreous luster

DOGTOOTH
SPAR

NAILHEAD
SPAR

CALCITE
Hexagonal/trigonal
H 3 • SG 2.71
One of the most abundant minerals, most calcium carbonate is massive, occurring as limestone or marble. It can also form outstanding crystals.

TRONA
Monoclinic • H 2½–3 • SG 2.1
Hydrated sodium carbonate, or trona, is gray, yellowish, or brown. It forms on the Earth's surface, especially in saline desert environments.

WITHERITE
Orthorhombic
H 3–3½ • SG 4.29
Reasonably uncommon, this carbonate of barium tends to be white or gray, and occurs in hydrothermal veins.

pearly luster

SMITHSONITE
Hexagonal/trigonal
H 4–4½ • SG 4.3–4.45
A zinc carbonate found in the upper oxidized zones of zinc ore deposits, smithsonite is also mined for zinc.

prismatic crystals

limestone groundmass

BARYTOCALCITE
Monoclinic • H 4 • SG 3.66–3.71
This barium calcium carbonate is white to yellowish, and often found in hydrothermal veins within limestone.

curved crystal faces

DOLOMITE
Hexagonal/trigonal
H 3½–4 • SG 2.85
Dolomite is a calcium magnesium carbonate widespread in altered limestones. Dolomite rock, formed exclusively of massive dolomite, is used as a building stone.

MAGNESITE
Hexagonal/trigonal • H 3–4 • SG 3.0–3.1
A magnesium carbonate, magnesite usually occurs as white to brownish dense masses. It is made into furnace bricks and magnesia cement.

needlelike crystals

STRONTIANITE
Orthorhombic • H 3½ • SG 3.78
This strontium carbonate is found in hydrothermal veins and limestones. The strontium is used in sugar refining and fireworks.

≫

**FLOS FERRI
("FLOWERS OF IRON")
ARAGONITE**

ARAGONITE
Orthorhombic
H 3½–4 • SG 2.94–2.95
Aragonite is calcium carbonate,
chemically identical to calcite, but
with a different crystal structure
and much less common.

BOTRYOIDAL SIDERITE

SIDERITE
Hexagonal/trigonal
H 4 • SG 3.96
A brown-colored carbonate of iron,
named from the Greek for iron, *sideros*,
siderite occurs in a variety of forms.

**RHOMBOHEDRAL
SIDERITE**

*short
prismatic
crystals*

PHOSGENITE
Tetragonal • H 2½–3 • SG 6.1
This rare lead carbonate chloride is
formed close to the Earth's surface by the
reaction of lead-rich minerals with water.

ARTINITE
Monoclinic • H 2½ • SG 2
A hydrated magnesium
carbonate hydroxide, artinite has
a distinctive habit, with sprays
of white, needle-shaped crystals.
It occurs in serpentine rocks.

**ARAGONITE
TWINNED CRYSTALS**

*glasslike
sheen*

HYDROZINCITE
Monoclinic
H 2–2½ • SG 4
Hydrozincite, or zinc
carbonate hydroxide, is pale
gray, white, pink, or yellowish.
It fluoresces bluish white
under ultraviolet light.

*patch of green
malachite
around margins*

limonite matrix

AZURITE
Monoclinic • H 3½–4 • SG 3.77–3.78
Azurite is a hydrous copper carbonate. Its
rich blue color and frequent association
with green malachite in hydrothermal
veins are distinctive features.

*botryoidal
habit*

LEADHILLITE
Monoclinic • H 2½–3 • SG 6.55
This lead sulfate carbonate hydroxide
usually occurs as well-formed crystals
in the oxidized zones of lead deposits.

CERUSSITE
Orthorhombic • H 3–3½ • SG 6.55
A carbonate of lead occurring where lead-bearing veins have been altered, cerussite is the most common lead ore after galena.

CLUSTER OF TWINNED CRYSTALS

CERUSSITE CRYSTALS

rhombohedral crystals

RHODOCHROSITE
Hexagonal/trigonal • H 3½–4 • SG 3.7
Gem-quality crystals of this manganese carbonate, in shades of rose-pink, can be found in the US, South Africa, and Peru.

ANKERITE
Hexagonal/trigonal
H 3½–4 • SG 2.97
Ankerite is a carbonate of calcium with lesser iron, magnesium, and manganese. It is sometimes found in gold-bearing quartz veins.

aurichalcite crystals

AURICHALCITE
Monoclinic • H 1–2 • SG 3.96
A blue- or green-colored zinc copper carbonate hydroxide, aurichalcite forms in the oxidized zones of zinc and copper deposits.

characteristic green coloring

MALACHITE ON CHRYSOCOLLA (SILICATE MINERAL)

associated azurite

MALACHITE
Monoclinic • H 3½–4 • SG 4
This striking green copper carbonate often occurs as rounded masses. It is used for ornamentation and as a source of copper.

BOTRYOIDAL MALACHITE

BORATES

Borates occur when metallic elements combine with the borate radical (BO_3). There are over 100 borate minerals, the most common being borax, kernite, ulexite, and colemanite. Borates tend to be pale colored, relatively soft, and have low specific gravity. Many occur in evaporite sequences where saline waters have dried out and minerals are then precipitated among layers of sedimentary rocks.

BORACITE
Orthorhombic
H 7–7½ • SG 3
Magnesium borate chloride crystals are pale green or white, and glassy. Boracite occurs in salt deposits, notably in Germany and Poland.

translucent, prismatic crystals

COLEMANITE
Monoclinic • H 4½ • SG 2.42
This hydrated calcium borate hydroxide forms when saline water evaporates. It was the main source of boron until the discovery of kernite.

BORAX
Monoclinic • H 2–2½ • SG 1.7
A chalky-white sodium borate hydrate, borax has many applications, including in medicines, laundry detergents, glasses, and textiles.

KERNITE
Monoclinic • H 2½–3 • SG 1.9
A colorless or white sodium borate hydrate, kernite has less water than borax. The two minerals occur together.

ULEXITE
Monoclinic • H 2½ • SG 1.96
A hydrated sodium calcium borate hydroxide, ulexite's white, fibrous crystals transmit light down their length. It has uses similar to borax.

HOWLITE
Monoclinic • H 3½ • SG 2.6
Howlite is calcium borosilicate hydroxide. It generally forms as chalky, rounded masses.

NITRATES

Nitrates are a small group of compounds where metallic elements combine with the nitrate radical (NO_2). These minerals are usually very soft and have low specific gravity. Many dissolve easily in water and they only rarely form as crystals. They are generally confined to arid regions, forming coatings on the land surface that often cover wide areas. Commercially, nitrates can be used as fertilizers or explosives.

NITRATITE
Hexagonal/trigonal • H 1½–2 • SG 2.27
Sodium nitrate typically occurs as crusts on the ground surface in arid regions, especially in Chile. It is white, gray, brown, or yellow.

SULFATES

Sulfates are composed of metals joined to the sulfate radical (SO_4). There are about 200 sulfates and most are rare. Many, such as the more common gypsum, form in evaporite deposits where minerals are precipitated from drying saline solutions. Others form as weathering products, or as primary minerals in hydrothermal veins. Many are economically important—barite is used to lubricate drills on oil rigs.

GYPSUM
Monoclinic • H 2 • SG 2.32
A widespread mineral, gypsum, or hydrated calcium sulfate, makes plaster of Paris when heated and mixed with water.

SATIN SPAR

RADIATING GYPSUM

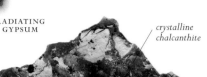

THENARDITE
Orthorhombic
H 2½–3 • SG 2.66
A pale gray or brownish mineral, thenardite is sodium sulfate. It is found on lava flows and around salt lakes.

prismatic crystal

ANGLESITE
Orthorhombic
H 2½–3 • SG 6.3–6.4
This lead sulfate comes in a variety of colors and forms. It is an alteration product of galena, the primary lead ore.

galena

crystalline chalcanthite

CHALCANTHITE
Triclinic
H 2½ • SG 2.28
Rich blue or green chalcanthite is hydrated copper sulfate. It forms through the oxidation of chalcopyrite and other copper sulfates.

radiating, hairlike crystals

rock groundmass

mass of needle-shaped brochantite crystals

LINARITE
Monoclinic
H 2½ • SG 5.3
Bright blue linarite is hydrous copper lead sulfate. It occurs in the oxidation zones of copper and lead ores.

CYANOTRICHITE
Orthorhombic • H 3 • SG 2.74–2.95
This hydrated copper aluminum sulfate is named after the Greek for "blue" and "hair," referring to its clusters of fine, blue crystals.

GLAUBERITE
Monoclinic
H 2½–3 • SG 2.8
Glauberite is sodium and calcium sulfate. Colorless, gray, or yellowish, it forms where saline water evaporates.

ALUNITE
Hexagonal/trigonal
H 3½–4 • SG 2.6–2.9
A hydrous sulfate of potassium and aluminum, alunite may be found at volcanic vents where rocks are altered by sulfur vapors.

CHROMATES

Chromate minerals form when metallic elements join with the chromate radical (CrO_4). They are rare minerals—crocoite is the only reasonably well-known chromate. They are generally brightly colored, and highly sought after by mineral collectors. Chromates often form when hydrothermal veins are altered by fluids.

RED CROCOITE

slender, elongate crystals with striations

adamantine luster

ORANGE CROCOITE

CROCOITE
Monoclinic • H 2½–3 • SG 6
Orange or red lead chromate forms in the oxidized zones of lead deposits. Good specimens come from Australia.

MELANTERITE
Monoclinic • H 2 • SG 1.9
White, green, or blue
melanterite is hydrated
iron sulfate. It is used in
purifying water supplies,
and as a fertilizer.

*prismatic
crystals*

EPSOMITE
Orthorhombic
H 2–2½ • SG 1.68
This hydrated magnesium
sulfate occurs in arid
regions and limestone cave
walls. It is the source of
Epsom salts.

JAROSITE
Hexagonal/trigonal
H 2½–3½ • SG 2.90–3.26
Jarosite is a hydrous sulfate of iron and
potassium, occurring as brown coatings
on pyrite and other iron minerals.

*scaly aggregates
of crystals*

CELESTINE
Orthorhombic • H 3–3½ • SG 3.96–3.98
Strontium sulfate is sought after, not only as
the main source of strontium, but also for the
beautiful, transparent, pale-colored crystals.

*iron oxide
groundmass*

COPIAPITE
Triclinic • H 2½–3 • SG 2.08–2.17
A yellow or green hydrated sulfate of
iron first described from Copiapó,
Chile, copiapite occurs where other
minerals are altered.

BROCHANTITE
Monoclinic
H 3½–4 • SG 3.97
A copper sulfate
hydroxide, brochantite
forms emerald-green
crystals, crusts, or masses.

ANHYDRITE
Orthorhombic
H 3–3½ • SG 2.98
A form of calcium sulfate,
anhydrite occurs alongside gypsum
but is less common. It alters to
gypsum in humid conditions.

BARITE
Orthorhombic
H 3–3½ • SG 4.5
The most common
barium mineral,
barium sulfate is
unusually heavy for a
pale-colored mineral.

POLYHALITE
Triclinic • H 3½ • SG 2.78
Polyhalite is hydrated potassium calcium
magnesium sulfate. Colorless, white,
pink, or red, it is widespread in many
marine salt deposits.

MOLYBDATES

Molybdates form when metals combine with
the molybdate radical (MoO_4). These minerals
are rare and tend to be dense and brightly
colored. Molybdate minerals occur in mineral
veins that have been altered by circulating water.
Wulfenite is the best-known molybdate mineral.
It is prized for its fine crystals and brilliant
orange or yellow colors.

TUNGSTATES

Tungstate minerals are compounds with metallic
elements joined to the tungstate radical (WO_4).
These minerals are rare and usually brittle and
dense. Some are dark colored and form as fine
crystals. Tungstates occur in hydrothermal veins
and pegmatites—very coarse-grained granitic
rocks where minerals form from fluids
permeating the rock.

HÜBNERITE
Monoclinic
H 4–4½ • SG 7.3
This manganese iron tungstate
is a major source of tungsten,
a metal used in steel alloys,
abrasives, and light bulbs.

*quartz
groundmass*

*hübnerite
crystal*

oily luster

*thin, square
wulfenite
crystal*

SCHEELITE
Tetragonal
H 4½–5 • SG 5.9–6.1
Mined for tungsten, this
calcium tungstate is found
in hydrothermal veins,
metamorphic and igneous
rocks, and alluvial sands.

*bipyramidal
scheelite crystal*

FERBERITE
Monoclinic
H 4–4½ • SG 7.5
Opaque black
ferberite is found in
hydrothermal veins and
granitic pegmatites. It
is iron tungstate, mined
as a source of tungsten.

WULFENITE
Tetragonal • H 2½–3 • SG 6.5–7.0
This lead molybdate is found in oxidized
zones of lead and molybdenum deposits.
It is a minor source of molybdenum.

PHOSPHATES

When metals combine with the phosphate radical (PO₄), phosphate minerals are formed. These minerals form a large group of over 200—however, many are very rare. Minerals in this group vary in hardness and specific gravity, and many are brightly colored. They usually form by the alteration of sulfide minerals, but some are primary minerals. A number of phosphates are rich in lead; others are radioactive.

HYDROXYLHERDERITE
Monoclinic • H 5–5½ • SG 2.95–3.01

Hydroxylherderite is calcium beryllium phosphate. It occurs as pale yellow or greenish crystals, with a glassy sheen, in granitic pegmatites.

DUFRÉNITE
Monoclinic • H 3½–4½ • SG 3.1–3.34

This hydrated phosphate of iron and calcium mainly occurs as green to black masses or crusts in altered veins and iron ores.

aggregate of xenotime crystals

XENOTIME
Tetragonal • H 4–5 • SG 4.4–5.1

Widely distributed yttrium phosphate is yellow-brown, gray, or greenish, and forms in igneous and metamorphic rocks.

META-AUTUNITE
Tetragonal
H 2–2½ • SG 3.05–3.2

Radioactive meta-autunite is a lemon-yellow or pale green hydrated phosphate of calcium and uranium. It occurs where uranium minerals are altered.

TURQUOISE
Triclinic • H 5–6 • SG 2.6–2.8

A sought after gem for thousands of years, this hydrated phosphate of copper and aluminum is found in altered igneous rocks.

PYROMORPHITE
Hexagonal • H 3½–4 • SG 6.5–7.1

A lead phosphate chloride, variably colored greenish, orange, yellowish or brownish, pyromorphite forms in the oxidized zones of lead deposits.

prismatic apatite crystal

APATITE
Hexagonal/monoclinic
H 5 • SG 3.1–3.2

Apatite is the group name used for three structurally identical calcium phosphate minerals: fluorapatite, chlorapatite, and hydroxylapatite.

tabular torbernite crystal

METATORBERNITE
Tetragonal • H 2–2½ • SG 3.22

This copper uranyl phosphate hydrate is related to meta-autunite and occurs in similar settings. It is distinguished by its green color.

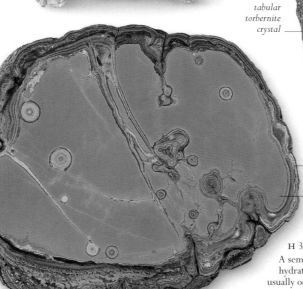

sliced nodule

waxy luster

VARISCITE
Orthorhombic
H 3½–4½ • SG 2.6–2.9

A semiprecious gemstone, this hydrated aluminum phosphate usually occurs as green, fine-grained masses in nodules, veins, or crusts.

TRIPLITE
Monoclinic • H 5–5½ • SG 3.5–3.9

Triplite is a phosphate of manganese, sometimes with iron and magnesium. It forms in granitic pegmatites.

WAVELLITE
Orthorhombic • H 3½–4 • SG 2.36

Wavellite is a rare aluminum phosphate hydroxide hydrate. Colorless, gray, or greenish, glassy, needlelike crystals form radiating aggregates on altered rock.

translucent mass of amblygonite

orange wavellite

AMBLYGONITE
Triclinic • H 5½–6 • SG 3.08

This rare lithium sodium aluminum fluophosphate mainly forms as masses, but crystals occur in Zimbabwe and Brazil.

VIVIANITE
Monoclinic • H 1½–2 • SG 2.68
Vivianite is hydrated iron phosphate. It commonly forms as clusters of dark prismatic crystals in altered iron deposits.

radiating acicular (needle-shaped) crystals

libethenite crystal

LIBETHENITE
Orthorhombic
H 4 • SG 4
Libethenite is a light to dark green copper phosphate hydroxide. It forms in the upper oxidized zone of copper deposits.

MONAZITE
Monoclinic • H 5–5½ • SG 4.6–5.4
Phosphate minerals containing either cerium, lanthanum, or neodymium are all referred to as monazite. They are mined for the various elements.

BRAZILIANITE
Monoclinic • H 5½ • SG 3
First discovered in Brazil, this sodium aluminum phosphate hydroxide is yellow or greenish and forms in cavities in granitic pegmatites.

LAZULITE
Monoclinic
H 5½–6 • SG 3.1
A relatively rare, semiprecious blue gemstone, this iron magnesium aluminum phosphate hydroxide occurs in metamorphic and igneous rocks.

bipyramidal crystal

VANADINITE
Hexagonal • H 3 • SG 6.88
Relatively rare lead vanadate chloride forms crystals in altered lead deposits. It is an important source of vanadium, used in steel alloys.

VANADATES

Vanadates are formed by the combination of metallic elements and the vanadate radical (VO_4). This group of minerals contains many rare examples, which tend to be dense and brightly colored. Vanadates often form when hydrothermal veins are altered by permeating fluids. Most vanadates have no commercial value; however, carnotite is an important source of uranium.

powdery crust on sandstone

CARNOTITE
Monoclinic • H 2 • SG 4.75
Generally occurring as powdery yellow crusts in uranium deposits, radioactive carnotite is a hydrated vanadate of potassium and uranium.

TYUYAMUNITE
Orthorhombic
H 2 • SG 3.3–3.6
This rare hydrated vanadate of calcium and uranium looks similar to carnotite, and also occurs in altered uranium deposits.

ADAMITE
Orthorhombic
H 3½ • SG 4.3–4.4
Zinc arsenate hydroxide occurs in altered arsenic and zinc deposits, sometimes as exceptional crystals.

ARSENATES

Arsenates are mostly rare minerals composed of metallic elements and the arsenate radical (AsO_3 or AsO_4). They generally have a fairly high specific gravity and low hardness. Many arsenates have bright colors—adamite is yellow or green, and clinoclase is green or blue. This group of minerals occurs in a variety of geological situations, but many arsenates occur in altered metal deposits.

ERYTHRITE
Monoclinic • H 1½–2½ • SG 3.18
Hydrated cobalt arsenate forms purple-pink crystals or coatings. Excellent examples occur in Canada and Morocco.

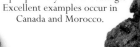

radiating clusters of clinoclase crystals

BAYLDONITE
Monoclinic
H 4½ • SG 5.6–5.7
This hydrated arsenate of copper, lead, and zinc is usually found as green or yellow crusts in altered hydrothermal veins.

olivenite crystals

OLIVENITE
Orthorhombic
H 3 • SG 4.4
Olivenite is copper arsenate hydroxide. It can be greenish, brownish, yellow, or gray and occurs in altered copper deposits.

quartz

CLINOCLASE
Monoclinic • H 2½–3 • SG 4.33
Clinoclase is a dark blue-green copper arsenate hydroxide that has a variety of forms in altered copper sulfide deposits.

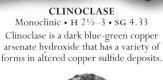

MIMETITE
Hexagonal • H 3½–4 • SG 7.0–7.3
Unusual, barrel-shaped crystals characterize this lead arsenate chloride, although other forms also exist. Mimetite occurs in altered lead deposits.

CHALCOPHYLLITE
Hexagonal/trigonal • H 2 • SG 2.7
Bright blue-green chalcophyllite is a copper aluminum arsenate sulfate hydrate. It forms in oxidized copper deposits.

SILICATES

Silicates are the most common and largest group of minerals. The fundamental building blocks are tetrahedra of silicon and oxygen (SiO_4) together with other elements. They are subdivided into six groups based on the arrangement of the silica tetrahedral. Some form as isolated tetrahedra (nesosilicates), some occur in pairs (sorosilicates), and others have a three-dimensional network of tetrahedra (tectosilicates). Some silicates form as chains of tetrahedra (inosilicates), while others form as sheets (phyllosilicates) or rings (cyclosilicates).

NESOSILICATES

HUMITE
Orthorhombic • H 6 • SG 3.24
A magnesium iron silicate fluohydroxide, humite generally occurs as yellow to orange granular masses in metamorphosed limestones and dolomites.

humite crust

NORBERGITE
Orthorhombic
H 6–6½ • SG 3.1–3.2
Norbergite mainly occurs as brownish yellow, white, or pink granular masses in metamorphic rocks. It is a magnesium silicate fluohydroxide.

DATOLITE
Monoclinic
H 5–5½ • SG 2.8–3.0
Datolite is a hydrous calcium boron silicate. Not very common, it is mainly found in veins or cavities in igneous rocks.

ANDRADITE
Cubic • H 6½–7 • SG 3.8
Yellowish green, brown, or black, andradite garnet is a calcium iron silicate. Cut gems are excellent at separating white light into colors.

MASSIVE DUMORTIERITE

DUMORTIERITE
Orthorhombic • H 8½ • SG 3.41
Dumortierite is a silicate of aluminum, iron, and boron. It usually forms fibrous aggregates of radiating crystals, but it can also be massive.

EUCLASE
Monoclinic
H 7½ • SG 3.05–3.10
Euclase is a beryllium aluminum silicate hydroxide. It may form white, colorless, green, or blue prismatic, striated crystals.

PYROPE
Cubic • H 7–7½ • SG 3.6
Pyrope garnet is a dark red magnesium aluminum silicate. It forms at high pressures in metamorphic and some igneous rocks.

KYANITE
Triclinic • H 5½–7 • SG 3.53–3.67
Kyanite is an aluminum silicate. Its bladed crystals in schist and gneiss formed at high pressures in the Earth.

rhombic crystal faces

ALMANDINE
Cubic • H 7–7½ • SG 4.3
The most common garnet, pinkish red almandine is an iron aluminum silicate. It is widely used as a gem.

green coloring due to vanadium

vitreous luster

red coloring due to iron

GREEN GROSSULAR

GROSSULAR
Cubic • H 6½–7 • SG 3.6
Grossular garnet is a silicate of calcium and aluminum that sometimes forms in marble. It comes in a wide range of colors.

RED GROSSULAR

OLIVINE
Orthorhombic
H 6½–7 • SG 3.27–4.32
Common in igneous rocks, nesosilicate minerals varying in composition from magnesium silicate to iron silicate are called olivine.

typical green color

translucent

wedge-shaped crystal

TITANITE TWINNED CRYSTALS

TITANITE
Monoclinic • H 5–5½ • SG 3.5–3.6
Variable in color, titanite is calcium titanium silicate. It is excellent at dispersing light—even better than diamond.

CRYSTALS IN MATRIX

pinkish brown topaz

TOPAZ
Orthorhombic • H 8 • SG 3.49–3.57
Topaz is aluminum silicate fluoride hydroxide. Crystal size is usually small, but a giant crystal weighing 596 lb (271 kg) is known from Brazil.

CHLORITOID
Monoclinic/triclinic • H 6½ • SG 3.6
Widespread in metamorphic and volcanic rocks, chloritoid is a dark green or black hydrous aluminum silicate of iron, magnesium, and manganese.

ANDALUSITE
Orthorhombic
H 6½–7½ • SG 3.13–3.16
Andalusite is an aluminum silicate. It occurs mainly in low-grade metamorphic rocks as coarse prismatic crystals with a square cross section.

prismatic crystal

ZIRCON
Tetragonal • H 7½ • SG 4.6–4.7
Zircon, or zirconium silicate, is a gem used extensively in jewelry. It is also the main source of the metal zirconium, used in nuclear reactors.

short, prismatic willemite crystal

WILLEMITE
Hexagonal/trigonal
H 5½ • SG 3.89–4.19
Willemite is zinc silicate. White, green, yellow, or reddish, and usually massive, it occurs in altered zinc deposits and metamorphosed limestone.

long, parallel, fibrous crystals

SILLIMANITE
Orthorhombic • H 6½–7½ • SG 3.23–3.27
Sillimanite is an aluminum silicate with long, slender crystals. Its composition is identical to andalusite, but it forms at higher temperatures and pressures.

EPIDOTE
Monoclinic
H 6–7 • SG 3.35–3.50
Epidote is an abundant mineral. Crystals of this hydrous calcium aluminum iron silicate are prismatic or tabular, green, and striated.

AXINITE
Triclinic
H 6–7 • SG 3.2–3.4
Axinite is a hydrous calcium iron manganese aluminum boron silicate, with axe-head-shaped crystals.

HEMIMORPHITE
Orthorhombic
H 4½–5 • SG 3.4–3.5
This hydrated silicate of zinc occurs in altered zinc deposits. It is very variable both in its color and form.

rounded crystal aggregates

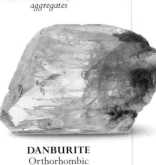

DANBURITE
Orthorhombic
H 7–7½ • SG 3
Danburite is a calcium boron silicate. Its variably colored crystals resemble topaz, but it can also be granular.

VESUVIANITE
Tetragonal/monoclinic
H 6–7 • SG 3.33–3.45
Vesuvianite is a hydrous calcium magnesium iron aluminum silicate, with fluorine. Green or yellow, it occurs in marble and igneous rocks.

»

CYCLOSILICATES

BENITOITE
Hexagonal
H 6–6½ • SG 3.64–3.68
This usually blue barium titanium silicate occurs in serpentinite and veins in schist. Gem-quality crystals come from California.

six-sided crystal

TOURMALINE
Hexagonal/trigonal
H 7–7½ • SG 3.0–3.2
Tourmaline is the name for a group of 11 hydrous boron silicate minerals with the same crystal structure but varying chemistry.

prismatic crystal

AQUAMARINE   EMERALD

BERYL
Hexagonal
H 6½–8 • SG 2.6–3.0
Beryllium aluminum silicate is both a source of beryllium and a gemstone. Gem varieties include emerald (green), sapphire (blue), and aquamarine (greenish blue).

MORGANITE
Hexagonal
H 7½–8 • SG 2.6–2.8
Morganite is a pink variety of beryl, colored by additional cesium or manganese. It forms tabular crystals in pegmatites.

SUGILITE
Hexagonal
H 5½–6½ • SG 2.7–2.8
This rare, hydrated silicate of potassium, sodium, iron, aluminum, lithium, and manganese occurs in metamorphic rock.

columnar, six-sided prismatic crystal

HELIODOR
Hexagonal
H 7½–8 • SG 2.6–2.8
Named after the Greek for "sun," heliodor is a yellow variety of beryl. Fine examples come from Russia.

rock groundmass

INOSILICATES

ACTINOLITE
Monoclinic
H 5–6 • SG 3.0–3.44
Actinolite is a more iron-rich, darker colored form of the amphibole tremolite. It is one of the asbestos minerals.

TREMOLITE
Monoclinic • H 5–6 • SG 2.9–3.2
A widespread amphibole, this hydrous silicate of calcium, magnesium, and iron forms in metamorphic rocks. It has been used as asbestos.

vitreous luster

radiating crystals

PECTOLITE
Triclinic • H 4½–5 • SG 2.74–2.88
This sodium calcium silicate hydroxide forms in cavities within basalt. It is common in the US, Canada, and England.

NEPHRITE
Monoclinic • H 6½ • SG 2.9–3.4
This very tough, cream to dark green form of the amphiboles tremolite and actinolite is commonly known as jade.

AEGIRINE
Monoclinic • H 6 • SG 3.55–3.60
This brown, green, or black pyroxene is a sodium iron silicate. It forms in metamorphic and dark igneous rocks.

HORNBLENDE
Monoclinic • H 5–6 • SG 3.28–3.41
Common in igneous and metamorphic rocks, dark amphibole or hornblende is a dark, hydrous silicate of calcium, magnesium, iron, and aluminum, with fluorine.

long, prismatic crystal

fibrous mass

striated crystal faces

WOLLASTONITE
Triclinic • H 4½–5 • SG 2.87–3.09
This calcium silicate, found in marble and other metamorphic rocks, is used in ceramics, paints, and as a replacement for asbestos.

RHODONITE
Triclinic
H 5½–6½ • SG 3.57–3.76
Rose-red or pink rhodonite is manganese calcium silicate, occurring as crystals, masses, and grains. It is popularly used in jewelry making.

slender prismatic crystal

SPODUMENE
Monoclinic • H 6½–7½ • SG 3.0–3.2
This pyroxene mineral is a lithium aluminum silicate. Some enormous crystals have been found; the largest weighed almost 100 tons.

prismatic diopside crystal

quartz

DIOPSIDE
Monoclinic • H 5½–6½ • SG 3.22–3.38
This pyroxene is a generally green-colored calcium magnesium silicate. It occurs in metamorphic and igneous rocks.

richterite crystal

RICHTERITE
Monoclinic
H 5–6 • SG 2.97–3.13
The amphibole richterite is a hydrous silicate of sodium, calcium magnesium, and iron. It occurs in metamorphosed limestones and igneous rocks.

PIGEONITE
Monoclinic
H 6 • SG 3.2–3.5
This brown to purplish black common pyroxene is a magnesium iron calcium silicate. It occurs in igneous rocks and in meteorites.

AUGITE
Monoclinic
H 5½–6 • SG 3.23–3.52
Augite is the most common pyroxene. This silicate of calcium, sodium, magnesium, iron, titanium, and aluminum occurs in igneous and metamorphic rocks.

ASTROPHYLLITE
Triclinic • H 3 • SG 3.3–3.4
This complex hydrous silicate of potassium, sodium, iron, manganese, and titanium occurs in gneiss and cavities in igneous rocks.

polished jadeite

JADEITE
Monoclinic
H 6–7 • SG 3.24
One of two carving materials commonly referred to as jade, this pyroxene mineral is a sodium aluminum iron silicate.

long prismatic crystal

rock groundmass

long, striated crystals

RIEBECKITE
Monoclinic • H 5 • SG 3.32–3.38
This amphibole is a hydrous sodium iron magnesium silicate, occurring in igneous rocks. The variety crocidolite, or "blue asbestos," occurs in metamorphosed ironstone.

PHYLLOSILICATES

spherical mass
of radiating
crystals

CLINOCHLORE
Monoclinic
H 2–2½ • SG 2.63–2.98
This hydrous silicate of iron,
magnesium, and aluminum forms
as green tabular crystals. It occurs
in a variety of rock types.

tabular
crystal

radiating
crystal groups

PREHNITE
Orthorhombic
H 6–6½ • SG 2.90–2.95
Prehnite is a hydrous silicate
of calcium and aluminum,
sometimes used as a gem. It
occurs in cavities in basalt.

OKENITE
Triclinic • H 4½–5 • SG 2.3
This hydrated calcium silicate
has fibrous or bladelike crystals,
usually white or tinted blue or
yellow. It occurs in basalt.

MUSCOVITE
Monoclinic
H 2½–4 • SG 2.77–2.88
Muscovite, or common mica,
is a potassium aluminum
aluminosilicate hydroxide, with
fluorine. It is very common in
metamorphic rocks and granite.

prismatic
crystal

PETALITE
Monoclinic
H 6–6½ • SG 2.3–2.5
Petalite is a lithium
aluminum silicate.
Crystals are usually
gray-white and in
aggregates. It is mined
for the lithium.

spherical crystal
aggregate

CAVANSITE
Orthorhombic • H 3–4 • SG 2.2–2.3
Cavansite is a hydrated calcium
vanadium silicate. It is blue
or greenish blue, and
occurs in cavities
in basalt.

PHLOGOPITE
Monoclinic
H 2–2½ • SG 2.76–2.90
The colorless, yellow, or
brown mica phlogopite is
a potassium magnesium
aluminosilicate hydroxide.

shiny tabular
crystal

SEPIOLITE
Orthorhombic • H 2–2½ • SG 2
This pale-colored clay mineral, a
hydrated magnesium silicate, usually
occurs as earthy masses in altered rocks.
It is used for ornamental carving.

typical blue
coloring

LEPIDOLITE
Monoclinic • H 2½–3 • SG 2.8–3.3
Lepidolite is the term used for mica minerals
that are potassium lithium aluminum
aluminosilicate hydroxides, with fluorine.

TECTOSILICATES

prismatic crystal

prismatic
crystal

CITRINE
Hexagonal/trigonal
H 7 • SG 2.7
Citrine is a yellow to
brownish variety of
quartz. It resembles
topaz and is often used
as a gemstone.

colorless
quartz

SMOKY QUARTZ
Hexagonal/trigonal • H 7 • SG 2.7
Smoky quartz is a brown variety
of quartz, or silicon dioxide. It
occurs in igneous rocks and
hydrothermal veins.

ROSE QUARTZ
Hexagonal/trigonal • H 7 • SG 2.7
Rose quartz is a prized, translucent pink
variety of quartz. Good crystals are rare;
more usually it forms massive aggregates.

MILKY QUARTZ
Hexagonal/trigonal
H 7 • SG 2.7
This very common, milky
white variety of quartz occurs
in all types of rocks and in
hydrothermal veins.

AMETHYST
Hexagonal/trigonal
H 7 • SG 2.7
Amethyst is a purple
variety of quartz, prized
since ancient times. It is
found in hydrothermal
veins and in lava cavities.

slender crystal

ZINNWALDITE
Monoclinic • H 2½–4 • SG 2.9–3.0
This brown, gray, or green mica is potassium lithium iron aluminum aluminosilicate hydroxide, with fluorine.

CHRYSOCOLLA
Orthorhombic
H 2–4 • SG 2.0–2.4
This blue or greenish blue hydrated silicate of copper and aluminum forms in altered copper deposits. Crystals are rare.

VERMICULITE
Monoclinic
H 1½ • SG 2.3
This green or yellow clay mineral often occurs where micas have been altered. It is a hydrated silicate of magnesium, iron, and aluminum.

GLAUCONITE
Monoclinic • H 2 • SG 2.4–2.95
A mica, glauconite is a potassium sodium magnesium aluminum iron aluminosilicate hydroxide. It occurs in marine sedimentary rocks.

tabular biotite crystal

CHRYSOTILE
Monoclinic • H 2½ • SG 2.53
Chrysotile is a hydrous silicate of magnesium, forming fibrous, silky white crystals in serpentinite rock. It is the most abundant of the asbestos form minerals.

BIOTITE
Monoclinic • H 2½–3 • SG 2.7–3.4
Biotite, or black mica, is a potassium iron magnesium aluminosilicate hydroxide, with fluorine. It is abundant in igneous and metamorphic rocks.

TALC
Triclinic/monoclinic
H 1 • SG 2.58–2.83
The softest of minerals, white, gray, or greenish talc is magnesium silicate hydroxide. Its many uses include toiletries, paint, and ceramics.

ALLOPHANE
Amorphous
H 3 • SG 2.8
A clay mineral, this aluminosilicate hydrate is an alteration product of feldspars and other minerals. It forms crusty masses.

PYROPHYLLITE
Triclinic/monoclinic
H 1–2 • SG 2.65–2.90
This aluminum silicate hydroxide, of variable form and color, occurs in low-grade metamorphic rocks. It has good insulating properties.

vitreous luster

prismatic crystals

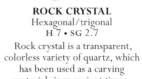

ROCK CRYSTAL
Hexagonal/trigonal
H 7 • SG 2.7
Rock crystal is a transparent, colorless variety of quartz, which has been used as a carving material since ancient times.

JASPER
Hexagonal/trigonal • H 7 • SG 2.7
Jasper is a variety of chalcedony, or microcrystalline quartz, and is used for jewelry. It is opaque, and impurities give a range of colors.

white quartz vein

AGATE
Hexagonal/trigonal
H 7 • SG 2.7
Forming in cavities in lavas, agate is a type of chalcedony. It is characterized by concentric color bands caused by impurities.

»

» TECTOSILICATES

translucent slab

alternating bands of black and white

CHALCEDONY
Hexagonal/trigonal
H 7 · SG 2.65
Chalcedony is microcrystalline quartz, or silicon dioxide. Pure chalcedony is white. It forms in veins and cavities of many rock types.

ONYX
Hexagonal/trigonal
H 7 · SG 2.7
Onyx is the striped, semiprecious variety of chalcedony. It is not particularly common; notable locations are India and South America.

CARNELIAN
Hexagonal/trigonal
H 7 · SG 2.7
Carnelian is a variety of chalcedony colored red to orange by iron oxide. The finest quality carnelian comes from India.

BLOODSTONE
Hexagonal/trigonal · H 7 · SG 2.7
Bloodstone is a variety of chalcedony colored dark green by traces of iron silicates. Flecks of red jasper throughout resemble blood.

CHRYSOPRASE
Hexagonal/trigonal · H 7 · SG 2.7
Chrysoprase is a variety of chalcedony containing nickel, which imparts a pale green color. It is the most valuable of the chalcedonies.

OPAL
Amorphous
H 5½–6½ · SG 1.9–2.3
Treasured opal is hydrated silicon dioxide, occurring as nodules, encrustations, or masses in most rock types. Impurities impart a variety of colors.

precious opal

streaks of yellow "potch" (common) opal

ironstone matrix

yellow cancrinite

prismatic crystal

CANCRINITE
Hexagonal/trigonal
H 5–6 · SG 2.42–2.51
The feldspathoid cancrinite is a variously colored hydrated sodium calcium aluminosilicate carbonate.

SCAPOLITE
Tetragonal
H 5½–6 · SG 2.50–2.78
The name scapolite encompasses a series of complex sodium calcium silicates, occurring mainly in metamorphic rocks. Scapolites have value as gems.

MICROCLINE
Triclinic
H 6–6½ · SG 2.55–2.63
A very common alkali feldspar, this potassium aluminosilicate is usually white or pinkish. The green variety is called amazonstone.

ANORTHITE
Triclinic
H 6–6½ · SG 2.74–2.76
This uncommon plagioclase feldspar is a calcium aluminosilicate. It forms pale-colored crystals, grains, or masses.

HEULANDITE
Monoclinic
H 3½–4 · SG 2.1–2.2
This zeolite is a hydrated sodium calcium aluminosilicate. It is used in petroleum refining as a molecular sieve.

SCOLECITE
Monoclinic
H 5 • SG 2.27
This zeolite is a hydrated calcium aluminosilicate. Usually colorless or white, it is common in igneous and metamorphic rocks.

long, slender, needlelike crystals

ANDESINE
Triclinic
H 6−6½ • SG 2.66−2.68
A plagioclase feldspar, andesine is a gray or white sodium calcium aluminosilicate. It is widespread in igneous rocks.

massive sodalite

SODALITE
Cubic
H 5½−6 • SG 2.14−2.40
The feldspathoid mineral sodalite is a sodium aluminum silicate chloride. Rare crystals have been found in Canada.

STILBITE
Monoclinic
H 3½−4 • SG 2.09−2.20
This widespread zeolite is a hydrated sodium calcium potassium aluminosilicate. It forms sheaflike crystals in a variety of rock types.

HARMOTOME
Monoclinic
H 4½ • SG 2.41−2.50
This widespread zeolite is a pale hydrated barium potassium aluminosilicate. It occurs in hydrothermal veins and volcanic rocks.

NATROLITE
Orthorhombic
H 5−5½ • SG 2.20−2.26
One of the most widespread zeolites, this hydrated sodium aluminosilicate occurs in cavities in basalts, and in hydrothermal veins.

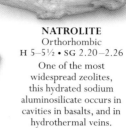

HYALOPHANE
Monoclinic
H 6−6½ • SG 2.6−2.8
A relatively rare barium feldspar, this potassium barium aluminosilicate is colorless, white, yellow, or pink.

ANALCIME
Cubic
H 5−5½ • SG 2.22−2.29
This pale zeolite, a hydrated sodium aluminosilicate, occurs in igneous, metamorphic, and some sedimentary rocks.

lazurite crystal

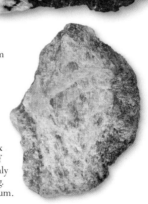

ALBITE
Triclinic
H 6−6½ • SG 2.60−2.63
Considered an alkali and plagioclase feldspar, albite is a pale-colored sodium aluminosilicate. It is an abundant mineral.

ANORTHOCLASE
Triclinic
H 6−6½ • SG 2.56−2.62
An alkali feldspar, anorthoclase is a sodium potassium aluminosilicate occurring as prismatic or tabular crystals.

calcite matrix

LAZURITE
Cubic
H 5−5½ • SG 2.4−2.5
The intense blue feldspathoid mineral lazurite is a sodium calcium aluminosilicate sulfate. It is the main mineral in lapis lazuli gems.

THOMSONITE
Orthorhombic
H 5−5½ • SG 2.25−2.40
A pale-colored zeolite, thomsonite is a hydrated aluminosilicate of sodium and calcium, widespread in cavities in basalts.

LAUMONTITE
Monoclinic
H 3−4 • SG 2.2−2.4
A widespread and common zeolite, laumontite is a hydrated calcium aluminosilicate. It occurs in igneous, metamorphic, and sedimentary rocks.

short, prismatic orthoclase crystal

POLLUCITE
Cubic
H 6½−7 • SG 2.7−3.0
A rare zeolite, this complex hydrated aluminosilicate of cesium and sodium commonly has other elements in it (e.g. calcium). It is a source of cesium.

ORTHOCLASE
Monoclinic • H 6−6½ • SG 2.55−2.63
Orthoclase alkali feldspar is a potassium aluminosilicate. It is a major component of many igneous and metamorphic rocks.

CHABAZITE
Hexagonal/trigonal
H 4−5 • SG 2.05−2.16
This common zeolite is a hydrated sodium calcium aluminosilicate. Crystals are colorless, white, yellow, or pink.

hairlike tufts of mesolite crystals

MESOLITE
Monoclinic • H 5 • SG 2.2−2.3
This white or colorless zeolite occurs in igneous and metamorphic rocks. It is a hydrated sodium calcium aluminosilicate.

HAÜYNE
Cubic • H 5½−6 • SG 2.5
A feldspathoid mineral, haüyne is a sodium calcium aluminosilicate, with sulfate and chlorine. It occurs mainly in silica-deficient volcanic rocks.

pseudocubic (almost cubic) rhombohedral crystal

ROCKS

Made of different mixtures of minerals, rocks form Earth's solid crust. The epitome of strength and solidity, they are actually in a state of constant change, with rock being destroyed and reformed over immense periods of time. Rocks are classified into three main groups, depending on how they form.

Igneous rocks can be formed by the cooling of magma deep underground, like this granite, or by volcanic eruption.

Sedimentary rocks such as red chalk are formed by the erosion of existing rocks and the recrystallization of their minerals.

Metamorphic rocks such as muscovite schist occur when changes in pressure, temperature, or both alter the mineralogy.

The world's oldest known rocks, from Canada's Northwest Territories, have existed for about four billion years. But most rocks are far younger than this. The chalk cliffs facing the English Channel date back to the Cretaceous Period, which ended 65 million years ago (see timeline, below), and the European Alps are younger still. Even in the Grand Canyon, the oldest rocks date back two billion years—less than half the lifetime of the planet as a whole. The reason for this is that Earth is tectonically active, with new rock being formed by the planet's internal heat. At the same time, existing rocks are broken down, in a endless cycle that started when Earth's crust first formed.

ROCK GROUPS

Rocks are classified into three overall groups—igneous, sedimentary, or metamorphic—reflecting the different ways in which they are formed. Igneous rock is created by volcanic heat, from molten magma in the Earth's mantle. The most common kind, a black crystalline rock called basalt, is produced by volcanic eruptions on Earth's surface. Basalt forms most of the seafloor. Volcanism can also create plutonic rock, which cools and solidifies beneath the surface in huge masses called batholiths. This is how most of the world's granite originates.

Sedimentary rocks form on Earth's surface. Their key feature is their layers, or strata, built up over long periods of time. Some sedimentary rocks—such as sandstone and shale—form when existing rocks erode, releasing particles that are then washed or blown away, forming rock elsewhere. Others, such as rock salt and rock gypsum, are created when saltwater evaporates, leaving behind its dissolved minerals, known as evaporite deposits. Sedimentary rock can also have a biological origin: chalk and limestone are formed from the microscopic skeletons of marine organisms, whereas coal is derived from the remains of plants, compressed over millions of years.

Metamorphism takes place deep beneath Earth's surface, when rocks are altered by heat, pressure, or both. For example, marble is created when limestone is heated by lava or by magma. Unlike limestone, marble is unlayered, and its fine texture enables it to be cut without splitting apart—a quality prized in sculpture. But if metamorphism is intense enough, rock turns back into molten magma. This completes the last link in the cycle, as solid rock is finally destroyed.

GRAND CANYON, COLORADO >
This view of the Grand Canyon shows the almost horizontal layering of sedimentary rocks and the effects of river erosion.

ROCK CYCLE

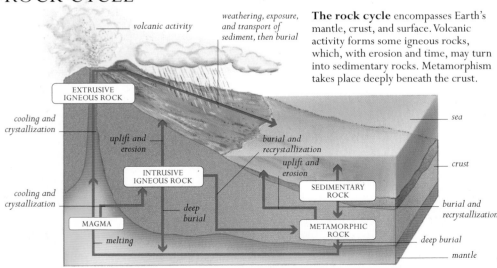

volcanic activity

weathering, exposure, and transport of sediment, then burial

The rock cycle encompasses Earth's mantle, crust, and surface. Volcanic activity forms some igneous rocks, which, with erosion and time, may turn into sedimentary rocks. Metamorphism takes place deeply beneath the crust.

EXTRUSIVE IGNEOUS ROCK

cooling and crystallization

uplift and erosion

burial and recrystallization

uplift and erosion

sea

crust

INTRUSIVE IGNEOUS ROCK

SEDIMENTARY ROCK

cooling and crystallization

MAGMA

deep burial

METAMORPHIC ROCK

burial and recrystallization

melting

deep burial

deep burial

mantle

GEOLOGICAL TIMELINE

Era	Period from modern to Precambrian strata	Epoch	Start (mya)
Cenozoic	Quarternary	Holocene	0.01
		Pleistocene	1.8
	Neogene	Pliocene	5.3
		Miocene	23
	Paleogene	Oligocene	34
		Eocene	55
		Paleocene	65
Mesozoic	Cretaceous		145
	Jurassic		199
	Triassic		251
Paleozoic	Permian		299
	Carboniferous (split into Lower Mississippian and Upper Pennsylvanian in USA)		359
	Devonian		416
	Silurian		433
	Ordovician		488
	Cambrian		542
Precambrian	Formation of the Earth		4,554

IGNEOUS ROCKS

Rocks that solidify from a molten state are called igneous and are broadly divided into extrusive (volcanic) rocks and intrusive rocks. Extrusive rocks form from lava on the Earth's surface, while intrusive rocks form underground, from magma. Lava and magma are rich in silica and metallic elements. As they cool, minerals such as feldspars form, along with quartz. Different combinations of these minerals make up many of the igneous rocks.

RHYOLITE
This fine-grained, pale-colored lava contains much quartz, mica, and feldspar. Often banded, it frequently includes visible phenocrysts (larger crystals).

vesicle (gas bubble cavity)

BASALT
A dark, fine-grained volcanic rock, basalt is the most common rock forming the oceanic crust.

VESICULAR BASALT
Dominated by the minerals plagioclase, pyroxene, and olivine, this lava is dark-colored, with numerous former gas cavities called vesicles.

dark, fine-grained rock

BANDED RHYOLITE
Similar in composition to granite, rhyolite also has tiny glass crystals because it forms from rapidly cooled lava. Banding shows the direction of magma flow.

AMYGDALOIDAL BASALT
Amygdaloidal means almond-shaped and refers to the gas bubbles found in basaltic lavas, which are typically infilled with secondary minerals such as zeolite, carbonate, and agate.

PAHOEHOE
Common in Hawaii, pahoehoe is named after the Hawaiian word *hoe*, meaning "to swirl." This refers to the ropy-textured, glassy surface of this basaltic lava.

PORPHYRITIC BASALT
This dark-colored rock has large crystals, typically of olivine or plagioclase, set in a fine-grained matrix.

PUMICE
Formed from frothy lava, pumice contains small crystals of feldspar in a glassy matrix. It is pale-colored and so highly porous that it floats.

PÉLÉ'S HAIR
Named after a Hawaiian goddess, and composed of numerous thin, brown, glassy strands, Pélé's hair is formed by lava sprays being blown in the wind.

IGNIMBRITE
A fine-grained, glassy, pale-colored volcanic tuff, ignimbrite often shows banding caused by the flow of lava when molten.

LITHIC TUFF
This rock contains fragments of previously formed rock in a fine, glassy matrix. Usually pale, lithic tuff is formed by violent volcanic eruptions.

SPINDLE BOMB
Molten, low-viscosity basaltic magma blasted through the air can assume aerodynamic forms such as this one. Such masses cool on the surface to form "bombs."

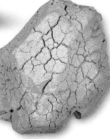

BREADCRUST BOMB
This variety of volcanic bomb is characterized by a cracked crust, due to continued expansion of the interior after the exterior has solidified.

AGGLOMERATE

Composed of relatively large rock fragments set in a finer matrix, agglomerate forms after volcanic explosions.

PITCHSTONE

This glassy, dense volcanic rock has variable composition and color, and a pitchlike, waxy, resinous sheen.

PORPHYRITIC TRACHYTE

This rock has a complex mineralogy of alkali feldspar, quartz, mica, pyroxene, and hornblende. Its matrix has large crystals.

PORPHYRITIC ANDESITE

Commonly composed of plagioclase feldspar, pyroxene, and amphibole, porphyritic andesite contains large crystals set in its fine-grained matrix.

DACITE

A light- to medium-colored rock with fine to coarse grains, dacite is mainly made of plagioclase feldspar and quartz, with pyroxene, biotite mica, and hornblende.

ANDESITE

Named after the Andes mountains, andesite is common in many of the world's subduction-related volcanic arcs—chains of islands or mountains formed by plate movement. It is typically about 60 percent silica by weight.

AMYGDALOIDAL ANDESITE

Brown, mold, purple, or red, this is a fine-grained volcanic rock containing infilled gas bubbles called amygdales.

SPILITE

This fine-grained, brownish, altered volcanic rock contains augite and plagioclase feldspar. It forms by alteration of basalt lava in contact with sea water.

RHOMB PORPHYRY

This igneous rock is characterized by large crystals of feldspar with a rhomb-shaped cross section set in its dark matrix.

TRACHYTE

Trachyte is a fine-grained volcanic group of rocks containing alkali feldspar and dark-colored mafic minerals such as biotite, hornblende, and pyroxene. These rocks are usually rough to the touch.

curved fracture

OBSIDIAN

Typically very dark, obsidian is formed by rapid cooling of highly viscous, hot, rhyolitic lava before individual minerals have had time to form crystals. This gives it a glassy texture. Obsidian has been used in jewelry since ancient times.

pale patch of devitrified glass

SNOWFLAKE OBSIDIAN

The "snowflakes" in this glassy, black volcanic rock, with a high silica content, are areas where minerals have crystallized from the glass.

patch of red garnet

GARNET PERIDOTITE

Peridotite is the main rock of the Earth's upper mantle. This dense, greenish variety is made of dark minerals including garnet, olivine, clinopyroxene, and orthopyroxene.

»

» IGNEOUS ROCKS

light plagioclase feldspar

GABBRO
This dark, plutonic rock contains plagioclase, pyroxene, and olivine. Coarse-grained, it forms from the slow cooling of basaltic magma at depth.

OLIVINE GABBRO
Gabbro is a dark, coarse-grained rock with a lot of pyroxene and plagioclase. Olivine occurs in significant amounts in this variety.

LAYERED GABBRO
Coarse-grained and dark-colored, this type of gabbro shows banding caused by the settling of minerals of different density in the magma.

MICROGRANITE
This granite—with fine to very fine grains, often with a porphyritic texture—occurs in sills and dykes, which are sheets of intrusive igneous rock.

PINK MICROGRANITE
WHITE MICROGRANITE

PORPHYRITIC GRANITE
Large crystals are set into the matrix of this medium-grained rock, which is composed mainly of quartz, mica, and feldspar.

BOJITE
A dark-colored igneous rock, bojite is a general term for a hornblende gabbro. It is coarse-grained and forms from magma.

KIMBERLITE
Dark-colored and coarse-grained, kimberlite is an ultramafic rock with a very low silica content. Of variable composition, it is the world's main source of diamonds.

black tourmaline

pink orthoclase feldspar

GRANODIORITE
This is the most common intrusive igneous rock in the continental crust. It is over 65 percent plagioclase.

PEGMATITE
Pegmatites are very coarse-grained rocks, formed from residual liquid magma after most of a granite intrusion has cooled and crystallized. Some pegmatites are an important source of gemstones.

LAMPROPHYRE
Lamprophyre's matrix is fine-grained and studded with distinct crystals of the water-bearing minerals mica and amphibole. It has a glistening texture.

NEPHELINE SYENITE
Coarse-grained, pale-colored, and composed of feldspar, mica, and hornblende, this rock also contains nepheline but no quartz.

SYENITE
Gray or pinkish, syenite is a plutonic rock that occurs in large intrusions. Coarse-grained, it has feldspar, mica, and hornblende, and little or no quartz.

black
biotite
mica

GRAPHIC GRANITE
This coarse-grained
rock contains quartz
and feldspar, whose
intergrowth results in a
graphic texture, vaguely
resembling runic writing.
It also contains mica.

gray quartz

HORNBLENDE GRANITE
Granite typically contains
quartz, feldspar, and mica.
This variety also contains
hornblende, a member of the
amphibole mineral group.

**PORPHYRITIC
GRANITE**
A pale-colored rock
composed of feldspar,
quartz, and mica,
porphyritic granite has
large, well-formed crystals
set into the matrix.

DOLERITE
Typically occurring
in sills and dykes,
dolerite is a dark,
medium-grained
rock composed of
plagioclase, pyroxene,
and iron oxides.

GRANITE
Coarse-grained and of varying color,
granite contains more than 10 percent
quartz. Typically a tough rock, resistant
to erosion, it is often used in building.

FELSITE
Fine-grained felsite forms
as sheetlike intrusions in
sills and dykes. Pale-colored,
it is composed mainly of
feldspar and quartz.

ADAMELLITE
This medium-grained
granite forms at depth with
crystals of quartz, mica,
and feldspar. One-third
to two-thirds of the total
feldspar is plagioclase.

ANORTHOSITE
Pale-colored,
anorthosite is composed
mainly of large crystals
of plagioclase feldspar.
It can also contain
olivine and augite.

DUNITE
Composed only of olivine, this
rock is dark green or brown, with a sugary
texture. It usually includes a little chromite.

light plagioclase
feldspar

green olivine
crystals

PERIDOTITE
A dark, dense rock
composed mainly of olivine
and pyroxene, peridotite is
coarse-grained and forms
slowly at great depth.

dark
pyroxene

LARVIKITE
A form of syenite,
larvikite is blue-black
and made of large
amounts of sodic feldspar
that can have a glittering
blue schiller effect.

DIORITE
Composed of plagioclase feldspar, with amphibole
and pyroxene, diorite contains little or no quartz. It
was used for ornamental purposes in ancient Egypt.

METAMORPHIC ROCKS

When existing rocks are subjected to heat or pressure or both within the Earth, they are transformed into different assemblages of minerals. Contact metamorphism occurs when intense, localized heat radiated by a body of igneous rocks causes the surrounding rocks to recrystallize. Regional metamorphism takes place over wide areas, at considerable depth, as a result of intense heat and pressure. The Earth's movements can result in rocks being pulverized by dynamic metamorphism.

GRANULITE
Formed at very high temperatures and pressures, granulite is dark, coarse-grained, and rich in pyroxene, garnet, mica, and feldspar.

GARNET SCHIST
The red garnets in this variety of schist indicate that it developed at relatively high temperatures and pressures deep within the continental crust.

CRENULATED SCHIST
Schist is characterized by its parallel planes of similarly oriented minerals within the rock. This corrugated pattern is called crenulation.

streaky pattern

MYLONITE
When rocks are pulverized deep within a fault zone, the rock dust and fragments produced form a fine-grained rock called mylonite.

SLATE
Dark and very fine-grained, slate is a compact rock with parallel cleavage planes. It is formed by low-pressure metamorphism.

KYANITE SCHIST
Mainly composed of feldspar, mica, and quartz, this schist also contains crystals of the blue mineral kyanite.

HALLEFLINTA
Originally volcanic tuff, rhyolite, or quartz porphyry, halleflinta is fine-grained, pale-colored, and rich in quartz. It is a type of hornfels.

SPOTTED SLATE
This is a dark, fine-grained rock characterized by black spots (porphyroblasts) of minerals, such as cordierite and andalusite.

BIOTITE SCHIST
Formed at relatively high pressures and temperatures, biotite schist contains feldspar, quartz, and a lot of biotite mica, which makes it dark.

MUSCOVITE SCHIST
A typical schist with pale, glittery muscovite mica, this rock also contains quartz and feldspar.

MIGMATITE
Formed at the highest temperatures and pressures, coarse-grained, folded migmatite contains bands of dark basaltic minerals and pale granitic minerals.

tubular structure

pink calcite

SKARN
Formed when carbonate rocks are altered by high-temperature contact metamorphism, skarn contains minerals rich in calcium, magnesium, and iron.

FULGURITE
When lightning strikes in deserts or on beaches, sand melts into small, tube-shaped structures. These structures are called sand fulgurites.

QUARTZITE
With a high percentage of quartz, this rock is harder than most metamorphic rocks. It is formed from sandstone altered at high temperature.

bladed crystal

CORDIERITE HORNFELS
Dark and splintery, cordierite hornfels forms from the heat of nearby igneous intrusions. It is fine- to medium-grained.

GNEISS
A medium- to coarse-grained rock formed at very high temperatures and pressures, gneiss is recognized by its alternating dark and light crystalline layers.

AUGEN GNEISS
Gneiss contains quartz, feldspar, and mica, often in parallel bands. Augen gneiss has crystals resembling lens-shaped eyes (*augen* in German).

GARNET HORNFELS
Hornfels is a tough, dark, flinty rock formed by heat next to an igneous intrusion. Reddish garnets occur in this particular variety.

CHIASTOLITE HORNFELS
Hornfels forms at very high temperatures near magma intrusions. Bladed crystals of pale chiastolite give this particular variety its name.

PYROXENE HORNFELS
Fine- to medium-grained, tough, and flinty, this variety of hornfels contains quartz, mica, and pyroxene. It forms next to igneous intrusions.

FOLDED GNEISS
Gneiss becomes plastic and folded at great depth. The dark bands are hornblende-rich and the paler ones are quartz and feldspar.

AMPHIBOLITE
Formed by moderate heat and varying pressure deep in the crust, this coarse-grained rock has an abundance of hornblende and plagioclase, as well as other minerals.

ECLOGITE
Made of two main minerals, green omphacite pyroxene and red garnet, eclogite is coarse-grained and forms at very high temperatures and pressures.

PHYLLITE
Phyllite is formed at lower temperatures and pressures than schist, but higher than slate. It is fine-grained and splits into slabs with a characteristic sheen.

GRANULAR GNEISS
This gneiss has a granular texture with equal-sized grains. It has dark bands of hornblende and biotite, and pale bands of quartz and feldspar.

marble fragment

MARBLE BRECCIA

clearly visible coarse-grain crystals

GRAY MARBLE

MARBLE
Formed by contact or regional metamorphism, marble is rich in calcite, often with colorful veins of other minerals. It is prized as a carving material.

GREEN MARBLE

blotchy texture

SERPENTINITE
Commonly banded, blotched, or streaked, serpentinite is a dense, but soft, metamorphic rock derived from peridotite. It is found in convergence zones between tectonic plates.

SEDIMENTARY ROCKS

These rocks result from deposition of sediment laid down by wind, water, and ice on the Earth's surface, and its subsequent burial. Sedimentary rocks are characterized by stratification or bedding, and they can contain fossils. They are broadly divided into clastic rocks—made up of fragments of rock and subdivided by grain size—and chemical and biochemical rocks, which are grouped according to their chemistry.

MILLET-SEED SANDSTONE
This rock is medium-grained, with a reddish iron oxide coating. Its well-rounded, equal-sized quartz grains have been shaped by the wind.

GREENSAND
Colored greenish by the silicate mineral glauconite, greensand is a quartz-rich sandstone formed in the ocean.

MICACEOUS SANDSTONE
Rich in quartz, this sandstone also contains glittery flakes of mica. It usually has medium-sized grains.

LIMONITIC SANDSTONE
This rock is colored red-brown or yellowish by the iron oxide mineral limonite, which coats its medium to fine quartz grains.

iron oxide gives red color

quartz grains colored by iron oxides

SANDSTONE

SANDSTONE
This typically occurs as stratified layers of sand-sized particles, held together by various mineral cements that impart different colors. Most are quartz-rich.

RED SANDSTONE

ROCK SALT
Formed of crystalline halite, rock salt is brownish and may contain clay. It is soluble and soft, and has a distinct taste.

ROCK GYPSUM
A crystalline rock associated with potash rock and created when salt water evaporates, rock gypsum is pale-colored, often fibrous, and very soft and soluble.

layered structure

BOULDER CLAY
Gray or brownish colored, boulder clay or till has a fine clay matrix filled with angular and rounded rock fragments.

CLAYSTONE
Of varying color, this very fine-grained rock is composed mainly of silicate clay minerals such as kaolinite—mostly derived from the weathering of feldspar.

TRAVERTINE
A pale-colored and often layered rock, travertine is virtually pure calcite. It is formed around hot springs and volcanic vents.

bands of hematite and chert

OOLITIC IRONSTONE
This rock is composed of small, rounded sedimentary grains (ooliths) of iron minerals, such as siderite, cemented by other iron minerals, as well as calcite and quartz.

LOESS
A clay with very fine, dustlike grains lifted by the wind from dry land surfaces, loess is crumbly and lacks obvious layering.

BANDED IRON FORMATION
This marine deposit has alternating bands of black hematite and red chert. It is one of the best ores of iron.

TUFA
This porous rock is formed by the precipitation of carbonate minerals from ambient temperature bodies of water, such as hot springs.

GRAY ORTHOQUARTZITE

ORTHOQUARTZITE
Rarely containing fossils, this rock is composed almost entirely of silica-cemented quartz grains. Orthoquartzite is also known as quartz arenite.

PINK ORTHOQUARTZITE

MANGANESE NODULE
Rich in valuable transition metals such as copper, manganese nodules are rounded, black concretions that form on the deep ocean floor.

rounded nodule

LIGNITE

JET

LIGNITE
Lignite is brown coal with a lower carbon content than bituminous coal. Jet is a hard, black, lustrous form of lignite, and will take a high polish.

ANTHRACITE
The purest form of coal, anthracite is black and shiny, with a vitreous surface. It has curved edges when broken.

BITUMINOUS COAL
This sedimentary rock has a lower carbon content than anthracite. Brittle and dull, it is the most abundant type of coal.

internal radiating structure

pale calcite filling cracks (septa)

SEPTARIAN NODULE
This type of concretion occurs in sedimentary rocks as individual, rounded masses, cemented by quartz or calcite. Internally they have septa (from the Latin *septum*, meaning "barrier").

PYRITE NODULE
Gray or black internally and brassy yellow outside, these nodules occur in shale and clay, and are composed entirely of pyrite.

»

NUMMULITIC LIMESTONE

A marine bryozoan, *Nummulites* is the main fossil in this rock. The cement is calcite, originally lime mud.

CORAL LIMESTONE

This rock is a mass of fossilized corals cemented by fine-grained calcite. It is gray to white or brownish.

fossilized crinoid, or sea lily, stem

CRINOIDAL LIMESTONE

Crinoids are echinoderms that are attached to the seabed by a flexible stem. Crinoidal limestone is a mass of broken stems cemented by hardened lime mud.

FRESHWATER LIMESTONE

This limestone is a pale, calcite-rich rock with some quartz and clay. It contains fossils of freshwater-dwelling organisms, which indicate where the rock formed.

flattened pisolith cemented by calcite

PISOLITIC LIMESTONE

This rock is made of pisoliths—pea-sized grains slightly larger than ooliths, often flattened, and loosely cemented by calcite.

OOLITIC LIMESTONE

This limestone is composed of ooliths—small, rounded, concentrically banded sedimentary grains rolled by seabed currents and cemented by carbonate mud.

BRYOZOAN LIMESTONE

This is a gray or reddish organic limestone, which has fossils of bryozoans in a matrix of hardened, calcite-rich mud.

LIMESTONE BRECCIA

Large, angular rock and quartz fragments cemented by calcite are typical of this rock, which forms at the base of cliffs.

FELDSPATHIC GRITSTONE

Coarse-grained and pale- to dark-colored, this gritstone contains a lot of quartz and up to 25 percent feldspar.

QUARTZ GRITSTONE

This gritstone is made of quartz with some feldspar and mica, all of coarse grain size.

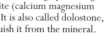

GRAYWACKE

This dark rock contains quartz, rock fragments, and feldspar, set in a mass of finer clay and chlorite. It forms in marine basins.

ARKOSE

Variable in color and medium- to coarse-grained, arkose is a sandstone with a high percentage of feldspar.

DOLOMITE

Often cream- or buff-colored, this rock contains a high percentage of dolomite (calcium magnesium carbonate). It is also called dolostone, to distinguish it from the mineral.

FOSSILIFEROUS SHALE

Fine-grained marine sedimentary rocks such as shale often contain large numbers of well-preserved fossils.

shale matrix

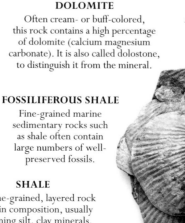

SHALE

This fine-grained, layered rock varies in composition, usually containing silt, clay minerals, organic materials, iron oxides, and minute crystals of minerals such as pyrite and gypsum.

POLYGENETIC CONGLOMERATE
The coarsest-grained sedimentary rock, polygenetic conglomerate has many different, rounded rock and mineral fragments in a fine matrix.

rounded quartz pebble

sandstone matrix

iron oxide gives red color

QUARTZ CONGLOMERATE
Varying in color, this rock typically has dirty white, pebble-sized quartz fragments set in a finer, darker matrix.

fine-grained matrix

RED CHALK

CHALK
Pure calcite, chalk is fine-grained, powdery, and easily crumbled. It is made of minute fossilized organisms, including coccoliths and radiolarians.

WHITE CHALK

SILTSTONE
This dark-colored rock has particles smaller than fine sand, but larger than clay grains—mainly of quartz. It also has organic material and calcite.

BRECCIA
This rock has large, angular fragments of rocks and minerals set in a fine matrix of sand or silt. It rarely forms in layers.

CHERT
A very fine-grained form of silica, chert occurs as bands and nodules in rocks such as limestone. It has many colors; red chert is called jasper.

MARL
Midway between clay and limestone in hardness, marl is a fine-grained, calcite-rich, layered rock. Chlorite and glauconite may give it a green color.

MUDSTONE
With a lot of clay and very fine quartz and feldspar fragments, mudstone differs from shale in that it forms as a block, with no layering.

FLINT
Usually found as nodules in chalk, flint is very hard, black, compact silica. It breaks leaving sharp, curved edges.

FOSSILS

Fossils are evidence of past life that has been buried and preserved in the rocks of Earth's crust. They give scientists important clues about how life has evolved, and they can also be used to date rocks, and create a timeline of the events that have shaped our world.

Over time the tissues of this plant have turned to charcoal. Only the outline remains, covered by a thin film of carbon.

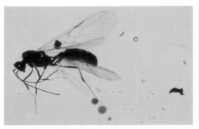

An insect became stuck on resin oozing from a tree. The resin turned into amber, perfectly preserving the creature.

This fish skeleton has been fossilized in shale. Here all the atoms of the original skeleton have been chemically replaced.

Life has existed on Earth for about 3.8 billion years. Initially, living things were small and soft-bodied, and left few obvious traces of their existence. But during the past billion years, life gradually changed. Organisms evolved hard body parts, which—given enough time—could fossilize. The significance of this change is hard to overstate. It turned the world's sedimentary rocks into a global data bank, teeming with an incredible array of fossilized species, arranged in the exact order in which they appeared. These fossils show the path evolution has followed. They also highlight mass extinctions, when enormous numbers of species have been wiped out in a relatively short space of time.

DEAD AND BURIED

Fossilization is a lottery, and only a small fraction of living organisms end up being preserved. On land, it is usually triggered by chance events—for example, when animals are overwhelmed by landslides or flash floods, or when they drown in lakes. Marine animals have a much better chance of being fossilized, because sediment routinely accumulates over their dead remains. Fine sediment can preserve soft bodies, but the best fossils are left by animals with hard body parts, such as shells or bones. After burial, the remains are slowly infiltrated by dissolved

minerals, which literally turns them to stone. Once formed, many fossils are destroyed by heat, pressure, or geological movements while still deep underground. But if a fossil survives all this, uplift may eventually bring it back to the surface, where erosion can release it from its parent rock (see panel, below). There, it has to be found before it eventually breaks apart.

These body fossils can be breathtaking objects, particularly when they are complete skeletons several yards long. However, they are not the only kind of fossilized remains. Rocks can also yield trace fossils, which are fossilized footprints, burrows, or other signs of animal activity. Trace fossils provide indirect but fascinating evidence of how animals lived: for example, dinosaur footprints can show how fast they moved, how they interacted in herds, and even how they put on weight as they grew.

Much further back in time, rocks sometimes contain chemical fossils—ancient carbon-based compounds that have been produced by biological processes. Although unspectacular, these chemical smudges are key evidence in the hunt for Earth's earliest living things.

SUDDEN DEATH >
Here, many trilobites of Ordovician age are fossilized together, suggesting the sudden burial of these animals by sediment.

INDEX FOSSILS

The geological timescale was largely established using fossils. Species which lived across a wide geographical range but only existed for a short time, are known as index fossils. They can be used to identify particular strata and link them from place to place—the presence of the same index fossil in different places shows that the strata were laid down at the same time. Index fossils therefore help geologists to date rocks and build a relative time sequence. Mesozoic ammonites (a group of extinct marine mollusks) are among the best index fossils—an ammonite zone may be as little as a million years.

HOW FOSSILS FORM

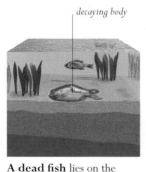

decaying body

A dead fish lies on the seabed, where its flesh may rot or be eaten. It must be buried rapidly to be preserved. As the mud turns into shale, it will compress and flatten the body.

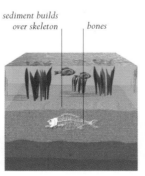

sediment builds over skeleton   *bones*

The fish skeleton has been covered by sediment. To become fossilized, the bones must undergo chemical changes, driven by heat and pressure, which replace the bone with other minerals.

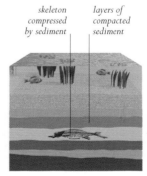

skeleton compressed by sediment   *layers of compacted sediment*

More sediment has accumulated on the seabed and compressed the lower layers. The composition of the bones changes further as they are enclosed between the layers of sediment.

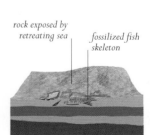

rock exposed by retreating sea   *fossilized fish skeleton*

Millions of years later the seabed sediment, now turned to rock, has been exposed by retreating seas. Weathering has further denuded the rock around the fossilized fish skeleton.

FOSSIL PLANTS

Plants are among the first organisms to appear in the fossil record. Algae are found in rocks from the Precambrian. Vascular plants (plants with tissues for conducting water and nutrients) evolved in the Silurian Period, and by Carboniferous times, Earth was colored green by vast, coal-forming swamp forests. Flowering plants developed later, in the Mesozoic Era.

EARLY LAND PLANT
Cooksonia hemisphaerica
Found in Silurian and Devonian rocks, *Cooksonia* was one of the earliest vascular plants. It had a stiff stem and leafless branches.

CALAMOPHYTON STEMS
Calamophyton primaevum
A primitive, leafless plant, probably related to ferns, *Calamophyton* is found in Devonian and Early Carboniferous rocks.

CLADOXYLON STEMS
Cladoxylon scoparium
Fossilized in Devonian and Carboniferous rocks, *Cladoxylon* was a low-growing plant, with a tough central stem and leafless, light-absorbing branches.

branch

tough stem

SEED FERN LEAF
Alethopteris serlii
A seed fern from Carboniferous and Permian strata, *Alethopteris* had compound pinnate fronds, consisting of thick, strongly veined leaflets.

CYCLOPTERIS LEAFLETS
Cyclopteris orbicularis
Oval leaflets from the seed fern *Neuropteris* are given the scientific name *Cyclopteris*. Its fossils belong to Carboniferous strata.

SEED FERN SEEDS
Trigonocarpus adamsi
The name *Trigonocarpus* is given to fossil seeds from seed ferns found in Carboniferous strata. Each seed has three ribs.

HORSETAIL FOLIAGE
Asterophyllites equisetiformis
Found in Carboniferous and Permian strata, *Asterophyllites* had needle-shaped leaves and a structure similar to that of modern horsetails.

CLIMBING HORSETAIL
Sphenophyllum emarginatum
Found fossilized in rocks of Devonian to Triassic age, this horsetail had wedge-shaped leaves and long, soft stems adapted for climbing.

SIGILLARIA STEM
Sigillaria aeveolaris
Found in Carboniferous and Permian rocks, *Sigillaria* was a giant relative of club mosses that grew to over 98 ft (30 m). It had a narrow stem, and its leaves grew in clumps.

vertical rib

LEPIDODENDRON ROOT
Stigmaria ficoides
Occurring in Carboniferous to Permian rocks, *Stigmaria* is the name for the fossil roots of the club moss relative, *Lepidodendron*.

CARBONIFEROUS FERN
Oligocarpia gothanii
This ground-hugging fern is found in Carboniferous and Permian strata. It occupied a wetland habitat.

SALVINIA RHIZOME
Salvinia formosa
A floating water fern from the tropics, *Salvinia* is found fossilized in rocks from Cretaceous to Recent strata.

pinnate (feather-like) leaf

CRETACEOUS FERN
Weichselia reticulata
Found fossilized in Cretaceous strata, *Weichselia* was similar to modern bracken and had fronds that were divided twice.

SEED FERN LEAFLETS
Dicrodium sp.
A seed fern from the Triassic Period, *Dicrodium* had pinnate leaves and its fronds were about 3 in (7.5 cm) long.

PALEOZOIC CONIFER
Lebachia piniformis
A cone-bearing plant from Carboniferous and Permian strata, *Lebachia* is an ancestor of modern conifers.

SECTION THROUGH CONE

CONIFER SEED CONES
Taxodium dubium
Found in Jurassic strata, *Taxodium* is related to modern cypress trees. It grew in damp habitats and had needlelike leaves.

COAST REDWOOD CONE
Sequoia dakotensis
Cones of the giant evergreen tree *Sequoia* have been found in Cretaceous and Recent rocks. Some living members of *Sequoia* are over 2,000 years old.

CRETACEOUS CONIFER
Glyptostrobus sp.
This conifer grew in swamps during the Cretaceous Period and into the Cenozoic Era. *Glyptostrobus* was an important coal-forming tree.

JURASSIC CONIFER
Araucaria mirabilis
This extinct monkey-puzzle tree, *Araucaria*, bore characteristic female cones with spiraly arranged scales attached to a central axis.

SUBFOSSIL TREE RESIN
Kauri pine amber
Amber is the hardened resin from pine trees, such as Kauri pines. First occurring in the Early Cretaceous, it often contains fossils of insects that perished on the fragrant, sticky resin.

seed

CARBONIFEROUS GYMNOSPERM
Cordaites sp.
An ancestor of the conifers, *Cordaites* grew during the Carboniferous and Permian Periods. It was a tree-sized plant that reproduced by seed.

GIGANTOPTERID LEAVES
Gigantopteris nicotianaefolia
A flowerless plant from Permian times, this species was so named because its leaves resembled those of tobacco plants.

PERMIAN GINKGO LEAVES
Psygmophyllum multipartitum
Still found in China, ginkgos first appeared in the Permian Period. The fan-shaped leaves can be identified in fossils of *Psygmophyllum*, a precursor of the modern-day ginkgo.

TRIASSIC GINKGO
Baiera munsteriana
Growing up to 6 in (15 cm) in length, the fan-shaped leaves of *Baiera* were split into separate ribs. In living ginkgos, the leaves are almost entire.

central axis

growth rings

PALM FRUIT
Nipa burtinii
Fossils of *Nipa* date from the Eocene Period onward. This palm holds its woody seeds in a 10 in (25 cm) ball-shaped fruit.

OAK TREE TRUNK
Quercus sp.
The well-known oak, *Quercus,* first occurs as fossils in Cretaceous strata. There are over 500 species of oak living today.

MAGNOLIA LEAF
Magnolia longipetiolata
One of the earliest genera of flowering plants, magnolias first appeared in the Cretaceous Period. Early insects fed on their nectar.

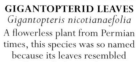
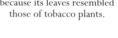

FOSSIL INVERTEBRATES

Invertebrates, animals without solid internal skeletons, are among the most common fossils found. They first appeared in the Precambrian, but it is only in Early Cambrian times that complex invertebrates such as trilobites become numerous in the fossil record. Fossils of invertebrates such as arthropods, mollusks, brachiopods, echinoderms, and corals are especially common, because they had hard external structures and lived in the sea, where most fossil-bearing rocks form.

ARCHAEOCYATHID
Metaldetes taylori
These reef-building organisms are known only from the Cambrian Period. *Metaldetes* had a cuplike structure, not unlike that of a coral.

corallite

STROMATOPOROID
Stromatopora concentrica
Found in rocks ranging from the Ordovician to the Permian, often in reef limestone, the fossil sponges in this group were made of porous, calcium-rich tubes.

CALCAREOUS SPONGE
Peronidella pistilliformis
Characterized by needlelike spicules (in the form of calcite) fused together, *Peronidella* is found in Triassic and Cretaceous rocks.

tubelike compartment

sheetlike structure

TREPOSTOME BRYOZOAN
Diplotrypa sp.
A bryozoan from Ordovician strata, *Diplotrypa* was a small invertebrate, not unlike coral, that lived in dome-shaped colonies.

CHEILOSTOME BRYOZOAN
Biflustra sp.
Found in Cenozoic rocks, this bryozoan genus is extant. It has minute compartments, which house zooids—the soft-bodied individuals of a colony.

LACE CORAL
Schizoretepora notopachys
A lace coral, *Schizoretepora* is found in Eocene to Pleistocene strata. It lived on rocky seabeds.

BRANCHING BRYOZOAN
Constellaria sp.
A bryozoan that built branching colonies on the seabed, *Constellaria* occurs in Ordovician strata.

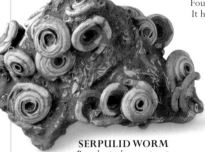

SERPULID WORM
Rotularia bognoriensis
Found in Jurassic to Eocene rocks, *Rotularia* is a genus of serpulid worm. Like all serpulids, each worm protected its soft body by producing coiled tubes made of calcium carbonate.

SPRIGGINA
Spriggina floundersi
A very early fossil found in Precambrian rocks, *Spriggina* had a long, wormlike body. Its classification is uncertain.

thecae housed soft-bodied individuals of the colony

network of branches

single, curved branch

"TUNING FORK" GRAPTOLITE
Didymograptus murchisoni
A graptolite (extinct colonial invertebrate) with two stipes (branches), *Didymograptus* is found in Ordovician rocks. It grew from ¾ to 23½ in (2–60 cm) in length.

BRANCHING GRAPTOLITE
Rhabdinopora socialis
Until recently, this graptolite was called *Dictyonema*. It had numerous thin, radiating stipes, and is from Ordovician strata.

SPIRAL GRAPTOLITE
Monograptus convolutus
A single branch, with thecae (cuplike structures) on one side, characterizes *Monograptus*, which is found in Early Silurian strata. *M. convolutus* had unusual coiling.

TABULATE CORAL
Catenipora sp.
A simple, tabulate coral with a chainlike structure, *Catenipora* lived in warm, shallow seas during Ordovician and Silurian times.

chainlike colonial structure

SCLERACTINIAN CORAL
Meandrina sp.
Shaped like the human brain, this colonial coral has ridges and valleys on its surface. First found in Eocene rocks, *Meandrina* still survives today.

RUGOSE CORAL
Goniophyllum pyramidale
A solitary coral occurring in Silurian rocks, *Goniophyllum* has a cone-shaped structure in which the polyp lived.

thick corallite wall

CAMBRIAN TRILOBITE
Paradoxides bohemicus
Some *Paradoxides* trilobites grew to nearly 3¼ ft (1 m) in length. This species had long spines on its body and is from Cambrian strata.

SILURIAN TRILOBITE
Dalmanites caudatus
Common in the Silurian Period, *Dalmanites* had a segmented thorax and a pointed tail spine.

DEVONIAN ENROLLED TRILOBITE
Phacops sp.
Characterized by compound eyes, *Phacops* is found in Devonian strata. Trilobites could roll up, like many modern arthropods.

ORDOVICIAN TRILOBITE
Eodalmanitina macrophtalma
An Ordovician trilobite, *Eodalmanitina* had large, crescent-shaped eyes. Its thorax was composed of 11 segments.

HORSESHOE CRAB RELATIVE
Euproops rotundatus
Related to horseshoe crabs, *Euproops* had a crescent-shaped headshield and long tail spine.

pincers

LOBSTER
Eryma leptodactylina
A fossil lobster from rocks of Jurassic and Cretaceous times, *Eryma* was 2¼ in (6 cm) long, and similar to modern species.

CRAB
Avitelmessus grapsoideus
Covered with many spines, this crab is from the Cretaceous Period. It grew to 10 in (25 cm) in width.

COCKROACH RELATIVE
Archimylacris eggintoni
A relative of cockroaches from the Carboniferous Period, *Archimylacris* had hind wings with a distinctive vein pattern.

SPIRIFERID BRACHIOPOD
Spiriferina walcotti
A common brachiopod from Triassic and Jurassic strata, *Spiriferina* had a rounded shell up to 1¼ in (3 cm) wide, with clearly visible growth lines.

growth line

ARTICULATE BRACHIOPOD
Leptaena rhomboidalis
A brachiopod from Ordovician, Silurian, and Devonian strata, *Leptaena* grew to about 2 in (5 cm) in width. Its shell had concentric and radial ribs.

RHYNCHONELLID BRACHIOPOD
Homeorhynchia acuta
Found in Early Jurassic strata, *Homeorhynchia* was a small brachiopod that grew to about ⅜ in (1 cm) in width.

SWAMP CLAM
Carbonicola pseudorobusta
Occurring in nonmarine rocks of the Carboniferous Period, *Carbonicola* had a tapered shell. Its fossils have been used in the relative dating of these rocks.

OYSTER
Gryphaea arcuata
This fossil oyster occurs in Triassic and Jurassic rocks. *Gryphaea* had one large, hooked valve and a smaller, flat one.

CLAM
Crassatella lamellosa
A small bivalve from Cretaceous to Miocene times, *Crassatella* had marked concentric growth lines on its shell.

MUSSEL RELATIVE
Ambonychia sp.
Found in Ordovician strata, *Ambonychia*, an early bivalve mollusk, grew up to 2½ in (6 cm) wide. Both valves had radial ribs on the surface.

SCALLOP
Pecten maximus
Found in Jurassic to Recent strata, this scallop was a bivalve mollusk that swam by flapping its valves, which were ribbed.

NAUTILOID
Vestinautilus cariniferous
This early *Nautilus* relative had very open coiling and little shell ornamentation. *Vestinautilus* is found in Carboniferous rocks.

ORDOVICIAN GASTROPOD
Murchisonia bilineata
A gastropod mollusk from Silurian to Permian strata, *Murchisonia* grew up to 2 in (5 cm) tall, with ridges around its whorls.

ROSTROCONCH
Conocardium sp.
Occurring in Devonian and Carboniferous strata, *Conocardium* resembled clams, but its shell had no functional hinge.

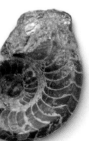

whorl with ridges

JURASSIC GASTROPOD
Pleurotomaria anglica
Found in Jurassic and Cretaceous rocks, this gastropod mollusk had a broad shell with a combination of radial and spiral patterns.

simple rib

CARBONIFEROUS AMMONOID
Goniatites crenistria
An ammonoid mollusk from Devonian and Carboniferous rocks, *Goniatites* had angular sutures where the chamber walls joined the shell.

DEVONIAN AMMONOID
Soliclymenia paradoxa
An Early Devonian ammonoid, *Soliclymenia* had thin ribs across the shell. Some species had unusual, triangular shells.

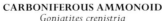

TRIASSIC AMMONOID
Ceratites nodosus
Found in Triassic strata, *Ceratites* was an ammonoid mollusk. Its heavily ornamented shell had open coiling and strong ribs.

AMMONITE
Mortoniceras rostratum
This ammonite mollusk from the Cretaceous Period grew up to 4 in (10 cm) in diameter, with ribs across the shell.

shell aperture

BELEMNITE
Pachyteuthis abbreviata
A squid relative from the
Jurassic, *Pachyteuthis* had
a calcite guard. It was
about 4 in (10 cm) long.

respiratory
structure

CYSTOID
Pseudocrinites bifasciatus
Characterized by
rhomboid-shaped respiratory
structures, *Pseudocrinites* lived
in Silurian and Devonian
times. It was attached to
the seabed by its stem.

DEVONIAN CRINOID
Cupressocrinites crassus
Up to 1¼ in (3 cm) in
diameter, this crinoid from
Devonian strata had a tall,
five-sided cup at the
end of its stem.

flexible arm

stem

JURASSIC CRINOID
Pentacrinites sp.
Named for its star-shaped
ossicles (stem segments),
Pentacrinites grew to over 3¼ ft
(1 m) in height. It is often found
attached to fossil wood.

densely
packed
branches

BRITTLESTAR
Lapworthura miltoni
An early fossil brittlestar,
Lapworthura is from
Ordovician and Silurian
strata. It grew to 4 in (10 cm)
in diameter. This species
had five relatively short,
thick arms.

tubercle
where spine
was attached

STARFISH
Tropidaster pectinatus
This extinct starfish from the Early
Jurassic was about 1 in (2.5 cm)
wide. It had five thick arms.

BLASTOID
Pentremites pyriformis
This echinoderm from
the blastoid group lived
during Carboniferous
times. It had long,
armlike structures that
were used for feeding.

SEA URCHIN
Hemicidaris intermedia
This common sea urchin from
Jurassic strata grew to around
1½ in (4 cm) in diameter. Its
many bumps (tubercles)
supported stout spines.

HEART URCHIN
Lovenia sp.
A heart-shaped, burrowing
sea urchin, *Lovenia* is known
from the Palaeocene and still
exists today. It is up to 2 in
(5 cm) in diameter.

FOSSIL VERTEBRATES

Fossilized vertebrate remains are not as common as
those of invertebrates, since many vertebrates lived on
land, where fewer fossils form, and they evolved much
later than the invertebrates. Fish were the earliest
vertebrates to evolve, some dating back to Cambrian
times. Their rapid evolution in the Silurian and Devonian
led to amphibians, which first appeared in the Devonian
Period. Dinosaurs flourished during the Mesozoic Era,
toward the end of which mammals began to diversify.

ZENASPID FISH
Zenaspis sp.
Found in Devonian
strata, *Zenaspis* had a
massive headshield. Up
to 10 in (25 cm) long,
its body was covered
in bony scales.

*finlike structure
aided movement*

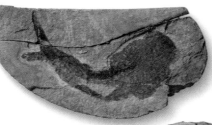

**EARLY FISHLIKE
VERTEBRATE**
Loganellia sp.
A primitive, jawless, flattened
"fish," *Loganellia* was covered
with toothlike scales. Up
to 4¾ in (12 cm) long, it is
found in Devonian rocks.

LOBE-FINNED FISH
Eusthenopteron foordi
The bones in the heavily built
fins of this Late Devonian
fish were similar to those in
the limbs of land-dwelling
vertebrates (tetrapods).

PSAMMOSTEID FISH
Drepanaspis sp.
A jawless, primitive fish,
Drepanaspis had a flattened
headshield. It is found only
in Devonian strata.

PLACODERM
Bothriolepis canadensis
A Devonian placoderm
(an extinct group of
jawless fish), *Bothriolepis*
had large head and
trunk-shields and
spinelike pectoral fins.

SHARK TOOTH
Carcharocles auriculatus
The serrated edges of
the teeth of this Cenozoic
shark could easily cut
through flesh.

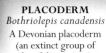

eye socket

*fine vertebral
column*

SHOAL OF DACE
Leuciscus pachecoi
Found in Miocene strata,
extinct species of *Leuciscus*
or dace resembled modern
bony fish. *L. pachecoi* grew
to 2½ in (6 cm) in length.

STINGRAY
Heliobatis radians
Found in Eocene strata,
Heliobatis was a primitive
stingray that grew to
about 12 in (30 cm)
in length and had a
skeleton of cartilage.

PRIMITIVE FROG
Rana pueyoi
Dating back to Miocene times,
Rana is a genus of frogs. *R. pueyoi* grew
to 6 in (15 cm) in length and shared
features such as long hind limbs
with modern frogs.

**SKULL OF
LARGE, PREDATORY
BONY FISH**
Xiphactinus sp.
A bony fish from Late Cretaceous
strata, *Xiphactinus* was
a marine predator with a
muscular body and large
front teeth.

*long, sharp
tooth*

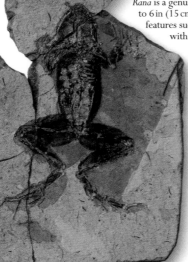

**DIPLOCAULID
AMPHIBIAN**
Diplocaulus magnicornis
A salamander-like amphibian
from the Permian Period,
Diplocaulus had protrusions on
the sides of its skull. It grew
up to 3¼ ft (1 m) in length.

DIMETRODON SKULL
Dimetrodon loomisi
Noted for a sail-like structure on its back, *Dimetrodon* was an early relative of mammals from the Permian Period. A high skull and short snout translated into a powerful bite.

DICYNODONT SKULL
Pelanomodon sp.
This tuskless herbivore was a dicynodont—a member of a group of mammal relatives that lived in the Permian and Triassic Periods.

eye socket

GIANT MONITOR LIZARD VERTEBRA
Varanus priscus
A huge monitor lizard, *Varanus priscus* grew to 23 ft (7 m) in length. It is found in rocks of Pleistocene age.

CYNODONT SKULL
Cynognathus crateronotus
A carnivore with a stout skull and large canine teeth, *Cynognathus* belonged to a group of mammal precursors called cynodonts. It is found in Triassic strata.

PLESIOSAUR FLIPPER
Cryptoclidus eurymerus
Growing up to 26 ft (8 m) in length, *Cryptoclidus* was a long-necked plesiosaur from the Jurassic Period.

MARINE TURTLE SKULL
Puppigerus crassicostata
Fossilized marine turtles are found in rocks ranging from the Mesozoic Era to Recent times. *Puppigerus* had a heavy shell and is found in Eocene strata.

EARLIEST BIRD
Archaeopteryx lithographica
Archaeopteryx was the earliest-known bird. Very few fossils have been found, from Jurassic strata in Germany.

GIANT GROUNDBIRD SKULL
Phorusrhacus inflatus
A carnivore up to 8¼ ft (2.5 m) tall, *Phorusrhacus* was a flightless bird with a powerful beak. It is found in Miocene rocks.

EARLY HORSE TEETH
Protorohippus sp.
A dog-sized, many-toed ancestor of the modern horse, *Protorohippus* is found in Eocene strata. It had low-crowned molars.

SABER-TOOTHED CAT SKULL
Smilodon sp.
Large, curved canine teeth were typical of *Smilodon*, which was the size of a tiger and lived in the Pleistocene Period.

vertebrae

EARLY ELEPHANT JAW
Phiomia serridens
Found in Eocene to Oligocene strata, *Phiomia* was 8¼ ft (2.5 m) tall. It had tusks in its upper jaw, and a small trunk.

low, sloping forehead

ANTHROPOID SKULL
Proconsul africanus
The first fossil anthropoid (apelike primate) to be found in Africa, *Proconsul* comes from Miocene strata.

NOTOUNGULATE
Toxodon platensis
Growing to about 8¾ ft (2.7 m), *Toxodon* had a sturdy body with a hippopotamus-like head. It lived from Pliocene to Pleistocene times.

PLATEOSAURUS SKULL
Plateosaurus sp.
A bulky plant eater from the Late Triassic, *Plateosaurus* grew to about 26 ft (8 m) in length and had a very small head.

bony spike

DIPLODOCUS TAIL VERTEBRA
Diplodocus longus
Found in Jurassic strata, *Diplodocus* was a giant plant eater that grew up to 89 ft (27 m) in length. Its tail was long and whiplike.

BRACHIOSAURUS THIGH BONE
Brachiosaurus sp.
A huge, plant-eating dinosaur that grew to 82 ft (25 m) in length, *Brachiosaurus* lived in the Jurassic and Cretaceous Periods.

COELOPHYSIS SKELETON
Coelophysis bauri
Fossils of *Coelophysis* are found in Triassic rocks. Only 9¾ ft (3 m) long, this carnivore had a birdlike skeleton.

PROCERATOSAURUS PARTIAL SKULL
Proceratosaurus bradleyi
A carnivore from Middle Jurassic strata in Gloucestershire, England, *Proceratosaurus* had a bony crest on its head.

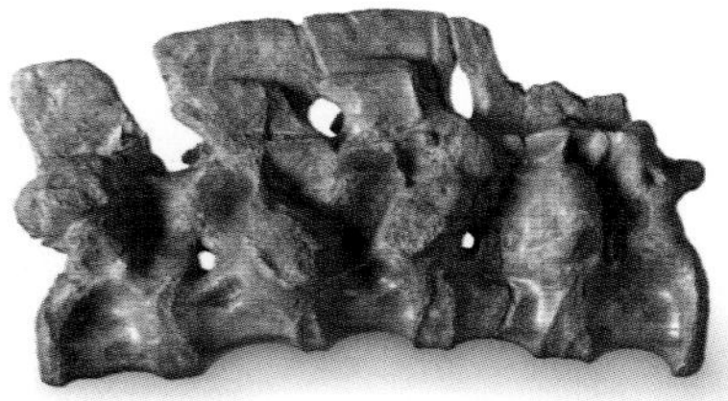

skull

long tail used for balance

MEGALOSAURUS SACRAL VERTEBRAE
Megalosaurus bucklandi
Found in Middle Jurassic strata, *Megalosaurus* was 30 ft (9 m) long, with a large head and strong hind limbs. It was carnivorous.

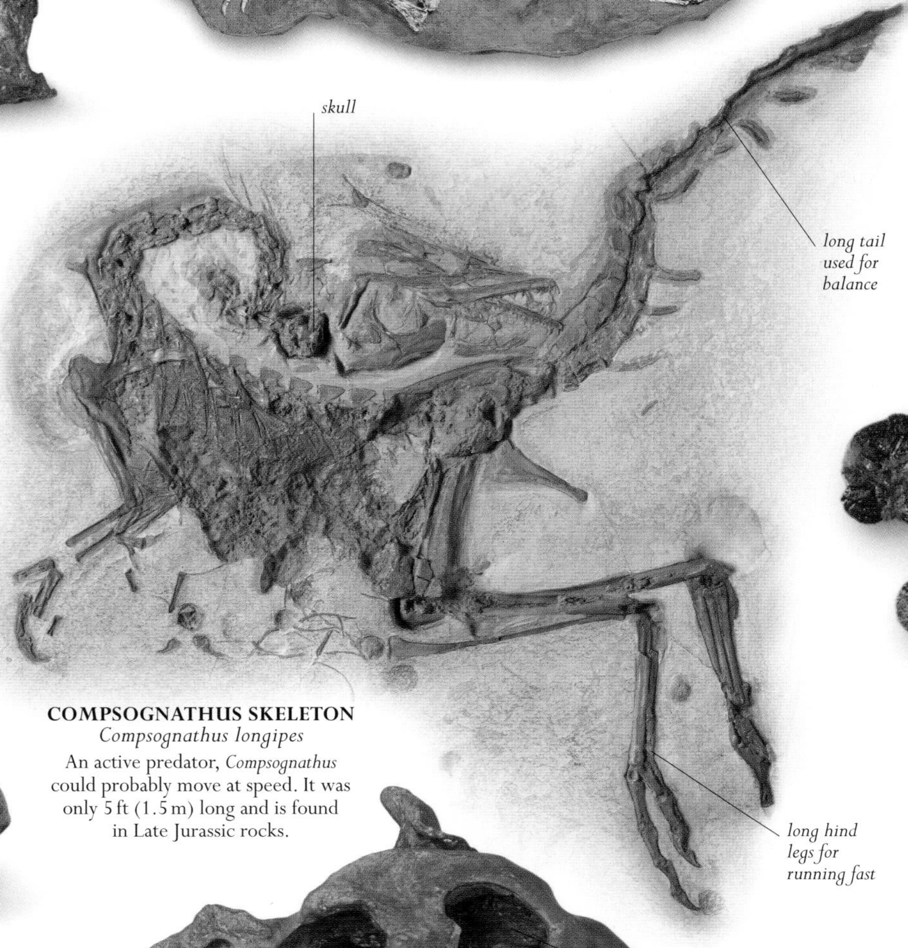

COMPSOGNATHUS SKELETON
Compsognathus longipes
An active predator, *Compsognathus* could probably move at speed. It was only 5 ft (1.5 m) long and is found in Late Jurassic rocks.

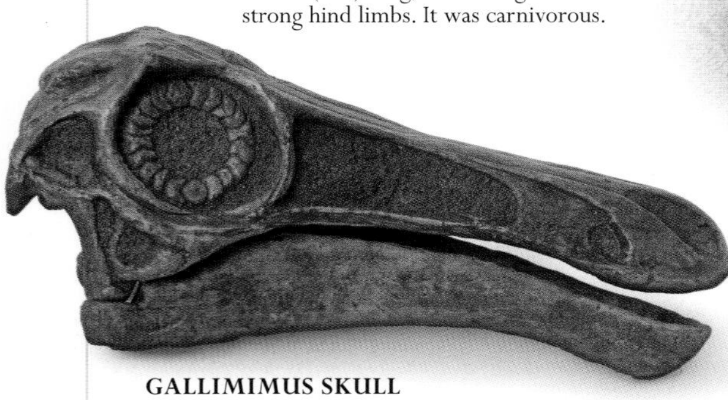

long hind legs for running fast

GALLIMIMUS SKULL
Gallimimus bullatus
Growing up to 20 ft (6 m) long, *Gallimimus* had a birdlike, beaked skull, and long neck and legs.

small brain cavity

strong, serrated teeth

ALBERTOSAURUS SKULL
Albertosaurus sp.
A predator and close relative of *Tyrannosaurus rex*, *Albertosaurus* grew to 26 ft (8 m) in length and is found in Late Cretaceous rocks.

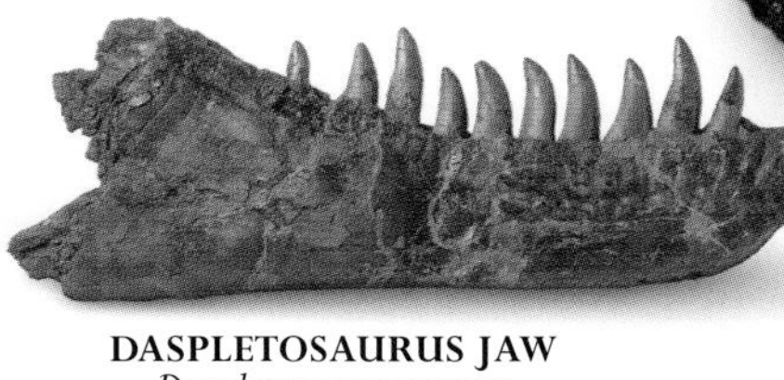

DASPLETOSAURUS JAW
Daspletosaurus torosus
This Cretaceous dinosaur had massive hind legs and small arms, and it grew to a length of 30 ft (9 m). It had a powerful jaw with the formidable teeth of a carnivore.

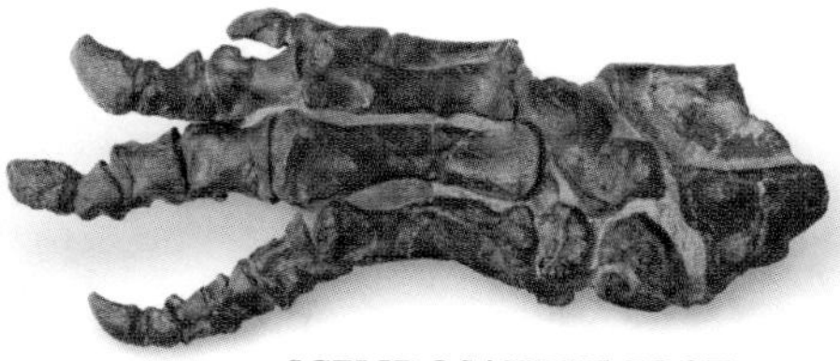

SCELIDOSAURUS FOOT
Scelidosaurus harrisonii
Found in Early Jurassic rocks, *Scelidosaurus* grew to 13 ft (4 m) in length and was covered with sharp, bony knobs. It had long toes and blunt claws.

STEGOSAURUS PLATE
Stegosaurus sp.
A Late Jurassic plant eater, *Stegosaurus* grew up to 30 ft (9 m) in length. It had two rows of huge bony plates running along its back.

ANKYLOSAURUS SKULL
Ankylosaurus magniventris
Found in Cretaceous rocks, *Ankylosaurus* was a heavily armored plant eater. It grew to about 20 ft (6 m) in length.

EUOPLOCEPHALUS TAIL CLUB
Euoplocephalus tutus
Growing to 23 ft (7 m) in length, *Euoplocephalus* lived in the Late Cretaceous. A bony club at the tip of its tail was probably used for defense.

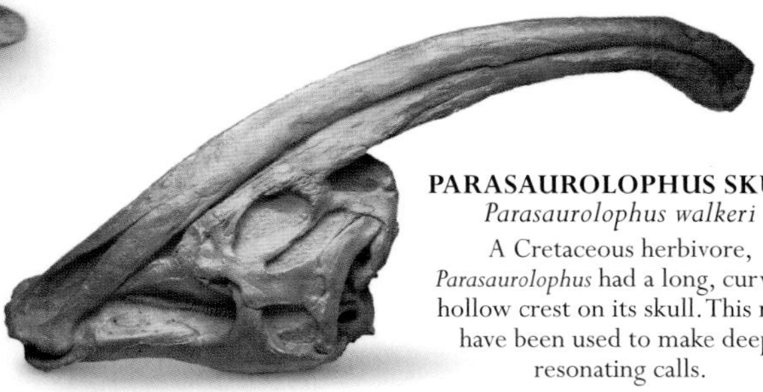

PARASAUROLOPHUS SKULL
Parasaurolophus walkeri
A Cretaceous herbivore, *Parasaurolophus* had a long, curved, hollow crest on its skull. This may have been used to make deep, resonating calls.

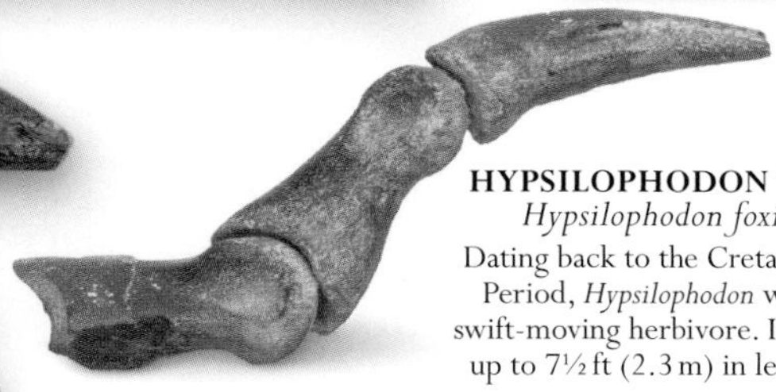

HYPSILOPHODON TOE
Hypsilophodon foxii
Dating back to the Cretaceous Period, *Hypsilophodon* was a swift-moving herbivore. It grew up to 7½ ft (2.3 m) in length.

PACHYCEPHALOSAURUS SKULL
Pachycephalosaurus wyomingensis
Living at the end of the Cretaceous Period, *Pachycephalosaurus* had a thick, domed skull and grew up to 16 ft (5 m) in length.

large nasal cavity

cheek horn

STEGOCERAS SKULL
Stegoceras validum
Found in Cretaceous strata, *Stegoceras* grew to 6½ ft (2 m) in length. Its small, serrated teeth indicate it was probably a herbivore.

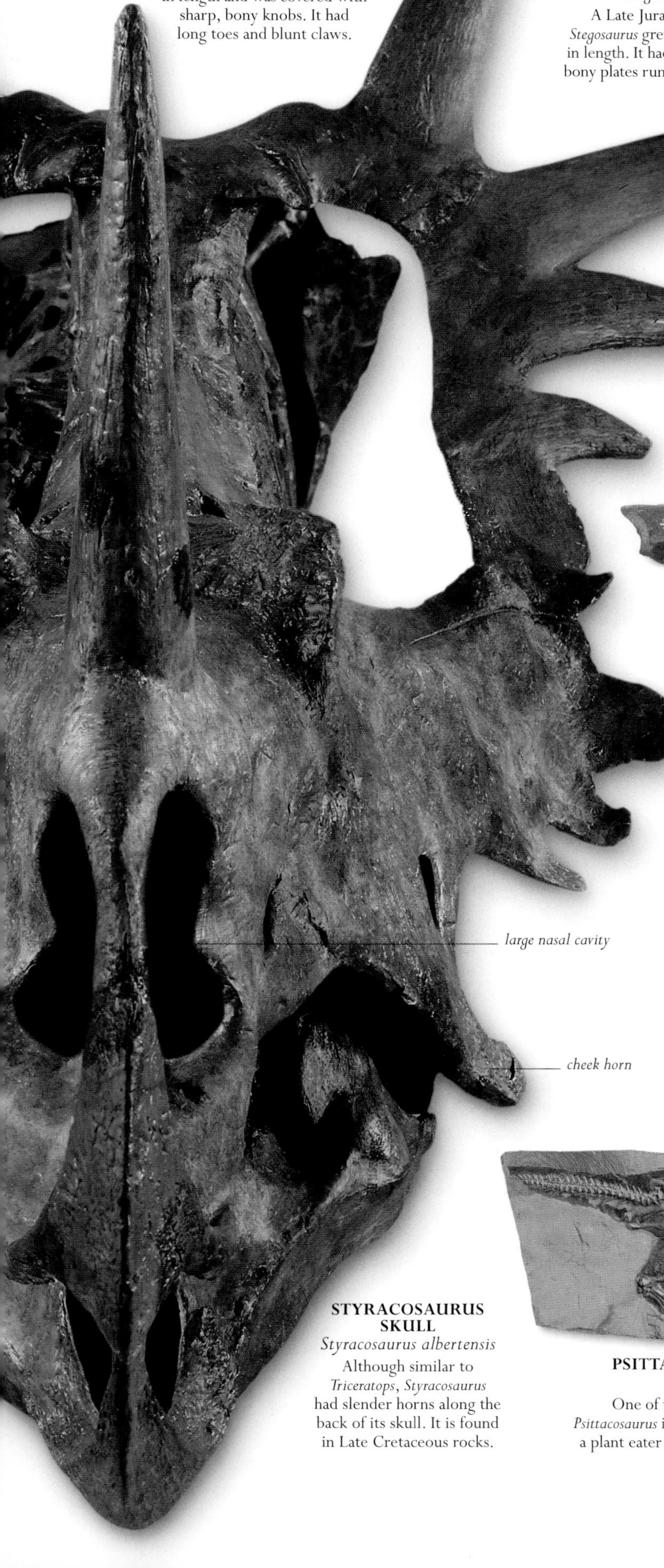

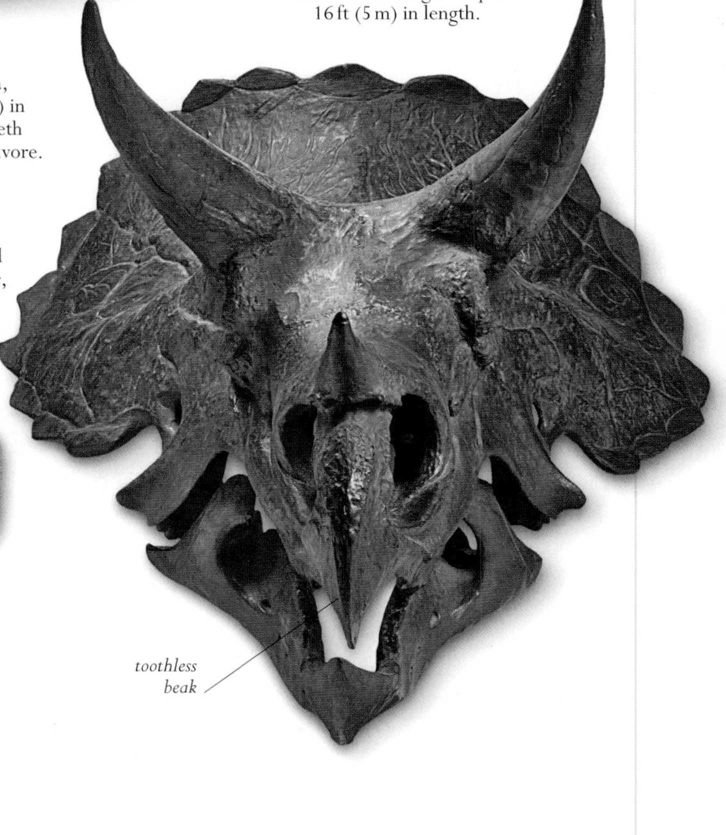

TRICERATOPS SKULL
Triceratops prorsus
A massive horned and plated skull characterized *Triceratops*, a plant eater from Late Cretaceous strata.

STYRACOSAURUS SKULL
Styracosaurus albertensis
Although similar to *Triceratops*, *Styracosaurus* had slender horns along the back of its skull. It is found in Late Cretaceous rocks.

PSITTACOSAURUS SKELETON
Psittacosaurus sp.
One of the earliest horned dinosaurs, *Psittacosaurus* is found in Cretaceous strata. It was a plant eater and grew to 6½ ft (2 m) in length.

toothless beak

EUOPLOCEPHALUS

Euoplocephalus belonged to a family of dinosaurs called the ankylosaurids, which were characterized by an armored head and bony plating on the body. This dinosaur grew to about 23 ft (7 m) in length and weighed around two tons. Its tail, body, and neck were covered by plates and bands of tough, leathery skin set with bony studs. Two rows of larger spikes ran along its back. Even its eyes were protected by bony eyelids. Fused bony knobs on the end of the long tail formed a tail club, which the dinosaur could swing at attacking predators. *Euoplocephalus* was a herbivore—its beaked mouth was ideally suited for grazing on vegetation in the thick forests of the Late Cretaceous. It may even have used the blunt hooves at the end of each toe to dig in the ground for roots and tubers. *Euoplocephalus* was probably a solitary dinosaur, though juveniles may have lived in herds.

SIZE 23 ft (7 m)
DATE Late Cretaceous
DISTRIBUTION North America
GROUP Ankylosaurids

> ARMORED HEAD
The head had a massive skull, with protective spikes at the back and a beaked mouth. The name *Euoplocephalus* means "well-armored head."

∨ NECK VERTEBRAE
Although the head was relatively small and the neck short, the neck vertebrae had to be strong to support the head's weighty, studded armor plating.

short shoulder blade _____

WALKING TANK >
Euoplocephalus had a wide, low-slung body, supported by short, stout limbs. In cross section it would have been almost round. Its small head was protected at the back by spikes, and its beaked mouth would have had small, ribbed teeth adapted for munching plants.

< STUDDED PLATES
One of the most important features of *Euoplocephalus* was its armor plating. This was made of tough plates of skin studded with ridged ovals of bone.

broad, rounded ribcage ___

∧ FRONT FOOT
This dinosaur's limbs were short, stocky, and strong. The front feet had short, sturdy toes, which helped support the considerable body weight.

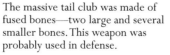

∧ TAIL CLUB
The massive tail club was made of fused bones—two large and several smaller bones. This weapon was probably used in defense.

∧ TAIL VERTEBRAE
About halfway along the tail, the typical tail vertebrae—armed with spikes—gave way to a fused, bony structure. This rigid structure would have supported the club at the end of the tail. The tail would have been quite muscular.

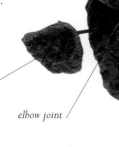

in life, bony studs were covered with scaly horn

elbow joint

bony studs and
spines along back

head spike

massive leg bones
supported the weight
of the armored body

large nasal cavity suggests
this dinosaur had a keen
sense of smell

hind feet had three
toes, each tipped
with a blunt hoof

MICROS LIFE

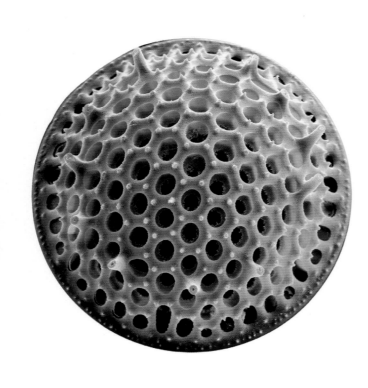

COPIC

Despite their tiny size, microscopic organisms dominate life on Earth. They were the first living things to evolve, and they underpin all the world's ecosystems, releasing and recycling nutrients that other forms of life need to survive. From the simplest prokaryotes to the most complex protists, they make up a constellation of varied and versatile life forms that largely goes unseen.

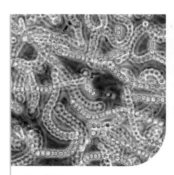

≫ 90
PROKARYOTES

Life at its most basic, prokaryotes have tiny cells without nuclei. Most live as individuals, though some form colonies such as filaments and chains. Prokaryotes include bacteria and archaea.

≫ 94
PROTISTS

Among the most numerous living things, protists are also some of the most diverse. Some have no fixed shape, but many have elaborate mineral skeletons or shells. They are typically single-celled.

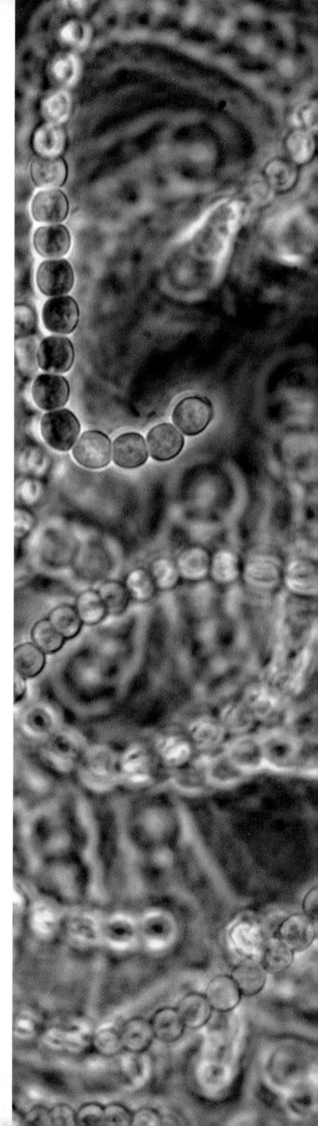

PROKARYOTES

An alien visiting Earth might conclude that its true masters were the prokaryotes—the archaea and the bacteria. They outnumber, and are more diverse than, the more complex life forms known as the eukaryotes, and thrive in every nook and cranny of the planet.

DOMAIN	ARCHAEA
KINGDOMS	5
CLASSES	9
ORDERS	18
FAMILIES	28
SPECIES	More than 2,000 known

DOMAIN	BACTERIA
KINGDOMS	28
CLASSES	49
ORDERS	About 79
FAMILIES	About 232
SPECIES	More than 8,000 known

Hydrothermal vents at the bottom of oceans are home to various thermophile archaea living at high temperatures.

Staphylococcus bacteria are common agents of food poisoning in humans, forming microscopic clusters in food.

DEBATE

CAULDRON OF LIFE

Many archaea are adapted for extremes; in hot water their fragile DNA is lagged by protective protein. Similiar protein braces the longer chromosomes of eukaryotes (fungi, plants, and animals). So perhaps what began as insulation in primeval pools evolved into a "scaffold" for the extra DNA needed for more complex life.

Prokaryotes are single-celled organisms and among the earliest forms of life. They also form one of life's two fundamental divisions, along with eukaryotes. While all living cells possess DNA, eukaryote cells also have a nucleus, and most also have mitochondria, which are energy-generating structures. Prokaryote cells have neither nucleus nor mitochondria. The archaea and bacteria are only distantly related to each other, having evolved from separate, as yet unknown genetic origins. Archaea cells are contained by chemically unique membranes overlain by a tough outer cell wall, and within the cell their DNA is often protein-covered. Bacteria cells have very different physical and chemical makeups, particularly in their cell wall. These properties make the typical archaea especially suited to harsh habitats, while bacteria are ubiquitous and thrive in every environment.

SMALLEST LIVING THINGS

All prokaryotes are tiny, their size measured in microns or micrometers (µm); 1 µm equals one-thousandth of a millimeter. A human hair is about 80 µm thick, but most prokaryotes are just 1–10 µm in length, and only photographed through an electron microscope. Yet they survive in almost every corner of the biosphere, from the outer atmosphere to deep within Earth's crust, and from ocean depths to inside the human body. For instance, the bacteria in your gut outnumber your own cells by about ten to one. Some prokaryotes thrive in boiling water or freezing ice, survive radiation, and even live off poisonous gases and corrosive acids. Most get nourishment from dead material, others from infesting living bodies. While some make food in darkness, by using the energy in minerals, others use photosynthesis, using the energy in light to convert carbon dioxide and water into food and oxygen. Although notorious for causing infectious diseases, prokaryotes are actually essential to human health, with humans depending on their gut bacteria to help break down food and even to manufacture essential nutrients. For almost four billion years, prokaryotes have profoundly influenced Earth's climate, rock formation, and evolution of other life.

MICROSCOPIC CYANOBACTERIA COLONY >
Although bacteria have single cells, the cells of some, such as cyanobacteria, can join together in spectacular, long filaments.

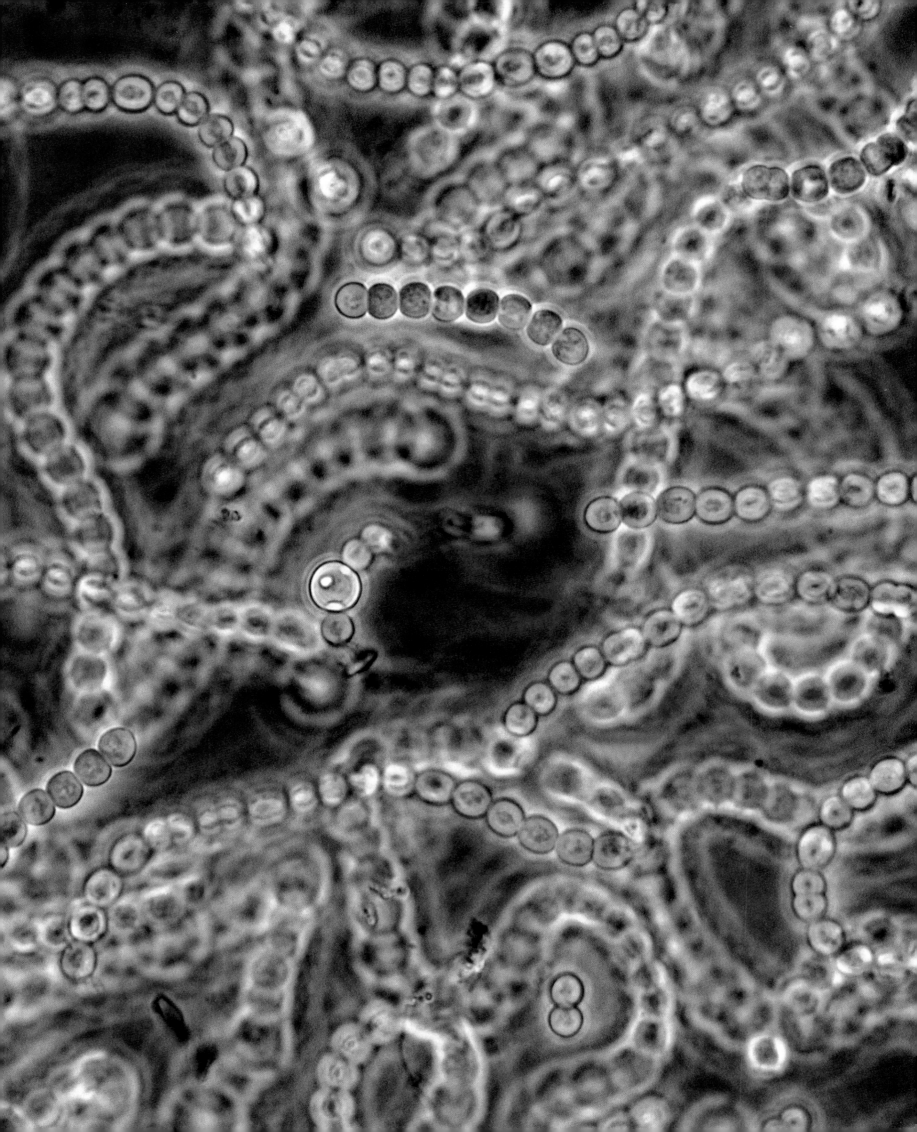

ARCHAEA

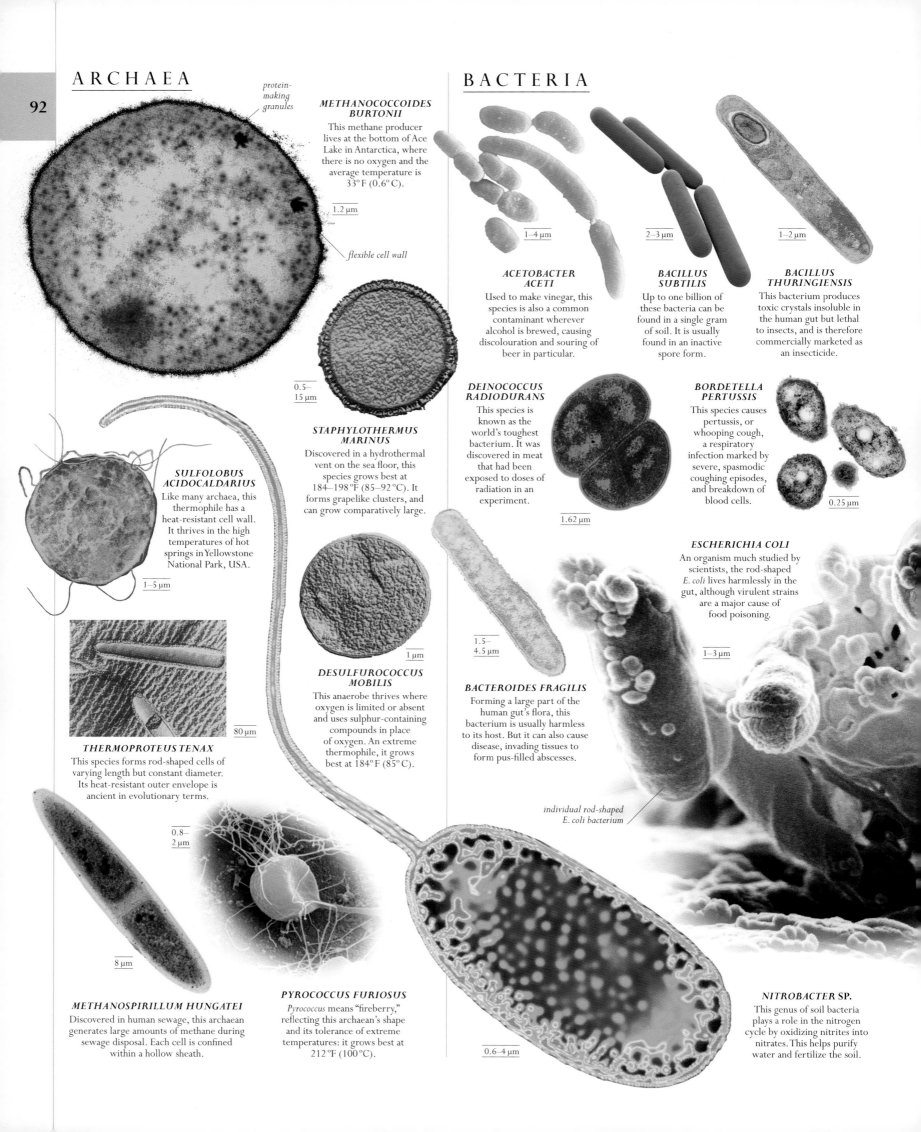

protein-making granules

flexible cell wall

METHANOCOCCOIDES BURTONII

This methane producer lives at the bottom of Ace Lake in Antarctica, where there is no oxygen and the average temperature is 33°F (0.6°C).

1.2 µm

SULFOLOBUS ACIDOCALDARIUS

Like many archaea, this thermophile has a heat-resistant cell wall. It thrives in the high temperatures of hot springs in Yellowstone National Park, USA.

1—5 µm

STAPHYLOTHERMUS MARINUS

Discovered in a hydrothermal vent on the sea floor, this species grows best at 184—198°F (85—92°C). It forms grapelike clusters, and can grow comparatively large.

0.5—15 µm

THERMOPROTEUS TENAX

This species forms rod-shaped cells of varying length but constant diameter. Its heat-resistant outer envelope is ancient in evolutionary terms.

80 µm

0.8—2 µm

DESULFUROCOCCUS MOBILIS

This anaerobe thrives where oxygen is limited or absent and uses sulphur-containing compounds in place of oxygen. An extreme thermophile, it grows best at 184°F (85°C).

1 µm

METHANOSPIRILLUM HUNGATEI

Discovered in human sewage, this archaean generates large amounts of methane during sewage disposal. Each cell is confined within a hollow sheath.

8 µm

PYROCOCCUS FURIOSUS

Pyrococcus means "fireberry," reflecting this archaean's shape and its tolerance of extreme temperatures: it grows best at 212°F (100°C).

BACTERIA

ACETOBACTER ACETI

Used to make vinegar, this species is also a common contaminant wherever alcohol is brewed, causing discolouration and souring of beer in particular.

1—4 µm

BACILLUS SUBTILIS

Up to one billion of these bacteria can be found in a single gram of soil. It is usually found in an inactive spore form.

2—3 µm

BACILLUS THURINGIENSIS

This bacterium produces toxic crystals insoluble in the human gut but lethal to insects, and is therefore commercially marketed as an insecticide.

1—2 µm

DEINOCOCCUS RADIODURANS

This species is known as the world's toughest bacterium. It was discovered in meat that had been exposed to doses of radiation in an experiment.

1.62 µm

BORDETELLA PERTUSSIS

This species causes pertussis, or whooping cough, a respiratory infection marked by severe, spasmodic coughing episodes, and breakdown of blood cells.

0.25 µm

ESCHERICHIA COLI

An organism much studied by scientists, the rod-shaped *E. coli* lives harmlessly in the gut, although virulent strains are a major cause of food poisoning.

1—3 µm

BACTEROIDES FRAGILIS

Forming a large part of the human gut's flora, this bacterium is usually harmless to its host. But it can also cause disease, invading tissues to form pus-filled abscesses.

1.5—4.5 µm

individual rod-shaped E. coli bacterium

0.6—4 µm

NITROBACTER SP.

This genus of soil bacteria plays a role in the nitrogen cycle by oxidizing nitrites into nitrates. This helps purify water and fertilize the soil.

cells dividing

NOSTOC SP.
This cyanobacteria forms gelatinous filaments in colonies that are hardy enough to survive from the poles to the tropics.

3–7 µm

DNA

2–3 µm

3–8 µm

4–8 µm

1–3 µm

CLOSTRIDIUM BOTULINUM
This bacterium thrives where oxygen is limited or absent. Living in soil, it produces the neurotoxins that cause botulism, but which also have medical and cosmetic uses.

CLOSTRIDIUM TETANI
A soil organism that can grow in dead tissue in wounds or burns, *C. tetani* produces tetanospasmin, the neurotoxin that causes tetanus.

SALMONELLA ENTERICA
Belonging to the same family as *E. coli*, some subspecies of *Salmonella* bacteria cause gastroenteritis; others cause typhoid fever.

SHIGELLA DYSENTERIAE
This gut bacteria produces shiga toxin, which can cause epidemic dysentery. As few as 10 bacteria are needed to cause an infection.

pigmented membranes for photosynthesis

STREPTOCOCCUS PNEUMONIAE
This bacterium, present inside humans, can cause pneumonia. In children and the elderly it is the leading cause of invasive infection that moves into all parts of the body.

0.9 µm

1.5– 6 µm

LACTOBACILLUS ACIDOPHILUS
Found in the intestine and vagina, this bacterium has nutritional and antimicrobial properties. It is used in probiotic drinks and supplements.

1 µm

STAPHYLOCOCCUS EPIDERMIDIS
This spherical bacterium can be found on the skin as part of normal flora, but can cause infections in immune-compromised patients.

1.5– 2 µm

flagellum helps propel organism

0.4– 0.5 µm

1–3 µm

VIBRIO CHOLERAE
Highly mobile curved rods with a single flagellum at one end, this species secretes a potent enterotoxin that causes cholera.

FUSOBACTERIUM NUCLEATUM
This species makes its home in the human mouth, and forms a major component of plaque. It can also cause premature birth.

PSYCHROBACTER URATIVORANS
This species is a psychrophile, or cryophile, which means it thrives at very low temperatures, thanks to natural antifreeze molecules in its cytoplasm within the cell membrane.

cell wall

NITROSOSPIRA SP.
Filling a vital ecological niche, this nitrifying soil bacteria oxidizes ammonia to form nitrites as part of the nitrogen cycle.

1 µm

numerous rods of bacteria grouped together

cell contents, or cytoplasm

1–3 µm

ENTEROCOCCUS FAECALIS
Normally a harmless inhabitant of the human digestive tract and vagina, this species can invade wounds and is resistant to many antibiotics.

0.5–1 µm

YERSINIA PESTIS
This species is the agent that causes bubonic plague. Transmitted to humans via fleas that live on rats, it causes up to 3,000 cases worldwide each year.

PROTISTS

From microscopic amoebae to giant kelp, the protists defy simple description, yet this informal grouping of eukaryotes included the first life forms to evolve that were more complex than the prokaryotes. It still produces most of Earth's food and oxygen.

DOMAIN	EUKARYOTA
KINGDOM	PROTISTA
CLADES	7
FAMILIES	About 778
SPECIES	About 70,500

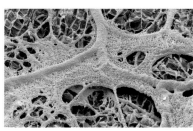

In the slime molds, which are a group of amoebas, thousands of individual nuclei coexist within one giant cell.

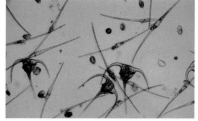

Many single-celled protists have extraordinary shapes, such as these grapple-like dinoflagellates.

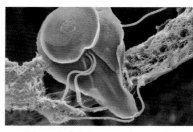

Some protoctists cause devastating disease, such as this *Giardia*, which infects intestines of humans and other animals.

DEBATE

MULTIPLE KINGDOMS

The kingdom Protista includes many organisms that are not classified as fungi, plants, or animals. And it includes many groups, ranging from single-celled amoebas to multicellular seaweeds, that are not closely related to one another. Many scientists think that Protista should be split into more than one kingdom.

Protists are mainly single-celled creatures that, unlike prokaryotes, have cell nuclei. Their basic cell nature separates them from the higher eukaryotes—plants, fungi, and animals—that later emerged from them. The protists include an incredible range of organisms with diverse lifestyles and ecological niches. Most are microscopic, ranging in size from 10 to 100 μm, and some are tiny enough to infest red blood cells. Others join together to produce multicellular colonies such as kelp, a seaweed that grows up to tens of feet in length, or the strange, funguslike slime molds, which form creeping puddles of slime that are essentially one giant cell. Typical protists include amoebae, which move around and capture food particles with their pseudopods (cell extensions), or plankton drifting in the sea, such as the beautiful diatoms, which have intricate silicon-based skeletons.

A HIDDEN REALM OF LIFE

Protists are among the most numerous creatures on Earth. Vast numbers of them live in the oceans and rivers, in sediments at the bottom of seas and lakes, and in the soil, while many spend all or part of their life cycles as parasites inside other organisms. They play a vital role in the planet's ecosphere, in particular as primary photosynthesizers, using the energy in light to convert carbon dioxide and water into food, while releasing oxygen into the atmosphere. They also act as predators and recyclers of matter. A few species are well known because they cause major diseases: the parasitic *Plasmodium* causes malaria, one of the biggest killers of humans; another parasite, *Trypanosoma brucei*, causes sleeping sickness. Also familiar are the dinoflagellates, a group of plankton organisms that can cause "red tides"—huge blooms of toxic creatures that kill fish and poison humans.

The taxonomy of the protists is complex because they don't form a natural kingdom. However, molecular and genetic analysis has allowed most protists to be placed in a few large clades, groups that each share a common ancestor: the amoebas and their relatives, the flagellates, the rhizarians, the alveolates, the heterokonts, and red and green algae.

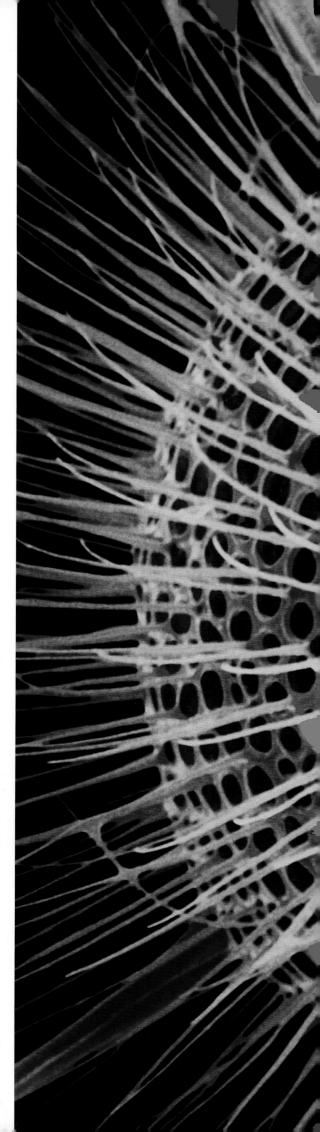

MICROSCOPIC DEATH STAR >
The spines of this rhizarian are covered with a slime that has oozed from within its cell and is used to capture prey.

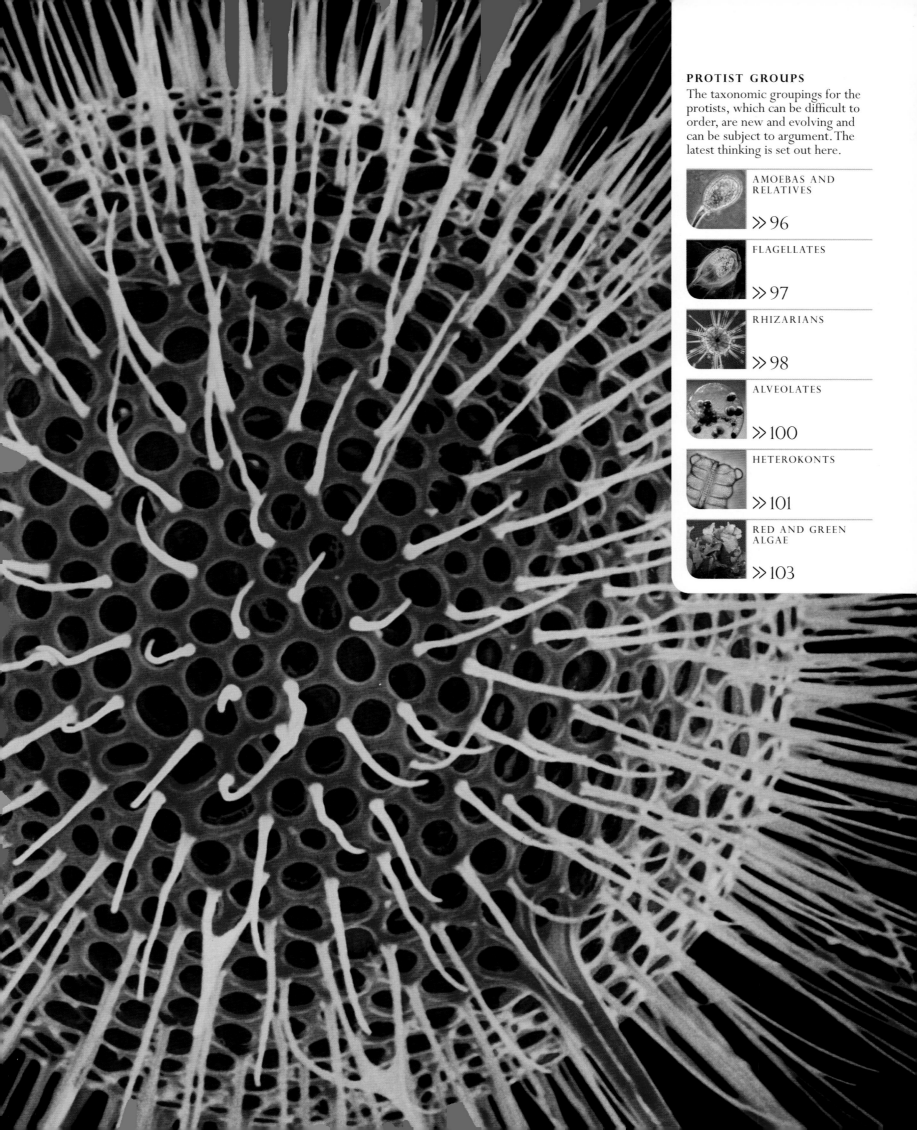

AMOEBAS AND RELATIVES

Two of the protist clades, the Amoebozoa and the Opisthokonta, have evolved different ways of moving around and obtaining food.

Amoebas of the clade Amoebozoa can change shape by oozing projections of their single cells called pseudopods. They use these "false feet" to creep forward and hunt smaller organisms, enveloping victims within a sac of fluid and digesting them while still alive. Some amoebas are giant cells, and are just visible to the naked eye. A few are gut parasites, causing amoebic dysentery in humans. One group, the slime molds, have a remarkable strategy for avoiding starvation. When prey runs low their cells are attracted to one another by a chemical distress signal, and aggregate into a tiny creeping slug. This then sprouts upward a number of stalks, which burst to scatter spores. Each spore hatches into a new amoeba ready to hunt on new ground.

ANIMAL-FUNGUS ORIGINS

Most protists of the clade Opisthokonta have developed a single whiplike flagellum for propulsion in open water. Early in the history of life, some of these protists may have given rise to animals, and the single flagellum is still seen in the tail of animal sperm today. Others, called nucleariids, lost their flagellum and actually reverted to the amoeba-like state. The nucleariids may be closely related to fungi, as fungi also lack flagella and fertilize one another without the involvement of free-swimming sperm.

DOMAIN	EUKARYOTA
KINGDOM	PROTISTA
CLADES	2
FAMILIES	About 50
SPECIES	About 4,000

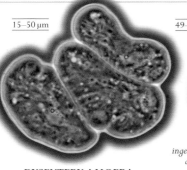

15–50 µm

DYSENTERY AMOEBA
Entamoeba histolytica
This parasitic amoeba lives in the human intestine and can cause amoebic dysentery. It can contain up to eight nuclei.

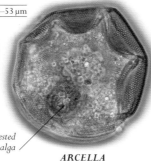

49–53 µm

ingested alga

ARCELLA BATHYSTOMA
This amoeba has a circular shell pitted with pores, domed on one side but sometimes with angular facets that develop into spines.

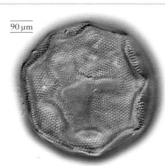

90 µm

ARCELLA GIBBOSA
This amoeba has a yellow or brown, domed, circular shell, with an aperture in the flat side and a series of regular depressions on the dome.

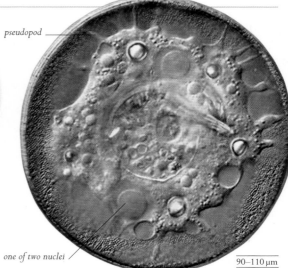

pseudopod

one of two nuclei

90–110 µm

ARCELLA DISCOIDES
This amoeba with two nuclei has a yellow-brown shell with a hole on one side through which pseudopods emerge.

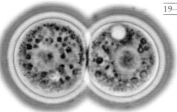

19–40 µm

PROTACANTHAMOEBA CALEDONICA
Originally identified in a Scottish estuary, close relatives of this amoeba have been found in the liver of tench in the Czech Republic.

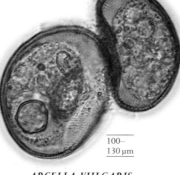

100–130 µm

ARCELLA VULGARIS
Mainly found in stagnant water and soil, this amoeba has a convex shell, with a hole that allows pseudopods to emerge.

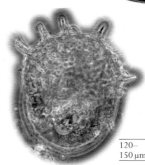

120–150 µm

CENTROPYXIS ACULEATA
This amoeba lives on algae in lakes and marshes. It uses sand and the cell walls of some algae to make a shell with four to six spines.

1.2–2.2 µm

SCALY NUCLEARIID
Pompholyxophrys ovuligera
Once assigned to the heliozoans clade, this flagellate of the Opisthokonta is covered in hollow scales or beads.

⅘ in
2 cm

STEMONITIS SP.
This "chocolate tube" or "pipe cleaner" slime mold begins as a mass of cell content with many nuclei from which many spore-bearing stalks sprout up.

pseudopod

180–230 µm

DIFFLUGIA PROTEIFORMIS
This pond-ooze amoeba constructs a shell by sticking together minuscule grains of sand and the cell walls of some algae.

FLAGELLATES

The whiplike propulsion of single-celled organisms has evolved in several groups of protists, some unrelated. But it predominates in the flagellates.

Flagellates are powerful swimming microbes that move by the whiplike action of one or more threads called flagella. Many are predators, preying on smaller organisms such as bacteria. However, unlike changeable amoebas, these flagellates have a fixed, rigid shape and direct food into a cellular "mouth" at the base of their flagella. Remarkably, however, some, such as *Euglena*, have a degree of behavioral versatility that provides them with plant- and animal-type nutrition, depending upon circumstance. When exposed to bright light these organisms can photosynthesize;

but in darkness their light-absorbing cell organelles, known as chloroplasts, shrivel and they revert to being a predator.

LIVING INSIDE ANIMALS

Many flagellates of this group lack the ordinary cellular features that would enable them to use oxygen in their respiration and survive inside the oxygen-poor environment of animal guts. Many are extraordinarily specialized, subsisting on the partially-digested food inside the abdomens of insects, but causing their host no harm. Others are parasites that cause devastating disease, in humans as well as other animals. One notorious group, the trypanosomes, are transmitted by biting insects and are responsible for sleeping sickness and leishmaniasis in the tropics.

DOMAIN	EUKARYOTA
KINGDOM	PROTISTA
CLADE	EXCAVATA
FAMILIES	40
SPECIES	About 2,500

DEBATE
EVOLUTION DOWN-SIZING

The flagellates inside termite guts lack mitochondria, the cell organelles that most other protists use to respire with oxygen. Some think these flagellates are very primitive protists. Others say they are more advanced organisms, and lost their mitochondria by evolving in an oxygen-poor environment.

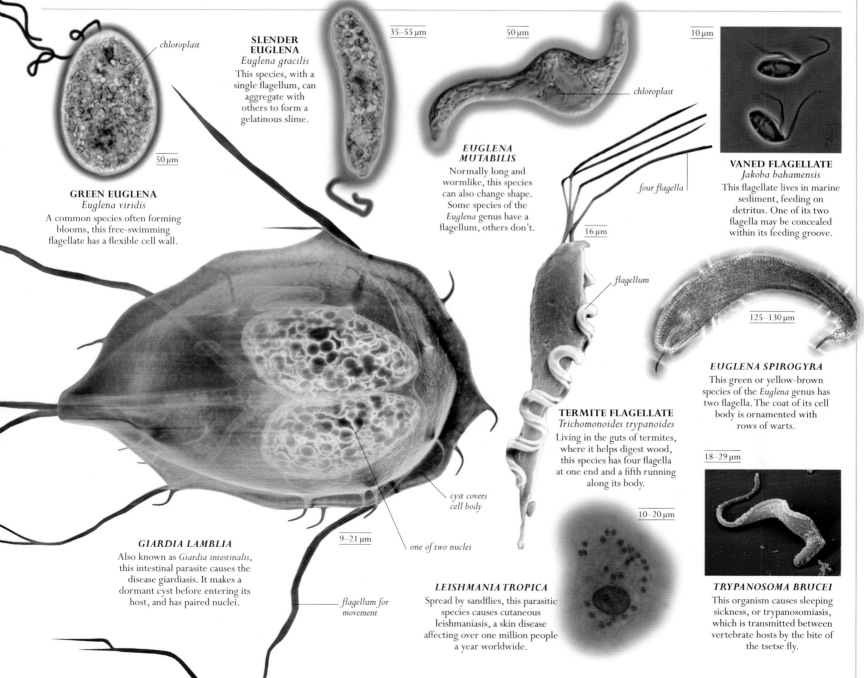

chloroplast

GREEN EUGLENA
Euglena viridis
A common species often forming blooms, this free-swimming flagellate has a flexible cell wall.

50 µm

SLENDER EUGLENA
Euglena gracilis
This species, with a single flagellum, can aggregate with others to form a gelatinous slime.

35–55 µm

50 µm

EUGLENA MUTABILIS
Normally long and wormlike, this species can also change shape. Some species of the *Euglena* genus have a flagellum, others don't.

chloroplast

10 µm

VANED FLAGELLATE
Jakoba bahamensis
This flagellate lives in marine sediment, feeding on detritus. One of its two flagella may be concealed within its feeding groove.

four flagella

16 µm

flagellum

125–130 µm

EUGLENA SPIROGYRA
This green or yellow-brown species of the *Euglena* genus has two flagella. The coat of its cell body is ornamented with rows of warts.

18–29 µm

cyst covers cell body

GIARDIA LAMBLIA
Also known as *Giardia intestinalis*, this intestinal parasite causes the disease giardiasis. It makes a dormant cyst before entering its host, and has paired nuclei.

9–21 µm

one of two nuclei

flagellum for movement

TERMITE FLAGELLATE
Trichomonoides trypanoides
Living in the guts of termites, where it helps digest wood, this species has four flagella at one end and a fifth running along its body.

10–20 µm

LEISHMANIA TROPICA
Spread by sandflies, this parasitic species causes cutaneous leishmaniasis, a skin disease affecting over one million people a year worldwide.

TRYPANOSOMA BRUCEI
This organism causes sleeping sickness, or trypanosomiasis, which is transmitted between vertebrate hosts by the bite of the tsetse fly.

RHIZARIANS

The clade Rhizaria includes two phyla that are among the most beautiful of all of the miniature protists: the radiolarians and the forams.

The unique, intricately sculptured shells of both phyla make them distinctive members of the microscopic world, and some have left an impressive fossil record. Radiolarians build glassy shells with silica, stripping the ocean surface of this otherwise abundant mineral. Their body radiates long pseudopods that poke through their shells like the rays of the sun, supplemented in some by silica-hardened spines. Radiolarians use their pseudopods for catching food, but some also harbor living algae, gaining nourishment from the sugars their photosynthetic partners make as they

drift in sunlit tropical oceans. Their silica dependence binds radiolarians to the sea, but their allies, the cercozoans, which include some amoebas, have also exploited soil and freshwater. Cercozoans typically retain the long pseudopods, but have shelled and shell-less forms, and some have flagella, depending on habitat demands.

The forams, or foraminiferans, have flourished in the oceans for hundreds of millions of years and their calcified shells litter the ocean floors to form layers of chalky sediment. Foram shells are so distinctive, even when fossilized, that geologists can use them to date deposits and identify hidden reservoirs of oil. When alive, each shell harbors a tiny amoeba that hunts with pseudopods—like the radiolarians—although some are big enough to catch animal larvae.

DOMAIN	EUKARYOTA
KINGDOM	PROTISTA
CLADE	RHIZARIA
FAMILIES	108
SPECIES	About 14,000

DEBATE
CONVERGING GIANTS

The pseudopods of radiolarians and forams mesh to form nets around their cells. This perhaps supports the idea that both forms descend from a common ancestor. But it is possible that this net evolved independently in the two groups, perhaps a result of both having large single cells.

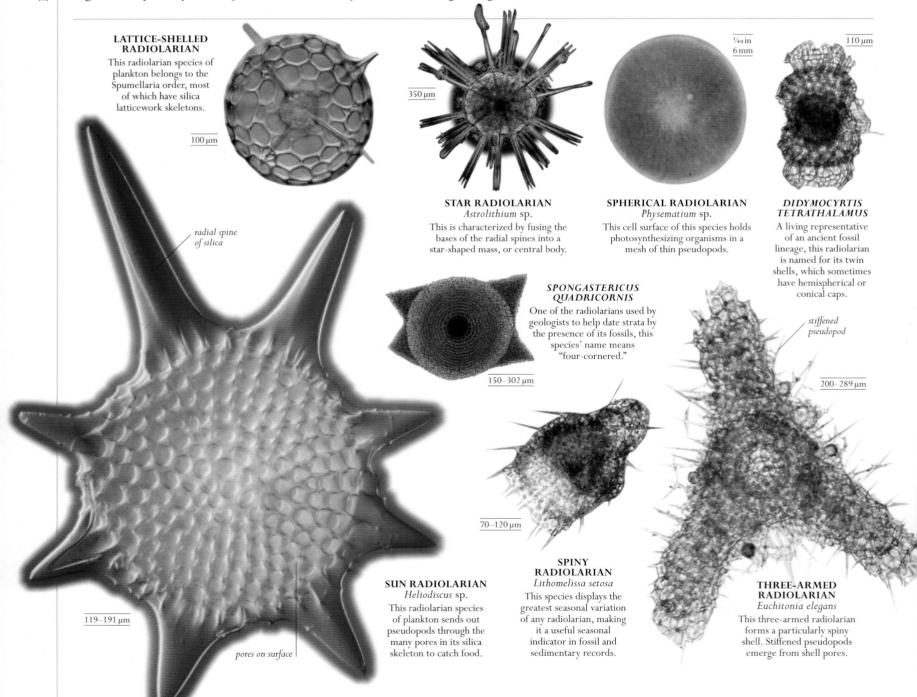

LATTICE-SHELLED RADIOLARIAN
This radiolarian species of plankton belongs to the Spumellaria order, most of which have silica latticework skeletons.

100 μm

radial spine of silica

STAR RADIOLARIAN
Astrolithium sp.
This is characterized by fusing the bases of the radial spines into a star-shaped mass, or central body.

350 μm

SPHERICAL RADIOLARIAN
Physematium sp.
This cell surface of this species holds photosynthesizing organisms in a mesh of thin pseudopods.

¹⁄₄₀ in
6 mm

DIDYMOCYRTIS TETRATHALAMUS
A living representative of an ancient fossil lineage, this radiolarian is named for its twin shells, which sometimes have hemispherical or conical caps.

110 μm

SPONGASTERICUS QUADRICORNIS
One of the radiolarians used by geologists to help date strata by the presence of its fossils, this species' name means "four-cornered."

150–302 μm

stiffened pseudopod

200–289 μm

SUN RADIOLARIAN
Heliodiscus sp.
This radiolarian species of plankton sends out pseudopods through the many pores in its silica skeleton to catch food.

119–191 μm

pores on surface

SPINY RADIOLARIAN
Lithomelissa setosa
This species displays the greatest seasonal variation of any radiolarian, making it a useful seasonal indicator in fossil and sedimentary records.

70–120 μm

THREE-ARMED RADIOLARIAN
Euchitonia elegans
This three-armed radiolarian forms a particularly spiny shell. Stiffened pseudopods emerge from shell pores.

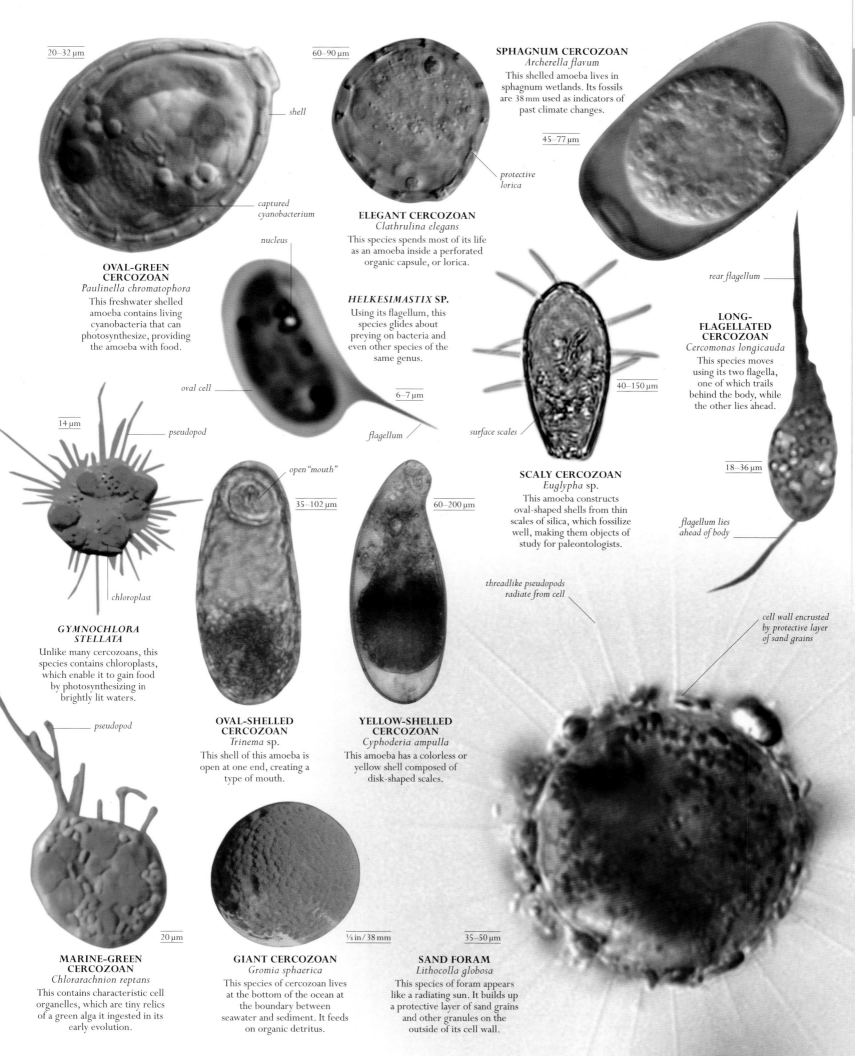

20–32 μm

shell

60–90 μm

SPHAGNUM CERCOZOAN
Archerella flavum
This shelled amoeba lives in
sphagnum wetlands. Its fossils
are 38 mm used as indicators of
past climate changes.

45–77 μm

*protective
lorica*

OVAL-GREEN
CERCOZOAN
Paulinella chromatophora
This freshwater shelled
amoeba contains living
cyanobacteria that can
photosynthesize, providing
the amoeba with food.

*captured
cyanobacterium*

ELEGANT CERCOZOAN
Clathrulina elegans
This species spends most of its life
as an amoeba inside a perforated
organic capsule, or lorica.

nucleus

HELKESIMASTIX SP.
Using its flagellum, this
species glides about
preying on bacteria and
even other species of the
same genus.

oval cell

flagellum

6–7 μm

rear flagellum

LONG-
FLAGELLATED
CERCOZOAN
Cercomonas longicauda
This species moves
using its two flagella,
one of which trails
behind the body, while
the other lies ahead.

14 μm

pseudopod

surface scales

40–150 μm

SCALY CERCOZOAN
Euglypha sp.
This amoeba constructs
oval-shaped shells from thin
scales of silica, which fossilize
well, making them objects of
study for paleontologists.

*flagellum lies
ahead of body*

18–36 μm

chloroplast

open "mouth"

35–102 μm

60–200 μm

GYMNOCHLORA
STELLATA
Unlike many cercozoans, this
species contains chloroplasts,
which enable it to gain food
by photosynthesizing in
brightly lit waters.

pseudopod

OVAL-SHELLED
CERCOZOAN
Trinema sp.
This shell of this amoeba is
open at one end, creating a
type of mouth.

YELLOW-SHELLED
CERCOZOAN
Cyphoderia ampulla
This amoeba has a colorless or
yellow shell composed of
disk-shaped scales.

*threadlike pseudopods
radiate from cell*

*cell wall encrusted
by protective layer
of sand grains*

20 μm

⅓ in / 38 mm

35–50 μm

MARINE-GREEN
CERCOZOAN
Chlorarachnion reptans
This contains characteristic cell
organelles, which are tiny relics
of a green alga it ingested in its
early evolution.

GIANT CERCOZOAN
Gromia sphaerica
This species of cercozoan lives
at the bottom of the ocean at
the boundary between
seawater and sediment. It feeds
on organic detritus.

SAND FORAM
Lithocolla globosa
This species of foram appears
like a radiating sun. It builds up
a protective layer of sand grains
and other granules on the
outside of its cell wall.

ALVEOLATES

Alveolates are united by possession of the same anatomical quirk—a fringe of tiny sacs around the cell called alveoli, after which the clade is named.

Three groups of superficially different but still single-celled protists are included in the alveolates: the dinoflagellates, the ciliates, and the apicomplexans. Predatory dinoflagellates swim in the ocean, using two whiplike flagella that emerge from grooves in their cell coat, running at right angles to one another. Some can discharge stinging barbs to immobilize their prey. Others release poisons, and sudden blooms of dinoflagellates are responsible for toxic red tides in parts of the world. A few are bioluminescent, sparkling with light when disturbed at night.

Most ciliates prowl on bacteria. Their smooth graceful motion is due to the coordinated beating of countless tiny hairs called cilia, which can cover their entire single-celled body. They also use these cilia to waft their food into a furrow that approximates to a mouth. Ciliates are virtually ubiquitous; some even live in the stomachs of mammalian herbivores, where they help digest the tough cellulose in plants.

In contrast, all apicomplexans are parasites. They are named for an arrangement of structures, called the apical complex, which helps them penetrate the living cells of animals, from which they derive their nourishment. Infamous among the apicomplexans are the malaria parasites, which penetrate the red blood cells of animals for food, destroying the cells in the process.

DOMAIN	EUKARYOTA
KINGDOM	PROTISTA
CLADE	ALVEOLATA
FAMILIES	222
SPECIES	About 20,000

DEBATE
PROTOZOANS

Early classifications placed food-eating microbes into different classes of the animal phylum Protozoa. Some microbes, such as the ciliates and the dinoflagellates, are now grouped together in modern protist taxonomy, although the exact nature of their relationships is debatable.

STALKED CILIATE
Vorticella sp.
This organism has an inverted bell-shaped body on a stalk, which is known to coil like a spring when stimulated.

50–160 μm

TRUMPET CILIATE
Stentor muelleri
This algae-feeding species, with a horn-shaped cell body, is large for a single-celled organism.

¹⁄₁₀ in
2–3 mm

SOIL CILIATE
Colpoda inflata
This normally kidney-shaped species plays an important role in soil ecology but is vulnerable to pesticides.

35–90 μm

POND CILIATE
Colpoda cucullus
Usually found in freshwater among decaying plants, this ciliate has food-containing organelles called vacuoles within its cell.

40–110 μm

GUT CILIATE
Balantidium coli
This is the only ciliate known to parasitize humans. Infection can cause intestinal ulcers, or serious intestinal infections.

50–130 μm

TOXOPLASMA GONDII
Passing between cats and other mammals (including humans), this parasite causes the disease toxoplasmosis, which is dangerous to the unborn human fetus.

6 μm

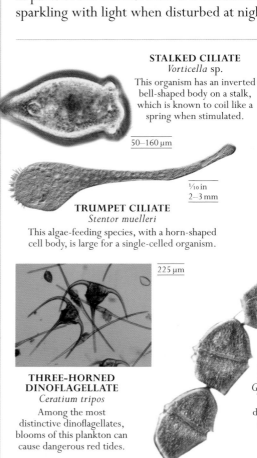

THREE-HORNED DINOFLAGELLATE
Ceratium tripos
Among the most distinctive dinoflagellates, blooms of this plankton can cause dangerous red tides.

225 μm

CHAIN-FORMING DINOFLAGELLATE
Gymnodinium catenatum
This species of dinoflagellate forms long, swimming chains of up to 32 cells.

38–50 μm

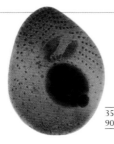

MAHOGANY-TIDE DINOFLAGELLATE
Karlodinium veneficum
This dinoflagellate plankton can undergo a population explosion leading to "mahogany tides" that are lethal to fish.

19–17 μm

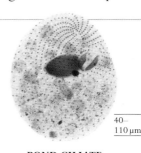

CRYPTOSPORIDIUM PARVUM
This apicomplexan can cause cryptosporidiosis, a diarrheal disease transmitted by ingestion of its spores via feces, usually in contaminated water.

4–6 μm

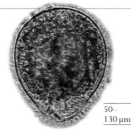

PLASMODIUM FALCIPARUM
The most deadly of the malaria-causing *Plasmodium* species, this apicomplexan parasite is responsible for the deaths of over one million people a year worldwide.

9–14 μm

organelle

200–2,000 μm

gas bag

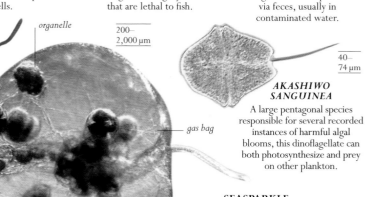

GYMNODINIUM SP.
Producing a neurotoxin, blooms of this dinoflagellate, found in both fresh- and saltwater, cause red tides and shellfish poisoning.

10–100 μm

KARENIA BREVIS
Formerly known as *Gymnodinium brevis*, this dinoflagellate plankton is a major cause of red tides in the Gulf of Mexico.

20–40 μm

AKASHIWO SANGUINEA
A large pentagonal species responsible for several recorded instances of harmful algal blooms, this dinoflagellate can both photosynthesize and prey on other plankton.

40–74 μm

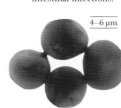

SEASPARKLE
Noctiluca scintillans
This bioluminescent, plankton-forming species has a gas bag allowing it to float just below the surface.

AMPHIDINIUM CARTERAE
This plankton-forming species causes ciguatera, a fish poisoning that can cause human fatalities if contaminated fish are consumed.

11–24 μm

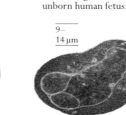

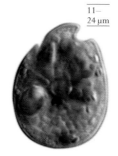

HETEROKONTS

Heterokonts include some types of algae, which are photosynthesizing protists growing in or near water that do not have true leaves or roots.

Heterokonts are mostly defined as having two different types of flagella on the sperm they use in reproduction. One flagellum is covered in tiny, bristly hairs, or mastigonemes, and the other is smooth and whiplike. Heterokonts include diatoms and brown algae, as well as water molds. Diatoms are single-celled algae with finely sculptured silica walls divided into two parts called valves. They predominate in phytoplankton, plankton communities of small organisms that drift in open waters and engage in photosynthesis. Here, near the surface, their pigments absorb energy from sunlight to make food. Diatoms contain green chlorophyll –as do plants—but also a brown pigment called fucoxanthin, which serves to broaden the spectrum of light that can be used, making their photosynthesis significantly more efficient.

Brown algae occupy coastal habitats around the world. They also make use of fucoxanthin, and have evolved into complex multicellular seaweeds that superficially resemble a plant. Rather than true roots and leaves, their body consists of a holdfast for clasping to rock and a creeping thallus that lacks the internal veins of a true leaf. Nevertheless, some brown seaweeds, such as kelps, can grow to extraordinary lengths and make extensive underwater forests on some offshore coastlines.

DOMAIN	EUKARYOTA
KINGDOM	PROTISTA
CLADE	HETEROKONTOPHYTA
FAMILIES	177
SPECIES	About 20,000

DEBATE
FROM ALGA TO MOULD

Water moulds grow and feed like mould, but unlike true moulds, which are fungi, have plantlike cell walls and heterokont flagella, while some also cause plant diseases. DNA analysis links them with diatoms and brown algae, so perhaps they evolved from algae that abandoned their chloroplasts and turned parasitic.

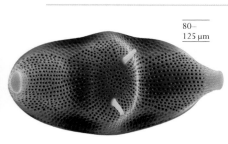

80– 125 µm

EPIPHYTIC DIATOM
Biddulphia sp.
The cause of brown film on fish tanks, in the wild this species grows on seaweed and rocks.

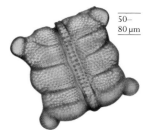

50– 80 µm

BARREL DIATOM
Biddulphia pulchella
This image clearly displays the different parts of a diatom: the two valves bound together by a narrow girdle.

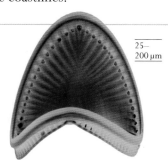

25– 200 µm

SADDLE DIATOM
Campylodiscus sp.
This genus of diatom shows a groove between the tubular lips, known as canals, around the edge of its valves.

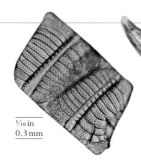

¹⁄₁₀ in
0.3 mm

COLONIAL DIATOM
Isthmia nervosa
This species grows on other algae, specifically seaweeds, forming branching colonies.

60– 240 µm

***GYROSIGMA* SP.**
The name of this diatom refers to the sigmoidal curve of its cell, meaning that it follows a very slight S-shape.

125 µm

SLIMY DIATOM
Lyrella lyra
This secretes a gluey slime from its central groove to help the cells glide along their host's surface.

18– 90 µm

POND DIATOM
Pinnularia sp.
Two chloroplasts are visible in this pen-shaped diatom, found in ponds and wet ground.

***STEPHANODISCUS* SP.**
With its disk shape, areolae (open circles), and a ring of spines, this diatom occurs individually or in chains.

12– 20 µm

disk-shaped girdle

spine

areola

10– 100 µm

200– 1,000 µm

URCHIN HELIOZOAN
Actinosphaerium sp.
This diatom resembles a sea urchin. It moves by shifting cell contents into its thin pseudopods.

GROOVED DIATOM
Diploneis sp.
The valves of this species of diatom appear as two heavily pronounced lips, known as canals, on either side of a slitlike groove called a raphe.

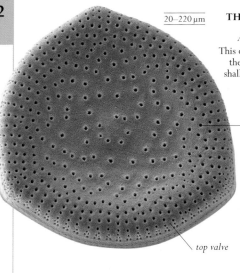

20–220 µm

THREE-CORNERED DIATOM
Actinoptychus sp.
This diatom can be found in the neritic zone on the shallow continental shelf.

pores allow intake of minerals for use in photosynthesis

top valve

⅜ in
0.9 cm

SPIDER'S-WEB DIATOM
Arachnoidiscus sp.
Strong radial ribs and web-patterned valves are features of this disk-shaped diatom. It can grow very large.

protective cell wall

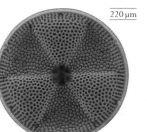

220 µm

SUN-RAY DIATOM
Actinoptychus heliopelta
A distinctive pattern of alternating raised and depressed sectors on its top and bottom valves characterize this diatom species.

20–220 µm

FIVE-ARMED DIATOM
Actinoptychus sp.
A five-arm appearance is produced by the alternating sectors on the valves of this species of the *Actinoptychus* genus.

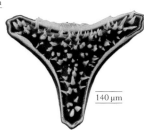

140 µm

TRICERATIUM SP.
More than 400 species of this genus of marine diatom are known. Diatoms of this genus are often, but not always, three sided.

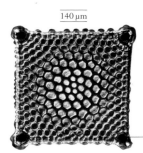

140 µm

TRICERATIUM FAVUS
The heavily silicified cell walls of this species can leave a trail of silica in the water as it moves. This can show how seawater penetrates into freshwater environments.

geometric cell wall

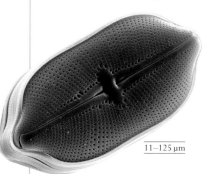

11–125 µm

BOAT-SHAPED DIATOM
Navicula sp.
This species belongs to the largest genus of diatom, with thousands of species known.

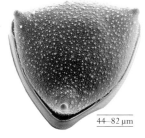

44–82 µm

DAPPLED DIATOM
Stictodiscus sp.
Seen under a scanning electron microscope, this diatom shows pores dappled on its valve surface, above a surrounding girdle.

fronds are leaf-like structures used in photosynthesis

air bladder on frond helps keep algae bouyant

shrublike fronds

6½ ft
2 m

24 in
60 cm

THONGWEED
Himanthalia elongata
During its reproductive phase, this brown seaweed of the Northern Hemisphere produces long straplike fronds, which are types of divided leaves.

13 ft
4 m

TOOTHED WRACK
Fucus serratus
This robust, shrubby, olive-brown seaweed grows on lower shores all round the North Atlantic.

SUGAR KELP
Saccharina latifolia
Found on rocky shores and at depths of 26–98 ft (8–30 m) in northern seas, this brown seaweed is also often known as sugar wrack.

6½–11 ft
2–3.5 m

CUVIE
Laminaria hyperborea
This algae is commercially important in iodine production. It grows at depths of 26–98 ft (8–30 m), mainly in the Northern Hemisphere.

3¼–9¾ ft
1–3 m

WIREWEED
Sargassum muticum
Originally from Japan, this invasive brown seaweed has now spread to Europe. It can grow as much as 4 in (10 cm) in a day.

12–39 in
30–100 cm

SEA OAK
Halidrys siliquosa
A large brown seaweed, common in rock pools in Europe, this species has a distinctive zigzag stem and air-filled pods known as air bladders.

RED AND GREEN ALGAE

Red and green algae are among the largest protists. While red algae have a high diversity in the sea, green algae do better in freshwater.

Although some are microscopic organisms, the most familiar red algae grow as multicellular seaweeds. They have evolved an extraordinary diversity of forms, ranging from rock-hugging encrustations that look more like lichens to bristly or leafy clumps. Like plants, most algae on land or in water use green chlorophyll during photosynthesis to catch the energy of sunlight to make food. But red algae have pigments that allow them to do the same thing in deeper coastal waters. In this environment a combination of pigments enables them to absorb the little blue light that penetrates the sea at such depth. This means that red algae can live much deeper in the oceans than the brown and green seaweeds.

FROM POND TO LAND

Green algae can be found in marine shallows, freshwater ponds, and streams, and form most of the photosynthesizing plankton there. They are also found on land, where some species form green mats that, for instance, coat the surfaces of tree trunks. Many share the same chlorophyll pigments that occur in land plants. While green algae typically produce sperm with two identical flagella, they also have spore-producing stages in their life cycle. Since simple plants have sperm- and spore-producing stages, the common ancestor of all true plants was a green alga.

DOMAIN	EUKARYOTA
KINGDOM	PROTISTA
CLADE	ARCHAEPLASTIDA
FAMILIES	181
SPECIES	About 10,000

DEBATE
CLOSE RELATIVES?

Plants conquered the land far better than any alga, but one group of green algae, the stoneworts, may be their cousins. Stoneworts are rooted in mud by cell filaments called rhizoids, similar to roots in simple plants. Their reproductive structures and branches are similar to those of plants too.

RED ALGAE

There are about 6,000 species of red algae. Unlike brown and green algae, they do not produce flagellated sperm. Instead, reproduction involves a fusion of cells more reminiscent of that in fungi. Some red algae become calcified, forming colonies. And, in spite of their name, red algae pigments make them red, olive, or gray.

20 in
50 cm

DULSE
Palmaria palmata
In North Atlantic coastal communities, this is a traditional source of protein and vitamins.

GLAUCOPHYTES

Glaucophytes are little more than bags of simple chloroplasts known as cyanelles. A freshwater group, they are sparsely but widely distributed around the world.

10–15 μm

CYANOPHORA PARADOXA
Like other glaucophytes, the cyanelles of this species may have descended from cyanobacteria that the species ingested early in its evolution.

⅜–6 in
1–15 cm

CORAL WEED
Corallina officinalis
This red seaweed is common in rock pools worldwide, forming branching, featherlike fronds.

frond

EYELASH WEED
Calliblepharis ciliata
This northern hemisphere seaweed has flat fronds, fringed with numerous "branchlets."

12 in
30 cm

12 in
30 cm

16 in
40 cm

AGARDHIELLA SUBULATA
Originally from the western Atlantic, Caribbean, and Gulf of Mexico, this fleshy red alga has invaded some parts of Europe.

3⅛ in
8 cm

SCHMITZIA HISCOCKIANA
Found in areas swept almost bare by the tide, this red seaweed is fleshy and gelatinous, with flattened branches and fingerlike projections.

brittle branch

2¾ in
7 cm

4–12 in
10–30 cm

GRACILARIA FOLIIFERA
Rising as slender dull purple stems, this red seaweed branches sparsely to give straplike fronds. It grows abundantly in shallow lagoons worldwide.

GRACILARIA BURSA-PASTORIS
A long, thin red seaweed with forked or alternating branches, this species is found from southern England to the Pacific and the Caribbean.

6¾ in
17 cm

MASTOCARPUS STELLATUS
This red seaweed has conspicuous papillae, which are reproductive bodies, on the fronds. It is found in the North Atlantic.

MAERL
Phymatolithon calcareum
This species, an encrusting coralline alga of the British Isles, is commercially dredged and ground up as a calcium-rich soil additive.

PTEROCLADIELLA CAPILLACEA
With feathery branches that taper toward the tip, this red seaweed, found in pools worldwide, often develops a Christmas tree appearance.

8 in
20 cm

flat, leaflike structure

6 in
15 cm

BUSH WEED
Ahnfeltia sp.
A source of the gelatinous substance agar, used in petri dishes, this Northern Hemisphere seaweed forms dense tufts of fronds.

6¾ in
17 cm

blades on frond

MELANAMANSIA FIMBRIFOLIA
Found in North America and Australia, this seaweed grows on sediment-covered reefs at depths of up to 180 ft (55 m).

SPOROLITHON PTYCHOIDES
This red algae forms a crust using calcareous deposits in its cell walls. It is found worldwide in rocky, current bottoms and coastal pools.

1/16 in
2.5 mm

CHONDRIA DASYPHYLLA
This species, found worldwide, has feathery fronds ending in clublike branchlets that sprout spore-bearing branchlets and urn-shaped fruit bodies.

4–8¾ in
10–21 cm

12 in
30 cm

CERAMIUM VIRGATUM
This small red seaweed grows worldwide on rocks and other seaweeds. From a tiny holdfast it grows into filament-like fronds with forked tips.

LEATHER WEED
Ptilophora leliaertii
First described in 2004 after it was discovered on a reef off the coast of South Africa, this red seaweed has feathery compound branches.

8 in
20 cm

14 in
35 cm

1/16–5/8 in
2–15 mm

3⅜ in
8.5 cm

GELIDIELLA ACEROSA
An important source of agar, this red seaweed, found in India, has a slender, cylindrical shoot, known as a stolon.

stolon

GELIDIUM PUSILLUM
Growing worldwide from an extensive creeping base that incorporates shells and small snails, this turf-forming red seaweed has flattened, leaflike fronds.

BROAD WEED
Lenormandiopsis nozawae
On both sides of this broad-bladed seaweed, found in temperate areas, are clusters of spore-bearing bodies, themselves home to tiny parasitic algae.

1⅛ in
3 cm

GELIDIELLA CALCICOLA
This small seaweed of the British Isles creeps over the surface of rock. Irregular branches project up and form tufts. It also grows on algae, particularly maerl.

12 in
30 cm

SEA BEECH
Delesseria sanguinea
Known for its
characteristic beechlike
leaves, this red seaweed
grows in the understory
of kelp forests in Europe.

LAURENCIA OBTUSA
This tropical species is
a source of halogenated
terpenoids—chemical
defenses against
grazing crabs and
urchins—which
also make useful
antifouling agents.

2¾–8¾ in
7–22 cm

fanlike branch

8¾ in
22 cm

IRISH MOSS
Chondrus crispus
Also known as carrageen
moss, this seaweed of the
British Isles is important
as a source of carrageen,
which is a gelling agent.

BLACK CARRAGEEN
Furcellaria lumbricalis
This Northern Hemisphere
seaweed has cylindrical,
brown-black fronds that
branch into masses of
fleshy fingers.

2–10 cm
¾–4 in

12 in
30 cm

tubular frond

POLYSIPHONIA LANOSA
Growing on other seaweeds in
tufts like pom-poms in the
Northern Hemisphere, this red
alga has branching filaments
made up of long tubular cells.

GREEN ALGAE

Green algae include desmids, which typically
have a single cell divided into two symmetrical
semi-cells. They also include multicellular green
seaweeds, such as sea lettuce, and pond weeds,
such as stonewort, a close relative of land plants.
Some species are also adapted to resist
dessication and thrive on land.

100–460 µm

cell divide

semi-cell

CLOSTERIUM SP.
This crescent-shaped desmid,
found worldwide, displays
semicells, each housing a
chloroplast, joined at the
isthmus, where the nucleus
is situated.

32–70 µm

PENIUM SP.
This desmid green alga, found in
North America, is symmetrically
divided into cylindrical semicells with
blunt, oval ends and girdle bands.

egg-bearing capsule

12–47 in
30–120 cm

long stem is plantlike

20 in
50 cm

stem made of elongated cells

prong

350 µm

MICRASTERIAS SP.
This desmid, found in
temperate areas, has
multiple spined arms,
tipped with prongs that
allow the semicells
to interlock.

4¾–23½ in
12–60 cm

SEA LETTUCE
Ulva lactuca
A popular food
worldwide, this green
alga has broad,
crumpled fronds,
anchored to rock by a
holdfast. It can also
form floating
communities.

NITELLA TRANSLUCENS
This algae is part of the
Charales order, the closest
algal relatives of land plants.
It thrives in wet, acidic
conditions near peat bogs.

COMMON STONEWORT
Chara vulgaris
Also known as skunkweed,
this Charales algae of the
Northern Hemisphere emits
an unpleasant smell.

¾–3⅛ in
2–8 cm

SEA GRAPE
Caulerpa lentillifera
This edible species is
cultivated in large
ponds and is a popular
food in the Philippines.
Very succulent, it can
be eaten raw, either on
its own or in salad.

PLANTS

>> 108
LIVERWORTS

The simplest true plants, liverworts are typically small and ribbonlike, or straggling, with tiny leaves. They do not have flowers, and spread by shedding spores instead of making seeds.

By harnessing the energy in sunlight, and using it to grow, plants play a central role in life on Earth. They generate food for animals and other organisms, and often form their habitats as well. Some plants are small and simple, but they also include giant conifers, and a dazzling wealth of flowering species that have evolved a remarkable array of shapes and strategies for survival.

≫ 110
MOSSES

Widespread in damp, cool, and shaded places, mosses are nonflowering plants that have wiry stems and small, spirally-arranged leaves. Many mosses grow on bare rocks and on trees.

≫ 112
FERNS AND RELATIVES

Ferns are the largest plants that reproduce with spores instead of seeds. Many are low-growing, but some form trees, which have trunks made of fibrous roots rather than wood.

≫ 116
CYCADS, GINKGOS, AND GNETOPHYTES

These plants do not have flowers, but they do form seeds. Before flowering plants evolved, cycads formed an important part of the world's vegetation.

≫ 118
CONIFERS

Although far fewer in species than flowering plants, conifers dominate the landscape in some parts of the world. All of them are trees or shrubs, and they typically form seeds in woody cones.

≫ 122
FLOWERING PLANTS

This is by far the largest group of living plants, making up the bulk of the world's vegetation. They all grow flowers—often inconspicuously—and spread by forming seeds.

LIVERWORTS

Found mainly in damp, shaded habitats, the liverworts are thought to be the simplest of all the existing groups of land plants. They come in two distinct forms: some are flat and ribbonlike; others are more like mosses, with slender stems flanked by tiny leaves.

DIVISION	MARCHANTIOPHYTA
CLASSES	3
ORDERS	13
FAMILIES	86
SPECIES	8,000

DEBATE
PLANTS APART

Liverworts are traditionally classified with mosses and hornworts in a group of ancient plants known as the bryophytes. All three have a number of things in common, but liverworts have some unique features that set them apart from the other two. They are the only land plants without stomata, or adjustable breathing pores, during their sporophyte (spore-producing) stage. Also, their rhizoids, or hairlike roots, each contain just one cell. As a result, they are now classified in a division of their own, the Marchantiophyta.

Ribbonlike liverworts are unlike all other plants. Instead of stems and leaves, they have a flat body, or thallus, that repeatedly forks as it grows. Many species have a glistening upper surface with deeply divided lobes. Medieval herbalists noted this liverlike shape, giving these plants their common name. Superficially, leafy liverworts are quite different, with trailing or spreading stems. They usually have two ranks of leaves, and a third row of much smaller leaves on their underside. Some kinds form loose mats in damp grass, but they also grow on rocks and on trees. They are much more numerous than ribbonlike liverworts, particularly in the tropics, where many species live epiphytically, creeping across the shaded leaves of rain forest trees.

Unlike many flowering plants, liverworts have no preset limits to their growth. Most are less than ¾ in (2 cm) high but may keep spreading for years, forming fragmented patches several yards (meters) across. This fragmentation gradually creates new plants. In addition, some species produce clusters of cells, called gemmae, in depressions on their upper surface. Scattered by raindrops, gemmae grow into new plants.

FORMING SPORES

Liverworts also reproduce by scattering spores, microscopic cells that form new plants. Spore formation requires damp conditions, because a liverwort's female cells must first be fertilized by swimming male sperm. In some species, male and female cells are produced by the same plant, but in many, the sexes are separate. Sperm cells often reach egg cells with the help of raindrops, which splash them from one plant to another.

Many liverworts produce sperm or egg cells in structures that look like tiny umbrellas or parasols. After fertilization has taken place, the spores develop and are then scattered into the air.

∨ **SEPARATE SEXES IN THE COMMON LIVERWORT**
These parasol-shaped structures produce male sex cells. Female sex cells are produced by separate plants.

gemma cup

female reproductive
structures with
star-shaped rays

SNAKEWORT
Conocephalum conicum
F: Conocephalaceae

Found worldwide on streamside
rocks and in other wet places, this
ribbon-shaped liverwort has a
glossy upper surface, flecked with
minute translucent air chambers.

CRESCENT-CUP LIVERWORT
Lunularia cruciata
F: Lunulariaceae

Common in gardens and greenhouses,
this ribbon-shaped liverwort is bright
green, with distinctive reproductive
structures, gemma cups, that
look like tiny fingernails.

COMMON LIVERWORT
Marchantia polymorpha
F: Marchantiaceae

In spring and summer this
ribbon-shaped liverwort
produces conspicuous
spore-forming structures
that look like tiny parasols.
It is widespread in damp
habitats and gardens.

notch between
two lobes

glossy
thallus

CRYSTALWORT
Riccia fluitans
F: Ricciaceae

This variable liverwort has two
forms—one grows on wet mud,
while the other floats in ponds. Both
have narrow ribbonlike stems that
fork frequently.

OVERLEAF PELLIA
Pellia epiphylla
F: Pelliaceae

Growing on wet peat and rocks,
this ribbon-shaped liverwort
often forms a tufted mat. It
produces black spore capsules
on slender white stalks.

leaves may be
green or red

leaves pressed
together,
enfolding stem

TAMARISK SCALEWORT
Frullania tamarisci
F: Frullaniaceae

Colored a distinctive purplish
brown, this leafy liverwort forms
mats on tree trunks and rocks.
Its main leaves have small concave
lobes that hold water.

GREATER WHIPWORT
Bazzania trilobata
F: Lepidoziaceae

A native of damp woodlands,
this mound-forming liverwort
has main leaves that overlap and
curve downward, giving its
stems a caterpillar-like shape.

WALL SCALEWORT
Porella platyphylla
F: Porellaceae

This densely branched, leafy
liverwort grows in a wide variety
of habitats, from woodlands to
walls. Its main leaves overlap like
a set of scales.

COMPRESSED FLAPWORT
Nardia compressa
F: Jungermanniaceae

A leafy liverwort growing by mountain
streams, this species often forms extensive
mats of overlapping stems. It has two main
ranks of rounded leaves, pressed together.

two ranks of
main leaves

EVEN SCALEWORT
Radula complanata
F: Radulaceae

Ranging in color from light green
to brown, this scaly-leaved liverwort
creeps over flat, brightly lit surfaces as
varied as tree trunks and coastal rocks.

VARIABLE-LEAVED CRESTWORT
Lophocolea heterophylla
F: Geocalycaceae

Found on decaying logs and
tree stumps, this translucent,
leafy liverwort has two
ranks of main leaves, with
rounded or toothed tips.
It produces black spore
capsules on white stalks.

GREATER FEATHERWORT
Plagiochila asplenioides
F: Plagiochilaceae

A delicate liverwort with translucent leaves, this species
forms spreading mosslike tufts. It grows on shaded rocks
and soil, particularly on chalk and limestone.

MOSSES

Mosses are nonflowering plants that typically form mats or cushion-shaped clumps. Despite their small size, they are remarkably resilient plants. They are able to grow in a wide range of habitats, from woodlands to deserts, and are found on every continent, including Antarctica.

DIVISION	BRYOPHYTA
CLASSES	8
ORDERS	26
FAMILIES	118
SPECIES	12,000

In the far north sphagnum mosses form raised bogs—here interspersed with gray patches of reindeer moss lichen.

Mosses have thin leaves that usually spiral around slender, wiry stems, and they reproduce by scattering spores. Like liverworts, they need moist conditions to grow. They can be extremely abundant in habitats that are always damp, and some species—particularly sphagnum mosses—form extensive blankets, dominating the ground in cold parts of the world. Others, however, are able to remain dormant during droughts. They look gray and lifeless, but become green again within minutes if it rains.

Mosses have a two-stage life-cycle—a feature shared by all other plants. In mosses, the dominant stage is the gametophyte, which produces male and female cells. After the female cells are fertilized, they produce the spore-forming stages or sporophytes, which remain attached to the parent plant. Most moss sporophytes have a capsule on the end of a long stalk. Once ripe—a process that can take many months—the capsule splits open, releasing up to 50 million spores into the air.

Because moss spores are so small and light, they are carried great distances by the slightest breeze. This makes mosses very good at dispersing and enables them to colonize all kinds of microenvironments, from crevices in the bark of trees to damp walls and roofs.

MOSS LOOKALIKES

Several plants commonly known as mosses do not belong to this group. Club mosses are more complex than true mosses, and have more robust stems and leaves. Living species are typically less than 20 in (50 cm) high, but in prehistoric times, club mosses included treelike forms up to 130 ft (40 m) high. Other moss lookalikes include reindeer moss—actually a form of lichen—and Spanish moss, a flowering plant in the bromeliad family that grows hanging from tree branches.

∨ **DEEP COVER**
The cool, moist climate in New Zealand's Fjordland National Park creates perfect conditions for a wide variety of mosses.

1¼ in / 3 cm

BLACK ROCK MOSS
Andreaea rupestris
F: Andreaeaceae
Widespread on mountains, this dark-colored moss grows on bare rock. Unlike most mosses, its capsules release their spores through four microscopic slits.

1¼ in / 3 cm

FIRE MOSS
Ceratodon purpureus
F: Ditrichaceae
Found worldwide, particularly on burned or disturbed ground, this low-growing moss is also common on roofs and walls. It produces a dense carpet of spore capsules in spring.

¾ in / 2 cm

COMMON POCKET MOSS
Fissidens taxifolius
F: Fissidentaceae
This widespread moss has short, spreading stems bearing two ranks of pointed leaves. It grows in shaded ground and on rocks.

VELVET FEATHER-MOSS
Brachythecium velutinum
F: Brachytheciaceae
With its much-branched stems, this abundant moss forms creeping mats on dead wood and in poorly drained grass. It has a worldwide range.

spore capsules with conical lids

4 in / 10 cm

6 in / 15 cm

WHITE FORK MOSS
Leucobryum glaucum
F: Dicranaceae
Growing in large, neatly rounded cushions, this woodland moss has a characteristic gray-green hue, turning almost white in dry weather.

2 in / 5 cm

MOUNTAIN FORK MOSS
Dicranum montanum
F: Dicranaceae
A moss that forms compact feathery mounds, this species has narrow leaves that curl up when dry, often breaking off to form new plants.

2 in / 5 cm

COMMON TAMARISK MOSS
Thuidium tamariscinum
F: Thuidiaceae
Its finely divided "fronds" make this woodland moss resemble a miniature fern. It grows on rotting wood and rocks in Europe and northern Asia.

1¼ in / 3 cm

PURPLE FORK MOSS
Grimmia pulvinata
F: Grimmiaceae
Growing on rocks, roofs, and walls, this widespread moss has leaves that end in long silvery hairs. Its spore capsules grow on curved stalks.

3¼ in / 8 cm

SWAN'S-NECK THYME MOSS
Mnium hornum
F: Mniaceae
Brilliant green in spring, this is a common woodland moss in North America and Europe. Its spore capsules have curved stalks resembling a swan's neck.

1¼ in / 3 cm

CYPRESS-LEAVED PLAIT MOSS
Hypnum cupressiforme
F: Hypnaceae
This highly variable mat-forming moss has crowded, overlapping leaves. Found worldwide, it is common on rocks and walls, and beneath trees.

4 in / 10 cm

OSTRICH-PLUME FEATHER MOSS
Ptilium crista-castrensis
F: Hypnaceae
Found mainly in northern forests, this moss has feather-shaped, symmetrically branched stems, often forming extensive patches below spruces and pines.

32 in / 80 cm

WILLOW MOSS
Fontinalis antipyretica
F: Fontinalaceae
A submerged freshwater species, willow moss trails from rocks in slow-flowing rivers and streams. It has three rows of keeled, dark green leaves.

spore capsules are horizontal when ripe

3⁄8 in / 1 cm

CAPE THREAD-MOSS
Orthodontium lineare
F: Orthodontiaceae
This moss from the Southern Hemisphere was introduced to Europe in the early 20th century and has become an invasive species.

10 in / 25 cm

BLUNT-LEAVED BOG MOSS
Sphagnum palustre
F: Sphagnaceae
Like its relatives, this peat-forming moss grows on wet ground, and holds large amounts of water. Each stem ends in a flat-topped rosette of small branches.

24 in / 60 cm

COMMON HAIR-CAP MOSS
Polytrichum commune
F: Polytrichaceae
This tall, tussock-forming moss is common in wet moorlands throughout the Northern Hemisphere. Its stems are stiff and unbranched, with narrow, pointed leaves.

½ in / 1.5 cm

CORD MOSS
Funaria hygrometrica
F: Funariaceae
One of the world's most widespread mosses, this mat-forming species is particularly common on disturbed ground. When ripe, its spore capsules have conspicuous orange stalks.

FERNS AND RELATIVES

Most ferns can be recognized by their graceful fronds, which unroll as they grow. Together with horsetails and whisk ferns, they make up a diverse and ancient group of nonflowering plants that reproduce by means of spores. Ferns grow in a wide variety of habitats, although the majority thrive where there is moisture and shade.

DIVISION	PTERIDOPHYTA
CLASSES	4
ORDERS	11
FAMILIES	37
SPECIES	12,000

DEBATE

LIVING FOSSILS

Horsetails have changed little in over 300 million years, although most living species are much smaller than the giant forms that existed in prehistoric times. One of the largest genera, called *Calamites*, grew up to 65 ft (20 m) high, and had massive ridged stems over 24 in (60 cm) across. Opinions differ about how horsetails should be classified. Some botanists rank them in a separate division, but there is increasing evidence that they have close links with ferns, and should be classified in the same clade, or evolutionary group.

Some ferns are small and easily overlooked, but the group also includes large treelike species that can be over 50 ft (15 m) high. Many are compact plants, with a single cluster of fronds, but others have creeping stems that sprout fronds as they grow. Bracken—one of the most widespread creeping species—is particularly vigorous. Over many years, it can produce clumps, or clones, over ½ mile (800 m) across. Ferns also include species that live in fresh water and a wide variety of epiphytic species, which live on other plants.

Fern fronds are often finely divided, and usually develop from tightly coiled buds, known as fiddleheads. Ferns form spores on the underside of the fronds, in structures that look like raised dots or lines. In many fern species, all the fronds can produce spores, but in some, the fronds are of two kinds: fertile fronds form spores, while sterile fronds concentrate on collecting sunlight to enable the fern to grow. When a fern spore germinates and starts developing, it produces an intermediate stage known as a gametophyte. Tiny, flat, and paper-thin, this eventually produces a new sporophyte, or spore-forming plant.

FERN RELATIVES

As well as true ferns, this group includes some closely related plants. The most distinctive of these are horsetails—upright plants with a rough texture, and hollow, cylindrical stems. Their stems are symmetrical, with whorls of slender branches attached to equally spaced joints. Horsetails get their texture from granules of silica, and in the past they were widely used for scouring pots and pans.

Whisk ferns and adder's tongues are small but intriguing plants, with twig-shaped stems or a single divided frond. Like true ferns and horsetails, they also reproduce by means of spores.

∨ **CHARACTERISTIC COIL**
The delicate tips of young fern fronds are protected inside tight coils. These unwind as they stretch up toward the light.

HORSETAILS

32 in/80 cm

COMMON HORSETAIL
Equisetum arvense
F: Equisetaceae

Widespread in the Northern Hemisphere, this horsetail can be a troublesome weed. Its black underground rhizomes sprout hollow shoots, bearing symmetrical rings of bright green branches.

TRUE FERNS

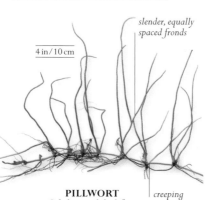

slender, equally spaced fronds

4 in/10 cm

PILLWORT
Pilularia globulifera
F: Marsileaceae

This marshland fern from western Europe forms clumps resembling tufts of grass. Its spores develop at ground level, in green capsules shaped like pills.

creeping stem

6 in/15 cm

WATER CLOVER
Marsilea quadrifolia
F: Marsileaceae

With its four-lobed fronds, this clump-forming aquatic fern looks deceptively like a flowering plant. It is widely distributed across the Northern Hemisphere.

5 ft/1.5 m

ROYAL FERN
Osmunda regalis
F: Osmundaceae

Often cultivated, this stately fern from the Northern Hemisphere has a rosette of spreading fronds, topped by much narrower fronds that bear its spores.

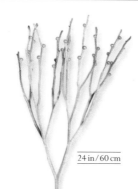

¾ in/2 cm

FLOATING FERN
Salvinia natans
F: Salviniaceae

Often forming a dense carpet, this water fern has small, oval leaves covered with water-repellent hairs. It is common throughout the tropics.

WHISK FERNS, MOONWORTS, AND ADDER'S TONGUES

24 in/60 cm

WHISK FERN
Psilotum nudum
F: Psilotaceae

A primitive relative of true ferns, this largely tropical plant has leafless brushlike stems, bearing berrylike capsules that produce its spores.

8 in/20 cm

ADDER'S-TONGUE FERN
Ophioglossum vulgatum
F: Ophioglossaceae

This unusual fern has a single oval frond, clasping a slender spore-producing stalk. It grows in grassy places throughout the Northern Hemisphere.

COMMON MOONWORT
Botrychium lunaria
F: Ophioglossaceae

Found worldwide in temperate regions, moonwort has a single frond. Spores are produced from rounded capsules, borne on a branched stalk.

8 in/20 cm

stem attached to short, fleshy rootstock

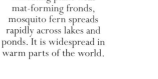

½ in/1.5 cm

MOSQUITO FERN
Azolla filiculoides
F: Azollaceae

A floating plant with mat-forming fronds, mosquito fern spreads rapidly across lakes and ponds. It is widespread in warm parts of the world.

3¼ ft/1 m

STICHERUS CUNNINGHAMII
F: Gleicheniaceae

This distinctive fern from New Zealand has a pencil-thin stem topped by a horizontal crown of narrow, radiating fronds. It spreads by creeping rhizomes.

33 ft
10 m

SILVER TREE FERN
Cyathea dealbata
F: Cyatheaceae

This tree fern gets its name from its fronds, which have silvery undersides. Native to New Zealand, it grows in open forest and scrub.

20 ft/6 m

fronds are evergreen in mild climates

TASMANIAN TREE FERN
Dicksonia antarctica
F: Dicksoniaceae

This sturdy-stemmed tree fern is widespread in Tasmania and southeast Australia. It needs shady conditions, and usually grows among trees.

BLACK TREE FERN
Cyathea medullaris
F: Cyatheaceae

60 ft
18 m

From New Zealand, this tall and slender tree fern has a sooty-black trunk and dark leaf stalks, contrasting with its bright green fronds.

SOFT TREE FERN
Cyathea smithii
F: Cyatheaceae

25 ft
8 m

Native to New Zealand and sub-Antarctic islands, this is the most southerly tree fern. It often has a collar of dead fronds hanging beneath its crown.

>>

SILVER BRAKE
Pteris argyraea
F: Pteridaceae
This shade-loving fern is named for its distinctive fronds, which have a silvery white central stripe. A pantropical species, it forms extensive clumps.

3¼ ft
1 m

24 in
60 cm

purple stems and veins

PAINTED BRAKE
Pteris tricolor
F: Pteridaceae
From Malaysia, painted brake has an unusual range of colors: its fronds are purple when young, becoming metallic green as they mature.

long terminal leaflet

20 in
50 cm

CLIFF BRAKE
Pellaea viridis
F: Pteridaceae
Native to South Africa, this drought-resistant fern has shining green fronds with wiry black stems. It grows in open woodlands on mountains.

BLACK MAIDENHAIR FERN
Adiantum capillus-veneris
F: Adiantaceae
Found in crevices on limestone, this widespread fern has bright green, translucent leaflets that contrast with its slender black stems.

12 in
30 cm

10 in / 25 cm

SILVER LIP FERN
Cheilanthes argentea
F: Adiantaceae
From eastern Asia, this small, evergreen fern has wedge-shaped fronds with black veins and distinctive silvery markings on their undersides.

LADDER BRAKE
Pteris vittata
F: Pteridaceae
Widespread in warm regions, this fern has upright or arching fronds with narrow linear leaflets. It usually grows on limestone and in alkaline soil.

3¼ ft
1 m

16 in
40 cm

NORTHERN BEECH FERN
Thelypteris phegopteris
F: Thelypteridaceae
Found as far north as Greenland, this compact fern grows in a variety of habitats from woodlands to rocky tundra.

30 in
75 cm

MOUNTAIN FERN
Oreopteris limbosperma
F: Thelypteridaceae
Found in damp habitats in acidic soil, this clump-forming European fern gives off a lemony smell if its fronds are bruised.

6½ ft
2 m

BRACKEN FERN
Pteridium aquilinum
F: Dennstaedtiaceae
Found on every continent except Antarctica, bracken fern spreads vigorously by growing underground rhizomes. It often dies back in winter, sprouting new fronds in spring.

purplish-black stalks

20 in
50 cm

SQUIRREL'S-FOOT FERN
Davallia trichomanoides
F: Davalliaceae
This epiphytic fern from Malaysia creeps over trees and other plants. Its rhizomes have furlike scales and tips that resemble a squirrel's feet.

6 ft
1.8 m

EUROPEAN CHAIN FERN
Woodwardia radicans
F: Blechnaceae
A luxuriant fern from damp, shaded habitats, this European species has large arching fronds that sometimes produce bulbils (small bulbs) at their tips.

30 in
75 cm

DEER FERN
Blechnum spicant
F: Blechnaceae
This evergreen fern has spreading sterile fronds, and upright fertile fronds with leaflets like narrow ribs. It is native to the temperate Northern Hemisphere.

SENSITIVE FERN
Onoclea sensibilis
F: Woodsiaceae
Originally from North America and eastern Asia, this wetland fern has sterile fronds that die down soon after the first frost.

winged midrib

toothed, pale green leaflets

24 in
60 cm

OAK FERN
Gymnocarpium dryopteris
F: Woodsiaceae
Delicate bright green fronds, sprouting separately, distinguish this northern fern. It grows in shaded screes as well as in woodlands.

16 in
40 cm

MALE FERN
Dryopteris filix-mas
F: Dryopteridaceae
One of the most common woodland ferns in Europe, this species has a crown of concentric fronds shaped like the tail of a shuttlecock.

4 ft
1.2 m

spore capsules on underside of fronds

HART'S-TONGUE FERN
Asplenium scolopendrium
F: Aspleniaceae
With its glossy strap-shaped fronds, this fern is often cultivated as an ornamental plant. It grows wild in Europe, western Asia, and North America.

24 in
60 cm

8 in
20 cm

MAIDENHAIR SPLEENWORT
Asplenium trichomanes
F: Aspleniaceae
Occurring from the tropics to sub-Arctic regions, this small clump-forming fern grows in rocky places. Its fronds have paired elliptical leaflets.

SOFT SHIELD FERN
Polystichum setiferum
F: Dryopteridaceae
Native to damp European woodlands, soft shield fern has feather-shaped fronds, with leaflets turned at an angle, instead of lying flat.

4 ft
1.2 m

BRITTLE BLADDER-FERN
Cystopteris fragilis
F: Woodsiaceae
Named for its rounded spore capsules borne on the underside of its fronds, this fern grows in temperate regions worldwide.

16 in
40 cm

6 in
15 cm

WALL RUE
Asplenium ruta-muraria
F: Aspleniaceae
Widespread across the Northern Hemisphere, this small tuft-forming fern grows on limestone rocks and on walls containing lime-rich mortar.

COMMON STAGHORN FERN
Platycerium bifurcatum
F: Polypodiaceae
Found in Indonesia and Australasia, this impressive epiphyte grows on tree-trunks. It has kidney-shaped, sterile fronds and spreading fertile fronds shaped like antlers.

35 in / 90 cm

fronds form cone-shaped crown

5 ft
1.5 m

OSTRICH FERN
Matteuccia struthiopteris
F: Woodsiaceae
Native to the Northern Hemisphere, this tall waterside fern has a symmetrical crown of sterile fronds in summer, followed by brown fertile fronds in the winter.

12 in
30 cm

COMMON POLYPODY
Polypodium vulgare
F: Polypodiaceae
Instead of forming clumps, this fern sprouts separate fronds along its creeping rhizome. It is common on rocks, trees, and leaf litter in northern temperate regions.

CYCADS, GINKGOS, AND GNETOPHYTES

Found mainly in warm parts of the world, the cycads, ginkgos, and gnetophytes are three ancient divisions of nonflowering plants with strikingly varied shapes. They range from climbers and low-growing shrubs to plants that are easily mistaken for palms.

DIVISION	CYCADOPHYTA
CLASS	1
ORDER	1
FAMILIES	3
SPECIES	304

DIVISION	GINKGOPHYTA
CLASS	1
ORDER	1
FAMILY	1
SPECIES	1

DIVISION	GNETOPHYTA
CLASS	1
ORDERS	3
FAMILIES	3
SPECIES	70

⌄ BUILT TO LAST

Like palms, cycads usually have a single crown of compound leaves, surrounding a central growing point, or apical meristem. Their tough leaves are able to withstand strong sunshine and drying winds.

These three groups are traditionally classed, along with conifers, as gymnosperms—a word that means "naked seed." Unlike flowering plants, gymnosperms form their seeds on exposed surfaces, which are typically specialized scales. Flowering plants produce their seeds in closed chambers, or ovaries.

Scientists have not yet resolved the exact relationship between cycads, ginkgos, and gnetophytes. Nor can they say where they stand in relation to conifers and flowering plants. At the cellular level, some features of gnetophytes suggest they are more closely related to conifers than to cycads and ginkgos, but they also have features in common with flowering plants.

Apart from the nature of their seeds, the three groups have little in common, and hardly ever grow in the same place. Cycads grow mainly in the tropics and subtropics, while the single surviving species of ginkgo comes from China. Gnetophytes are more diverse, including trees and climbers from the tropics, densely branched shrubs from dry habitats, and the bizarre *Welwitschia*, which grows only in the Namib Desert of southwest Africa.

VARYING FORTUNES

Cycads have a long history, dating back nearly 300 million years. Once an important part of the world's vegetation, they have gradually lost ground in the face of competition from flowering plants. Today, nearly a quarter of cycad species are endangered, under threat from illegal plant collection and habitat change.

Gnetophytes face fewer problems, and the ginkgo has seen a dramatic improvement in its status. For centuries it was preserved by Buddhist monks, who grew it in temple gardens long after it had vanished in the wild. Introduced to Europe in the 1700s, it proved easy to grow and highly tolerant of polluted air. It is now planted worldwide in parks and on city streets.

CYCADS

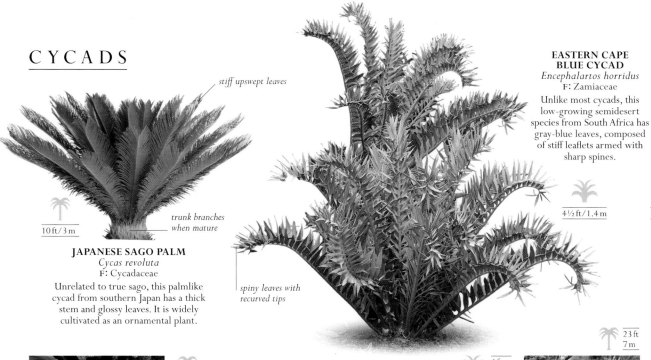

stiff upswept leaves

10 ft / 3 m

trunk branches when mature

JAPANESE SAGO PALM
Cycas revoluta
F: Cycadaceae

Unrelated to true sago, this palmlike cycad from southern Japan has a thick stem and glossy leaves. It is widely cultivated as an ornamental plant.

spiny leaves with recurved tips

EASTERN CAPE BLUE CYCAD
Encephalartos horridus
F: Zamiaceae

Unlike most cycads, this low-growing semidesert species from South Africa has gray-blue leaves, composed of stiff leaflets armed with sharp spines.

4½ ft / 1.4 m

20 ft / 6 m

PRICKLY CYCAD
Encephalartos altensteinii
F: Zamiaceae

Found near South Africa's eastern coast, this tall, subtropical cycad is named for its spiny leaves. Its yellow cones produce bright red seeds.

6 ft / 1.8 m

MEXICAN FERN PALM
Dioon edule
F: Zamiaceae

A cycad rather than a true palm, this slow-growing plant from eastern Mexico has oval seed cones up to 12 in (30 cm) long.

6½ ft / 2 m

MEXICAN HORNCONE
Ceratozamia mexicana
F: Zamiaceae

This robust cycad from eastern Mexico has a large crown of spreading leaves and gray-green cones with distinctive horns on their scales.

4 ft 1.2 m

COONTIE
Zamia pumila
F: Zamiaceae

Once used as a source of starch, this dwarf cycad from the Caribbean has short, half-buried stems and upright, reddish brown cones.

23 ft 7 m

MACROZAMIA MOOREI
F: Zamiaceae

One of Australia's tallest cycads, this palmlike plant has giant seed cones up to 36 in 990 cm) long. It grows in dry woodlands.

10 ft / 3 m

BURRAWONG
Macrozamia communis
F: Zamiaceae

Native to Australia's southeast coast, this cycad has large cones with red, fleshy seeds. It often grows in dense stands.

GNETOPHYTES

curled leaves

LONGLEAF JOINT-FIR
Ephedra trifurca
F: Ephedraceae

Also known as Mexican tea, this mound-forming plant has a dense mass of leafless stems. It grows in deserts in Mexico and the southern United States.

6½ ft / 2 m

MELINJO
Gnetum gnemon
F: Gnetaceae

50 ft / 15 m

A nonflowering tree from southeast Asia and the Pacific, melinjo has evergreen leaves and nutlike seeds. Both are used in cooking.

3¼ ft / 1 m

CHINESE EPHEDRA
Ephedra sinica
F: Ephedraceae

This straggling plant from eastern Asia contains powerful alkaloids and has a long history of use as a medicinal herb.

male cones attached to central stem

3¼ ft / 1 m

GINKGO

98 ft / 30 m

MAIDENHAIR TREE
Ginkgo biloba
F: Ginkgoaceae

This distinctive tree is easily recognized by its fan-shaped leaves, which turn bright yellow in fall. Originally restricted to southern China, it is now cultivated worldwide.

EDIBLE KERNEL

FLESHY FRUIT

WELWITSCHIA
Welwitschia mirabilis
F: Welwitschiaceae

Endemic to the Namib Desert, this extremely long-lived plant has a single pair of strap-shaped leaves. Over the centuries they crack and split, creating a tangled mound of foliage.

CONIFERS

Conifers evolved over 300 million years ago, long before the world's first broad-leaved trees. With their tough, waxy leaves, conifers thrive in harsh climates. They are less diverse than broad-leaved trees, but they dominate forests on cold mountains, and in the far north.

DIVISION	PINOPHYTA
CLASS	1
ORDER	1
FAMILIES	7
SPECIES	630

Between areas of open tundra, conifers grow far to the north of the Arctic Circle, forming the largest forest on Earth.

Although relatively few in species, conifers include the world's tallest trees, as well as the heaviest, the longest-lived, and some of the most widespread. Conifers are traditionally placed in a group known as gymnosperms, along with cycads, ginkgos, and gnetophytes. Unlike broad-leaved trees, they do not have flowers and produce pollen and seeds in cones.

LEAVES AND CONES

Most conifers are evergreen, with highly resinous leaves that are good at withstanding cold winds and strong sunshine. Pines have slender needles, arranged singly or in bundles, while other conifers often have linear leaves or flat scales. A few conifers—including larches and some redwoods—are deciduous, with soft leaves that are shed each year.

Conifers have two types of cones, which normally grow on the same tree. Male cones produce pollen. Small, soft, and often very numerous, they typically appear in spring, and wither soon after shedding their pollen into the air. The larger female cones contain one or more seeds. They can take several years to develop, and usually become hard and woody when mature.

The cones of some species, including firs and cedars, release their seeds by slowly disintegrating on the tree. Pine cones remain intact, and often remain on the tree long after they are ripe. Most of them shed their seeds in dry weather, when their scales open wide, but some hold on to their seeds until they are scorched. This is a specialized adaptation that allows pines to recolonize areas that have been burned by forest fires.

A few groups of conifers, the yews, junipers, and podocarps, have small berrylike cones with fleshy scales. Their seeds are spread by birds, which scatter them as they feed.

∨ **CONIFERS IN CONTROL**
Conifers often form extensive single-species stands. These pines are growing in California's Yosemite National Park in the USA.

COLORADO BLUE SPRUCE
Picea pungens
F: Pinaceae

Bright blue-gray leaves, with prickly tips, make this a popular ornamental tree. Native to western North America, it typically grows on mountains.

115 ft / 35 m

CONE

FOUR-SIDED RIGID LEAVES

GRAND FIR
Abies grandis
F: Pinaceae

The grand fir is a tall, fast-growing tree that is native to western North America. Its leaves have an orangelike scent when crushed.

165 ft / 50 m

CALIFORNIA RED FIR
Abies magnifica
F: Pinaceae

This drought-resistant fir is found on dry mountain slopes. It has leaves that curve upward and upright cones up to 8 in (20 cm) long.

130 ft / 40 m

EUROPEAN SILVER FIR
Abies alba
F: Pinaceae

Named after the silver bands on the underside of its leaves, this fir has upright resinous cones. They disintegrate in order to scatter their seeds.

130 ft / 40 m

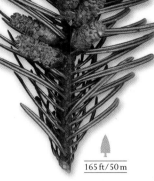

119

PLANTS • CONIFERS

CAUCASIAN FIR
Abies nordmanniana
F: Pinaceae

Widely grown as a Christmas tree, this fir is Europe's tallest conifer. It originally comes from mountains in the Black Sea region.

165 ft / 50 m

SITKA SPRUCE
Picea sitchensis
F: Pinaceae

Thriving in cold, wet conditions, the sitka spruce is often grown as a forestry tree. It comes from the coastal belt of western North America.

undersides of leaves have white bands

cylindrical cones with toothed scales

50 ft / 15 m

SINGLE-LEAF PINYON
Pinus monophylla
F: Pinaceae

Unique among pines, this low-growing tree has leaves that grow singly, rather than in clusters. It comes from rocky slopes in Mexico and the American Southwest.

165 ft / 50 m

NORWAY SPRUCE
Picea abies
F: Pinaceae

An important lumber tree, this fast-growing conifer has spiky leaves and cylindrical cones. Its natural range includes most of northern Europe.

SCOTCH PINE
Pinus sylvestris
F: Pinaceae

Ranging from the British Isles to China, the Scotch pine is the world's most widespread conifer. When mature, it has peeling bark and often an asymmetrical shape.

115 ft / 35 m

LONG, THICK LEAVES

65 ft / 20 m

CONE WITH EDIBLE SEEDS

LODGEPOLE PINE
Pinus contorta
F: Pinaceae

Growing in coastal dunes and bogs, this North American pine has paired leaves and prickly cones. The cones release their seeds after being scorched by fire.

33 ft / 10 m

STONE PINE
Pinus pinea
F: Pinaceae

Prized for its edible seeds (pine nuts), this Mediterranean pine has large oval cones, and an elegant umbrella-shaped outline when mature.

100 ft / 30 m

male pollen-producing cones

stiff, sharp needles

AROLLA PINE
Pinus cembra
F: Pinaceae

Found on European mountains, this slow-growing tree produces small cones that fall intact, like those of other pines. Its seeds are then dispersed by birds.

65 ft / 20 m

MARITIME PINE
Pinus pinaster
F: Pinaceae

Native to the western Mediterranean, the maritime pine grows rapidly on poor, sandy soil. It has glossy brown cones that are up to 8 in (20 cm) long.

115 ft / 35 m

AUSTRIAN PINE
Pinus nigra
F: Pinaceae

Tall, rangy, and open-crowned, this pine has long leaves arranged in pairs. Despite its name, it is widespread throughout Europe, typically growing on limestone.

130 ft / 40 m

CHINESE PINE
Pinus tabuliformis
F: Pinaceae

With age, this oriental pine develops a distinctive flat, spreading crown. It grows on mountains, and produces small egg-shaped cones.

80 ft / 25 m

MONTEREY PINE
Pinus radiata
F: Pinaceae

Originally restricted to a small area of California, this fast-growing pine is now widely planted for lumber, particularly in the Southern Hemisphere.

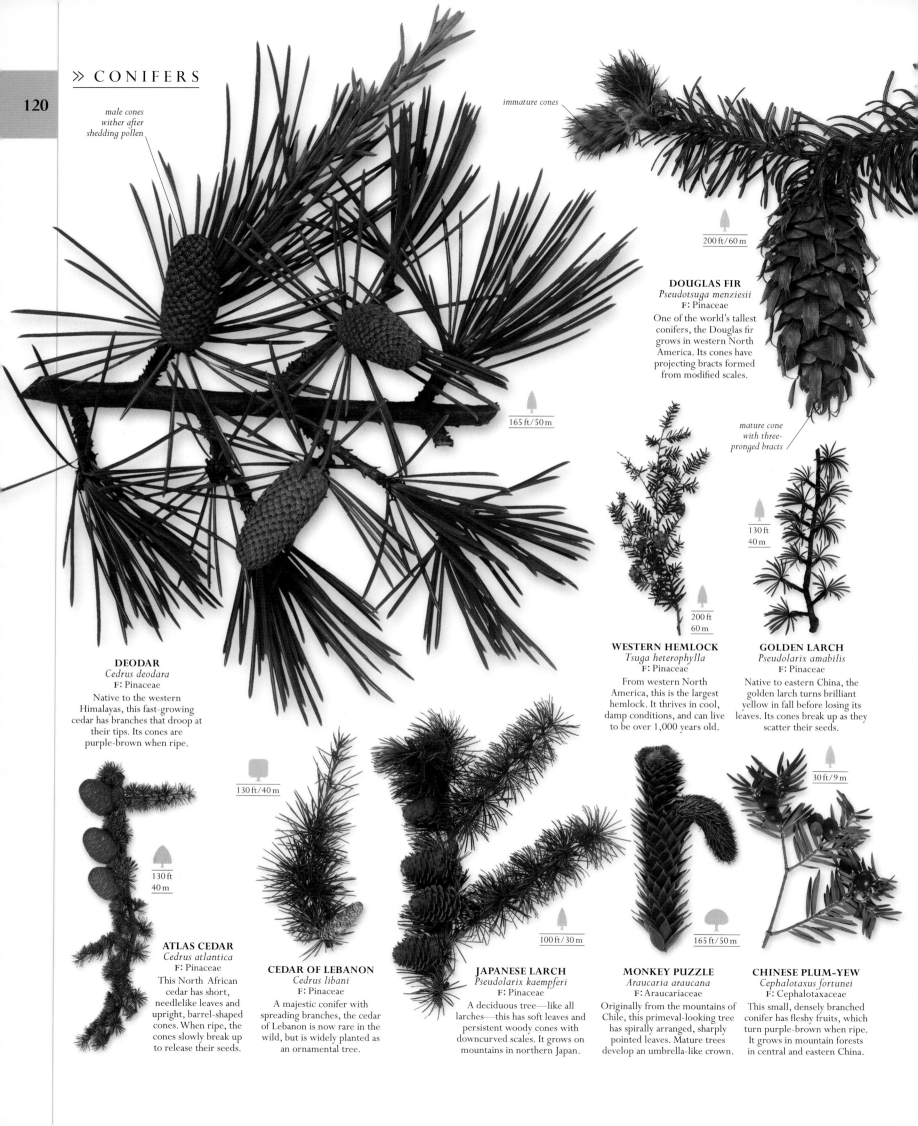

male cones wither after shedding pollen

immature cones

200 ft / 60 m

DOUGLAS FIR
Pseudotsuga menziesii
F: Pinaceae
One of the world's tallest conifers, the Douglas fir grows in western North America. Its cones have projecting bracts formed from modified scales.

165 ft / 50 m

mature cone with three-pronged bracts

130 ft
40 m

DEODAR
Cedrus deodara
F: Pinaceae
Native to the western Himalayas, this fast-growing cedar has branches that droop at their tips. Its cones are purple-brown when ripe.

200 ft
60 m

WESTERN HEMLOCK
Tsuga heterophylla
F: Pinaceae
From western North America, this is the largest hemlock. It thrives in cool, damp conditions, and can live to be over 1,000 years old.

GOLDEN LARCH
Pseudolarix amabilis
F: Pinaceae
Native to eastern China, the golden larch turns brilliant yellow in fall before losing its leaves. Its cones break up as they scatter their seeds.

130 ft / 40 m

130 ft
40 m

ATLAS CEDAR
Cedrus atlantica
F: Pinaceae
This North African cedar has short, needlelike leaves and upright, barrel-shaped cones. When ripe, the cones slowly break up to release their seeds.

CEDAR OF LEBANON
Cedrus libani
F: Pinaceae
A majestic conifer with spreading branches, the cedar of Lebanon is now rare in the wild, but is widely planted as an ornamental tree.

100 ft / 30 m

JAPANESE LARCH
Pseudolarix kaempferi
F: Pinaceae
A deciduous tree—like all larches—this has soft leaves and persistent woody cones with downcurved scales. It grows on mountains in northern Japan.

165 ft / 50 m

MONKEY PUZZLE
Araucaria araucana
F: Araucariaceae
Originally from the mountains of Chile, this primeval-looking tree has spirally arranged, sharply pointed leaves. Mature trees develop an umbrella-like crown.

30 ft / 9 m

CHINESE PLUM-YEW
Cephalotaxus fortunei
F: Cephalotaxaceae
This small, densely branched conifer has fleshy fruits, which turn purple-brown when ripe. It grows in mountain forests in central and eastern China.

JAPANESE UMBRELLA PINE
Sciadopitys verticillata
F: Cupressaceae
As a wild tree, this pinelike conifer is restricted to the mountains of Japan. It has slender leaves arranged like the spokes of an umbrella.

80 ft / 25 m

FRAGILE RIPE CONE

LAWSON CYPRESS
Chamaecyparis lawsoniana
F: Cupressaceae
Like other cypresses, this has small cones and sprays of tiny scalelike leaves. Originally from western North America, it exists in many cultivated forms.

165 ft / 50 m

98 ft
30 m

JAPANESE CEDAR
Cryptomeria japonica
F: Cupressaceae
A cypress rather than a true cedar, this tree has slender leaves and small rounded cones. It grows on mountains in China and Japan.

WESTERN JUNIPER
Juniperus occidentalis
F: Cupressaceae
This long-lived tree grows on rocky mountain slopes in the western United States. Like other junipers, it produces seeds inside berrylike cones.

65 ft
20 m

80 ft / 25 m

CHINESE JUNIPER
Juniperus chinensis
F: Cupressaceae
Widespread in temperate eastern Asia, this shrub or small tree has prickly leaves when young, and scalelike leaves when mature.

linear leaves in opposite ranks

360 ft
110 m

COAST REDWOOD
Sequoia sempervirens
F: Cupressaceae
Native to the coast of northern California, this is the world's tallest tree. Mature trees have soaring trunks and relatively sparse branches, and can be over 1,000 years old.

330 ft / 100 m

GIANT SEQUOIA
Sequoiadendron giganteum
F: Cupressaceae
This California redwood is the world's most massive tree. The largest living specimens weigh over 5,000 tons, and have fireproof bark up to 2 ft (60 cm) thick.

RIPENING CONE

130 ft / 40 m

DAWN REDWOOD
Metasequoia glyptostroboides
F: Cupressaceae
Native to central China, this deciduous redwood is extremely rare in the wild. Until the 1940s it was thought to be extinct, being known only from fossils.

80 ft / 25 m

MONTEREY CYPRESS
Cupressus macrocarpa
F: Cupressaceae
Although widely cultivated, the wild Monterey cypress is restricted to a small region of California's coast. Mature trees typically have an irregular spreading shape.

arils turn red when ripe

200 ft / 60 m

TAIWANIA
Taiwania cryptomerioides
F: Cupressaceae
One of Asia's largest conifers, this tropical species has a trunk up to 10 ft (3 m) thick. It has spiny-tipped leaves and small, rounded cones.

130 ft / 40 m

SWAMP CYPRESS
Taxodium distichum
F: Cupressaceae
Also known as the bald cypress, this deciduous conifer grows in swamps in the southeastern United States. Its trunk often has a buttressed base.

165 ft / 50 m

WESTERN RED CEDAR
Thuja plicata
F: Cupressaceae
This large tree grows in northwestern North America. It has flat sprays of small scalelike leaves, and its wood is highly rot-resistant, even when dead.

98 ft / 30 m

CALIFORNIA NUTMEG
Torreya californica
F: Taxaceae
This rare conifer is restricted to canyons and mountains in California. It is unrelated to the true nutmeg, although it has nutlike seeds.

65 ft / 20 m

EUROPEAN YEW
Taxus baccata
F: Taxaceae
This long-lived tree produces seeds in fleshy arils—modified cone scales. Found wild in Europe and southwest Asia, it is commonly planted.

FLOWERING PLANTS

With over 250,000 species, flowering plants—or angiosperms—are by far the largest group of plants, as well as the most diverse. They play a vital role in most land-based ecosystems, producing food and shelter for animals and many other living things.

DIVISION	ANGIOSPERMAE
CLADES	3
ORDERS	40
FAMILIES	603
SPECIES	About 255,000

Wind-pollinated flowers such as hazel catkins shed clouds of pollen grains into the air. They are rarely brightly colored.

Animal-pollinated flowers are usually conspicuous, and have sticky pollen grains. Hummingbirds transfer pollen as they feed.

Fleshy fruits evolved to attract animals. Wild cucumbers are eaten by antelopes, which scatter the seeds in their droppings.

Dry fruits often burst open when their seeds are ripe. This willowherb is scattering its fluffy seeds into the wind.

The first flowering plants evolved 140 million years ago, which makes them relative latecomers to life on Earth. Since then, they have become the dominant forms of plant life. The smallest flowering plants are not much bigger than a pinhead, but this group also includes all broadleaved trees, and a variety of other species, from cacti and grasses to orchids and palms.

Flowering plants share several key features that help make them successful. Foremost among these are flowers—actually a collection of highly modified leaves. In most flowers, the outermost layers are the sepals, then the petals. These surround the male stamens, which make pollen, and the female carpels, which collect pollen arriving from similar flowers, so that their female ovules can develop into seeds. Some flowers spread their pollen through the air, but many more use animals to transfer their pollen. In these showy species, flowers lure animals to sip the sugary nectar in the innermost part of the flower, brushing past the pollen on the way. These plants mostly attract insects, but birds and bats are also important pollinators.

STRATEGIES FOR SPREADING

Flowering plants are not unique in making seeds, but they are the only plants that produce them in fruits. A fruit develops from a flower's ovaries—the closed chambers that house the seeds, close to the flower's central stem. Fruits have a double function: to protect the seeds and to help them spread. Fleshy fruits do this by attracting animals, which feed and scatter the seeds in their excrement. Dry fruits work in a variety of ways. Some burst open when their seeds are ripe. Others have hooks that latch onto clothing, skin, or fur. Still more drift in water or through the air. Many flowering plants spread by the way they grow, often producing new plants from creeping stems. This can create vast, interconnected clones. The largest examples, formed by North American aspen trees, stretch over 100 acres (40 ha), and may be 10,000 years old.

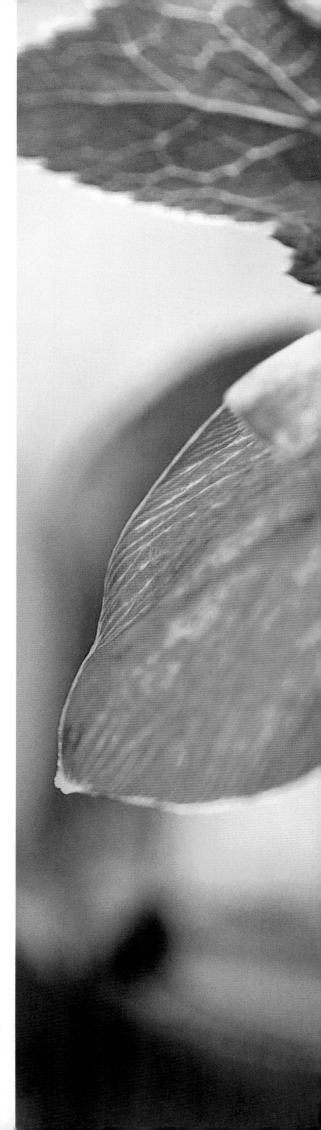

ATTRACTING ATTENTION >
Unlike most flowers, this hellebore has colorful sepals, and small green petals on the inside. It is pollinated by bees.

BASAL ANGIOSPERMS

Of the more than 50 orders of flowering plants, or angiosperms, a few, known as basal angiosperms, evolved early and still exist today.

Outwardly, basal, or primitive, angiosperms have little in common, and grow in widely scattered parts of the world. They include trees, shrubs, and climbers, as well as submerged and emergent water plants. Some have large, eye-catching flowers, while in others the flowers are small or insignificant. Most are pollinated by insects, although hornworts bloom underwater and produce drifting pollen grains.

Fossil and genetic evidence shows that basal angiosperms evolved at different times. The very first angiosperms appeared about 140 million years ago. These were probably the Amborellales, an order that exists in the form of a single species of shrub, found on only one island in the South Pacific. Later, the Nymphaeales appeared. Aquatic plants with showy flowers, they include more than 70 species of water lilies, found all over the globe. Two other orders then followed: the Austrobaileyales, with nearly 100 species of woody plants—perennials with a hard durable stem made of wood—found mainly in the tropics; and the Chloranthales, a small order of about 60 species, from the tropical Americas, eastern Asia, and the Pacific. The Ceratophyllales, or hornworts, may have been the last to split off. Meanwhile, the main line of flowering plants underwent a massive radiation, or diversification, leading to the vast majority of species alive today.

DIVISION	ANGIOSPERMAE
GROUPS	BASAL ANGIOSPERMS
ORDERS	5
FAMILIES	9
SPECIES	251

CHLORANTHALES

There are four genera in the family Chloranthaceae, which is the sole family in the order Chloranthales. Members of this primitive order are aromatic trees and shrubs with pairs of toothed leaves and inconspicuous flowers that have no petals.

SARCANDRA GLABRA
F: Chloranthaceae
With a range of medicinal uses, this evergreen inhabits damp ground, especially wooded streambanks, in southeast Asia, China, and Japan.

24 in/60 cm

berry clusters in winter

AMBORELLALES

6½ ft
2 m

This order of primitive evergreen shrub contains a single family, which has a single genus with only one species, *Amborella trichopoda*. It bears small flowers, males and females being produced on separate plants, and red berries, each containing a single seed.

AMBORELLA TRICHOPODA
F: Amborellaceae
Found only in rain forests on the island of New Caledonia in the South Pacific, this is a primitive flowering plant, threatened by habitat destruction.

AUSTROBAILEYALES

The order Austrobaileyales is made up of only four families. Members of this order are trees, shrubs, and climbers. The flowers of most species are single and have many corolla petals. Perhaps the best known species is star anise, a fragrant spice.

STEM, LEAVES, AND FLOWERS

60 ft/18 m

FRUIT

STAR ANISE
Illicium verum
F: Illiciaceae
Widely used for flavoring, star anise fruits are woody and star-shaped. This woodland species is native to China and Vietnam.

many-petalled single flower

CERATOPHYLLALES

An order of submerged freshwater plants, Ceratophyllales is not closely related to any other order. Its members belong to a single family, Ceratophyllaceae, and are characterized by dissected whorled leaves, spiny one-seeded fruits, separate male and female flowers, and a lack of roots.

3¼ ft
1 m

RIGID HORNWORT
Ceratophyllum demersum
F: Ceratophyllaceae
This is a submerged species with no roots, inhabiting ponds and ditches in non-arctic Europe. It has tiny flowers and whorls of leaves.

50 ft/15 m

AUSTROBAILEYA SCANDENS
F: Austrobaileyaceae
The flowers of this rare primitive climber, found only in rain forests in Queensland, Australia, smell like rotting fish to attract pollinating flies.

NYMPHAEALES

This is a primitive order of aquatic plants, with floating, submerged, or, more rarely, emergent leaves. The order Nymphaeales are the water lilies, which often grown ornamentally for their showy flowers. Nymphaeales are cultivated throughout the world.

WATER LILY FLOWER

10 ft / 3 m

RIMMED LEAVES

AMAZON WATER LILY
Victoria amazonica
F: Nymphaeaceae
Native to deep Amazonian backwaters, this water lily has enormous round leaves with upright rims. The flowers open at night.

star-shaped semidouble flower

125

6½ ft / 2 m

FANWORT
Cabomba caroliniana
F: Cabombaceae
From still, freshwater habitats in the central and southeastern United States, fanwort has submerged and floating leaves. It can be invasive.

NYMPHAEA 'SUNRISE'
F: Nymphaeaceae
With large, showy, fragrant flowers and mid-green floating foliage, this hybrid, thought to be of American origin, is one of the largest flowering water lilies.

20 in
50 cm

WHITE WATER LILY
Nymphaea alba
F: Nymphaeaceae
This European species has fragrant flowers and its fruits ripen under water, releasing floating seeds. It inhabits lakes, ponds, and slow-flowing rivers.

5 ft / 1.5 m

starlike, pure white flowers

CHINESE FOXNUT
Euryale ferox
F: Nymphaeaceae
This inhabitant of deep, slow-flowing water and still water in parts of Asia has prickly leaves, flowers, and one-seeded berries.

5 ft / 1.5 m

sharp prickles protect leaves above surface from animals

stems, seeds, and fruits are all edible

bright purple flower

grass-green to olive-green shoots

∨ FLOATING FLOWERS
Water lily flowers are hermaphrodite—they have both male and female organs. However, they do not normally pollinate themselves, as the female reproductive parts mature before the male ones start producing pollen. This time lag increases the chances that the flower is fertilized by incoming pollen, rather than by pollen that the flower makes itself.

stamen

petal

∧ FLOWER
The flowers have numerous white petals and bright yellow stamens, and can grow to 6 in (15 cm) in diameter.

sepal

seeds

< OVARY SECTION
The female reproductive organs (consisting of ovules, ovary, and stigma) are fused together. Cavities contain ovules, which turn into seeds when fertilized.

∨ LEAVES
The water lily has large leaves, up to 12 in (30 cm) across. Unlike most leaves, their stomata (breathing pores) are on the upper surface, which has a water-repellent coating.

furled
young leaf

aerenchyma

∧ STEM SECTION
Stems have longitudinal air spaces (aerenchyma), as well as structural tissue, creating buoyancy and allowing oxygen to circulate.

WHITE WATER LILY
Nymphaea alba

One of about 50 wild species of water lilies, this handsome plant lives in still or slow-flowing water, casting a deep shade with its rounded, glossy leaves. It grows in water up to 5 ft (1.5 m) deep, and flowers in mid- to late summer, producing a succession of pure white blooms. Each flower lasts for three to four days, opening in the morning and closing by late afternoon. They attract pollinating beetles, which often spend the night inside the flowers before being released at dawn. White water lilies are useful to water animals: pond snails glue their eggs to the underside of the leaves, and fishes hide beneath them, avoiding predatory birds. After the flowers are pollinated, they produce buoyant seeds. These float for several weeks, before sinking into the mud.

SIZE Leaf diameter 4–12 in (10–30 cm)
HABITAT Ponds, streams, lakes
DISTRIBUTION Europe
LEAF TYPE Simple, orbicular with a basal notch

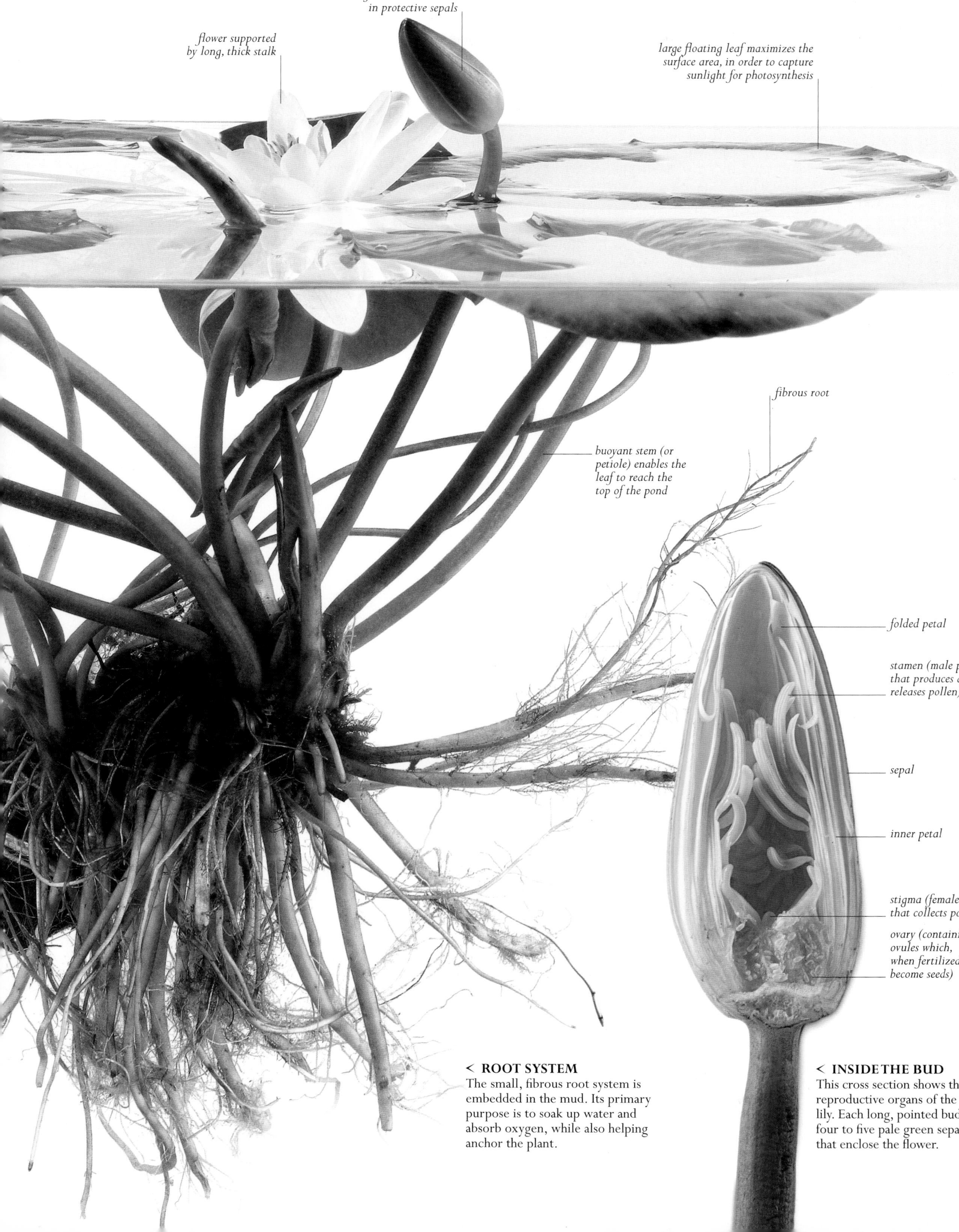

flower supported
by long, thick stalk

flower bud enclosed
in protective sepals

large floating leaf maximizes the
surface area, in order to capture
sunlight for photosynthesis

fibrous root

buoyant stem (or
petiole) enables the
leaf to reach the
top of the pond

folded petal

stamen (male part
that produces and
releases pollen)

sepal

inner petal

stigma (female part
that collects pollen)

ovary (containing
ovules which,
when fertilized,
become seeds)

< ROOT SYSTEM
The small, fibrous root system is
embedded in the mud. Its primary
purpose is to soak up water and
absorb oxygen, while also helping
anchor the plant.

< INSIDE THE BUD
This cross section shows the
reproductive organs of the water
lily. Each long, pointed bud has
four to five pale green sepals
that enclose the flower.

MAGNOLIIDS

The magnoliids are a large group of primitive flowering plants that are botanically between basal angiosperms and monocots.

Found in tropical and temperate regions, the magnoliids form a major plant group, that evolved early in the history of flowering plants. They take their name from the magnolia family, which is one of the largest in the group. Plants in this family nearly all have woody stems, and some kinds grow into large trees. However, the group also includes herbaceous or nonwoody species, both as freestanding plants and as climbers.

Some herbaceous magnoliids have highly specialized flowers. In birthworts and pipevines, for example, the flower is shaped like a flaring tube, lined with backward-pointing hairs. Flowers like these act as temporary traps for pollinating flies, which are attracted by a powerful scent. But these are an exception. Most magnoliid flowers are structurally simple, with numerous spirally arranged parts, attached separately to a central stem. Instead of sepals and petals (see p.122) the flowers have a single layer of flaps that are similar in color, size known as tepals. The fossil record shows that similar flowers existed more than 100 million years ago.

COMMON CHARACTERISTICS

Magnoliids are grouped together mainly by genetic evidence, but they also share features that can be seen with a microscope, or with the naked eye. At a microscopic level, their pollen grains have a single pore. This small but significant feature links them with monocots, but distinguishes them from eudicots—by far the largest group of flowering plants—which have three-pored pollen grains.

Most magnoliids have leaves with smooth margins, and a network of branching veins. Their fruits may be soft and fleshy, or hard and conelike, and can contain one or several seeds. Species with fleshy fruits use animals to disperse their seeds: many of them are swallowed whole and then scattered by birds. In prehistoric times, wild avocados may have been dispersed by giant ground sloths. Now that ground sloths are extinct, avocados depend on human cultivation to disperse their seeds and ensure survival.

DIVISION	ANGIOSPERMAE
CLADE	MAGNOLIIDAE
ORDERS	4
FAMILIES	20
SPECIES	7,100

CANELLALES

There are two families in the order Canellales: the Canellaceae and the Winteraceae. These are aromatic trees and shrubs with leathery entire leaves. The flowers in most species have both male and female parts and the fruit is a berry. The leaves and bark of some species can be used medicinally. The Winteraceae are a primitive family—its members have woody stems that contain no water-conducting vessels.

WINTER'S BARK
Drimys winteri
F: Winteraceae

Native to the coastal rain forest of Chile and Argentina, *Drimys winteri* has aromatic bark and leaves and fragrant flowers.

16 ft
5 m

PIPERALES

The order Piperales includes herbaceous plants, climbers, trees, and shrubs and is widely distributed in tropical regions. The stems have scattered bundles of vascular tissue, a characteristic of monocots. Members of the family Piperaceae have tiny flowers that lack petals and are clustered in spikes. Many species are aromatic.

3¼ ft
1 m

BIRTHWORT
Aristolochia clematitis
F: Aristolochiaceae

A foul-smelling and poisonous perennial, birthwort was grown for medicinal use. It is native to Europe, growing in damp places.

4 in
10 cm

ASARABACCA
Asarum europaeum
F: Aristolochiaceae

This European woodland creeper is also called wild ginger. Its glossy evergreen leaves hide insignificant flowers.

13 ft
4 m

BLACK PEPPER
Piper nigrum
F: Piperaceae

An evergreen climber from shady habitats in India and Sri Lanka, this species is widely cultivated for its aromatic fruit.

fruits, or peppercorns, used dried

MAGNOLIALES

Consisting almost exclusively of trees and shrubs, the Magnoliales are a primitive order, widely distributed in the fossil record. Although very variable, most have simple alternately arranged leaves and flowers with both male and female parts. There are six families in the order, of which the Magnoliaceae are the best known, being widely grown in gardens for their spectacular flowers.

tepals protect stamen

65 ft
20 m

100 ft
30 m

TULIP TREE
Liriodendron tulipifera
F: Magnoliaceae

A native of eastern North America, the tulip tree grows in woodlands. The leaves turn yellow before dropping in fall.

MAGNOLIA
Magnolia campbellii
F: Magnoliaceae

The flowers of this deciduous species appear in early spring, before the leaves. Magnolia grows in mountain forests in China, India, and Nepal.

LAURALES

The order Laurales consists of seven families of trees, shrubs, and woody climbing plants. A few genera grow in temperate parts of the world, but most are found in tropical and subtropical regions. Their classification is based on genetic analysis rather than morphological characteristics. Many plants are aromatic and are used in perfumery, cooking, and medicine. Others provide lumber or are grown ornamentally.

BAY LAUREL
Laurus nobilis
F: Lauraceae
The aromatic leaves of this species are used for flavoring. It grows in woods, scrub, and rocky habitats around the Mediterranean.

unripe fruit

50 ft/15 m

SASSAFRAS
Sassafras albidum
F: Lauraceae
This suckering deciduous tree of woodlands in eastern North America has aromatic leaves that are brightly colored in fall.

80 ft
25 m

FRUIT

CINNAMON STICK

CINNAMON
Cinnamomum verum
F: Lauraceae
The spice cinnamon comes from the aromatic bark of this species, found in lowland forests in Sri Lanka.

60 ft
18 m

LEAVES

CAROLINA ALLSPICE
Calycanthus floridus
F: Calycanthaceae
Native to woods and streamsides in the southeastern United States, Carolina allspice has aromatic leaves and bark and large, fragrant flowers.

60 ft
18 m

CALIFORNIA LAUREL
Umbellularia californica
F: Lauraceae
Sometimes called "headache tree," as scent from the crushed leaves can cause headaches, this species from the United States flowers in winter.

8 ft/2.5 m

LEAF CLUSTER

EDIBLE FRUIT

60 ft
18 m

AVOCADO TREE
Persea americana
F: Lauraceae
Probably originating in southern Mexico, in well-drained parts of rain forests, the avocado tree is widely cultivated, for its edible pear-shaped fruit.

white to dark pink flowers open early in spring

SCENTED FLOWERS

EVERGREEN LEAVES

65 ft
20 m

YLANG-YLANG
Cananga odorata
F: Annonaceae
Oil from the fragrant flowers of ylang-ylang are used in perfumery. This evergreen tree comes from parts of Asia and Australia.

26 ft
8 m

SWEETSOP
Annona squamosa
F: Annonaceae
Believed to have Caribbean origins, sweetsop, or sugar or custard apple, is widely cultivated. The fruit has edible flesh, which looks and tastes like custard.

GLOSSY LEAVES

mace

SEED, OR NUTMEG, WITH MACE COVERING

60 ft
18 m

NUTMEG
Myristica fragrans
F: Myristicaceae
The spices nutmeg and mace both come from the seeds of *Myristica fragrans*, a native of the Molucca—or Spice—Islands of Indonesia.

26 ft
8 m

LEAVES

EDIBLE FRUIT

PAWPAW
Asimina triloba
F: Annonaceae
This deciduous tree grows in damp woods in the eastern United States. Singly borne flowers produce edible fruit.

MONOCOTYLEDONS

Defined by their unique internal anatomy, monocots include grasses and palms, plus lilies, orchids, and many other ornamental plants.

Early in the evolution of flowering plants, their family tree developed two major branches. A smaller but still substantial branch became the monocotyledons (monocots). The larger branch became the eudicotyledons (eudicots). Monocots get their name from their seeds, which have a single prepacked seed-leaf, or cotyledon. There is no other foolproof way of recognizing a monocot, but there are some strong clues. Most have long narrow leaves with parallel veins, and the flower parts (petals, stamens, and so on) are arranged in threes or multiples of three, and not fours or fives as in most eudicots, and the pollen grains only have a single opening. In many monocot flowers, such as tulips, the sepals and petals are almost identical and known as tepals, but many of the magnoliids share this feature as well.

Below ground, monocots usually have a cluster of fibrous roots, instead of a main taproot with smaller roots leading off it. Another key characteristic—revealed under a microscope—involves the structure of their stems. They have scattered bundles of vascular tissue—the specialized cells that carry water and sap—unlike eudicots, which have theirs in concentric rings. This makes monocot stems more flexible than those of a typical eudicot, but it also makes it harder for them to evolve into trees. Treelike monocots, which principally include palms, have a very different way of growing from typical broad-leaved trees and conifers. Their trunks grow taller, but not thicker, and are usually topped by a single rosette of leaves.

SURVIVAL STRATEGIES

Smaller monocots vary greatly, from climbers to acqautic plants that survive harsh periods as underground storage organs, such as bulbs or tubers. Grasses thrive on the grazing effects of animals, and are the only plant family that forms an entire habitat: grassland. In the tropics, many monocots are epiphytic, growing high up in trees. This group includes bromeliads, and many orchids—by far the largest monocot family, with over 25,000 species worldwide.

DIVISION	ANGIOSPERMAE
CLADE	MONOCOTYLEDONEAE
ORDERS	11
FAMILIES	81
SPECIES	58,000

DEBATE

DID MONOCOTS HAVE AQUATIC ORIGINS?

Living monocots include many freshwater plants, and also some of the few flowering plants that live in the sea. According to a longstanding theory, monocots may have originally evolved in fresh water, before diversifying and taking up life on land. This would account for the long, thin leaves found in many species, and also the internal structure of their stems (see left). Having spread to land, some terrestrial monocots then came full circle and evolved into aquatic forms—duckweeds, for example, originated in this way.

ACORALES

Only one genus and as few as two species make up the order Acorales. Called sweet flags, these waterside and wetland plants have a fleshy flower spike of small flowers and were once classified as aroids (see right). Botanists now believe they represent the earliest branch of the monocot family tree, and may hold clues as to what the first monocots looked like.

3¼ ft / 1 m

SWEET FLAG
Acorus calamus
F: Acoraceae
Once cut to strew on floors because of its fresh citrus scent, this waterside plant grows across the Northern Hemisphere.

ALISMATALES

This order includes many common aquatic plants as well as the mainly land-based aroid family (Araceae). Aroids, which can be dramatic in appearance, have a distinctive reproductive anatomy consisting of a fleshy spike of tiny flowers called a spadix and a leaflike surround called a spathe. The other families in the order include many freshwater species, as well as several families of seagrasses.

WATER PLANTAIN
Alisma plantago-aquatica
F: Alismataceae
Found in waterside habitats, this plant is common across the Northern Hemisphere. It has white, pink, or purple flowers that last only a day.

OVAL FLOATING LEAF

3¼ ft 1 m

BRANCHED FLOWER CLUSTER

WATER HAWTHORN
Aponogeton distachyos
F: Aponogetonaceae
Native to South Africa, this plant has become widely naturalized elsewhere. Its vanilla-scented flowers open just above the water surface.

3¼ ft 1 m

elliptical to oval leaves

male flowers

FLOWER STEM

ARROW-SHAPED LEAF

female flowers

3¼ ft / 1 m

ARROWHEAD
Sagittaria sagittifolia
F: Alismataceae
This European wetland plant grows arrowhead-shaped leaves above water, ribbonlike ones below water, and sometimes floating leaves too.

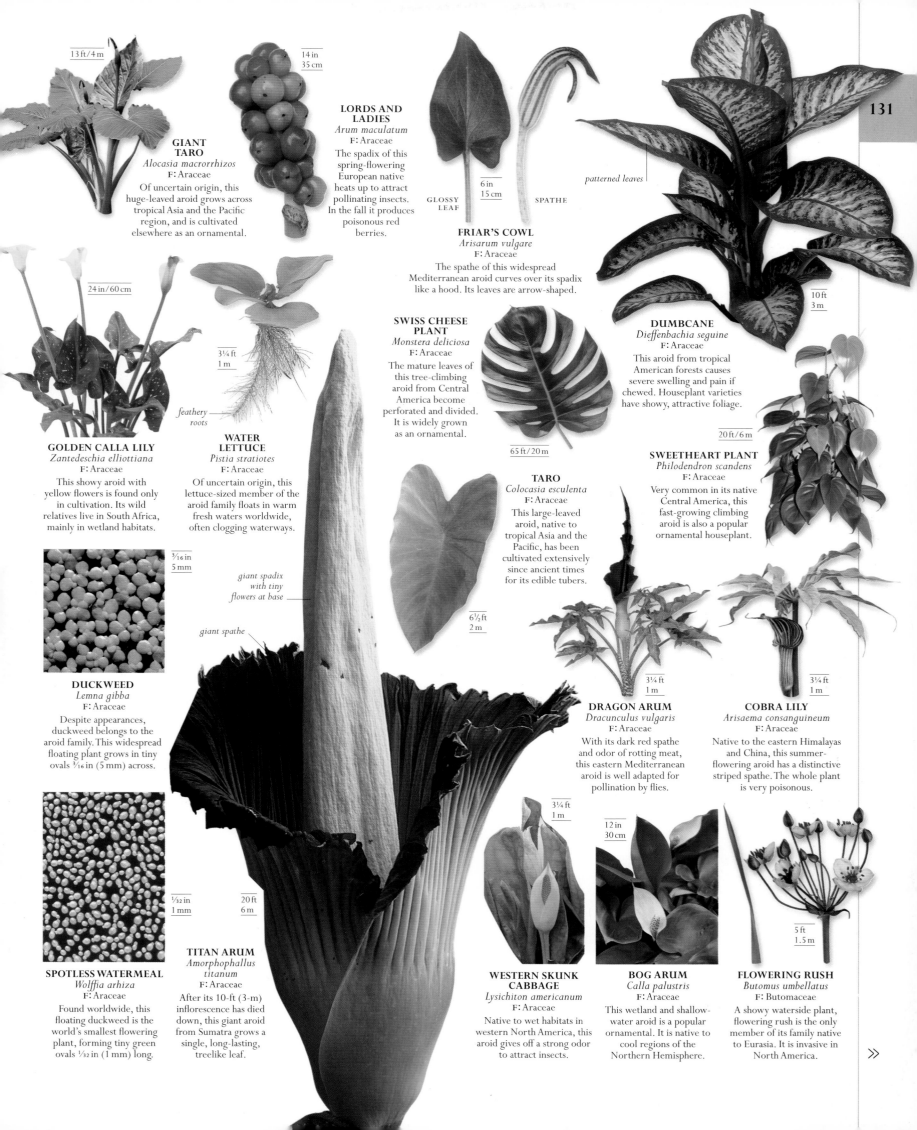

GIANT TARO
Alocasia macrorrhizos
F: Araceae
Of uncertain origin, this huge-leaved aroid grows across tropical Asia and the Pacific region, and is cultivated elsewhere as an ornamental.

13 ft/4 m

14 in 35 cm

LORDS AND LADIES
Arum maculatum
F: Araceae
The spadix of this spring-flowering European native heats up to attract pollinating insects. In the fall it produces poisonous red berries.

GLOSSY LEAF

6 in 15 cm

SPATHE

patterned leaves

FRIAR'S COWL
Arisarum vulgare
F: Araceae
The spathe of this widespread Mediterranean aroid curves over its spadix like a hood. Its leaves are arrow-shaped.

SWISS CHEESE PLANT
Monstera deliciosa
F: Araceae
The mature leaves of this tree-climbing aroid from Central America become perforated and divided. It is widely grown as an ornamental.

65 ft/20 m

DUMBCANE
Dieffenbachia seguine
F: Araceae
This aroid from tropical American forests causes severe swelling and pain if chewed. Houseplant varieties have showy, attractive foliage.

10 ft 3 m

24 in/60 cm

GOLDEN CALLA LILY
Zantedeschia elliottiana
F: Araceae
This showy aroid with yellow flowers is found only in cultivation. Its wild relatives live in South Africa, mainly in wetland habitats.

3¼ ft 1 m

feathery roots

WATER LETTUCE
Pistia stratiotes
F: Araceae
Of uncertain origin, this lettuce-sized member of the aroid family floats in warm fresh waters worldwide, often clogging waterways.

TARO
Colocasia esculenta
F: Araceae
This large-leaved aroid, native to tropical Asia and the Pacific, has been cultivated extensively since ancient times for its edible tubers.

6½ ft 2 m

20 ft/6 m

SWEETHEART PLANT
Philodendron scandens
F: Araceae
Very common in its native Central America, this fast-growing climbing aroid is also a popular ornamental houseplant.

³⁄₁₆ in 5 mm

giant spadix with tiny flowers at base

giant spathe

DUCKWEED
Lemna gibba
F: Araceae
Despite appearances, duckweed belongs to the aroid family. This widespread floating plant grows in tiny ovals ³⁄₁₆ in (5 mm) across.

DRAGON ARUM
Dracunculus vulgaris
F: Araceae
With its dark red spathe and odor of rotting meat, this eastern Mediterranean aroid is well adapted for pollination by flies.

3¼ ft 1 m

COBRA LILY
Arisaema consanguineum
F: Araceae
Native to the eastern Himalayas and China, this summer-flowering aroid has a distinctive striped spathe. The whole plant is very poisonous.

3¼ ft 1 m

¹⁄₃₂ in 1 mm

20 ft 6 m

SPOTLESS WATERMEAL
Wolffia arhiza
F: Araceae
Found worldwide, this floating duckweed is the world's smallest flowering plant, forming tiny green ovals ¹⁄₃₂ in (1 mm) long.

TITAN ARUM
Amorphophallus titanum
F: Araceae
After its 10-ft (3-m) inflorescence has died down, this giant aroid from Sumatra grows a single, long-lasting, treelike leaf.

3¼ ft 1 m

WESTERN SKUNK CABBAGE
Lysichiton americanum
F: Araceae
Native to wet habitats in western North America, this aroid gives off a strong odor to attract insects.

12 in 30 cm

BOG ARUM
Calla palustris
F: Araceae
This wetland and shallow-water aroid is a popular ornamental. It is native to cool regions of the Northern Hemisphere.

5 ft 1.5 m

FLOWERING RUSH
Butomus umbellatus
F: Butomaceae
A showy waterside plant, flowering rush is the only member of its family native to Eurasia. It is invasive in North America.

»

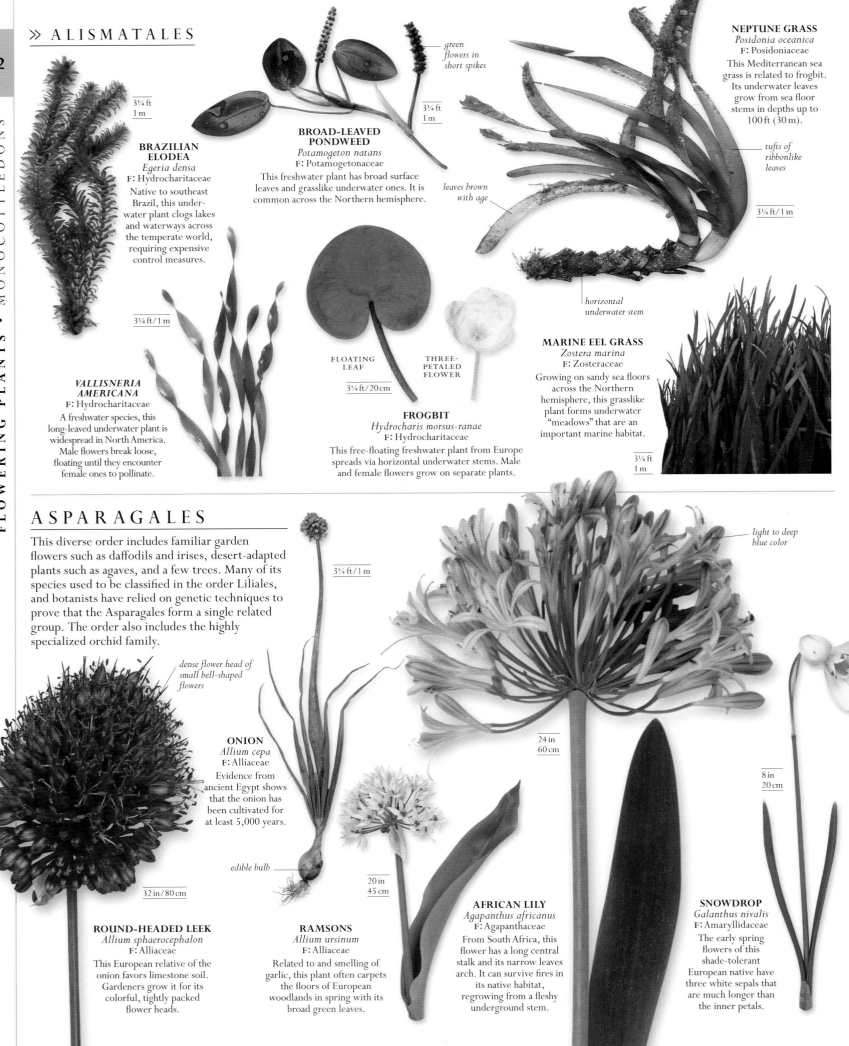

3¼ ft
1 m

BRAZILIAN ELODEA
Egeria densa
F: Hydrocharitaceae
Native to southeast Brazil, this under-water plant clogs lakes and waterways across the temperate world, requiring expensive control measures.

green flowers in short spikes

3¼ ft
1 m

BROAD-LEAVED PONDWEED
Potamogeton natans
F: Potamogetonaceae
This freshwater plant has broad surface leaves and grasslike underwater ones. It is common across the Northern hemisphere.

NEPTUNE GRASS
Posidonia oceanica
F: Posidoniaceae
This Mediterranean sea grass is related to frogbit. Its underwater leaves grow from sea floor stems in depths up to 100 ft (30 m).

tufts of ribbonlike leaves

leaves brown with age

3¼ ft/1 m

3¼ ft/1 m

VALLISNERIA AMERICANA
F: Hydrocharitaceae
A freshwater species, this long-leaved underwater plant is widespread in North America. Male flowers break loose, floating until they encounter female ones to pollinate.

FLOATING LEAF

3¼ ft/20 cm

THREE-PETALED FLOWER

FROGBIT
Hydrocharis morsus-ranae
F: Hydrocharitaceae
This free-floating freshwater plant from Europe spreads via horizontal underwater stems. Male and female flowers grow on separate plants.

horizontal underwater stem

MARINE EEL GRASS
Zostera marina
F: Zosteraceae
Growing on sandy sea floors across the Northern hemisphere, this grasslike plant forms underwater "meadows" that are an important marine habitat.

3¼ ft
1 m

ASPARAGALES

This diverse order includes familiar garden flowers such as daffodils and irises, desert-adapted plants such as agaves, and a few trees. Many of its species used to be classified in the order Liliales, and botanists have relied on genetic techniques to prove that the Asparagales form a single related group. The order also includes the highly specialized orchid family.

dense flower head of small bell-shaped flowers

3¼ ft/1 m

light to deep blue color

24 in
60 cm

ONION
Allium cepa
F: Alliaceae
Evidence from ancient Egypt shows that the onion has been cultivated for at least 5,000 years.

edible bulb

8 in
20 cm

32 in/80 cm

ROUND-HEADED LEEK
Allium sphaerocephalon
F: Alliaceae
This European relative of the onion favors limestone soil. Gardeners grow it for its colorful, tightly packed flower heads.

20 in
45 cm

RAMSONS
Allium ursinum
F: Alliaceae
Related to and smelling of garlic, this plant often carpets the floors of European woodlands in spring with its broad green leaves.

AFRICAN LILY
Agapanthus africanus
F: Agapanthaceae
From South Africa, this flower has a long central stalk and its narrow leaves arch. It can survive fires in its native habitat, regrowing from a fleshy underground stem.

SNOWDROP
Galanthus nivalis
F: Amaryllidaceae
The early spring flowers of this shade-tolerant European native have three white sepals that are much longer than the inner petals.

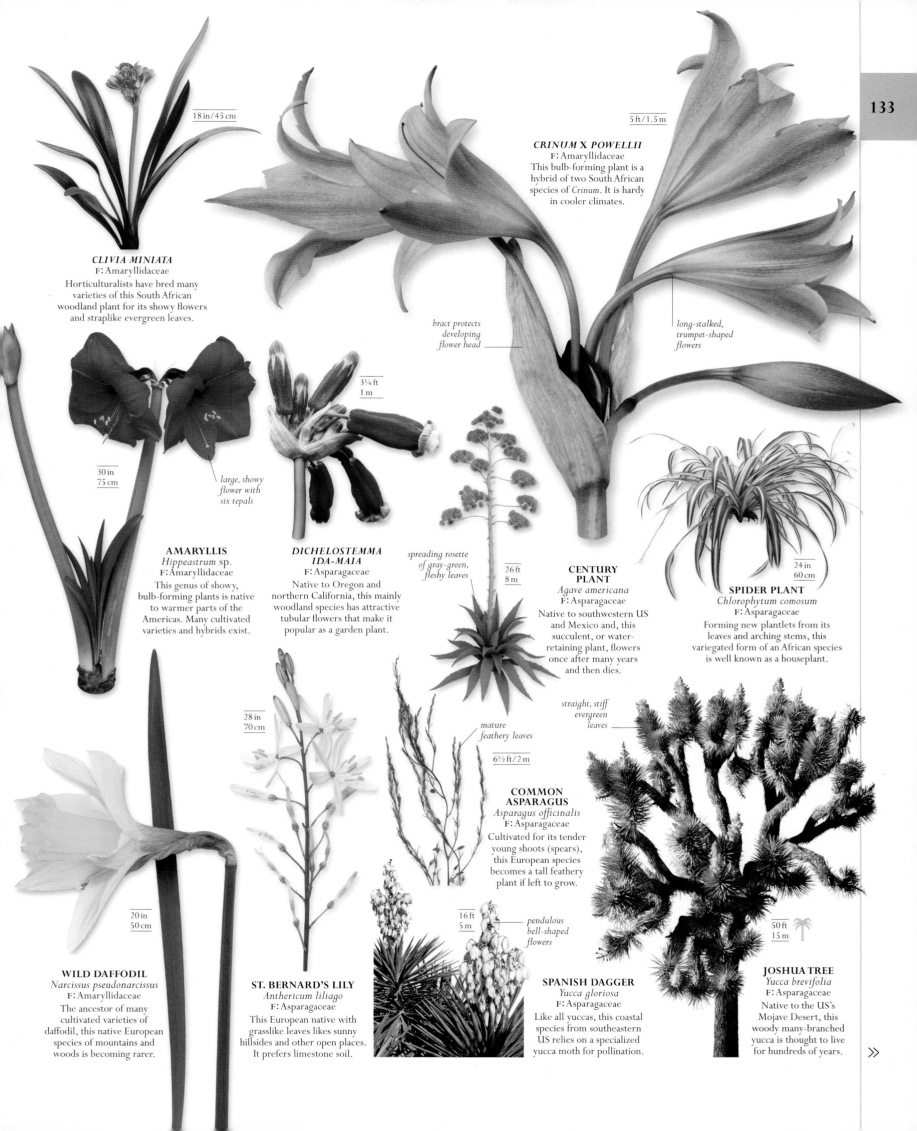

CLIVIA MINIATA
F: Amaryllidaceae
Horticulturalists have bred many
varieties of this South African
woodland plant for its showy flowers
and straplike evergreen leaves.

18 in/45 cm

CRINUM X POWELLII
F: Amaryllidaceae
This bulb-forming plant is a
hybrid of two South African
species of *Crinum*. It is hardy
in cooler climates.

5 ft/1.5 m

bract protects
developing
flower head

long-stalked,
trumpet-shaped
flowers

30 in
75 cm

large, showy
flower with
six tepals

AMARYLLIS
Hippeastrum sp.
F: Amaryllidaceae
This genus of showy,
bulb-forming plants is native
to warmer parts of the
Americas. Many cultivated
varieties and hybrids exist.

3¼ ft
1 m

**DICHELOSTEMMA
IDA-MAIA**
F: Asparagaceae
Native to Oregon and
northern California, this mainly
woodland species has attractive
tubular flowers that make it
popular as a garden plant.

spreading rosette
of gray-green,
fleshy leaves

26 ft
8 m

**CENTURY
PLANT**
Agave americana
F: Asparagaceae
Native to southwestern US
and Mexico and, this succulent, or water-
retaining plant, flowers
once after many years
and then dies.

24 in
60 cm

SPIDER PLANT
Chlorophytum comosum
F: Asparagaceae
Forming new plantlets from its
leaves and arching stems, this
variegated form of an African species
is well known as a houseplant.

straight, stiff
evergreen
leaves

28 in
70 cm

mature
feathery leaves

6½ ft/2 m

**COMMON
ASPARAGUS**
Asparagus officinalis
F: Asparagaceae
Cultivated for its tender
young shoots (spears),
this European species
becomes a tall feathery
plant if left to grow.

20 in
50 cm

WILD DAFFODIL
Narcissus pseudonarcissus
F: Amaryllidaceae
The ancestor of many
cultivated varieties of
daffodil, this native European
species of mountains and
woods is becoming rarer.

ST. BERNARD'S LILY
Anthericum liliago
F: Asparagaceae
This European native with
grasslike leaves likes sunny
hillsides and other open places.
It prefers limestone soil.

16 ft
5 m

pendulous
bell-shaped
flowers

SPANISH DAGGER
Yucca gloriosa
F: Asparagaceae
Like all yuccas, this coastal
species from southeastern
US relies on a specialized
yucca moth for pollination.

50 ft
15 m

JOSHUA TREE
Yucca brevifolia
F: Asparagaceae
Native to the US's
Mojave Desert, this
woody many-branched
yucca is thought to live
for hundreds of years.

FLOWERING PLANTS • MONOCOTYLEDONS

flower stem holds up to 15 flowers

10 in
25 cm

SOLOMON'S SEAL
Polygonatum multiflorum
F: Asparagaceae
A mainly woodland plant, this Eurasian species has leafy arched stems and tube-shaped scentless flowers.

28 in
70 cm

3¼ ft / 1 m

BUTCHER'S BROOM
Ruscus aculeatus
F: Asparagaceae
Flowers and fruits grow straight out of the "leaves" (actually flattened stems) of this shrubby European plant.

LILY-OF-THE-VALLEY
Convallaria majalis
F: Asparagaceae
A sweetly scented plant from temperate Eurasia, lily-of-the-valley has poisonous red berries.

CAST-IRON PLANT
Aspidistra elatior
F: Asparagaceae
Native to Japan, this woodland plant is now popular elsewhere. Its small purple flowers bloom close to the soil.

24 in
60 cm

5 ft
1.5 m

SEA SQUILL
Drimia maritima
F: Asparagaceae
Native to Mediterranean coasts, this large-bulbed plant grows a tall spike of white flowers in late summer after its leaves have withered.

bell-shaped flower

12 in
30 cm

STAR-OF-BETHLEHEM
Ornithogalum angustifolium
F: Asparagaceae
This widespread European plant closes its flowers quickly in dull weather. Its narrow grooved leaves have a central white stripe.

24 in
60 cm

MOTHER-IN-LAW'S TONGUE
Sansevieria trifasciata
F: Asparagaceae
This tropical west African species with stiff patterned leaves is a popular houseplant. It is also grown for its fibers.

20 in
50 cm

PORTUGUESE SQUILL
Scilla peruviana
F: Asparagaceae
From southwestern Europe, this showy bulb-forming perennial has long broad leaves growing from the plant's base.

12 in
30 cm

APHYLLANTHES
Aphyllanthes monspeliensis
F: Asparagaceae
When not flowering, this Mediterranean species looks like a clump of rushes, with many slender stems that are almost leafless.

18 in
45 cm

spike of up to 50 fragrant flowers

12 in
30 cm

HYACINTH
Hyacinthus orientalis
'Blue Jacket'
F: Asparagaceae
This is one of many fragrant, colorful varieties derived from a southwest Asian species over many centuries.

CAPE COWSLIP
Lachenalia aloides
F: Asparagaceae
Native to the southwest tip of Africa, this bulb-forming species grows only one or two strap-shaped leaves.

BLUE HOSTA
Hosta 'Halycon'
F: Asparagaceae
This bluish-leaved hybrid
is one of many garden
varieties of *Hosta*, a genus
of shade-tolerant plants
native to Northeast Asia.

18 in
45 cm

18 in
45 cm

ENGLISH BLUEBELL
*Hyacinthoides
non-scripta*
F: Asparagaceae
Native to western Europe, this
bulb-forming plant sometimes
carpets whole woodland floors
with blue flowers
in spring.

SMALL CAMAS
Camassia quamash
F: Asparagaceae
The edible bulbs of this
spring-flowering
meadowland plant from
western North America
were a staple food for
American Indians.

3¼ ft
1 m

DRAGON TREE
Dracaena draco
F: Asparagaceae
Rare in its native Canary
Islands, this many-
branched tree survives as a
popular ornamental. Its
red sap was prized as
"dragon's blood."

40 ft
12 m

24 in
60 cm

TASSEL HYACINTH
Muscari comosum
F: Asparagaceae
The flowerheads of this
Mediterranean species
have fertile and sterile
flowers, the latter forming
the purplish "tassel" at
the top.

32 in
80 cm

COMMON FRINGE LILY
Thysanotus tuberosus
F: Asparagaceae
Native to southeast Australia,
this narrow-leaved plant has
fringe-petaled flowers that
each open for only one day.

65 ft
20 m

CABBAGE TREE
Cordyline australis
F: Asparagaceae
This ornamental tree, shown here
as a young plant, bears clusters of
scented flowers. It was once an
important food source for the
Maori people of New Zealand.

**FIELD
GLADIOLUS**
Gladiolus italicus
F: Iridaceae
A Mediterranean
member of the iris
family, field gladiolus
flowers between
March and June.

3¼ ft
1 m

BEARDED IRIS
Iris x *germanica*
F: Iridaceae
This fragrant hybrid iris,
common in gardens,
may have originated in
southeastern Europe but
is now naturalized across
the continent.

3¼ ft
1 m

MONTBRETIA
Crocosmia x
crocosmiiflora
F: Iridaceae
A 19th-century hybrid
created from two
South African species, this
popular garden plant is
invasive in some areas.

24 in
60 cm

3¼ ft
1 m

**PALE YELLOW-EYED
GRASS**
Sisyrinchium striatum
F: Iridaceae
Native to Chile
and Argentina, this
representative of a large
New World genus has
pale flowers marked
with fine purple stripes.

12 in
30 cm

KEW'S FREESIA
Freesia x *kewensis*
F: Iridaceae
Derived from native
South African
ancestors, this freesia
is a sweet-smelling
garden hybrid in the
Northern Hemisphere.

24 in
60 cm

24 in
60 cm

**GRAND CHRISTMAS
BELLS**
Blandfordia grandiflora
F: Blandfordiaceae
Nectar-sipping birds seek out
this large flower of coastal
eastern Australia. Its leaves
are narrow and grasslike.

SAFFRON CROCUS
Crocus sativus
F: Iridaceae
The dried stigmas of this
cultivated Mediterranean
crocus have long been
sold as saffron, used
for coloring and
flavoring food.

18 in
45 cm

BARBARY NUT
Gynandriris sisyrinchium
F: Iridaceae
This wild Mediterranean iris grows
a corm (a bulblike underground
stem) to help it survive unfavorable
conditions.

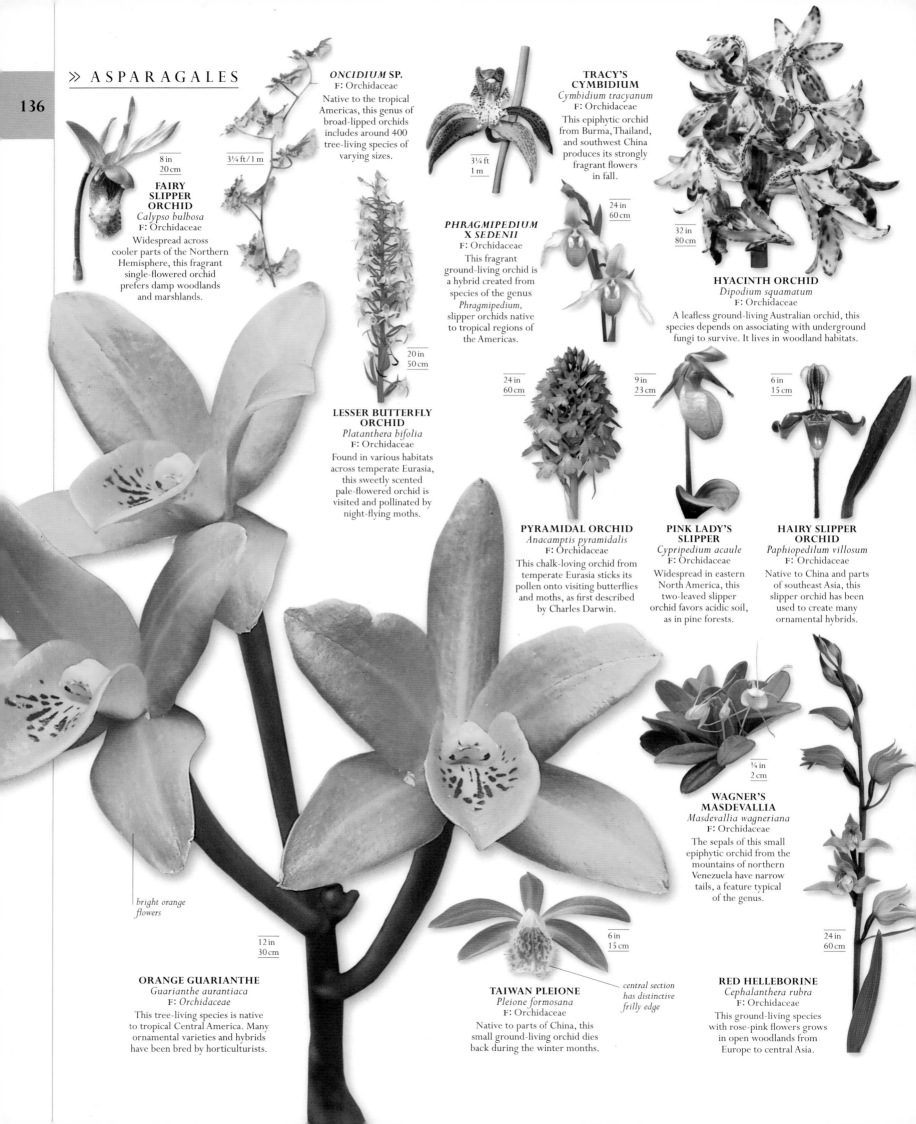

FAIRY SLIPPER ORCHID
Calypso bulbosa
F: Orchidaceae
Widespread across cooler parts of the Northern Hemisphere, this fragrant single-flowered orchid prefers damp woodlands and marshlands.

8 in
20 cm

3¼ ft / 1 m

ONCIDIUM SP.
F: Orchidaceae
Native to the tropical Americas, this genus of broad-lipped orchids includes around 400 tree-living species of varying sizes.

PHRAGMIPEDIUM X SEDENII
F: Orchidaceae
This fragrant ground-living orchid is a hybrid created from species of the genus *Phragmipedium*, slipper orchids native to tropical regions of the Americas.

3¼ ft
1 m

TRACY'S CYMBIDIUM
Cymbidium tracyanum
F: Orchidaceae
This epiphytic orchid from Burma, Thailand, and southwest China produces its strongly fragrant flowers in fall.

24 in
60 cm

32 in
80 cm

HYACINTH ORCHID
Dipodium squamatum
F: Orchidaceae
A leafless ground-living Australian orchid, this species depends on associating with underground fungi to survive. It lives in woodland habitats.

20 in
50 cm

LESSER BUTTERFLY ORCHID
Platanthera bifolia
F: Orchidaceae
Found in various habitats across temperate Eurasia, this sweetly scented pale-flowered orchid is visited and pollinated by night-flying moths.

24 in
60 cm

9 in
23 cm

6 in
15 cm

PYRAMIDAL ORCHID
Anacamptis pyramidalis
F: Orchidaceae
This chalk-loving orchid from temperate Eurasia sticks its pollen onto visiting butterflies and moths, as first described by Charles Darwin.

PINK LADY'S SLIPPER
Cypripedium acaule
F: Orchidaceae
Widespread in eastern North America, this two-leaved slipper orchid favors acidic soil, as in pine forests.

HAIRY SLIPPER ORCHID
Paphiopedilum villosum
F: Orchidaceae
Native to China and parts of southeast Asia, this slipper orchid has been used to create many ornamental hybrids.

¾ in
2 cm

WAGNER'S MASDEVALLIA
Masdevallia wagneriana
F: Orchidaceae
The sepals of this small epiphytic orchid from the mountains of northern Venezuela have narrow tails, a feature typical of the genus.

bright orange flowers

12 in
30 cm

ORANGE GUARIANTHE
Guarianthe aurantiaca
F: Orchidaceae
This tree-living species is native to tropical Central America. Many ornamental varieties and hybrids have been bred by horticulturists.

6 in
15 cm

TAIWAN PLEIONE
Pleione formosana
F: Orchidaceae
Native to parts of China, this small ground-living orchid dies back during the winter months.

central section has distinctive frilly edge

24 in
60 cm

RED HELLEBORINE
Cephalanthera rubra
F: Orchidaceae
This ground-living species with rose-pink flowers grows in open woodlands from Europe to central Asia.

VIOLET BIRD'S-NEST ORCHID
Limodorum abortivum
F: Orchidaceae
This European orchid lacks green leaves so its root tips obtain food through *Russula* fungi found alongside it.

32 in/80 cm

DENDROBIUM SP.
F: Orchidaceae
Including over 1,000 species of diverse shapes, and sizes, this genus of epiphytic orchids is found from southeast Asia to New Zealand.

6½ ft
2 m

12 in
30 cm

35 in
90 cm

TONGUE ORCHID
Serapias lingua
F: Orchidaceae
The lip of this unusual Mediterranean orchid's flower hangs down like a tongue, acting as a landing platform for visiting insects.

LIZARD ORCHID
Himantoglossum hircinum
F: Orchidaceae
The long lips of Europe's largest wild orchid have a fanciful resemblance to tiny lizards, resulting in its common name.

LIPPEROSE MOTH ORCHID
Phalaenopsis 'Lipperose'
F: Orchidaceae
This cultivated hybrid is one of many developed using the genus *Phalaenopsis* from southeast Asia with its broad flattened flowers.

3¼ ft
1 m

three sepals and two petals on each flower

24 in
60 cm

3¼ ft
1 m

24 in
60 cm

drooping sepals

12 in
30 cm

MILITARY ORCHID
Orchis militaris
F: Orchidaceae
This chalk-loving Eurasian orchid may have acquired its name because the individual flowers resemble helmeted human figures.

COMMON DONKEY ORCHID
Diuris corymbosa
F: Orchidaceae
Native to southern Australia, this orchid gets its common name from the shape of its side petals, which resemble donkeys' ears.

SEED POD

LEATHERY EVERGREEN LEAVES

ROTHCHILD'S VANDA
Vanda 'Rothschildiana'
F: Orchidaceae
This cultivated hybrid was created by crossing species of the genus *Vanda*, which are tree-living orchids native to tropical Asia.

FLAT-LEAVED VANILLA ORCHID
Vanilla planifolia
F: Orchidaceae
Native to Mexico and Central America, this climbing orchid is the source of the flavoring vanilla.

GREENHOOD ORCHID
Pterostylis sp.
F: Orchidaceae
In this mainly Australian genus of ground-living orchids, the upper sepal often forms a hood over the center of the flower.

lanceolate leaves on each stalk

AUTUMN LADY'S-TRESSES
Spiranthes spiralis
F: Orchidaceae
This small, mainly grassland orchid from temperate Eurasia is notable for its flowers being arranged spirally up the flower stalk.

8 in
20 cm

flower resembles a fat bumblebee

sepal forms protective hood over flower

stamen carries pollen

flower attracts insects

3¼ ft/1 m

16 in
40 cm

20 in
50 cm

DARK RED HELLEBORINE
Epipactis atrorubens
F: Orchidaceae
The long roots of this fragrant Eurasian orchid help it grow successfully even within the cracks of limestone cliffs.

14 in
35 cm

NUN'S ORCHID
Phaius tankervilleae
F: Orchidaceae
This fragrant ground-living orchid is widely cultivated. Its native range is tropical and subtropical southeast Asia and the South Pacific.

flower lip closes when touched by an insect

BEE ORCHID
Ophrys apifera
F: Orchidaceae
Although in a genus whose flowers attract pollinating male insects by mimicking females, this Eurasian orchid in practice often pollinates itself.

∨ **PLANT EXPANSION**
Dinema polybulbon grows all over its chosen
support structure, whether it be a rock or
another plant, and forms a dense mat of
stolons—the horizontal stems that reach
out from the base of the plant.

*broad whitish petal
forms a lip, which acts
as a flag to attract
pollinating insects*

*outer three sepals
enclose the flower bud*

MANY-BULBED ORCHID
Dinema polybulbon

This diminutive orchid is the only species in the genus *Dinema*. Rambling and prolific, it is an epiphytic or lithophytic plant, meaning that it grows on trees or rocks, using them for support. It gains its nutrients from the air, animal detritus, or other plant remains, and absorbs water from rain or mist. Moisture collects on the orchid's leaves, which are thick and waxy, and have reduced stomata—all features that help reduce water loss. The leaves funnel water into swollen, upright stems known as pseudobulbs, which are in fact water-storage organs that evolved to keep the orchid hydrated during tropical dry seasons. The pseudobulbs are attached to horizontal stems, which are in turn connected to both underground and aerial roots. These absorb dissolved nutrients through an outer layer of cells called velamen. During winter, each pseudobulb produces a single flower. Hardy in frost to 14°F (−10°C), this species is a favorite of orchid collectors.

SIZE Height 3 in (7.5 cm)
HABITAT Humid, mixed forests
DISTRIBUTION Mexico, Central America, Jamaica, Cuba
LEAF TYPE Parallel veins

flattened leaf tip has a notch

PSEUDOBULB >
These small, oval, bulblike structures, sparsely distributed on a horizontal stem, store water to be used when surface water is scarce.

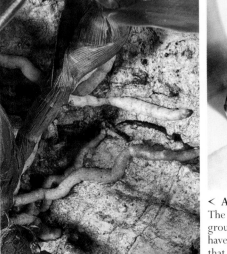

∧ FLOWER
The pseudobulb produces a small, sweet-smelling flower with narrow yellow-brown sepals, purple petals, and a spreading white lip.

< POLLEN SACS
Pollen grains are contained in two sacs. A sticky mass glues the sacs onto the pollinator, which carries the pollen to another plant.

< AERIAL ROOTS
The orchid's aerial roots grow above ground. The older parts of the root have a protective layer of dead cells that act like blotting paper, helping absorb and retain water.

» ASPARAGALES

spikes of white or cream flowers

26 ft
8 m

GRASS TREE
Xanthorrhoea australis
F: Xanthorrhoeaceae
Over time, this bushfire-resistant Australian plant eventually grows into a solidly branched tree. Its tiny flowers are borne on tall spikes.

20 ft
6 m

NEW ZEALAND FLAX
Phormium tenax
F: Hemerocallidaceae
This common New Zealand species grows large leaves at ground level and a tall flower stalk bearing tubular reddish flowers.

5 ft
1.5 m

DAY LILY
Hemerocallis fulva
F: Hemerocallidaceae
Each flower of this ornamental species lasts one day. Native to East Asia, it is widely naturalized in North America.

COMMON ASPHODEL
Asphodelus aestivus
F: Asphodelaceae
This common narrow-leaved Mediterranean plant produces many white or yellow flowers on a tall stem.

5 ft
1.5 m

sap from leaves has healing properties

3¼ ft
1 m

MEDICINAL ALOE
Aloe vera
F: Asphodelaceae
This spiny succulent adapted to dry areas has been cultivated since ancient times for its reputed medicinal properties.

orange-red flower spike fades to yellow near base

5 ft
1.5 m

TORCH LILY
Kniphofia uvaria
F: Asphodelaceae
Native to the southwest tip of Africa, this multiflowered plant with long narrow leaves is now a popular ornamental specimen.

buds at top of spike open last

18 in
45 cm

LILIALES

Many plants once considered part of the lily family, including onions and hyacinths, have recently been moved into a completely separate order, the Asparagales (see pp.132–39). The trimmed-down Liliales includes the true lilies, their close relatives the tulips, and various other plants including alstroemerias. Many are colorful bulb-forming species while some are climbers.

18 in
45 cm

WILD TULIP
Tulipa sylvestris
F: Liliaceae
Still widespread in its native Europe, this yellow-flowered tulip, a relative of the cultivated garden tulip grows wild in meadows, rocky hillsides, as well as open woodland.

40 cm
16 in

50 cm
20 in

HERB PARIS
Paris quadrifolia
F: Liliaceae
Found in ancient woodland in temperate Eurasia, this four-leaved plant has a single green flower that develops into a black inedible fruit.

SNAKESHEAD FRITILLARY
Fritillaria meleagris
F: Liliaceae
This bulb-forming European species has distinctive chequered flowers. It grows wild in damp meadows but is more common in cultivation.

DIOSCOREALES

This relatively small order is dominated by the yams, a family of mainly tropical climbing plants. Several species of yam have been cultivated since ancient times for their large edible tubers. The order also contains a few lilylike plants such as bog asphodel.

BOG ASPHODEL
Narthecium ossifragum
F: Nartheciaceae
This European plant lives in nutrient-poor upland habitats. As its seeds develop, the flowerhead becomes a fiery orange color.

16 ft
5 m

edible tuber

YAM
Dioscorea sp
F: Dioscoreaceae
Widely cultivated in the tropics, yams grow large, starch-rich tubers. Plants have stems that trail or climb and heart-shaped leaves.

13 ft/4 m

BLACK BRYONY
Tamus communis
F: Dioscoreaceae
This poisonous European climbing plant of the yam family is named for its black underground tuber. Its globular fruits turn bright red when ripe.

unripe fruits are green

twining stems of heart-shaped leaves

MADONNA LILY
Lilium candidum
F: Liliaceae
Often encountered in Christian art as a symbol
of purity, this widely cultivated lily is native to
the eastern Mediterranean.

3¼ ft
1 m

*female stigma
receives the
pollen*

*male stamens
carry the pollen*

*leaves arranged
spirally up stem*

GAGEA RETICULARIS
F: Liliaceae
This small bulb-forming plant
grows in open habitats across much
of temperate Asia, as well as
southeast Europe and North Africa.

6 in
15 cm

4 ft
1.2 m

TURK'S CAP LILY
Lilium martagon
F: Liliaceae
This widespread Eurasian
lily has the reflexed
(backward-bending) tepals
that form the flower shape
characteristic of the species.

GIANT HIMALAYAN LILY
Cardiocrinum giganteum
F: Liliaceae
Found from the Himalayas to
China, this giant lily grows for
several seasons before flowering,
after which the main plant dies.

10 ft
3 m

FLAME LILY
Gloriosa superba
F: Colchicaceae
This striking plant from
South African woodlands is
a climber, clambering
upwards with the aid
of tendrils.

6½ ft / 2 m

FLOWERHEADS

6 in
15 cm

**MEADOW
SAFFRON**
Colchicum autumnale
F: Colchicaceae
The crocuslike flowers
of this European native
appear in the autumn,
months before its
leaves. Despite being
poisonous, it is
cultivated widely.

**TYPICAL
LEAF**

*flower with
six tepals*

*leaf twists from stem
so underside appears
uppermost*

*up to 40 tubular
flowers*

CALDAS' BOMAREA
Bomarea caldasii
F: Alstroemeriaceae
This showy-flowered
climbing plant is a close
relative of alstroemerias. It
is native to South America.

13 ft
4 m

PERUVIAN LILY
Alstroemeria sp.
F: Alstroemeriaceae
This lilylike South American
genus includes many popular
ornamentals. The leaves twist
during growth to become
anatomically upside-down.

4 ft
1.2 m

» LILIALES

CHILEAN BELLFLOWER
Lapageria rosea
F: Philesiaceae
Cultivated for its spectacular bell-shaped flowers, this is native to humid woodlands of Chile.

30 ft
10 m

50 ft
15 m

BERRIES

FLOWER CLUSTER

SMILAX
Smilax aspera
F: Smilacaceae
Native to the Mediterranean and southwest Asia, this climber has male and female flowers on separate plants.

5 ft
1.5 m

FALSE HELLEBORE
Veratrum sp.
F: Melanthiaceae
Native to the Northern Hemisphere, this genus of poisonous plants produces branched flowerheads of often greenish flowers.

PANDANALES

Found mainly in the tropics, this order includes over 1,300 species of trees, shrubs, climbers, and smaller plants. Many look superficially like palms, except that they have simpler, strap-shaped leaves. Around half of them belong to the single genus *Pandanus* (screw pines).

MULTI-SEEDED FRUIT

SCREW PINE
Pandanus tectorius
F: Pandanaceae
No relation of real pines, this tropical coastal tree has long been a vital source of materials for Pacific Island cultures.

MATURE TREE

60 ft
18 m

ARECALES

This order is made up of one large family, the palm trees. Growing typically from one central bud, palms vary from small shrubs to towering trees. There are also many species of slender climbing rattans. The huge leaves of palm trees are either feather-shaped or fan-shaped. Although they may conjure up an image of deserts or sandy beaches, most species actually live in tropical rain forests.

100 ft
30 m

65 ft
20 m

SUGAR PALM
Arenga pinnata
F: Arecaceae
This native of India and southeast Asia grows showy yellow flowerheads. It yields products including sugar and fibers.

DATE PALM
Phoenix dactylifera
F: Arecaceae
Seen here as a young tree, this cultivated palm from the Middle East grows as separate male and female trees.

100 ft
30 m

80 ft
25 m

pinnate leaf

COCONUT PALM
Cocos nucifera
F: Arecaceae
Now widely cultivated, this species probably originated in the western Pacific, colonizing new islands via its floating fruits.

MATURE TREE

SINGLE-SEEDED FRUIT

MATURE TREE

RIPE FRUIT

BETEL NUT PALM
Areca catechu
F: Arecaceae
Cultivated for its seeds, which release a psychoactive substance when chewed, this palm originates from southeast Asia.

DESERT FAN PALM
Washingtonia filifera
F: Arecaceae
Dead leaves below this palm's crown provide shelter for desert birds and insects of its native southwestern United States.

60 ft
18 m

PETTICOAT PALM
Copernicia macroglossa
F: Arecaceae
Native to Cuba, this relatively small palm is named for the skirt of dead leaves that it retains below its crown.

23 ft
7 m

PINNATE
LEAF

65 ft
20 m

RIPE FRUIT

AFRICAN OIL PALM
Elaeis guineensis
F: Arecaceae
A plant of humid tropical lowlands, this palm is extensively cultivated beyond its native Africa for the oils contained in its fruits.

20 ft
6 m

BOTTLE PALM
Hyophorbe lagenicaulis
F: Arecaceae
This swollen-based palm comes from tiny Round Island near Mauritius. It is widely grown as an ornamental.

50 ft
15 m

SAGO PALM
Metroxylon sagu
F: Arecaceae
Shown here as a young plant, this swamp-living palm probably originated in New Guinea but now grows throughout southeast Asia.

65 ft
20 m

CHUSAN PALM
Trachycarpus fortunei
F: Arecaceae
From China, this cold-tolerant species has separate male and female plants. The females produce round blue-black fruits.

SEED

100 ft
30 m

COCO DE MER
Lodoicea maldivica
F: Arecaceae
The seeds of this palm from the Seychelles are the biggest in the plant kingdom. They take six years to mature.

80 ft
25 m

ROYAL PALM
Roystonea regia
F: Arecaceae
Avenues of this handsome smooth-trunked species are common in the tropics. Native to Caribbean islands, it thrives on rich soils.

characteristic bulge

30 ft
10 m

RAFFIA PALM
Raphia farinifera
F: Arecaceae
At 65 ft (20 m) long, the leaves of this African and Madagascan palm are the largest of any tree.

65 ft
20 m

PALMYRA PALM
Borassus flabellifer
F: Arecaceae
This tall-trunked species from southern Asia favors drier habitats. It is cultivated for its fruit and sugar-yielding sap.

fan-shaped leaf up to 3¼ ft (1 m) across

30 ft
10 m

BRAZILIAN WAX PALM
Copernicia prunifera
F: Arecaceae
This palm from central and northeast Brazil has brown flowers. The wax coating on its leaves is harvested commercially.

10 ft
3 m

EUROPEAN FAN PALM
Chamaerops humilis
F: Arecaceae
Often trunkless in the wild, cultivated specimens grow several short trunks from a single base. It originates from Mediterranean countries.

80 ft
25 m

CHILEAN WINE PALM
Jubaea spectabilis
F: Arecaceae
Also known as the coquito palm, this massive-trunked, cold-tolerant palm is native only to central Chile, where it is now a protected species.

COMMELINALES

This order includes a variety of mainly low-growing plants, typically from the warmer regions of the world. Many have attractive three-petaled blue flowers (reduced to two petals in some species), making them popular as ornamentals. Their leaves and stems are also often fleshy, with a sticky sap that hardens on exposure to air.

BLUE SPIDERWORT
Commelina coelestis
F: Commelinaceae
Sometimes planted as ground cover, this sprawling plant is native to Mexico and Central America.

24 in
60 cm

QUEEN'S SPIDERWORT
Dichorisandra reginae
F: Commelinaceae
A popular garden flower grown in warmer regions, this tropical woodland species is from Peru. It has blue flowers with white centers.

10 cm
4 in

TURTLE VINE
Callisia repens
F: Commelinaceae
This succulent creeping plant from forest edges in the American tropics spreads by rooting from stems.

TAHITIAN BRIDAL VEIL
Tripogandra multiflora
F: Commelinaceae
This native of Central and South America has weak stems and leaves with purple undersides.

2¼ ft
70 cm

trailing stems

6 in
15 cm

INCH PLANT
Tradescantia zebrina
F: Commelinaceae
With striped, succulent foliage, this ground-living species from the tropical Americas is also a popular houseplant.

12 in
30 cm

POALES

Grasses dominate the order Poales. Their wind-pollinated flowers have no need of showy petals. Sedges and rushes also belong in this order, as do the bromeliads, which are mainly epiphytic, meaning they live on but do not feed off other plants. Many produce brightly colored flower spikes that emerge from rosettes of tough, broad leaves.

18 in
45 cm

SLENDER CLUB-RUSH
Isolepis cernua
F: Cyperaceae
With narrow green stems and silvery flowerheads, this widespread temperate sedge is also dubbed "fiber-optics grass."

12 in
30 cm

PINK QUILL
Tillandsia cyanea
F: Bromeliaceae
Native to Ecuador, pink quill grows on trees, in forests between 2,000 ft and 3,300 ft (600–1,000 m) above sea level.

12 in
30 cm

SCARLET STAR
Guzmania lingulata
F: Bromeliaceae
This tree-living bromeliad has a wide native range from Central America to Brazil. It is a popular ornamental species.

arching stalk of flower cluster

long, narrow leaves

5 ft
1.5 m

cluster of separate flowers

12 in
30 cm

PINEAPPLE
Ananas comosus
F: Bromeliaceae
Christopher Columbus first introduced Europeans to this cultivated bromeliad with its large fruit. It probably originates from Brazil.

12 in
30 cm

red bracts surround flowers

BIRD'S-NEST BROMELIAD
Nidularium innocentii
F: Bromeliaceae
The small flowers of this Brazilian bromeliad nestle within the chamber formed by the surrounding colored bracts. *Nidularium* means "little nest."

spiny-leaved rosettes

vaselike rosette of leaves collects rainwater

14 in
35 cm

DYER'S TILLANDSIA
Tillandsia dyeriana
F: Bromeliaceae
This epiphytic bromeliad is endangered in the wild due to destruction of its native mangrove forests in Ecuador.

LORENTZ'S BROMELIAD
Deuterocohnia lorentziana
F: Bromeliaceae
Native to the high, dry Andes of Argentina, this is a ground-dwelling bromeliad.

10 in
25 cm

BLUSHING BROMELIAD
Neoregelia carolinae
F: Bromeliaceae
At flowering time, the central leaves of this Brazilian bromeliad turn crimson, and it produces blue or violet flowers.

10 ft / 3 m

TALL KANGAROO PAW
Anigozanthos flavidus
F: Haemodoraceae

From sandy areas of southwestern Australia, this species gets its common name from the appearance of its woolly-haired flower buds.

yellow blotch on uppermost petal of each flower

18 in
45 cm

3¼ ft / 1 m

flowers in compact conical cluster

spike of male flowers

COMMON WATER HYACINTH
Eichhornia crassipes
F: Pontederiaceae

Native to Amazonia, this floating plant is a major tropical pest, but can be used to absorb pollution.

swollen leaf stalk

PICKEREL WEED
Pontederia cordata
F: Pontederiaceae

Fast-growing and with striking blue flowerheads, this water-edge species is common across eastern North America and has become invasive elsewhere.

PENDULOUS SEDGE
Carex pendula
F: Cyperaceae

Like other true sedges (genus *Carex*), this European species has triangular stems and separate male and female flower clusters.

4½ ft
1.4 m

6½ ft
2 m

16 ft / 5 m

spike of female flowers

crimson central leaves

giant spike of more than 3,000 flowers

CHINESE WATER CHESTNUT
Eleocharis dulcis
F: Cyperaceae

Native to eastern Asia, this wetland sedge has tall, leafless stems. It is cultivated for its edible underwater tubers.

PAPYRUS SEDGE
Cyperus papyrus
F: Cyperaceae

This tall African perennial is a wetland plant. The Ancient Egyptians wrote on a paperlike material called papyrus made from its leaves.

evergreen variegated foliage

3¼ ft
1 m

yellow bracts around small white flowers

33 ft
10 m

stiff, spiny leaves

ZEBRA PLANT
Aechmea chantinii
F: Bromeliaceae

This large spiny-leaved epiphytic bromeliad comes from the rain forests of South America. It is pollinated by hummingbirds.

HAHN'S CATOPSIS
Catopsis hahnii
F: Bromeliaceae

This epiphytic bromeliad is native to the cloud forests of southern Mexico and Central America.

QUEEN OF THE ANDES
Puya raimondii
F: Bromeliaceae

Native to the central Andes, the world's largest bromeliad produces a single colossal flower spike after many years of growth and then dies.

gray-green leaves with waxy texture

20 in
50 cm

rosette of strap-shaped leaves

SOFT RUSH
Juncus effusus
F: Juncaceae

This very widespread rush thrives in damp infertile soil. Its cylindrical stems are filled with spongy tissue, or pith.

5 ft
1.5 m

4½ ft
1.3 m

24 in
60 cm

LARGE QUAKING GRASS
Briza maxima
F: Poaceae

This annual grass from Europe gets its name from the way its fine-stalked flower heads shake at the slightest breeze.

6 ft
1.8 m

FALSE OAT-GRASS
Arrhenatherum elatius
F: Poaceae

Common across its native Europe, this tall-flowered wild relative of the oat has now spread to many other parts of the world.

24 in
60 cm

SAND COUCH
Elytrigia juncea
F: Poaceae

This tough grass grows on the seaward side of European sand dunes. Its roots, or rhizomes, help to hold the windblown sand together to form dunes.

CRESTED DOG'S-TAIL
Cynosurus cristatus
F: Poaceae

Low-growing apart from its flower stalks, this perennial grass from Europe and west Asia resists trampling and is often used for lawns.

30 in
75 cm

COCK'S-FOOT
Dactylis glomerata
F: Poaceae

Often grown in hay fields and pastures, this common grass with distinctive tufty flower heads is native to Eurasia and North Africa.

8¼ ft
2.5 m

young seeds

JOB'S TEARS
Coix lacryma-jobi
F: Poaceae

This cereal grass is native to tropical Asia but is widely grown elsewhere. The seeds of one variety are traded as beads for jewelry.

14 in
35 cm

DRIED SEED HEAD

20 ft
6 m

TALL, BROAD LEAF

HARE'S TAIL
Lagurus ovatus
F: Poaceae

The distinctive soft-haired flower heads of this mainly coastal Mediterranean grass make it popular in dried flower arrangements.

GIANT REED
Arundo donax
F: Poaceae

Originally from central Asia, this outsize marshland grass has been widely planted elsewhere. Its woody stems have many traditional uses.

EVERGREEN LEAVES

15 ft
5 m

BLACK BAMBOO
Phyllostachys nigra
F: Poaceae

Like all the bamboos, this black-stemmed species has hard woody stems. It is native to east and South China. Bamboos are the fastest growing woody plant, growing up to 24 in (60 cm) a day.

awns protect seeds

THREE-AWNED GOAT GRASS
Aegilops neglecta
F: Poaceae

A relative of wheat, this low-growing, drought-resistant annual grass is native to Eurasia and parts of Africa.

14 in
35 cm

FINGER-THICK CANE

COMMON REED
Phragmites australis
F: Poaceae

Widespread in both temperate and tropical regions, this shallow-water grass can colonize large areas via its creeping horizontal stems.

20 ft
6 m

3¼ ft
1 m

SWEET VERNAL GRASS
Anthoxanthum odoratum
F: Poaceae

Native to Eurasia, this early-flowering grass contains a chemical called coumarin that makes it smell pleasantly of fresh-mown hay.

4 ft
1.2 m

MARRAM GRASS
Ammophila arenaria
F: Poaceae

This tough grass from Europe flourishes on sand dunes, which its long underground stems and roots help to stabilize.

6 ft
1.8 m

OAT
Avena sativa
F: Poaceae

This cultivated grass, grown widely as a cereal crop for livestock and human consumption, thrives in cool, wet climates.

awn

32 in
80 cm

BARLEY
Hordeum vulgare
F: Poaceae

Originating in the ancient Near East, this cultivated cereal grass is notable for the long hairlike bristles (awns) on its flower heads.

3¼ ft
1 m

BREAD WHEAT
Triticum aestivum
F: Poaceae

The world's top-tonnage cereal, this species originated in the ancient Near East. It is a hybrid of both the wild and the earlier cultivated wheats.

16 in
40 cm

CREEPING BENT-GRASS
Agrostis stolonifera
F: Poaceae

This common perennial grass with feathery flower heads is found worldwide. It spreads using creeping horizontal stems, or stolons.

9¾ ft
3 m

CITRONELLA
Cymbopogon nardus
F: Poaceae

A type of lemongrass, citronella originates from tropical Asia. It yields an oil used in perfumes and for repelling insects.

PERENNIAL RYE-GRASS
Lolium perenne
F: Poaceae

Much used in pastures, lawns, and sports fields, this common Eurasian grass has now spread around the world.

6 ft
1.8 m

RICE
Oryza sativa
F: Poaceae

Native to east Asia, this major grain crop of warmer regions is usually cultivated in shallow water or in flood-prone areas.

single stem

10 ft
3 m

MAIZE
Zea mays
F: Poaceae

This major food crop was first cultivated in ancient Mexico. It has male or female flower heads—edible corn is the female flower.

20 ft
6 m

SUGAR CANE
Saccharum officinarum
F: Poaceae

Possibly related to maize, this tropical cultivated grass may have originated in New Guinea. Sugar is extracted from its thick stems.

»

9¾ ft
3 m

VETIVER
Chrysopogon zizanioides
F: Poaceae

Native to India, this tropical grass is widely cultivated for its fragrant oil and to bind soils to prevent erosion.

36 in
90 cm

32 in
80 cm

YORKSHIRE FOG
Holcus lanatus
F: Poaceae

A common European grass found in damp meadows, Yorkshire fog has leaves with soft velvety hairs.

dense white flower heads

PAMPAS GRASS
Cortaderia selloana
F: Poaceae

This tall grass from southern South America is a popular ornamental species. It has become invasive in some areas.

9¾ ft
3 m

» POALES

GIANT FEATHER GRASS
Stipa gigantea
F: Poaceae
Gardeners cultivate this tall grass for its showy flowerheads, which persist until winter. It is native to Spain and Portugal.

8 ft
2.5 m

female flower

male flower heads

5 ft
1.5 m

BRANCHED BUR-REED
Sparganium erectum
F: Sparganiaceae
Separate ball-shaped clusters of male and female flowers develop on the same flower stalk in this widespread wetland species of the Northern Hemisphere.

10 ft
3 m

CATTAIL
Typha latifolia
F: Typhaceae
This common Northern Hemisphere wetland plant has a distinctive cigar-shaped female flowerhead. The male flowers grow above it in a tuft.

YELLOW-EYED GRASS
Xyris sp.
F: Xyridaceae
This genus of grasslike plants, widespread in warmer regions of the world, bears small yellow flowers on slender stems.

ZINGIBERALES

A feature of this mainly tropical order is that many species grow giant leaves at the end of stalks. Although there are no true woody trees, some species such as the banana plant grow very large. Many Zingiberales have showy flowers and foliage and have become ornamentals. The ginger family, Zingiberaceae, is the largest in the order, includes several other important spice plants beside ginger itself.

16 in
40 cm

LOVELY PRAYER PLANT
Ctenanthe amabilis
F: Marantaceae
Native to forest floors in the Brazilian tropics, this warmth-loving plant requires high humidity if grown in cultivation.

3¼ ft
1 m

ETERNAL FLAME
Calathea crocata
F: Marantaceae
Related to *Ctenanthe* and *Maranta*, this Brazilian species has more striking flowers than its relatives although living in similar forest habitats.

24 in
60 cm

SPIRAL FLAG
Chamaecostus igneus
F: Costaceae
This orange-flowered relative of the ginger family is native to tropical eastern. Brazil, and is also grown as an ornamental.

12 in
30 cm

SILVER-VEINED PRAYER PLANT
Maranta leuconeura
F: Marantaceae
This Brazilian forest plant folds its leaves together at night to conserve moisture. Cultivated varieties have strikingly patterned foliage.

40 ft/12 m

ENSET
Ensete ventricosum
F: Musaceae
This African-relative of the banana has long been cultivated in Ethiopia for its nutritious rootstock and stem, not its (inedible) fruit.

30 ft
9 m

BANANA
Musa acuminata
F: Musaceae
The seedless hybrid of Asian wild ancestors, cultivated bananas grow sterile male flowers at the end of their fruiting branches.

6½ ft
2 m

INDIAN SHOT
Canna indica
F: Cannaceae
This South American plant has unusual flowers in which some of the "petals" are actually modified pollen-producing stamens. There are many cultivated varieties.

24 in/60 cm

each yellow bract protects four or five tiny flowers

CHINESE YELLOW BANANA
Musella lasiocarpa
F: Musaceae
Possibly extinct in the wild, this mountain species from China produces a yellow flower spike that can last for months.

VIOLET-
STRIPED
WHITE
FLOWER

LANCEOLATE
LEAF

SEED
POD

5.5 m
18 ft

CARDAMOM
Elettaria cardamomum
F: Zingiberaceae
This tropical plant is
native to forests in
South India and Sri
Lanka, but cultivated
elsewhere. Its seeds are
the spice cardamom.

WHITE
FLOWER

5 ft
1.5 m

LESSER
GALANGAL
Alpinia officinarum
F: Zingiberaceae
Native to South China,
this relative of ginger
develops a swollen
underground stem and is
similarly used as a spice.

RHIZOME AND
ROOTS

AROMATIC
GINGER
Kaempferia galanga
F: Zingiberaceae
Native to tropical Asia,
this short-stemmed
species develops small
flowers and is grown as
an ornamental.

12 in
30 cm

paddle-shaped leaves,
green at tip, yellow
at leaf stem

symmetrical
fan-shaped tree

FLESHY
RHIZOME

3¼ ft / 1 m

TURMERIC
Curcuma longa
F: Zingiberaceae
Native to S.E. Asia, this large-leaved
plant develops a swollen
underground stem from which the
yellow spice turmeric is obtained.

STEM WITH
FLOWER HEAD

EDIBLE
RHIZOME

3¼ ft
1 m

GINGER
Zingiber officinale
F: Zingiberaceae
The spice called ginger is
the underground stem of
a leafy Southeast Asian
plant that is now known
only from cultivation.

50 ft
15 m

TRAVELLER'S TREE
Ravenala madagascariensis
F: Strelitziaceae
Native to open forests in
Madagascar, this relative
of *Strelitzia* is pollinated
by lemurs in its
natural habitat.

6½ ft
2 m

BIRD-OF-PARADISE
Strelitzia reginae
F: Strelitziaceae
Orange-and-blue bird-pollinated
flowers open one at a time from a
beaklike sheath in this South
African plant.

purple flowers
develop in summer

10 in
25 cm

HUME'S ROSCOEA
Roscoea humeana
F: Zingiberaceae
This relative of the
ginger plant grows
unusual orchidlike
flowers. Its native
habitat is the mountains
of Southwest China.

12 ft
4 m

HELICONIA
STRICTA
F: Heliconiaceae
In its natural habitat, this
large-leaved tropical plant
from northern areas of
South America is
pollinated by nectar-
sipping hummingbirds.

EUDICOTYLEDONS

About three-quarters of all the world's current flowering plants are classified in the eudicotyledon grouping, which evolved over 125 million years ago.

Eudicots get their name from the cotyledons, or seed-leaves, in the seed before it germinates. Unlike monocots, which have a single seed-leaf, eudicots have two. They include an enormous variety of plants, from agricultural weeds to towering rain forest trees, and they are of huge economic importance, as well as being prized as garden flowers. Many eudicots are annuals, with short life spans measured in months or even weeks. Others are biennials and perennials, which live from two years to a century or more.

Despite their great diversity, eudicots have many physical features in common. Their leaves often have a netlike tracery of veins, unlike the parallel veins of monocots, and their stems have well-developed vascular systems, arranged in rings, for transporting water and sap. As well as getting taller, woody species thicken and strengthen their stems as they grow. This

"secondary growth" is absent in most monocots, which is why eudicots make up most of the world's flowering shrubs and trees. Below ground level, most eudicots have a taproot, with smaller roots branching off from it.

Eudicot flowers usually have parts in fours or fives, rather than the threes found in monocots, with sepals and petals that look different, both in color and in shape. Each species has its own distinctive kind of pollen, the grains always have three pores; monocots pollen grains only have a single opening.

A VITAL ROLE

The fossil pollen record suggests that the eudicots diverged from other flowering plants some 125 million years ago. Since then, they have colonized every land habitat, although they are less common as water plants and absent from the sea. Their importance to animals is impossible to overestimate. Countless species—with the notable exception of grazing animals, which feed on grasses (monocots)—depend on them for food and shelter, and many act as pollinators, or help to spread seeds.

DIVISION	ANGIOSPERMAE
CLADE	EUDICOTYLEDONAE
ORDERS	38
FAMILIES	307
SPECIES	182,227

DEBATE
FOUR-WAY SPLIT

For many years, all flowering plants were classified in two groups—the monocotyledons (monocots) and the dicotyledons (dicots)—based on the number of their seed-leaves. However, DNA analysis, combined with palynology—the study of pollen—has shown that this classification does not fully reflect plant evolution. As a result, flowering plants are now split into four overall groups: basal angiosperms, magnoliids, monocots, and the eudicots, or true dicots.

BUXALES

Containing two families, the order Buxales is distributed in temperate, tropical, and subtropical regions. The majority of the species are trees and shrubs, characterized by simple evergreen leaves without leaflike stalks, or stipules, at their bases. Many species are grown ornamentally, and boxwood is used for carving.

PROTEALES

Of the Proteales, the largest family, Proteaceae, includes Southern Hemisphere evergreen trees and shrubs. The Platanaceae are deciduous trees found in the Northern Hemisphere. Nelumbonaceae are aquatic plants from Asia, Australia, and North America.

5 ft
1.5 m

SWEET BOX
Sarcococca hookeriana
F: Buxaceae
This species bears clusters of small fragrant flowers in winter, and is found in shady habitats in western China.

16 ft
5 m

BOXWOOD
Buxus sempervirens
F: Buxaceae
Native to woods and scrub on rocky calcareous hillsides, this evergreen is widely cultivated in gardens and used for topiary.

LOTUS LEAF

SEED HEADS

SEED POD SEEN FROM ABOVE

3 ft/1 m

SACRED LOTUS
Nelumbo nucifera
F: Nelumbonaceae
Growing in shallow freshwater habitats in parts of Asia and Australia, sacred lotus carries large, fragrant flowers on long stalks above the water.

50 ft
15 m

SAW BANKSIA
Banksia serrata
F: Proteaceae
Saw Banksia grows in woodlands and scrub in Australia. Its bark is fire-resistant, enabling it to survive bush fires.

KING PROTEA
Protea cynaroides
F: Proteaceae
This species is native to hillsides and scrub in South Africa. Clusters of small flowers are enclosed by petal-like bracts.

6½ ft
2 m

outer bract

collection of flowers at center of flowerhead

FLOWERING PLANTS · EUDICOTYLEDONS

simple, evergreen leaf

23 ft / 7 m

SILKY OAK
Grevillea robusta
F: Proteaceae
This fast-growing species
from the Australian rainforest
bears one-sided clusters of
flowers, which have brightly
colored leaflike sepals in
place of petals.

sepals

115 ft
35 m

**KAHILI
FLOWER**
Grevillea banksii
F: Proteaceae
Cultivated as an
ornamental species for its
striking "bottle-brush"
flowers, this native of
Australia grows in woods
and open habitats.

10 ft
3 m

WARATAH
Telopea speciosissima
F: Proteaceae
From dry woodlands in New South
Wales, Australia, common waratah
is also known as Sydney waratah.

6½ ft / 2 m

6½ ft / 2 m

MOUNTAIN DEVIL
Lambertia formosa
F: Proteaceae
From coastal and montane heathlands
and forest in New South Wales,
Australia, this species has pink-tinged
bracts surrounding the flower clusters.

**RED PINCUSHION
PROTEA**
Leucospermum cordifolium
F: Proteaceae
This South African species
is noted for its brightly
colored, spherical
flowerheads. It grows in
acidic soil.

2 m / 6½ ft

CONE BUSH
Isopogon anemonifolius
F: Proteaceae
Spherical flowerheads above
feathery leaves characterize this
native of dry woodlands and
heathlands. Cone bush is found
in New South Wales, Australia.

*fruit takes
about six
months to ripen*

*thick, stiff-
textured, broad
leaf, similar
to maple*

100 ft / 30 m

*leafy spike of red, or
occasionally yellow
or white flowers*

30 ft
10 m

UNRIPE
NUTS

FLOWER
SPIKE

CHILEAN FIRE BUSH
Embothrium coccineum
F: Proteaceae
From forest and open habitats in
southern Chile, this species is cultivated
for its flame-colored flowers, and thrives
in sheltered gardens.

**MACADAMIA
NUT**
Macadamia integrifolia
F: Proteaceae
Macadamia is native
to coastal rainforest in
Australia and is
cultivated for its
edible nuts.

65 ft
20 m

LONDON PLANETREE
Platanus x hispanica
F: Platanaceae
Planted in London since the 17th
century, this deciduous hybrid arose
from the cross-fertilization of two
species of *Platanus* in Spain. Tolerant
of pollution, it is widely planted in
urban parks and streets.

RANUNCULALES

Annual and perennial herbaceous species, woody and herbaceous climbers, shrubs, and trees are all included in the order Ranunculales, named after the buttercup family, Ranunculaceae, which contains the greatest number of species in the order. Many of the genera are familiar ornamental garden plants, including clematis species, columbines, poppies, delphiniums, and anemones.

10 ft
3 m

oval, toothed leaf

oblong berry

EVERGREEN
LEAF

BRIGHT RED
BERRIES

BARBERRY
Berberis vulgaris
F: Berberidaceae

A European species of hedges and scrub, barberry has distinctive triple spines, hanging flower clusters, and red berries.

12 in
30 cm

SACRED BAMBOO
Nandina domestica
F: Berberidaceae

This species grows in mountain valleys in India, China, and Japan. It is also known as heavenly bamboo.

6½ ft
2 m

CHOCOLATE VINE
Akebia quinata
F: Lardizabalaceae

Bearing sprays of scented flowers in spring, chocolate vine inhabits forest edges in China, Korea, and Japan.

BISHOP'S HAT
Epimedium davidii
F: Berberidaceae

This evergreen from western China grows in woods and scrub. The young leaves are coppery, later turning green.

flower cluster

20 in
50 cm

twining stem

33 ft
10 m

20 ft
6 m

MOONSEED
Menispermum canadense
F: Menispermaceae

Although the fruits of this climber resemble black grapes, they are extremely poisonous. The species inhabits woodland and banks of streams in Canada and USA.

13 ft
4 m

fragrant flower

CAROLINA SNAILSEED
Cocculus carolinus
F: Menispermaceae

This climbing woodland species from southeast USA has tiny flowers; the males and females are borne on separate plants.

30 in
75 cm

LEONTICE
Leontice leontopetalum
F: Berberidaceae

Native to cultivated ground and dry hillsides in North Africa and eastern Mediterranean countries, leontice grows from a tuber.

16 in/40 cm

MAY APPLE
Podophyllum peltatum
F: Berberidaceae

A native of North America, May apple—also known as American mandrake—grows in open woodland.

33 ft/10 m

JAPANESE STAUNTON VINE
Stauntonia hexaphylla
F: Lardizabalaceae

Native to woodland in Japan and South Korea, this vigorous evergreen climber has woody stems and fragrant flowers.

PINNATE
LEAF

CREAMY-
COLOURED
FLOWER

CLIMBING CORYDALIS
Ceratocapnos claviculata
F: Papaveraceae

This scrambling European species grows in woods and shady habitats on acid soils, supporting itself through its leaf tendrils.

YELLOW HORNED POPPY
Glaucium flavum
F: Papaveraceae
Native to much of Europe and western Asia, yellow horned poppy usually inhabits coastal shingle. Its long curved fruits are distinctive.

36 in
90 cm

GREATER CELANDINE
Chelidonium majus
F: Papaveraceae
Formerly grown by herbalists, greater celandine is native to Europe and northern Asia, growing in woodland, scrub, and rocky places.

36 in
90 cm

YELLOW CORYDALIS
Corydalis lutea
F: Papaveraceae
This European species grows on walls and in stony and rocky places. It spreads vigorously by seed.

12 in
30 cm

4 ft
1.2 m

BLEEDING HEART
Dicentra spectabilis
F: Papaveraceae
Named for its heart-shaped flowers, it grows in damp woodland edges in Siberia, northern China, and Korea.

spurred petal

PINNATE LEAF

18 in
45 cm

CALIFORNIA POPPY
Eschscholzia californica
F: Papaveraceae
A species of open habitats in western USA and Mexico, California poppy is grown in for its brightly colored flowers.

12 in
30 cm

WELSH POPPY
Meconopsis cambrica
F: Papaveraceae
Inhabiting shady, rocky places in hilly areas and often grown in gardens, Welsh poppy is native to western Europe.

YELLOW OR ORANGE FLOWER

12 in
30 cm

COMMON FUMITORY
Fumaria officinalis
F: Papaveraceae
Found in most of Europe on cultivated and waste ground, this species usually grows on light soils.

pollen produced by a ring of dark central anthers

MATILIJA POPPY
Romneya coulteri
F: Papaveraceae
A species of scrub and grassland in California and Mexico, this poppy has fragrant flowers and is often grown in gardens.

8 in
20 cm

24 in
60 cm

cut stem exudes poisonous latex

COMMON POPPY
Papaver rhoeas
F: Papaveraceae
Arable and waste ground are home to this poppy, native to Europe, North Africa, and parts of Asia. It is used as a symbol to commemorate World War I.

OPIUM POPPY
Papaver somniferum
F: Papaveraceae
Grown as the source of opium, heroin, and poppy seed, this inhabits cultivated and disturbed ground in Eurasia.

20 in
50 cm

6½ ft
2 m

SICKLEFRUIT HYPECOUM
Hypecoum imberbe
F: Papaveraceae
This native of southern Europe grows in cultivated and waste ground and on walls.

FLOWERING PLANTS • EUDICOTYLEDONS

SEEDHEAD

cup-shaped flower

CLIMBING STEMS

FIVE-PETALED FLOWER

WINTER ACONITE
Eranthis hyemalis
F: Ranunculaceae
This tuberous perennial is found in damp woodland and shady places in central Europe. It flowers in late winter (hence its common name) and early spring.

6 in
15 cm

TRAVELER'S JOY
Clematis vitalba
F: Ranunculaceae
Also known as "old man's beard" for its gray feathery fruit, traveler's joy is a woody climber found in wood edges and hedgerows in Europe and North Africa.

100 ft
30 m

16 in
40 cm

PHEASANT'S EYE
Adonis annua
F: Ranunculaceae
An increasingly uncommon annual species, pheasant's eye is found in cultivated and waste ground in southern Europe and southwest Asia.

3 ft
1 m

MEADOW BUTTERCUP
Ranunculus acris
F: Ranunculaceae
Found in much of Europe and temperate regions of western Asia, this perennial species is found in damp grassland.

20 in
50 cm

3 ft
1 m

FIELD LOVE-IN-A-MIST
Nigella arvensis
F: Ranunculaceae
This annual species is native to central and southern Europe, North Africa, and southwest Asia. It inhabits cultivated and disturbed ground.

ROCKET LARKSPUR
Consolida ambigua
F: Ranunculaceae
Growing on cultivated and disturbed ground on light soils, rocket larkspur is an annual native to the Mediterranean region.

THIN-STEMMED FOLIAGE

24 in
60 cm

KINGCUP
Caltha palustris
F: Ranunculaceae
Kingcup, or marsh marigold, inhabits marshes, ditches, wet woods, and wet grassland in much of Europe, Asia, and North America.

20 in
45 cm

CROWN ANEMONE
Anemone coronaria
F: Ranunculaceae
Mediterranean countries are home to crown anemone, a tuberous perennial species which grows on stony hillsides, roadsides, and cultivated ground.

6½ ft
2 m

5 ft
1.5 m

5 ft
1.5 m

flowers form on long spikes

stem, leaves, and root are toxic

12 in
30 cm

PASQUE FLOWER
Anemone pulsatilla
F: Ranunculaceae
Chalk slopes with short turf are the home of pasque flower, a perennial species native to central Europe and western Asia.

SCARLET LARKSPUR
Delphinium cardinale
F: Ranunculaceae
Also known as cardinal larkspur where it grows in California, USA, and Baja California, Mexico, this short-lived perennial is found on dry hillsides.

COMMON MEADOW RUE
Thalictrum flavum
F: Ranunculaceae
This perennial species grows in wet meadows and fens beside freshwater in Europe and temperate regions of Asia.

MONKSHOOD
Aconitum napellus
F: Ranunculaceae
Named for its helmeted flowers, monkshood inhabits damp woods and streamsides in Europe. It is an extremely poisonous perennial species.

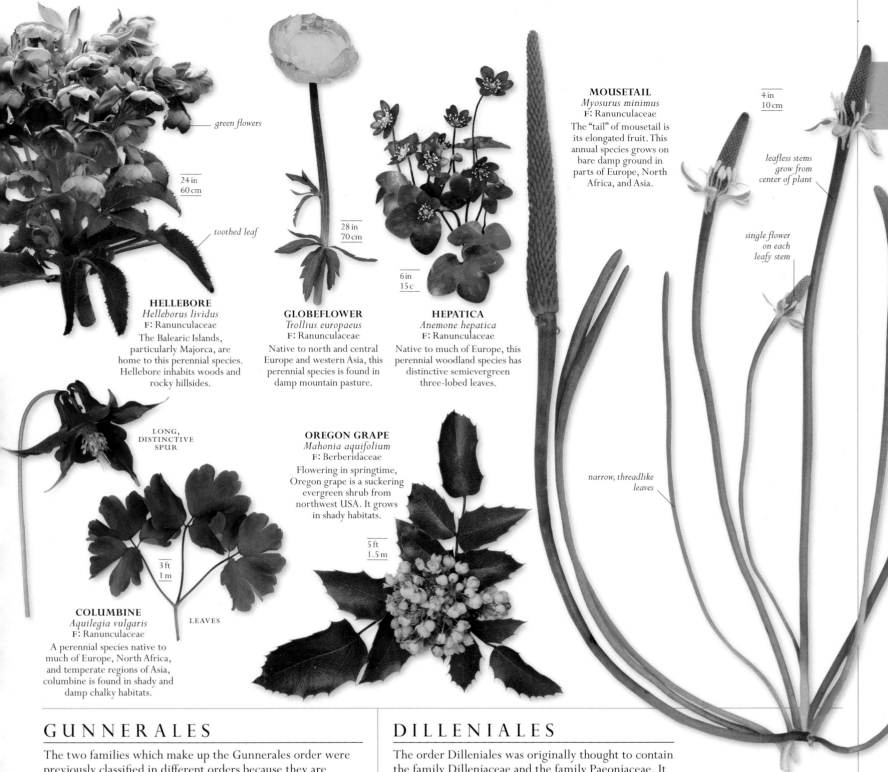

green flowers

24 in
60 cm

toothed leaf

HELLEBORE
Helleborus lividus
F: Ranunculaceae
The Balearic Islands,
particularly Majorca, are
home to this perennial species.
Hellebore inhabits woods and
rocky hillsides.

28 in
70 cm

GLOBEFLOWER
Trollius europaeus
F: Ranunculaceae
Native to north and central
Europe and western Asia, this
perennial species is found in
damp mountain pasture.

6 in
15 c

HEPATICA
Anemone hepatica
F: Ranunculaceae
Native to much of Europe, this
perennial woodland species has
distinctive semievergreen
three-lobed leaves.

MOUSETAIL
Myosurus minimus
F: Ranunculaceae
The "tail" of mousetail is
its elongated fruit. This
annual species grows on
bare damp ground in
parts of Europe, North
Africa, and Asia.

4 in
10 cm

*leafless stems
grow from
center of plant*

*single flower
on each
leafy stem*

LONG,
DISTINCTIVE
SPUR

OREGON GRAPE
Mahonia aquifolium
F: Berberidaceae
Flowering in springtime,
Oregon grape is a suckering
evergreen shrub from
northwest USA. It grows
in shady habitats.

5 ft
1.5 m

*narrow, threadlike
leaves*

3 ft
1 m

LEAVES

COLUMBINE
Aquilegia vulgaris
F: Ranunculaceae
A perennial species native to
much of Europe, North Africa,
and temperate regions of Asia,
columbine is found in shady and
damp chalky habitats.

GUNNERALES

The two families which make up the Gunnerales order were
previously classified in different orders because they are
visually very different. However, genetic analysis has recently
shown the two families to be closely related. The family
Gunneraceae consists of a single genus of large herbaceous
plants growing in damp habitats, whereas Myrothamnaceae
species are inhabitants of African deserts. *Gunnera* species are
often grown in gardens as ornamental plants.

8 ft
2.5 m

GUNNERA
Gunnera manicata
F: Gunneraceae
Massive leaves and tall
flower spikes characterize
this perennial species. It is
found beside fresh water in
Brazil and Colombia.

DILLENIALES

The order Dilleniales was originally thought to contain
the family Dilleniaceae and the family Paeoniaceae. It
now only contains the Dilleniaceae, a family of trees,
shrubs, and climbers found in tropical regions. Most
species have alternately arranged leaves and bisexual
flowers (with male and female parts) with five sepals,
five petals, and numerous stamens. Some produce dry
fruit that open to shed their seeds; others have berries.

7 m
23 ft

20 ft
6 m

GOLDEN
GUINEA VINE
Hibbertia scandens
F: Dilleniaceae
A vigorous climber or scrambler, this
evergreen shrub is found in Australia,
often growing near the coast.

SIMPOH AIR
Dillenia suffruticosa
F: Dilleniaceae
A large and vigorous
evergreen shrub endemic to
Malaysia, Sumatra, and
Borneo, simpoh air is found
on swampy ground and
forest edges.

CARYOPHYLLALES

An extremely diverse order, the Caryophyllales include trees, shrubs, climbers, succulents, and herbaceous species, and range from carnations to cacti. Most members exist in dry conditions, but many have developed various special adaptations enabling them to survive difficult environments. Perhaps the most extreme example of this is the ability of carnivorous species to trap and digest insects for extra nutrients.

⅜ in / 10 mm

CONOPHYTUM MINUTUM
F: Aizoaceae
A minute, clump-forming species, with fleshy pebble-like leaves, *Conophytum minutum* inhabits semidesert areas in South Africa.

1 in / 2.5 cm

WARTY TIGER JAWS
Faucaria tuberculosa
F: Aizoaceae
Native to semidesert areas of South Africa this species has leaves that resemble an open mouth, hence the name.

3¼ ft / 1 m

PURPLE DEWPLANT
Disphyma crassifolium
F: Aizoaceae
This procumbent species—with stems that trail along the ground—has fleshy leaves and daisylike flowers. Native to South Africa, Australia, and New Zealand, it lives on saline soils.

many-petaled yellow or light pink flower

fleshy leaf

4 in / 10 cm

SCHWANTESIA RUEDEBUSCHII
F: Aizoaceae
A clump-forming species with unequal pairs of keeled fleshy leaves, this succulent grows on hillsides in Namibia and South Africa.

16 in / 40 cm

LAMPRANTHUS SP.
F: Aizoaceae
This species of succulent plants from South Africa grows in semidesert areas, particularly near the coast. The numerous daisylike flowers are brightly colored.

1¼ in / 3 cm

TITANOPSIS CALCAREA
F: Aizoaceae
Inhabiting semidesert areas of South Africa, this clump-forming succulent has fleshy leaves. It flowers in late summer and autumn.

HOTTENTOT FIG
Carpobrotus edulis
F: Aizoaceae
This sprawling, fleshy South African species with showy flowers and edible figlike fruit grows in open dry habitats, where it can be invasive.

10 ft / 3 m

LIVING STONES
Lithops aucampiae
F: Aizoaceae
Growing among pebbles in semidesert regions of South Africa, this dwarf clump-forming species has fleshy leaves.

1¼ in / 3 cm

yellow flowers

carmine-red flowers

⅜ in / 10 mm

leaves

ICE PLANT
Mesembryanthemum crystallinum
F: Aizoaceae
Popularly named ice plant for the tiny glistening swellings covering the entire plant, this species inhabits saline areas in parts of Africa, Europe, and western Asia.

3¼ ft / 1 m

FAIRY ELEPHANT'S FEET
Frithia pulchra
F: Aizoaceae
A minute succulent, *Frithia pulchra* inhabits temperate open ground in South Africa. The barrel-shaped fruits open after rain to release their seeds.

GIBBAEUM VELUTINUM
F: Aizoaceae
Pairs of unequal-sized fleshy leaves, joined near their bases, characterize this mat-forming species. It is found in semidesert areas in South Africa.

3 in / 8 cm

LOVE-LIES-BLEEDING
Amaranthus caudatus
F: Amaranthaceae
Believed to originate in South America, this edible species of various habitats has been used as a source of food since ancient times.

8 ft / 2.5 m

35 in
90 cm

POLPALA
Aerva lanata
F: Amaranthaceae
With catkinlike flower clusters, this perennial native from tropical regions of Asia and Africa grows on dry and waste ground.

ACHYRANTHES BIDENTATA
F: Amaranthaceae
Forest edges, streamsides, and moist shady places in China, Japan, India, and Nepal are home to this species.

30 in
75 cm

ANNUAL SEABLITE
Suaeda maritima
F: Amaranthaceae
Mainly coastal, inhabiting salt marshes in Europe, but growing inland in parts of Asia and North America, annual seablite has green leaves that turn red.

12 in
30 cm

panicles of tiny flowers

5 ft
1.5 m

HEART-SHAPED LEAF

TALL, LEAFY STEM

SEA BEET
Beta vulgaris
F: Amaranthaceae
The wild ancestor of garden beets, sea beet is a fleshy inhabitant of bare ground by the sea in parts of Europe, North Africa, and Asia.

3¼ ft
1 m

SEA PURSLANE
Halimione portulacoides
F: Amaranthaceae
Salt marshes, especially edges of tidal channels and pools, are home to sea purslane, a sprawling silvery species from parts of Europe, Africa, and Asia.

12 in
30 cm

COMMON GLASSWORT
Salicornia europaea
F: Amaranthaceae
Common glasswort is a bushy species of muddy salt marshes in western Europe. Its succulent stems are sometimes cooked as a vegetable.

white, pink, or lilac flower

PURPLE FLOWER SPIKE

EDIBLE SALAD LEAF

MEXICAN TEA
Dysphania ambrosioides
F: Amaranthaceae
A short-lived inhabitant of cultivated and waste ground in tropical America, this aromatic species is used as a seasoning and in tea.

3¼ ft
1 m

RED ORACHE
Atriplex hortensis
F: Amaranthaceae
This spinachlike species with edible leaves grows on seashores and in saline (salty) habitats, and is originally from Asia.

4 ft
1.2 m

paired fleshy leaf

24 in
60 cm

PLUMED COCKSCOMB
Celosia argentea
F: Amaranthaceae
This showy plant lives on dry slopes and stony ground in tropical regions of Africa, Asia, and America. It is often grown in gardens.

»

PARODIA GRAESSNERI
F: Cactaceae

This cactus has a spherical stem and funnel-shaped flowers. It inhabits mountainous parts of South America.

6 in
15 cm

spines protect slow-growing plant

ERIOSYCE SUBGIBBOSA
F: Cactaceae

This spherical species grows in dry stony places, often at the coast in its native Chile.

12 in
30 cm

unstalked flower

5 ft
1.5 m

PERUVIAN OLD MAN CACTUS
Espostoa lanata
F: Cactaceae

Identified by long white hairs on the columnar stem, this is a slow-growing species from hilly areas of Peru and southern Ecuador.

16 in
40 cm

BARREL CACTUS
Echinocactus sp.
F: Cactaceae

Native to northern Argentina, this is a barrel-shaped species of stony ground and rocky hillsides.

flesh in leaves stores water

4 in/10 cm

23½ in
60 cm

CLEISTOCACTUS BROOKEI
F: Cactaceae

With semierect or spreading fleshy single stems, this cactus species inhabits mountainous areas of Bolivia.

23½ in/60 cm

39 ft
12 m

OLD MAN CACTUS
Cephalocereus senilis
F: Cactaceae

Long white hairs on the stem give this cactus its common name. It is native to rocky areas in Mexico.

REBUTIA HELIOSA
F: Cactaceae

A clump-forming species with brightly colored flowers, this native of Bolivia grows in partially shaded mountainous habitats.

GOLDEN COLUMN
Weberbauerocereus johnsonii
F: Cactaceae

This is a tall species from Peru, where it inhabits sandy soil.

LEUCHTENBERGIA PRINCIPIS
F: Cactaceae

The stem of this species is either spherical or short and cylindrical, and bears fragrant flowers. It inhabits hilly regions of northern Mexico.

20 ft
6 m

52 ft
16 m

arms, or branches, develop to assist fertilization

13 ft/4 m

SAGUARO CACTUS
Carnegiea gigantea
F: Cactaceae

Living for up to 150 years in desert areas of Arizona, California, and in Mexico, saguaro cactus is an extremely tall species.

MISTLETOE CACTUS
Rhipsalis baccifera
F: Cactaceae

Mistletoe cactus is epiphytic—grows on another plant—and lives in tropical America, Madagascar, Sri Lanka, and tropical Africa.

3¼ ft
1 m

16 ft
5 m

49 ft
15 m

PACHYCEREUS PRINGLEI
F: Cactaceae

This very tall branching treelike species inhabits semidesert areas of Mexico. Its flowers open at night.

HARRISIA JUSBERTII
F: Cactaceae

Of uncertain origin, but probably native to hilly areas of Argentina or Paraguay, *Harrisia jusbertii* is a columnar night-flowering cactus.

PACHYCEREUS SCHOTTII
F: Cactaceae

A tall slow-growing species from southern Arizona, this species has unpleasant-smelling flowers which open at night.

6 in
15 cm

14 in
35 cm

12 in / 30 cm

flower bud

RAT'S-TAIL CACTUS
Aporocactus flagelliformis
F: Cactaceae
Living on trees or rocks in wooded areas of Mexico, rat's-tail cactus has trailing fleshy stems and colorful flowers.

5 ft
1.5 m

MATUCANA INTERTEXTA
F: Cactaceae
Mountainous regions of Peru and Bolivia are home to this clump-forming species with a spherical or short cylindrical stem.

ASTROPHYTUM ORNATUM
F: Cactaceae
A spherical or columnar stem with long brownish-yellow protective spines characterizes this species. It grows in arid areas of Mexico.

CLARET CUP HEDGEHOG
Echinocereus triglochidiatus
F: Cactaceae
Growing in deserts, scrub, and on rocky slopes in the southern US and northern Mexico, this variable cactus is pollinated by hummingbirds.

8 in
20 cm

8 in
20 cm

4 in
10 cm

photosynthesis takes place through stems

STENOCACTUS MULTICOSTATUS
F: Cactaceae
This spherical species inhabits shady places in lowland areas of northeast Mexico. The funnel-shaped flowers have faintly striped petals.

GYMNOCALYCIUM HORSTII
F: Cactaceae
This spherical clump-forming cactus grows on rocky hillsides in Argentina, Uruguay, Paraguay, and parts of Brazil.

6 in
15 cm

TURK'S CAP CACTUS
Melocactus salvadorensis
F: Cactaceae
With a characteristic flower-bearing structure on top of a spherical stem when mature, turk's cap cactus inhabits open rocky ground in northeast Brazil.

OLD LADY CACTUS
Mammillaria hahniana
F: Cactaceae
Growing in semidesert regions, old lady cactus is a Mexican species with grayish hairs on the spherical stem.

newly formed prickly pear fruit

16 ft / 5 m

CRAB CACTUS
Schlumbergera truncata
F: Cactaceae
An epiphytic winter-flowering species of tropical rain forest in southeast Brazil, crab cactus is often grown as a houseplant.

12 in
30 cm

cluster of spines

BRIGHT YELLOW FLOWERS

PADDLELIKE GREEN STEM

8 in
20 cm

GLORY OF TEXAS
Thelocactus bicolor
F: Cactaceae
Native to arid regions of Texas and northeast Mexico, this is a spherical species.

PRICKLY PEAR
Opuntia ficus-indica
F: Cactaceae
With flattened jointed stems and edible egg-shaped fruit, prickly pear is a native of rocky hillsides and dry ground in Mexico, and is often naturalized elsewhere.

∨ **SPIKY HAIRY STAR**
Viewed from above, *Astrophytum ornatum* resembles a star. The flower is surrounded by groups of spines on the eight ribs, showing the cross bands of white, woolly stem hairs.

spines grow from a small hollow, or areole, surrounded by white hair

CACTUS
Astrophytum ornatum

Commonly known as the Monk's Hood Cactus, this cactus's genus name means "star plant" in ancient Greek. *Astrophytum ornatum* was first collected in 1827 by Thomas Coulter, an Irish doctor and botanist, who sent it to Professor de Candolle at the Botanic Garden of Geneva. When de Candolle unpacked the cactus he thought it was covered in fungus, but discovered that the white spots were actually tufts of hair or trichomes. These woolly scales may help the cactus collect water and protect it against the sun, but they also provide camouflage. *Astrophytum ornatum* is the most densely covered of all the *Astrophytum* cacti, and also the spiniest. This cactus is now rare in the wild.

SIZE 4 ft (1.2 m)
HABITAT Hot, arid zones
DISTRIBUTION Mexico
LEAF TYPE Spines

spines protect the slow-growing plant from being eaten

pith, used for water storage

fibrous roots

< ROOT SYSTEM
Fibrous and shallow roots help the cactus absorb water from a large area, essential in an environment where after brief showers only an inch or two of soil is wet.

OUTER PETALS >
The numerous, narrow outer petals of the flower are pale yellow and have brown tips. The flower itself grows to 4¼ in (11 cm) in diameter.

< ˅ STEM HAIRS
This type of cactus exhibits woolly, white tufts of trichomes arranged in bands across the ribs. This stem hair is usually dense in young plants and more sparse in older plants.

petal

stamen (male part that produces and releases pollen)

stigma

style (female part that connects the stigma to the ovary)

ovary

> FLOWER SECTION
The yellow petals have a broad oblong shape with a slightly serrated apex. The flower has a yellow carpel (the female reproductive organs, made up of the stigma, style, and ovary) and yellow stamens.

˄ OVARY
Located below all the other reproductive parts of the flower, the ovary contains ovules that develop into seeds.

˄ STIGMA
The flower has a single stigma with seven to twelve lobes. This is the part of the flower used to catch pollen. The stigma is about ½ in (1.5 cm) long.

FLOWERING PLANTS • EUDICOTYLEDONS

23½ in
60 cm

DEPTFORD PINK
Dianthus armeria
F: Caryophyllaceae
Inhabiting dry grassland,
especially on light sandy soil
in much of Europe, Deptford
pink has starry flowers with
toothed petals.

3¼ ft
1 m

CORNCOCKLE
Agrostemma githago
F: Caryophyllaceae
A native of eastern
Mediterranean countries,
corncockle was formerly a
widespread cornfield weed,
but is now much rarer.

23½ in
60 cm

COW BASIL
Vaccaria pyramidata
F: Caryophyllaceae
From cultivated and waste ground
in parts of Europe and Asia, cow
basil has pinkish flowers and bluish
green leaves.

10 in
25 cm

SEA SANDWORT
Honckenya peploides
F: Caryophyllaceae
A fleshy prostrate species, sea
sandwort is found on coastal
sand and pepple in North
America, Europe, and Asia.

12 in
30 cm

**THYME-LEAVED
SANDWORT**
Arenaria serpyllifolia
F: Caryophyllaceae
Found in bare or disturbed
ground in Europe, temperate
Asia, and North America, this
species gets its common
name from its tiny leaves.

numerous pink, or
occasionally white,
flowers

long flower
stalk

ALPINE CAMPION
Lychnis alpina
F: Caryophyllaceae
This native of the European Alps,
Pyrenees, and subarctic areas of
North America, Europe, and west
Asia grows on mineral-rich rocks.

8 in
20 cm

12 in
30 cm

32 in
80 cm

FIVE-
PETALED
FLOWER

LANCEOLATE
LEAVES

four or five small green
leaves per shoot

3¼ ft / 1 m

RAGGED ROBIN
Lychnis flos-cuculi
F: Caryophyllaceae
Divided petals give ragged robin
its name. It is a European
species found in marshy fields
and damp ground.

FIELD MOUSE-EAR
Cerastium arvense
F: Caryophyllaceae
A native of dry grassland in North
America, Europe, North Africa,
and western temperate Asia.

32 in / 80 cm

calyx

central cushion has
long tap root

MOSS CAMPION
Silene acaulis
F: Caryophyllaceae
Cushion-forming and mosslike,
this grows on mountains in North
America, western, central, and
northern Europe, as well as Asia.

BLADDER CAMPION
Silene vulgaris
F: Caryophyllaceae
Named for its inflated calyx, Bladder
campion inhabits open grassy places
in Europe, North Africa, and
temperate regions of Asia.

capsule containing flowers

20 in
50 cm

CHILDING PINK
Petrorhagia nanteuilii
F: Caryophyllaceae
The flowers of childing pink open one at a time. This species inhabits dry grassy places in sandy soil in much of Europe.

3¼ ft
1 m

SOAPWORT
Saponaria officinalis
F: Caryophyllaceae
Once used to make soap, this species is found along streams and damp ground in Europe and Asia.

20 in
50 cm

COMMON CHICKWEED
Stellaria media
F: Caryophyllaceae
This sprawling species is found on cultivated and open ground worldwide and is sometimes used as a salad vegetable.

ROUND-LEAVED SUNDEW
Drosera rotundifolia
F: Droseraceae
Inhabiting open boggy moorland and meadows in North America, Europe, and northern Asia, this insectivorous species has leaves with numerous sticky hairs covered in insect-dissolving enzymes.

leaf hairs trap insects

4 in
10 cm

flowers surrounded by bracts

brightly colored bract

26 ft / 8 m

trap triggered by hairs on leaf surface

when trap closes "teeth" prevent insect escaping

VENUS FLY-TRAP
Dionaea muscipula
F: Droseraceae
An insectivorous perennial species from coastal bogs in North and South Carolina, Venus fly-trap has hinged, two-lobed leaves.

4 in
10 cm

BOUGAINVILLEA
Bougainvillea glabra
F: Nyctaginaceae
What seem to be petals on bougainvillea are actually bracts. This slow-growing evergreen climber from Brazil is widely cultivated.

12 in / 30 cm

SEA HEATH
Frankenia laevis
F: Frankeniaceae
A mat-forming species, sea heath grows in bare sandy ground in the drier parts of salt marshes in Europe and western Asia.

lid closes on trapped insects

rim secretes sugar to attract insects

FOUR O'CLOCK FLOWER
Mirabilis jalapa
F: Nyctaginaceae
From dry open habitats in tropical Central and South America, this species has fragrant flowers which open during the late afternoon.

3¼ ft / 1 m

fluid in funnel digests insect

4¾ in
12 cm

PITCHER PLANT
Nepenthes vogelii
F: Nepenthaceae
Found in the montane and submontane forests of Borneo, this insectivorous species grows on tree moss.

FLOWERING PLANTS • EUDICOTYLEDONS

12 in / 30 cm

ERIOGONUM UMBELLATUM
F: Polygonaceae
Inhabiting well-drained montane forest and scrub in northern and western US and Canada, this spreading mat-forming species has long-lasting flowers.

6½ ft / 2 m

ST. CATHERINE'S LACE
Eriogonum giganteum
F: Polygonaceae
With clusters of tiny long-lasting flowers much visited by butterflies, St. Catherine's lace is native to arid areas of the Southwest US.

BLACK BINDWEED
Fallopia convolvulus
F: Polygonaceae
Growing on waste and cultivated ground, black bindweed is native to much of Europe, North Africa, and temperate Asia.

3¼ ft / 1 m

WAVY-EDGED LEAF

3¼ ft 1 m

SEED STALK

23½ in 60 cm

23½ in 60 cm

COMMON BUCKWHEAT
Fagopyrum esculentum
F: Polygonaceae
Buckwheat originated in temperate areas of Asia, and is cultivated for its seeds, which are made into flour and also fed to birds.

CURLED DOCK
Rumex crispus
F: Polygonaceae
Curled dock inhabits grassy places, waste ground, and coastal pebbles in Europe and most of Africa. It can be an invasive species.

KNOTGRASS
Polygonum aviculare
F: Polygonaceae
This spreading species of open cultivated and bare ground in Europe and Asia grows on coasts and inland.

small, green flowers

reddish flowers pollinated by wind

SEED STALK

ROUND LEAF

fleshy, kidney-shaped leaves

12 in 30 cm

35 in 90 cm

39 ft 12 m

9¾ ft 3 m

MOUNTAIN SORREL
Oxyria digyna
F: Polygonaceae
Inhabiting damp ledges and streamsides on mountains in Arctic and temperate regions of the Northern Hemisphere, mountain sorrel is often tinged red.

COMMON BISTORT
Persicaria bistorta
F: Polygonaceae
Inhabiting grassy areas in much of Europe and central Asia, common bistort has dense cylindrical flower spikes.

CORAL VINE
Antigonon leptopus
F: Polygonaceae
This fast-growing species climbs by means of tendrils. It lives in tropical forest and scrub in Mexico.

POKEWEED
Phytolacca americana
F: Phytolaccaceae
With poisonous blackberrylike fruit this unpleasant-smelling perennial is native to open and shady parts of eastern North America and Mexico.

8¼ ft 2.5 m

CHINESE RHUBARB
Rheum palmatum
F: Polygonaceae
With enormous roots and large, poisonous leaves, Chinese rhubarb inhabits streamsides and damp ground in mountainous areas in China and Tibet.

large leaf supported by strong stem

12 in 30 cm

2 in 5 cm

SPRING BEAUTY
Claytonia perfoliata
F: Portulacaceae
Also known as miner's lettuce, this native of cultivated and bare ground in western North America, Mexico, and Cuba has two fused leaves beneath the flowers.

TALINUM OKANOGANENSE
F: Portulacaceae
The flowers of this prostrate species open in the afternoon. Dry grassland and scrub in western North America are its home.

LEWISIA BRACHYCALYX
F: Portulacaceae
Native to damp stony meadows in mountainous areas of southwestern US, this species has a basal rosette of fleshy leaves.

CEYLON LEADWORT
Plumbago zeylanica
F: Plumbaginaceae
Inhabiting open areas and thickets, Ceylon leadwort is native to tropical regions of Africa, the Middle East, and southwestern Asia.

STATICE
Limonium sinuatum
F: Plumbaginaceae
Statice has winged branching stems and is found in rocky and sandy coastal habitats and in saline areas inland in Mediterranean countries.

16 in / 40 cm

20 in / 50 cm

3¼ in / 8 cm

SUMMER PURSLANE
Portulaca oleracea
F: Portulacaceae
Inhabiting many parts of the world including much of Europe, China, and Japan, purslane is an edible fleshy species of disturbed ground.

SEA PINK
Armeria maritima
F: Plumbaginaceae
In much of Western Europe, thrift, or sea pink, inhabits coastal rocks and cliffs, salt marshes, and mountains. It is a cushion-forming species.

5 ft / 1.5 m

10 in / 25 cm

TAMARISK
Tamarix gallica
F: Tamaricaceae
Usually coastal, tamarisk also grows in saline areas inland. It originates from southern Europe, North Africa, and the Canary Islands, and is planted elsewhere.

9¾ ft / 3 m

CALANDRINIA FELTONII
F: Portulacaceae
Now found in the Falkland Islands, but also in gardens, this species is from open rocky or grassy areas in Argentina.

12 in / 30 cm

165

JOJOBA
Simmondsia chinensis
F: Simmondsiaceae
Deserts in Arizona and California and Mexico are home to jojoba, a species often cultivated for its oil.

6½ ft / 2 m

SANTALALES

Found mainly in tropical and subtropical regions, the order Santalales includes many species of parasitic and semiparasitic plants, such as mistletoe, which grow attached to other plants from which they obtain water and nutrients. Although it is DNA analysis that determines their classification, most of the order have seeds lacking an outer covering. Santalales classification is still being debated.

SANDALWOOD
Santalum album
F: Santalaceae
Cultivated for its wood and fragrant oil, this semiparasitic species grows in dry rocky areas in parts of Asia.

30 ft / 9 m

FIRE TREE
Nuytsia floribunda
F: Loranthaceae
This species is semiparasitic, deriving moisture and nutrients from the roots of surrounding plants in woodland in southwestern Australia.

MISTLETOE
Viscum album
F: Santalaceae
Mistletoe forms spherical masses on tree branches. A semiparasitic species with white berries, it inhabits much of Europe, North Africa, and Asia.

berries can be poisonous

33 ft / 10 m

3¼ ft / 1 m

OSYRIS
Osyris alba
F: Santalaceae
This broomlike semiparasitic species with fragrant flowers is native to southern Europe, North Africa, and southwest Asia, inhabiting dry rocky places.

4 ft / 1.2 m

FLOWERING PLANTS · EUDICOTYLEDONS

SAXIFRAGALES

The order Saxifragales is named after saxifrages. The Latin name *saxifraga* literally means "rock breaker," because these plants often grow in cracks in rocks and walls. Other prominent members of the order are currants, the *Ribes* species, hydrangeas, and the stonecrops *Crassula*. Stonecrops are succulents, or water-retaining plants, adapted to dry conditions. There are over 1,000 species and many are grown as indoor plants.

MEXICAN FIRECRACKER
Echeveria setosa
F: Crassulaceae
Named for its brightly colored flowers, this Mexican species has rosettes of succulent leaves densely covered with white hairs.

2 in
5 cm

6 in
15 cm

COMMON HOUSELEEK
Sempervivum tectorum
F: Crassulaceae
Originating from mountains in central Europe, houseleek is often planted on roofs and walls. This plant forms dense mats of succulent rosettes.

20 in
50 cm

CRASSULA DECEPTOR
F: Crassulaceae
Bearing sweetly scented flowers, this variable species is from South Africa. A white powdery coating protects the stems from sun damage and drought.

SAUCER PLANT
Aeonium tabuliforme
F: Crassulaceae
This native of hillsides and coastal cliffs on Tenerife, Canary Islands, forms a flat succulent rosette. After flowering the plant dies.

24 in / 60 cm

ORPINE
Sedum telephium
F: Crassulaceae
Rocky terrain, woods, and hedgebanks are the habitats of orpine, a species native to much of Europe, temperate Asia, and North America.

24 in
60 cm

NAVELWORT
Umbilicus rupestris
F: Crassulaceae
This species has round fleshy leaves, each with a central dimple. Navelwort inhabits rocks and walls in much of Europe.

20 in
50 cm

16 in
40 cm

FLAMING KATY
Kalanchoe blossfeldiana
F: Crassulaceae
Arid areas of Madagascar are home to this bushy species with glossy succulent leaves and brightly colored flowers.

16 in
40 cm

ROSEROOT
Sedum rosea
F: Crassulaceae
Inhabiting mountain rocks and coastal cliffs, roseroot is a fleshy species found in the Arctic and alpine regions in Europe, North America, and Asia.

BLACK CURRANT
Ribes nigrum
F: Grossulariaceae
This species grows in damp woods in much of Europe and central Asia. Black currant is cultivated for its flavorful fruit.

6½ ft
2 m

6½ ft / 2 m

BUFFALO CURRANT
Ribes odoratum
F: Grossulariaceae
Scented flowers and thornless stems characterize buffalo currant, an inhabitant of rocky and sandy areas in central USA.

3¼ ft / 1 m

WESTERN WATER MILFOIL
Myriophyllum hippuroides
F: Haloragaceae
This aquatic species has finely divided leaves. It is native to freshwater habitats in western North America.

PARROTIOPSIS
Parrotiopsis
jacquemontiana
F: Hamamelidaceae
Instead of petals, the
flowers of this forest
species from the
western Himalayas have
white bracts, or
modified leaves.

20 ft / 6 m

40 ft
12 m

CHINESE SWEETGUM
Liquidambar formosana
F: Hamamelidaceae
A deciduous tree of damp
woodland in China and
Taiwan, this species is noted
for its brightly colored
autumn foliage.

BUDS AND
FLOWERS

LEAVES CHANGING
COLOR

**VIRGINIAN WITCH
HAZEL**
Hamamelis virginiana
F: Hamamelidaceae
The flowers are fragrant and the
leaves turn yellow in autumn. It
inhabits woodland in eastern
North America.

13 ft / 4 m

CHINESE ASTILBE
Astilbe chinensis var. *taquetii*
F: Saxifragaceae
With distinctive plume-like
clusters of flowers, this
moisture-loving plant grows in
damp woodland and by streams
in China, Korea, and Siberia.

3¼ ft / 1 m

50 ft / 15 m

PERSIAN IRONWOOD
Parrotia persica
F: Hamamelidaceae
Native to forest in the Caucasus region and
northern Iran, Persian ironwood is a
winter-flowering deciduous tree with
brilliantly colored autumn leaves.

**MOTHER
OF THOUSANDS**
Saxifraga stolonifera
F: Saxifragaceae
Native to shady habitats
in China and Japan, this
perennial produces
stolons, stemlike
clusters that can root to
form new plants.

12 in
30 cm

**YELLOW
SAXIFRAGE**
Saxifraga aizoides
F: Saxifragaceae
Streamsides and wet
stony areas of mountains
are home to this
mat-forming species of
Europe, North America,
and western Asia.

8 in
20 cm

ALUMROOT
Heuchera americana
F: Saxifragaceae
Inhabiting rocky
woodland in North
America, alumroot
has glossy leaves
which are mottled
when young.

24 in / 60 cm

*flowers form
in a flattish
topped cluster*

ELEPHANT'S EARS
Bergenia stracheyi
F: Saxifragaceae
An inhabitant of damp woodland and
meadows in the western Himalayas and
Afghanistan, elephant's ears has fragrant
flowers and large glossy leaves.

12 in
30 cm

*large glossy
leaves*

**OPPOSITE-LEAVED
GOLDEN-SAXIFRAGE**
Chrysosplenium oppositifolium
F: Saxifragaceae
The procumbent stems of this
species form extensive patches
in damp shady habitats in
western and central Europe.

6 in / 15 cm

COMMON PEONY
Paeonia officinalis
F: Paeoniaceae
A herbaceous species
found in woodland,
meadows, and scrub
in parts of Europe,
common peony is
known for its
showy flowers.

28 in
70 cm

28 in / 70 cm

**PIGGYBACK
PLANT**
Tolmiea menziesii
F: Saxifragaceae
Young plantlets growing on
leaf bases give piggyback
plant its name. It is a hairy
perennial species found in
damp shady places in
North America.

VITALES

This order consists of a single family, the Vitaceae or grape family. It contains 14 genera and 850 species, including the important grapevine and the ornamental Virginia creeper. Members of the Vitaceae are mainly native to the tropics or warm temperate regions. Mostly vines or lianas. they usually have swollen nodes (where leaves fork from the stem) and tendrils for climbing. Their flowers are usually held in flat-topped clusters.

foliage turns red in the fall

GRAPEVINE
Vitis vinifera
F: Vitaceae

115 ft / 35 m

Humans have used grapes to make wine, food, and medicine since neolithic times. Very widespread, the grapevine is native to the Mediterranean, Europe, and Asia.

fruit smaller than domesticated varieties

CRIMSON GLORY VINE
Vitis coignetiae
F: Vitaceae
Cultivated for its giant, dimpled leaves (12 in / 30 cm across) and fall color, this deciduous climber is native to temperate Asia.

VIRGINIA CREEPER
Parthenocissus quinquefolia
F: Vitaceae
This prolific climber from eastern and northern America clings onto smooth surfaces using the adhesive pads on its small, forked tendrils.

98 ft / 30 m

49 ft / 15 m

GERANIALES

The Geraniales has only four plant families. The *Geraniaceae* or geranium family is the largest, with 800 species within seven genera, of which the genus *Geranium* or cranesbills has 260 species. The genus *Pelargonium* has 280 species including the garden plants commonly known as geraniums. *Melianthaceae* is another family of the Geraniales; its members are trees and shrubs native to tropical and southern Africa.

8¼ ft / 2.5 m

GIANT HONEYBUSH
Melianthus major
F: Melianthaceae
Nectar drips from the bronze flower spikes of this South African native. Touching the leaves causes them to give off a strong odor.

12 in / 30 cm

23½ in / 60 cm

8 in / 20 cm

20 in / 50 cm

HERB ROBERT
Geranium robertianum
F: Geraniaceae
Widespread in the Northern Hemisphere, this sprawling species has red stems and long flower stalks. It has a strong, mousy smell.

APPLE PELARGONIUM
Pelargonium odoratissimum
F: Geraniaceae
This perennial originates from South Africa and has spreading flower stalks. It is cultivated for "oil of geranium," which has a strong scent of apple and rose.

MEADOW CRANESBILL
Geranium pratense
F: Geraniaceae
Preferring grasslands on chalky soil, this perennial is native to Europe and Asia. It is the main food source for the northern brown argus butterfly.

ROCK STORKSBILL
Erodium petraeum
F: Geraniaceae
Storksbills take their name from the shape of their seeds. This Mediterranean perennial is cultivated in rock gardens.

MYRTALES

The order Myrtales is found throughout the tropics and warmer regions. Of the 14 families, the Myrtaceae (myrtle family) is the biggest with over 5,650 species. They tend to produce essential oils, such as myrtle, clove, and eucalyptus. The Lythraceae (loosestrife and pomegranate family) have petals that emerge from a floral tube (calyx) that includes petals and the stamen. Most of the Onagraceae (willowherb or evening primrose family) have four colored sepals and petals.

SWAMP LOOSESTRIFE
Decodon verticillatus
F: Lythraceae
Native to northeastern America, this shrub grows in swamps. It has stems with up to six sides, its leaves grow in whorls of three and it has ⅓ in (1 cm) flowers.

8¼ ft / 2.5 m

PURPLE LOOSESTRIFE
Lythrum salicaria
F: Lythraceae
Numerous purple-red, square stems grow from the woody, creeping rootstock of this perennial. Native to Europe, Asia, southeastern Australia, and northwestern Africa, it is invasive.

5 ft / 1.5 m

CRAPE MYRTLE
Lagerstroemia indica
F: Lythraceae
Blooming for up to 120 days, this tree is native to China, Korea, and Japan. Its bark is smooth, mottled, pinkish gray, and is shed each year.

20 ft / 6 m

HENNA
Lawsonia inerma
F: Lythraceae
Native to N. Africa and the Middle East, henna's leaves make a reddish-brown dye and its fragrant flowers yield essential oils.

20 ft / 6 m

triangular leaves

WATER CHESTNUT
Trapa natans
F: Lythraceae
This floating plant is from Europe and Asia. Within a nut with four hornlike barbed spines is the edible starchy seed.

30 in / 75 cm

multiseeded fruit develops from flower

POMEGRANATE
Punica granatum
F: Lythraceae
This spiny, shrubby tree, from southwestern Asia, is widely cultivated in the Mediterranean for its pulpy fruit, containing numerous seeds.

23 ft / 7 m

CIGAR FLOWER
Cuphea ignea
F: Lythraceae
A densely branched, perennial shrub, cigar flower is a garden and house ornamental. Native to Mexico and the West Indies, its fruit is paperlike capsules.

35 in / 90 cm

RANGOON CREEPER
Quisqualis indica
F: Combretaceae
This climber from tropical Asia has clusters of tubular red flowers. Its ellipsoidal fruit, each with five wings, taste of almonds.

59 ft / 18 m

INDIAN ALMOND
Terminalia catappa
F: Combretaceae
Found on coasts of the Indo-Pacific oceans, this tree has horizontal branches. Its corky fruit, dispersed by water, contains an edible, almond-tasting nut.

98 ft / 30 m

flowers have five or six petals

ROSE GRAPE
Medinilla magnifica
F: Melastomataceae
This ornamental from the Philippines is an epiphyte, growing on trees. Its leaves have the longitudinal veins characteristic of the melastomes.

20 ft / 6 m

BRAZILIAN SPIDER FLOWER
Tibouchina urvilleana
F: Melastomataceae
In warm regions this Brazilian ornamental blooms for most of the year. Its velvety leaves have red edges and three to five obvious longitudinal veins.

16 ft / 5 m

VIRGINIA MEADOW BEAUTY
Rhexia virginica
F: Melastomataceae
This hairy, perennial herb grows in wetlands of the eastern US. It has square stems and stalkless leaves with toothed margins.

23½ in / 60 cm

stamens form bottlebrush-shaped flowers

TONGHI BOTTLEBRUSH
Callistemon subulatus
F: Myrtaceae

A spreading shrub with small woody fruit containing hundreds of seeds. It is found mainly in New South Wales and Victoria, Australia.

9¾ ft
3 m

GREEN BOTTLEBRUSH
Callistemon viridiflorus
F: Myrtaceae

Resisting snow, frost, and drought, this sprawling shrub from southern Australia has leaves with sharp, pointed ends. It attracts birds and butterflies.

9¾ ft
3 m

SCARLET KUNZEA
Kunzea baxteri
F: Myrtaceae

Unlike *Callistemon* bottlebrushes these woody shrubs from the south coast of West Australia lose their five sepals and petals, leaving a one-celled fruit.

9¾ ft
3 m

RED GUM
Eucalyptus camaldulensis
F: Myrtaceae

Widespread in Australia, with smooth, pale bark and blue-green leaves, this tree has durable wood and its nectar makes good honey.

130 ft
40 m

TASMANIAN SNOW GUM
Eucalyptus coccifera
F: Myrtaceae

With gray-white bark, peeling in long strips to reveal a creamy-white trunk, this Tasmanian native has rounded, unstalked, opposite leaves.

16 ft
5 m

COMMON MYRTLE
Myrtus communis
F: Myrtaceae

The aromatic leaf of myrtle yields essential oils. Native to the Mediterranean, myrtle has fragrant flowers and produces blue-black berries.

82 ft
25 m

elliptical leaves smell of peppermint

39 ft
12 m

ALLSPICE
Pimenta dioica
F: Myrtaceae

From the Caribbean, southern Mexico, and Central America, this unisexual tree has small white flowers that form brown berrylike pitted fruit. Unripe fruit is dried and ground to make "allspice."

120 ft
36 m

CIDER GUM
Eucalyptus gunnii
F: Myrtaceae

This hardy Tasmanian tree forms juvenile leaves that are round and silvery, maturing to scythe-shaped and blue-gray. It has red-brown peeling bark.

39 ft
12 m

URN-FRUITED GUM
Eucalyptus urnigera
F: Myrtaceae

This tree from southeast Tasmania has urn-shaped fruit, blue-gray juvenile leaves, and white flowers in clusters of three with numerous stamens.

49 ft
15 m

82 ft
25 m

66 ft
20 m

CHILEAN MYRTLE
Luma apiculata
F: Myrtaceae

This slow-growing tree has a contorted trunk and smooth gray-orange peeling bark. Its fruit is a black berry.

POHUTAKAWA
Metrosideros excelsa
F: Myrtaceae

This tree produces red flowers in December. It is native to North Island, New Zealand, clinging to rocky cliffs with its long, hanging roots.

CAJUPUT
Melaleuca cajuputi
F: Myrtaceae

Native to the islands of Indonesia, this tree has aromatic leaves that contain an emerald-colored medicinal oil.

39 ft
12 m

CLOVE
Syzygium aromaticum
F: Myrtaceae

The dried flower buds of this native of Indonesia and the Philippines are used as a spice. The flowers are cream-colored with red stamens and purple single-seeded berries.

EVERGREEN LEAVES

26 ft
8 m

EDIBLE FRUIT

GUAVA
Psidium guajava
F: Myrtaceae

This tree from southern Mexico has flaky, coppery bark. The fruit has a sweet, musky smell when ripe, and contains numerous hard, yellow seeds.

3¼ ft
1 m

5 ft
1.5 m

9¾ ft
3 m

5 ft
1.5 m

large, purple-
pink flower

GODETIA
Clarkia amoena
F: Onagraceae
This annual is native to coastal
hills of western North America.
Cultivated in gardens, its flower
has four broad petals and forms
a dry capsule, containing
numerous seeds.

FUCHSIA FULGENS
F: Onagraceae
Clinging to rocks as a
lithophyte or trees as an
epiphyte with its long,
tuberous roots, this
deciduous perennial is
native to the mountains
of Mexico.

HARDY FUCHSIA
Fuchsia magellanica
F: Onagraceae
This widespread, deciduous,
ornamental perennial
originates from Chile,
where it usually has its roots
in water. It produces
black berries.

**ROSEBAY
WILLOWHERB**
Chamerion angustifolium
F: Onagraceae
This herbaceous perennial,
widespread throughout the
temperate northern
hemisphere, spreads rapidly
using horizontal stems. Its
fruit capsule splits four
ways to release seeds.

20 in
50 cm

5 ft
1.5 m

**SHOWY EVENING
PRIMROSE**
Oenothera speciosa
F: Onagraceae
This herbaceous, smooth-
stemmed perennial is from the
southeastern US and Mexico. Its
white flowers become pink with
age and close in full sun.

EVENING PRIMROSE
Oenothera biennis
F: Onagraceae
This biennial from eastern
North America has bluish green
leaves in a basal rosette and
erect stems bearing flowers that
open at night. The seed oil has
medicinal properties.

stigma has four
white lobes

6½ ft
2 m

28 in
70 cm

ENCHANTER'S NIGHTSHADE
Circaea lutetiana
F: Onagraceae
This woodland plant is distributed through
the temperate northern hemisphere. Flowers
of two sepals and two notched petals form
small, round, hairy fruit.

characteristic
four petals

**GREAT
WILLOW HERB**
Epilobium hirsutum
F: Onagraceae
This very hairy perennial
forms extensive patches with
its horizontal stems. It is native
to most of North America,
Europe, and parts of Asia.

SEEDBOX
Ludwigia alternifolia
F: Onagraceae
Preferring to grow in swamps, this plant
is native to Missouri. It has box-shaped
seedpods borne on angled stems.

3¼ ft
1 m

CELASTRALES

This is a diverse order with few visibly
distinguishing characteristics, apart from
small flowers and an obvious nectar disc.
The Celastrales contains about 1,300
species, within 100 genera. The divisions
within the order are probably still not
completely resolved but 93 of the genera
are in the family Celastraceae (staff vine
or bittersweet) of vines or shrubby trees.

SPINDLE TREE
Euonymus europaeus
F: Celastraceae
Native to Europe, the
wood of this tree was
used to make
spindles—spikes used
to spin thread. Its fruit
ripens to red and split
into four to reveal
orange-coated,
poisonous seeds.

nectar
disc

20 ft
6 m

elliptical leaves

**GRASS OF
PARNASSUS**
Parnassia palustris
F: Parnassiaceae
This rosette plant grows
in bogs in the northern
temperate zone. Each
flower is borne on a stem,
clasped by a stalkless leaf.

12 in
30 cm

CUCURBITALES

Including trees, shrubs, herbs, and climbers, the Cucurbitales are mainly found in tropical areas. There are about 2,300 species contained within seven families. The Begonia family has 1,400 species, 130 of which are horticultural plants, and the Gourd family has 825 species, including some important food species such as squash and pumpkins. Both these families have monoecious (has both male and female flowers) species.

13 ft
4 m

EDIBLE FLOWER

NUTRITIOUS FRUIT

WHITE BRYONY
Bryonia dioica
F: Cucurbitaceae
Common in central and southern Europe in hedgerows on calcium carbonate-rich soils, this is a vigorous climber with an enormous, tuberous root.

26 ft
8 m

HAIRY LEAVES

large, spherical fruit narrows above middle

fruit stem

CUCUMBER
Cucumis sativus
F: Cucurbitaceae
The cucumber has tubular yellow flowers, and originated in tropical Asia. Cultivated for 3,000 years, there are many varieties including the gherkin.

IMMATURE FRUIT

3¼ ft
1 m

LARGE LEAF

SQUASH
Cucurbita pepo
F: Cucurbitaceae
With five-sided prickly stems and large yellow-orange flowers, pumpkins, marrows, and zucchini are all varieties of *Cucurbita pepo* from Central America.

15 ft/4.5 m

FRUIT   LEAF SPRIG

SMOOTH LOOFAH
Luffa cylindrica
F: Cucurbitaceae
This climbing, annual vine is native to South America. The fruit grows to 24 in (60 cm) long and contains edible, white flesh, and a fibrous structure.

CALABASH
Lagenaria siceraria
F: Cucurbitaceae
This vine produces whitish crinkly flowers. Its edible, hard-shelled fruit floats in the sea for months and is used as a container.

16 ft
5 m

3¼ ft
1 m

SQUIRTING CUCUMBER
Ecballium elaterium
F: Cucurbitaceae
This Mediterranean plant has fruits that fill up with a thick, sticky liquid as they ripen and explode their contents as far as 20 ft (6 m).

33 ft/10 m

WATERMELON
Citrullus lanatus
F: Cucurbitaceae
With yellow-green flowers, this annual, prostrate vine originated in South Africa and its fruit is cultivated throughout warmer climates.

16 in/40 cm

BEGONIA LISTADA
F: Begoniaceae
Native to Brazil, this species sprawls, has thick, velvety leaves with red undersides, and white pinkish flowers. It was first described in 1981.

poisonous squirting fruit

leaves are up to 4 in (10 cm) long

BALSAM PEAR
Momordica charantia
F: Cucurbitaceae
Also known as bitter gourd, this tropical vine produces bitter-tasting fruit that splits three ways to release red-brown seeds encased in scarlet arils, or coats.

16 ft
5 m

150 ft
45 m

TETRAMELES NUDIFLORA
F: Datiscaceae
This Indo-Malayan tree has a thick, heavily-fluted, cylindrical trunk. There are small, cream male and female flowers on different trees that form seed capsules.

FABALES

Found all over the world, except for Antarctica, the Fabales have compound leaves with tiny outgrowths (stiplues) at the base and seed pods that open when mature. Called legumes, these plants have nodules swellings that contain bacteria on their roots that help them fix nitrogen from the air into soil. The Fabaceae, the largest of the four families, includes plants like the pea, overleaf, with flowers that have a large upper petal with adjacent smaller petals.

50 cm
20 in

— pinnate leaves

SEED POD

HERBACEOUS PLANT

PEANUT
Arachis hypogaea
F: Fabaceae

Native to Central and South America, this legume has yellow, reddish veined pealike flowers. Its seed pod matures in the ground, producing peanuts.

5 ft / 1.5 m

SENSITIVE PLANT
Mimosa pudica
F: Fabaceae

This prickly, Brazilian shrub's leaves close when touched. Its flowerheads (³⁄₈ in / 1 cm) usually have pink flowers and long pink-purple stamens.

GOLDEN SHOWER TREE
Cassia fistula
F: Fabaceae

This slender tree from southeast Asia has pendulous pealike flowers and pinnate leaves with three to eight pairs of leaflets. It contains poisonous seeds.

66 ft
20 m

DECIDUOUS LEAF

32 in
80 cm

GLOSSY LEAVES

FLOWER RACEME

POD HAS EDIBLE SEEDS

66 ft
20 m

MIMOSA
Acacia dealbata
F: Fabaceae

This tree has fragrant flowers and blue-green bark, aging to black. Preferring mountain gullies, it is native to Southwest Australia.

SAINFOIN
Onobrychis viciifolia
F: Fabaceae

This fodder crop from Eurasia has green, oval, pinnate leaves with 6–14 pairs of leaflets. Densely packed pink flowers appear on a main stem.

130 ft
40 m

WEST INDIAN LOCUST TREE
Hymenaea courbaril
F: Fabaceae

This hardwood tree has a straight, thick trunk and stems of white purplish flowers with large petals and long stamens. Orange gum oozes from the trunk.

4 ft
1.2 m

LEAF SPRIG

flowers are clusters of stamens with no petals

23½ in
60 cm

KIDNEY VETCH
Anthyllis vulneraria
F: Fabaceae

A climber found in dry, often chalky, grasslands, with silky-haired, spreading stems. The cream to red flowers are in clusters with a ruff of thick downy sepals.

66 ft
20 m

SEED PODS

TAMARIND
Tamarindus indica
F: Fabaceae

This evergreen tree with drooping branches from East Africa and Asia has yellow-orange flowers on stems and edible pulp inside the seed pod.

FLAT WATTLE
Acacia glaucoptera
F: Fabaceae

This spreading, erect shrub of Southwest Australia has narrow, black, twisted seed pods and yellow, globular flowers.

oblong leaf with 20–30 pairs of leaflets

39 ft / 12 m

SILK TREE
Albizia julibrissin
F: Fabaceae

This deciduous tree is native to southwest Asia. It has dark green bark, striped vertically with age, and is fast-growing but short lived.

»

35 in
90 cm

WILD LICORICE
Astragalus glycyphyllos
F: Fabaceae

With similar leaves to true licorice, this perennial herb from northwest Europe can be used as tea and has a curved pod.

3¼ ft
1 m

BLUE FALSE INDIGO
Baptisia australis
F: Fabaceae

The sap of this herbaceous perennial, native to woods and streamsides in the eastern US, turns purple when exposed to air, and is used as an indigo dye substitute.

EVERGREEN
LEAVES

OBLONG,
GLOSSY
LEAVES

SPONGY
SEED

130 ft/40 m

BLACK BEAN TREE
Castanospermum australe
F: Fabaceae

This lumber tree, with orange-red flowers carried on stems, has large, woody seedpods. The seeds are poisonous.

33 ft
10 m

SEEDS

CAROB TREE
Ceratonia siliqua
F: Fabaceae

PULPY
PODS

This Mediterranean tree has a thick trunk, dense foliage, and small, dull-green flowers. Pulp from the seedpods is a chocolate substitute.

JUDAS TREE
Cercis siliquastrum
F: Fabaceae

This deciduous, ornamental and lumber tree from the eastern Mediterranean flowers profusely in the spring, then later develops flat 4-in (10-cm) seedpods.

33 ft
10 m

young seedpod

20 in
50 cm

CHICK PEA
Cicer arietinum
F: Fabaceae

One of the earliest cultivated vegetables from the Middle East, its seedpod contains one to three seeds, which can be cooked to make hummus.

5 ft
1.5 m

5 ft/1.5 m

GOAT'S-RUE
Galega officinalis
F: Fabaceae

Naturalized in temperate areas, this perennial, with long, cylindrical red-brown seedpods, is believed by some to improve lactation and reduce fevers and diabetes.

9¾ ft
3 m

MOUNT ETNA BROOM
Genista aetnensis
F: Fabaceae

Native to Sicily, around the slopes of Mount Etna, and Sardinia, this small tree has few leaves; the green shoots responsible for photosynthesis.

DORYCNIUM
Dorycnium hirsutum
F: Fabaceae

This Mediterranean perennial forms gray-green shrubby mounds. Its red-brown cylindrical pods can be mistaken for small berries.

20 in
50 cm

GORSE
Ulex parviflorus
F: Fabaceae

This dense, spiny, perennial shrub from the Meditteranean has yellow leaves and short, brown-black pods. Fire and brush clearance aid seed germination.

PURPLISH
PINK
FLOWER

POD
CONTAINS
10–15
SEEDS

33 ft/10 m

**BROAD-LEAVED
EVERLASTING PEA**
Lathyrus latifolius
F: Fabaceae

Distributed through central Europe, this vigorous, climbing perennial has winged stems and its flowers are white or shades of pink. Its seeds (peas) form in long pods.

**COMMON BIRD'S
FOOT TREFOIL**
Lotus corniculatus
F: Fabaceae

Found in pastureland in Europe, Asia, and Africa, it has yellow flowers and five lobed leaves, three of which sit above the others. Its seedpods resemble a bird's foot.

12 in
30 cm

5 ft/1.5 m

COMMON VETCH
Vicia sativa
F: Fabaceae

This scrambling, widespread annual is native to Europe and the Mediterranean. An animal feed crop, its flowers are normally paired and its tendrils can be branched.

PEA FLOWERS

12 in 30 cm

TWISTED SEEDPODS

CROWN VETCH
Coronilla varia
F: Fabaceae

This rapidly spreading
perennial of central southern
Europe has thick leaves and
deep roots, good for binding
soil and controlling erosion.

3¼ ft 1 m

6½ ft 2 m

BROOM
Cytisus scoparius
F: Fabaeae

Found in northwest Europe's
heaths, this shrub has strongly
scented flowers and many slender,
ridged, and angled branches
traditionally used for brooms.

HORSESHOE VETCH
Hippocrepis comosa
F: Fabaceae

This prostrate perennial, native
to England, Holland, and Germany, is an
important food plant for blue butterfly
caterpillars. The flowers develop into
a clawlike seedpods.

ALFALFA
Medicago sativa
F: Fabaceae

This deeply rooting perennial grows
on chalk grassland throughout the
US, southwest Asia, and Europe.
It is a useful animal-feed crop and
has medicinal properties.

32 in 80 cm

4 ft 1.2 m

flowers densely packed

20 in 50 cm

RED CLOVER
Trifolium pratense
F: Fabaceae

A food crop for grazing
animals, this perennial
has long-stalked basal
(primitive) leaves, and an
oblong fruit pod hidden
within its flower head.

COMPOUND LEAVES

BLUE-VIOLET FLOWERS

3¼ ft 1 m

ROOT

LICORICE
Glycyrrhiza glabra
F: Fabaceae

This feathery perennial has smooth,
small, oblong pods and deep roots.
Extract from the roots is 50 times
sweeter than sugar and is medicinal.

strong stem

12 ft 3.7 m

25 ft 7.5 m

narrow leaflet

SCARLET RUNNER BEAN
Phaseolus coccineus
F: Fabaceae

Native to Central American
mountains, the shoot of this
perennial twists clockwise
around its support. Its seeds
(beans) are variously colored.

COMMON LABURNUM
Laburnum anagyroides
F: Fabaceae

This very poisonous tree
is native to central and
southern Europe.
Clusters of brown hairy
pods (3 in (7.5 cm))
contain toxic black seeds.

82 ft / 25 m

BLACK LOCUST
Robinia pseudoacacia
F: Fabaceae

This deciduous, vigorous tree has
poisonous roots and bark, producing
suckers and flat brown seedpods. It
originates from the southeast US.

16 in 40 cm

NICEAN MILKWORT
Polygala nicaeensis
F: Polygalaceae

This Mediterranean perennial
species has flowers with two
sepals and three joined petals,
one of which is fringed, and a
small capsule as a fruit.

GARDEN LUPIN
Lupinus polyphyllus
F: Fabaceae

Originating from
western North America
and naturalized
throughout Europe,
this popular
ornamental has hairy,
black pods and
undersides of leaves. Its
flowers are fragrant.

FAGALES

Some of the world's best-known trees belong to this order. Fagales contains the beech genus, *Fagus*, the Betulaceae or birch family, and the Juglandaceae or walnut family, as well as five other families. These trees dominate the woodlands in which they grow. They have simple leaves and small, unisexual flowers (either male or female) that are pollinated by wind.

65 ft
20 m

fruit and seed develop from catkins

FOREST OAK
Casuarina torulosa
F: Casuarinaceae

The timber of this Australian tree is prized by wood-turners. Its pendulous branches carry long, needlelike leaves and seeds in small, warty cones.

EUROPEAN HAZELNUT
Corylus avellana
F: Betulaceae

This European shrubby tree is harvested for its edible nuts. The distinctive male catkins appear very early in spring.

female flowers

100 ft
30 m

catkins, or cylindrical clusters of male flowers

80 ft
25 m

JAPANESE HOP HORNBEAM
Ostrya japonica
F: Betulaceae

This species from eastern Asia has gray-brown scaly bark. Its seeds, or nuts, are enclosed in husks, like hops, and hang in long clusters.

100 ft
30 m

RED ALDER
Alnus rubra
F: Betulaceae

A native of western North America, this tree prefers damp slopes and margins of watercourses. It has pale gray bark that becomes red when bruised or scraped.

30 ft
10 m

80 ft
25 m

EUROPEAN HORNBEAM
Carpinus betulus
F: Betulaceae

Frequenting fencerows, this European native has small, green, male and female catkin flowers that develop into nuts with three-lobed bracts.

bracts

male catkins

leaf bud

100 ft
30 m

ENGLISH WALNUT
Juglans regia
F: Juglandaceae

A species from the mountains of central Asia, the walnut tree is valued for fruit and lumber. It has smooth gray bark and male catkins up to 10 cm (4 in) long.

LEAF SPRIG WITH UNRIPE FRUIT

100 ft
30 m

RIPE WALNUT

RIPE PECAN

PINNATE LEAVES

PECAN
Carya illinoensis
F: Juglandaceae

The pecan is native to North America and cultivated for its edible nuts. The nuts are enclosed in a husk that splits four ways when ripe.

130 ft
40 m

PLATYCARYA STROBILACEA
F: Juglandaceae

Native to eastern Asia, this tree has clusters of erect male catkins and a fruit, like a conifer cone, with winged seeds.

50 ft
15 m

BUTTERNUT
Juglans cinerea
F: Juglandaceae

This deciduous tree, native to North America, has yellow-green catkins. Small bunches of egg-shaped nuts contain sweet seeds.

100 ft
30 m

SILVER BIRCH
Betula ermanii
F: Betulaceae

This tree has peeling, whitish bark and a weeping habit. A species from Europe and northern Asia, it prefers light, sandy soils.

toothed leaves

spiny bur encases nuts

EUROPEAN CHESTNUT
Castanea sativa
F: Fagaceae

Cultivated for 3,000 years for its nuts, this tree is from southeastern Europe and western Asia. Its catkins contain male flowers in the upper part and female flowers below.

AMERICAN BEECH
Fagus grandifolia
F: Fagaceae

This broadly spreading tree from North America has gray-brown bark. It has glossy deciduous leaves and paired fruits (beechnuts) in husks.

100 ft
30 m

deciduous leaves

SCALY ACORN CUP

CATKIN SPRIG

115 ft
35 m

ENGLISH OAK
Quercus robur
F: Fagaceae
Long-lived, with valuable lumber, this tree is common in western Europe, especially Britain. It bears male flowers, in the form of catkins, and long-stalked acorns.

SCARLET OAK
Quercus coccinea
F: Fagaceae
The leaves of this North American tree turn a characteristic deep red in fall. This oak bears long, yellow-green male catkin flowers and its acorns have a glossy cup.

80 ft
25 m

AUTUMN LEAF

SPRING COLOR

30 ft
10 m

KERMES OAK
Quercus coccifera
F: Fagaceae
An evergreen, shrubby tree from the Mediterranean area, this species has hollylike leaves that are bronze when young, and yellow-brown male catkins.

50 ft
15 m

JAPANESE STONE OAK
Lithocarpus edulis
F: Fagaceae
This Japanese evergreen tree has upright creamy-white catkins, which have female flowers at the base and males above. The edible acorns ripen over two years.

80 ft
25 m

RAULI
Nothofagus nervosa
F: Nothofagaceae
A timber tree, rauli is native to Argentina and Chile. The young leaves are bronze. Greenish female flowers grow in clusters and form bristly husks enclosing small nuts.

SWEET GALE
Myrica gale
F: Myricaceae
This sweet-smelling shrub, traditionally used as an insect repellent, grows in peat bogs in northern temperate zones. It has either male or female reddish catkins.

nut clusters form in leaf axils

6½ ft
2 m

MALPIGHIALES

One of the largest and most diverse orders of mostly tropical plants, Malpighiales contains about 16,000 species. The plants in this order are genetically simialr but physically different. Well-known families are the euphorbias, which have unisexual flowers (either male or female), and others that include species such as the passionflower, willow, and viola. Many of the families in this order are hardly known outside their native regions.

POISONOUS FRUIT

PAIRED LEAVES

32 in
80 cm

TUTSAN
Hypericum androsaemum
F: Clusiaceae
This small shrub from western Europe has reddish, two-ridged stems. Its aromatic leaves have medicinal uses, but its berries are poisonous.

32 in
80 cm

PERFORATE ST. JOHN'S-WORT
Hypericum perforatum
F: Clusiaceae
This European perennial is a common plant of banks, fields, and roadsides. It has round stems with ridges on either side, leaves with translucent dots, and many-seeded capsules.

CEYLON IRONWOOD
Mesua ferrea
F: Clusiaceae
This tree yields heavy lumber, hence its common name. It has large flowers with four white petals and bright red young leaves.

100 ft
30 m

CANDLE-NUT TREE
Aleurites moluccana
F: Euphorbiaceae
Oil from the nuts of this tropical tree was used to make candles. It has stems of small, creamy flowers and its variable leaves are gray-green when young.

TRILOBED LEAF
80 ft
25 m

NUT CONTAINING HARD SEED

SEGMENTED LEAF

FLOWER HEAD

CASTOR OIL PLANT
Ricinus communis
F: Euphorbiaceae
Castor oil is extracted from the poisionous seeds of this species from tropical Africa. With upright flower heads, the upper female flowers have red stigmas and lower males yellow anthers.

13 ft
4 m

130 ft
40 m

RUBBER TREE
Hevea brasiliensis
F: Euphorbiaceae
A native of Brazil, this tree is famous for the milky sap (latex) under its bark used for rubber. It has leaves with three leaflets and pungent yellow flowers.

16 ft
5 m

CASSAVA
Manihot esculenta
F: Euphorbiaceae
Originating from South America, its tuberous roots are used to make tapioca. It has small flower clusters that grow on secondary branches.

13 ft
4 m

STICKY LEAVES

OIL-YIELDING SEEDS

JATROPHA GOSSYPIFOLIA
F: Euphorbiaceae
This invasive, toxic plant, banned in Australia, is native to tropical America. It has sticky leaves, watery sap, and flowers with purple bracts.

16 in
40 cm

DOG'S MERCURY
Mercurialis perennis
F: Euphorbiaceae
This downy perennial has a single upright stem. Plants are either male or female and produce tiny green flowers on long thin stalks.

4 ft
1.2 m

LARGE MEDITERRANEAN SPURGE
Euphorbia characias
F: Euphorbiaceae
This ornamental, Mediterranean perennial has upright, purplish woolly stems, bare and smooth at the base. The fruit, or seed capsule, is hairy and berrylike.

16 in
40 cm

PETTY SPURGE
Euphorbia peplus
F: Euphorbiaceae
This poisonous, weedy annual, native to Europe, North Africa, and west Asia, has dividing stems and flower stalks with three branches.

CROWN OF THORNS
Euphorbia milii
F: Euphorbiaceae
This semisucculent, climbing shrub from Madagascar has very spiny stems. Its leaves are mainly on the new shoots.

6 ft
1.8 m

8 in
20 cm

LIVING BASEBALL
Euphorbia obesa
F: Euphorbiaceae
Rare in the wild, this ball-shaped succulent originates from the Great Karoo, South Africa. Its tiny flowers grow from "eyes" at its top.

20 ft
6 m

CROTON
Croton tiglium
F: Euphorbiaceae
Used in Chinese medicine, this tree from Southeast Asia has malodorous leaves and s male and female flowers that form seed capsules.

12 in
30 cm

SAUSAGE SPURGE
Monadenium guentheri
F: Euphorbiaceae
This evergreen succulent from tropical Africa has white flower bracts with purple markings. It has fat sickle-shaped leaves mainly on young growth.

24 in
60 cm

PERENNIAL FLAX
Linum perenne
F: Linaceae
This slender, often spreading perennial is native to Europe, especially the Alps and Britain. Its flowers form buds at intervals at the top of the stem.

20 ft
6 m

COCA
Erythroxylum coca
F: Erythroxylaceae
The leaves of this evergreen shrub from northwestern South America yield cocaine. It has gray bark and clusters of small yellowish white flowers.

FIVE-LOBED LEAF AND FLOWER

65 ft
20 m

SEED-CONTAINING BERRY

leaves have twining tendril at base

BLUE PASSION FLOWER
Passiflora caerulea
F: Passifloraceae
This ornamental South American vine has scented flowers associated with Christian symbolism. Flowers have five sepals and petals, five stamens, and three purple stigmas.

AYAHUASCA
Banisteriopsis caapi
F: Malpighiaceae

This woody vine, native to the Amazon, is traditionally used to make a sacred, medicinal drink. It produces stalks of pink flowers and winged seedpods.

33 ft / 10 m

RED MANGROVE
Rhizophora mangle
F: Rhizophoraceae

Found throughout the tropics, especially in swampy salt marshes, the red mangrove has prop roots and its seeds germinate before leaving their parent tree.

80 ft / 25 m

IDESIA
Idesia polycarpa
F: Salicaceae

A mountain tree from east Asia, this has smooth gray bark. Small yellow-green, fragrant flowers and purple-red berries form.

70 ft / 21 m

RAFFLESIA
Rafflesia arnoldii
F: Rafflesiaceae

Rafflesia is a parasite on southeast Asian rain forest vines. Its 3¼ ft-(1m-) foul-smelling flowers are the world's largest. The smell attracts flies.

24 in / 60 cm

WHITE WILLOW
Salix alba
F: Salicaceae

This waterside tree is native to Europe and Asia. Trees have either male or female catkins, never both. The bark is a source of salicin, the active ingredient of aspirin.

80 ft / 25 m

flowers are up to 5 in (12 cm) across

50 ft / 15 m

GIANT GRANADILLA
Passiflora quadrangularis
F: Passifloraceae

This passion flower is a perennial native of South America. It bears huge, oblong fruit and has four-sided stems.

PUSSY WILLOW
Salix caprea
F: Salicaceae

This shrubby tree, native to Europe and Asia, has hairy, oval, toothed, alternate leaves. Female catkins form capsules, releasing cottony seeds.

40 ft / 12 m

ASPEN
Populus tremula
F: Salicaceae

Native to Europe and Asia, the gray bark of young trees has diamond-shaped scars. Flattened stalks cause the leaves of aspen to quiver.

80 ft / 25 m

fragrant flowers may be white, red, or purple

WHITE POPLAR
Populus alba
F: Salicaceae

This deciduous tree, native to cenral Europe and central Asia, tolerates water and salt. It is unisexual—most trees are female. Catkins form capsules that release fluffy seeds.

100 ft / 30 m

AZARA MICROPHYLLA
F: Salicaceae

This evergreen tree from Argentina and Chile has small, vanilla-scented flowers with yellow stamens. Each leaf has a circular growth (stipule) at the base.

33 ft / 10 m

WILD PANSY
Viola tricolor
F: Violaceae

This sprawling, short-lived perennial is native to Europe, growing in grassy habitats with neutral to acid soils. Used in herbal remedies, it is also known as heart's-ease.

12 in / 30 cm

SHRUB VIOLET
Hybanthus floribundus
F: Violaceae

This Australian woody perennial accumulates nickel. Shrub violet has bluish petals with a yellow patch and small, dark green oval leaves.

4 ft / 1.2 m

FLOWERING PLANTS • EUDICOTYLEDONS

OXALIDALES

This order contains about 2,300 species, grouped within six families. Of these, the Cephalotaceae contains only one species – the carnivorous albany pitcher plant. The Cunoniaceae are woody plants that produce woody fruit capsules containing small seeds. The wood sorrel family (Oxalidaceae), is the largest with 560 species within eight genera. They are characterized by divided leaves that exhibit or open out by day and then close up at night.

ALBANY PITCHER PLANT
Cephalotus follicularis
F: Cephalotaceae
This carnivorous plant is native to the coast of southwest Australia. It has basal, oval leaves, and a liquid-filled pitcher that traps prey but it is not a member of the pitcher family (see p.192).

8 in
20 cm

flowers have five petals

13 ft
4 m

CHRISTMAS BUSH
Ceratopetalum gummiferum
F: Cunoniaceae
From coastal eastern Australia, this shrub produces insignificant white flowers in spring. The pink and red sepals enlarge in winter, enclosing the fruit.

14 in
35 cm

PINK SORREL
Oxalis articulata
F: Oxalidaceae
Growing from a swollen rhizome, or root stalk, this plant from South America forms leafy mounds topped with clusters of flowers. These produce seed capsules that explode.

FLOWER CLUSTER

40 ft
12 m

BLACK WATTLE
Callicoma serratifolia
F: Cunoniaceae
The wood of this shrubby tree from the coast of New South Wales, Australia, was used by early settlers to make wattle-and-daub shelters. Its young foliage is bronze.

LEAFLET CLUSTER

RIPE FRUIT

FLOWER SPRIG

STAR FRUIT
Averrhoa carambola
F: Oxalidaceae
This bushy tree is native to Southeast Asia is widely cultivated for its edible star-shaped fruit. It flowers four times a year.

50 ft/15 m

ROSALES

This order of plants contains nine families, among them the Rosaceae (rose family), Cannabaceae (hemp family), Moraceae (mulberry family), Rhamnaceae (buckthorn family), Ulmaceae (elm family), and Urticaceae (nettle family). Members of the Rosales are often grown for fruit or other products. Plants in this order tend to have five sepals and numerous stamens. Mostly insect-pollinated, they are often thorny or hairy.

oil-rich berries

SEA BUCKTHORN
Hippophae rhamnoides
F: Elaeagnaceae
Widespread through Asia and Europe, this tree bears yellow flowers before the leaves. The small, bright-orange berries are rich in vitamin C.

33 ft
10 m

OLEASTER
Elaeagnus angustifolia
F: Elaeagnaceae

Native to west Asia, this spreading, deciduous tree has spiny shoots, covered in silver scales. Its egg-shaped, yellow-red fruit is edible.

20 ft/6 m

HEMP
Cannabis sativa
F: Cannabaceae

Originating from central and west Asia, this perennial is a source of cannabis from its leaves, fiber for making rope, and oil is extracted from its seeds.

6½ ft 2 m

HOP
Humulus lupulus
F: Cannabaceae

This climbing perennial is widespread throughout the temperate Northern Hemisphere. The conelike female flower or hop is used to flavor and preserve beer.

23 ft/7 m

FEMALE FLOWER

MALE FLOWER

BLACK MULBERRY
Morus nigra
F: Moraceae

This deciduous, spreading tree, widely cultivated for its rich-flavored fruit, is native to the Middle East. It has bumpy, fissured orange bark.

43 ft 13 m

young fruit

JACKFRUIT
Artocarpus heterophyllus
F: Moraceae

From lowland Southeast Asia, this tree yields the largest tree-borne fruit, weighing up to 66lb (30 kg) and 36 in (90 cm) long when ripe.

65 ft 20 m

GLOSSY LEAF

FLESHY FRUIT

COMMON FIG
Ficus carica
F: Moraceae

This fig is native to southwest Asia and the eastern Mediterranean. As with all figs, flowers are inside a bud and pollinated by a special wasp that enters the bud.

33 ft 10 m

SACRED FIG
Ficus religiosa
F: Moraceae

Under this sacred tree, Buddha is said to have gained enlightenment. Native to Southeast Asia, its flowers, then its fruit, are enclosed within purple, flecked figs.

100 ft 30 m

EARLY FLOWER STAGE

LEAF AND FLOWER SPRIG

PAPER MULBERRY
Broussonetia papyrifera
F: Moraceae

Fine paper is made from the inner bark of this tree from Japan and Taiwan. It produces large quantities of pollen.

50 ft/15 m

LEAF SPRIG

OSAGE ORANGE
Maclura pomifera
F: Moraceae

Native to North America, this tree is used for hedging and its roots and wood were valued by American Indians.

65 ft 20 m

UNRIPE FRUIT

JUJUBE
Ziziphus jujube
F: Rhamnaceae

This thorny, shrubby tree is widely cultivated for its fruit in China and India. The immature, smooth, oval, green pitted fruit tastes like apple.

33 ft/10 m

COMMON BUCKTHORN
Rhamnus cathartica
F: Rhamnaceae

A shrubby, invasive tree that has spine-tipped shoots, tiny yellow-green flowers, and black berries. It is native to Europe, Asia, and Africa.

26 ft/8 m

NEW JERSEY TEA
Ceanothus americanus
F: Rhamnaceae

This bush from the northeastern US has purple, three-lobed capsules, containing seeds. Its red roots and hairy leaves have been used to make tea.

30 in 75 cm

FLOWERING PLANTS • EUDICOTYLEDONS

CHINESE PHOTINIA
Photinia serratifolia
F: Rosaceae
A plant of Chinese forests, this tree is commonly cultivated for ornamental purposes. Its dense wood is used in furniture making.

26 ft/8 m

SILVERWEED
Potentilla anserina
F: Rosaceae
This silky-haired, creeping perennial grows in wasteland, pastures, and dunes in the Americas, Europe, Asia, Australia, and New Zealand.

32 in
80 cm

SNOWY MESPILUS
Amelanchier lamarckii
F: Rosaceae
Frothy clusters of star-shaped flowers in spring are followed by dark red berries in summer in this showy tree from eastern North America.

40 ft/12 m

LEAVES AND EDIBLE FRUIT

SPRING FLOWERS

WILD CHERRY
Prunus avium
F: Rosaceae
The wild ancestor of orchard cherries grows in woods and hedges in Europe, Asia, and North Africa and is naturalized in North America.

80 ft
25 m

SUMMER FLOWER

EDIBLE FRUIT

STRAWBERRY
Fragaria vesca
F: Rosaceae
The wild strawberry is a perennial in North American and European woodlands. Tiny edible fruit form from the swollen flower receptacle.

12 in
30 cm

THREE-LOBED LEAF

ORCHARD PLUM
Prunus domestica
F: Rosaceae
These trees originated as a cross between the cherry plum of China and the European blackthorn, but without the sharp thorns of the latter.

40 ft
12 m

PRAIRIE CRAB APPLE
Malus ioensis
F: Rosaceae
One of several crab apples native to North America, this species is cultivated for its somewhat tart fruit.

35 ft/11 m

double flower garden variety

32 in
80 cm

JAPANESE ROSE
Rosa rugosa
F: Rosaceae
Tolerant of salt spray, this rose from east Asia is often planted as a hedge near the sea. It has prickly stems and pink flowers with wrinkled petals.

6½ ft/2 m

WHITE OR PINK FLOWERS

BRIGHT RED HIPS

10 ft
3 m

DOG ROSE
Rosa canina
F: Rosaceae
The thorny, arching branches of this rose are familiar in hedgerows in Europe and North Africa. This plant is also naturalized in North America.

APOTHECARY'S ROSE
Rosa gallica var. *officinalis*
F: Rosaceae
A popular garden plant, this European parent of many floribunda and hybrid tea roses has a long pedigree. "Attar of roses" is a fragrant oil distilled from its petals.

fragrant, deep pink flowers

MOUNTAIN AVENS
Dryas octopetala
F: Rosaceae
The flowers of this Arctic and mountain undershrub turn to follow the sun to warm their centers and attract pollinating insects.

20 in
50 cm

OAKLIKE
LEAVES

EIGHT-PETALED
FLOWER

COTONEASTER
Cotoneaster horizontalis
F: Rosaceae
This creeping Chinese shrub is commonly grown in gardens for its flowers and fruit, and occasionally escapes. The semievergreen leaves form flattened clumps.

3¼ ft
1 m

FIRETHORN
Pyracantha rogersiana
F: Rosaceae
This thorny evergreen shrub from eastern China belongs to a genus often planted for their attractive but inedible, orange, berrylike fruit.

10 ft / 3 m

WHITE OR PINK
FLOWERS

EDIBLE
FRUIT

BLACKBERRY
Rubus fruticosus
F: Rosaceae
Familiar in Europe as scrambling hedgerow bushes bearing edible fruit in fall, blackberries form a group of very closely related "microspecies."

8 ft
2.5 m

WILD SERVICE TREE
Sorbus torminalis
F: Rosaceae
This rare tree grows in ancient woods in Europe, Asia Minor, and North Africa. Its name is derived from the brew, *cerevisia*, made from its fruit.

80 ft
25 m

AMERICAN MOUNTAIN ASH
Sorbus americana
F: Rosaceae
This deciduous woodland tree is native to eastern North America. Its orange berries last into winter, providing food for thrushes and jays.

many small flowers

40 ft
12 m

MEDLAR
Mespilus germanica
F: Rosaceae
Originally from central and southern Europe, medlar produces a hard, yellowish brown fruit. This turns soft and edible in fall, then decays.

FIVE-PETALED
WHITE FLOWER

20 ft / 6 m

FRUIT WITH REMAINS
OF SEPALS

WILLOW-LEAVED PEAR
Pyrus salicifolia
F: Rosaceae
Cultivated for its pendulous, silvery foliage, not its inedible fruit, this Middle Eastern tree is endangered in the wild in Turkey.

five-petaled white flowers

40 ft
12 m

SALAD BURNET
Sanguisorba minor
F: Rosaceae
Native from Europe to Iran, and introduced in North America, this perennial of lime-rich grassland has edible leaves, hence its name.

TOOTHED
LEAFLETS

FLOWER HEADS

24 in
60 cm

lobed leaf

showy five-petaled flowers

WATER AVENS
Geum rivale
F: Rosaceae
This downy perennial grows in damp places in North America, Europe, and Asia Minor. Hooked hairs on the fruit latch onto animals for dispersal.

24 in
60 cm

NODDING
SUMMER
FLOWER

BURRLIKE
FRUIT

LADY'S MANTLE
Alchemilla vulgaris
F: Rosaceae
This name encompasses several closely related grassland species from Europe, Asia, and eastern North America. Their leaves are said to resemble a trailing gown.

24 in
60 cm

red fruit with single stem

COMMON HAWTHORN
Crataegus monogyna
F: Rosaceae
Occurring in woods and hedges in the wild from Europe to Afghanistan, this densely branched small tree produces massed white blossom in spring, followed by deep red fruit. It is also cultivated in gardens.

RED FRUIT,
OR HAWS

FRAGRANT
BLOSSOM

33 ft
10 m

»

» ROSALES

HACKBERRY
Celtis occidentalis
F: Ulmaceae

Native to North America, hackberry has bright green, elmlike leaves, and red fruits eaten by many birds and mammals.

red berry

130 ft
40 m

120 ft
36 m

JAPANESE ZELKOVA
Zelkova serrata
F: Ulmaceae

In the USA elms killed by Dutch Elm Disease are sometimes replaced by this valuable Asian timber tree. The Japanese grow it in bonsai form.

100 ft
30 m

toothed leaf

EUROPEAN FIELD ELM
Ulmus minor
F: Ulmaceae

This tree used to be a major feature in European landscapes, but was decimated by Dutch Elm Disease.

FRIENDSHIP PLANT
Pilea involucrata
F: Urticaceae

Several *Pilea* species make valuable houseplants, including this species from Central and South America with strongly veined leaves.

1 ft
30 cm

STINGING NETTLE
Urtica dioica
F: Urticaceae

Stinging hairs discourage animals from eating nettle leaves. This species grows in disturbed ground in Europe, Asia, North Africa, and North America.

6½ ft / 2 m

BRASSICALES

Many members of the order Brassicales have bitter or fragrant oils in their leaves, stems, or swollen roots. Although these oils evolved to deter herbivores, they make many species pleasantly edible for humans, and have culinary, perfumery, or herbal uses. The Brassicaceae, or cabbage family, is the most important group, with 3,300 species.

WILD CABBAGE
Brassica oleracea
F: Brassicaceae

Humans have grown this western European plant for millennia. Cauliflower, broccoli, and brussels sprouts are cultivated forms of the same species.

STALKLESS BLUE-GREEN LEAF

FLOWERHEAD

3¼ ft
1 m

male flowers

30 ft
10 m

PAPAYA
Carica papaya
F: Caricaceae

This South American treelike plant produces yellow flowers. The female flowers develop into large orange-fleshed fruits of the same name.

30 ft / 10 m

HORSERADISH TREE LEAF

FLOWER STEM

HORSERADISH TREE
Moringa oleifera
F: Moringaceae

This tropical Asian tree has corky gray bark and fernlike leaves. A condiment can be made from its crushed roots.

NASTURTIUM
Tropaeolum majus
F: Tropaeolaceae

A colorful annual from Central and South America, this is a popular garden plant. The flowers and leaves can be eaten in salads.

5 ft
1.5 m

CAPER
Capparis spinosa
F: Capparaceae

A perennial spiny bush, caper is native to the Mediterranean. Its flower buds are used salted and pickled for culinary use.

20 in / 50 cm

COMMON MIGNONETTE
Reseda odorata
F: Resedaceae

Originally from North Africa, it is grown in gardens across southern Europe. An oil from its fragrant flowers is used in perfumery.

24 in
60 cm

WALLFLOWER
Erysium cheiri
F: Brassicaceae

Probably native to cliffs and meadows in the eastern Mediterranean, this species has been cultivated throughout Europe since medieval times.

CORALROOT BITTERCRESS
Cardamine bulbifera
F: Brassicaceae
A plant of central European beechwoods, this creeping perennial occurs from the British Isles eastward to the Caucasus and Asia Minor.

28 in
70 cm

AUBRIETA
Aubrietia deltoides
F: Brassicaceae
Originating from the Aegean region, aubrieta grows readily on walls throughout the warmer parts of Europe.

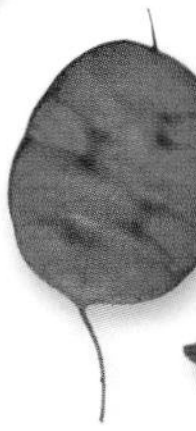

12 in
30 cm

HONESTY
SEED POD

HONESTY
Lunaria annua
F: Brassicaceae
Native in the wild in southeast Europe, honesty is often grown in gardens. Its seed pods are used in dried flower arrangements.

4 ft
1.2 m

5 ft
1.5 m

HONESTY
FLOWER
STEM

HORSERADISH
Armoracia rusticana
F: Brassicaceae
The pungent taste of the roots, which are used for horseradish sauce, evolved to deter herbivores from eating this Eurasian perennial.

FLOWER
STEM

EDIBLE
LEAVES

3¼ ft
1 m

WINTERCRESS
Barbarea vulgaris
F: Brassicaceae
Native throughout Europe, wintercress was once cultivated for winter salads, and introduced for this purpose to North America, Australia, and New Zealand.

12 in
30 cm

SWEET ALISON
Lobularia maritima
F: Brassicaceae
This Mediterranean annual is widely cultivated for its sweet-scented flowers. "Alison" from *alyssum* in Greek refers to its reputation for curing madness.

16 in
40 cm

lobed leaves
at base of stem

scented purple,
red, or white
flowers

FLOWERHEAD

four-petaled
flower

SHEPHERD'S PURSE
Capsella bursa-pastoris
F: Brassicaceae
Purse-shaped pods that split to release massed seeds have helped this species spread around the world from its native Mediterranean.

32 in/80 cm

rounded
leaves
are gray
beneath

24 in
60 cm

24 in
60 cm

DISTINCTIVE
LEAF

WILD RADISH
FLOWER STEM

LEAF
STEM

SEA KALE
Crambe maritima
F: Brassicaceae
This cabbagelike perennial grows on gravel beaches and seacliffs around Eurasia. Its spherical fruits, which float for many days, are dispersed by the sea.

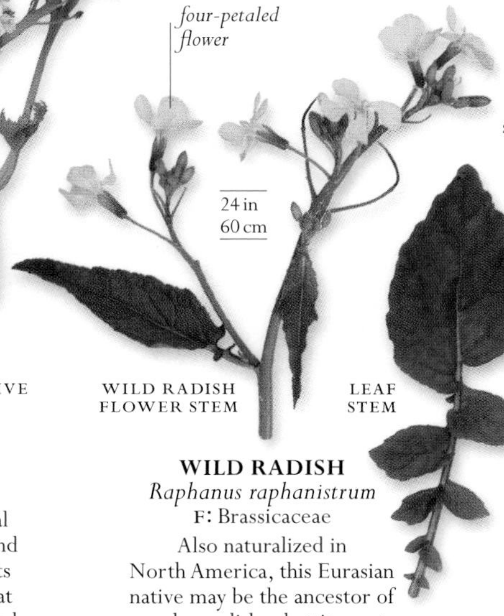

WILD RADISH
Raphanus raphanistrum
F: Brassicaceae
Also naturalized in North America, this Eurasian native may be the ancestor of garden radishes but its root is not round.

WATERCRESS
Nasturtium officinale
F: Brassicaceae
This wild Eurasian perennial grows near water and is also cultivated in tanks. Its young shoots and peppery leaves, rich in vitamin C, are used in salads.

24 in
60 cm

HOARY STOCK
Matthiola incana
F: Brassicaceae
Native to coastal rocks in southwest Europe and west Asia, stock is much cultivated. "Ten Weeks Stock" is a short-lived garden variety.

MALVALES

The Malvales is a moderately large order, mainly of shrubs and trees, found with greatest variety in tropical and warm temperate regions, but extending into cooler parts. It includes two main families: the rockrose family (Cistaceae), mostly from the Northern Hemisphere; and the more widespread mallow family (Malvaceae) of herbs, shrubs, and massive trees. Among the smaller families are the Bixaceae, which has just five species.

33 ft/10 m

spiny fruit

ANNATTO
Bixa orellana
F: Bixaceae
The food dye annatto comes from the spiny fruit of this pink-flowered shrub or small tree, native to tropical America.

FLOWERING SPRIG

3¼ ft/1 m

HAIRY ROCKROSE
Cistus incanus
F: Cistaceae
Several scientific names have been given to this widespread Mediterranean shrub, because its leaves vary greatly in hairiness and size.

13 ft/4 m

CALIFORNIAN FLANNELBUSH
Fremontodendron californicum
F: Malvaceae
The flowers of this spreading shrub produce showy masses in early summer. It lives high up in the granite mountains of California.

COMMON ROCKROSE
Helianthemum nummularium
F: Cistaceae
Found on sunny banks throughout most of Europe, this low shrub prefers lime-rich soil. Different subspecies grow in Europe's mountains.

20 in
50 cm

five-petaled rose-red flowers

CHINESE HIBISCUS
Hibiscus rosa-sinensis
F: Malvaceae
The tropical Chinese hibiscus, or China rose, is one of several *Hibiscus* species cultivated especially for their showy blossoms.

OBLONG LEATHERY LEAVES

RIBBED POD

40 ft
12 m

COCOA
Theobroma cacao
F: Malvaceae
Originally from Brazilian rain forests, this tree is cultivated around the tropics. Cocoa comes from the seeds, known as beans, inside its fruit pods.

15 ft/4.5 m

toothed leaves

6½ ft/2 m

strongly scented flowers

FRUITING STEM

32 in
80 cm

CAFFEINE-CONTAINING SEED

82 ft/25 m

NUTRITIOUS BAOBAB FRUIT

MATURE TREE

GLOSSY OVAL LEAVES

82 ft
25 m

MEZEREON
Daphne mezereum
F: Thymelaeaceae
Damp woods and shady gorges are the typical habitat of this deciduous shrub, which is found across most of Europe.

MUSK MALLOW
Malva moschata
F: Malvaceae
Native to North Africa and southern Europe, and a garden plant farther north, this tall perennial grows in grassy and bushy places.

COLA NUT TREE
Cola nitida
F: Malvaceae
The caffeine-rich seeds (called cola nuts) from the pods of this West African tree are chewed for their stimulant effects.

BAOBAB TREE
Adansonia digitata
F: Malvaceae
This massive African tree is called an upside-down tree because, when leafless, its branches look like roots. It can live 3,000 years.

pollen is attached to filaments on central stamen tube

TRAILING ABUTILON
Abutilon megapotamicum
F: Malvaceae
Native to Argentina, Brazil, and Uruguay, this spreading shrub is also popular as a colorful ornamental in warm, sunny gardens.

GLOSSY LEAVES

6 ft
1.8 m

SPINY FRUIT

130 ft
40 m

DURIAN
Durio zibethinus
F: Malvaceae
The spiny fruit of this Asian rain forest tree smell like sweaty socks. This attracts animals to eat the fruit then disperse the seeds in their droppings.

120 ft
36 m

AMERICAN LIME
Tilia americana
F: Malvaceae
This medium to large deciduous tree adds to fall colors in woodlands across northeastern America. It often grows alongside sugar maple.

6 ft
1.8 m

FLOWER SPRIG

LEAF CLUSTERS

230 ft
70 m

DOWN-FILLED POD

KAPOK
Ceiba pentandra
F: Malvaceae
Massive kapok trees grow wild in West Africa, and Central and South America. Fiber from their fruit, or pods, is used as a stuffing for toys.

OPEN BOLL

WILD HOLLYHOCK
Alcea pallida
F: Malvaceae
This tall evergreen perennial, a close relative of garden hollyhocks, comes from the eastern Mediterranean. It grows in rocky places and scrubland.

5 ft
1.5 m

UPLAND COTTON
Gossypium hirsutum
F: Malvaceae
This Central American shrub is the most common cultivated cotton species. The cotton fibers protect seeds inside the fruit, or boll.

SAPINDALES

The Sapindales is a large and important order, mostly of trees, shrubs, and woody vines, often with divided leaves. It includes many dominant woodland species and commercially important species such as citrus fruit. More than half its members belong to two families: the maple family (Sapindaceae), with around 1,900 species; and the rue family (Rutaceae), which mostly originate in Australia and South Africa, and has 1,700 species.

MANGO
Mangifera indica
F: Anacardiaceae
A native of Asia, mango is one of the most widely cultivated fruit in the tropical world, and is rich in vitamin A.

EVERGREEN LEAVES

59 ft / 18 m

RIPE FRUIT

16 ft / 5 m

CASHEW
Anacardium occidentale
F: Anacardiaceae
Originally from Venezuela and Brazil, this shrubby tree was taken to Asia and Africa in the 1500s, and cultivated for its nuts.

39 ft / 12 m

33 ft / 10 m

STAG'S HORN SUMACH
Rhus typhina
F: Anacardiaceae
This deciduous shrub or small tree grows on forest edges and wasteland in northeastern America. It has spiky clusters of red berries.

SMOKE BUSH
Cotinus coggygria
F: Anacardiaceae
Finely branched clusters of pale green flowers give this bush a smoky appearance. It is found in southern Europe and Asia.

» SAPINDALES

CRABWOOD
Carapa guianensis
F: Meliaceae

180 ft
55 m

The dark wood of this tropical South American tree is sometimes sold as Brazilian mahogany. Soap is made from its seeds.

NEEM TREE
Azadirachta indica
F: Meliaceae

130 ft
40 m

RIPE FRUIT

PINNATE LEAVES

Valued for its wood, medicinal oils, and edible shoots, this tree grows throughout the Old World tropics. Indian farmers produce insecticides from the oil and leaves.

CHINESE MAHOGANY
Toona sinensis
F: Meliaceae

80 ft
25 m

Chinese people eat the leaves of this east Asian tree as a vegetable. Furniture is made from its hard, reddish wood.

SCENTED BORONIA
Boronia megastigma
F: Rutaceae

6½ ft
2 m

This erect shrub of wet sandy sites in western Australia has bell-like flowers, brownish on the outside and golden-green inside.

FRINGED RUE
Ruta chalepensis
F: Rutaceae

24 in
60 cm

From rocky habitats in southern Europe and southwestern Asia, this is thought to be the rue mentioned in the Bible.

PEPPER AND SALT
Eriostemon spicatus
F: Rutaceae

5 ft
1.5 m

Widespread in sandy and gravely places in southwestern Australia, this low shrub has narrow leaves and pink, white, or bluish flowers.

SYDNEY ROCKROSE
Boronia serrulata
F: Rutaceae

5 ft
1.5 m

This small shrub, in a genus confined to Australia, grows in coastal heaths near Sydney. It has bright pink, cup-shaped flowers.

glossy rounded leaves

MEXICAN ORANGE
Choisya ternata
F: Rutaceae

6½ ft
2 m

Originating from Mexico but common in gardens elsewhere, this irregular, bushy, evergreen shrub has branched clusters of sweet-scented, white flowers.

HOP TREE
Ptelea trifoliata
F: Rutaceae

20 ft
6 m

This small tree native to northeastern America is grown for ornament. The flowers of the male tree are smaller than those of the female tree

SALMON CORREA
Correa pulchella
F: Rutaceae

3¼ ft / 1 m

A native of South Australia, this small shrub is grown in gardens for its delicate, pendulous, tubular flowers.

COMMON PRICKLY ASH
Zanthoxylum americanum
F: Rutaceae

33 ft
10 m

This spiny North American tree grows as far north as Quebec, Canada. American Indians chewed its bark to help soothe toothaches.

SKIMMIA
Skimmia japonica
F: Rutaceae

6½ ft / 2 m

Much planted in gardens, parks, and amenity areas, this evergreen, aromatic shrub from east Asia produces red berries in late summer.

JAPANESE BITTER ORANGE
Poncirus trifoliata
F: Rutaceae

26 ft / 8 m

STEM

GOLF BALL-SIZED FRUIT

The small, inedible yellow fruit of this spiny shrub resemble oranges with a downy skin. They have several medicinal uses.

LEMON
Citrus limon
F: Rutaceae

20 ft
6 m

EVERGREEN LEAVES

RIPENING FRUIT

The Romans spread this evergreen tree throughout Europe. Thought to originate in Asia, it is widely cultivated for its fruit.

SEVILLE ORANGE
Citrus aurantium
F: Rutaceae

unripe fruit

30 ft
9 m

Unlike the sweet orange eaten raw (*C. sinensis*), this species' bitter fruit is only good for cooking. Both originate from tropical Asia.

JAPANESE MAPLE
Acer palmatum
F: Sapindaceae
Centuries of breeding have produced many cultivated varieties of this native Japanese tree, with varying leaf shapes and spectacular fall colors.

40 ft
12 m

SYCAMORE
Acer pseudoplatanus
F: Sapindaceae
Native to mountain woods in Europe and Asia, sycamore is widely planted elsewhere. Its winged seeds spread in the wind.

FLOWERING STEM

WINGED SEEDS

100 ft
30 m

SUGAR MAPLE
Acer saccharum
F: Sapindaceae
This tree is native to the northeastern US and southeastern Canada. Maple syrup is made by boiling sap collected from the tree in spring.

115 ft
35 m

GOLDEN-RAIN TREE
Koelreuteria paniculata
F: Sapindaceae
This showy tree from east Asia is much planted in temperate regions for its cascading yellow blossoms and distinctive bladderlike seedpods.

40 ft
12 m

CONKER IN SPIKY SHELL

100 ft
30 m

young flowers have yellow "eye"

100 ft
30 m

WHITE SPRING FLOWERS

LYCHEE
Litchi chinensis
F: Sapindaceae
Probably originating in southern China, this tree is cultivated for its fruit. Their sweet flesh is encased within a tough shell.

HORSE CHESTNUT
Aesculus hippocastanum
F: Sapindaceae
Native to southeast Europe, this tree is often planted for shade along city streets. Its fruit resembles chestnuts but are hard and inedible.

130 ft
40 m

BRIGHT GREEN LEAVES

EDIBLE NUTS

JAVA ALMOND
Canarium indicum
F: Burseraceae
Native to rain forests in the Pacific islands, this is one of the region's most useful trees, providing wood, oil, and edible nuts.

toothed leaves

26 ft
8 m

65 ft
20 m

redder eye of older flower

26 ft
8 m

PINNATE LEAVES

FLOWER SPIKE

26 ft
8 m

YELLOWHORN
Xanthoceras sorbifolium
F: Sapindaceae
This small tree grows wild in northern China. Its scientific name records its leaves, which resemble those of rowan (*Sorbus*).

TREE OF HEAVEN
Ailanthus altissima
F: Simaroubaceae
Hailing from China, this slightly rank-smelling tree is planted along city streets because it endures pollution and most soil types.

BITTERWOOD
Quassia amara
F: Simaroubaceae
Boiled extracts from the bark and wood of this tree are used to make tonics against malaria in tropical America.

FRANKINCENSE
Boswellia sacra
F: Burseraceae
Frankincense, a gum resin used in incense and to fix perfumes, oozes as milky juice from cuts in trunks of these Arabian trees.

CORNALES

Although the scope of the Cornales varies between different classification systems, it is used here as a small order of just five or six families. The main family is the dogwood family (Cornaceae), a loose taxonomic grouping of shrubs and small trees found in temperate zones and on tropical mountains. It also includes the hydrangea family (Hydrangeaceae), represented by several popular garden plants.

HYDRANGEA
Hydrangea macrophylla
F: Hydrangeaceae
Native to China and Japan, this showy shrub is cultivated for its rounded clusters of rose, lavender, blue, and even white flowers.

5 ft / 1.5 m

FLOWERING DOGWOOD
Cornus florida
F: Cornaceae
40 ft
12 m
The showy white "petals" of this North American dogwood tree are bracts (specialized leaves) around clusters of tiny flowers.

DOVE TREE
Davidia involucrata
F: Cornaceae
This small flowering tree from southwestern China is impressive in bloom, with flower heads ¾ in (2 cm) across, surrounded by creamy bracts.

80 ft
25 m

MOCK ORANGE
Philadelphus sp.
F: Hydrangeaceae
Around 65 species of this shrubby genus, named for their orange-blossom fragrance, are native to western North America, Asia, and Mexico.

15 ft
4.5 m

ERICALES

The Ericales is an important order, both economically and as a one of the world's main herbaceous groups. It includes the acid-soil loving heather family (Ericaceae), with over 4,000 species of flowering shrubs; and the primrose family (Primulaceae), found mostly in the mountain habitats of northern temperate regions; and carnivorous plants in the pitcher plant family (Sarraceniaceae).

CORALBERRY
Ardisia crenata
F: Myrsinaceae
A greenhouse favorite for its long-lasting, bright berries, this Asian shrub has become an invasive pest in Hawaii, Florida, and Texas.

6 ft
1.8 m

JACOB'S LADDER
Polemonium caeruleum
F: Polemoniaceae
This tall perennial has cup-shaped lavender or white flowers. It grows in rocky and grassy places in North America and northern and central Europe.

3¼ ft
1 m

PERENNIAL PHLOX
Phlox paniculata
F: Polemoniaceae
A pyramidal cluster of trumpet-shaped, pink or lavender flowers top the stem of this perennial, native to open woods in the southeastern US.

3¼ ft / 1 m

HEATHER
Calluna vulgaris
F: Ericaceae
This evergreen shrub, with spikes of pale purple flowers, dominates vast areas of moorland in northern Europe and eastern North America.

3¼ ft
1 m

DORSET HEATH
Erica species
F: Ericaceae
One of several, similar spiky *Ericae*, this species is known from southern England (including Dorset), as well as western Ireland and France.

24 in
60 cm

STRAWBERRY TREE
Arbutus unedo
F: Ericaceae
Although related to heathers, the warty, red fruit of this evergreen Mediterranean tree look remarkably like strawberries but taste insipid.

40 ft
12 m

scarlet, bell-shaped flowers

50 ft / 15 m

TREE RHODODENDRON
Rhododendron arboreum
F: Ericaceae
There are about 850 species of *Rhododendron*, most with leathery, evergreen leaves and showy flowers. Many, like this one, come from the Himalayas.

WHITE, BELL-SHAPED FLOWERS

16 ft
5 m

12 in
30 cm

HIMALAYAN PIERIS
Pieris formosa
F: Ericaceae
Formosa means "beautiful," a reference to the hanging clusters of white, urn-shaped flowers on this Asian bush or small tree.

ACIDIC BERRIES

COWBERRY
Vaccinium vitis-idaea
F: Ericaceae
This neat heather, with leathery leaves and glossy red berries in fall, is found across North America, northern Europe, and Asia.

thick trunk covered in small leaves

6 in
15 cm

CHECKERBERRY
Gaultheria procumbens
F: Ericaceae
This creeping, aromatic shrub forms patches beneath oaks and conifers in eastern North America. Its red fruit lasts through the winter.

10 ft
3 m

SWEET PEPPER BUSH
Clethra alnifolia
F: Clethraceae
This deciduous shrub grows in wet forests and bogs in eastern North America. Its leaves turn yellow or orange in fall.

DECIDUOUS LEAVES

ORANGE BERRIES

65 ft
20 m

PERSIMMON
Diospyros virginiana
F: Ebenaceae
This native North American tree has yellowish white, bell-shaped flowers and produces globular orange fruit, about 1½ in (4 cm) across.

6½ ft / 2 m

65 ft
20 m

INDIAN BALSAM
Impatiens glandulifera
F: Balsaminaceae
Exploding fruit pods, which shoot their seeds outward, have helped this Himalayan species become established by rivers in most of Europe.

33 ft
10 m

FRUIT

black seeds

165 ft
50 m

ROUND, WOODY FRUIT

BOOJUM TREE
Fouquieria columnaris
F: Fouquieriaceae
This strange tree is mainly confined to Baja, California, US. Its stem and upright, spiny branches are green and can photosynthesize.

green flesh

OPEN FRUIT

KIWI
Actinidia chinensis
F: Actinidiaceae
Also known as Chinese gooseberry, this woody climber was introduced from China to New Zealand, where the berries were sold as "kiwi fruit."

BRAZIL NUT
Bertholletia excelsa
F: Lecythidaceae
Brazil "nuts" are actually seeds inside this South American tree's hard, cannonball-like fruit. In the wild, agoutis (large rodents) open the fallen fruit.

MULTIPLE "NUTS" INSIDE

∧ **INSECT TRAP**
Insects caught in the pitcher find it very difficult to escape. The leaf digests the soft parts, but the skeletons remain.

FLOWER ∨
The flowers have five sepals, five yellow petals (which redden as they mature), and a whitish, umbrella-like style. They hang upside down.

∧ **"WINDOWS"**
Insects fly toward the bright translucent "windows" or areoles on the upper part of the pitcher, but then fall into the tube.

SEED CAPSULE ∨
The broad, rough capsules crack open and scatter many tiny knobbly seeds, which are about ⅛ in (3 mm) long.

RHIZOME ∨
The plant's domed pitchers rise up from a horizontally branching, underground stem known as a rhizome.

ROOTS >
The fibrous roots are 8–12 in (20–30 cm) long and grow along the length of the rhizome.

HOODED PITCHER PLANT
Sarracenia minor

This insectivorous plant is highly adapted to living in nutrient-poor, acidic environments, digesting insects to obtain the phosphorus and nitrogen from their bodies. Insects are attracted by nectar into the pitcher, the long thin tube of modified leaves from which the plant gets its name. The interior of the pitcher is covered in a slippery waxy coating and downward-pointing hairs that become coarser nearer the bottom of the tube. The insects fall in and, unable to get out, eventually die from exhaustion. Digestive glands in the tube walls exude enzymes which break down the insects, resulting in a liquid slurry that is then absorbed into the leaf, providing the pitcher plant with valuable nutrients.

petal

ovary forms seed capsules when fertilized

bowl-shaped style from which insects collect pollen

SIZE Up to 12 in (30 cm) tall
HABITAT Upland savanna, wet pinewood, bogs
DISTRIBUTION Southeast USA
LEAF TYPE Hooded, narrow, hollow cone or ascidium

∨ **IN BLOOM**
The yellow, odorless flowers are produced from late March to mid-May. Bees are the main pollinator, and will prefer this species over other nearby plants.

∨ **HOODS**
The concave hoods prevent rainwater from filling up the pitcher and also shade the opening, making it harder for trapped insects to find the exit.

nectar glands at the back of the hood

reddish purple colors of the hood attract insects to the pitcher

∨ **CARNIVORE AT RISK**
The hooded pitcher plant is now rare and a threatened species due to habitat loss. It is found in North Carolina, South Carolina, Georgia, and Florida. There are two varieties: *S. minor* grows to 12 in (30 cm), while *S. minor* var. *okefenokeensis* grows up to 4 ft (1.2 m) and is found only at Okefenokee Swamp in Georgia.

15 ft
4.5 m

30 ft / 10 m

GRANTHAM'S CAMELLIA
Camellia granthamiana
F: Theaceae
One of around 80 *Camellia* species,
all from Asia, this endangered tree
from China was discovered in 1955.

40 ft / 12 m

EPAULETTE TREE
Pterostyrax hispidus
F: Styracaceae
Native to east Asia, this
large-leaved, deciduous tree is
sometimes cultivated for its
hanging clusters of fragrant,
creamy-white flowers.

SILKY CAMELLIA
Stewartia malacodendron
F: Theaceae
Named after its silkily
haired young stems, this
deciduous shrub or small
tree grows in woodland in
eastern USA.

56 ft / 17 m

TEA
Camellia sinensis
F: Theaceae
The dried fermented leaves and buds of
this evergreen Asian shrub are used to
make tea. Their astringent tannins deter
grazing animals.

*protruding red
stamen*

18 in / 45 cm

14 in
35 cm

*rough, hairy,
lanceolate leaf*

26 ft / 8 m

3¼ ft
1 m

**PINKEY
PITCHER PLANT**
Sarracenia x stevensii
F: Sarraceniaceae
In pitcher plants, like this
hybrid, insects, attracted by
nectar, fall into pitcher-
shaped leaves, where they
get trapped and are digested
by enzymes.

**YELLOW
PITCHER PLANT**
Sarracenia flava
F: Sarraceniaceae
Pitcher plants trap
and digest insectsto
supplement their
nutrient-poor diet.
This species is native
to southeast USA.

**PARROT
PITCHER PLANT**
Sarracenia psittacina
F: Sarraceniaceae
Sarracenia species grow
wild only in eastern North
America. The pitchers of this
species, found from Florida
to Louisiana, form
horizontal traps.

SPANISH CHERRY
Mimusops elengi
F: Sapotaceae
This evergreen Indian tree is planted in
tropical countries for its fragrant flowers.
Its durable wood is used in shipbuilding
and construction.

3¼ ft / 1 m

*hood resembles
cobra ready to strike*

COBRA PLANT
Darlingtonia californica
F: Sarraceniaceae
The only species of
Darlingtonia, this carnivorous
plant grows in coastal bogs
and beside mountain streams
in western USA.

20 in / 50 cm

3¼ ft / 1 m

TANSY PHACELIA
Phacelia tanacetifolia
F: Boraginaceae
Mostly annuals of dry places,
phacelias flower prolifically after
rain. More than 80 members of
this genus are found in California.

VIPER'S-BUGLOSS
Echium vulgare
F: Boraginaceae
Widespread in grassy places in
Europe and temperate Asia, this
rough-haired biennial has pinkish
flower buds but deep blue flowers.

CREEPING JENNY
Lysimachia nummularia
F: Primulaceae
From Sweden eastward to the Caucasus, this creeping perennial with rounded leaves grows in damp grassy places, shady hedgerows, and streamsides.

flower with five petals

24 in
60 cm

COMMON COMFREY
Symphytum officinale
F: Boraginaceae
Symphytum means "growing together," named because this perennial of damp ground throughout Eurasia was traditionally used in the treatment of injuries.

4 ft
1.2 m

WATER FORGET-ME-NOT
Myosotis scorpioides
F: Boraginaceae
This surface-breaking plant of streams and ponds is widely distributed in Eurasia, and introduced to the Americas and New Zealand.

28 in
70 cm

PURPLE GROMWELL
Buglossoides purpureocaerulea
F: Boraginaceae
Upright flower stems grow from creeping ground stems in this woodland perennial of chalky soils in southern Europe and southwest Asia.

24 in/60 cm

paired oval leaves with rounded tips

COMMON LUNGWORT
Pulmonaria officinalis
F: Boraginaceae
With blotched leaves supposedly resembling unhealthy lungs, and once thought to be efficacious against tuberculosis, this perennial grows in shady places throughout central Europe.

12 in
30 cm

BORAGE
Borago officinalis
F: Boraginaceae
Stiff hairs cover the stem and leaves of this southern European wayside plant, also cultivated for ornament and its oil-rich seeds.

24 in
60 cm

SCARLET PIMPERNEL
Anagallis arvensis
F: Primulaceae
Just as likely to have deep blue as scarlet flowers, this sprawling plant grows in disturbed ground virtually worldwide, except the tropics.

12 in
30 cm

NORTHERN PRIMROSE
Primula scandinavica
F: Primulaceae
Growing on mountain slopes in Norway and Sweden, this resembles the more common European birdseye primrose, but is much smaller.

6 in
15 cm

MOSQUITO BILLS
Dodecatheon hendersonii
F: Primulaceae
Dodecatheon flowers are called "shooting stars" because of their rocket or dart shapes. This western North American species has magenta to white flowers.

12 in
30 cm

HAIRY ANDROSACE
Androsace villosa
F: Primulaceae
In a genus of arctic-alpines, this perennial of the mountains of southern Europe is distinguished by its silky leaves and dense white flowerheads.

3 in
7 cm

HARDY CYCLAMEN
Cyclamen hederifolium
F: Primulaceae
Cyclamen leaves grow from underground stems; their tubular autumn flowers have back-folded petals. This species grows in bushy places in southern Europe.

4 in
10 cm

COWSLIP
Primula veris
F: Primulaceae
A typical primrose, with a head of up to 30 nodding flowers, cowslip grows in lime-rich meadows in S. Europe and temperate Asia.

30 cm
12 in

WILD PRIMROSE
Primula sp.
F: Primulaceae
Supposedly the *prima rosa* or "first flower" of spring, this perennial grows in woodland clearings and hedgebanks in western and southern Europe.

6 in
15 cm

GARRYALES

Any classification system raises anomalies, and the order Garryales is one of these. Once placed in the Cornales, modern chemical investigation separated this order of two families and around 20 species. The silky tassel family (Garryaceae) includes two genera: *Garrya* from North America and *Aucuba* from Eastern Asia. The Eucommiaceae covers just one species: *Eucommia ulmoides*, a tree of Chinese mountain forests.

16 ft/5 m

SPOTTED LAUREL
Aucuba japonica
F: Garryaceae
Some cultivars of this ornamental Japanese shrub grow yellow-blotched leaves. Plants are either male, with upright flower spikes, or female with small clustered flowerheads.

16 ft/5 m

SILK-TASSEL BUSH
Garrya elliptica
F: Garryaceae
Growing in coastal scrub in California and Oregon, this small tree has showy, gray-green male catkins and shorter, silvery female catkins.

GENTIANALES

The order name celebrates the gentians of mountains and gardens, but the Gentianaceae is a small family. There are five families, the largest is the madder family (Rubiaceae), with over 13,000 species, including tropical shrubs such as coffee. The oleander family (Apocyanaceae), strychnine family (Loganiaceae), and jessamine family (Gelsemiaceae) complete the order.

3¼ ft
1 m

MADAGASCAR PERIWINKLE
Catharanthus roseus
F: Apocynaceae

The leaves of this garden ornamental, endangered in its native Madagascar, contain tiny amounts of alkaloids used to treat childhood leukemias. This use has saved it from extinction.

50 ft
15 m

OVAL LEAF

FIVE-PETALED FLOWER

COMMON FRANGIPANI
Plumeria rubra
F: Apocynaceae

Native from Mexico to the Caribbean islands, the flowers of this ornamental tree are fragrant at night and attract sphinx moths for pollination.

6 in/15 cm

BEAD PLANT
Nertera granadensis
F: Rubiaceae

Named for its small, beadlike fruits, this perennial with tiny, green flowers is native to Australia, New Zealand, and South America.

5 ft
1.5 m

13 ft/4 m

GREATER PERIWINKLE
Vinca major
F: Apocynaceae

Long, rooting stems of this evergreen perennial creep across woodland floors in southern and central Europe and North Africa.

OLEANDER
Nerium oleander
F: Apocynaceae

All parts of this evergreen shrub are poisonous. It grows by streams from the Mediterranean to China, producing clusters of fragrant pink flowers.

30 in
75 cm

BUTTERFLY WEED
Asclepias tuberosa
F: Apocynaceae

American Indians chewed roots as a cure for pleurisy. This perennial grows in fields and on roadsides in eastern North America.

succulent, leafless stem

10 ft
3 m

CLEAVERS
Galium aparine
F: Rubiaceae

Prickles on the stem and leaf margins attach to animal fur, dispersing parts of this agricultural weed of Europe and western Asia.

glossy, evergreen leaf

CROSSWORT
Cruciata laevipes
F: Rubiaceae

With stem leaves in crosslike whorls and clustered, honey-scented flowers, this perennial grows in grassy places across Europe and Asia.

24 in
60 cm

50 ft
15 m

QUININE TREE
Cinchona calisaya
F: Rubiaceae

The antimalaria drug quinine comes from the bark of this South American tree. Seeds were smuggled to Asia in the mid 19th century for cultivation there.

33 ft/10 m

COFFEE SHRUB
Coffea arabica
F: Rubiaceae

Native to Ethiopia but transported elsewhere for cultivation, this evergreen shrub produces crimson drupes (stoned fruit) within which are two seeds (coffee beans).

6½ ft
2 m

CAPE JASMINE
Gardenia augusta
F: Rubiaceae

Originating in Asia, this evergreen shrub has fragrant, waxy, tubular white flowers that turn yellow with age. These ripen to berrylike fruits.

fruits red when ripe

STRYCHNINE
Strychnos nux-vomica
F: Loganiaceae
The poison strychnine comes
from seeds of this evergreen
tree from southeast Asia.

EGG-SHAPED
LEAVES

82 ft
25 m

STRYCHNINE
SEEDS

PERSIAN VIOLET
Exacum affine
F: Gentianaceae
Grown as an indoor ornamental but native to
Socotra Island in the Indian Ocean, this evergreen
biennial has purple flowers with yellow centers.

12 in
30 cm

SPRING GENTIAN
Gentiana verna
F: Gentianaceae
With ground-level leaf
rosettes and deep blue,
tubular flowers, this perennial
grows in Arctic and mountain
regions of Europe and Asia.

20 in
50 cm

5 in
12 cm

COMMON
CENTAURY
Centaurium erythraea
F: Gentianaceae
Grassy places and
dunes from Europe to
southwest Asia are
home to this annual
with funnel-shaped,
five-lobed
pink flowers.

12 in
30 cm

flower's center emits
stench of rotting meat

rosette of
oval leaves

CARRION FLOWER
Stapelia gettleffii
F: Apocynaceae
Carrion-flowers are succulent,
spiny plants from southern Africa.
The stench of their barred or
blotched flowers attracts carrion
flies as pollinators.

hairy-edged
petals

MADAGASCAR
JASMINE
Stephanotis floribunda
F: Gentianaceae
Originally from Madagascar,
this woody, twining vine is
a popular greenhouse plant
with leathery leaves and
clusters of fragrant, waxy,
white flowers.

20 ft
6 m

LAMIALES

Modern taxonomy has expanded the
scope of this order to include up to 21
families, typically with tubular flowers
and unequal petal lobes. The largest
are the mint family (Lamiaceae, often
known by its older name Labiatae) and
the figwort family (Scrophulariaceae),
each with 5,000–6,000 species. Others
include the olive family (Oleaceae) and
plantain family (Plantaginaceae).

1.5 m
5 ft

STOUT
FLOWER
SPIKE

CRINKLY
LOBED
LEAF

3¼ ft
1 m

BEAR'S
BREECHES
Acanthus mollis
F: Acanthaceae
This robust perennial of rocky places in
the western Mediterranean has spikes of
purple-veined white flowers with a
three-lobed lower lip.

dark green leaf
with creamy veins

20 in
50 cm

SHRIMP BUSH
Justicia brandegeana
F: Acanthaceae
This popular ornamental
from tropical America
and the Caribbean
has white, tubular,
two-lipped flowers
surrounded by a reddish,
shrimp-shaped leaves.

ZEBRA PLANT
Aphelandra squarrosa
F: Acanthaceae
Originating from coastal Brazilian
forests, this popular houseplant has
pale-veined leaves and spikes of
yellow flowers surrounded
by yellow specialized leaves.

BLACK
MANGROVE
Avicennia germinans
F: Acanthaceae
This tree forms thickets beside tidal
estuaries along tropical Atlantic
coasts. Pointed fruits, falling into
the mud, sprout new plants.

50 ft
15 m

»

BUGLE
Ajuga reptans
F: Lamiaceae

12 in / 30 cm

Growing around Europe, North Africa, and Southwest Asia, this woodland and meadow perennial produces flower stems from far-creeping, rooting runners.

tubular flower with large three-lobed lip

oval leaves, often bronzy underneath

SAGE
Salvia officinalis
F: Lamiaceae

24 in / 60 cm

This grayish shrub from Spain, southern France, and the Balkans is widely cultivated for its pungent leaves, used to flavour foods.

ROSEMARY
Rosmarinus officinalis
F: Lamiaceae

6½ ft / 2 m

From dry Mediterranean habitats, the oils in the leaves of this shrub reduce potential water loss from transpiration. Rosemary is a well-known culinary herb.

LATE SUMMER FLOWERS

BASIL
Ocimum basilicum
F: Lamiaceae

EDIBLE LEAVES

32 in / 80 cm

Also known as sweet basil, this is an annual from India and Iran. In cultivation, its pungent leaves are used as a culinary herb.

COMMON LAVENDER
Lavandula angustifolia
F: Lamiaceae

This is another evergreen shrub of dry Mediterranean scrub habitats. The water-saving oils in its leaves are valued in perfumery.

32 in / 80 cm

DENSELY PACKED FLOWERS

BETONY
Betonica officinalis
F: Lamiaceae

32 in / 80 cm

A common hedgerow and grassland herb in most of Europe, the Caucasus, and North Africa, betony has reddish purple or white flowers.

CHASTE TREE
Vitex agnus-castus
F: Lamiaceae

20 ft / 6 m

Chaste tree, from damp places in southern Europe, was once thought to preserve chastity. It is used to regulate hormonal functions in alternative herbal treatments.

JERUSALEM SAGE
Phlomis fruticosa
F: Lamiaceae

5 ft / 1.5 m

Native to dry rocky habitats in the eastern Mediterranean, this evergreen shrub with gray-felted leaves is a widely grown garden ornamental.

WILD BERGAMOT
Monarda fistulosa
F: Lamiaceae

LANCEOLATE LEAF

CLUSTER OF FLOWERS

4 ft / 1.2 m

A minty tea is made from the grayish leaves of this showy American perennial of dry fields and thickets from New England to Texas.

branched cluster of pinkish purple flowers

3¼ ft / 1 m

WILD MARJORAM
Origanum vulgare
F: Lamiaceae

A perennial aromatic herb of grassy and stony habitats in southern Europe and southwest Asia, it is grown for many culinary uses.

stalked, oval leaf

ROUND-LEAVED MINT BUSH
Prostanthera rotundifolia
F: Lamiaceae

10 ft / 3 m

This aromatic Australian shrub with circular leaves and pink to purple spring flowers grows in open forests from Queensland to Tasmania.

COMMON THYME
Thymus vulgaris
F: Lamiaceae

1¼ ft / 40 cm

Native to dry rocky places in the western Mediterranean, this densely branched shrub is used as a culinary herb, and an ingredient in perfumes and soaps.

WATER MINT
Mentha aquatica
F: Lamiaceae

3¼ ft / 1 m

Growing part-submerged in ponds and ditches in Europe, Africa, and Southwest Asia, this herb is one parent of the hybrid peppermint.

FLOWER
SPRIG

EGG-SHAPED
FRUIT

OLIVE
Olea europaea
F: Oleaceae
The fleshy fruits of this evergreen
Mediterranean tree contain 40 percent
unsaturated oils. Before they are eaten,
the olives must be pickled in brine.

50 ft / 15 m

heart-
shaped leaf

10 ft / 3 m

65 ft / 20 m

GOLDEN BELL
Forsythia suspensa
F: Oleaceae
Also called weeping forsythia,
this deciduous shrub has
hollow, pendulous stems and
yellow flowers. It is native to
China and perhaps Japan.

MANNA ASH
Fraxinus ornus
F: Oleaceae
This unusual ash from
southern Europe has dense
clusters of white flowers.
Manna, a sugary gum from its
bark, is used medicinally.

leaves up
to 5 in
(12 cm) long

23 ft / 7 m

12 in
30 cm

LILAC
Syringa vulgaris
F: Oleaceae
Gardeners have developed
many hybrids and varieties
of this showy, deciduous
tree, found wild on scrubby
hillside in southeast Europe.

dense, pyramid-
shaped, cluster of
scented flowers

TRUMPET-
SHAPED
FLOWER

ROSETTE
OF STICKY
LEAVES

**FALSE AFRICAN
VIOLET**
Streptocarpus saxorum
F: Gesneriaceae
Originating from Kenya and
Tanzania, this evergreen
perennial has whorls of
small, hairy, almost succulent
leaves and five-lobed,
trumpet-shaped flowers.

40 ft
12 m

COMMON JASMINE
Jasminum officinale
F: Oleaceae
Cultivated for its showy
scented flowers, this
scrambling, deciduous
shrub from the Caucasus
to China twists counter-
clockwise as it climbs.

**COMMON
BUTTERWORT**
Pinguicula vulgaris
F: Lentibulariaceae
This bog plant from
northern Europe,
Asia, and North
America gets
nutrients by trapping
and digesting insects
on its sticky leaves.

7 in
18 cm

40 ft
12 m

6 in
15 cm

AFRICAN VIOLET
Saintpaulia tongwensis
F: Gesneriaceae
Popular houseplants, African violets
live in rain forests in tropical East
Africa. Most of the 20 species are
endangered in the wild.

60 ft / 18 m

BEANLIKE
PODS

LONG
FLOWER
CLUSTER

INDIAN BEAN TREE
Catalpa bignonioides
F: Bignoniaceae
Despite its common name,
this woodland tree grows in
southern USA. It is planted
farther north and in Europe for
its showy flowers.

24 in
60 cm

HARDY GLOXINIA
Incarvillea delavayi
F: Bignoniaceae
A hardy garden perennial
with rosy-purple trumpet
flowers, this herbaceous plant
grows wild in mountain
grasslands in India, Tibet,
and China.

50 ft
15 m

JACARANDA
Jacaranda mimosifolia
F: Bignoniaceae
Originally from Argentina
and Brazil, this tropical tree
with hanging clusters of
lavender-colored flowers is
widely planted for shade
and ornament.

TRUMPET VINE
Campsis x tagliabuana
F: Bignoniaceae
A garden hybrid between
a North American and an
Asian species, this vigorous
climbing shrub has clusters
of orange-red, trumpet-
shaped flowers.

>>

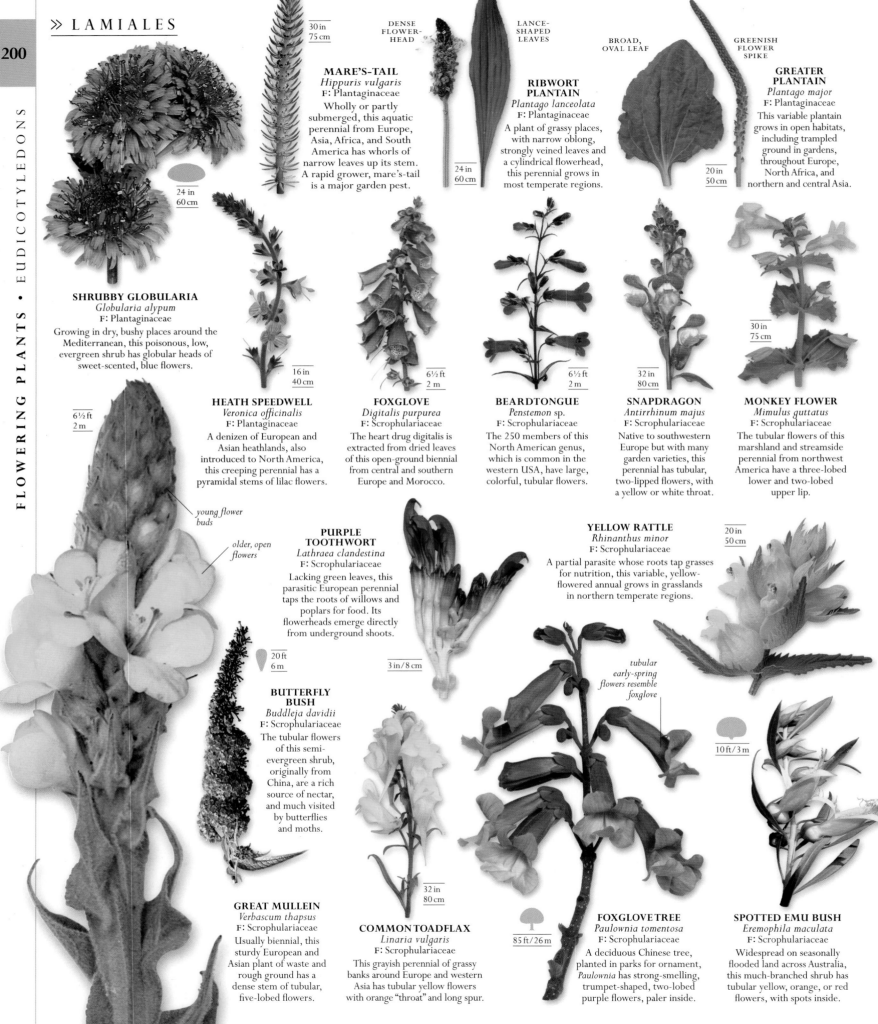

MARE'S-TAIL
Hippuris vulgaris
F: Plantaginaceae
Wholly or partly submerged, this aquatic perennial from Europe, Asia, Africa, and South America has whorls of narrow leaves up its stem. A rapid grower, mare's-tail is a major garden pest.

DENSE FLOWER-HEAD

LANCE-SHAPED LEAVES

30 in / 75 cm

24 in / 60 cm

RIBWORT PLANTAIN
Plantago lanceolata
F: Plantaginaceae
A plant of grassy places, with narrow oblong, strongly veined leaves and a cylindrical flowerhead, this perennial grows in most temperate regions.

BROAD, OVAL LEAF

GREENISH FLOWER SPIKE

GREATER PLANTAIN
Plantago major
F: Plantaginaceae
This variable plantain grows in open habitats, including trampled ground in gardens, throughout Europe, North Africa, and northern and central Asia.

20 in / 50 cm

SHRUBBY GLOBULARIA
Globularia alypum
F: Plantaginaceae
Growing in dry, bushy places around the Mediterranean, this poisonous, low, evergreen shrub has globular heads of sweet-scented, blue flowers.

24 in / 60 cm

6½ ft / 2 m

HEATH SPEEDWELL
Veronica officinalis
F: Plantaginaceae
A denizen of European and Asian heathlands, also introduced to North America, this creeping perennial has a pyramidal stems of lilac flowers.

16 in / 40 cm

FOXGLOVE
Digitalis purpurea
F: Scrophulariaceae
The heart drug digitalis is extracted from dried leaves of this open-ground biennial from central and southern Europe and Morocco.

6½ ft / 2 m

BEARDTONGUE
Penstemon sp.
F: Scrophulariaceae
The 250 members of this North American genus, which is common in the western USA, have large, colorful, tubular flowers.

6½ ft / 2 m

SNAPDRAGON
Antirrhinum majus
F: Scrophulariaceae
Native to southwestern Europe but with many garden varieties, this perennial has tubular, two-lipped flowers, with a yellow or white throat.

32 in / 80 cm

MONKEY FLOWER
Mimulus guttatus
F: Scrophulariaceae
The tubular flowers of this marshland and streamside perennial from northwest America have a three-lobed lower and two-lobed upper lip.

30 in / 75 cm

young flower buds

older, open flowers

PURPLE TOOTHWORT
Lathraea clandestina
F: Scrophulariaceae
Lacking green leaves, this parasitic European perennial taps the roots of willows and poplars for food. Its flowerheads emerge directly from underground shoots.

3 in / 8 cm

YELLOW RATTLE
Rhinanthus minor
F: Scrophulariaceae
A partial parasite whose roots tap grasses for nutrition, this variable, yellow-flowered annual grows in grasslands in northern temperate regions.

20 in / 50 cm

BUTTERFLY BUSH
Buddleja davidii
F: Scrophulariaceae
The tubular flowers of this semi-evergreen shrub, originally from China, are a rich source of nectar, and much visited by butterflies and moths.

20 ft / 6 m

tubular early-spring flowers resemble foxglove

10 ft / 3 m

GREAT MULLEIN
Verbascum thapsus
F: Scrophulariaceae
Usually biennial, this sturdy European and Asian plant of waste and rough ground has a dense stem of tubular, five-lobed flowers.

COMMON TOADFLAX
Linaria vulgaris
F: Scrophulariaceae
This grayish perennial of grassy banks around Europe and western Asia has tubular yellow flowers with orange "throat" and long spur.

32 in / 80 cm

85 ft / 26 m

FOXGLOVE TREE
Paulownia tomentosa
F: Scrophulariaceae
A deciduous Chinese tree, planted in parks for ornament, *Paulownia* has strong-smelling, trumpet-shaped, two-lobed purple flowers, paler inside.

SPOTTED EMU BUSH
Eremophila maculata
F: Scrophulariaceae
Widespread on seasonally flooded land across Australia, this much-branched shrub has tubular yellow, orange, or red flowers, with spots inside.

NEW ZEALAND LILAC
Veronica hulkeana
F: Plantaginaceae

This evergreen shrub, also known as a hebe, is a popular garden plant. It has showy stems of lilac flowers, growing wild on cliffs on the eastern side of New Zealand's South Island.

20 in
50 cm

HEBE "RED EDGE"
Veronica sp.
F: Plantaginaceae

Often used in landscaping schemes, this hardy hebe cultivar is thought to be a cross between *Veronica albicans* and *Veronica pimeleoides*.

24 in
60 cm

flowers attractive to butterflies for pollination

opposite, oval shiny leaves

BODINIER'S BEAUTYBERRY
Callicarpa bodinieri
F: Verbenaceae

Related to American beautyberry, this Chinese ornamental shrub is grown in gardens for its decorative purple berries, which are bitter but not poisonous.

10 ft
3 m

LANTANA
Lantana camara
F: Verbenaceae

The tubular flowers of this prickly shrub open yellow or orange then turn red. Native to the Americas, it is one of the most invasive weeds in warm regions.

5 ft
1.5 m

VERVAIN
Verbena officinalis
F: Verbenaceae

Widely distributed in rough grassland in temperate and tropical regions, this stiffly hairy perennial has slender spikes of two-lipped, lilac flowers.

3¼ ft
1 m

FLAMING GLORYBOWER
Clerodendrum splendens
F: Verbenaceae

This African vine, with clusters of tubular, red flowers, twines around trees in its native forest habitat, or trellises in gardens.

12 ft
3.7 m

SOLANALES

The potato family (Solanaceae), an economically important family of up to 4,000 species, dominates the order Solanales. Many species contain poisonous alkaloids. The bindweed family (Convolvulaceae) includes tropical climbers and low-growing herbs. Other families within this order are the Hydrolaceae, with one genus found in the Americas; five African trees in the Montiniaceae; and a pan-tropical herb in the Sphenocleaceae.

GREAT BINDWEED
Convolvulus sylvaticus
F: Convolvulaceae

A plant of hedgerows and waste ground, even in cities, and native to the Mediterranean, this large-flowered perennial spreads by far-creeping rhizomes (underground plant stems).

10 ft
3 m

trumpet-shaped blue flowers with white or yellow centre

three-lobed leaf

13 ft
4 m

MORNING GLORY
Ipomoea tricolor
F: Convolvulaceae

This twining herbaceous climber from tropical America has lobed leaves and funnel-shaped flowers that open only in the morning.

COMMON DODDER
Cuscuta epithymum
F: Convolvulaceae

24 in
60 cm

An annual forming a dense network of threadlike, twining stems, parasitic dodder taps into gorse, heather, and other species for food.

≫ SOLANALES

SPINY FRUIT CAPSULE

1.5 m
5 ft

FUNNEL-SHAPED FLOWER

THORN-APPLE
Datura stramonium
F: Solanaceae
Rank-smelling and highly poisonous, this widespread annual weed is thought to originate from America.

32 in
80 cm

HENBANE
Hyoscyamus niger
F: Solanaceae
Poisonous, narcotic, and evil-smelling, this annual herb inhabits disturbed ground, especially where cattle graze, in Europe, Asia, and North Africa.

LEAFY FLOWER HEAD

5 ft/1.5 m

GLOSSY BLACK BERRY

DEADLY NIGHTSHADE
Atropa belladonna
F: Solanaceae
Packed with powerful narcotics and truly deadly, this tall, perennial herb grows in woods and scrub in Europe, North Africa, and western Asia.

soft, succulent berry containing many seeds

7 ft/2 m

TOMATO
Solanum lypopersicum
F: Solanaceae
Probably bred originally from yellow-fruited cherry tomatoes from Peru and Ecuador, this short-lived perennial is cultivated for its fleshy, edible berries.

30 ft/10 m

RED ANGELS' TRUMPETS
Brugmansia sanguinea
F: Solanaceae
This evergreen tree from western South America, with poisonous foliage, is grown in warm, sunny gardens for its showy, pendulous, tubular flowers.

24 in/60 cm

CHINESE LANTERN
Physalis alkekengi
F: Solanaceae
From southern Europe despite its name, this relative of cape gooseberry is mainly grown for ornament. A lantern-shaped flower conceals its berry.

APIALES

The order Apiales is dominated by the highly distinctive and economically important carrot family (Apiaceae, often known by the older name Umbelliferae), with at least 3,500 species. The Araliaceae, consisting of ivies and ginseng, is another large family, while the parchment-bark family, Pittosporaceae, of evergreen shrubs and trees is medium-sized. Seven smaller families, with few species, complete the order.

2 m
6½ ft

WILD ANGELICA
Angelica sylvestris
F: Apiaceae
A typical "umbellifer" this plant has with 15–40 umbrella-like spokes supporting its flower-bearing stem. Angelica grows in grassy places in Europe and temperate Asia.

PALMATE LEAF

36 in/90 cm

LATE-SUMMER FLOWERS

ASTRANTIA
Astrantia major
F: Apiaceae
A plant of alpine meadows and open woods throughout Europe, astrantia has showy pinkish bracts, or specialized leaves, under its heads of small flowers.

6½ ft
2 m

HOGWEED
Heracleum sphondylium
F: Apiaceae
With hollow, hairy stems and umbrella-shaped flower heads with 10–20 stems, hogweed grows in roadsides from eastern Europe to Britain.

16 ft/5 m

UMBELS OF YELLOW FLOWERS

FINELY DIVIDED LEAF

GIANT FENNEL
Ferula communis
F: Apiaceae
This robust, giant perennial, with pungent-smelling, hollow stems, grows in grassy and meadowy places around the Mediterranean, Asia, and North Africa.

COMPOUND UMBEL

8 ft
2.5 m

LACY LEAF

HEMLOCK
Conium maculatum
F: Apiaceae
A biennial of wet places in Europe and Asia, widely introduced elsewhere, hemlock has hollow, purple-spotted stems. All parts are poisonous.

3¼ ft
1 m

FEATHERY LEAF

UMBEL OF WHITE FLOWERS

DEVELOPING FRUIT STAGE

WILD CARROT
Daucus carota
F: Apiaceae
The wild ancestor of cultivated carrots has just slightly swollen tap-roots. It grows throughout Europe, temperate Asia, and North Africa.

HOT PEPPER
Capsicum frutescens
F: Solanaceae

4 ft
1.2 m

Much cultivated in tropical Asia and equatorial America, peppers are South American natives. The fruit of this small shrub are chilies.

TOBACCO
Nicotiana tabacum
F: Solanaceae

10 ft
3 m

Originally South American, this herb is the most common of two species grown for smoking tobacco. The leaves contain nicotine.

POTATO
Solanum tuberosum
F: Solanaceae

3¼ ft
1 m

FLOWERS AND PINNATE LEAVES

EDIBLE TUBERS

Potatoes are a major food crop. The swollen tubers turn green (and poisonous) in sunlight. The plant was bred from South American ancestors.

berrylike red fruit

COMMON HOLLY
Ilex aquifolium
F: Aquifoliaceae

80 ft / 24 m

An evergreen, woodland understorey shrub or small tree from Europe, North Africa, and northwestern Asia, holly has spiny leaves to discourage grazers.

egg-shaped flower head

33 ft
10 m

KOHUHU
Pittosporum tenuifolium
F: Pittosporaceae

Growing wild in mountain forests on both islands of New Zealand, this evergreen tree has almost black branches and honey-scented spring flowers.

24 in / 60 cm

leathery, spiny leaves

32 in / 80 cm

SEA HOLLY
Eryngium maritimum
F: Apiaceae

Leathery leaves help this sturdy perennial reduce water loss and repel salt spray in sand dunes around Europe, North Africa, and southeastern Asia.

IVY
Hedera helix
F: Araliaceae

100 ft
30 m

An evergreen shrub from Europe and southwestern Asia, ivy climbs up other plants or forms creeping carpets in woods.

ORIENTAL GINSENG
Panax ginseng
F: Araliaceae

The root of this Asian herb is used as a traditional remedy. Its species name, *Panax*, comes from the Greek word *panacea*, meaning a cure-all for diseases.

AQUIFOLIALES

The mainly tropical holly family (Aquifoliaceae) includes trees and shrubs with characteristically toothed leaves, and is the main representative of the Aquifoliales order. Other members of this small and recent order are an odd family of twining herbs, the Cardiopteridaceae; three Asian shrubs in the Helwingiaceae; four South American shrubs and trees in the Phyllonomaceae; and a family of tropical trees, the Stemonuraceae.

ASTERALES

Around 13 families make up the Asterales. The largest is the daisy family (Asteraceae) with some 25,000 species. Typically, each flowerlike head (capitula) has many individual flowers, called florets, surrounded by showy rays. Some of the bellflower family (Campanulaceae) show similar characteristics. This order includes the bogbean (Menyanthaceae) and the fan-flower (Goodeniaceae) families, and several smaller families.

ray florets

disk florets

5 ft
1.5 m

MICHAELMAS DAISY
Symphotrichum novi-belgii
F: Asteraceae

Previously classified as a species of *Aster*, this showy, variable garden plant is a native of eastern North America.

bract

4 ft
1.25 m

SALSIFY
Tragopogon porrifolius
F: Asteraceae

A biennial herb of grassy places around the Mediterranean, salsify has lilac or reddish purple flowerheads, surrounded by longer, pointed bracts.

COMMON RAGWORT
Senecio jacobaea
F: Asteraceae

Toxic to farm animals and avoided by rabbits, this native perennial from Europe and western Asia has invaded grasslands almost worldwide.

5 ft
1.5 m

capitula of disk florets, which mature into seed

lobed ray florets may be yellow, orange, or maroon

11½ ft
3.5 m

oval, toothed leaf

SUNFLOWER
Helianthus annuus
F: Asteraceae

Probably Mexican in origin, this tall annual is grown as an ornamental as well as commercially for its seeds, which contain 27–40 percent polyunsaturated oil and 13–20 percent protein.

3¼ ft
1 m

CORNFLOWER
Centaurea cyanus
F: Asteraceae

Probably native to southern Europe and western Asia, cornflower seeds were spread with corn, making it an agricultural weed before herbicides.

SMALL WHITE FLOWER HEAD

SERRATED LEAF

OXEYE DAISY
Leucanthemum vulgare
F: Asteraceae

One of the most common white-flowered daisies of Europe and western Asia, this variable patch-forming perennial quickly seeds into disturbed ground.

32 in
75 cm

tall, robust stem

5 in
12 cm

DAISY
Bellis perennis
F: Asteraceae

Originally from Europe and western Asia, this short perennial of grazed grasslands and lawns has been transported almost worldwide. It is commonly considered a weed.

UPRIGHT OR HORIZONTAL LEAF

12 in
30 cm

COMMON DANDELION
Taraxacum officinale
F: Asteraceae

The name for this familiar weed covers an amazing diversity of minutely differing forms, with perhaps 1,000 recognized "microspecies" in Europe and Asia.

YELLOW FLOWERHEAD

PINK OR PURPLE RAY FLORET

4 ft
1.2 m

PURPLE CONEFLOWER
Echinacea purpurea
F: Asteraceae

Named for its cone-shaped central disk floret, this ornamental perennial from eastern North America is cultivated for herbal remedies against colds and 'flu.

ROUGH, TOOTHED LEAF

MILK THISTLE
Silybum marianum
F: Asteraceae
This weakly spiny biennial, with white-blotched leaves, grows in waste and cultivated ground around S. Europe, N. Africa, and W. Asia.

8 ft
2.5 m

CANADIAN GOLDENROD
Solidago canadensis
F: Asteraceae
Native to N. America but much grown in gardens, this downy perennial has yellow flowerheads borne in curved, spreading flower spikes.

8 ft
2.5 m

4 ft
1.2 m

GLOBE THISTLE
Echinops bannaticus
F: Asteraceae
Native to grassy places in S.E. Europe and W. Asia, this hairy-stemmed perennial has a spherical head of bluish, tubular "disk" florets.

SPINY HEAD

6½ ft
2 m

GLOBE ARTICHOKE
Cynara cardunculus
F: Asteraceae
A widely grown Mediterranean vegetable, unknown in the wild, this is probably a variety of cardoon, a wasteground perennial in that region.

DEEPLY CUT LEAF

6½ ft
2 m

RHUBARB-LIKE LEAF

BUTTERBUR
Petasites hybridus
F: Asteraceae
Plants of butterbur have stems with either male or female flowers. It grows in wet meadows and beside streams throughout Europe.

EARLY SPRING FLOWER

20 in
50 cm

COTTONWEED
Otanthus maritimus
F: Asteraceae
This shrubby perennial grows in coastal habitats in southern Europe, North Africa, and southwest Asia. The common name derives from its woolly stems and leaves.

6 ft
1.8 m

DENSE BLAZING STAR
Liatris spicata
F: Asteraceae
A plant of damp ground in eastern USA, this plant is named after its dense, elongated, spike of rose-purple flowerheads.

SPEAR THISTLE
Cirsium vulgare
F: Asteraceae
Common in disturbed ground in Europe and western Asia, this robust perennial with spine-tipped leaves has been spread worldwide by agriculture.

5 ft
1.5 m

outer ray florets

20 in
50 cm

leaves hairy on both sides

24 in
60 cm

YARROW
Achillea millefolium
F: Asteraceae
Found in grassy places throughout Europe and western Asia, this feathery-leaved perennial has been introduced to North America, Australia, and New Zealand.

6½ ft
2 m

BLUE OR LAVENDER FLOWERS

OBLONG ROSETTE LEAF

6½ ft
2 m

CHICORY
Cichorium intybus
F: Asteraceae
Introduced almost worldwide from Europe, western Asia, and North Africa, it is cultivated for salad leaves and, from its roots, as a coffee substitute.

FRENCH TARRAGON
Artemisia dracunculus
F: Asteraceae
Leaves of this bushy herb from southeast Europe, Asia, and North America are used to flavor fish and other foods.

COTTON LAVENDER
Santolina chamaecyparissus
F: Asteraceae
Widely grown in gardens, this strongly aromatic, dwarf, evergreen shrub with gray-hairy leaves is native to rocky places around the Mediterranean.

24 in
60 cm

GARDEN MARIGOLD
Calendula officinalis
F: Asteraceae
This species has been cultivated so long that its origins are unknown. Marigold extract from the flowers is used to treat skin problems.

20 in
50 cm

GARDEN GAZANIA
Gazania sp.
F: Asteraceae
Seventeen species of *Gazania* grow in southern Africa, most in dry habitats. Cultivars like these can survive in gardens with little watering.

STRAND GAZANIA
Gazania rigens
F: Asteraceae
This sprawling, mat-forming perennial often flowers prolifically on dunes and rocky outcrops along the southern Cape coast of South Africa.

8 in / 20 cm

PERENNIAL LEAF

4 ft
1.2 m

FLOWER SPIKE

GREAT BLUE LOBELIA
Lobelia siphilitica
F: Campanulaceae

A showy American perennial of woods and meadows from New England to Alabama, this plant was once thought to cure syphilis.

tubular flowers

16 in
40 cm

ROYAL BLUEBELL
Wahlenbergia gloriosa
F: Campanulaceae

The state plant of the Australian Capital Territory, this erect perennial with deep blue flowers grows in mountain grasslands in South Australia.

28 in/70 cm

BELL-SHAPED FLOWER

OVAL LEAVES

BALLOON FLOWER
Platycodon grandiflorum
F: Campanulaceae

The only member of its genus, this Asian perennial with blue or white, bell-shaped flowers also comes in many garden varieties.

5 ft
1.5 m

BUCKBEAN
Menyanthes trifoliata
F: Menyanthaceae

This far-creeping perennial produces its beanlike fruits in bogs and shallow water in North America, Greenland, northern Europe, and Asia.

CLUSTERED FLOWERHEAD

20 in
50 cm

ROUND-HEADED RAMPION
Phyteuma orbiculare
F: Campanulaceae

With dark blue flowers in rounded heads, this unbranched perennial grows in chalk grasslands from southern England to Greece.

LANCEOLATE LEAF

3¼ ft
1 m

NETTLE-LEAVED BELLFLOWER
Campanula trachelium
F: Campanulaceae

Found in fencerows around Europe, Iran, and North Africa, this roughly hairy perennial has nettlelike leaves and blue, bell-shaped flowers.

ONE-SIDED FLOWER

6 in
15 cm

FLESHY LEAF

SWAMPWEED
Selliera radicans
F: Goodeniaceae

This creeping herb is found on coastal sandflats around Chile, Australia, and New Zealand — where it also grows alongside mountain streams.

5 ft
1.5 m

FRINGED WATERLILY
Nymphoides peltata
F: Menyanthaceae

Unlike true water lilies, this aquatic perennial has smaller yellow flowers with five fringed petals. It is found from Europe to Japan.

DIPSACALES

The order Dipsacales is found throughout the world, but especially in the Northern Hemisphere. It is generally characterized by small flowers, often in compact heads, and includes many popular ornamental plants. The valerian family (Valerianaceae) is the most diverse, with around 350 species, but the order also includes the scabious and teasel family (Dipsacaceae), honeysuckle family (Caprifoliaceae), and muskroot family (Adoxaceae).

6 in
15 cm

MUSKROOT
Adoxa moschatellina
F: Adoxaceae

Inflorescences of this delicate woodland perennial from North America, Europe, and Asia have five flowers at right angles (with one facing upward).

32 in
80 cm

RED VALERIAN
Centranthus ruber
F: Valerianaceae

Butterflies appreciate the flowers of this grayish perennial. It is found on coastal rocks and old walls around the Mediterranean and western Asia, and is naturalized elsewhere.

lobed leaves

shiny, red berries

40 ft/12 m

FRUIT

FLOWER

ELDERBERRY
Sambucus nigra
F: Adoxaceae

A shrub or small tree with lobed leaves and flat-topped flowerheads, elderberry grows in Europe, western Asia, and northern Africa.

6½ ft
2 m

EUROPEAN CRANBERRYBUSH
Viburnum opulus
F: Adoxaceae

Native across Europe to Asia, this deciduous shrub has flattened flowerheads with large, sterile outer flowers, around smaller, fertile ones.

13 ft
4 m

GARDEN VALERIAN
Valeriana officinalis
F: Valerianaceae

This perennial is found in grassy places from northern Europe to Japan. It has flattened heads of pale pink, five-petaled flowers, which are darker when in bud.

SCENTED FLOWERS

PERENNIAL LEAVES

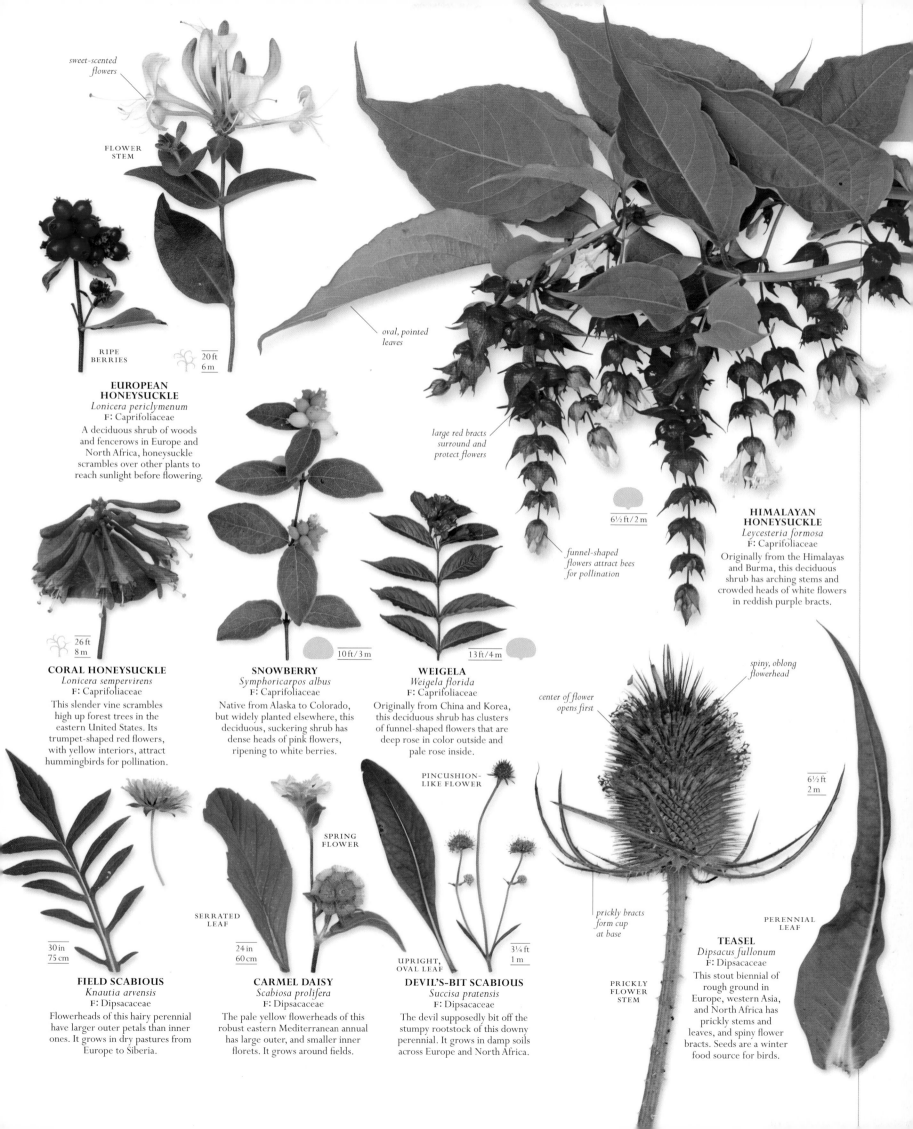

sweet-scented flowers

FLOWER STEM

RIPE BERRIES

20 ft 6 m

EUROPEAN HONEYSUCKLE
Lonicera periclymenum
F: Caprifoliaceae

A deciduous shrub of woods and fencerows in Europe and North Africa, honeysuckle scrambles over other plants to reach sunlight before flowering.

oval, pointed leaves

large red bracts surround and protect flowers

6½ ft / 2 m

funnel-shaped flowers attract bees for pollination

HIMALAYAN HONEYSUCKLE
Leycesteria formosa
F: Caprifoliaceae

Originally from the Himalayas and Burma, this deciduous shrub has arching stems and crowded heads of white flowers in reddish purple bracts.

26 ft 8 m

CORAL HONEYSUCKLE
Lonicera sempervirens
F: Caprifoliaceae

This slender vine scrambles high up forest trees in the eastern United States. Its trumpet-shaped red flowers, with yellow interiors, attract hummingbirds for pollination.

10 ft / 3 m

SNOWBERRY
Symphoricarpos albus
F: Caprifoliaceae

Native from Alaska to Colorado, but widely planted elsewhere, this deciduous, suckering shrub has dense heads of pink flowers, ripening to white berries.

13 ft / 4 m

WEIGELA
Weigela florida
F: Caprifoliaceae

Originally from China and Korea, this deciduous shrub has clusters of funnel-shaped flowers that are deep rose in color outside and pale rose inside.

spiny, oblong flowerhead

center of flower opens first

6½ ft 2 m

PINCUSHION-LIKE FLOWER

SPRING FLOWER

SERRATED LEAF

30 in 75 cm

24 in 60 cm

UPRIGHT, OVAL LEAF

prickly bracts form cup at base

3¼ ft 1 m

PERENNIAL LEAF

PRICKLY FLOWER STEM

FIELD SCABIOUS
Knautia arvensis
F: Dipsacaceae

Flowerheads of this hairy perennial have larger outer petals than inner ones. It grows in dry pastures from Europe to Siberia.

CARMEL DAISY
Scabiosa prolifera
F: Dipsacaceae

The pale yellow flowerheads of this robust eastern Mediterranean annual has large outer, and smaller inner florets. It grows around fields.

DEVIL'S-BIT SCABIOUS
Succisa pratensis
F: Dipsacaceae

The devil supposedly bit off the stumpy rootstock of this downy perennial. It grows in damp soils across Europe and North Africa.

TEASEL
Dipsacus fullonum
F: Dipsacaceae

This stout biennial of rough ground in Europe, western Asia, and North Africa has prickly stems and leaves, and spiny flower bracts. Seeds are a winter food source for birds.

FUNGI

Ranging from mushrooms to microscopic molds, fungi were once classified as plants. Today, they are recognized as a distinct kingdom of living things. Growing in or through their food, they digest organic matter, and often become visible only when they reproduce. Fungi are both allies and enemies to other forms of life: vital recyclers and mutually benefiting partners, they also include parasites and pathogens.

MUSHROOMS

As well as mushrooms, this large group includes bracket fungi, puffballs, and many other species. Their fruitbodies vary in shape, but all make their spores on microscopic cells called basidia.

SAC FUNGI

These fungi produce their spores in microscopic sacs, which often form a felty layer on their fruitbodies. Many are cup-shaped, but they also include truffles, morels, and single-celled yeasts.

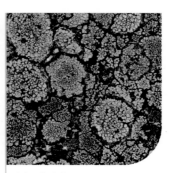

LICHENS

Living partnerships between fungi and algae, lichens colonize all kinds of bare surfaces. Some form flat crusts, while others are like tiny plants. Most grow slowly, and are exceptionally long-lived.

MUSHROOMS

The phylum Basidiomycota includes the majority of what are commonly called mushrooms and toadstools. Found largely in temperate woodland, nearly all share the ability to form sexual spores externally on special cells called basidia.

PHYLUM	BASIDIOMYCOTA
CLASSES	3
ORDERS	52
FAMILIES	177
SPECIES	About 32,000

These typical mushrooms show the cap with radiating gills on the underside, supported by the central stem.

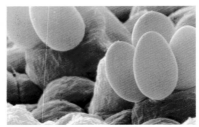

The microscopic spores of the field mushroom are here shown still attached to the basidia, or reproductive cells.

This stinkhorn is covered with a dark spore mass, which rapidly liquifies and emits a foul odor, usually of rotting flesh.

DEBATE
COLORFUL MYSTERY

Many mushrooms have bright colors: red, violet, blue, and green. Scientists have little idea what advantage these colors give to the mushroom. Unlike flowers, they do not need bright colors to attract pollinators. Red and orange pigments possibly prevent damage from sunlight, while others may act as a warning to predators.

Few fungi are as diverse in form as the Basidiomycota. Their fruitbodies (the parts above ground in which spores are produced) have evolved to provide varied and efficient mechanisms for spore dispersal. These fruitbodies often take the form of a stem, cap, and gills, but many are simple, crustlike sheets, and others are more complex, bracketlike forms. More exotic still are the totally enclosed, spherical types such as the puffballs and the earthstars. Then there are the beautiful, coral-like structures of the club and coral fungi, and the often weird, animal-like shapes of the stinkhorns. Each structure supports and produces special spore-producing cells called basidia, and—depending on the fungus concerned—uses a variety of mechanisms to disperse these spores. For example, gilled fungi forcibly eject their spores, to be carried off by air currents. The puffballs, however, depend on a combination of wind and raindrops to puff out their spores. Stinkhorns, on the other hand, use foul odors and bright colors to attract animals, such as insects and other invertebrates, to eat their spore mass and to pass the spores unharmed through their digestive tracts. In this way the spores are scattered as far as the animals travel before excreting them.

FUNGAL FEEDING

The main body of the fungus is usually underground, formed of fine fungal threads called hyphae, which make up a mycelium (fungal body). These threads permeate the substance on which the fungus grows, be it soil, old leaf litter, fallen timber, living plant tissues, or even dead and decaying animals. The mycelium may spread over a large area, and in many fungi it also forms a mutually beneficial relationship with the roots of plants. In this relationship—called a mycorrhiza—the mycelium wraps around and penetrates the plant's roots, providing the fungus with access to carbohydrates from the plant. In return, the plant gains from the mycelium's greater ability to absorb water and mineral nutrients. Other mushrooms break down dead organic matter, or devour living organisms for food—both processes help renew the soil and to improve the growing conditions for other living things.

MUSHROOMS AMONG FALLEN LEAVES >
Many fungi feed on dead and decaying plant tissues and may spread in huge numbers where nutrients and moisture occur.

AGARICALES

Many of the most familiar mushrooms and toadstools belong to this order. They include fungi with fleshy, not woody, fruitbodies (the body supporting the spore-forming cells). Many have caps and stems with gills; some also have pores. Other forms include bird's-nest, brackets, crusts, trufflelike, and puffballs. Most live in leaf litter, soil, or on wood (causing a white rot); others are parasitic or live in association with plant roots (mycorrhizal).

FIELD MUSHROOM
Agaricus campestris
F: Agaricaceae
Common in meadows of North America and Eurasia; this fungus has a rounded cap; bright pink gills turning brown with maturity; and a short stem with a small ring.

1½–4 in
4–10 cm

2–4 in
5–10 cm

CULTIVATED MUSHROOM
Agaricus bisporus
F: Agaricaceae
Selling in millions, this familiar fungus is common worldwide. Its cap varies from white to dark brown with scales.

slightly fibrous surface

THE PRINCE
Agaricus augustus
F: Agaricaceae
A large, orange-brown, scaly cap and a white, woolly stem with a floppy ring characterize this common species from North America and Eurasia.

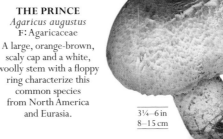

3¼–6 in
8–15 cm

2–6 in
5–15 cm

FRECKLED DAPPERLING
Lepiota aspera
F: Agaricaceae
From woods and gardens in North America and Eurasia; this uncommon fungus's brown cap has pyramidal scales that rub off. The brown stem has a ring.

2–4¾ in
5–12 cm

YELLOW STAINER
Agaricus xanthodermus
F: Agaricaceae
This species has a flattened cap center, chrome-yellow stains on the stem base, and an unpleasant odor. It is common in western North America and Eurasia.

2¾–6 in
7–15 cm

HORSE MUSHROOM
Agaricus arvensis
F: Agaricaceae
Common in parks in North America and Eurasia; this species stains dull brassy yellow. The pendent ring on its stem has a cogwheellike pattern on its underside.

⅜–1½ in
1–4 cm

STINKING DAPPERLING
Lepiota cristata
F: Agaricaceae
A rubberlike smell is a feature of this species of *Lepiota* (most common in woods and meadows) from North America and Eurasia.

yellow stains

MATURE CAP

¾–2¼ in
2–6 cm

1¼–3¼ in
3–8 cm

YOUNG CAP

ring

1½–4½ in
4–11 cm

BLUSHING DAPPERLING
Leucoagaricus badhamii
F: Agaricaceae
Found in Eurasia; this rare fungus grows in woods and gardens with rich soils. White when young, it bruises blood-red when handled, and then blackens.

LAWYER'S WIG
Coprinus comatus
F: Agaricaceae
This fungus is common in disturbed soils and pathsides in North America and Eurasia. Its tall, shaggy cap is unmistakable, and dissolves into inky fluid.

2–3¼ in
5–8 cm

ORANGE-GIRDLED DAPPERLING
Lepiota ignivolvata
F: Agaricaceae
Found in Eurasia; this uncommon species has an orange-margined ring set low down on the club-shaped, white stem.

2–6 in
5–15 cm

WHITE DAPPERLING
Leucoagaricus leucothites
F: Agaricaceae
Common in meadows and grassy roadsides in North America and Eurasia; this fungus's ivory fruitbody turns gray-brown with age. Its gills are white to pale pink.

SHAGGY PARASOL
Chlorophyllum rhacodes
F: Agaricaceae
Common in North America and Eurasia; this fungus has a shaggy, brown cap; thick, double-edged ring; reddening flesh; and swollen stem base.

8–20 in
20–50 cm

⅜–1¼ in
1–3 cm

⅜–2 in
1–5 cm

PLANTPOT DAPPERLING
Leucocoprinus birnbaumii
F: Agaricaceae
Common worldwide; this species grows in soils of potted plants. Its features include a delicate golden-yellow cap and a slender stem with a ring.

INNER LAYER

OUTER SKIN

GRAY PUFFBALL
Bovista plumbea
F: Agaricaceae
This ball-shaped fungus is smooth when young. As it matures, its skin flakes to reveal a papery inner layer. It is common in North America and Eurasia.

GIANT PUFFBALL
Calvatia gigantea
F: Agaricaceae
Common in hedges, fields, and gardens in N. America and Eurasia; this mushroom's large, smooth, white fruitbody with a white to yellow interior is unmistakable.

bulbous stem base

4–12 in
10–30 cm

PARASOL
Macrolepiota procera
F: Agaricaceae
Common in meadows in North America and Eurasia; this tall species has a scaly cap and snakeskin-patterned stem with a thick ring.

¾–1¾ in
2–4.5 cm

MEADOW PUFFBALL
Lycoperdon pratense
F: Agaricaceae
This short-stemmed species has an internal membrane separating the spore-mass from the stem. It is common in meadows throughout Eurasia.

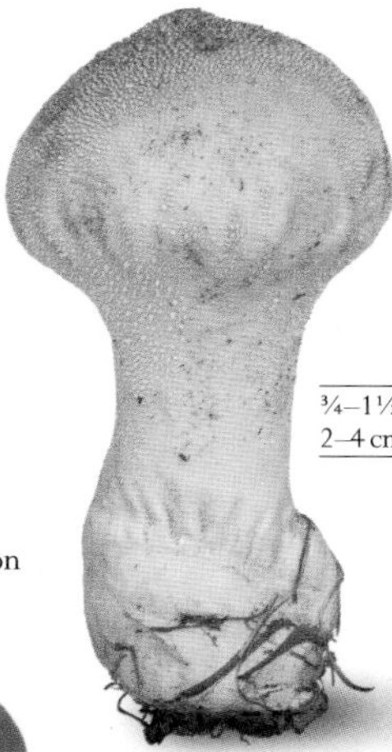

¾–1½ in
2–4 cm

COMMON PUFFBALL
Lycoperdon perlatum
F: Agaricaceae
The most common white puffball, this species is found in North America and Eurasia. Its granular spines rub off to leave circular scars in a regular pattern.

⅜–1¼ in
1–3 cm

SPINY PUFFBALL
Lycoperdon echinatum
F: Agaricaceae
This species is found in beech woods in North America and Eurasia. It has groups of long spines that meet at their tips. Its spores are purplish brown.

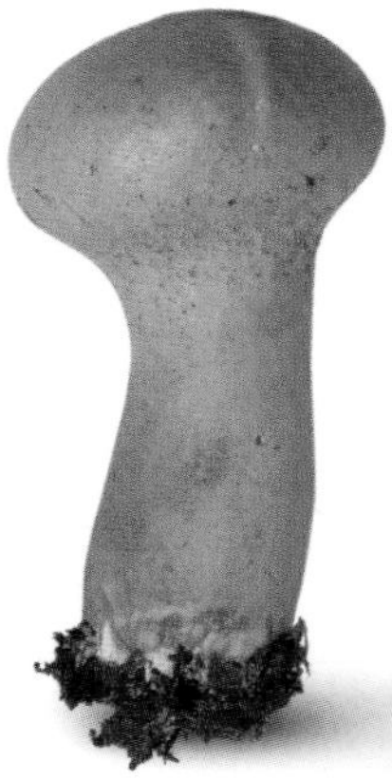

2–4 in
5–10 cm

PESTLE PUFFBALL
Lycoperdon excipuliforme
F: Agaricaceae
One of the tallest puffballs, this fungus has a buff-white stem that is left behind once the spores are released. It is found in Eurasia.

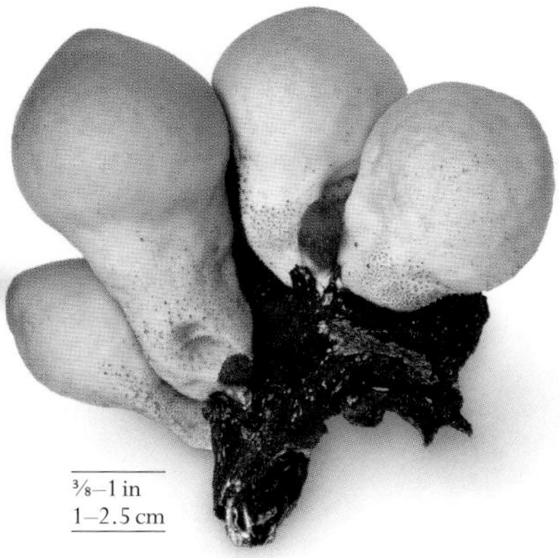

⅜–1 in
1–2.5 cm

STUMP PUFFBALL
Lycoperdon pyriforme
F: Agaricaceae
Common in North America and Eurasia; this pear-shaped species grows on wood. It has prominent cords at its base, and is firm when young.

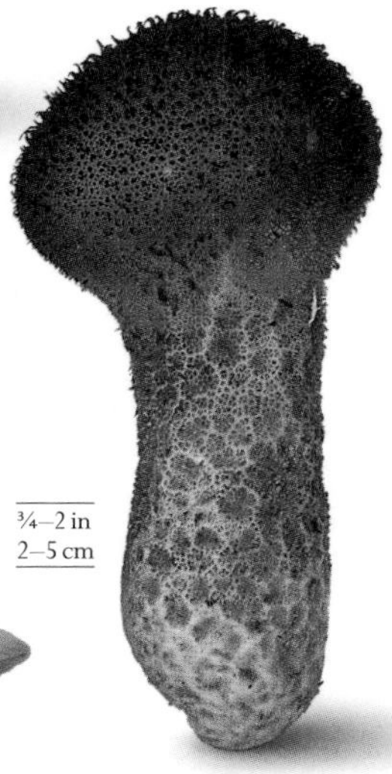

¾–2 in
2–5 cm

STINKING PUFFBALL
Lycoperdon foetidum
F: Agaricaceae
Common in acidic meadows and woods in Eurasia; this yellowish brown puffball has dark brown spines. Its flesh has an unpleasant smell.

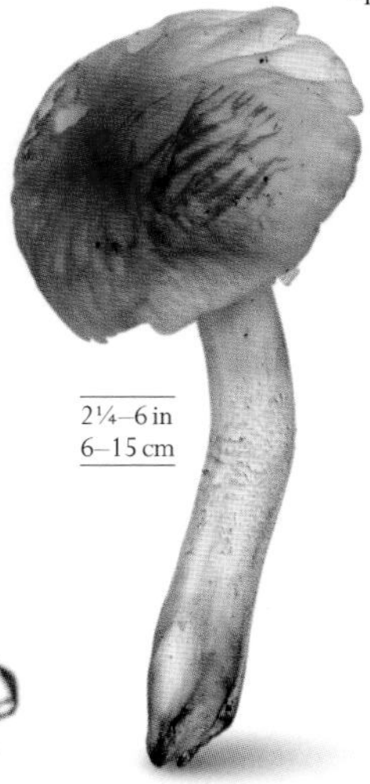

2¼–6 in
6–15 cm

POPLAR FIELDCAP
Agrocybe cylindracea
F: Bolbitiaceae
Found in Eurasia; this rare species grows on poplar wood. The cap cracks when dry and the stem has a ring.

⅜–1½ in
1–3.5 cm

COMMON FIELDCAP
Agrocybe pediades
F: Bolbitiaceae
Usually found in lawns in Eurasia; this fungus's smooth, yellowish cap and slender stem have a mealy smell and no veil.

1¼–2¾ in
3–7 cm

SPRING FIELDCAP
Agrocybe praecox
F: Bolbitiaceae
Common in North America and Eurasia; during spring, this fungus's cap may have marginal veil fragments. The stem may have a fragile ring.

2¼–4 in
6–10 cm

WINTER STALKBALL
Tulostoma brumale
F: Agaricaceae
Typically found in sandy soils in dune slacks in North America and Eurasia; this fungus has tiny, whitish yellow, round heads on slender, pale brown stalks.

1¼–2 in
3–5 cm

SANDY STILTBALL
Battarrea digueti
F: Tulostomataceae
Found in very dry, sandy soils in North America; this tall, stalked species emerges from a leathery "egg." Its brown cap contains the spores.

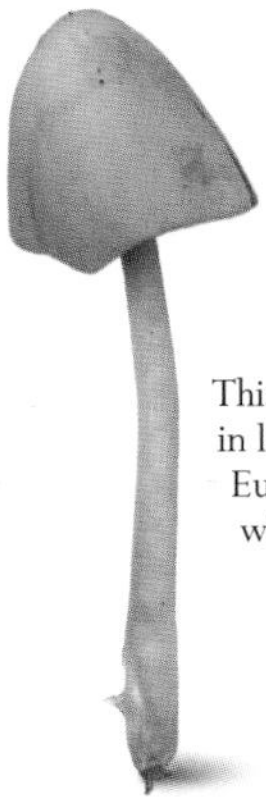

⅜–½ in
1–1.5 cm

MILKY CONECAP
Conocybe apala
F: Bolbitiaceae
This species is frequently found in lawns in North America and Eurasia; Its conical, ivory cap with orange-brown gills sits on a tall stem.

⅜–1¼ in
1–3 cm

YELLOW FIELDCAP
Bolbitius vitellinus
F: Bolbitiaceae
Growing in meadows, this fungus is common in North America and Eurasia; It has a delicate, sticky cap that lasts only a day or so.

»

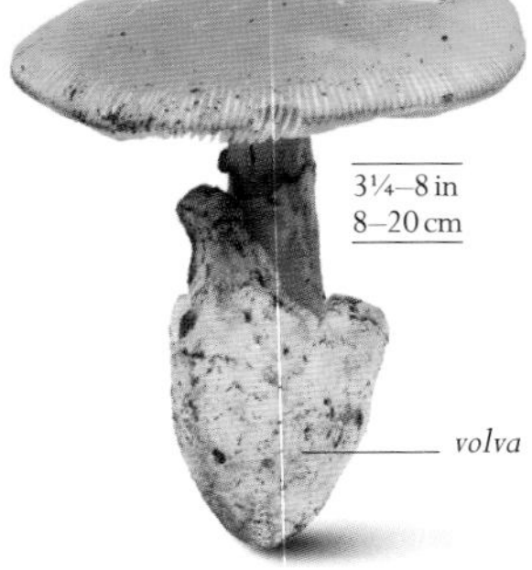
volva

3¼–8 in
8–20 cm

CAESAR'S MUSHROOM
Amanita caesarea
F: Amanitaceae

Found under oaks, mainly from the Mediterranean to central Eurasia; this mushroom has an orange cap and stem, and a white volva.

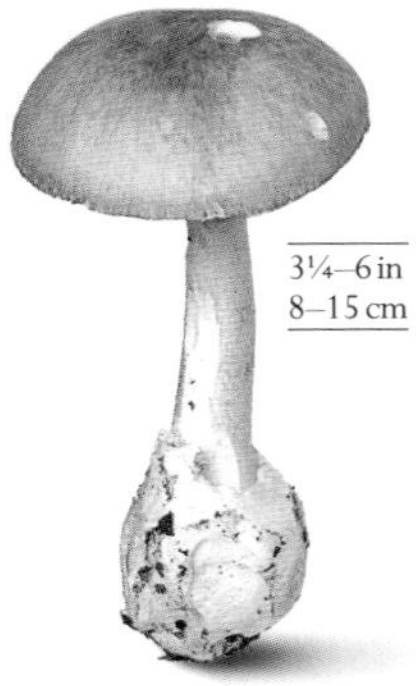

3¼–6 in
8–15 cm

DEATH CAP
Amanita phalloides
F: Amanitaceae

Common in parts of N. America and Eurasia; this species has a fragile ring, large volva, and sickly smell. Its cap may be greenish, yellowish, or white.

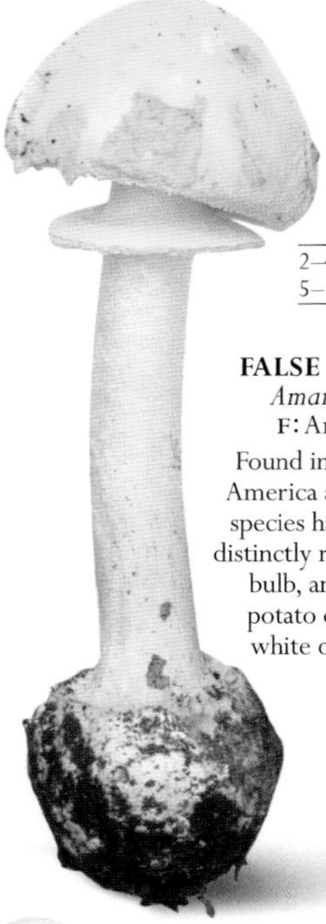

2–4 in
5–10 cm

FALSE DEATHCAP
Amanita citrina
F: Amanitaceae

Found in eastern North America and Eurasia; this species has a stem with a distinctly rimmed, rounded bulb, and flesh with a potato odor. Its cap is white or pale yellow.

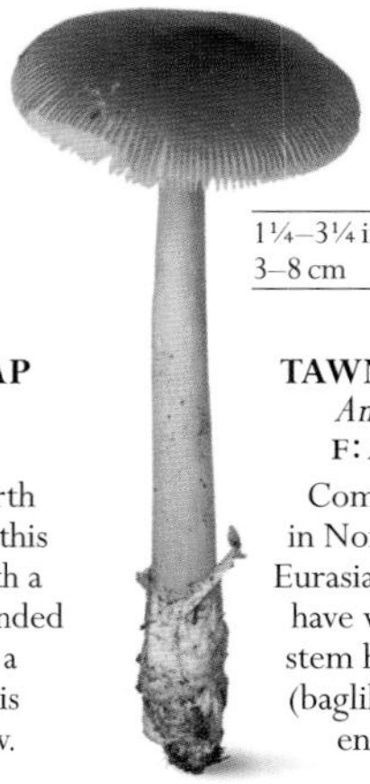

1¼–3¼ in
3–8 cm

TAWNY GRISETTE
Amanita fulva
F: Amanitaceae

Common in woods in North America and Eurasia; this fungus may have veil patches. The stem has a white volva (baglike remains of an enclosing veil).

buff warts
rounded bulb

2¼–7 in
6–18 cm

BLUSHER
Amanita rubescens
F: Amanitaceae

This species is common in mixed woods in N. America and Eurasia; Its cap varies from cream to brown, and its flesh bruises pinkish red.

2¼–6 in
6–15 cm

FLY AGARIC
Amanita muscaria
F: Amanitaceae

Common in North America and Eurasia; this distinctive mushroom is especially common under birches. Its white warts may wash off after rain.

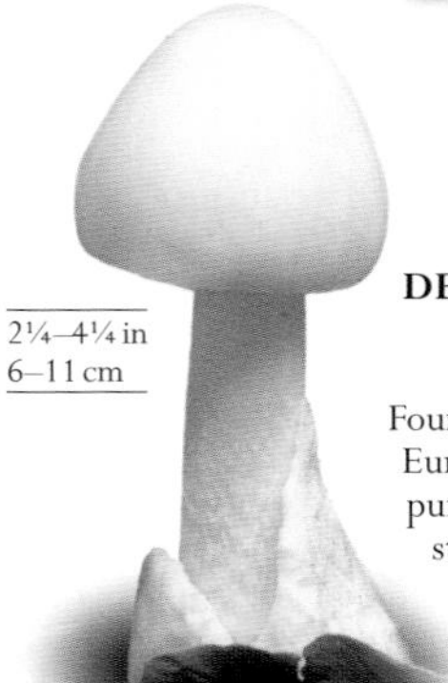

2¼–4¼ in
6–11 cm

DESTROYING ANGEL
Amanita virosa
F: Amanitaceae

Found throughout northern Eurasia; this rare fungus is pure white with a slightly sticky, bell-shaped cap and a white volva.

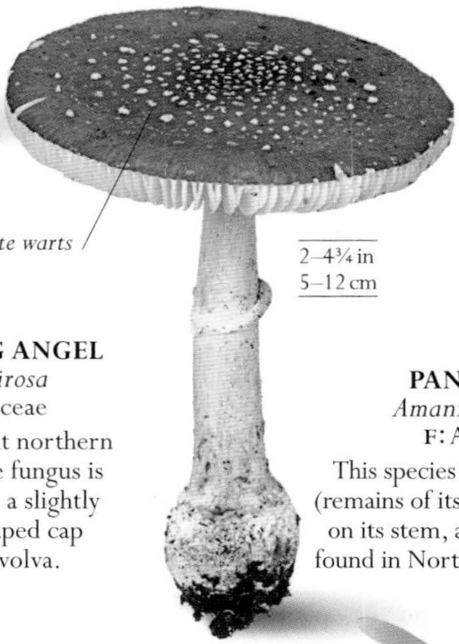
pure white warts

2–4¾ in
5–12 cm

PANTHER CAP
Amanita pantherina
F: Amanitaceae

This species has pure white warts (remains of its protective veil), a ring on its stem, and a ridged bulb. It is found in North America and Eurasia.

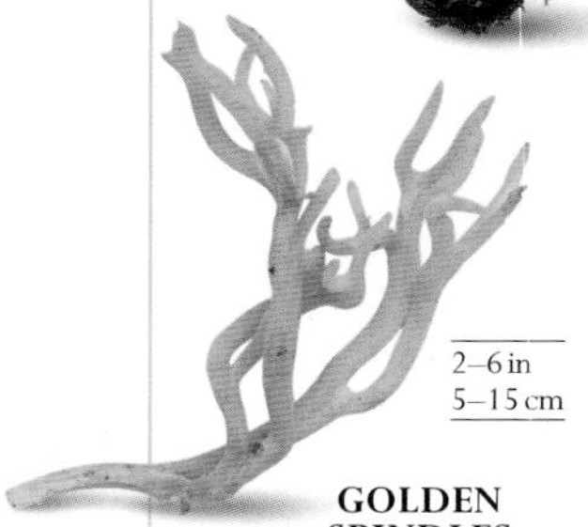

2–6 in
5–15 cm

GOLDEN SPINDLES
Clavulinopsis corniculata
F: Clavariaceae

This golden-yellow club with antlerlike branches is relatively common in acidic, unimproved meadows and grassy clearings in woods in Eurasia.

YELLOW CLUB
Clavulinopsis helvola
F: Clavariaceae

Common in grassland in North America and Eurasia; this is one of several yellow clubs. It can only be reliably identified by its spiny spores.

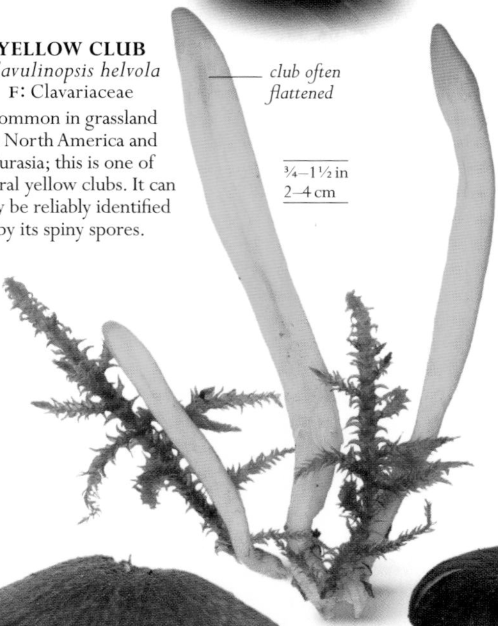
club often flattened

¾–1½ in
2–4 cm

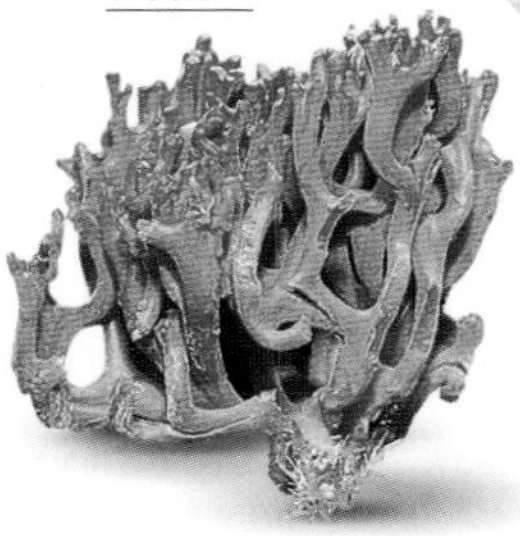

1½–3¼ in
4–8 cm

VIOLET CORAL
Clavaria zollingeri
F: Clavariaceae

A rare species found in mossy meadows and woodlands in North America and Eurasia; this species has violet, coral-like branches.

1¼–4 in
3–10 cm

WHITE SPINDLES
Clavaria fragilis
F: Clavariaceae

This fungus forms dense clusters of pure white, simple clubs in woods and meadows in Eurasia.

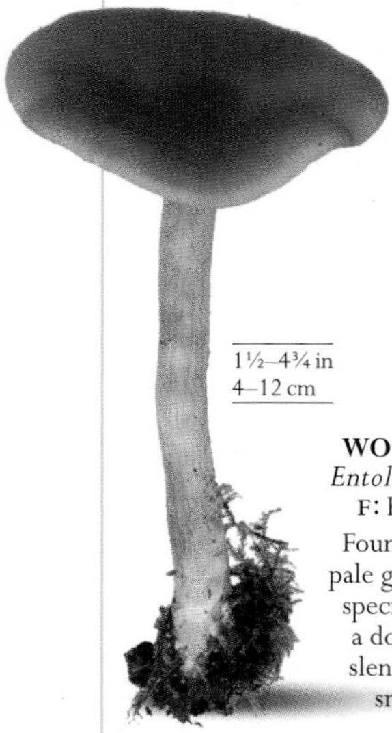

1½–4¾ in
4–12 cm

WOOD PINKGILL
Entoloma rhodopolium
F: Entolomataceae

Found in Eurasia; this pale gray to gray-brown species has a cap with a domed center and slender stem. It may smell of bleach.

⅜–1¼ in
1–3 cm

MOUSEPEE PINKGILL
Entoloma incanum
F: Entolomataceae

Smelling of mice, this pinkgill has a hollow stem and a bright grass-green cap that fades to brown with age. It grows in unimproved meadows in Eurasia.

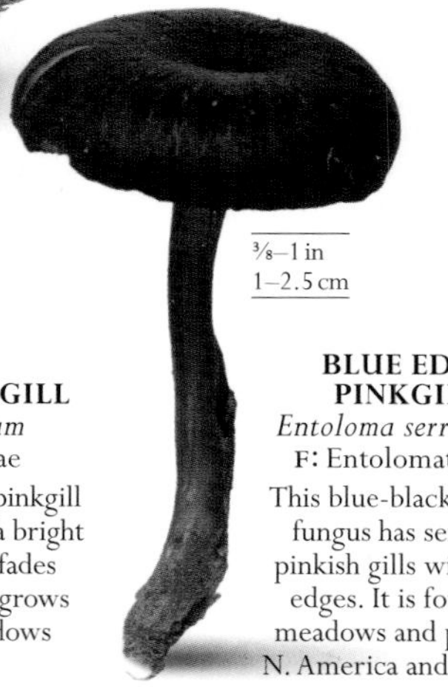

⅜–1 in
1–2.5 cm

BLUE EDGE PINKGILL
Entoloma serrulatum
F: Entolomataceae

This blue-black capped fungus has serrated pinkish gills with dark edges. It is found in meadows and parks in N. America and Eurasia.

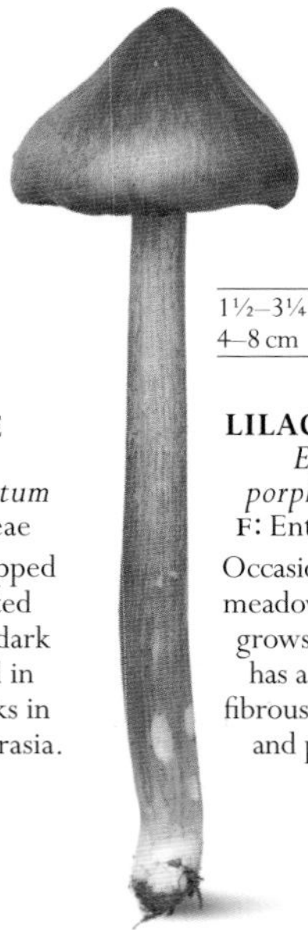

1½–3¼ in
4–8 cm

LILAC PINKGILL
Entoloma porphyrophaeum
F: Entolomataceae

Occasionally found in meadows, this fungus grows in Eurasia. It has a gray-purple fibrous cap and stem, and pinkish gills.

1¼–3½ in
3–9 cm

THE MILLER
Clitopilus prunulus
F: Entolomataceae

Common in mixed woods in North America and Eurasia; this fungus has a cap that changes shape with age, starting convex then becoming depressed.

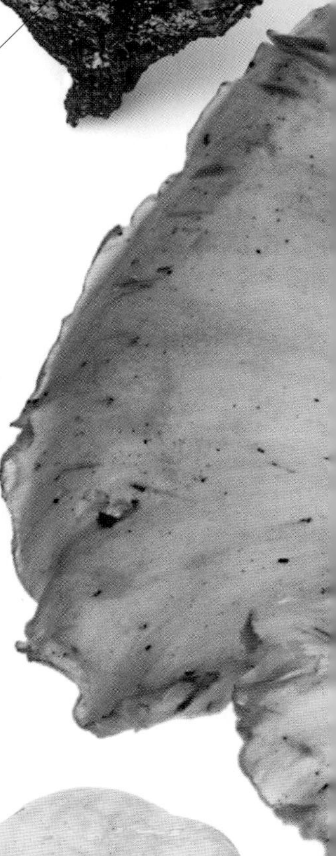

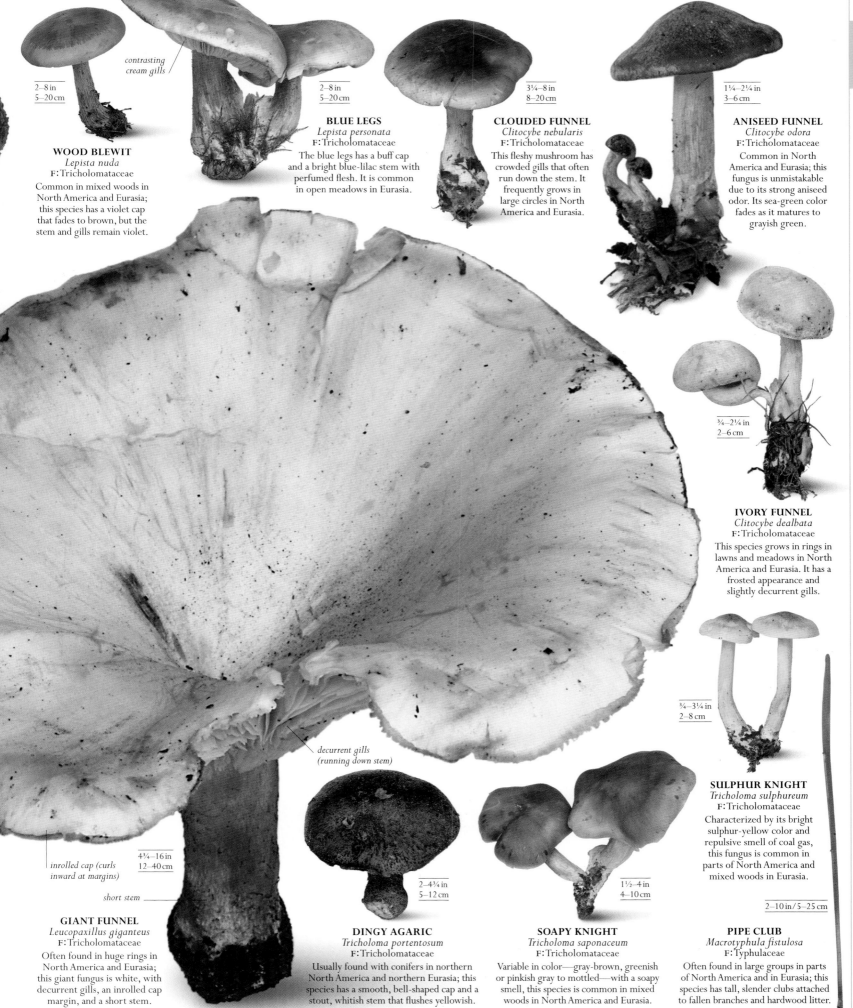

WOOD BLEWIT
Lepista nuda
F: Tricholomataceae
Common in mixed woods in North America and Eurasia; this species has a violet cap that fades to brown, but the stem and gills remain violet.

2–8 in
5–20 cm

contrasting cream gills

BLUE LEGS
Lepista personata
F: Tricholomataceae
The blue legs has a buff cap and a bright blue-lilac stem with perfumed flesh. It is common in open meadows in Eurasia.

2–8 in
5–20 cm

CLOUDED FUNNEL
Clitocybe nebularis
F: Tricholomataceae
This fleshy mushroom has crowded gills that often run down the stem. It frequently grows in large circles in North America and Eurasia.

3¼–8 in
8–20 cm

ANISEED FUNNEL
Clitocybe odora
F: Tricholomataceae
Common in North America and Eurasia; this fungus is unmistakable due to its strong aniseed odor. Its sea-green color fades as it matures to grayish green.

1¼–2¼ in
3–6 cm

IVORY FUNNEL
Clitocybe dealbata
F: Tricholomataceae
This species grows in rings in lawns and meadows in North America and Eurasia. It has a frosted appearance and slightly decurrent gills.

¾–2¼ in
2–6 cm

GIANT FUNNEL
Leucopaxillus giganteus
F: Tricholomataceae
Often found in huge rings in North America and Eurasia; this giant fungus is white, with decurrent gills, an inrolled cap margin, and a short stem.

decurrent gills (running down stem)

inrolled cap (curls inward at margins)

short stem

4¾–16 in
12–40 cm

DINGY AGARIC
Tricholoma portentosum
F: Tricholomataceae
Usually found with conifers in northern North America and northern Eurasia; this species has a smooth, bell-shaped cap and a stout, whitish stem that flushes yellowish.

2–4¾ in
5–12 cm

SOAPY KNIGHT
Tricholoma saponaceum
F: Tricholomataceae
Variable in color—gray-brown, greenish or pinkish gray to mottled—with a soapy smell, this species is common in mixed woods in North America and Eurasia.

1½–4 in
4–10 cm

SULPHUR KNIGHT
Tricholoma sulphureum
F: Tricholomataceae
Characterized by its bright sulphur-yellow color and repulsive smell of coal gas, this fungus is common in parts of North America and mixed woods in Eurasia.

¾–3¼ in
2–8 cm

PIPE CLUB
Macrotyphula fistulosa
F: Typhulaceae
Often found in large groups in parts of North America and in Eurasia; this species has tall, slender clubs attached to fallen branches and hardwood litter.

2–10 in / 5–25 cm

>>

FLY AGARIC
Amanita muscaria

The fly agaric is arguably the most famous of all fungi, and it appears in illustrations in children's books all over the world. The bright scarlet cap, usually scattered with white spots of veil tissue, make it one of the easiest fungi to identify. Found originally throughout most of Europe, northern Asia, and North America, it now grows wherever humans have planted the host trees—usually birch—with which the fungus forms a mutually beneficial relationship. It is now commonly found in parts of Africa, India, and Australasia. All parts of the fly agaric are poisonous, although rarely fatal.

large floppy ring is easily torn

SIZE Cap diameter 2¼–6 in (6–15 cm)
HABITAT Birch and pine woods
DISTRIBUTION Almost worldwide
SPORE COLOR White

yellow flesh under outer skin

gills are pure white

white warts loosely attached to cap

∨ SLICE THROUGH CAP
In cross section, the yellowish orange flesh immediately beneath the red skin of the cap is revealed. The gills, or lamellae, are bright white, in contrast.

stem flesh

crowded gills

∨ SCALES
The white or pale yellow warty scales are all that remain of an enveloping veil of tissue that once enclosed the young cap.

red skin peels off easily

< FAMILIAR FUNGUS
The fly agaric changes shape dramatically as it grows and expands, but the color and key characters usually remain constant. The cap may fade to yellow if there has been persistent rain. Its common name refers to an old folk remedy—the red skin of the cap was used in a saucer of milk to poison annoying house flies.

STEM BASE >
The base is swollen and clublike, with warty ridges encircling the upper part of the bulb. Threads of mycelium (fungal tissue) connect to fine tree roots below the soil.

∧ GILLS
The fly agaric's radiating gills are of different lengths, some stopping well short of the stem. This allows them to fill the available space, ensuring that spore production is maximized.

GROWTH STAGES OF A FLY AGARIC

veil warts

warts dispersed

partial veil beneath cap

veil starting to tear

ring now formed

fully exposed gills

upturned cap

∧ BABY BUTTON
The young button is completely enclosed in a warty universal veil.

∧ BROKEN VEIL
The stem has started to grow, and the universal veil on the cap has split apart.

∧ GROWING CAP
The cap expands, with the partial veil still attached underneath.

∧ GILLS APPEAR
The partial veil starts to tear, forming the ring and exposing the gills above.

∧ SPORE DISPERSAL
Spores grow on the exposed gills and are then released into the air.

∧ OLD AGE
At the end of its life the cap coloration fades and the cap may turn upward.

THE GYPSY
Cortinarius caperatus
F: Cortinariaceae
From North America and northern Eurasia; this species, found in conifer woods, has a frosted cap with a white veil and a stem with a distinct sheathing ring.

2–4¾ in
5–12 cm

DAPPLED WEBCAP
Cortinarius bolaris
F: Cortinariaceae
Common in beech woods in Eurasia; this fungus has tiny reddish brown scales on its cap and stem. The flesh stains yellow-orange in the stem.

1¼–2¼ in
3–6 cm

2–3¼ in
5–8 cm

raised point at cap center

2¼–4 in
6–10 cm

BLUE AND BUFF WEBCAP
Cortinarius malachius
F: Cortinariaceae
Found under pines in northern Eurasia; this uncommon species has a pale brown cap with a lilac tint, and a stem with bands of white veil.

2–4 in
5–10 cm

ELEGANT WEBCAP
Cortinarius elegantissimus
F: Cortinariaceae
Found throughout Eurasia in beech woods in chalky soil; this fungus has a vivid orange-yellow cap and a yellow stem with a large, marginate bulb.

fine scales and fibers cover cap surface

SCALY WEBCAP
Cortinarius pholideus
F: Cortinariaceae
This uncommon species is found under birches in Eurasia. It is easily recognized by its scaly, brown cap and stem, and violet gills and stem apex when young.

1¼–3¼ in
3–8 cm

PEARLY WEBCAP
Cortinarius alboviolaceus
F: Cortinariaceae
This species is common in mixed woods in North America and Eurasia. Its silvery white fruitbody has a lilac tint. The gills mature to cinnamon-brown.

scaly ring

2¼–4 in
6–10 cm

RED AND OLIVE WEBCAP
Cortinarius rufoolivaceus
F: Cortinariaceae
Found in beech woods throughout Eurasia, this uncommon species has a distinctively copper-colored cap—with a pink or olive-green margin. The bulbous stem is violet to greenish yellow.

2¼–4 in
6–10 cm

ORANGE WEBCAP
Cortinarius mucosus
F: Cortinariaceae
Restricted to North America and northern Eurasia; this species is common in pine woods. Its orange-brown cap and stout, white stem are very slimy. The gills are rust-brown.

minute scales on cap

ring zone

2¼–6 in
6–15 cm

RED BANDED WEBCAP
Cortinarius armillatus
F: Cortinariaceae
The cinnabar-red bands of veil on the club-shaped stem distinguish this common species of birch woods in North America and Eurasia.

2–4¾ in
5–12 cm

club-shaped stem

SPLENDID WEBCAP
Cortinarius splendens
F: Cortinariaceae
This rare species has a golden-yellow cap, yellow flesh, a sulphur-yellow veil, and a bulbous stem. It mainly occurs in beech woods in Eurasia.

1¼–2¾ in
3–7 cm

⅜–1¼ in
1–3 cm

1½–4 in
4–10 cm

BITTER BIGFOOT WEBCAP
Cortinarius sodagnitus
F: Cortinariaceae
Found mainly in southern England and Mediterranean Eurasia; this uncommon species occurs in beech woods on chalky soil. The bulb on the stem is large.

PIXIE WEBCAP
Cortinarius flexipes
F: Cortinariaceae
An odor of pelargoniums and delicate white hairs on the pointed cap characterize this species. It is common under birches in North America and northern Eurasia.

VIOLET WEBCAP
Cortinarius violaceus
F: Cortinariaceae
Found in mixed woodland throughout North America and Eurasia; this rare species has a distinctive intensely violet-colored cap and stem.

club-shaped stem

BIRCH WEBCAP
Cortinarius triumphans
F: Cortinariaceae
Common under birches in Eurasia; this fungus is yellow-orange overall, with prominent girdles of yellow veil on the stem.

3¼–6 in
8–15 cm

CAP (UPPER SIDE)   CAP (UNDERSIDE)

PEELING OYSTERLING
Crepidotus mollis
F: Crepidotaceae

This fungus occurs in North America and Eurasia; Its small, fan-shaped, pale cap has a gelatinous skin, which peels easily. The stem is absent or very short.

¾–2¾ in
2–7 cm

VARIABLE OYSTERLING
Crepidotus variabilis
F: Crepidotaceae

One of several similar species in North America and Eurasia; its cap is dry with fine fibers. It usually lacks a stem.

³⁄₁₆–1¼ in
0.5–3 cm

TINY PIXIE CAP
Galerina calyptrata
F: Hymenogastraceae

Found in Eurasia; this fungus is one of many species only identifiable under a microscope. Its rounded cap is orange-brown and striated.

⅛–⁵⁄₁₆ in
0.3–0.8 cm

CAP NOT FULLY EXPANDED

ROOTING POISONPIE
Hebeloma radicosum
F: Hymenogastraceae

The deeply rooting stem with a large ring, buff cap with flat scales, and strong smell of marzipan characterize this Eurasian species.

2–4¾ in
5–12 cm

attached to wood

DEADLY GALERINA
Galerina marginata
F: Hymenogastraceae

This species has an orange-brown cap and a small ring on the stem. It grows on fallen wood in North America and Eurasia.

⅜–2 in
1–5 cm

POISONPIE
Hebeloma crustuliniforme
F: Hymenogastraceae

From North America and Eurasia; this species smells strongly of radish. Its cap is ivory-white to buff, and slimy when wet. In damp weather, the gills exude droplets.

1½–3½ in
4–9 cm

SPLIT FIBERCAP
Inocybe rimosa
F: Inocybaceae

Common in mixed woodland in North America and Eurasia; its pointed fibrous cap is straw-yellow and the stem is tall and slender.

1¼–2¾ in
3–7 cm

red band

STAR FIBERCAP
Inocybe asterospora
F: Inocybaceae

A flattened bulb at the stem's base and star-shaped spores distinguish this small, brown fibrous-capped species from Eurasia.

1¼–2¾ in
3–7 cm

radial fibers

WHITE FIBERCAP
Inocybe geophylla
F: Inocybaceae

This is one of the most common *Inocybe* species in woods in North America and Eurasia; It has a conical, silky cap and a slender stem.

⅜–1½ in
1–4 cm

WHITE FORM

VIOLET FORM

TORN FIBERCAP
Inocybe lacera
F: Inocybaceae

Found in North America and Eurasia; this is distinguished by its cylindrical spores. It has a scaly fibrous cap and a slender brown stem.

⅜–1¾ in
1–4.5 cm

stout stem

DEADLY FIBERCAP
Inocybe erubescens
F: Inocybaceae

Found in mixed woodland in chalk in Eurasia; this uncommon species has a fibrous cap that discolors with age, and stout stem that bruises red.

1¼–3½ in
3–9 cm

LILAC LEG FIBERCAP
Inocybe griseolilacina
F: Inocybaceae

A common species in beech woods in Eurasia; it has a scaly cap and a pale violet, slender stem.

⁵⁄₁₆–1½ in
0.8–4 cm

³⁄₈–1½ in
1–4 cm

green stem apex

SCARLET WAXCAP
Hygrocybe coccinea
F: Hygrophoraceae
The scarlet, waxy cap, gills, and stem make this a very striking fungus of unimproved grasslands in North America and Eurasia.

½–2¼ in
1.5–6 cm

SNOWY WAXCAP
Hygrocybe virginea
F: Hygrophoraceae
The most common waxcap in all types of grassland in Eurasia; its waxy cap and slender stem are translucent. It has decurrent gills.

½–2 in
1.5–5 cm

PARROT WAXCAP
Hygrocybe psittacina
F: Hygrophoraceae
This fungus occurs in North America and Eurasia; its slimy cap is vivid green to orange and the upper part of the stem is bright green.

fleshy, waxy, pink gills

1¼–2¾ in
3–7 cm

PINK WAXCAP
Hygrocybe calyptriformis
F: Hygrophoraceae
This highly recognizable but rare fungus has a pointed pink cap, fragile pale stem, and waxy gills. Found in Eurasia, it grows in unimproved meadows.

³⁄₈–2 in
1–5 cm

BLACKENING WAXCAP
Hygrocybe conica
F: Hygrophoraceae
Very common in grasslands and woodlands in N. America and Eurasia; this species has a conical reddish orange cap and fibrous stem that blacken with age or bruising.

1–2¼ in
2.5–6 cm

gills are broadly attached

MEADOW WAXCAP
Hygrocybe pratensis
F: Hygrophoraceae
This is one of the larger meadow-occurring waxcaps, and is found in North America and Eurasia. It has a stout stem and a fleshy cap with gills running down the stem.

1½–4¾ in
4–12 cm

pale yellow margin

1½–3¼ in
4–8 cm

CLUB FOOT
Ampulloclitocybe clavipes
F: Hygrophoraceae
Common in mixed woods in late fall, this species from North America and Eurasia has decurrent gills and a swollen, spongy base.

½–2¾ in
1.5–7 cm

GOLDEN WAXCAP
Hygrocybe chlorophana
F: Hygrophoraceae
Found in Eurasia; this is the most common waxcap found in meadows. It has a bright yellow-orange, slightly sticky cap.

CRIMSON WAXCAP
Hygrocybe punicea
F: Hygrophoraceae
Found in unimproved grassland in North America and Eurasia; this uncommon species is the largest blood-red waxcap. Its stem is dry and fibrous with a white base.

IVORY WOODWAX
Hygrophorus eburneus
F: Hygrophoraceae
This species is found in beech woods in North America and Eurasia; it has a sticky cap and stem, thick gills, and a flowery smell.

1¼–3¼ in
3–8 cm

¾–2 in
2–5 cm

³⁄₈–2 in
1–5 cm

AMETHYST DECEIVER
Laccaria amethystina
F: Hydnangiaceae
This highly recognizable, slender fungus from North America and Eurasia has an intense violet coloration when fresh, and powdery spores dusting the gills.

CONIFERCONE CAP
Baeospora myosura
F: Cyphellaceae
Found in North America and Eurasia; this is one of a few fungi growing mostly on pinecones. It has crowded, narrow gills and a velvety stem.

³⁄₁₆–¾ in
0.5–2 cm

pinecone

1¼–2 in
3–5 cm

DECEIVER
Laccaria laccata
F: Hydnangiaceae
Variable color—from brick-red to flesh-pink—with a dry cap and thick gills; this species is abundant in temperate woods in North America and Eurasia.

HERALD OF WINTER
Hygrophorus hypothejus
F: Hygrophoraceae
Appearing in pine woods after a frost in North America and Eurasia; this sticky fungus has an olive-brown cap and yellow stem.

cap frequently splits as it ages

ST. GEORGE'S MUSHROOM
Calocybe gambosa
F: Lyophyllaceae
This ivory-buff species from Eurasia often grows in fairy rings at woodland margins in late spring. The mushroom smells strongly of newly ground meal.

¾–1½ in
1–4 cm

3/16–½ in
0.5–1.5 cm

1¼–4¾ in
3–12 cm

PINK DOMECAP
Calocybe carnea
F: Lyophyllaceae
This species grows in short grass in North America and Eurasia; its smooth cap and fibrous stem are rose-pink and the gills are white.

2–4 in
5–10 cm

SILKY PIGGYBACK
Asterophora parasitica
F: Lyophyllaceae
One of two well-known parasitic species to live on rotten fruitbodies of brittle-caps, this fungus from N. America and Eurasia has a rounded silky cap.

CLUSTERED DOMECAP
Lyophyllum decastes
F: Lyophyllaceae
Common on roadsides, pathsides, and disturbed soil in North America and Eurasia; this fungus forms clumps of tough caps with stout stems.

HORSEHAIR PARACHUTE
Marasmius androsaceus
F: Marasmiaceae
Found in Eurasia; this fungus has a distinctive black hairlike stem, while its radially grooved cap is pale pinkish brown.

⅛–⅜ in
0.3–1 cm

HAIRY PARACHUTE
Crinipellis scabella
F: Marasmiaceae
Found on dead stems of grasses, this recognizable fungus is from Eurasia. It has a small cap and a long, slender stem covered with dense, bristling brown hair.

3/16–½ in
0.5–1.5 cm

3/16–⅜ in
0.5–1 cm

cap depressed at center

radial groove

FAIRY RING CHAMPIGNON
Marasmius oreades
F: Marasmiaceae
This classic fairy-ring mushroom is common in open grass in North America and Eurasia; it has fleshy, beige caps with thick gills.

⅜–2 in
1–5 cm

3/16–¾ in
0.5–2 cm

COMMON BIRD'S NEST
Crucibulum laeve
F: Nidulariaceae
Forming miniature nests with spore-filled, egglike cases, this common species may be hard to spot in woody debris. It occurs in North America and Eurasia.

3/16–⅜ in
0.5–1 cm

wiry stem

GARLIC PARACHUTE
Marasmius alliaceus
F: Marasmiaceae
This species grows in beech woods in Eurasia; it has a tall, slender, blackish stem and a strong odor of rancid garlic.

½–1½ in
1.5–4 cm

COLLARED PARACHUTE
Marasmius rotula
F: Marasmiaceae
Like little parachutes, this mushroom's rounded, depressed caps are radially grooved. Its tough stems are attached to woody debris. It occurs in North America and Eurasia.

FLUTED BIRD'S NEST
Cyathus striatus
F: Nidulariaceae
This fungus is recognized by its hairy, brown, and tall fluted nests containing 10–15 egglike cases. It occurs in woody debris across North America and Eurasia, but is not common.

»

SILVERLEAF FUNGUS
Chondrostereum purpureum
F: Cyphellaceae
Common on cherry and plum trees, it is found in North America and Eurasia. Violet below when young, it darkens to purple-brown with age.

¾–2 in
2–5 cm

wavy margin

purple underside

SAFFRONDROP BONNET
Mycena crocata
F: Mycenaceae
Common in woods with chalky soils in Eurasia; this species exudes a bright saffron-orange juice when it is broken.

⅜–1¼ in
1–3 cm

COMMON BONNET
Mycena galericulata
F: Mycenaceae
Abundant in temperate woods in North America and Eurasia; this fungus is variable in color. Its pinkish gray gills have linking veins and cross-ridges.

⅜–2¼ in
1–6 cm

⅛–⅜ in
0.3–1 cm

ORANGE BONNET
Mycena acicula
F: Mycenaceae
Common in leaf litter and debris in broadleaf woods in North America and Eurasia; this fungus has a tiny, translucent cap that is striped almost to the center.

³⁄₁₆–1 in
0.5–2.5 cm

YELLOWLEG BONNET
Mycena epipterygia
F: Mycenaceae
Occurring in North America and Eurasia; this fungus grows in acidic soils in woods and meadows. Its cap and stem have a sticky, peelable layer.

⅜–1½ in
1–4 cm

domed cap center

1¼–2¼ in
3–6 cm

BLACKEDGE BONNET
Mycena pelianthina
F: Mycenaceae
Found in Eurasia; this fungus is characterized by a strong smell of radish. It has purplish black-edged gills and a pale lilac to gray-brown cap.

CLUSTERED BONNET
Mycena inclinata
F: Mycenaceae
Found in dense clusters on wood in North America and Eurasia; this fungus has a cap with toothed margins and a strong soapy smell.

ROSY BONNET
Mycena rosea
F: Mycenaceae
Found in beech woods in Eurasia; this common species has a robust pink cap and stem, and smells strongly of radish.

¾–2¼ in
2–6 cm

OLIVE OYSTERLING
Panellus serotinus
F: Mycenaceae
From North America and Eurasia; this mushroom fruits in winter, often near water on hardwood tree trunks. Its cap is slimy when wet.

1¼–4 in
3–10 cm

club-shaped stem

1½–3¼ in
4–8 cm

SPINDLE TOUGHSHANK
Gymnopus fusipes
F: Omphalotaceae
Abundant in Eurasian oak woods from early summer onward; this fungus forms large clumps of tough fruitbodies on tree roots.

1–2¼ in
2.5–6 cm

WOOD WOOLLYFOOT
Gymnopus peronatus
F: Omphalotaceae
The common name of this species from Eurasia refers to the stiff, fuzzy hairs at the base of the stem.

2–6 in
5–15 cm

1½–4 in
4–10 cm

SPOTTED TOUGHSHANK
Rhodocollybia maculata
F: Omphalotaceae
Common in mixed woods across North America and Eurasia; this fungus has a white cap, stem, and gills. The crowded gills stain rust-red with age.

1¼–2¼ in
3–6 cm

BUTTER CAP
Rhodocollybia butyracea
F: Omphalotaceae
This fungus is abundant in woods in North America and Eurasia. Varying from blackish or reddish brown to dark ocher, its cap is oily to the touch.

JACK O'LANTERN
Omphalotus illudens
F: Omphalotaceae
Found across North America and Eurasia; this bright orange, poisonous fungus is famous for gills that glow in the dark with an eerie greenish light.

HONEY FUNGUS
Armillaria gallica
F: Physalacriaceae
Found in woods across Eurasia; this species is usually found on the ground. It is a weak parasite on the trees around which it grows.

1¼–4 in
3–10 cm

1–4 in
2.5–10 cm

WRINKLED PEACH
Rhodotus palmatus
F: Physalacriaceae
Found on fallen logs of mainly elm, this uncommon species of North America and Eurasia has an unusual, peach-pink, wrinkled cap and a fruity odor.

⅜–2¼ in
1–6 cm

VELVET SHANK
Flammulina velutipes
F: Physalacriaceae
Growing throughout winter in North America and Eurasia; this fungus is characterized by its often sticky cap and velvety stem.

¾–6 in
2–15 cm

PORCELAIN FUNGUS
Oudemansiella mucida
F: Physalacriaceae
Usually found on beech logs in Eurasia; this fungus has gray-white caps that are slimy when wet. The tough stem has a thin ring.

1–4 in
2.5–10 cm

ROOTING SHANK
Xerula radicata
F: Physalacriaceae
Found across North America and Eurasia; this deep-rooted fungus has a stiff, tall stem. The cap is slimy when wet, and the gills are widely spaced.

BRANCHING OYSTER
Pleurotus cornucopiae
F: Pleurotaceae
Growing in clusters on fallen logs, usually of elm, this fungus has trumpet-shaped caps with gills running down the short, frequently branching stems. It occurs in North America and Eurasia.

1½–4¾ in
4–12 cm

caps often overlap

2¼–8 in
6–20 cm

tiny glistening granules

OYSTER MUSHROOM
Pleurotus ostreatus
F: Pleurotaceae
From North America and Eurasia; this species is found on trees and logs. Its shelflike caps vary from blue-green to pale buff. The stem is almost absent.

white veil breaks into patches

GLISTENING INKCAP
Coprinellus micaceus
F: Psathyrellaceae
Usually found in clusters, this mushroom's rounded, grooved caps are dusted with micalike flakes of veil. It is common on wood in North America and Eurasia.

¾–1¼ in
2–3 cm

3/16–⅜ in
0.5–1 cm

1–3 in
2.5–7.5 cm

FAIRIES' BONNETS
Coprinellus disseminatus
F: Psathyrellaceae
Often found in clusters on rotted stumps in North America and Eurasia; it has tiny umbrella-like caps with deeply fluted grooves. The gills are blackish when mature.

WEEPING WIDOW
Lacrymaria lacrymabunda
F: Psathyrellaceae
Common on pathsides and disturbed soil in North America and Eurasia; it is named for its black gills that weep droplets from their edges.

½–2¾ in
1.5–7 cm

5/16–1½ in
0.8–4 cm

1–3¼ in
2.5–8 cm

2–3¼ in
5–8 cm

PALE BRITTLESTEM
Psathyrella candolleana
F: Psathyrellaceae
Usually found in clusters in woody debris in early summer; it has fragile, slender stems and pale or buff caps. It occurs in North America and Eurasia.

CLUSTERED BRITTLESTEM
Psathyrella multipedata
F: Psathyrellaceae
The densely clustered stems of this species are joined at the base. It is usually found in open grass across Eurasia.

COMMON INKCAP
Coprinopsis atramentaria
F: Psathyrellaceae
Occurring in North America and Eurasia; this species has egg-shaped caps that dissolve to a blackish fluid as the spores are released.

MAGPIE INKCAP
Coprinopsis picacea
F: Psathyrellaceae
Found on chalky soils in Eurasian woodlands; this uncommon species has white, fluffy scales that contrast with the dark gray-brown cap.

»

1½–4 in
4–10 cm

DEER SHIELD
Pluteus cervinus
F: Pluteaceae
Found in North America and Eurasia; this species is variable in color, with a radially fibrous cap, often dome-centered; pink gills not attached to the stem; and a fibrous stem.

⅜–2¼ in
1–6 cm

YELLOW SHIELD
Pluteus chrysophaeus
F: Pluteaceae
A golden to greenish yellow cap, yellow gills turning pink, and a whitish stem characterize this species. Found in Eurasia, it grows on decayed wood.

1–3¼ in
2.5–8 cm

WILLOW SHIELD
Pluteus salicinus
F: Pluteaceae
Common in broad-leafed woodlands in Eurasia; this species can be identified by the bluish gray staining at the base of its slender stem.

4–10 in
10–25 cm

BEEFSTEAK FUNGUS
Fistulina hepatica
F: Schizophyllaceae
Resembling a fleshy steak, with bloodlike red juice, it is particularly common in warmer parts of North America and Eurasia.

fine silky-hairy surface

4–10 in
10–25 cm

SILKY ROSEGILL
Volvariella bombycina
F: Pluteaceae
Growing on deciduous trees in North America and Eurasia; this rare species has a white to pale lemon cap and a whitish stem with a baglike thin veil enclosing the base of the stem.

⅜–2 in
1–5 cm

CAP (UNDERSIDE)

COMMON PORECRUST
Schizophyllum commune
F: Schizophyllaceae
Found in North America and Eurasia; this fan-shaped fungus is identified by the gill-like spore surface on the underside of the cap.

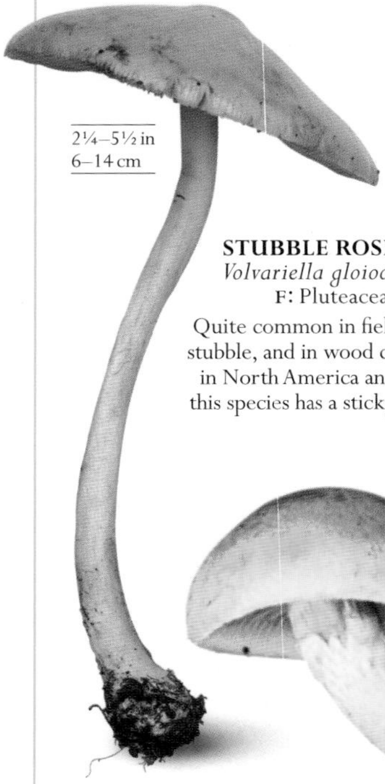

CAP
(UPPER SIDE)

greenish yellow gills

orange cap center

2¼–5½ in
6–14 cm

STUBBLE ROSEGILL
Volvariella gloiocephala
F: Pluteaceae
Quite common in fields and old stubble, and in wood chip mulch, in North America and Eurasia; this species has a sticky, gray cap.

1¼–2¾ in
3–7 cm

CONIFER TUFT
Hypholoma capnoides
F: Strophariaceae
Found in North America and Eurasia; this uncommon species occurs on conifer wood. Its whitish gills mature to grayish lilac.

1¼–2¾ in
3–7 cm

SULPHUR TUFT
Hypholoma fasciculare
F: Strophariaceae
Abundant in temperate woods in North America and Eurasia; this species has greenish yellow gills that mature to dark purple, and are easy to recognize in the field.

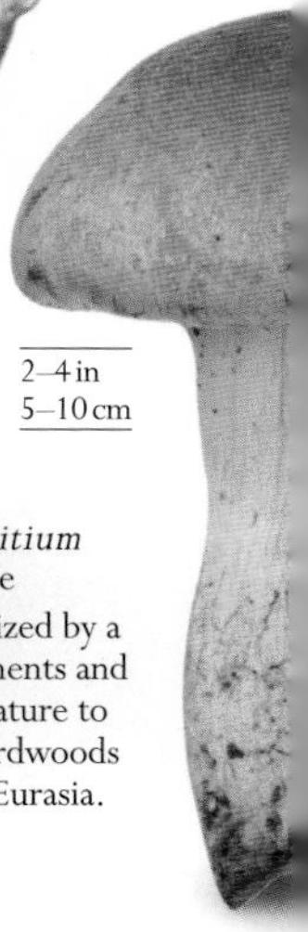

2–4 in
5–10 cm

BRICK TUFT
Hypholoma sublateritium
F: Strophariaceae
This species is characterized by a fleshy cap with veil fragments and pale yellow gills that mature to lavender. It grows on hardwoods in North America and Eurasia.

SHEATHED WOODTUFT
Kuehneromyces mutabilis
F: Strophariaceae
Often confused with the deadly *Galerina marginata*, this species can be identified by its sticky cap, scaly stem, and brown gills. It occurs in North America and Eurasia.

¾–2¾ in
2–7 cm

ring on stem

½–2¼ in
1.5–6 cm

REDLEAD ROUNDHEAD
Leratiomyces ceres
F: Strophariaceae
Previously known as *Stropharia aurantiaca*, this species has a bright red cap and reddish flushed stem. It grows among wood chips in North America and Eurasia.

ALDER SCALYCAP
Pholiota alnicola
F: Strophariaceae
Found in Eurasia; this species is characterized by a sticky cap and tufted habit. Despite its common name, it is usually seen at the base of birches.

1¼–2¾ in
3–7 cm

1¼–4¾ in
3–12 cm

GOLDEN SCALYCAP
Pholiota aurivella
F: Strophariaceae
This fairly common species is found on beech logs or trunks in North America and Eurasia. Its sticky golden caps have dark orange-brown scales.

2–6 in
5–15 cm

DUNG ROUNDHEAD
Stropharia semiglobata
F: Strophariaceae
Occurring in animal dung or in grass where animals graze; it is found in North America and Eurasia. It has a sticky cap and a slender stem with a ring.

³⁄₁₆–1½ in
0.5–4 cm

SHAGGY SCALYCAP
Pholiota squarrosa
F: Strophariaceae
Growing in North America and Eurasia; this fungus has a distinctive dry, sharply scaly cap and stem; pale yellow gills; and a smell of corn or radish.

scaly, furry surface

LIBERTY CAP
Psilocybe semilanceata
F: Strophariaceae
This well-known agaric has a conical, dull yellow cap with a distinct point. Common in meadows in late fall, it occurs in North America and Eurasia.

³⁄₁₆–¾ in
0.5–2 cm

BLUE-GREEN SLIMEHEAD
Stropharia cyanea
F: Strophariaceae
This fungus has a blue-green cap that fades to yellow. It is found in North America and Eurasia.

1¼–2¾ in
3–7 cm

cap margin with veil remnants

2¼–6 in
6–15 cm

1½–2¾ in
4–7 cm

WHITELACED SHANK
Megacollybia platyphylla
F: Incertae sedis
Found in North America and Eurasia; this species has a radially fibrous, pale gray-brown cap; widely spaced, deep gills; and a stem base with rootlike strands.

COMMON CAVALIER
Melanoleuca polioleuca
F: Incertae sedis
Common in grass in Eurasia; this species has a gray-brown cap with white gills. Its stem has blackish flesh at the base.

PLUMS AND CUSTARD
Tricholomopsis rutilans
F: Incertae sedis
Common on pine stumps; this species has a reddish purple cap and stem, and golden-yellow gills. It grows in North America and Eurasia.

2–6 in
5–15 cm

SPECTACULAR RUSTGILL
Gymnopilus junonius
F: Incertae sedis
Occurring in Eurasia; this fungus is found in clumps, usually at the base of a tree. It has a dry cap with crowded, shallow yellowish gills.

³⁄₈–1½ in
1–4 cm

PETTICOAT MOTTLEGILL
Panaeolus papilionaceus
F: Incertae sedis
The minute toothed remains of the veil at the cap margin and mottled black gills characterize this species. It is found in North America and Eurasia.

root attaches to conifer wood

2–4 in
5–10 cm

2–8 in
5–20 cm

³⁄₈–2¼ in
1–6 cm

EGGHEAD MOTTLEGILL
Panaeolus semiovatus
F: Incertae sedis
Found in animal dung, this species occurs in North America and Eurasia; it is characterized by a sticky gray cap and a ring around its tall stem.

1¼–2¾ in
3–7 cm

GOBLET
Pseudoclitocybe cyathiformis
F: Incertae sedis
This distinctive species has a very dark coloration and a tall, fibrous stem. Found in Eurasia, it is common in late fall and winter.

¾–3¼ in
2–8 cm

CINNABAR POWDERCAP
Cystodermella cinnabarina
F: Incertae sedis
Found in North America and Eurasia; this species has a brick-red cap and pale cream gills. The cap and stem have a granular surface.

TROOPING FUNNEL
Infundibulicybe geotropa
F: Incertae sedis
This fungus has a funnel-shaped, fleshy cap; a tall stem; and a pale leather-brown color. It occurs in Eurasia.

BOLETALES

This order contains fleshy fungi and includes those with both pored and gilled fruitbodies, most with cap and stem, but some are crusts, puffballs, or trufflelike. The majority live in association with trees (mycorrhizal), but some feed on dead wood and cause a brown rot, while others are parasitic. The spore-producing layer—the hymenium—is easily loosened from the flesh.

orange-brown cap

1½–6 in
4–15 cm

cylindrical stem

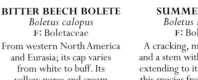

2¼–5½ in
6–14 cm

2¾–6 in
7–15 cm

BAY BOLETE
Boletus badius
F: Boletaceae
A common species found among conifers or beech trees in North America and Eurasia; its color varies from orange-brown to bay.

BITTER BEECH BOLETE
Boletus calopus
F: Boletaceae
From western North America and Eurasia; its cap varies from white to buff. Its yellow pores and cream flesh bruise blue.

SUMMER BOLETE
Boletus reticulatus
F: Boletaceae
A cracking, matte, brown cap and a stem with a fine white net extending to its base distinguish this species from eastern North America and Eurasia.

wrinkled cap surface

4–10 in
10–25 cm

PENNY BUN
Boletus edulis
F: Boletaceae
Found worldwide; this species is identified by the fine white markings on its stem, the unchanging cream flesh, and the white pores that mature to yellow.

PARASITIC BOLETE
Boletus parasiticus
F: Boletaceae
This small mushroom from North America and Eurasia grows only on the common earthball, causing its host to become hollow.

sticky cap

black scales on face

1¼–2 in
3–5 cm

2¼–6 in
6–15 cm

¾–2¾ in
2–7 cm

PEPPERY BOLETE
Chalciporus piperatus
F: Boletaceae
Common under conifers and also with fly agaric under birches in North America and Eurasia; it has cinnamon pores and yellow flesh.

BITTER BOLETE
Tylopilus felleus
F: Boletaceae
This species is found in North America and Eurasia; characterized by pores that turn pink with age and a strongly netted stem.

2–4 in
5–10 cm

common earthball

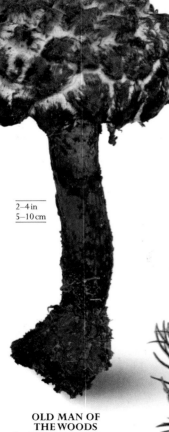

OLD MAN OF THE WOODS
Strobilomyces strobilaceus
F: Boletaceae
Found in North America and Eurasia; this rare species is unmistakable with black woolly scales on its cap and stem, and white tubes.

3¼–6 in
8–15 cm

ORANGE BIRCH BOLETE
Leccinum versipelle
F: Boletaceae
Found in Eurasia; this species has a yellow-orange fleshy cap. The stem has black, woolly flecks and the flesh stains lilac-black.

2¼–6 in
6–15 cm

BROWN BIRCH BOLETE
Leccinum scabrum
F: Boletaceae
One of many similar species, this fungus is from North America and Eurasia. Its flesh may flush pink when cut, and the cap is sticky when wet.

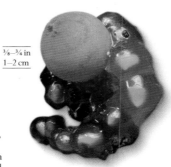

⅜–¾ in
1–2 cm

STALKED PUFFBALL-IN-ASPIC
Calostoma cinnabarinum
F: Calostomataceae
From North America; this species' bright cinnabar-red ball on a stem emerges from a gelatinous layer.

2–39 in
5–100 cm

WET ROT
Coniophora puteana
F: Coniophoraceae
Found worldwide; this fungus forms a brown sheet of tissue, often warty or wrinkled, on wet wood. It causes serious damage to buildings.

2–3½ in
5–9 cm

BAROMETER EARTHSTAR
Astraeus hygrometricus
F: Diplocystidiaceae
Frequent in North America and Eurasia; its starlike arms peel back to reveal the inner spore-filled ball, but close again in dry weather.

CORNFLOWER BOLETE
Gyroporus cyanescens
F: Gyroporaceae
Growing in acidic soil in eastern North America and Eurasia; this uncommon species is distinguished by its brittle, hollow stem.

2–3¼ in
5–8 cm

ROSY SPIKE
Gomphidius roseus
F: Gomphidiaceae
Growing in association with the bolete *Suillus bovinus* under pines in Eurasia; this fungus has a slimy, rose-pink cap with grayish gills.

½–2 in
1.5–5 cm

COPPER SPIKE
Chroogomphus rutilus
F: Gomphidiaceae
Common under pines in western North America and Eurasia; this species has a pointed, coppery brown cap.

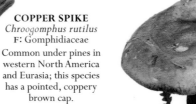

2¼–6 in
6–15 cm

BROWN ROLLRIM
Paxillus involutus
F: Paxillaceae
Common in mixed woods in North America and Eurasia; the downy, inrolled cap margin and soft yellow gills staining brown characterize this species.

1½–3¼ in
4–8 cm

¾–3¼ in
2–8 cm

FALSE CHANTERELLE
Hygrophoropsis aurantiaca
F: Hygrophoropsidaceae
Sometimes mistaken for a chanterelle, this species from North America and Eurasia is distinguished by its crowded, multiforked, soft gills.

1½–4 in
4–10 cm

COMMON EARTHBALL
Scleroderma citrinum
F: Sclerodermataceae
This potatolike fungus is common in damp woodland in North America and Eurasia; its thick skin has dark scales and a spore-filled black interior.

¾–2 in
2–5 cm

POTATO EARTHBALL
Scleroderma bovista
F: Sclerodermataceae
Common in woods in North America and Eurasia; its smooth outer skin cracks into a fine mosaic. The inner purple-black spore-mass dries brownish.

DYEBALL
Pisolithus arhizus
F: Sclerodermataceae
This species is found worldwide in poor, sandy soils associated with pines. The egglike spore sacs in its interior are embedded in blackish jelly.

2–4 in
5–10 cm

4–12 in
10–30 cm

VELVET ROLLRIM
Tapinella atrotomentosa
F: Tapinellaceae
Common on pine stumps in North America and Eurasia; it has a cap with an inrolled margin; soft, thick gills; and a velvety stem.

cap skin peels off

1½–4 in
4–10 cm

dots can weep milky latex

WEEPING BOLETE
Suillus granulatus
F: Suillaceae
Common under pines in North America and Eurasia; its ringless stem is covered with glandular dots.

2–4 in
5–10 cm

SLIPPERY JACK
Suillus luteus
F: Suillaceae
Found on pines in North America and Eurasia; this fungus has a slimy cap and yellow pores. The large ring on its stem has a lavender underside.

1¼–2¾ in
3–7 cm

BOVINE BOLETE
Suillus bovinus
F: Suillaceae
Common under pines in North America and Eurasia; this species has a sticky cap and angular, irregular pores.

yellow-brown cap surface is dry and rough

2–4 in
5–10 cm

LARCH BOLETE
Suillus grevillei
F: Suillaceae
Confined to larch woods in North America and Eurasia; this species is a yellow-orange to brick-red color. The stem has an ascending ring.

2¾–5 in
7–13 cm

VELVET BOLETE
Suillus variegatus
F: Suillaceae
Frequent under pines in Eurasia; this species has a felty-scaly cap, with dark cinnamon-brown pores, and a ringless stem.

CANTHARELLALES

The species of the Cantharellales order may look like agarics (members of the Agaricales order), but differ in several important respects. They may have fleshy fruitbodies with a cap and stem, but they lack true gills, having instead a smooth, wrinkled, or folded gill-like spore-producing surface on the underside. The spores are smooth and usually white to cream.

CRESTED CORAL
Clavulina coralloides
F: Clavulinaceae
Very common in woods in North America and Eurasia; this fungus forms a coral-like mass of white branches, each dividing into fine points.

1¼–3¼ in
3–8 cm

⅜–2¼ in
1–6 cm

TRUMPET CHANTERELLE
Craterellus tubaeformis
F: Cantharellaceae
Often found in large troops in mixed woods in North America and Eurasia; this fungus has a variable color and shallow, blunt veins rather than gills.

³⁄₁₆–¾ in
0.5–2 cm

HORN OF PLENTY
Craterellus cornucopioides
F: Cantharellaceae
A distinctive species found throughout Eurasia; this thin trumpet grows in troops in beech leaf litter. It produces a white spore deposit.

cap often depressed at center

2–6 in
5–15 cm

inrolled margin

¾–4¾ in
2–12 cm

CHANTERELLE
Cantharellus cibarius
F: Cantharellaceae
Found in North America and Eurasia; this species has blunt-edged gills with numerous cross-veins and a smell of apricots.

tapered stem

WOOD HEDGEHOG
Hydnum repandum
F: Hydnaceae
Occurring in North America and Eurasia; this dusky orange species is irregular in shape, with tiny spines underneath the cap.

irregular, often lobed cap

GEASTRALES

The Geastrales, or earthstars, share the common character of a thick outer layer—the peridium—which splits apart and peels back to form starlike arms. These reveal a central spore sac, like a puffball, from which the dark brown, warted spores exit via a pore at the apex. Earthstars are found in leaf litter and bare, sandy soil—for example, in sand dunes.

STRIATE EARTHSTAR
Geastrum striatum
F: Geastraceae
Found in Eurasia; this is one of the smaller earthstars. Its pale gray spore sac is stalked and has a sharply pointed, striated opening.

1¼–2½ in
3–6.5 cm

collar below spore sac

COLLARED EARTHSTAR
Geastrum triplex
F: Geastraceae
Growing in North America and Eurasia; this is the most common earthstar. Its rays crack to leave a cuplike collar around the spore sac.

1½–4¾ in
4–12 cm

1¼–2¼ in
3–6 cm

2–3¼ in
5–8 cm

2¾–6 in
7–15 cm

SESSILE EARTHSTAR
Geastrum fimbriatum
F: Geastraceae
Found in North America and Eurasia; this fungus's globe-shaped, pale brown fruitbody splits open with 5–9 arms. Its grayish spore sac has a fringed opening.

ARCHED EARTHSTAR
Geastrum fornicatum
F: Geastraceae
This earthstar's arms stay attached to a disc of tissue in the soil, and its stalked spore sac has a prominent opening. It occurs in North America and Eurasia.

PEPPER POT
Myriostoma coliforme
F: Geastraceae
Found in dry, sandy soils in North America and Eurasia; this rare species has a distinctive, large spore sac with several pore openings.

GOMPHALES

Although some species were included with the chanterelles in Cantharellales, DNA analysis of the Gomphales fungi suggests that they are more closely related to the stinkhorn fungi in the Phallales. They often form large fruitbodies, varying in shape from simple clubs (in genus *Clavariadelphus*) to trumpet- or chanterelle-like structures with a complex spore-producing surface (in genus *Gomphus*).

¾–2¼ in
2–6 cm

GIANT CLUB
Clavariadelphus pistillaris
F: Clavariadelphaceae

A rare species of North America and Eurasia; this fungus forms a large, swollen club, with a smooth to slightly wrinkled surface, which bruises purple-brown.

2–4 in
5–10 cm

SCALY VASE CHANTERELLE
Gomphus floccosus
F: Gomphaceae

Common in North America; this resembles a fleshy, trumpet-shaped vase with scales at the top and wrinkled "gills" on its undersurface.

stains green with age

½–1½ in
1.5–4 cm

1¼–3¼ in
3–8 cm

UPRIGHT CORAL
Ramaria stricta
F: Gomphaceae

A fairly common species in North America and Eurasia; it is always attached to decaying wood or wood-chip mulch. The branches are pale brown, bruising reddish.

2¾–6 in
7–15 cm

ROSSO CORAL
Ramaria botrytis
F: Gomphaceae

Found in beech woods in North America and Eurasia; this uncommon coral fungus has pinkish white branches with deep red tips.

GREENING CORAL
Ramaria abietina
F: Gomphaceae

Found in conifer woods in North America and Eurasia; this yellow-olive fungus has densely packed branches that bruise green.

GLOEOPHYLLALES

This is an order of wood-decay fungi that is characterized by the ability to produce a brown rot of wood. The order Gloeophyllales has a single family, the Gloeophyllaceae, which includes the genus *Gloeophyllum*. Some well-known bracket fungi on conifer wood are members of this genus.

ANISE MAZEGILL
Gloeophyllum odoratum
F: Gloeophyllaceae

This fungus, which grows on decayed conifer wood, is found in North America and Eurasia; it has irregular brackets with yellow pores and an aniseed odor.

2–8 in
5–20 cm

HYMENOCHAETALES

This group contains a number of diverse types of fungi including some crust fungi, polypores, such as *Inonotus* and *Phellinus*, as well as several agaric species, such as *Rickenella*. The Hymenochaetales are defined through molecular studies and have few uniting physical characteristics. Many feed on wood and may cause a white rot of wood.

⅜–2¼ in
1–6 cm

OAK CURTAIN CRUST
Hymenochaete rubiginosa
F: Hymenochaetaceae

Found across Eurasia; this fungus forms overlapping brackets and crusts mainly on fallen oak. Its tough fruitbodies have concentric markings.

4–16 in
10–40 cm

thick, pale margin

WILLOW BRACKET
Phellinus igniarius
F: Hymenochaetaceae

This gray to almost black perennial bracket from North America and Eurasia grows over many years. It is hoof-shaped and extremely woody.

DACRYMYCETALES

This order forms very simple, rounded or branched gelatinous fruitbodies, usually with bright orange coloration. Smooth or wrinkled, the Dacrymycetales have unusual spore-producing cells (basidia), which typically have two stout stalks (sterigmata), each bearing a spore. They mainly feed on dead wood.

1¼–3¼ in
3–8 cm

ALDER BRACKET
Inonotus radiatus
F: Hymenochaetaceae

This deep reddish brown fungus with a paler margin often forms vertical chains of brackets on alder and other trees in North America and Eurasia.

YELLOW STAGSHORN
Calocera viscosa
F: Dacrymycetaceae

Attached to conifer wood, this is found in North America and Eurasia; its clubs usually divide into branches with a gelatinous, rubbery texture.

3⁄16–1½ in
0.5–4 cm

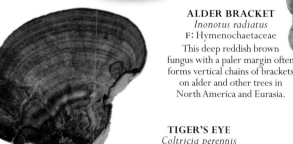

¾–4 in
2–10 cm

TIGER'S EYE
Coltricia perennis
F: Hymenochaetaceae

Frequently found in acidic meadows of North America and Eurasia; this fungus's goblet-shaped, thin fruitbodies are concentrically zoned.

⅛–⅜ in
0.3–1 cm

ORANGE MOSSCAP
Rickenella fibula
F: Incertae sedis in Rickenella clade

This tiny species is common in mossy grasslands of North America and Eurasia; its bright orange cap is radially marked with a darker center.

POLYPORALES

The Polyporales form a large group of diverse fungi. Most of these are polypores—wood decomposers whose spores are formed in tubes (somewhat like the tubes of the boletes) or sometimes on spines. Most lack fully developed stems and grow shelf-, bracket-, or crust-like fruitbodies on wood, but some have more or less central stems and grow at the base of trees. A few Polyporales appear to grow from soil.

UPPER SURFACE

4–12 in
10–30 cm

UNDERSIDE

OAK MAZEGILL
Daedalea quercina
F: Fomitopsidaceae

Found on fallen oaks in North America and Eurasia; this perennial species has tough brackets with long mazelike pores on the underside.

RAZORSTROP FUNGUS
Piptoporus betulinus
F: Fomitopsidaceae

This large, kidney-shaped bracket is pale brown to white. A damaging parasite of birches, it is found in North America and Eurasia.

2–12 in
5–30 cm

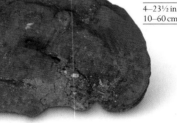

4–23½ in
10–60 cm

ARTIST'S BRACKET
Ganoderma applanatum
F: Ganodermataceae

Found in North America and Eurasia; this woody perennial bracket grows for many years to reach huge sizes. Its fallen spores are a rich cinnamon-brown.

bracket grows
in layers

6–12 in
15–30 cm

DYER'S MAZEGILL
Phaeolus schweinitzii
F: Fomitopsidaceae

Usually found at the base of conifers in North America and Eurasia; this large, furry, cushionlike bracket is used for the manufacture of dye.

hairy surface

4–20 in
10–50 cm

CHICKEN OF THE WOODS
Laetiporus sulphureus
F: Fomitopsidaceae

These large brackets, found in North America and Eurasia; usually grow on oaks and sometimes on other trees.

LACQUERED BRACKET
Ganoderma lucidum
F: Ganodermataceae

This deep reddish to purple-brown bracket with a shiny surface may have a long, lateral stem. It is found in North America and Eurasia.

4–12 in
10–30 cm

GIANT POLYPORE
Meripilus giganteus
F: Meripilaceae

One of the largest polypores, its overlapping brackets are thick and fleshy. It grows around beech and other trees in North America and Eurasia.

4–10 in
10–25 cm

RED BELTED POLYPORE
Fomitopsis pinicola
F: Fomitopsidaceae

Occurring in North America and Eurasia; this hoof-shaped, woody bracket is usually found on pines, and sometimes on birches.

margin is wavy
and lobed

4–20 in
10–50 cm

surface
bruises black

4–8 in
10–20 cm

ZONED ROSETTE
Podoscypha multizonata
F: Meruliaceae

Growing in soil from buried oak roots, this rare species occurs in Eurasia; it has densely packed lobes that form a circular mass.

JELLY ROT
Phlebia tremellosa
F: Meruliaceae

Found on logs in Eurasia; this species has a pale, velvety upper side and a yellow-to-orange underside with dense ridges.

1½–6 in
4–15 cm

1¼–2¾ in
3–7 cm

SMOKY BRACKET
Bjerkandera adusta
F: Meruliaceae

This common bracket, found in North America and Eurasia; can be identified by the ash-gray pore surface on its underside.

TINDER BRACKET
Fomes fomentarius
F: Polyporaceae
2–12 in
5–30 cm
This gray-brown, hoof-shaped, perennial bracket grows on birches and other deciduous trees. It is found in North America and Eurasia.

DRYAD'S SADDLE
Polyporus squamosus
F: Polyporaceae
4–23½ in
10–60 cm
Found in early summer in North America and Eurasia; these circular or fan-shaped brackets have concentric scales and pores on the underside.

BLUSHING BRACKET
Daedaleopsis confragosa
F: Polyporaceae
3¼–6 in
8–15 cm
One of the most common brackets, especially on willows, in N. America and Eurasia; this semicircular species has cream pores that bruise pinkish red.

BIRCH MAZEGILL
Lenzites betulina
F: Polyporaceae
Usually found on birches in North America and Eurasia; this tough, leathery species has pores that can be extremely elongated in gill-like ridges.

UPPER SURFACE

1¼–4 in
3–10 cm

UNDERSIDE

funnel-shaped cap

BLACK FOOTED POLYPORE
Polyporus badius
F: Polyporaceae
2–8 in
5–20 cm
This fungus has a funnel-shaped, leathery cap and its stem is black at the base. It grows on fallen beech logs in North America and Eurasia.

velvety stem base

TUBEROUS POLYPORE
Polyporus tuberaster
F: Polyporaceae
2–8 in
5–20 cm
Growing on fallen branches in North America and Eurasia; this species may root into the ground and form a large, tuberous mass.

WINTER POLYPORE
Polyporus brumalis
F: Polyporaceae
1¼–3¼ in
3–8 cm
This small species grows on fallen branches in North America and Eurasia; it has somewhat large, decurrent (running down the stem) pores. The stem is central or off-center.

CINNABAR BRACKET
Pycnoporus cinnabarinus
F: Polyporaceae
1¼–4 in
3–10 cm
Growing on dead deciduous wood in North and America and Eurasia; this rare species has a bright reddish orange, leathery, annual bracket.

PURPLEPORE BRACKET
Trichaptum abietinum
F: Polyporaceae
¾–1½ in
2–4 cm
Found on fallen conifers in North America and Eurasia; this fan-shaped bracket has concentric zones of pale gray, often tinged green by algae, with a purple margin.

HAIRY BRACKET
Trametes hirsuta
F: Polyporaceae
2–4¾ in
5–12 cm
This semicircular bracket is covered in minute hairs. It grows on dead deciduous wood or sometimes on old gorse stems in North America and Eurasia.

TURKEYTAIL
Trametes versicolor
F: Polyporaceae
¾–2¾ in
2–7 cm
Found in North America and Eurasia in a variety of colors, this species has brackets with concentric zones of different colors and a white pore surface.

fleshy, bright yellow-orange bracket

LUMPY BRACKET
Trametes gibbosa
F: Polyporaceae
This cream-colored bracket is often stained green from algae. Its pores may be long and mazelike. It grows on fallen deciduous logs in North America and Eurasia.

small-lobed bracket

HEN OF THE WOODS
Grifola frondosa
F: Sparassidaceae
¾–2¼ in
2–6 cm
Found in North America and Eurasia; this species grows at the base of oak, often where lightening has struck. It has a dense cluster of small brackets.

basal stem

WOOD CAULIFLOWER
Sparassis crispa
F: Sparassidaceae
4–16 in
10–40 cm
Growing at the base of conifers in North America and Eurasia; this species has cream lobes that are flattened and fleshy like a cauliflower.

4–12 in
10–30 cm

RUSSULALES

The best-known genuses within this order are *Russula* and *Lactarius*, which, although resembling typical mushrooms, are not related at all to the true Agaricales. Apart from the cap-and-stem shapes, Russulales produce fruitbodies in a wide range of forms. Most Russulales have spores with warts that stain blue-black in iodine solutions. When cut, the *Lactarius* species produce a latex, which may be white to colored.

1¼–2¼ in
3–6 cm

LIVER MILKCAP
Lactarius hepaticus
F: Russulaceae
Associated with pines in Eurasia; this mushroom has a smooth cap with a grooved margin. The gills bleed a white latex that stains yellow when cut.

1½–3¼ in
4–8 cm

OAKBUG MILKCAP
Lactarius quietus
F: Russulaceae
Common under oaks in Eurasia; its reddish brown cap has darker zones, and the flesh has a sweet, oily smell.

UGLY MILKCAP
Lactarius turpis
F: Russulaceae
A common species of birch woods in North America and Eurasia; it is olive-green to almost black in color and has a slimy cap.

2–6 in
5–15 cm

BEECH MILKCAP
Lactarius blennius
F: Russulaceae
This is a common species growing with beech in Eurasia. Slimy when wet, the gray-green cap has spots around the margin.

1½–3½ in
4–9 cm

depressed cap center

2–6 in
5–15 cm

SAFFRON MILKCAP
Lactarius deliciosus
F: Russulaceae
Found with pines in North America and Eurasia; this fungus has an orange-zoned and spotted cap that stains green with age. The flesh bleeds orange-red latex.

grooved cap margin

1¼–2¼ in
3–6 cm

2–6 in
5–15 cm

WOOLLY MILKCAP
Lactarius torminosus
F: Russulaceae
Common under birches in North America and Eurasia; this mushroom has a dark pink, hairy cap that is strikingly zoned.

CURRY MILKCAP
Lactarius camphoratus
F: Russulaceae
As the fruitbody of this fungus dries, it gives off a smell of curry, which persists for many weeks. It is found in North America and Eurasia.

hairy and inrolled margin

2¼–4 in
6–10 cm

SOOTY MILKCAP
Lactarius fuliginosus
F: Russulaceae
Found in deciduous woods in Eurasia; this uncommon species has a dark brown cap and stem that bleed a white latex that rapidly turns pink.

3¼–8 in
8–20 cm

PEPPERY MILKCAP
Lactarius piperatus
F: Russulaceae
This uncommon species from mixed woods in North America and Eurasia has a funnel-shaped cap with extraordinarily crowded, narrow gills. It exudes white latex when cut.

dry, smooth cap surface

2–6 in
5–15 cm

2–4 in
5–10 cm

CHARCOAL BURNER
Russula cyanoxantha
F: Russulaceae

Growing in mixed woods in North America and Eurasia; this mushroom's cap varies from purple-lilac to uniformly green. The gills are forked, flexible, and oily.

BLOODY BRITTLEGILL
Russula sanguinaria
F: Russulaceae

Associated with pines in North America and Eurasia; this species has a scarlet cap and red-streaked stem. Its spores are pale ocher.

1¼–2¾ in
3–7 cm

BEECHWOOD SICKENER
Russula nobilis
F: Russulaceae

Exclusively found with beech in Eurasia; this species has a scarlet cap and bluish white gills.

1½–3½ in
4–9 cm

GREEN BRITTLEGILL
Russula aeruginea
F: Russulaceae

Common under birches in North America and Eurasia; this mushroom's pale olive to grass-green cap has tiny rust-colored spots. Its spores are pale cream.

2–4 in
5–10 cm

YELLOW SWAMP BRITTLEGILL
Russula claroflava
F: Russulaceae

A common species on moss in boggy birch woods in North America and Eurasia; its cap, yellowish cream gills, and white stem all bruise gray-black.

2–4¾ in
5–12 cm

OCHER BRITTLEGILL
Russula ochroleuca
F: Russulaceae

One of the most common species found in Eurasia; it is identifiable by its matte yellow-ocher or greenish yellow cap and white gills.

1½–4 in
4–10 cm

PRIMROSE BRITTLEGILL
Russula sardonia
F: Russulaceae

Found in Eurasia; this species grows in association with pine. Variable in color, from purple to green or yellow, it has a fruity odor.

2¼–6 in
6–15 cm

CRAB BRITTLEGILL
Russula xerampelina
F: Russulaceae

Found in North America and Eurasia; this fungus is one of many related species that can only be identified under a microscope or by habitat.

dry and mattee cap surface

1¼–3¼ in
3–8 cm

SICKENER
Russula emetica
F: Russulaceae

Growing in wet pine woods in North America adn Eurasia; the sickener has a bright scarlet cap that contrasts with its pure white gills and stem.

1½–4¾ in
4–12 cm

gills may have red edges

ROSY BRITTLEGILL
Russula rosea
F: Russulaceae

Found in Eurasia; this fungus has a carmine-red, dry, hard cap that fades rapidly. Its stem may also be red and the flesh has a smell similar to cedar.

stem often flushes red

3¼–6 in
8–15 cm

STINKING BRITTLEGILL
Russula foetens
F: Russulaceae

This large, orange-brown fungus with a grooved, lumpy cap margin grows in North America and Eurasia. It has a sour, rancid odor.

EARPICK FUNGUS
Auriscalpium vulgare
F: Auriscalpiaceae

Growing on pinecones in North America and Eurasia; this unique mushroom looks like a bent-over spoon with minute spines hanging down from the small, furry cap.

³⁄₁₆–¾ in
0.5–2 cm

¾–2¼ in
2–6 cm

HAIRY CURTAIN CRUST
Stereum hirsutum
F: Stereaceae

Found in North America and Eurasia; this mushroom varies in shape from fully crustlike to small overlapping brackets. It is hairy above and smooth below.

4–20 in
10–50 cm

BLEEDING BROADLEAF CRUST
Stereum rugosum
F: Stereaceae

Growing in North America and Eurasia; this species forms small crusts, and occasionally small brackets on wood. Its upper surface bleeds red where cut.

2–10 in
5–25 cm

ROOT ROT
Heterobasidion annosum
F: Bondarzewiaceae

Usually parasitic on conifers in North America and Eurasia; this species has a pale brown crust that darkens with age.

4–16 in
10–40 cm

CORAL TOOTH
Hericium coralloides
F: Hericiaceae

Usually found on beech trees in North America and Eurasia; this endangered species has white fruitbody branches with pendent spines on the lower surfaces.

AURICULARIALES

Although often grouped with other jelly fungi, the Auriculariales are separated by their unusual basidia (spore-producing cells). These vary in shape but all are partitioned by dividing membranes into four divisions, with each division producing a spore.

WITCHES' BUTTER
Exidia glandulosa
F: Auriculariaceae
Found in temperate North America and Eurasia, frequently on hardwood trees; it resembles a wrinkled mass of gelatinous tar. It shrivels when dry to a hard, black mass.

¾–4 in
2–10 cm

⅜–3¼ in
1–8 cm

broad, thick, jellylike bracket

TRIPE FUNGUS
Auricularia mesenterica
F: Auriculariaceae
Found in Eurasia; this species is common on dead wood, especially elm. It resembles a small bracket fungus from the top, and has a wrinkled, rubbery, grayish purple underside.

1½–6 in
4–15 cm

1½–4¾ in
4–12 cm

JELLY EAR
Auricularia auricula-judae
F: Auriculariaceae
Common on dead wood of deciduous trees in North America and Eurasia; this fungus has thin, elastic "ears" that are velvety outside and wrinkled inside.

JELLY TOOTH
Pseudohydnum gelatinosum
F: Incertae sedis
This species from North America and Eurasia has a pale translucent gray to pale brown coloration. It is occasionally found on conifer stumps.

soft, peglike spines on underside, on which spores are formed

THELEPHORALES

This diverse group includes bracket fungi, crust fungi, earthfans, and toothed fungi. Many of these have tough, leathery flesh, and commonly feature knobbed or spiny spores. The group was only identified as a result of molecular studies since the Thelephorales have few physical features in common.

DRAB TOOTH
Bankera fuligineoalba
F: Bankeraceae
An uncommon species of conifer woods in Eurasia; this fungus has a short, stout stem. Its cap has an underside covered with tiny grayish white spines.

2–4 in
5–10 cm

thick, fleshy scales

1½–5½ in
4–14 cm

BITTER TOOTH
Sarcodon scabrosus
F: Bankeraceae
Found in mixed woods in North America and Eurasia; this rare species has a centrally depressed cap with irregular scales. The spines beneath are pale buff.

1¼–4 in
3–10 cm

stout, velvety stem

BLACK TOOTH
Phellodon niger
F: Bankeraceae
Growing in mixed woods in North America and Eurasia; this uncommon species smells of fenugreek when dry. The irregularly shaped cap is gray to purplish black with gray spines underneath.

blue-green stem base

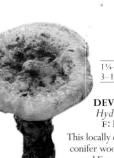

1¼–6 in
3–15 cm

DEVIL'S TOOTH
Hydnellum peckii
F: Bankeraceae
This locally common species from conifer woods in North America and Eurasia has a flattened, knobby, woolly cap that often exudes blood-red droplets. The underside has pale brown spines.

ragged margin

1½–4 in
4–10 cm

EARTHFAN
Thelephora terrestris
F: Thelephoraceae
Quite common in woods or in meadows in North America and Eurasia; this fungus grows on soil or woody debris. Its fan-shaped fruitbodies overlap, forming clumps with paler, fringed margins.

PHALLALES

Named for the phallic shape of many of the species in this group, such as the stinkhorns, this order also contains some false truffles. The stinkhorns "hatch" from an egglike structure, often in just a few hours.

spores on inside of cage

4 in
10 cm

cage bursts from "egg"

RED CAGE FUNGUS
Clathrus ruber
F: Clathraceae
Found in parks and gardens in Eurasia; this rare species has a red cage with black, foul-smelling spores. The cage hatches from a small, pale "egg."

1–5½ in
2.5–14 cm

DEVIL'S FINGERS
Clathrus archeri
F: Clathraceae
Introduced from Australasia, this species is found mainly in southern Eurasia, where it remains rare; its red arms emerge from a white "egg" and have blackish spores that smell fetid.

EXOBASIDIALES

This small group consists mainly of gall-forming plant parasites whose spore-producing cells form a layer on the leaf surface. Some cause disease in cultivated plants of the *Vaccinium* genus, which includes the common blueberry.

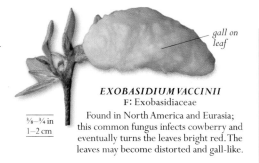

gall on leaf

EXOBASIDIUM VACCINII
F: Exobasidiaceae

Found in North America and Eurasia; this common fungus infects cowberry and eventually turns the leaves bright red. The leaves may become distorted and gall-like.

³⁄₈–³⁄₄ in
1–2 cm

UROCYSTIDIALES

This order contains some well-known smut fungi, in particular species of the genus *Urocystis*. These are parasites of flowering plants such as anemone, onions, wheat, and rye, and often cause serious injury to the host plant.

¹⁄₁₆–⁵⁄₃₂ in
2–4 mm

black, powdery spores on leaf surface

ANEMONE SMUT
Urocystis anemones
F: Urocystidaceae

This smut fungus from North America and Eurasia forms dark brown, powdery, raised pustules on the leaves of anemones and some other plants.

foul-smelling spore mass on cap

hollow, spongy stem

DOG STINKHORN
Mutinus caninus
F: Phallaceae

This common stinkhorn is found in mixed woods in North America and Eurasia; its tip, covered in greenish black spores, is joined to the spongy stem that emerges from a white "egg."

³⁄₈–4³⁄₄ in
1–12 cm

2–8 in
5–20 cm

white skirt drops down from cap

PHALLUS MERULINUS
F: Phallaceae

This tropical species is mainly found in Australasia; it hatches from a white "egg." There are many similar species, some with brightly colored "skirts."

large, white "egg"

2–8 in
5–20 cm

STINKHORN
Phallus impudicus
F: Phallaceae

Common in mixed woods, this fungus is found in Eurasia; its spore-covered, honeycomb-like cap hatches in a few hours from an "egg." Its foul odor is often evident many yards away.

PUCCINIALES

One of the largest orders of fungi with over 7,000 species, the rust fungi include numerous serious parasites of crop plants. They can have very complex life cycles with multiple hosts and produce different types of spores at different stages in their life.

black, powdery spores in yellow spots on lower leaf surface

RASPBERRY YELLOW RUST
Phragmidium rubi-idaei
F: Phragmidiaceae

This rust from North America and Eurasia causes pustules to form on the upper surface of leaves. It survives winter with the help of black spores on the underside of leaves.

orange rust damaging rose stem

ROSE RUST
Phragmidium tuberculatum
F: Phragmidiaceae

This common rust is found in North America and Eurasia; it causes orange pustules on the undersides of leaves and on distorted stems. The pustules turn black in late summer.

ALEXANDERS RUST
Puccinia smyrnii
F: Pucciniaceae

Found across Eurasia; this common rust fungus forms raised plaques or warts on the leaves of Alexanders (*Smyrnium olusatrum*).

yellow warts of rust fungus

leaf surface spotted with rust fungus

raised blister of rust fungus

HOLLYHOCK RUST
Puccinia malvacearum
F: Pucciniaceae

This serious pest of hollyhocks is found in North America and Eurasia; it covers the leaves with small pustules. Older leaves die and fall off.

PUCCINIA ALLII
F: Pucciniaceae

Common on onions, garlic, and leeks in North America and Eurasia; this species forms pustules on infected leaves. These break open to release dustlike airborne spores.

rounded bright orange-yellow pustules erupt through leaf surface

powdery pustules on leaf

HYPERICUM RUST
Melampsora hypericorum
F: Melampsoraceae

This common species from Eurasia is visible as scattered, raised pustules on the undersides of *Hypericum* leaves.

FUCHSIA RUST
Pucciniastrum epilobii
F: Pucciniastraceae

A parasite of fuchsias and fireweed in Eurasia; this species infects the leaves and forms pustules on the undersides of the leaves.

SAC FUNGI

The Ascomycota, or sac fungi, are fungi that produce their spores in little sacs called asci located on the fruitbody—the part of the fungus that projects above ground. They are the largest group of fungi, and include many cup- and saucer-shaped species.

PHYLUM	ASCOMYCOTA
CLASSES	7
ORDERS	56
FAMILIES	226
SPECIES	About 33,000

Many sac fungi display vivid colors, although it is uncertain what biological function such bright colors may have.

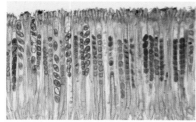

Spore-producing asci are shown here under a microscope. Arranged in dense layers, the asci contain eight spores each.

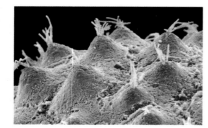

Many species form their asci in special protective chambers called perithecia; from here they discharge their spores.

Ranging in size from the microscopic to around 8 in (20 cm) in height, the sac fungi occupy a wide variety of habitats, growing on dead, dying, and living tissue, and floating in fresh and saltwater environments. Many species are parasitic, and include some of the most serious crop pests. Others form mutually beneficial relationships, called mycorrhizae, with plants. Sac fungi include some of the most important fungi in the history of medicine, such as the source of penicillin, while others are serious pathogens—*Pneumocystis jirovecii*, for instance, can cause lung infections in people with a weak immune system. This phylum also contains the yeasts, which, being vital components in the production of alcohol and bread, have played a pivotal role in human history.

The Ascomycota display a wide range of fruitbody shapes, including cup-shaped, club-shaped, potato-like, simple crusts or sheets, pimplelike, coral-like, shield-shaped, or stalked with a spongelike cap. Depending on the type of fruitbody, the spore-producing asci may grow externally on a special fertile layer or be contained internally. Not all species have a sexual stage; in fact, many use asexual methods of reproduction. Most yeasts grow and rapidly colonize new areas as a result of asexual division, or budding, where a small bud forms on the outside of a yeast cell, then separates, becoming a new cell.

THE CUP FUNGUS

The common name of cup fungus refers to one of the most conspicuous shapes of fruitbody associated with the Ascomycota. Its open top (which is sometimes disk- or saucer-shaped) allows wind and rain to scatter the spores that line its inner surface. In some varieties, water is absorbed by the asci, which contain the spores, so that pressure builds up and forcibly ejects the spores up to 12 in (30 cm) from the fruitbody. Close examination of surfaces such as rotting logs, fallen branches, or leaves may reveal a fascinating world of almost microscopic cups. In the larger cup species, disturbance of the cup can produce such a vigorous ejection of spores that their release is not only visible as a faint cloud of spores, but also audible.

ORANGE PEEL CUP >
The orange peel fungus is a good example of the simple cup-shaped fruitbody adopted by many fungi species.

HYPOCREALES

The fungi in this order are usually distinguished by their brightly colored, spore-producing structures. These are usually yellow, orange, or red. The Hypocreales are often parasitic on other fungi and also on insects. The best known among them is the genus *Cordyceps* with club- or branch-like fruitbodies (bodies that support the spore-forming cells). Some species have medicinal uses.

1¼–2¼ in
3–6 cm

2–5 in
5–13 cm

false truffle parasitized by truffleclub

SCARLET CATERPILLAR CLUB
Cordyceps militaris
F: Cordycipitaceae
Found in North America and Eurasia; this fungus is parasitic on moth pupae. The head of its club bears tiny spore-producing structures.

SNAKETONGUE TRUFFLECLUB
Elaphocordyceps ophioglossoides
F: Ophiocordycipitaceae
Parasitic on buried false truffles, this species is found in North America and Eurasia; it forms yellowclubs with an elongated, greenish black head.

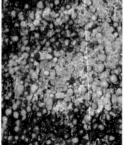

rounded, cushionlike fruitbodies

purple ergot on flower heads

infected bolete fruitbody

½ in
1.5 cm

8–12 in
20–30 cm

CORAL SPOT
Nectria cinnabarina
F: Nectriaceae
Abundant on damp wood in North America and Eurasia; this fungus forms pink "pimples" when sexually immature and red-brown clusters when mature.

ERGOT
Claviceps purpurea
F: Clavicipitaceae
This species has caused outbreaks of mass poisoning. Found in North America and Eurasia, it is parasitic on grass and cereal crops.

BOLETE EATER
Hypomyces chrysospermus
F: Hypocreaceae
This common mold, found in N. America and Eurasia on bolete, turns bright golden yellow, with a fluffy texture.

XYLARIALES

Members of this order often have their spore-producing cells in chambers, which are embedded in a woody growth called stroma. Although many species live on wood, some also occur on animal dung, fruit, leaves, and soil, or are associated with insects. The order includes many economically important plant parasites.

¾–3¼ in
2–8 cm

tips covered in powdery spores

⅜–½ in
1–1.5 cm

NAIL FUNGUS
Poronia punctata
F: Xylariaceae
Found in horse dung in North America and Eurasia; this species is in decline. Its flattened disc has numerous tiny holes from which spores are released.

⅜–1½ in
1–4 cm

CANDLESTUFF FUNGUS
Xylaria hypoxylon
F: Xylariaceae
Common on dead wood in North America and Eurasia; this species resembles a snuffed-out candle with a velvety black stem.

DEAD MAN'S FINGERS
Xylaria polymorpha
F: Xylariaceae
Growing on dead wood in North America and Eurasia; this fungus forms brittle black clubs with a rough surface, tiny pores, and thick, white flesh.

stemless fruitbody has hard, brittle surface

dead trunk of ash tree

¾–4 in
2–10 cm

¼–⅜ in
0.5–1 cm

BEECH WOODWART
Hypoxylon fragiforme
F: Xylariaceae
Occurring in clusters on beech logs in North America and Europe; it forms hard, rounded fruitbodies with tiny spore-releasing chambers.

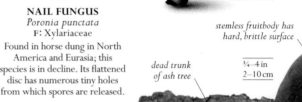

CRAMP BALLS
Daldinia concentrica
F: Xylariaceae
Found in North America and Eurasia; this fungus has rounded fruitbodies that reveal concentric whitish zones when cut in half. They eject black spores from their outer layer.

ERYSIPHALES

Parasitic on leaves and fruit of flowering plants, the Erysiphales are powdery mildews. The hyphae (filaments) of their mycelium (vegetative part of the fungus) penetrate the cells of the host plant and take up nutrients.

mildew patches

affected apple leaf

feltlike white mycelium covers leaf surface

OAK POWDERY MILDEW
Erysiphe alphitoides
F: Erysiphaceae
Growing on oaks in North America and Eurasia; this fungus covers young leaves, causing them to shrivel and blacken.

APPLE POWDERY MILDEW
Podosphaera leucotricha
F: Erysiphaceae
Common on apple leaves in North America and Eurasia; this mildew appears first on the underside of leaves as whitish patches, before spreading rapidly.

POWDERY MILDEW
Golovinomyces cichoracearum
F: Erysiphaceae
Found in North America and Eurasia; this species appears on members of Asteraceae (the sunflower family), causing patches on leaves, which eventually die.

CAPNODIALES

Commonly called sooty molds, these sac fungi are frequently found on leaves. They feed on the honeydew excreted by insects or on liquid exuded from the leaves. Some cause skin problems in humans.

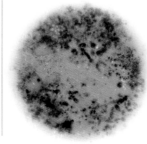

CLADOSPORIUM CLADOSPORIOIDES
F: Davidiellaceae
This mold, common in North America and Eurasia, grows on damp bathroom walls. It may cause an allergic reaction in some people.

HELOTIALES

The fungi of this order are distinguished by their disc- or cup-shaped fruitbodies, unlike some similar cup fungi. Their saclike spore-producing cells, or asci, do not have an apical lid (a flap through which they open). Most members live on nutrient-rich soil, dead logs, and other organic matter. This order also includes some of the most damaging plant parasites.

³⁄₁₆–½ in
0.5–1.5 cm

BROWN CUP
Rutstroemia firma
F: Rutstroemiaceae

Consisting of a light brown cup and a narrow stem, this fungus grows on fallen branches, especially of oaks, in Europe. It turns the host wood black.

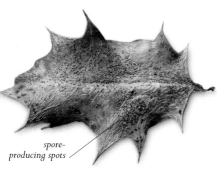

spore-producing spots

TROCHILA ILICINA
F: Dermateaceae

Abundant on fallen holly leaves, this fungus is found in North America and Eurasia. It forms spore-producing spots on the leaf surface.

merged black spots

ROSE BLACK SPOT
Diplocarpon rosae
F: Dermateaceae

Commonly found on rose leaves in North America and Eurasia; this fungus causes black spots, which merge together to cause large black patches.

1¼–2¾ in
3–7 cm

spore-producing inner surface

smooth surface of cup

SCALY EARTHTONGUE
Geoglossum fallax
F: Geoglossaceae

Growing in meadows in North America and Eurasia; this uncommon fungus is one of several species with blackish, flattened clubs, which can only be separated using a microscope.

¹⁄₁₆–³⁄₈ in
0.2–1 cm

³⁄₁₆–1½ in
0.5–4 cm

BLACK BULGAR
Bulgaria inquinans
F: Bulgariaceae

Found in North America and Eurasia; this species has a brown outer surface. Its spore-producing inner surface is smooth, black, and rubbery.

³⁄₁₆–1¼ in
0.5–3 cm

ANEMONE CUP
Dumontinia tuberosa
F: Sclerotiniaceae

Parasitic on the tubers of anemones, this common Eurasian species has a long, black stem and a simple brown cup.

BEECH JELLYDISC
Neobulgaria pura
F: Helotiaceae

Frequently found on fallen beech logs in Eurasia; this translucent jellydisc varies from pale pink to pale lilac. The discs are often distorted from crowding together.

³⁄₁₆–1¼ in
0.5–3 cm

buff-colored pustules on infected fruit

APPLE BROWN ROT
Monilia fructigena
F: Sclerotiniaceae

Very common in Eurasia; this fungus occurs primarily on apples and pears, although it is also found on *Prunus* sp., causing brown fruit rot.

¹⁄₃₂–⅛ in
1–3 mm

LEMON DISC
Bisporella citrina
F: Helotiaceae

Very common on dead hardwood in Eurasia; this species occurs in clusters. Its golden-yellow discs sometimes cover entire branches.

⅛–⅜ in
0.3–1 cm

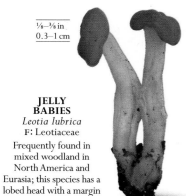

JELLY BABIES
Leotia lubrica
F: Leotiaceae

Frequently found in mixed woodland in North America and Eurasia; this species has a lobed head with a margin that rolls back on itself.

¹⁄₁₆–⅜ in
0.2–1 cm

GREEN ELFCUP
Chlorociboria aeruginascens
F: Helotiaceae

Found in North America and Eurasia; the blue-green mature cups are somewhat uncommon, but the green stains they cause on fallen oak logs are easily visible.

³⁄₁₆–¾ in
0.5–2 cm

LARGE PURPLE DROP
Ascocoryne cylichnium
F: Helotiaceae

Quite common on fallen beech logs in Eurasia; this species attaches itself centrally to the wood and forms gelatinous, irregular discs when sexually mature.

BOG BEACON
Mitrula paludosa
F: Helotiaceae

This species, found in North America and Eurasia, grows on plant remains in shallow water in spring and early summer. It has a rounded to tongue-shaped head.

PEZIZALES

Members of this order produce spores inside saclike structures, or asci, which typically open by rupturing to form an operculum (terminal lid) and eject the spores. The order includes a number of species of economic importance, such as morels, truffles, and desert truffles.

3/16–3/4 in
0.5–2 cm

COMMON EARTHCUP
Geopora arenicola
F: Pyronemataceae
Common in Eurasia; this fungus is difficult to spot since it lies buried in sandy soil. It has a smooth, spore-producing inner surface.

tall orange cups

2–4 in
5–10 cm

HARE'S EAR
Otidea onotica
F: Pyronemataceae
This common species is often found in clusters in broadleaf forests in North America and Eurasia; its tall cups are split down on one side.

3/16–3/8 in
0.5–1 cm

COMMON EYELASH
Scutellinia scutellata
F: Pyronemataceae
This is one of many similar species; its fruitbody is a scarlet cup with a fringe of black hairs. This fungus is common on wet, rotten wood in North America and Eurasia.

3/16–1/2 in
0.5–1.5 cm

TOOTHED CUP
Tarzetta cupularis
F: Pyronemataceae
A common species in alkaline soil in woods of North America and Eurasia; its gobletlike cup has a short stem.

dark brown, wrinkled cap

ORANGE PEEL FUNGUS
Aleuria aurantia
F: Pyronemataceae
Often found along gravely dirt tracks in North America and Eurasia; this is an unmistakable species, with thin, orange caps.

irregular ridges and pits

3/4–4 in
2–10 cm

1 1/2–4 in
4–10 cm

2–6 in
5–15 cm

FALSE MOREL
Gyromitra esculenta
F: Discinaceae
This poisonous species is found throughout North America and Eurasia; it usually grows under conifers in spring. The shiny brown cap looks like a wrinkled brain.

BLEACH CUP
Disciotis venosa
F: Morchellaceae
Growing in spring in damp woodland in North Americ and Eurasia; this short-stemmed species has a chlorinelike smell. Its inner surface is brown and wrinkled and the outer surface is pale.

2–8 in
5–20 cm

HALF-FREE MOREL
Morchella semilibera
F: Morchellaceae
This hollow morel looks like a dark, ridged thimble on a scaly, pale stem. It is common in spring in mixed woods in North America and Eurasia.

smooth cap surface

2–6 in
5–15 cm

hollow stem

THIMBLE MOREL
Verpa conica
F: Morchellaceae
An uncommon species, this fungus grows in woods and hedges in chalky soils in North America and Eurasia; its smooth, thimblelike cap sits on a hollow stem.

2–4 in
5–10 cm

2–6 in
5–15 cm

BLACK MOREL
Morchella elata
F: Morchellaceae
Common in woods during spring in North America and Eurasia; this species has a pinkish buff to black cap, with cross-connected black ridges, and a hollow stem.

MOREL
Morchella esculenta
F: Morchellaceae
This prized species grows in spring in calcareous woodland in North America and Eurasia; it has a spongelike, hollow cap and a hollow stem.

1/2–2 3/4 in
1.5–7 cm

1 1/4–4 in
3–10 cm

1–3 in
2.5–7.5 cm

BLISTERED CUP
Peziza vesiculosa
F: Pezizaceae
Common in North America and Eurasia; this fungus typically occurs in clusters in compost, straw, or manure. Its brittle cups can have ragged edges.

CELLAR CUP
Peziza cerea
F: Pezizaceae
Found across North America and Eurasia; this fungus often occurs on damp brickwork. It is dark ocher on the inside and pale outside.

BAY CUP
Peziza badia
F: Pezizaceae
One of many similar species, this common fungus of North American and Eurasian woods has a spore-producing inner surface. Its brown cup develops an olive tint with age.

**¾–2¾ in
2–7 cm**

**¾–3¼ in
2–8 cm**

PERIGORD TRUFFLE
Tuber melanosporum
F: Tuberaceae
This highly prized truffle grows underground around oaks in the Mediterranean region; the truffles are found using dogs or pigs.

WHITE TRUFFLE
Tuber magnatum
F: Tuberaceae
Prized in Italy and France, this expensive truffle of alkaline soil in southern Europe can be cultivated on suitable host trees such as oak and poplar, by inoculation.

SUMMER TRUFFLE
Tuber aestivum
F: Tuberaceae
Found across southern and central Europe; this highly valued truffle grows underground, near a variety of broadleaf trees.

¾–2 in / 2–5 cm

**⅜–3¼ in
1–8 cm**

SCARLET ELFCUP
Sarcoscypha austriaca
F: Sarcoscyphaceae
Found from winter to early spring on fallen branches in North America and Eurasia; this species has a scarlet cup contrasting with its pale outer surface.

**¾–2¼ in
2–6 cm**

**2–6 in
5–15 cm**

WHITE SADDLE
Helvella crispa
F: Helvellaceae
Possibly poisonous, this species is common in mixed woods in North America and Eurasia; its thin, saddlelike cap sits on a fragile, ribbed stem.

ELFIN SADDLE
Helvella lacunosa
F: Helvellaceae
Found in mixed woods in North America and Eurasia; this common species has a lobed, dark cap on a gray, fluted, and columned stem.

EUROTIALES

Better known for the blue and green molds, this order includes the fungi *Penicillium* (famous for producing penicillin, the first discovered antibiotic) and *Aspergillus* (a significant cause of disease in humans).

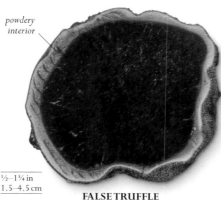

powdery interior

**½–1¾ in
1.5–4.5 cm**

FALSE TRUFFLE
Elaphomyces granulatus
F: Elaphomycetaceae
Common in sandy soil below conifers in North America and Eurasia; this reddish brown truffle has a roughened surface and purple-black inner spore mass.

TAPHRINALES

This order contains many plant parasites, with most species in the genus *Taphrina*. All species have two growth states: in the saprophytic state, they are yeastlike and propagate by budding; but in the parasitic state, they emerge through plant tissues, causing distorted leaves and galls.

BIRCH BESOM
Taphrina betulina
F: Taphrinaceae
Found across Eurasia; this common species causes witches' broom, a disease of birch trees that causes clumps of narrow twigs to grow at branch ends.

**8–37 in
20–95 cm**

PLEOSPORALES

Typical members of this order develop their asci within a flasklike fruiting body. The asci have two wall layers: at maturity, the inner wall protrudes beyond the outer wall, ejecting the spores. Many species grow on plants; some form lichens.

dark cones

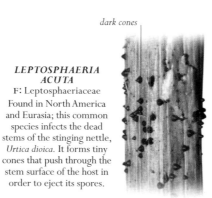

LEPTOSPHAERIA ACUTA
F: Leptosphaeriaceae
Found in North America and Eurasia; this common species infects the dead stems of the stinging nettle, *Urtica dioica*. It forms tiny cones that push through the stem surface of the host in order to eject its spores.

RHYTISMATALES

Commonly called tar spot fungi, the species in this order infect plant matter, such as leaves, twigs, bark, female conifer cones, and occasionally berries. Many species attack the needles of conifers, causing needle drop. The tar spot of maple leaves is perhaps the most frequently seen.

brown spots on upper leaf surface

**⅜–¾ in
1–2 cm**

**1–2 in
2.5–5 cm**

TAR SPOT
Rhytisma acerinum
F: Rhytismataceae
Abundant on acer or maple trees in North America and Eurasia; this fungus causes irregular spots, with paler yellow margins, which disfigure the leaves.

YELLOW FAN
Spathularia flavida
F: Cudoniaceae
This common species grows in wet, mossy conifer woods of N. America and Eurasia; it has flattened, pale to darker yellow, rubbery heads.

PEACH LEAF CURL
Taphrina deformans
F: Taphrinaceae
This fungus infects most varieties of peach and nectarine trees in North America and Eurasia. The affected leaves curl and crinkle, and frequently turn reddish purple.

**5½–16 in
14–40 cm**

red patches caused by infection

PEAR SCAB
Venturia pyrina
F: Venturiaceae
Found in pear orchards in North America and Eurasia; this common parasite causes the fruit to distort, discolor, and even drop before it is ripe.

dark, sunken spots

BOEREMIA HEDERICOLA
F: Didymellaceae
Found in North America and Eurasia; this species causes circular white lesions on the leaves of the ivy plant, which turn brown and die.

LICHENS

From outcrops of rock exposed to the sea to desert areas where the only place they can grow is inside the rock itself, lichens survive in the harshest parts of the world. They are some of nature's pioneers, creating foundations on which others can live.

PHYLA	ASCOMYCOTA BASIDIOMYCOTA
CLASSES	10
ORDERS	15
FAMILIES	40
SPECIES	About 18,000

Asexual soredia are bundles of fungal hyphae (filaments) and algal cells. Here they are budding off and awaiting dispersal.

Asexual isidia are tiny peglike cells on the edge of the lichen, which break off to form new colonies of lichen.

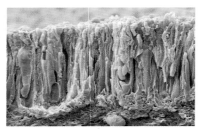

Lichen cells are shown here in microscopic cross section. Spore-producing asci rise from algal cells.

DEBATE
UNLIKELY PARTNERS?

Lichen evolution remains a mystery. Scientists are still trying to understand how and why fungi and algae came to live together. It may have started as an attack by one on the other, which developed into a partnership. Not all lichen associations are necessarily mutually beneficial—some may be parasitic rather than a partnership.

A lichen is not a single organism, but a composite consisting of a green alga or cyanobacterium and a fungus, living together in a mutually beneficial association. The algal partner provides nutrients through photosynthesis, while the fungal partner aids the alga by retaining water and capturing mineral nutrients. The fungus is normally a member of the Ascomycota (the sac fungi), or, more rarely, of the Basidiomycota (mushrooms)—lichen classification reflects the type of fungus involved. Typically, the fungus surrounds the photosynthetic cells of the alga, enclosing them within special fungus tissues unique to lichens. It appears that although neither partner is capable of surviving on its own, together they can endure the most extreme conditions. Lichens have even been found some 250 miles (400 km) from the South Pole, but they also grow in familiar places, such as dry-stone walls, rocks, and bark.

Lichens broadly fall into three types, based on shape: foliose lichens, which have leaves; crustose lichens, which form a crust; and fruticose lichens, which have branches. However, there are some that defy this categorization, such as filamentous (hairlike) and gelatinous (water-absorbent) lichens.

REPRODUCTION

Many lichens reproduce sexually by means of spores. These are produced by the fungal half of the partnership and are usually formed in special cup- or disklike structures called apothecia. These spores, once ejected, must land next to a suitable algal partner if they are to form another lichen and survive. Other lichens produce spores inside special chambers called perithecia, which are like microscopic volcanoes, releasing their spores through a hole at the top. Alternatively, lichens may reproduce asexually by budding or breaking off specialized parts of their body. These soredia or isidia contain a mix of fungal and algal cells, which, if they land in a suitable habitat, will go on to form new lichen colonies. Rocky shorelines in North America are home to vast colonies of lichens several miles in length. These will have taken hundreds or even thousands of years to spread to such a huge size.

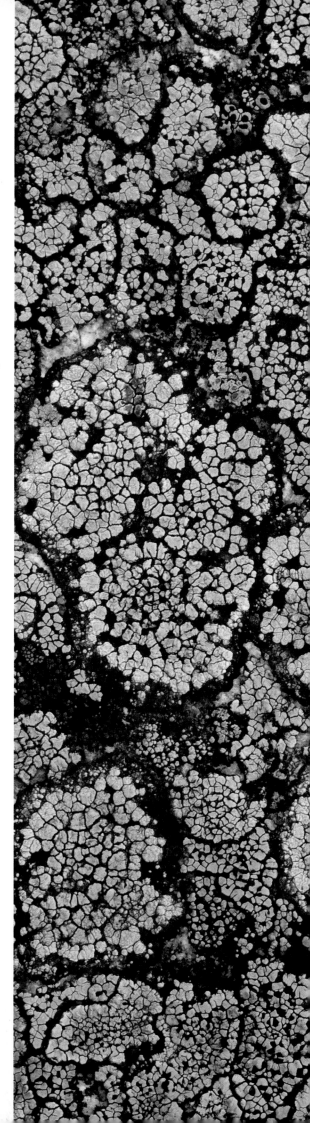

SPREADING MAP LICHEN >
Typical of lichens, *Rhizocarpon* is able to colonize harsh environments such as dry, exposed rock surfaces.

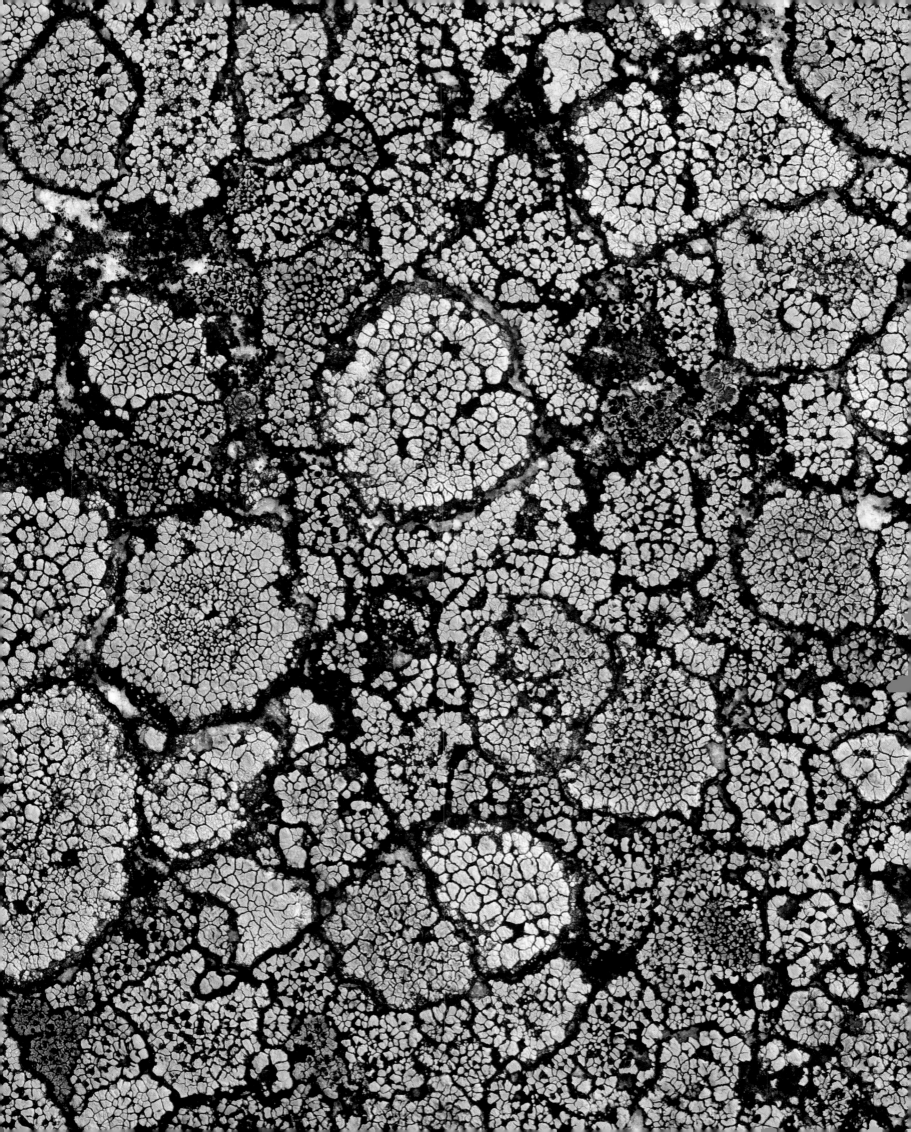

ORANGE LICHEN
Caloplaca verruculifera
F: Teloschistaceae

This lichen has radiating lobes with apothecia (spore-producing discs) at the center. It is found on rocks in coastal areas in North America and Eurasia, often near bird perches.

2–4 in
5–10 cm

GOLDEN-EYE LICHEN
Teloschistes chrysophthalmus
F: Teloschistaceae

A critically endangered species in America, Eurasia, and the tropics; this lichen grows on shrubs and small trees in old orchards and hedgerows. Its branched lobes produce large, orange discs.

1–3 in
2.5–7.5 cm

COMMON WALL LICHEN
Xanthoria parientina
F: Teloschistaceae

Growing on trees, walls, and roofs in North America, Eurasia, Africa, and Australia; this lichen has rounded, yellow-orange lobes.

1–3 in
2.5–7.5 cm

rounded lobes that lift at edges

2–6 in / 5–15 cm

BEARD LICHEN
Usnea filipendula
F: Parmeliaceae

Found mainly in northern regions; this species forms green-gray, pendent clumps on trees. Spiny apothecia form at the tips.

REINDEER MOSS
Flavocetraria nivalis
F: Parmeliaceae

Found in mountain meadows and upland moors in North America and Eurasia; this species has brownish, flattened leafy fronds with spiny margins.

HAMMERED SHIELD LICHEN
Parmelia sulcata
F: Parmeliaceae

This lichen's flattened lobes are gray-green, with rounded tips and powdery, reproductive structures on the surface. It is commonly found on trees in North America and Eurasia.

1–3 in
2.5–7.5 cm

HOARY ROSETTE LICHEN
Physcia aipolia
F: Physciaceae

Growing on tree bark in the Americas and Eurasia; this gray to brownish gray species forms rough patches with a lobed margin. It has black apothecia.

1–3 in
2.5–7.5 cm

1–3 in
2.5–7.5 cm

1–3 in
2.5–7.5 cm

POWDER-HEADED TUBE-LICHEN
Hypogymnia tubulosa
F: Parmeliaceae

Common on twigs and tree trunks, this lichen has lobes that are gray-green above and dark underneath. It is found in North America and Eurasia.

HOODED TUBE-LICHEN
Hypogymnia physodes
F: Parmeliaceae

Found worldwide on trees, rocks, and walls, this lichen has pale gray-green lobes with wavy margins. The rare apothecia are red-brown with gray margins.

⅜–2 in
1–5 cm

DEVIL'S MATCHSTICK
Cladonia floerkeana
F: Cladoniaceae

This species is common in peaty soil in North America and Eurasia; it forms a crust of greenish gray scales from which stalks emerge, tipped with scarlet apothecia.

CARTILAGE LICHEN
Ramalina fraxinea
F: Ramalinaceae

Growing on trees in North America and Eurasia; this species forms flattened, gray-green branches dotted with apothecia.

1–5 in
2.5–12.5 cm

CORAL LICHEN
Sphaerophorus globosus
F: Sphaerophoraceae

Found on rocks in mountainous areas of North America and northern Eurasia; this fungus forms dense cushions of pinkish brown branches with globular apothecia.

1–4 in
2.5–10 cm

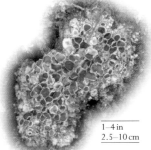

BLACK SHIELDS
Tephromela atra
F: Mycoblastaceae

This lichen has pale gray crustose lobes that look like dried oatmeal. Its apothecia are black. The lichen is found on exposed rocks in North America and Eurasia.

1–4 in
2.5–10 cm

1–4 in / 2.5–10 cm

STONEWALL RIM-LICHEN
Lecanora muralis
F: Lecanoraceae

This lichen often grows on concrete and on rocks. Its gray-green lobes radiate outward. It is found in North America and Eurasia.

1–4 in
2.5–10 cm

REINDEER LICHEN
Cladonia portentosa
F: Cladoniaceae

One of several reindeer lichens, this species is common in meadows and moorland in N. America and Eurasia. Its thin, hollow branches repeatedly divide.

BLACK TAR LICHEN
Verrucaria maura
F: Verrucariaceae

2–20 in
5–50 cm

Occurring on rocks along seacoasts in North America and Eurasia; this species has a dark gray, cracked crust that contains apothecia.

DOG LICHEN
Peltigera praetextata
F: Peltigeraceae

8–12 in
20–30 cm

Found on rocks in North America and Eurasia; this lichen has large gray-black lobes, with paler margins and reddish brown apothecia.

BLISTERED JELLY LICHEN
Collema furfuraceum
F: Collemataceae

1–2 in
2.5–5 cm

This lichen has flat, gelatinous, wrinkled lobes. It is found on rocks and trees in areas with high rainfall in North America and Eurasia.

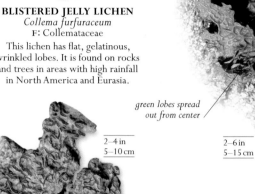

green lobes spread out from center

TREE LUNGWORT
Lobaria pulmonaria
F: Lobariaceae

2–6 in
5–15 cm

Mainly found on tree bark in coastal areas in North America, Eurasia, and Africa; this species is declining due to habitat loss. The branching lobes are pale orange underneath.

ROCK TRIPE
Lasallia pustulata
F: Umbilicariaceae

2–8 in
5–20 cm

This North American and Eurasian lichen forms colonies on nutrient-rich rocks in coastal or upland areas. It has a gray-brown upper surface with many oval pustules.

PETALLED ROCK-TRIPE
Umbilicaria polyphylla
F: Umbilicariaceae

1–3 in
2.5–7.5 cm

A common species on rocks in mountains in North America and Eurasia; it has smooth, broad lobes that are dark brown above and black beneath.

2–4 in
5–10 cm

COMMON SCRIPT LICHEN
Graphis scripta
F: Graphidaceae

black, slitlike openings release spores

Often seen on tree bark in North America and Eurasia; this lichen forms a thin, gray-green crust with slitlike openings.

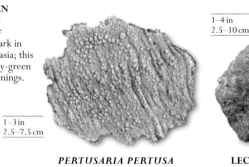

PERTUSARIA PERTUSA
F: Pertusariaceae

Common on tree bark in North America and Eurasia; this lichen forms gray crusts with a pale margin. The crusts are covered with groups of warts with tiny openings.

1–4 in
2.5–10 cm

LECIDEA LICHEN
Lecidea fuscoatra
F: Lecideaceae

This species is common on siliceous rocks and old brick walls in North America and Eurasia. It forms a gray, cracked crust, with sunken, black apothecia.

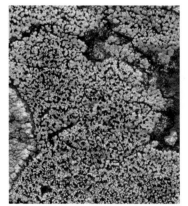

MAP LICHEN
Rhizocarpon geographicum
F: Rhizocarpaceae

3/16–2½ in
5–65 mm

Common on rocks in mountains in northern regions and Antarctica; this lichen forms a flat patch bordered by a black line of spores. Groups of these lichens have a patchwork appearance.

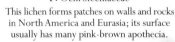

2–8 in
5–20 cm

1–3 in
2.5–7.5 cm

CRAB-EYE LICHEN
Ochrolechia parella
F: Ochrolechiaceae

This lichen forms patches on walls and rocks in North America and Eurasia; its surface usually has many pink-brown apothecia.

dome-shaped apothecia

1–5 in
2.5–12.5 cm

BAEOMYCES RUFUS
F: Baeomycetaceae

Forming gray-green crusts on sandy soil and rocks, this lichen has brown ball-like apothecia on stalks a few millimeters high. It is found in North America and Eurasia.

ANIMA

LS

Animals make up the largest kingdom of living things. Driven by the need to eat food, and to escape being eaten, they are uniquely responsive to the world around them. The vast majority are invertebrates, but mammals and other chordates often outclass them in size, strength, and speed.

≫ 248
INVERTEBRATES

Hugely varied in shape, invertebrates also have a wide range of lifestyles. Insects make up the largest group, but others include jellyfish, worms, and animals protected by hard shells.

≫ 318
CHORDATES

Most of the world's large animals are chordates. Externally, they can be covered in fur, feathers, or overlapping scales, but internally nearly all have a backbone, which makes up part of a bony skeleton.

INVERTEBRATES

With nearly two million species identified, animals make up the largest kingdom of living things on Earth. The vast majority of them are invertebrates—animals without a backbone. Invertebrates are extraordinarily varied; many are microscopic, but the largest are over 33 ft (10 m) long.

Invertebrates were the first animals to evolve. Initially, they were small, soft-bodied, and aquatic—features that many living invertebrates still share. During the Cambrian Period, which ended about 540 million years ago, invertebrates underwent a spectacular burst of evolution, developing a huge range of body forms and some very different ways of life. This evolutionary explosion produced almost all the major phyla, or groups, of invertebrates that exist today.

GREAT DIVERSITY

There is no such thing as a typical invertebrate, and many of the phyla have very little in common. The simplest kinds do not have heads or brains, and usually rely on internal fluid pressure to keep their shape. At the other extreme, arthropods have well-developed nervous systems and elaborate sense organs, including complex eyes. Crucially, they also have an external body case, or exoskeleton, with legs that bend at flexible joints. This particular body plan has proved to be remarkably

successful, allowing arthropods to invade every natural habitat in water, on land, and in the air. Invertebrates also include animals with shells, and ones that are reinforced by mineral crystals or hard plates. However, unlike vertebrates, they never have internal skeletons made of bone.

SPLIT LIVES

Most invertebrates begin life as eggs. Some look like miniature versions of their parents when they hatch, but many start life with a very different body form. These larvae change shape, food sources, and feeding methods as they grow up. For example, sea urchins have drifting larvae, which filter food from the sea, while the adults scrape algae from rocks. The change in shape—metamorphosis—can be gradual, or it can take place abruptly, with the young animal's body being broken down, and an adult body being assembled in its place. Metamorphosis enables invertebrates to exploit more than one food source, and it also helps them spread, often over great distances.

SPONGES
Among the simplest animals, sponges have sievelike bodies, with an internal skeleton of mineral crystals. Classified as the phylum Porifera, there are about 15,000 species.

ARTHROPODS
The phylum Arthropoda is the biggest in the animal kingdom, with over a million species identified. It includes insects, crustaceans, arachnids, centipedes, and millipedes.

INVERTEBRATE TREE

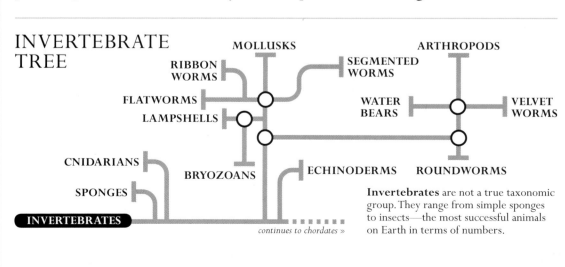

MOLLUSKS

RIBBON WORMS

FLATWORMS

LAMPSHELLS

CNIDARIANS

SPONGES

INVERTEBRATES

BRYOZOANS

SEGMENTED WORMS

ECHINODERMS

ARTHROPODS

WATER BEARS

VELVET WORMS

ROUNDWORMS

Invertebrates are not a true taxonomic group. They range from simple sponges to insects—the most successful animals on Earth in terms of numbers.

continues to chordates »

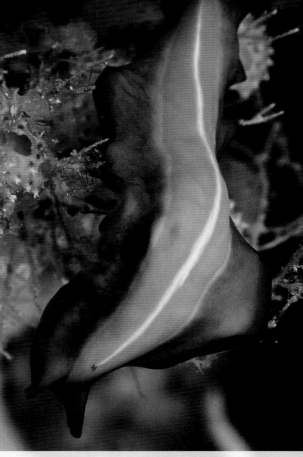

CNIDARIANS
Members of the phylum Cnidaria are soft-bodied invertebrates that kill their prey with stinging cells. Of 11,000 known species, the vast majority are marine.

FLATWORMS
Numbering about 20,000 species, the phylum Platyhelminthes contains animals with flat, paper-thin bodies, and a distinguishable head and tail.

SEGMENTED WORMS
With about 15,000 species, the phylum Annelida contains worms whose sinuous bodies are divided into ringlike segments. It includes earthworms and leeches.

CRUSTACEANS
Mainly aquatic, crustaceans are arthropods that breathe with gills. Classified as the subphylum Crustacea, there are over 50,000 species, including crabs and lobsters.

MOLLUSKS
One of the most diverse invertebrate groups, the phylum Mollusca contains about 110,000 species. They include gastropods, bivalves, and cephalopods.

ECHINODERMS
Recognizable by their five-fold symmetry, members of the phylum Echinodermata have skeletons of small, chalky plates set in their skin. There are about 7,000 species.

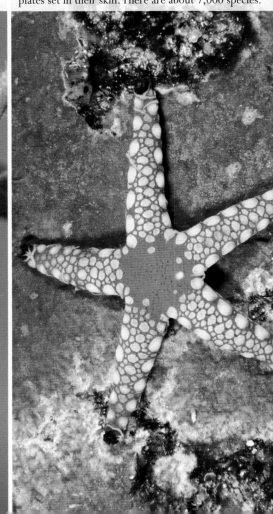

SPONGES

Simple, mostly marine animals, adult sponges live permanently attached to rocks, corals, and shipwrecks. A few species live in freshwater.

The Porifera vary in size and shape, from thin sheets to huge barrels, but all have the same basic structure: different types of specialized cells, but no organs. A system of water canals branches through the sponge, and water is drawn into them through tiny pores in the sponge's surface. Special chambers or cells lining the water canals trap and engulf bacteria from plankton for food, and waste water exits through large openings called osculae.

Many sponges are classically "spongy," others are rock hard, soft, or even slimy depending on the supporting skeleton, which is made up of tiny spicules of silicon dioxide or calcium carbonate. Spicules vary in shape and number between species and can be used in identification.

PHYLUM	PORIFERA
CLASSES	3
ORDERS	24
FAMILIES	127
SPECIES	About 15,000

3¼ ft
1 m

BLUE SPONGE
Haliclona sp.
F: Chalinidae
One of very few blue-colored sponges, this species commonly grows on the tops of corals and rocks off northern Borneo.

CALCAREOUS SPONGES

The skeleton of Calcarea is composed of densely packed, mostly starlike, calcium carbonate spicules, each with three or four pointed rays. Variable in shape and crunchy to the touch, most species are small and lobe- or tube-shaped.

3¼ in
8 cm

LEMON SPONGE
Leucetta chagosensis
F: Leucettidae
Growing on steep coral reefs in the western Pacific, this saclike sponge provides a splash of bright color.

⅜–1½ in
1–4 cm

CLATHRINA CLATHRUS
F: Clathrinidae
Made up of many tubes, each only a few millimeters wide, this northeastern Atlantic sponge is a distinctive yellow color.

spicules surrounding osculum

1¾–2 in / 2–5 cm

PURSE SPONGE
Sycon ciliatum
F: Sycettidae
This simple, hollow sponge of northeast Atlantic coasts has spiky calcareous spicules surrounding its osculum.

tiny pores allow water entry

4 in
10 cm

3¼ in
8 cm

LEUCONIA NIVEA
F: Baeriidae
The shape of this northeastern Atlantic sponge varies from lobes and cushions to crusts. It grows in areas of high water movement.

RED PURSE SPONGE
Grantessa sp.
F: Heteropiidae
Shaped like a small gourd, this delicate sponge grows between coral heads on shallow reefs in Malaysia and Indonesia.

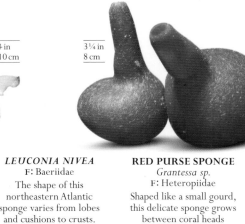

DEMOSPONGES

Ninety-five percent of all Demospongiae fall into this group. Although they are varied in appearance, most have a skeleton made up of scattered silicon dioxide spicules and a flexible organic collagen called spongin. A few encrusting species have no skeleton and some have only spongin.

3¼ ft
1 m

BROWN TUBE SPONGE
Agelas tubulata
F: Agelasidae
This sponge consists of uneven brown tubes arranged in clusters. It is common on deep reefs in the Caribbean and Bahamas.

14 in
35 cm

MEDITERRANEAN BATH SPONGE
Spongia officinalis adriatica
F: Spongiidae
This sponge has a pliable elastic skeleton, which retains its shape after it has been cleaned and dried. It is used for washing.

GOLF BALL SPONGE
Paratetilla bacca
F: Tetillidae
One of many tropical ball-shaped sponges, this species grows on sheltered coral reefs in the western Pacific.

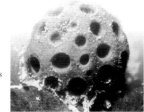

4¾ in / 12 cm

2–4 in
5–10 cm

ELEPHANT HIDE SPONGE
Pachymatisma johnstonia
F: Geodiidae
Mounds of this tough sponge can cover large areas of rocks and wrecks in clear coastal waters of the northeast Atlantic.

12–16 in
30–40 cm

RED TREE SPONGE
Negombata magnifica
F: Podospongiidae
Attempts to farm this beautiful Red Sea sponge are meeting with some success. It contains chemicals of potential medical importance.

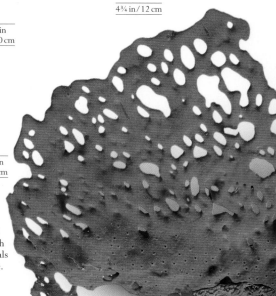

BORING SPONGE
Cliona celata
F: Clionaidae
Although it appears as yellow lumps, much of this European sponge remains hidden as strands growing through shells and calcareous rocks.

20 in
50 cm

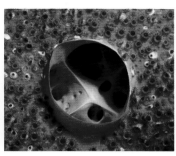

BORING SPONGE
Cliona delitrix
F: Clionaidae
Water leaves this Caribbean sponge through large holes called oscula (shown here). This sponge bores through corals by secreting acid.

6–12 in
15–30 cm

YELLOW FINGER SPONGE
Callyspongia ramosa
F: Callyspongiidae
The bright color of this tropical Pacific sponge is due to the chemicals it contains. Sponge extracts are used by the pharmaceutical industry.

12–16 in
30–40 cm

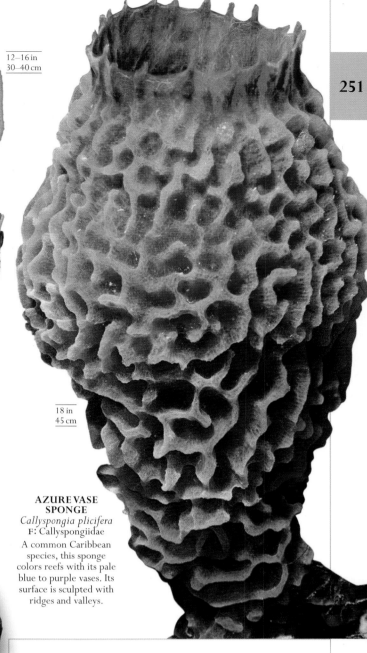

18 in
45 cm

AZURE VASE SPONGE
Callyspongia plicifera
F: Callyspongiidae
A common Caribbean species, this sponge colors reefs with its pale blue to purple vases. Its surface is sculpted with ridges and valleys.

BREADCRUMB SPONGE
Halichondria panicea
F: Halichondriidae
This northeastern Atlantic sponge grows as crusts on rocky shores and in shallow water. Its color comes from symbiotic algae.

12 in
30 cm

SPIRASTRELLA CUNCTATRIX
F: Spirastrellidae
One of many colorful encrusting sponges found in the ocean, this species covers inshore rocks in the Mediterranean and North Atlantic.

³⁄₈–³⁄₄ in
1–2 cm

NIPHATES SP.
F: Niphatidae
Like many sponges, this tropical species grows as irregular lumps and crusts. Its surface is covered in small projections called conuli.

2–4 in
5–10 cm

PINK VASE SPONGE
Niphates digitalis
F: Niphatidae
The shape of this Caribbean reef sponge varies from a tube to an open vase. Small anemones often grow on its rough surface.

12 in
30 cm

BARREL SPONGE
Xestospongia testudinaria
F: Petrosiidae
Small fish and many invertebrates live on and in this huge Indo-Pacific sponge, which grows big enough for a person to fit inside.

6½ ft
2 m

GLASS SPONGES

Hexactinellida are a small group of sponges lives mostly in the deep ocean, where some form reeflike mounds up to 65 ft (20m) high. The six-rayed, starlike, silica spicules that make up the skeleton are usually fused together as a rigid lattice. After death, the skeleton remains as a ghostly outline.

13½ in
35 cm

VENUS'S FLOWER BASKET
Euplectella aspergillum
F: Euplectellidae
Growing in tropical oceans below about 500 ft (150m), this sponge was collected and displayed by Victorians for its delicate silica skeleton.

rigid lattice of silica spicules

4 in
10 cm

ELK HORN SPONGE
Axinella damicornis
F: Axinellidae
This common bright yellow Mediterranean sponge also grows on steep rock faces on Irish coasts and on the west coast of Great Britain.

APLYSINA ARCHERI
F: Aplysinidae
The graceful long tubes of this sponge sway gently with the current on the Caribbean reefs where it grows.

2½–6½ ft
0.8–2 m

DEEP SEA GLASS SPONGE
Hyalonema sieboldi
F: Hyalonematidae
This glass sponge lives in the deep sea, and is kept clear of the mud by its long thin stalk of silica spicules.

19–32 in
48.5–80.5 cm

CNIDARIANS

This phylum includes jellyfish, corals, and anemones. They have stinging tentacles to catch live prey, which they digest in a simple saclike gut.

All cnidarians are aquatic and most are marine. They have two body plans: a free-swimming bell-like shape called a medusa—as seen in jellyfish—and a static polyp, typical of anemones. Neither medusae nor polyps have a head or front end. Their tentacles encircle a single gut opening, used to take in food and to eliminate waste.

A cnidarian's nervous system consists of a simple network of fibers; there is no brain. This means that the animal's behavior is usually simple. Although they are carnivorous, cnidarians cannot actively pursue their prey, with the possible exception of the box jellyfish. Instead, most species wait for swimming animals to blunder into the reach of their tentacles.

STINGS

Cnidarians' outer skins—and in some species the inner skin as well—are peppered with tiny stinging capsules of a kind that is unique to animals of

this phylum. These stinging organs are called cnidocysts (or cnidocytes), and it is from these that cnidarians get their name. Cnidocysts are concentrated on the tentacles, and are triggered by touch or chemical signals, both when the animal comes into contact with potential prey and when it is under attack. Each cnidocyst contains a microscopic sac of venom and discharges a tiny coiled "harpoon" to inject the venom into flesh. Some stings can penetrate human skin and inflict severe pain, but the vast majority of cnidarians are harmless to humans.

ALTERNATING LIFE CYCLE

Many cnidarians have a life cycle that alternates between medusa and polyp—usually with one of these forms dominating, although in some groups one or the other form is missing. The free-swimming medusa is normally the sexual stage. In most species fertilization is external: sperm and eggs are released in open water, forming planktonic larvae that resemble tiny flatworms. These settle in order to grow into polyps. Specialized polyps then produce new medusae to complete the cycle.

PHYLUM	CNIDARIA
CLASSES	4
ORDERS	22
FAMILIES	278
SPECIES	11,300

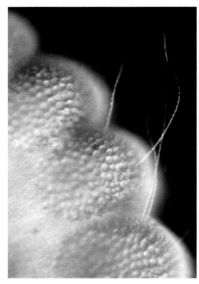

This microscopic view shows cnidocysts (stinging cells) that have been triggered to discharge their venom-laden harpoons.

BOX JELLYFISH

Members of the class Cubozoa are found in tropical and subtropical waters. They differ from true jellyfish in having a greater ability to control the direction and speed of their movement, rather than drifting with the currents. A flapping skirt at the bottom of the bell powers them at considerable speed. Box jellyfish have eyes, set in clusters on the sides of their transparent bell. They can see enough to steer past obstacles and to spot prey.

bell controls speed

1–9¾ ft
0.3–3 m

SEA WASP
Chironex fleckeri
F: Chirodropidae
This Indo-Pacific species—the largest box jellyfish—delivers a particularly painful sting and has caused human fatalities.

TRUE JELLYFISH

The familiar jellyfish bell is the medusa stage of the Scyphozoa life cycle. The polyp is reduced, or in some deepwater forms, missing. Jellyfish polyps undergo strobilation, a process that produces new tiny medusae by budding. Rhizostome jellyfish lack a fringe of tentacles around the bell.

UPSIDE-DOWN JELLYFISH
Cassiopea andromeda
F: Cassiopeidae
Superficially resembling an anemone, this Indo-Pacific rhizostome jellyfish lives at the bottom of lagoons, mouth upward, with its bell pulsating to circulate water.

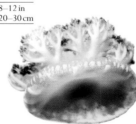

8–12 in
20–30 cm

4–8 in
10–20 cm

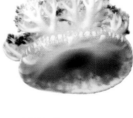

18–28 in
45–70 cm

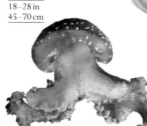

DEEP SEA CORONATE JELLYFISH
Periphylla periphylla
F: Periphyllidae
This is one of many little-known deepwater species in a group of jellyfish characterized by a bell with an encircling groove.

WHITE-SPOTTED JELLYFISH
Phyllorhiza punctata
F: Mastigiidae
This rhizostome jellyfish is native to the western Pacific, but has been introduced to North America, where it may threaten commercial fisheries.

1½ in
4 cm

STALKED JELLY FISH
Haliclystus auricula
F: Lucernariidae
Although it lives more like a static polyp, this curious North Atlantic cnidarian, a possible cousin of box jellyfish, is in fact a medusa.

5½–6½ in
14–16 cm

8–16 in
20–40 cm

MOON JELLYFISH
Aurelia aurita
F: Ulmaridae
This globally widespread genus has four long "arms" as well as small marginal tentacles. It swarms to breed near coasts and polyps settle in estuaries.

SPOTTED LAGOON JELLYFISH
Mastigias papua
F: Mastigiidae
Like other rhizostome jellyfish, this algae-containing species traps planktonic food with mucus. It enters South Pacific lagoons, including those of oceanic islands.

HYDROZOANS

Most animals of the class Hydrozoa live as branching sea-dwelling colonies forming miniature polyp-bearing forests. Some anchor themselves to surfaces by creeping horizontal stems, called stolons. Colonies are supported by a horny translucent sheath and some of the polyps produce sexual medusae. Freshwater hydra lack medusae: instead, sex organs develop directly on solitary polyps.

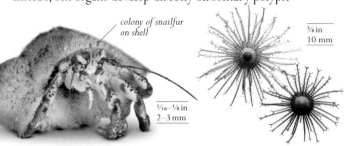

colony of snailfur on shell

SNAILFUR
Hydractinia echinata
F: Hydractiniidae

One of a family of colonial spiny hydrozoans, this northeast Atlantic species grows on the surface of snail shells containing hermit crabs.

1/16–1/8 in
2–3 mm

3/8 in
10 mm

BLUE BUTTONS
Porpita porpita
F: Porpitidae
O: Anthoathecata

This colonial, jellyfish-like hydrozoan of tropical oceans is sometimes regarded as a highly modified individual polyp.

2–4 in
5–10 cm

PURPLE LACE "CORAL" HYDROID
Distichopora violacea
F: Stylasteridae

Like some other hydrozoans, colonies of this Indo-Pacific lace coral have polyps specialized for different functions, such as feeding and stinging.

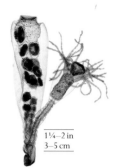

1 1/4–2 in
3–5 cm

COMMON OBELIA
Obelia geniculata
F: Campanulariidae

This is a worldwide hydrozoan, abundant on intertidal seaweeds. It grows in zigzag colonies of cup-encased polyps that sprout from creeping horizontal stems.

PORTUGUESE MAN OF WAR
Physalia physalis
F: Physaliidae

Resembling a jellyfish, this is really an ocean-going colony of hydrozoans. Stinging tentacles and specialized polyps are suspended from a gas-filled float.

16 in
40 cm

HULA SKIRT SIPHONPHORE
Physophora hydrostatica
F: Physophoridae

This widespread free-floating hydrozoan colony has a smaller float than its relative, the Portuguese man of war, and carries prominent swimming bells.

3/8 in
10 mm

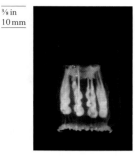

EIGHT-RIBBED HYDROMEDUSA
Melicertum octocostatum
F: Melicertidae

This hydrozoan from the North Atlantic and North Pacific oceans belongs to a family related to those with cup-encased polyps. It is known mainly from medusae.

gas-filled float

33–165 ft
10–50 m

3/8 in
10 mm

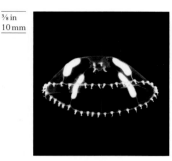

CUP HYDROMEDUSA
Phialella quadrata
F: Phialellidae

Like related *Obelia*, branching colonies of this globally widespread hydrozoan release free-living medusae, which reproduce sexually.

16–23 1/2 in
4–6 cm

FEATHER HYDROID
Aglaophenia cupressina
F: Aglaopheniidae

Feather hydroids and the closely related sea ferns belong to a group of colonial hydrozoans with cup-encased polyps. This is an Indo-Pacific species.

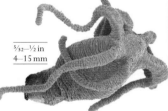

5/32–1/2 in
4–15 mm

COMMON HYDRA
Hydra vulgaris
F: Hydridae

Lacking the medusa stage, hydras have small freshwater polyps that can reproduce asexually by budding. This is a globally widespread cold-water species, shown here in enhanced color.

3/4–1 in
2–2.5 cm

FRESHWATER JELLYFISH
Craspedacusta sowerbyi
F: Olindiasidae

Sporadic in freshwater ponds, lakes, and streams worldwide, minute polyps of this animal develop into dominant medusae.

1 1/2–6 1/2 ft
0.5–2 m

LION'S MANE JELLYFISH
Cyanea capillata
F: Cyaneidae

A large species of Arctic waters, this jellyfish has many tentacles arranged in dense groups. It has a powerful sting and its prey includes fish.

umbrella-like bell

6–12 in
15–30 cm

BLUE JELLYFISH
Cyanea lamarckii
F: Cyaneidae

A smaller cousin of the lion's mane jellyfish, this North Atlantic species has a milder sting. It preys on ctenophores and other medusae.

oral arm

4–8 in
10–20 cm

PINK-HEARTED HYDROID
Tubularia sp.
F: Tubulariidae

Tubularia hydrozoans have polyps on long stems with two whorls of tentacles, one around the polyp base and one around its mouth.

16–20 in
40–50 cm

FIRE "CORAL" HYDROID
Millepora sp.
F: Milleporidae

Like others of its family, this fiercely stinging colonial hydrozoan has a calcified skeleton and is a reef-builder. It is distantly related to true stony corals.

ANEMONES AND CORALS

Unlike other cnidarians, the Anthozoa lack the free-swimming medusa stage altogether and most remain static throughout their life cycle. The polyps, many of which have a flowerlike appearance, produce sperm and eggs. Members of this class include solitary anemones, colonial sea pens, soft corals, and tropical reef-forming stony corals.

8–12 in
20–30 cm

slender red or yellow lobe

3¼–5 ft
1–1.5 m

polyp cluster

COMMON SEA FAN
Gorgonia ventalina
F: Gorgoniidae
Sea fans, such as this Caribbean species, have upright fanlike colonies supported by a central axis, which is stiffened with a horny substance called gorgonin.

8–12 in
20–30 cm

3½–20 ft
1–6 m

white polyps

TOADSTOOL LEATHER CORAL
Sarcophyton trocheliophorum
F: Alcyoniidae
This soft coral can form enormous leathery colonies on tropical reefs of the Indo-Pacific. It is fast-growing, nourished by photosynthetic algae.

4–6 in
10–15 cm

SEA STRAWBERRY
Gersemia rubiformis.
F: Nephtheidae
This stalked soft coral forms bright pink lumpy colonies. It occurs in northern parts of the Pacific and Atlantic Oceans.

CARNATION CORAL
Dendronephthya sp.
F: Nephtheidae
Typical of a family of stalked soft corals with polyps arranged in clusters, this colorful species occurs on Indo-Pacific tropical reefs.

20–40 in
50–100 cm

4–8 in
10–20 cm

flexible body

16–20 in
40–50 cm

RED DEAD-MAN'S FINGERS
Alcyonium glomeratum
F: Alcyoniidae
This is a more slender, upright ally of dead-man's fingers, found on sheltered rocky European coastlines. The color varies from red to yellow.

COMMON DEAD-MAN'S FINGERS
Alcyonium digitatum
F: Alcyoniidae
A typical soft coral, this thickly lobed European species has colonies of polyps, which are carried on a fleshy mass unsupported by a rigid skeleton.

RED CORAL
Corallium rubrum
F: Coralliidae
Colonies of this Mediterranean sea fan (not a true coral) are supported by a skeleton of tiny, enmeshed calcified needles. It is valued for jewelry.

14–16 in
35–40 cm

KIDNEY SEA PEN
Sarcoptilus grandis
F: Pennatulidae
Widespread in temperate waters, this sea pen has kidney-shaped branches that occur in rows along each side of its body.

ORANGE SEA PEN
Ptilosarcus gurneyi
F: Pennatulidae
One of many brightly colored sea pens, this species from the Pacific coast of North America retracts into its burrow when disturbed by predators.

1½–6½ ft
0.5–2 m

calcareous tubes

6–12 in
15–30 cm

tentacle

20–40 in
50–100 cm

20–40 in
50–100 cm

WHITE SEA WHIP
Junceella fragilis
F: Ellisellidae
Sea whips are threadlike relatives of sea fans, with their supporting horny axis reinforced with calcium. This is an Indonesian reef species.

RED SEA WHIP
Ellisella sp.
F: Ellisellidae
Ellisella sea whips form two-prong branching colonies, some of which may form dense underwater bushes. They occur in tropical and temperate waters.

ORGAN PIPE CORAL
Tubipora musica
F: Tubiporidae
This Indo-Pacific soft coral has polyps in upright, calcified tubes, which are connected to the colony by a rootlike network at the base.

BLUE CORAL
Heliopora coerulea
F: Helioporidae
Despite its hard, calcified skeleton, this species is more closely related to soft coral than stony coral. It is the only member of its order.

4–8 in
10–20 cm

FLOWERPOT CORAL
Goniopora columna
F: Poritidae
This species has daisylike polyps
that are greatly lengthened when
fully extended. It is an
Indo-Pacific relative of the
lobe coral.

13–16 ft
4–5 m

LOBE CORAL
Porites lobata
F: Poritidae
One of the most common
reef-builders of the Indo-Pacific,
this coral forms large encrusting
colonies in places buffeted by
strong wave action.

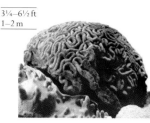

3¼–6½ ft
1–2 m

LOBED BRAIN CORAL
Lobophyllia sp.
F: Mussidae
Massive colonies of this brain
coral—either flattened or
dome-shaped—occur on reefs
in tropical waters of
the Indo-Pacific.

4–4¾ in
10–12 cm

DAHLIA ANEMONE
Urticina felina
F: Actiniidae
Sticky swellings on this anemone
may carry so much debris that,
with tentacles retracted, it
resembles a small pile of gravel.
It occurs around the North Pole.

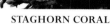

3¼–9¾ ft
1–3 m

STAGHORN CORAL
Acropora sp.
F: Acroporidae
The branching stag horn corals
are among the largest tropical
reef-builders. Nourished by
photosynthetic algae, these
species are fast growing.

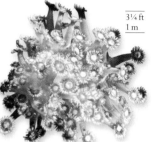

3¼ ft
1 m

DAISY CORAL
Goniopora sp.
F: Poritidae
Long-reaching polyps of
Goniopora stony corals on tropical
reefs typically have 24 tentacles.
They are among the most
flowerlike of corals.

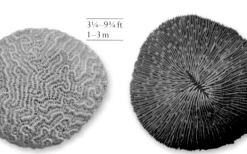

3¼–9¾ ft
1–3 m

**LARGE-GROOVED
BRAIN CORAL**
Colpophyllia sp.
F: Faviidae
The hemispherical brainlike
structure of this coral is typical of
its family. It is a tropical reef-builder
that contains photosynthetic algae.

4–8 in
10–20 cm

MUSHROOM CORAL
Fungia fungites
F: Fungiidae
This is one of many tropical
mushroom corals that are not
reef-building, but live among
other species as solitary polyps
crawling over the ocean floor.

20–40 in
50–100 cm

**MAGNIFICENT
SEA ANEMONE**
Heteractis magnifica
F: Stichodactylida
This giant anemone is from
Indo-Pacific reefs, where it lives in
close partnership with a variety of
fish species, including clown fish.

300–600 ft / 100–200 m

**ATLANTIC COLD
WATER CORAL**
Lophelia pertusa
F: Caryophylliidae
Unlike many deep sea corals
lacking nutrient-supplying
algae, this North Atlantic
species forms extensive reefs,
although only very slowly.

1–6 in
2.5–15 cm

PLUMOSE ANEMONE
Metridium senile
F: Metridiidae
This is a globally widespread
member of a family of
anemones that is characterized
by a fuzzy mass of tentacles. It
can split to form genetically
identical populations.

2–2¾ in
5–7 cm

**SNAKELOCKS
ANEMONE**
Anemonia viridis
F: Actiniidae
The conspicuously long
tentacles of this common
European intertidal anemone
rarely retract, even when
exposed at low tide.

⅜–½ in
10–15 mm

DEVONSHIRE CUP CORAL
Caryophyllia smithii
F: Caryophylliidae
This northeastern Atlantic species
belongs to a family containing cold
water corals, some with large,
anemone-like polyps. Barnacles
frequently attach to it.

4–6 in
10–15 cm

COMMON TUBE ANEMONE
Cerianthus membranaceus
F: Cerianthidae
Tube anemones burrow through
sediment, using unique nonstinging
cnidocysts to build feltlike tubes from
mucus. This species occurs on
European offshore mud.

20–40 in
50–100 m

**ATLANTIC
BLACK CORAL**
Antipathes sp.
F: Antipathidae
Mostly found in deep waters,
black corals consist of spiny polyp
colonies encased in slender,
horny exoskeletons.

FLATWORMS

Structurally among the simplest of animals, flatworms live wherever a moist habitat can provide their paper-thin bodies with oxygen and food.

Flatworms belong to the major phylum known as Platyhelminthes, and are found in various forms in the ocean, in freshwater ponds, or even within the bodies of other animals. Superficially resembling leeches, flatworms are in fact far simpler animals. With no blood system and no organs for breathing, they use their entire body surface to absorb oxygen from moisture outside. The smallest flatworms, which have no gut, also absorb their food in this way. In other species the gut has one opening but branches frequently, so that digested food

can still reach all tissues, even without a blood stream to circulate the nutrients. Some free-living flatworms are detritus-eaters, gliding along on a bed of microscopic beating hairs. Others prey on other invertebrates.

INTERNAL PARASITES

Tapeworms and flukes are parasites. Inside the host animal, their flattened, gutless bodies are perfectly suited for absorbing nutrients. Many have complex ways of passing from host to host and may infect more than one type of animal. They enter the body of the host in infected food or sometimes by penetrating the skin. Once inside, a parasite may go deeper into the body—burrowing through the gut wall to lodge in vital organs.

PHYLUM	PLATYHELMINTHES
CLASSES	5
ORDERS	33
FAMILIES	About 400
SPECIES	About 20,000

DEBATE
A NEW PHYLUM?

Traditionally, the tiny, gutless, and brainless marine animals called acoelans are viewed as flatworms. Controversial studies suggest that they belong in a new phylum of their own, which would make them the first with a "head" and a "tail," in contrast with radial (round) cnidarians.

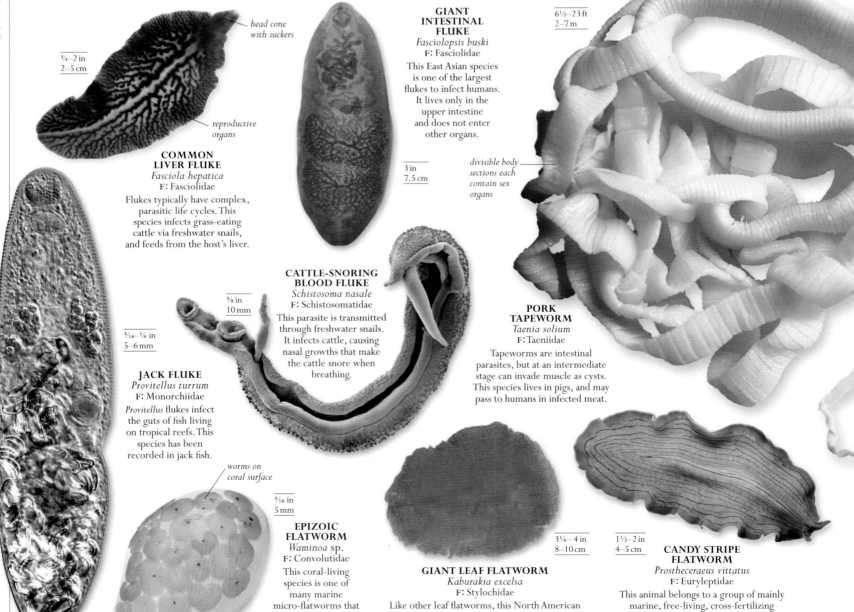

COMMON LIVER FLUKE
Fasciola hepatica
F: Fasciolidae
Flukes typically have complex, parasitic life cycles. This species infects grass-eating cattle via freshwater snails, and feeds from the host's liver.

¾–2 in
2–5 cm

head cone with suckers

reproductive organs

GIANT INTESTINAL FLUKE
Fasciolopsis buski
F: Fasciolidae
This East Asian species is one of the largest flukes to infect humans. It lives only in the upper intestine and does not enter other organs.

3 in
7.5 cm

6½–23 ft
2–7 m

divisible body sections each contain sex organs

CATTLE-SNORING BLOOD FLUKE
Schistosoma nasale
F: Schistosomatidae
This parasite is transmitted through freshwater snails. It infects cattle, causing nasal growths that make the cattle snore when breathing.

⅜ in
10 mm

JACK FLUKE
Provitellus turrum
F: Monorchiidae
Provitellus flukes infect the guts of fish living on tropical reefs. This species has been recorded in jack fish.

³⁄₁₆–¼ in
5–6 mm

PORK TAPEWORM
Taenia solium
F: Taeniidae
Tapeworms are intestinal parasites, but at an intermediate stage can invade muscle as cysts. This species lives in pigs, and may pass to humans in infected meat.

worms on coral surface

EPIZOIC FLATWORM
Waminoa sp.
F: Convolutidae
This coral-living species is one of many marine micro-flatworms that resemble planktonic larvae of cnidarians.

³⁄₁₆ in
5 mm

GIANT LEAF FLATWORM
Kaburakia excelsa
F: Stylochidae
Like other leaf flatworms, this North American intertidal species is mainly carnivorous. It smothers its prey with a mouth-bearing extension of its gut.

3¼–4 in
8–10 cm

1½–2 in
4–5 cm

CANDY STRIPE FLATWORM
Prostheceraeus vittatus
F: Euryleptidae
This animal belongs to a group of mainly marine, free-living, cross-fertilizing hermaphrodites called polyclads. It is an Atlantic Ocean species.

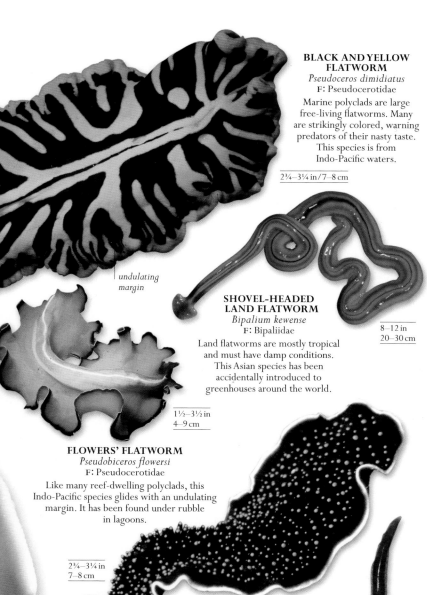

BLACK AND YELLOW FLATWORM
Pseudoceros dimidiatus
F: Pseudocerotidae
Marine polyclads are large free-living flatworms. Many are strikingly colored, warning predators of their nasty taste. This species is from Indo-Pacific waters.

2¾–3¼ in / 7–8 cm

undulating margin

SHOVEL-HEADED LAND FLATWORM
Bipalium kewense
F: Bipaliidae
Land flatworms are mostly tropical and must have damp conditions. This Asian species has been accidentally introduced to greenhouses around the world.

8–12 in
20–30 cm

FLOWERS' FLATWORM
Pseudobiceros flowersi
F: Pseudocerotidae
Like many reef-dwelling polyclads, this Indo-Pacific species glides with an undulating margin. It has been found under rubble in lagoons.

2¾–3¼ in
7–8 cm

GOLD-SPECKLED FLATWORM
Thysanozoon nigropapillosum
F: Pseudocerotidae
Many *Thysanozoon* polyclads are coated with pimply swellings, which are yellow-tipped on this otherwise velvety black species of the Indo-Pacific.

DRAB FRESHWATER FLATWORM
Dugesia lugubris
F: Planariidae
Triclads are flatworms with a triple-branched gut. This European species is one of many triclads found in freshwater habitats; others are marine.

½–¾ in
1.5–2 cm

NEW ZEALAND LAND FLATWORM
Arthurdendyus triangulatus
F: Geoplanidae
This large soil-living species is native to New Zealand, but has invaded Europe. It preys on earthworms.

BROWN FRESHWATER FLATWORM
Dugesia tigrina
F: Planariidae
Native to freshwater habitats of North America, this species has been introduced to Europe.

⅜–½ in
1–1.5 cm

STREAM FLATWORM
Dugesia gonocephala
F: Planariidae
Many of the freshwater *Dugesia* flatworms, such as this European species of running water, have earlike flaps for detecting water currents.

¾–1¼ in
2–3 cm

ROUNDWORMS

The simple, cylindrical roundworms are remarkably successful—surviving almost everywhere, resisting drought, and reproducing rapidly.

The phylum Nematoda is ubiquitous. There can be millions of roundworms in a square metre of soil, and they also live in freshwater or marine habitats. Many are parasites. These worms, or nematodes, can be extraordinarily prolific, producing hundreds of thousands of eggs per day. When environmental conditions deteriorate they can survive heat, frost, or drought by enclosing themselves in a cyst and becoming dormant.

Roundworms have muscle-lined body cavity and two openings to their gut—the mouth and anus. Their cylindrical bodies are coated in a tough layer, called a cuticle—similar to that of arthropods—which they moult periodically as they grow.

PHYLUM	NEMATODA
CLASSES	2
ORDERS	12
FAMILIES	About 160
SPECIES	About 20,000

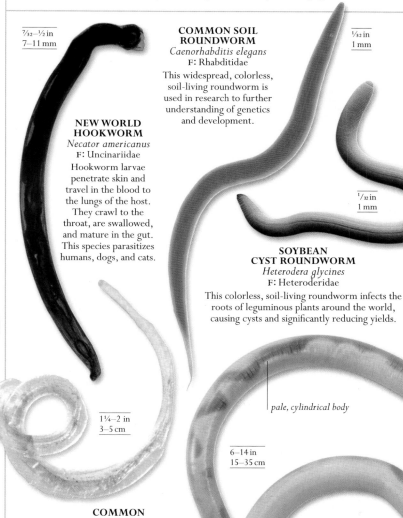

7/32–½ in
7–11 mm

COMMON SOIL ROUNDWORM
Caenorhabditis elegans
F: Rhabditidae
This widespread, colorless, soil-living roundworm is used in research to further understanding of genetics and development.

1/32 in
1 mm

NEW WORLD HOOKWORM
Necator americanus
F: Uncinariidae
Hookworm larvae penetrate skin and travel in the blood to the lungs of the host. They crawl to the throat, are swallowed, and mature in the gut. This species parasitizes humans, dogs, and cats.

1/32 in
1 mm

4–6½ in
10–17 cm

SOYBEAN CYST ROUNDWORM
Heterodera glycines
F: Heteroderidae
This colorless, soil-living roundworm infects the roots of leguminous plants around the world, causing cysts and significantly reducing yields.

pale, cylindrical body

1¼–2 in
3–5 cm

6–14 in
15–35 cm

COMMON WHIPWORM
Trichuris trichiura
F: Trichuridae
Like many other gut parasites, the mainly tropical whipworm infects humans who eat food contaminated with feces. It completes its life cycle in the intestines.

LARGE INTESTINAL ROUNDWORM
Ascaris sp.
F: Acarididae
A common parasite of humans in regions of poor sanitation, this species enters the intestine in contaminated food before infecting the lungs.

SEGMENTED WORMS

With more complex muscle and organ systems than flatworms, many species of this phylum, called Annelida, are accomplished swimmers or burrowers.

Segmented worms include earthworms, ragworms, and leeches. Their blood circulates in vessels and they have a firm sac of fluid (the coelom) running the length of their body, which keeps motion of the gut separate from motion of the body wall. The coelum is split into sections, with each one corresponding to a body segment. Each body segment contains a set of muscles, and coordination of these muscle sets can send a wave of contraction down the length of the body, or make it flex to and fro. This makes many annelids highly mobile both on land and in the water.

Marine segmented worms—the predatory ragworms and their filter-feeding relatives —carry bundles of bristles along their body, typically borne on little paddles that help with swimming, burrowing, or even walking. This group of bristled, segmented worms are called polychaetes or bristle worms.

Land-living earthworms are more sparsely bristled detritus-feeders, and are important recyclers of dead vegetation and aerators of soil. Many leeches are specialized further, and carry suckers to extract blood from a host. Their saliva contains chemicals that stop blood clotting. Other leeches are predatory. Both earthworms and leeches have a saddle-shaped, glandular structure around their bodies called a clitellum, which they use to make their egg cocoons.

PHYLUM	ANNELIDA
CLASSES	4
ORDERS	8
FAMILIES	About 130
SPECIES	About 15,000

SLUDGE WORM
Tubifex sp.
F: Naididae
This widely distributed worm can be seen in mud polluted by sewage, the front end buried, the rear end wiggling to extract oxygen.

¾–2¾ in
2–7 cm

MEGADRILE EARTHWORM
Glossoscolex sp.
F: Glossoscolecidae
Glossoscolex are large earthworms from tropical C. and S. America. Many are found in rain forest habitats.

20 in
50 cm

TIGER WORM
Eisenia foetida
F: Lumbricidae
This European inhabitant of rotting vegetation secretes defensive pungent fluid and, like other earthworms, has a "saddle" for producing cocoons of eggs.

4–6 in
10–15 cm

COMMON EARTHWORM
Lumbricus terrestris
F: Lumbricidae
This earthworm—native to Europe but introduced elsewhere—drags leaves into its burrow at night as a source of food.

6–10 in
15–25 cm

clitellum or "saddle"

CHRISTMAS TREE TUBE WORM
Spirobranchus giganteus
F: Serpulidae
Characterized by its spiral whorls of tentacles, used for filter feeding and extracting oxygen, this species is widespread on tropical reefs.

1½–2¾ in
4–7 cm

VELVET WORMS

These soft-bodied worms, cousins of arthropods, lumber slowly along dark forest floors like giant caterpillars, but they are extraordinary hunters.

These animals have the body of an earthworm and the multiple limbs of a millipede, but belong in a phylum of their own: Onychophora. Velvet worms are rarely seen in their warm native rain

forests of tropical America, Africa, and Australasia; they shun the open, preferring to hide in crevices and leaf litter. They come out at night or after rainfall to hunt other invertebrates. Velvet worms capture their prey in a unique way: they immobilize their victims by spraying them with sticky slime that is produced by glands that open through pores straddling the mouth.

PHYLUM	ONYCHOPHORA
CLASS	1
ORDER	1
FAMILIES	2
SPECIES	About 200

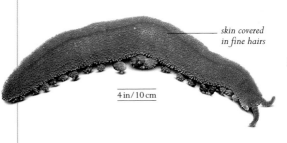

skin covered in fine hairs

4 in/10 cm

SOUTHERN AFRICAN VELVET WORM
Peripatopsis moseleyi
F: Peripatopsidae
This species belongs to a family with a globally southern distribution.

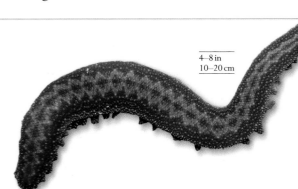

4–8 in
10–20 cm

CARIBBEAN VELVET WORM
Epiperipatus broadwayi
F: Peripatidae
This species is one of a family of equatorial velvet worms, which typically have more legs than those of the southern continents.

FIRE WORM
Hermodice carunculata
F: Amphinomidae

A tropical Atlantic offshore species, this polychaete preys on corals, sucking out the soft flesh from the hard skeleton. Its body extensions have painfully irritating bristles.

2¼–12 in
6–30 cm

SEA MOUSE
Aphrodita aculeata
F: Aphroditidae

This mud-burrowing scale worm is found in the shallow waters of North Europe. Its scales are covered with a hairy coat.

4–8 in
10–20 cm

1–1¼ in/2.5–3 cm

SEA CUCUMBER SCALE WORM
Gastrolepidia clavigera
F: Polynoidea

This Indo-Pacific polychaete has flattened back scales. It is parasitic on sea cucumbers.

4¾–10 in
12–25 cm

LUGWORM
Arenicola marina
F: Arenicolidae

An earthwormlike polychaete, the lugworm lives in burrows on beaches and mud-flats, where it ingests sediment and feeds on detritus.

3¼–13 ft
1–4 m

HONEYCOMB WORM
Sabellaria alveolata
F: Sabellariidae

This tubeworm builds its tubes from sand and shell fragments. Dense populations form honeycomb-like reefs in the Atlantic and Mediterranean.

2–6 in
5–15 cm

GREEN PADDLE WORM
Eulalia viridis
F: Phyllodocidae

Paddleworms are active carnivores with leaf-shaped flaps on their body extensions. This European species lives among intertidal rocks and kelp.

PACIFIC FEATHERDUSTER WORM
Sabellastarte sanctijosephi
F: Sabellidae

A tropical Indo-Pacific tubeworm, this species is common along coastlines, including coral reefs and tidal pools.

3¼–4 in
8–10 cm

10–16 in/25–40 cm

KING RAGWORM
Alitta virens
F: Nereididae

Ragworms, close relatives of paddleworms, have two-pronged body extensions. This Atlantic mud-burrowing species can deliver a painful bite.

2–2¾ in
5–7 cm

RED TUBE WORM
Serpula vermicularis
F: Serpulidae

Serpulid tubeworms build hard chalky tubes. Like most serpulids, this widespread species has tentacles modified to plug the tube after retraction.

1/16–1/8 in
2–3 mm

NORTHERN SPIRAL TUBE WORM
Spirorbis borealis
F: Serpulidae

The small coiled tube of this tubeworm is cemented to brown *Fucus* seaweeds and kelps along North Atlantic coastlines.

GIANT TUBE WORM
Riftia pachyptila
F: Siboglinidae

This tubeworm lives in the hot sulfur-rich darkness of ocean-floor volcanic vents in the Pacific. The red plume harbors bacteria that make its food using chemicals from these vents.

body used as anchor

6½–7¾ ft
2–2.4 m

WATER BEARS

Visible only through a microscope, stubby-legged, agile water bears share their aquatic community with microbes and far simpler invertebrates.

Tiny water bears, or tardigrades (meaning "slow-stepped") clamber through miniature forests of water weed, four pairs of short legs clinging on with clawed feet. Most are less than a millimeter long. They abound among moss or algae, where many use their needlelike jaws to pierce the cells of this vegetation and suck its sap. Many water bear species are known only as females, which reproduce asexually by spawning offspring from unfertilized eggs. If its habitat dries up, a water bear is able to survive by going into a kind of suspended animation called cryptobiosis, shrivelling into a husk, sometimes for years at a time, until rainfall revives it.

PHYLUM	TARDIGRADA
CLASSES	3
ORDERS	5
FAMILIES	20
SPECIES	About 1,000

MOSS WATER BEAR
Echiniscus sp.
F: Echiniscidae

Many water bears live in moss, but their ability to survive in a dried state has helped a number of species to disperse worldwide.

0.25 mm

clawed feet

SEAWEED WATER BEAR
Echiniscoides sigismundi
F: Echiniscoididae

One of many little-known marine water bears, this species has been recorded among seaweed in coastal locations around the world.

0.25 mm

ARTHROPODS

Jointed legs and flexible armor have helped a phylum that includes winged insects and underwater crustaceans to evolve unrivaled diversity.

There are more species of arthropods known to scientists than all the other animal phyla added together—and doubtless many more await discovery. They have an extraordinary range of lifestyles including grazing, predation, filtering particles of food in water, and drinking fluids such as nectar or blood.

Arthropods are coated by an exoskeleton made of a tough material called chitin. It is sufficiently flexible around the joints to allow for mobility, but resists stretching. As arthropods grow, they must periodically molt their exoskeleton, which is replaced by a slightly larger one. The exoskeleton serves as protective armor but also helps reduce water loss in very dry habitats.

BODY PARTS

Arthropods evolved from a segmented ancestor, perhaps like an annelid worm. Body segmentation persists in all arthropods, but is especially obvious in millipedes and centipedes. In other groups, various segments have fused together into discrete body sections. Insects are divided into a sensory head, a muscular thorax with legs and wings, and an abdomen containing most of the internal organs. In arachnids and some crustaceans, the head and thorax are fused into one single section.

GETTING OXYGEN

Aquatic arthropods, such as crustaceans, breathe with gills. The bodies of most land arthropods—insects and myriapods—are permeated by a network of microscopic, air-filled tubes called tracheae. These open out through pores known as spiracles on the sides of the body, usually one pair per body section. Small muscles inside the spiracles regulate the flow of air by means of valves. In this way, oxygen can seep directly to all cells of the body and does not need to be transported in blood. Some arachnids breathe with tracheae; others with leafy chambers in their abdomen that evolved from the gills of their aquatic ancestors. Most use a combination of the two.

PHYLUM	ARTHROPODA
CLASSES	14
ORDERS	69
FAMILIES	About 2,650
SPECIES	About 1,230,000

CAMOUFLAGE OR MIMICRY

Many arthropods have adapted to their environment by blending imperceptibly into it. Stick insects, for example, look so like twigs that they are very difficult for predators to spot. In contrast, arthropods, such as wasps, have evolved bright colors that warn others they are distasteful or dangerous. The hornet moth is completely harmless but avoids predation because it looks and sounds exactly like a hornet. Despite their similar appearance, other anatomical features separate them into different families. The hornet moth belongs to the family Lepidoptera and the hornet to the family Hymenoptera.

MILLIPEDES AND CENTIPEDES

Millipedes and centipedes form a group of multi-segmented arthropods called myriapods. Millipedes typically have two legs per body segment; centipedes have only one per segment. Whereas millipedes are vegetarian, centipedes are predatory carnivores.

shiny brown exoskeleton

head

1½–2 in
4–5 cm

stout sensory antenna

BROWN GIANT PILL MILLIPEDE
Zephronia sp.
F: Sphaerotheriidae
Giant pill millipedes, such as this Bornean species, differ from their smaller northern hemisphere relatives in that they have 13 body sections rather than 12.

2–3 mm / 1/16–1/8 in

AMERICAN SHORT-HEADED MILLIPEDE
Brachycybe sp.
F: Andrognathidae
This is a North American representative of a family of small, flat millipedes that live in rotting wood and leaf litter.

3/8–3/4 in
1–2 cm

BRISTLY MILLIPEDE
Polyxenus lagurus
F: Polyxenidae
One of a family of millipedes armed with defensive bristles, this tiny species of the Northern Hemisphere lives under the bark of trees and in leaf litter.

BLACK GIANT PILL MILLIPEDE
Zoosphaerium sp.
F: Sphaerotheriidae
One of many southern hemisphere giant pill millipedes, this species is native to Madagascar.

1¼–1½ in
3–4 cm

woodlouselike body

WHITE-RIMMED PILL MILLIPEDE
Glomeris marginata
F: Glomeridae
Pill millipedes, such as this European species, have fewer body segments than other millipedes and can roll into a ball when disturbed.

frontal shield

¼–¾ in
0.6–2 cm

BLACK SNAKE MILLIPEDE
Tachypodoiulus niger
F: Julidae

½–1½ in
1.5–4 cm

Unlike its close relatives, this distinctive, white-legged millipede from western Europe spends much time above ground—even climbing trees and walls.

½–1¼ in
1.5–3 cm

AMERICAN GIANT MILLIPEDE
Narceus americanus
F: Spirobolidae

3–5 in
7.5–13 cm

This large, Atlantic coast millipede belongs to a family of mainly American, cylindrical millipedes. Like its relatives, it releases noxious chemicals in defense.

BROWN SNAKE MILLIPEDE
Julus scandinavius
F: Julidae

A member of a large family of cylindrical millipedes with ring-encased segments, this species occurs in European deciduous woodland, especially on acid soils.

8–11 in/20–28 cm

AFRICAN GIANT MILLIPEDE
Archispirostreptus gigas
F: Spirostreptidae

One of the largest of all millipedes, this species occurs in tropical Africa and is one of many species that release irritating chemicals to defend themselves.

BORING MILLIPEDE
Polyzonium germanicum
F: Polyzoniidae

This somewhat primitive millipede has a scattered distribution in Europe. It lives in woodland and resembles beech bud scales when coiled.

EASTERN FLAT-BACKED MILLIPEDE
Polydesmus complanatus
F: Polydesmidae

Polydesmids have projections on their exoskeleton that make them look flat-backed. This eastern European species is a fast runner.

½–1¼ in
1.5–3 cm

YELLOW EARTH CENTIPEDE
Geophilus flavus
F: Geophilidae

Eyeless geophilids have more segments (and therefore legs) than other centipedes. This European soil-dweller has been introduced to the Americas and Australia.

³⁄₁₆–¾ in/0.5–1.8 cm

¾–1½ in
2–3.5 cm

1½–2¼ in
4–6 cm

TANZANIAN FLAT-BACKED MILLIPEDE
Coromus diaphorus
F: Oxydesmidae

The shiny pitted surface typical of many of the eyeless flat-backed millipedes is especially prominent in this species from the tropics.

¾–1¼ in
2–3 cm

1–2 in
2.5–5 cm

BANDED STONE CENTIPEDE
Lithobius variegatus
F: Lithobiidae

Once thought to be confined to Britain, this centipede is now known in continental Europe, where native populations have unbanded legs.

¾–1¼ in/2–3 cm

BROWN STONE CENTIPEDE
Lithobus forficatus
F: Lithobiidae

Found sheltering beneath bark and rock, stone centipedes characteristically have 15 body segments. This globally widespread species frequents forest, garden, and seashore.

HOUSE CENTIPEDE
Scutigera coleoptrata
F: Scutigeridae

An unmistakable centipede, this long-legged animal with compound eyes is one of the fastest invertebrate runners. Native to the Mediterranean region, it has been introduced elsewhere.

poison fangs

TIGER GIANT CENTIPEDE
Scolopendra hardwickei
F: Scolopendridae

Many of the giant *Scolopendra* centipedes have vibrant warning colors. This Indian species is one of several with a tiger pattern.

8–10 in
20–25 cm

one pair of jointed legs per body segment

4–6 in
10–15 cm

BLUE-LEGGED CENTIPEDE
Ethmostigmus trigonopodus
F: Scolopendridae

This close relative of the giant *Scolopendra* centipedes is widespread throughout Africa and is one of several species with blue-tinted legs.

ARACHNIDS

A class within the phylum arthropoda, arachnids include predatory spiders and scorpions, as well as mites and blood-sucking ticks.

Arachnids and related horseshoe crabs are chelicerates—arthropods named for their clawlike mouthparts (chelicerae). The head and thorax of chelicerates are fused into one body section, which carries the sensory organs, brain, and four pairs of walking legs. Unlike other arthropods, chelicerates lack antennae.

EFFECTIVE PREDATORS

Scorpions, spiders, and their relatives are land-living predators that have evolved quick ways of immobilizing and killing prey. Spiders use their chelicerae as fangs to inject venom and many first ensnare prey by spinning webs. Scorpions poison their prey with the sting in their tail. Between their legs and chelicerae, arachnids also have a pair of limblike pedipalps. These are modified into grasping pincers in scorpions or sperm-transferring clubs in male spiders.

MICROSCOPIC DIVERSITY

Many mites are too small to see with the naked eye. They abound in almost every habitat, where they scavenge on detritus, prey on other tiny invertebrates, or live as parasites. Some lead innocuous lives in skin follicles, feathers, or fur; others cause disease or allergy. One group —the ticks—are blood-suckers and can spread disease-causing microbes.

PHYLUM	ARTHROPODA
CLASS	ARACHNIDA
ORDERS	13
FAMILIES	650
SPECIES	65,000

This female wasp spider waits at the center of her dewy orb web to ensnare any insect prey that might fly in.

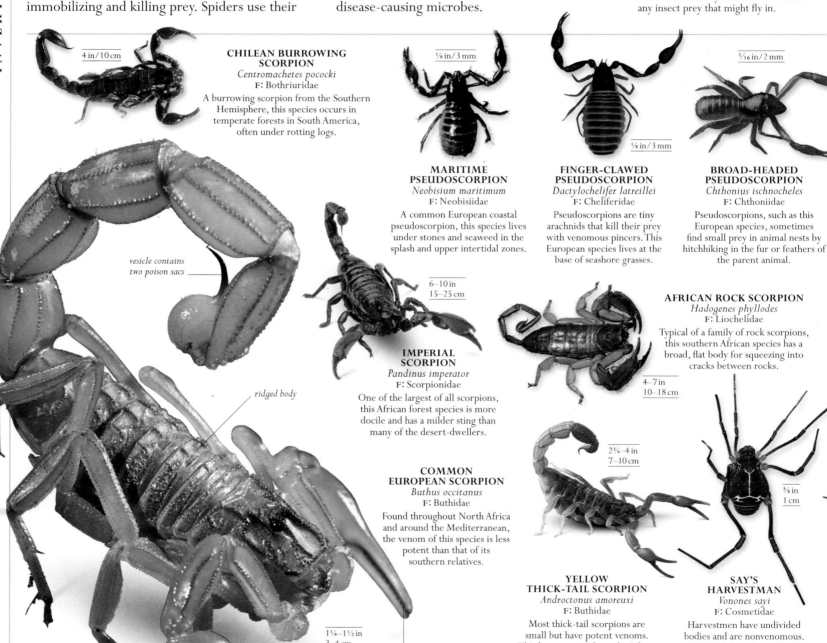

CHILEAN BURROWING SCORPION
Centromachetes pococki
F: Bothriuridae
A burrowing scorpion from the Southern Hemisphere, this species occurs in temperate forests in South America, often under rotting logs.

4 in/10 cm

MARITIME PSEUDOSCORPION
Neobisium maritimum
F: Neobisiidae
A common European coastal pseudoscorpion, this species lives under stones and seaweed in the splash and upper intertidal zones.

⅛ in/3 mm

FINGER-CLAWED PSEUDOSCORPION
Dactylochelifer latreillei
F: Cheliferidae
Pseudoscorpions are tiny arachnids that kill their prey with venomous pincers. This European species lives at the base of seashore grasses.

⅛ in/3 mm

BROAD-HEADED PSEUDOSCORPION
Chthonius ischnocheles
F: Chthoniidae
Pseudoscorpions, such as this European species, sometimes find small prey in animal nests by hitchhiking in the fur or feathers of the parent animal.

¹/₁₆ in/2 mm

vesicle contains two poison sacs

ridged body

IMPERIAL SCORPION
Pandinus imperator
F: Scorpionidae
One of the largest of all scorpions, this African forest species is more docile and has a milder sting than many of the desert-dwellers.

6–10 in
15–25 cm

AFRICAN ROCK SCORPION
Hadogenes phyllodes
F: Liochelidae
Typical of a family of rock scorpions, this southern African species has a broad, flat body for squeezing into cracks between rocks.

4–7 in
10–18 cm

COMMON EUROPEAN SCORPION
Buthus occitanus
F: Buthidae
Found throughout North Africa and around the Mediterranean, the venom of this species is less potent than that of its southern relatives.

YELLOW THICK-TAIL SCORPION
Androctonus amoreuxi
F: Buthidae
Most thick-tail scorpions are small but have potent venoms. This large species from the Sahara and Middle East has caused human fatalities.

2¾–4 in
7–10 cm

SAY'S HARVESTMAN
Vonones sayi
F: Cosmetidae
Harvestmen have undivided bodies and are nonvenomous. Many, including this American species, produce distasteful chemicals to deter predators.

⅜ in
1 cm

1¼–1½ in
3–4 cm

palps modified as pincers

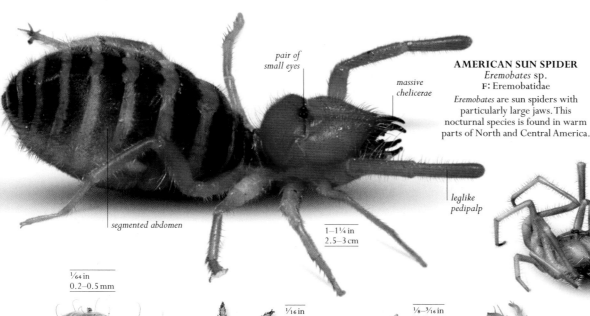

*pair of
small eyes*

*massive
chelicerae*

*leglike
pedipalp*

segmented abdomen

1–1¼ in
2.5–3 cm

AMERICAN SUN SPIDER
Eremobates sp.
F: Eremobatidae
Eremobates are sun spiders with
particularly large jaws. This
nocturnal species is found in warm
parts of North and Central America.

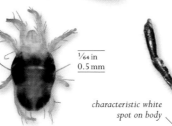

3¼–4 in
8–10 cm

PAINTED SUN SPIDER
Metasolpuga picta
F: Solpugidae
Sun spiders, also called camel
spiders, are fast-running, desert-living
relatives of true spiders. This is
a diurnal species from Africa.

2–3¼ in
5–8 cm

WIND SPIDER
Galeodes arabs
F: Galeodidae
A common species in the largest
genus of sun spiders, this Middle
Eastern species is named for its
ability to tolerate sand storms.

1/64 in
0.2–0.5 mm

FLOUR MITE
Acarus siro
F: Acaridae
This is an important pest
that feeds on stored cereal
products. Like many other
species, its presence
can cause allergic
reactions in humans.

1/16 in
2 mm

CHIGGER MITE
Neotrombicula autumnalis
F: Trombiculidae
Adults of chigger mites
are vegetarian, but their
larvae feed on the skin of
other animals, including
humans. Their chewing
causes intense irritation.

1/8–3/16 in
3–5 mm

COMMON VELVET MITE
Trombidium holosericeum
F: Trombidiidae
This is a widespread Eurasian
species of velvet mite. When
young, it lives parasitically off
other arthropods, but becomes
predatory when mature.

1/64 in
0.5 mm

TWO-SPOT
SPIDER MITE
Tetranychus urticae
F: Tetranychidae
Spider mites are a family
of mites that suck the sap
of plants. They weaken
the plant and can transmit
viral diseases.

5/16–1/2 in
8–12 mm

*characteristic white
spot on body*

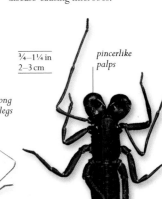

LONE STAR TICK
Amblyomma americanum
F: Ixodidae
Like other blood-sucking
ticks, this species—common
in woodlands in the USA—
can transmit a number of
disease-causing microbes.

1/32–1/16 in
1–2 mm

VARROA MITE
Varroa cerana
F: Varroidae
Parasitic on honeybees,
young of this mite feed on
bee larvae. When mature,
they attach to adult bees and
spread to other hives.

1/64 in
0.5 mm

MANGE MITE
Sarcoptes scabiei
F: Sarcoptidae
This tiny mite burrows into the
skin of various mammal species,
completing its life cycle there.
It causes scabies in humans and
mange in carnivores.

1/64–1/32 in
0.5–1 mm

CHICKEN MITE
Dermanyssus gallinae
F: Dermanyssidae
A blood-sucking parasite of
poultry, this mite lives—and
completes its life cycle—in
crevices away from its host,
but emerges at night to feed.

1/64 in
0.5 mm

PERSIAN FOWL TICK
Argas persicus
F: Argasidae
A blood-sucking parasite of fowl,
including domestic chickens, this
oval-shaped, soft-bodied tick can
spread disease between birds and
can cause paralysis.

3/8–1/2 in
1–1.5 cm

SPINY
HARVESTMAN
Discocyrtus sp.
F: Gonyleptidae
This South American
harvestman has spiny hind legs
as a possible defense against
predators. It lives under stones
and logs in forests.

HORNED HARVESTMAN
Phalangium opilio
F: Phalangiidae
A common harvestman of Eurasia
and North America, the male of
this species has a jaw ornamented
with projecting horns.

5/32–11/32 in / 4–9 mm

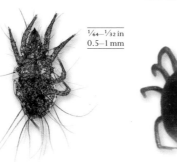

1¼–1½ in
3–4 cm

*very long
front legs*

WHIP SPIDER
Phrynus sp.
F: Phrynidae
Whip spiders are not true
spiders. They have long,
whiplike front legs and pincers
for grabbing prey, but lack
venom. All are tropical.

3/4–1¼ in
2–3 cm

*pincerlike
palps*

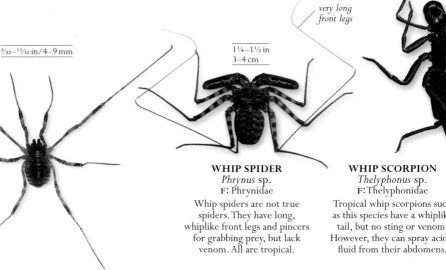

WHIP SCORPION
Thelyphonus sp.
F: Thelyphonidae
Tropical whip scorpions such
as this species have a whiplike
tail, but no sting or venom.
However, they can spray acidic
fluid from their abdomens.

FUNNEL-WEB SPIDER
Atrax robustus
F: Hexathelidae
Females of this aggressive
Australian spider live in
burrows with funnel-like,
silk-lined entrances. The
dangerous bite most
frequently comes from males
wandering for mates.

¾–2¾ in
2–7 cm

**MEXICAN RED-KNEED
TARANTULA**
Brachypelma smithi
F: Theraphosidae
This is one of many
theraphosids: large-bodied
spiders popularly called
bird-eaters or tarantulas.
They prey on large insects,
rarely on small vertebrates.

2–3 in
5–7.5 cm

CHACO TARANTULA
Acanthoscurria insubtilis
F: Theraphosidae
Many large theraphosids,
such as this South American
species, live in disused rodent
burrows. Despite their
popular name, they do
not eat birds.

2–3 in
5–7.5 cm

GOLIATH TARANTULA
Theraphosa blondi
F: Theraphosidae
O: Araneae
Like many of its relatives,
this South American bird-
eater—the largest of all
spiders—discharges
irritating hairs
when threatened.

pedipalps

4¾–5½ in
12–14 cm

*forward-pointing
chelicerae carrying fangs*

eight small eyes

*hairy brown
body*

**NORTH AMERICAN
TRAPDOOR SPIDER**
Ummidia audouini
F: Ctenizidae
This N. American spider builds
a cork-like trapdoor set with
trip-lines. Stumbling prey alert
the predator waiting in a
silk-lined burrow beneath.

1–2 cm
⅜–¾ in

SIX-EYED SPIDER
Oonops domesticus
F: Oonopidae
This tiny, pink, six-eyed
spider occurs in warm
parts of Eurasia, but
further north—including
in Britain—it is found
only in houses.

2 mm
1⁄16 in

WOODLOUSE SPIDER
Dysdera crocata
F: Dysderidae
This nocturnal European
spider has huge fangs that
catch and break into the hard
exoskeletons of woodlice. It
occurs in damp places where
woodlice are abundant.

⅜–½ in
1–1.2 cm

*spinnerets
extrude silk*

¼–½ in
0.6–1.6 cm

*white bands
on legs*

**LADYBIRD
SPIDER**
Eresus kollari
F: Eresidae
Only males of this Eurasian
spider have the eponymous
ladybird pattern. It catches
prey from within silk-
lined burrows on
heathland slopes.

**DWARF
SPIDER**
Gonatium sp.
F: Linyphiidae
Typical of a very large
family of tiny "money
spiders," this Northern
Hemisphere species builds
sheet webs and can travel
on the wind by ballooning
on silken threads.

⅛ in
3 mm

⅛–¼ in
3–6 mm

**NORTHERN
SPITTING SPIDER**
Scytodes thoracica
F: Scytodidae
Spitting spiders are sluggish
spiders that spray gummy,
poisonous fluid to immobilize
prey before biting. Its range
spans the Northern Hemisphere.

5⁄32–½ in
4–13 mm

EUROPEAN GARDEN SPIDER
Araneus diadematus
F: Araneidae
A weaver of orb webs from the Northern
Hemisphere, this spider of woodland,
heath, and gardens is marked by a white
cross on a variable background color.

⅜ in
1.2 cm

⁷⁄32–⅜ in
7–10 mm

**DADDY LONG-LEGS
SPIDER**
Pholcus phalangioides
F: Pholcidae
Spindly-legged pholcids vibrate
their webs when disturbed.
Many are cave dwellers, but
this cosmopolitan species
inhabits houses. Females carry
eggs in their jaws.

egg sac

CAVE SPIDER
Meta menardi
F: Araneidae
Many spiders of the
genus *Meta*, such as this
European species,
inhabit caves, where
they suspend their
drop-shaped egg sacs.

1⁄16–1 1⁄32 in
2–9 mm

**CRAB-LIKE SPINY
ORB-WEAVER**
Gasteracantha cancriformis
F: Araneidae
One of many American
orb-weavers with defensive
spines, this species inhabits
the southern USA and the
Caribbean. Colors of the
body and spines vary.

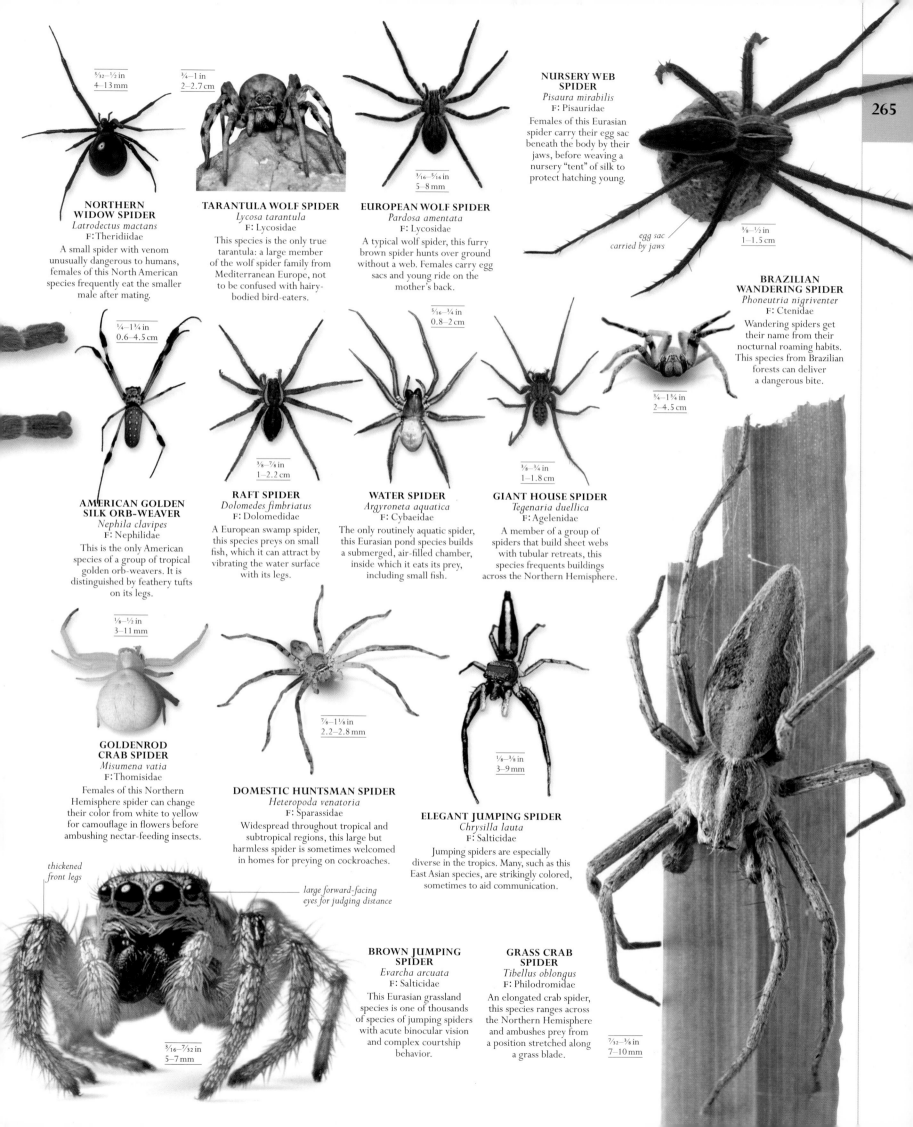

NORTHERN WIDOW SPIDER
Latrodectus mactans
F: Theridiidae
A small spider with venom unusually dangerous to humans, females of this North American species frequently eat the smaller male after mating.

⁵⁄₃₂–½ in
4–13 mm

TARANTULA WOLF SPIDER
Lycosa tarantula
F: Lycosidae
This species is the only true tarantula: a large member of the wolf spider family from Mediterranean Europe, not to be confused with hairy-bodied bird-eaters.

¾–1 in
2–2.7 cm

EUROPEAN WOLF SPIDER
Pardosa amentata
F: Lycosidae
A typical wolf spider, this furry brown spider hunts over ground without a web. Females carry egg sacs and young ride on the mother's back.

³⁄₁₆–⁵⁄₁₆ in
5–8 mm

NURSERY WEB SPIDER
Pisaura mirabilis
F: Pisauridae
Females of this Eurasian spider carry their egg sac beneath the body by their jaws, before weaving a nursery "tent" of silk to protect hatching young.

egg sac carried by jaws

⅜–½ in
1–1.5 cm

BRAZILIAN WANDERING SPIDER
Phoneutria nigriventer
F: Ctenidae
Wandering spiders get their name from their nocturnal roaming habits. This species from Brazilian forests can deliver a dangerous bite.

¾–1¾ in
2–4.5 cm

AMERICAN GOLDEN SILK ORB-WEAVER
Nephila clavipes
F: Nephilidae
This is the only American species of a group of tropical golden orb-weavers. It is distinguished by feathery tufts on its legs.

¼–1¾ in
0.6–4.5 cm

RAFT SPIDER
Dolomedes fimbriatus
F: Dolomedidae
A European swamp spider, this species preys on small fish, which it can attract by vibrating the water surface with its legs.

⅜–⅞ in
1–2.2 cm

WATER SPIDER
Argyroneta aquatica
F: Cybaeidae
The only routinely aquatic spider, this Eurasian pond species builds a submerged, air-filled chamber, inside which it eats its prey, including small fish.

⁵⁄₁₆–¾ in
0.8–2 cm

GIANT HOUSE SPIDER
Tegenaria duellica
F: Agelenidae
A member of a group of spiders that build sheet webs with tubular retreats, this species frequents buildings across the Northern Hemisphere.

⅜–¾ in
1–1.8 cm

GOLDENROD CRAB SPIDER
Misumena vatia
F: Thomisidae
Females of this Northern Hemisphere spider can change their color from white to yellow for camouflage in flowers before ambushing nectar-feeding insects.

⅛–½ in
3–11 mm

DOMESTIC HUNTSMAN SPIDER
Heteropoda venatoria
F: Sparassidae
Widespread throughout tropical and subtropical regions, this large but harmless spider is sometimes welcomed in homes for preying on cockroaches.

⅞–1⅛ in
2.2–2.8 mm

ELEGANT JUMPING SPIDER
Chrysilla lauta
F: Salticidae
Jumping spiders are especially diverse in the tropics. Many, such as this East Asian species, are strikingly colored, sometimes to aid communication.

⅛–⅜ in
3–9 mm

thickened front legs

large forward-facing eyes for judging distance

BROWN JUMPING SPIDER
Evarcha arcuata
F: Salticidae
This Eurasian grassland species is one of thousands of species of jumping spiders with acute binocular vision and complex courtship behavior.

³⁄₁₆–⁷⁄₃₂ in
5–7 mm

GRASS CRAB SPIDER
Tibellus oblongus
F: Philodromidae
An elongated crab spider, this species ranges across the Northern Hemisphere and ambushes prey from a position stretched along a grass blade.

⁷⁄₃₂–⅜ in
7–10 mm

MEXICAN RED-KNEED TARANTULA
Brachypelma smithi

With its stout, furry body, the Mexican red-kneed tarantula might seem more mammal-like than spiderlike. The female can live up to 30 years, which is unusual for an invertebrate, but the life span of the male is only up to six years. *Brachypelma smithi* is one of a group of spiders that gets the nickname of bird-eater because of size, and although most of their diet consists of fellow arthropods, they can and do take down small mammals and reptiles when the opportunity arises. In its native Mexico this species lives in burrows in earth banks, where it is safe to molt and lay eggs, and from which it ambushes its prey. Habitat destruction is now a threat, and, because the species makes a popular pet, it is routinely bred in captivity.

SIZE Body length 2–3 in (5–7.5 cm)
HABITAT Tropical deciduous forests
DISTRIBUTION Mexico
DIET Mostly insects

legs are covered in specialized hairs sensitive to air movements and touch

padded foot

< EYES
Like most spiders, the tarantula has eight simple eyes arranged at the front of the head. Even so, it has poor vision, and relies more on touch than sight to sense its surroundings and the presence of prey.

< JOINTS
Like all arthropods, tarantulas have jointed legs. Each leg consists of seven tubular sections of exoskeleton that connect via flexible joints. Muscles run through these joints to move the sections.

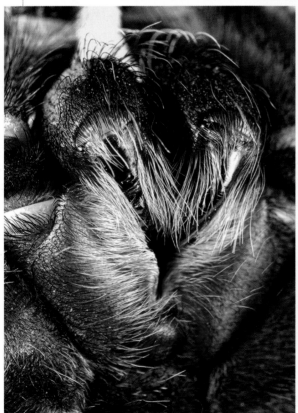

< FOOT
At the tip of each foot are two claws used for grip as the tarantula moves. As in other hunting spiders, there are also pads of tiny hairs for extra purchase on smooth surfaces.

^ VENOMOUS FANGS
Tarantula fangs hinge forward when attacking prey—unlike most other spiders, whose fangs are angled toward each other. Both fangs inject venom from muscular sacs within the head to paralyze the victim.

< SPINNERETS
Glands in the abdomen produce silk in liquid form. Using its hind legs, the tarantula pulls the silk from tubes called spinnerets. The silk solidifies into threads used to make egg sacs and line the spider's burrow.

dark abdomen contains
most of the animal's
vital internal organs

orange-red
patella, or knee

∧ HAIRY WEAPONS
Like many tropical American
tarantulas, this spider defends
itself by rubbing its hind legs
against its abdomen to release
the barbed, stinging hairs that
grow there. These urticating
hairs are tiny and light, and
designed to float in a cloud
around a predator's face,
getting in its eyes, nose, and
mouth. They cause intense
irritation when lodged in skin.

palps are used for
feeling, groping prey,
and transferring
sperm to the female

UNDERSIDE >
The head and thorax are
fused into one body part,
which carries the legs and
mouthparts. The abdomen
has openings for breathing
and reproduction, and two
pairs of spinnerets at the rear.

SEA SPIDERS

These fragile-looking marine animals live among seaweeds in shallow seas and coral reefs in tropical waters. The largest species live in the deep ocean.

Sea spiders of the class Pycnogonida are not true spiders, and are so remarkably different from other arthropods that some scientists think they belong to an ancient lineage that is not closely related to any groups alive today. Others think they are distant cousins of the arachnids. Most sea spiders are small—less than ³⁄₈ inch (1 cm) in body length. They have three or four pairs of legs and the head and thorax are fused together. Instead of clawlike mouthparts they have a stabbing proboscis, which they use like a hypodermic needle to suck fluids from their invertebrate prey. Their spindly body shape means they do not need gills; they rely instead on oxygen seeping directly into all their cells through their body surface.

PHYLUM	ARTHROPODA
CLASS	PYCNOGONIDA
ORDER	1
FAMILIES	8
SPECIES	About 1,000

³⁄₄ in / 2 cm

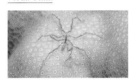
⁵⁄₁₆ in / 8 mm

³⁄₁₆ in / 5 mm

GIANT SEA SPIDER
Colossendeis megalonyx
F: Colossendeidae
One of the largest of all sea spiders, this is a deep-sea animal of sub-Antarctic waters, with a leg span of 28 in (70 cm).

SPINY SEA SPIDER
Endeis spinosa
F: Endeidae
This sea spider—found around European coasts but possibly ranging elsewhere—has an especially spindly body, with a long cylindrical proboscis.

FAT SEA SPIDER
Pycnogonum littorale
F: Pycnogonidae
Unlike most other sea spiders, this European species has a thick body and comparatively short, curved, clawed legs. It feeds on anemones.

YELLOW-KNEED SEA SPIDER
Unknown sp.
F: Callipallenidae
Some species of sea spiders—such as this species of Australian reefs—are strikingly colored, often as camouflage against their colorful surroundings.

⁵⁄₁₆ in / 8 mm

GRACEFUL SEA SPIDER
Nymphon gracile
F: Nymphonidae
One of the most common sea spiders of the Northeast Atlantic, this species occurs in the intertidal zone and in offshore shallows.

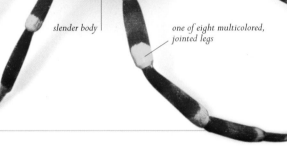
2–4 in
5–10 cm

slender body

one of eight multicolored, jointed legs

HORSESHOE CRABS

The Merostomata are a small class of sea-dwelling relatives of spiders and scorpions. Of prehistoric origin, they are widely regarded as "living fossils."

Horseshoe crabs were far more diverse in prehistory, when animals like them may well have been the first chelicerates on Earth. Their clawlike mouthparts and lack of antennae indicate that, despite their hard carapace, horseshoe crabs are most closely related to arachnids. The leaflike gills on the underside of their abdomens are the forerunners of similar internal structures—the so-called book lungs—that arachnids use to breathe on land. The pedipalps are used as a fifth pair of legs, which is one pair more than in spiders. Horseshoe crabs grub around in muddy ocean waters to catch their prey. They come ashore in large numbers to breed, laying their spawn in sand.

PHYLUM	ARTHROPODA
CLASS	MEROSTOMATA
ORDER	1
FAMILY	1
SPECIES	4

16–24 in / 40–60 cm

JAPANESE HORSESHOE CRAB
Tachypleus tridentatus
F: Limulidae
This horseshoe crab spawns in sandy shores of Eastern Asia. Habitat destruction and pollution have depleted its numbers in some parts of it's range.

fused head and thorax covered by carapace

long spinelike tail

spiny abdomen

ATLANTIC HORSESHOE CRAB
Limulus polyphemus
F: Limulidae
This species of the Northwest Atlantic Ocean gathers in spring, breeding in swarms along American coastlines—especially around the Gulf of Mexico.

16–24 in
40–60 cm

CRUSTACEANS

Most crustaceans are aquatic, breathe with gills, and have limbs for crawling or swimming. A few are sedentary as adults, parasitic, or land living.

The basic body plan of a crustacean consists of a head, a thorax, and an abdomen, although in many groups the head and thorax are fused together. In crabs, lobsters, and shrimps a frontal flap grows backward to cover the entire head and thorax with a shelllike carapace. Crustaceans are the only arthropods that have two pairs of antennae and primitively two-pronged limbs. Limbs on the thorax are usually used for locomotion, but in some species they are modified as pincers for feeding or defense. Many crustaceans also have well-developed abdominal limbs, which are often

used for brooding young. Gills tie most crustaceans to water, although woodlice and some crabs have modified breathing organs that enable them to live in damp places on land.

Being permanently underwater, where their weight is supported, allows crustaceans to develop thick, heavy exoskeletons. Many species have exoskeletons that are hardened with minerals, which are reabsorbed between molts when the outer coating of the body is shed. With their bodies buoyed up by water, many crustaceans have become bigger than their land-based relatives. The world's largest arthropod is the deep-sea Japanese spider crab (leg span 13ft/4m). At the other extreme, crustaceans make up the bulk of animal life in plankton, either as tiny larvae or as an adult shrimp and krill.

PHYLUM	ARTHROPODA
SUBPHYLUM	CRUSTACEA
ORDERS	42
FAMILIES	About 850
SPECIES	About 50,000

DEBATE
CRUSTACEAN ANCESTRY

Crustaceans have many unique features, including double antennae and two-pronged limbs, that suggest they are exclusive descendants of a single common ancestor. However, recent DNA analysis suggests that insects are also descended from the crustacean group.

WATER FLEAS AND RELATIVES

The class Branchiopoda are primarily freshwater crustaceans, which make up part of the plankton that flourish in short-lived pools, surviving long periods of drought as eggs. Their thoraxes carry leafy limbs for breathing and filter feeding. Water fleas are typically encased in a transparent carapace.

⅜–½ in
1–1.5 cm

BRINE SHRIMP
Artemia salina
F: Artemiidae
This soft-bodied, stalk-eyed animal is found worldwide in salt pools and swims upside-down. Its hard-shelled eggs can resist years of drought.

¹⁄₁₆–³⁄₁₆ in
2–5mm

LARGE WATER FLEA
Daphnia magna
F: Daphniidae
This North American water flea, like its relatives, can brood eggs inside its carapace. These hatch without fertilization, rapidly populating ponds as a result.

¹⁄₁₆ in
1.5mm

MARINE WATER FLEA
Evadne nordmanni
F: Podonidae
Most water fleas—named for their jerky swimming—occur in stagnant freshwater pools. This, however, is a saltwater species of oceanic plankton.

VERNAL POOL TADPOLE SHRIMP
Lepidurus packardi
F: Triopsidae
Tadpole shrimps are ancient, bottom-living crustaceans of temporary freshwater pools. Relatives of this Californian species have changed little in 220 million years of evolution.

2 in
5 cm

— two tails

BARNACLES AND COPEPODS

Like other marine crustaceans, the class Maxillopoda start life as tiny planktonic larvae. Barnacle larvae cement themselves, head down, on rocks. Most copepods stay free-swimming, although some become parasites.

³⁄₁₆–½ in
0.5–1.5 cm

COMMON ACORN BARNACLE
Semibalanus balanoides
F: Archaeobalanidae
Sensitive to drying out, this intertidal species is most abundant at the base of the barnacle zone on exposed North Atlantic rocky coastlines.

2–4 in/5–10 cm

GIANT ACORN BARNACLE
Balanus nubilus
F: Balanidae
The world's largest barnacle, this species occurs attached to rocks below the intertidal level along the Pacific North American coast.

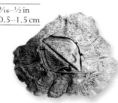

¾ in
1.8 cm

SEA LOUSE
Caligus sp.
F: Caligidae
A representative of a group of copepods that are parasitic on marine fishes, this crustacean attacks salmon and related species.

¾–1¼ in
2–3 cm

ASIAN ACORN BARNACLE
Tetraclita squamosa
F: Tetraclitidae
This acorn barnacle lives in the intertidal zone on Indo-Pacific shores. Recent research suggests it actually consists of several similar species.

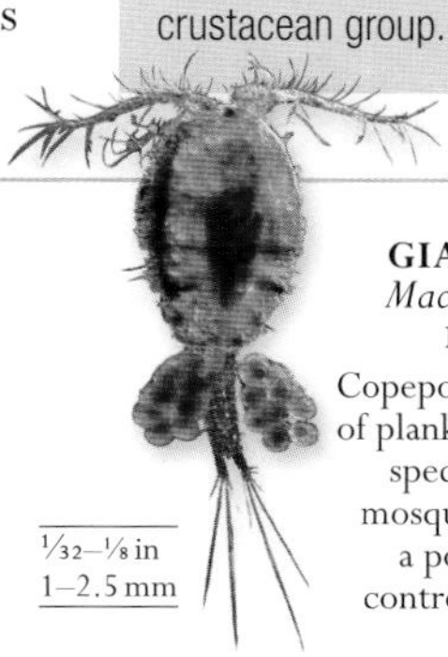

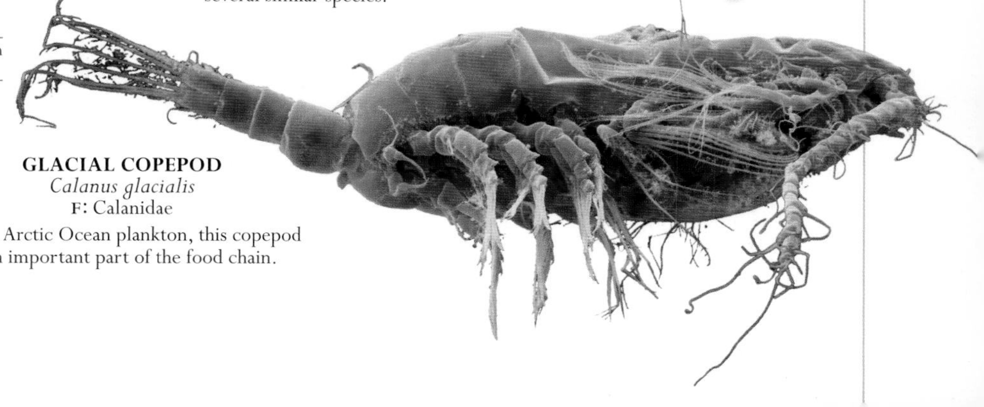

⅛–³⁄₁₆ in
3–5 mm

GLACIAL COPEPOD
Calanus glacialis
F: Calanidae
Within Arctic Ocean plankton, this copepod is an important part of the food chain.

¹⁄₃₂–⅛ in
1–2.5 mm

GIANT COPEPOD
Macrocyclops albidus
F: Cyclopidae
Copepods are tiny predators of plankton; this widespread species even preys on mosquito larvae, offering a potential means of controlling these insects.

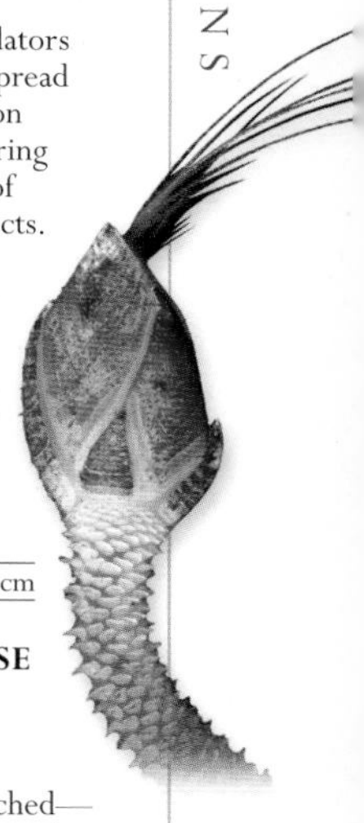

DEEP SEA GOOSE BARNACLE
Neolepas sp.
F: Scalpellidae
This goose barnacle ally lives around volcanic vents on the ocean floor, where it filter-feeds on other organisms, including bacteria.

2–4 in/5–10 cm

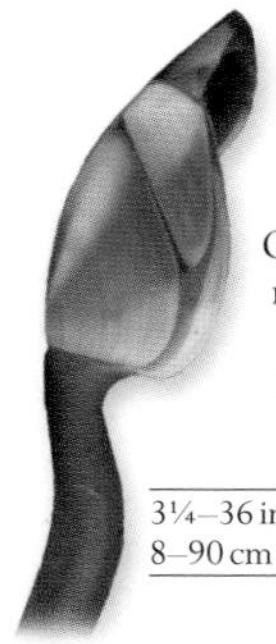

COMMON GOOSE BARNACLE
Lepas anatifera
F: Lepadidae
Goose barnacles are attached—mostly to oceanic flotsam—by a flexible stalk. This species occurs in temperate waters of the Northeast Atlantic.

3¼–36 in
8–90 cm

COMMON FISH LOUSE
Argulus sp.
F: Argulidae
This is a flattened, fast-swimming crustacean with an oval carapace. It uses suckers to attach to fish and suck their blood.

³⁄₁₆–⅜ in
0.5–1 cm

SEED SHRIMPS

Seed shrimps belong to the class Ostracoda. The entire body of a seed shrimp is enclosed within a two-valve hinged carapace, with just the limbs poking through. If threatened, the animal can shut itself inside. These small crustaceans crawl through vegetation in marine and freshwater habitats; some use their antennae to swim.

¾–1¼ in
2–3 cm

GIANT SWIMMING SEED SHRIMP
Gigantocypris sp.
F: Cypridinidae
Most seed shrimps are tiny crustaceans with two-valved carapaces; this is a large, deep-sea species with big eyes for hunting bioluminescent prey.

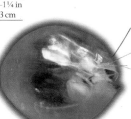

¹⁄₆₄–¹⁄₁₆ in
0.5–2 mm

CRAWLING SEED SHRIMP
Cypris sp.
F: Cyprididae
This widespread freshwater crustacean belongs to a group of small, hard-shelled seed shrimps that crawl through detritus.

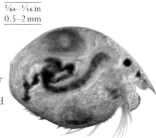

CRABS AND RELATIVES

The Malacostraca class is the most diverse crustacean group. The basic features are head, thorax, and an abdomen with multiple limbs. Two large orders within this class include the decapods, which have a carapace that curves around the fused head-thorax to contain a gill cavity, and the isopods (woodlice and relatives), which lack a carapace and are the largest group of land-living crustaceans.

ANTARCTIC KRILL
Euphausia superba
F: Euphausiidae
Swarms of these plankton-feeding crustaceans are critical components of the Southern Ocean food chains that support whales, seals, and seabirds.

1½–2¼ in
4–6 cm

RELICT OPOSSUM SHRIMP
Mysis relicta
F: Mysidae
Translucent, feathery-legged opossum shrimps carry their larvae in a brood pouch. Most live in coastal waters, but this species is found in freshwaters in the Northern Hemisphere.

⅜–¾ in
1–1.8 cm

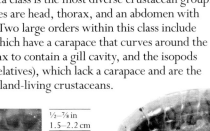

½–⅞ in
1.5–2.2 cm

⅜–¾ in
1–2 cm

COMMON FRESHWATER SHRIMP
Gammarus pulex
F: Gammaridae
This northern European amphipod, abundant in freshwater streams, belongs to a family of detritus-eating freshwater shrimps. Related species inhabit brackish waters.

COMMON SANDHOPPER
Orchestia gammarellus
F: Talitridae
One of many amphipods (sideways-flattened crustaceans), this is a European intertidal sandhopper—so called because it can jump by flipping its abdomen.

SPINY SKELETON SHRIMP
Caprella acanthifera
F: Caprellidae
Like other skeleton shrimps, this is a slender, slow-moving, predatory amphipod with few legs. It clings to seaweeds in European rockpools.

½ in / 13 mm

abdomen

COMMON WATER SKATER
Asellus aquaticus
F: Asellidae
A common European member of a family of freshwater woodlice relatives, this species crawls among detritus in stagnant water.

⅜–½ in / 1–1.5 cm

¾–1¼ in / 2–3 cm

4–4¾ in
10–12 cm

tail fin used for swimming

GIANT DEEPSEA ISOPOD
Bathynomus giganteus
F: Cirolanidae
A giant marine relative of woodlice, this species crawls along ocean beds, scavenging on dead animals and occasionally taking live prey.

7½–14 in
19–36 cm

segmented exoskeleton

BLACK-HEADED WOODLOUSE
Porcellio spinicornis
F: Porcellionidae
Often living alongside humans, especially in lime-rich habitats, this distinctly marked woodlouse is native to Europe but has invaded North America.

COMMON SEA SLATER
Ligia oceanica
F: Ligiidae
A large coastal woodlouse, this European species lives in rock crevices above the intertidal zone. It feeds on detritus.

⅜–¾ in
1–1.8 cm

COMMON PILL WOODLOUSE
Armadillidium vulgare
F: Armadillidiidae
Pill woodlice are characterized by their ability to roll into a ball when disturbed. This species is widespread across Eurasia and introduced elsewhere.

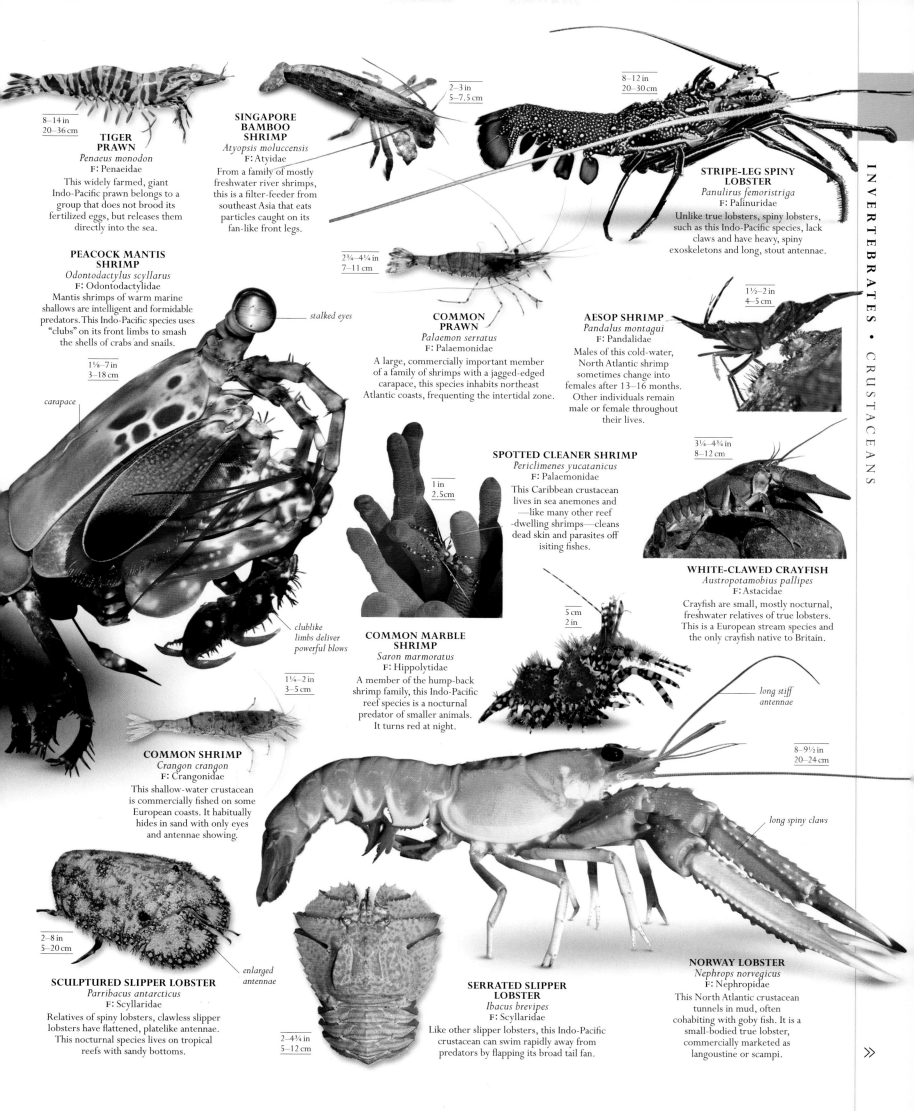

TIGER PRAWN
Penaeus monodon
F: Penaeidae
This widely farmed, giant Indo-Pacific prawn belongs to a group that does not brood its fertilized eggs, but releases them directly into the sea.

8–14 in
20–36 cm

SINGAPORE BAMBOO SHRIMP
Atyopsis moluccensis
F: Atyidae
From a family of mostly freshwater river shrimps, this is a filter-feeder from southeast Asia that eats particles caught on its fan-like front legs.

2–3 in
5–7.5 cm

8–12 in
20–30 cm

STRIPE-LEG SPINY LOBSTER
Panulirus femoristriga
F: Palinuridae
Unlike true lobsters, spiny lobsters, such as this Indo-Pacific species, lack claws and have heavy, spiny exoskeletons and long, stout antennae.

PEACOCK MANTIS SHRIMP
Odontodactylus scyllarus
F: Odontodactylidae
Mantis shrimps of warm marine shallows are intelligent and formidable predators. This Indo-Pacific species uses "clubs" on its front limbs to smash the shells of crabs and snails.

1⅛–7 in
3–18 cm

stalked eyes

carapace

2¾–4¼ in
7–11 cm

COMMON PRAWN
Palaemon serratus
F: Palaemonidae
A large, commercially important member of a family of shrimps with a jagged-edged carapace, this species inhabits northeast Atlantic coasts, frequenting the intertidal zone.

AESOP SHRIMP
Pandalus montagui
F: Pandalidae
Males of this cold-water, North Atlantic shrimp sometimes change into females after 13–16 months. Other individuals remain male or female throughout their lives.

1½–2 in
4–5 cm

SPOTTED CLEANER SHRIMP
Periclimenes yucatanicus
F: Palaemonidae
This Caribbean crustacean lives in sea anemones and —like many other reef-dwelling shrimps—cleans dead skin and parasites off isiting fishes.

1 in
2.5cm

3¼–4¾ in
8–12 cm

WHITE-CLAWED CRAYFISH
Austropotamobius pallipes
F: Astacidae
Crayfish are small, mostly nocturnal, freshwater relatives of true lobsters. This is a European stream species and the only crayfish native to Britain.

clublike limbs deliver powerful blows

1¼–2 in
3–5 cm

5 cm
2 in

COMMON MARBLE SHRIMP
Saron marmoratus
F: Hippolytidae
A member of the hump-back shrimp family, this Indo-Pacific reef species is a nocturnal predator of smaller animals. It turns red at night.

long stiff antennae

COMMON SHRIMP
Crangon crangon
F: Crangonidae
This shallow-water crustacean is commercially fished on some European coasts. It habitually hides in sand with only eyes and antennae showing.

8–9½ in
20–24 cm

long spiny claws

2–8 in
5–20 cm

SCULPTURED SLIPPER LOBSTER
Parribacus antarcticus
F: Scyllaridae
Relatives of spiny lobsters, clawless slipper lobsters have flattened, platelike antennae. This nocturnal species lives on tropical reefs with sandy bottoms.

enlarged antennae

2–4¾ in
5–12 cm

SERRATED SLIPPER LOBSTER
Ibacus brevipes
F: Scyllaridae
Like other slipper lobsters, this Indo-Pacific crustacean can swim rapidly away from predators by flapping its broad tail fan.

NORWAY LOBSTER
Nephrops norvegicus
F: Nephropidae
This North Atlantic crustacean tunnels in mud, often cohabiting with goby fish. It is a small-bodied true lobster, commercially marketed as langoustine or scampi.

»

12–16 in
30–40 cm

ROBBER CRAB
Birgus latro
F: Coenobitidae
The largest land-living arthropod,
this hermit crab is a relative of
squat lobsters and inhabits
Indo-Pacific island forests,
eating coconuts using
massive pincers.

⅜ in
1 cm

PINK SQUAT LOBSTER
Lauriea siagiani
F: Galatheidae
Many tropical squat lobsters are
associated with specific reef
organisms. This tiny, hairy
Indonesian species lives on
Xestospongia vase sponges.

**BLUE-STRIPED
SQUAT LOBSTER**
Galathea strigosa
F: Galatheidae
Slender-clawed squat lobsters
such as this European species are
decapods (ten-legged), but the last
leg pair is reduced, making them
appear eight-legged.

2¾–3½ in
7–9 cm

unequal-sized claws
used for signaling

anemone
attached
to shell

¾ in / 2 cm

ANEMONE PORCELAIN CRAB
Petrolisthes ohshimai
F: Porcellanidae
Porcelain Crabs are tiny, eight-legged
decapods more closely related to
squat lobsters than true crabs. This
Indo-Pacific species lives in giant
Stichodactyla anemones.

⁵⁄₁₆–½ in / 8–12 mm

antenna

**INDO-PACIFIC
PEA CRAB**
Pinnotheres sp.
F: Pinnotheridae
Tiny pea crabs complete their life
cycle on or in the bodies of other
marine invertebrates. This Philippine
species lives on cup corals.

1½ in
4 cm

RED REEF HERMIT CRAB
Paguristes cadenati
F: Diogenidae
Hermit crabs inhabit discarded
snail shells that accommodate
their soft coiled abdomen. This
species lives on Indo-Pacific and
eastern Atlantic reefs.

5–8 in
13–20 cm

**WHITE-SPOTTED
HERMIT CRAB**
Dardanus megistos
F: Diogenidae
This crustacean of the eastern Atlantic
and Indo-Pacific coasts is a "left-handed"
hermit crab, with an enlarged left claw.

2¼–4 in
6–10 cm

ANEMONE HERMIT CRAB
Dardanus pedunculatus
F: Diogenidae
This hermit crab of Indo-Pacific reefs
always carries a *Calliactis* anemone on
its shell, which shares its food and
provides camouflaged protection.

VELVET SWIMMING CRAB
Necora puber
F: Portunidae
Swimming crabs have paddlelike rear legs. This aggressive, red-eyed species is common at low shore level on rocky coasts in the northeast Atlantic.

2–2½ in
5–6.5 cm

BLUE SWIMMING CRAB
Portunus pelagicus
F: Portunidae
This Indo-Pacific swimming crab prefers sandy or muddy coastlines, with juveniles entering the intertidal zone. Like its relatives, it preys on other invertebrates.

2–2¾ in
5–7 cm

shell-like carapace

2–4 in
5–10 cm

EDIBLE CRAB
Cancer pagurus
F: Cancridae
The most commercially important European crab, this offshore species has a distinctive wide "piecrust" carapace and can live for more than 20 years.

SPOTTED CORAL CRAB
Carpilius maculatus
F: Carpiliidae
One of three related, brightly colored, coral-living crabs with many fossil relatives, this large species has been found in the Indian and Pacific oceans.

1¾–3½ in
4.5–9 cm

WARTY BOX CRAB
Calappa hepatica
F: Calappidae
A tortoiselike crab, this Indo-Pacific, sand-burrowing species habitually shields its face with its claws, earning it the alternative name of shame-face crab.

1½–2¼ in
4–6 cm

1½–2 in
4–5 cm

SPONGE CRAB
Dromia personata
F: Dromiidae
This Atlantic crab often carries sponge pieces to hide itself from predators. Its reduced hind legs are like those of the related, primitive hermit crabs.

PAINTED PEBBLE CRAB
Leucosia anatum
F: Leucosiidae
Belonging to a family of small, mostly diamond-shaped crabs with long claws, this colourful species occurs in the Indian Ocean.

¾–1¼ in
2–3 cm

12–16 in
30–40 cm

JAPANESE SPIDER CRAB
Macrocheira kaempferi
F: Inachidae
With legs spanning 13 ft (4 m), this is the world's biggest arthropod. It occurs in northwest Pacific waters and reputedly lives for up to 100 years.

3¼–4 in
8–10 cm

CHRISTMAS ISLAND RED CRAB
Gecarcoidea natalis
F: Gecarcinidae
This land-living crab lives in burrows only in the forests of Christmas Island. Huge numbers march in an annual migration to the sea for spawning.

⅜–1⅛ in
1–3 cm

PANAMIC ARROW CRAB
Stenorhynchus debilis
F: Inachidae
Arrow crabs are small, stalk-eyed, spiky-headed spider crabs. This eastern Pacific species is a reef-dwelling scavenger.

EUROPEAN FRESHWATER CRAB
Potamon potamios
F: Potamidae
A member of a large family of Eurasian crabs from alkaline freshwaters, this southern European species spends much of its time on land.

1½–2 in
4–5 cm

eyes on stalks

⅜–¾ in
1–2 cm

ORANGE FIDDLER CRAB
Uca vocans
F: Ocypodidae
Fiddler crabs are burrowing beach crabs with an enlarged signaling claw in males. This species lives on muddy western Pacific shores.

RED SEA GHOST CRAB
Ocypode saratan
F: Ocypodidae
Ghost crabs, such as this Indo-Pacific species, are pale relatives of fiddler crabs. Both groups are stalk-eyed, fast-running beach dwellers.

¾–1½ in / 2–4 cm

furry claws

2–2¼ in
5–6 cm

CHINESE MITTEN CRAB
Eriocheir sinensis
F: Varunidae
Native to eastern Asia, this furry-clawed, burrowing river crab was introduced to North America and Europe, where it has become a pest.

INSECTS

Insects first appeared on land more than 400 million years ago and today they account for more species than any other class on the planet.

Insects have evolved diverse lifestyles and although most are terrestrial, there are also numerous freshwater species but almost no marine. They have a number of key features, such as a small size, an efficient nervous system, high reproductive rates, and—in many cases—the power of flight, which have led to their success.

Insects include (among others) beetles, flies, butterflies, moths, ants, bees, and true bugs. Yet for all their diversity, insects are remarkably similar. Evolution has modified the basic insect anatomy many times over to produce a multitude

of variants based on three major body regions—the head, thorax, and abdomen. The head, made up of six fused segments, houses the brain and carries the major sensory organs: compound eyes, secondary light-receptive organs called ocelli, and the antennae. The mouthparts are modified according to diet, allowing the sucking of liquids or the chewing of solid foods.

The thorax is made up of three segments, each bearing a pair of legs. The posterior two thoracic segments usually each bear a pair of wings. The legs, which are each made up of a number of segments, can be greatly modified to serve a variety of functions from walking and running to jumping, digging, or swimming. The abdomen, which is usually made up of 11 segments, contains the digestive and reproductive organs.

PHYLUM	ARTHROPODA
CLASS	INSECTA
ORDERS	30
FAMILIES	About 1,000
SPECIES	About 1,000,000

DEBATE
HOW MANY SPECIES?

The actual number of insect species likely to exist far exceeds the number so far described, and many more are discovered each year. Estimates hover around 2 million. However, research based on sampling in species-rich rain forests suggests that there could be as many as 30 million species.

SILVERFISH

These primitive wingless insects of the order Thysanura have lengthy bodies that can be covered with scales. The head has a pair of long antennae and small eyes. The abdominal segments have small appendages (styles).

long front legs held forward

front segment of thorax

⅜–½ in
1–1.5 cm

½ in / 1.2 cm

SILVERFISH
Lepisma saccharina
F: Lepismatidae
This common domestic species can sometimes be a nuisance in kitchens, where it feeds on tiny scraps of dropped food.

FIREBRAT
Thermobia domestica
F: Lepismatidae
This insect is found worldwide and lives under stones and in leaf litter. Indoors, it prefers warm conditions and can be a pest in bakeries.

large triangular front wings

pale abdomen

¾–1 in
1.7–2.5 cm

MAYFLIES

Ephemeroptera are soft-bodied insects with slender legs and two pairs of wings. The head has a pair of short antennae and large, compound eyes. The end of the abdomen has two or three long tail filaments. The life cycle is dominated by the aquatic nymphal stages—the nonfeeding adults live for only a few hours or days.

⁷⁄₃₂–⅜ in
7–11 mm

⁵⁄₁₆–½ in
8–12 mm

½–¾ in
1.2–1.8 cm

MAYFLY
Ephemera danica
F: Ephemeridae
Breeding in rivers and lakes with silty bottoms, this large mayfly is widespread in Europe. Adults have long antennae and three tail filaments.

POND OLIVE
Cloeon dipterum
F: Baetidae
A widespread European species, this mayfly breeds in a range of habitats from ponds and ditches to troughs and water butts.

BLUE-WINGED OLIVE
Ephemerella ignita
F: Ephemerellidae
Adults of this northern European species have three tail filaments. The males' rounded eyes have two parts, the larger upper portion for spotting females.

SIPHLONURUS LACUSTRIS
F: Siphlonuridae
Very common in upland lakes in northern Europe, this summer mayfly has two long tails and greenish gray wings. The hind wings are small.

DAMSELFLIES AND DRAGONFLIES

These insects in the order Odonata have a distinctive long body with a mobile head and large eyes, giving good all-round vision. The adults have two pairs of similarly sized wings and are fast-flying hunters; the nymphs capture prey underwater using their specialized mouthparts. Dragonflies are robust with rounded heads, whereas damselflies are more slender with broader heads and widely separated eyes.

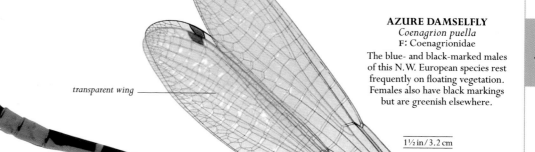

transparent wing

AZURE DAMSELFLY
Coenagrion puella
F: Coenagrionidae

The blue- and black-marked males of this N.W. European species rest frequently on floating vegetation. Females also have black markings but are greenish elsewhere.

1½ in / 3.2 cm

segmented abdomen

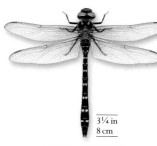

3¼ in
8 cm

TWIN-SPOTTED SPIKETAIL
Cordulegaster maculata
F: Cordulegastridae

This dragonfly is found in the eastern USA and southeastern Canada, where it prefers clean streams in wooded habitats.

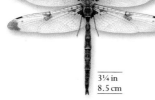

3¼ in
8.5 cm

PRINCE BASKETTAIL
Epitheca princeps
F: Corduliidae

A widespread North American species, the prince baskettail can be seen patrolling ponds, lakes, creeks, and rivers from dawn until dusk.

1½ in
3.2 cm

EMERALD DAMSELFLY
Lestes sponsa
F: Lestidae

Common in a wide band across Europe and Asia, this damselfly is found near still or slow-moving, well-vegetated water.

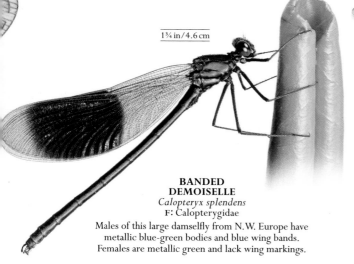

1¾ in / 4.6 cm

BANDED DEMOISELLE
Calopteryx splendens
F: Calopterygidae

Males of this large damselfly from N.W. Europe have metallic blue-green bodies and blue wing bands. Females are metallic green and lack wing markings.

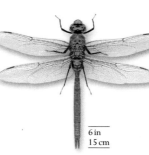

2½ in
6 cm

PLAINS CLUBTAIL
Gomphus externus
F: Gomphidae

Flying on warm sunny days and breeding in slow-moving, muddy streams and rivers, this dragonfly is widespread in the USA.

6 in
15 cm

COMET DARNER
Anax longipes
F: Aeshnidae

Occurring from Brazil to Massachusetts, the comet darner can be found over lakes and large ponds and has a steady, regular flight pattern.

3 in
7.6 cm

SWIFT RIVER CRUISER
Macromia illinoiensis
F: Macromiidae

This North American species patrols gravelly or rocky streams and rivers, but can also be seen flying away from water along roads or tracks.

wing reddish
at base

red or dark-
orange abdomen

3 in
7.6 cm

three long tail filaments

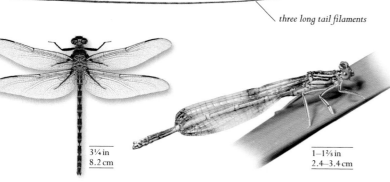

3¼ in
8.2 cm

GRAY PETALTAIL
Tachopteryx thoreyi
F: Petaluridae

A large dragonfly found in damp broad-leaved forests of the east coast of North America, this species breeds in boggy areas and seeps.

1–1⅓ in
2.4–3.4 cm

WHITE-LEGGED DAMSELFLY
Platycnemis pennipes
F: Platycnemididae

This damselfly from central Europe breeds in slow-moving, weedy canals and rivers. The tibia of the hind leg is expanded and appears slightly feathery.

1½–1¾ in
3.5–4.1 cm

BROAD-BODIED CHASER
Libellula depressa
F: Libellulidae

This central European species breeds in ditches and ponds. The abdomen of the male is blue above whereas that of the female is yellowish brown.

FLAME SKIMMER
Libellula saturata
F: Libellulidae

Favoring warm ponds, streams, and even hot springs, this species is common in the southwestern USA.

STONEFLIES

Members of the order Plecoptera are soft, slender-bodied insects with a pair of thin tail filaments and two pairs of wings. The nymphal stages are aquatic.

¾–1 in / 2–2.8 cm

PERLA BIPUNCTATA
F: Perlidae

The males of this species, which favors stony streams in upland regions, have much shorter wings and can be half the size of the females.

¹¹⁄₃₂–½ in
0.9–1.3 cm

YELLOW SALLY
Isoperla grammatica
F: Perlodidae

Particularly common in limestone areas, this species prefers clean, gravel-bottomed streams and stony lakes. Males are shorter than females.

STICK AND LEAF INSECTS

These slow-moving herbivorous insects of the order Phasmatodea have stick- or leaflike bodies that may be smooth or spiny. Many are well camouflaged to avoid predators.

TWO-STRIPED STICK INSECT
Anisomorpha buprestoides
F: Phasmatidae

Found in the southern USA, this species can eject an acidic, defensive liquid from glands in the thorax.

1½–2½ in
4.2–6.8 cm

JUNGLE NYMPH STICK INSECT
Heteropteryx dilatata
F: Heteropterygidae

This impressive species is found in Malaysia. Females are flightless and green; males are smaller, winged, and brownish.

6 in
15.5 cm

JAVANESE LEAF INSECT
Phyllium bioculatum
F: Phylliidae

Females of this southeast Asian insect are large, winged, and leaflike. The males are smaller, more slender, and brown.

2¾–3¾ in
7–9.4 cm

large, fan-shaped hind wing

flattened, leaflike abdomen

EARWIGS

These slender, slightly flattened, scavenging insects belong to the order Dermaptera. They have short front wings with large fan-shaped hind wings folded beneath. The flexible abdomen ends in a pair of multipurpose forceps.

½ in
1.4 cm

COMMON EARWIG
Forficula auricularia
F: Forficulidae

This species can be found under bark and in leaf litter. The female cares for her eggs and feeds the young nymphs.

¾ in
1.8 cm

TAWNY EARWIG
Labidura riparia
F: Labiduridae

The tawny earwig, the largest European species, is especially common along sandy river banks and in coastal areas.

PRAYING MANTIDS

These predatory insects of the order Mantodea have a triangular, highly mobile head with large eyes. The enlarged, spined front legs are specially modified for snatching prey. A pair of toughened front wings protects the larger membranous hind wings folded beneath.

large compound eyes on a triangular head

extended prothorax

leaflike front wing

2–3 in
5–7.4 cm

spiny, enlarged front femur

head crest

2¼ in
6 cm

COMMON PRAYING MANTIS
Mantis religiosa
F: Mantidae

The strike of a praying mantis lasts a fraction of a second, and impales the prey on the sharp spines of the front legs.

CONEHEAD MANTID
Empusa pennata
F: Empusidae

A slender mantid with a distinctive head crest, this southern European species eats small flies and can be green or brown.

1¼–2¼ in
3–6 cm

ORCHID MANTIS
Hymenopus coronatus
F: Hymenopodidae

With its flower-mimicking color and legs resembling petals, this southeast Asian mantis is able to hide among foliage and ambush small insects.

CRICKETS AND GRASSHOPPERS

Orthoptera are mainly herbivorous. They have two pairs of wings, although some can be short-winged or wingless. The hind legs are often large and used for jumping. They sing by rubbing their front wings together or their hind legs on a wing edge.

½–1 in
1.4–2.4 cm

COMMON FIELD GRASSHOPPER
Chorthippus brunneus
F: Acrididae

The common field grasshopper is typically found on short, dry, grazed grassland, where it is most active on sunny days.

2 in
5 cm

WELLINGTON TREE WETA
Hemideina crassidens
F: Stenopelmatidae

Native to New Zealand, this nocturnal insect lives in rotten wood and tree stumps. It eats plant material as well as small insects.

¾ in
2 cm

AFRICAN CAVE CRICKET
Phaeophilacris geertsi
F: Rhaphidophoridae

This central African species is an omnivorous scavenger. Its long antennae are an adaptation to life in dark microhabitats.

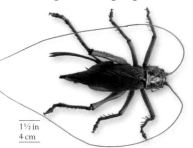

1½ in
4 cm

LEAF-ROLLING CRICKET
Gryllacris subdebilis
F: Gryllacrididae

Found in Australia, this species has relatively long wings and very long antennae, which may be up to three times the length of the body.

1½–2¼ in/4–6 cm

DESERT LOCUST
Schistocerca gregaria
F: Acrididae

Crowding of nymphs after rain stimulates this African insect to transform from a solitary into a gregarious form. It swarms in billions, devastating crops.

MOLE CRICKET
Gryllotalpa gryllotalpa
F: Gryllotalpidae

This species from Europe uses its strong front legs to burrow, just like a miniature mole. It is found in meadows and banks near rivers, where the soil is damp and sandy.

1½–2 in
4–4.5 cm

OAK BUSH CRICKET
Meconema thalassinum
F: Tettigoniidae

Found on a range of broadleaved trees, this European cricket feeds on small insects after dark. The female has a long, curved ovipositor.

½–¾ in
1.8–2 cm

ovipositor for laying eggs

warty surface

HOUSE CRICKET
Acheta domestica
F: Gryllidae

This nocturnal species makes an attractive chirping song. Originally from southwest Asia and North Africa, it has spread into Europe.

1in
2.4 cm

1¼ in / 2.8 cm

COMMON BLACK CRICKET
Gryllus bimaculatus
F: Gryllidae

This cricket is widespread in southern Europe, parts of Africa, and Asia and lives on the ground under wood or debris.

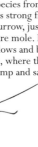

bright red markings

2¼–3¼ in
6–8 cm

FOAMING GRASSHOPPER
Dictyophorus spumans
F: Pyrgomorphidae

Bright colors advertise that this South African grasshopper is toxic to predators. It can also produce a noxious foam from its thoracic glands.

COCKROACHES

Members of the order Blattodea are scavenging insects, with a flattened oval body. The downward-pointing head is often largely concealed by a shieldlike pronotum and there are usually two pairs of wings. The tip of the abdomen has a pair of sensory prongs or cerci.

2–3¼ in
5–8 cm

MADAGASCAN HISSING COCKROACH
Gromphadorhina portentosa
F: Blaberidae

This large, wingless cockroach is reared worldwide as a pet. The male has prominent bumps on the thorax for male-to-male combat.

short terminal cerci

NYMPH

1¾ in
4.4 cm

TERMITES

These nest-building social insects of the order Isoptera live in colonies with different castes: reproductives (the kings and queens), workers, and soldiers. Workers are generally pale and wingless; reproductives have wings that are shed after a nuptial fight; and soldiers have large heads and jaws.

SUBTERRANEAN TERMITE
Coptotermes formosanus
F: Termitidae

Native to southern China, Taiwan, and Japan, this invasive species has now spread to other parts of the world, where it is a serious pest.

¼–⁷⁄₃₂ in
6–7 mm

AMERICAN COCKROACH
Periplaneta americana
F: Blattidae

This species, which was originally from Africa, is now found worldwide. It lives on ships and in food warehouses.

light brown pronotum

¹¹⁄₃₂–⁷⁄₁₀ in
0.8–1.3 cm

DUSKY COCKROACH
Ectobius lapponicus
F: Blattellidae

Small and fast running, this European species can be found among leaf litter and occasionally in foliage. It has been introduced into the USA.

⁹⁄₁₀ in
2.4 cm

PACIFIC DAMPWOOD TERMITE
Zootermopsis angusticollis
F: Termopsidae

This termite, found along the Pacific coastal states of North America, nests in and feeds on decayed, fungus-infected wood.

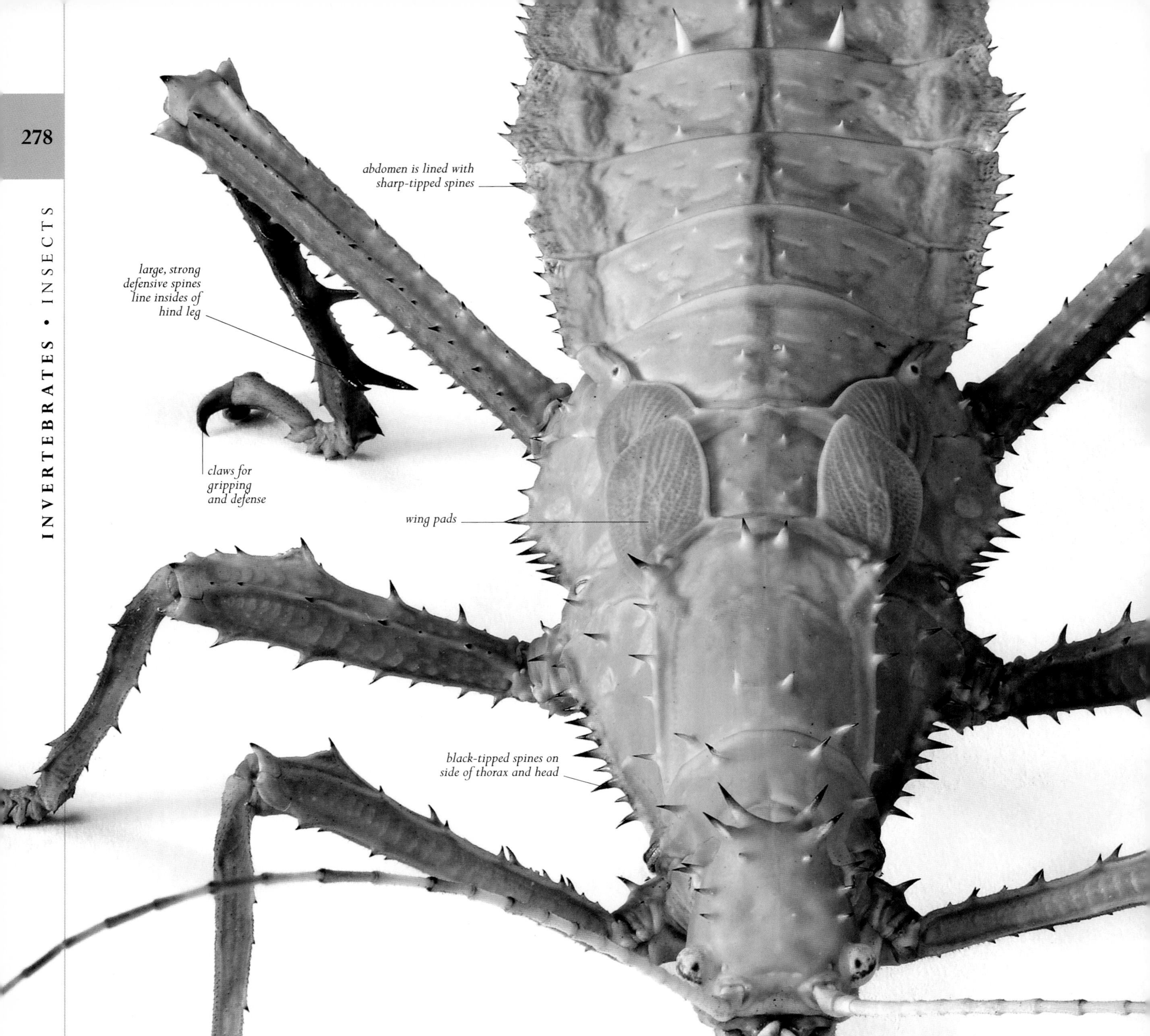

abdomen is lined with
sharp-tipped spines

large, strong
defensive spines
line insides of
hind leg

claws for
gripping
and defense

wing pads

black-tipped spines on
side of thorax and head

∧ **ADOLESCENT**
The small, nonoverlapping
wing pads show that this
female is still a nymph—
not yet sexually mature.
At the next molt, when it
sheds its skin, the nymph will
become an adult, acquiring
short, stubby wings and a
functional ovipositor. After
mating, the abdomen will
swell as the eggs develop.

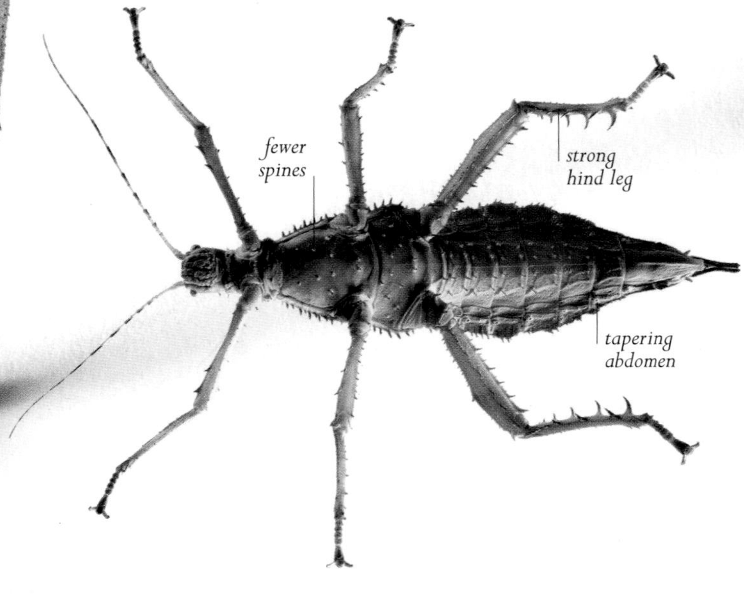

fewer
spines

strong
hind leg

tapering
abdomen

< **UNDERSIDE**
The female's underside is
dark green and, although
it has fewer spines than
the top surface, it is well
protected by the spiny legs.

JUNGLE NYMPH STICK INSECT
Heteropteryx dilatata

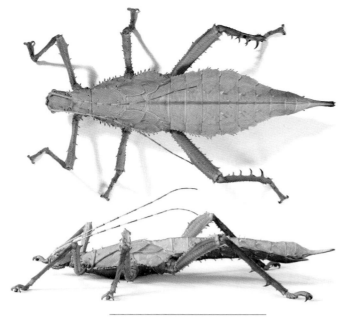

This species is also known as the Malayan jungle nymph. Females are large, with bright green coloration on their upper surface and darker green below. The adult male is much smaller, more slender, and darker in color. Both have wings, but the female is flightless. Nymphs and adults feed on the foliage of a range of different plants, including durian, guava, and mango. Mature females with eggs can be very aggressive. If disturbed, they make a loud hissing sound using their short wings, and splay their strong, spiny hind legs in a defensive posture. If attacked, they will kick out. This lively species has become a popular pet worldwide.

SIZE Up to 6 in (15.5 cm)
HABITAT Tropical forest
DISTRIBUTION Malaysia
DIET Foliage of various plants

MOUTHPARTS >
At rest, the jaws are concealed behind two pairs of food-handling appendages called palps. These are covered with sensory organs, which enable the insect to taste the surface of the leaves it feeds on.

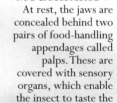

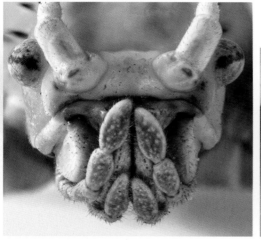

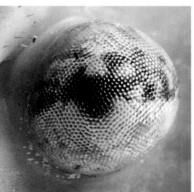

< COMPOUND EYE
Compound eyes—typical of insects—do not need to be as acute as those of many predatory arthropods, but they do need to be able to detect movement and potential enemies.

SEGMENTED BODY >
The tough, spiny plates that make up the body segments are joined by soft membranes that allow flexibility.

∧ OVIPOSITOR
The female—which can lay up to 150 eggs in her lifetime—uses her ovipositor to lay her large eggs one at a time, concealed in any suitable substrate such as leaf litter.

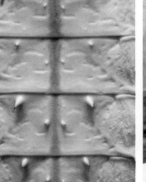
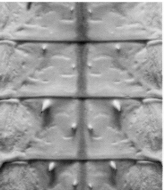

long, segmented antennae sense the immediate environment, including air movements

∧> FOOT
The foot is made up of a number of short tarsal segments and a long, spiny terminal segment, which bears a pair of sharp, curved claws.

TRUE BUGS

Abundant and widespread in terrestrial and aquatic habitats, members of the order Hemiptera range from minute, wingless insects to giant water bugs capable of catching fish and frogs. The mouthparts are used for piercing and sucking up liquids such as plant sap, dissolved prey tissues, or blood. Many species are plant pests and some transmit disease.

$\frac{1}{32}$–$\frac{1}{16}$ in
1–2 mm

GREENHOUSE WHITEFLY
Trialeurodes vaporariorum
F: Aleyrodidae
This small mothlike bug is found in temperate regions worldwide and can be a serious pest of greenhouse crops.

$\frac{3}{16}$ in
5 mm

AMERICAN LUPIN APHID
Macrosiphum albifrons
F: Aphididae
Aphids such as this North American species can quickly infest plants because females can produce many offspring without fertilization.

distinct, dark marginal area

$\frac{5}{16}$–$\frac{3}{8}$ in
8–10 mm

SPITTLE BUG
Aphrophora alni
F: Aphrophoridae
Commonly occurring on a wide range of trees and shrubs across Europe, this froghopper can vary in appearance from pale to dark brown.

$\frac{3}{8}$–$\frac{1}{2}$ in
1–1.2 cm

FROGHOPPER
Cecopis vulnerata
F: Cercopidae
The nymphs of this conspicuously colored European species live communally underground in a protective, frothy mass and feed on plant root sap.

INDIAN CICADA
Angamiana aetherea
F: Cicadidae
This cicada is found in India. As with all cicadas, males produce loud songs both for courtship and to signal aggression.

$1\frac{1}{4}$–$1\frac{1}{2}$ in
3.5–4 cm

pale area at base of hind wing

$\frac{7}{8}$–1 in
2.2–2.6 cm

HIMALAYAN CICADA
Pycna repanda
F: Cicadidae
Known in North India, this species is also found in parts of China and Nepal, where it prefers cool, alpine, deciduous forests.

$\frac{1}{4}$–$\frac{5}{16}$ in
6–8 mm

HORSE CHESTNUT SCALE
Pulvinaria regalis
F: Coccidae
Although commonly found on the bark of horse chestnut trees, this European scale insect will also attack a range of other deciduous species.

$\frac{1}{2}$ in
1.3 cm

LEAFHOPPER
Ledra aurita
F: Cicadellidae
Mottled coloration helps this flat-bodied leafhopper from northern Europe to blend with the lichen-covered bark of oak trees in its habitat.

$\frac{7}{32}$–$\frac{11}{32}$ in
7–9 mm

CICADELLA VIRIDIS
F: Cicadellidae
Found feeding on grasses and sedges in wet, boggy, or marshy areas in Europe and Asia, this leafhopper can also be found near garden ponds.

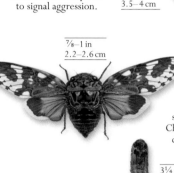

$3\frac{1}{4}$ in / 8 cm

large eyespot

PEANUT-HEADED BUG
Fulgora laternaria
F: Fulgoridae
This bug is found in Central and South America and the West Indies. The bulbous head was once thought to glow.

$\frac{3}{8}$–$\frac{1}{2}$ in
1–1.2 cm

THORN BUG
Umbonia crassicornis
F: Membracidae
Found in Central and South America, almost the whole body is concealed under the enlarged, thornlike pronotum.

elongated head

WART-HEADED BUG
Phrictus quinquepartitus
F: Fulgoridae
Also known as the dragon-headed bug, this species can be found in Costa Rica, Panama, Colombia, and parts of Brazil.

$2\frac{1}{4}$ in / 5.5 cm

$\frac{1}{16}$–$\frac{1}{8}$ in
2–3 mm

ASH PLANT LOUSE
Psyllopsis fraxini
F: Psyllidae
Found commonly on ash trees, the nymphs of this species cause red, swollen galls to form at the edges of leaves where they feed.

distinctive marking across forewings

brightly colored hind wings

$\frac{1}{8}$–$\frac{5}{32}$ in
3–4 mm

THISTLE LACE BUG
Tingis cardui
F: Tingidae
This bug occurs in western Europe and feeds on spear, musk, and marsh thistles. The body is covered in powdery wax.

EUROPEAN TORTOISE BUG
Eurygaster maura
F: Scutelleridae
This bug feeds on a wide range of grasses but can sometimes be a minor pest of cereal crops.

³⁄₈–½ in
1–1.2 cm

HAWTHORN SHIELD BUG
Acanthosoma haemorrhoidale
F: Acanthosomatidae
This attractive European species feeds on the buds and berries of hawthorn and occasionally other deciduous trees such as hazel.

½ in
1.3 cm

COMMON FLOWER BUG
Anthocoris nemorum
F: Anthocoridae
This predatory bug can be found on a wide range of plants. Despite its small size it can pierce human skin.

⁵⁄₃₂ in
4 mm

BIRCH BARK BUG
Aradus betulae
F: Aradidae
The flat body of this European bug allows it to live under the bark of birch trees, where it feeds on fungi.

³⁄₁₆ in
5 mm

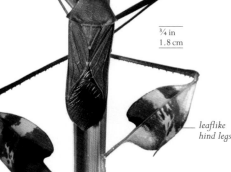

long antennae

leaflike hind legs

¾ in
1.8 cm

LEAF-FOOTED BUG
Bitta flavolineata
F: Coreidae
Found in parts of Central and South America, the leg expansions of this herbivorous bug may provide camouflage and protect it from enemies.

⁵⁄₃₂–³⁄₁₆ in
4–5mm

BEDBUG
Cimex lectularius
F: Cimicidae
This widespread species feeds on the blood of humans and other warm-blooded mammals. These wingless, flattened bugs are active after dark.

single sharp claw

strong front leg

hair-fringed hind leg

pair of appendages acts as breathing siphon

3¼–4 in
8–10 cm

GIANT WATER BUG
Lethocerus grandis
F: Belostomatidae
Using its strong front legs and toxic saliva, this widespread bug can overpower larger vertebrate prey such as frogs and fish.

³⁄₈–½ in
1–1.2 cm

COMMON POND SKATER
Gerris lacustris
F: Gerridae
Widespread and immediately recognizable as it darts about on the surface of water, the pond skater locates prey by the ripples it sends out.

½ in/1.2 cm

WATER BOATMAN
Corixa punctata
F: Corixidae
Using its powerful, oarlike rear legs for swimming, this common European species feeds on algae and detritus in ponds.

³⁄₈ in/1.1 cm

TOAD BUG
Nerthra grandicollis
F: Gelastocoridae
The warty surface and drab colouring of this African bug allow it to hide in mud and debris, where it ambushes insect prey.

½ in
1.5 cm

SAUCER BUG
Ilyocoris cimicoides
F: Naucoridae
The saucer bug traps air from the surface under its folded wings and hunts prey in the shallow margins of lakes and slow rivers. This species is from Europe.

½ in
1.2 cm

WATER MEASURER
Hydrometra stagnorum
F: Hydrometridae
This slow-moving European bug lives around the margins of ponds, lakes, and rivers and feeds on small insects and crustaceans.

¼ in
6 mm

COMMON GREEN CAPSID
Lygocoris pabulinus
F: Miridae
This widespread bug can be a serious pest of a range of plants, including fruit crops such as raspberries, pears, and apples.

»

» TRUE BUGS

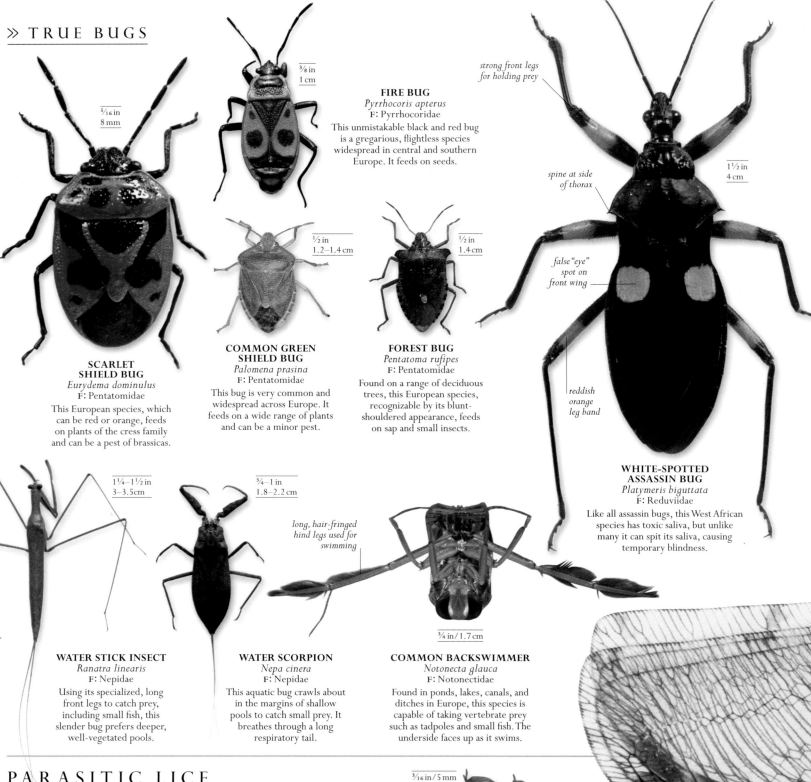

5/16 in
8 mm

3/8 in
1 cm

FIRE BUG
Pyrrhocoris apterus
F: Pyrrhocoridae
This unmistakable black and red bug
is a gregarious, flightless species
widespread in central and southern
Europe. It feeds on seeds.

strong front legs
for holding prey

1/2 in
1.2–1.4 cm

1/2 in
1.4 cm

spine at side
of thorax

1 1/2 in
4 cm

false "eye"
spot on
front wing

**SCARLET
SHIELD BUG**
Eurydema dominulus
F: Pentatomidae
This European species, which
can be red or orange, feeds
on plants of the cress family
and can be a pest of brassicas.

**COMMON GREEN
SHIELD BUG**
Palomena prasina
F: Pentatomidae
This bug is very common and
widespread across Europe. It
feeds on a wide range of plants
and can be a minor pest.

FOREST BUG
Pentatoma rufipes
F: Pentatomidae
Found on a range of deciduous
trees, this European species,
recognizable by its blunt-
shouldered appearance, feeds
on sap and small insects.

reddish
orange
leg band

**WHITE-SPOTTED
ASSASSIN BUG**
Platymeris biguttata
F: Reduviidae
Like all assassin bugs, this West African
species has toxic saliva, but unlike
many it can spit its saliva, causing
temporary blindness.

1 1/4–1 1/2 in
3–3.5cm

3/4–1 in
1.8–2.2 cm

long, hair-fringed
hind legs used for
swimming

3/4 in / 1.7 cm

WATER STICK INSECT
Ranatra linearis
F: Nepidae
Using its specialized, long
front legs to catch prey,
including small fish, this
slender bug prefers deeper,
well-vegetated pools.

WATER SCORPION
Nepa cinera
F: Nepidae
This aquatic bug crawls about
in the margins of shallow
pools to catch small prey. It
breathes through a long
respiratory tail.

COMMON BACKSWIMMER
Notonecta glauca
F: Notonectidae
Found in ponds, lakes, canals, and
ditches in Europe, this species is
capable of taking vertebrate prey
such as tadpoles and small fish. The
underside faces up as it swims.

PARASITIC LICE

These wingless insects of the order Phthiraptera are ectoparasites,
living on the bodies of birds and mammals. Their mouthparts are
modified for chewing skin fragments or sucking blood, and their
legs for gripping tightly to hair or feathers.

3/16 in / 5 mm

1/16–1/8 in
2–3 mm

**HUMAN
BODY LOUSE**
Pediculus humanus humanus
F: Pediculidae
This subspecies may have
evolved from the head louse
after the invention of clothing,
to which it glues its eggs.
It transmits disease.

1/16–1/8 in
2–3 mm

**CHICKEN
BODY LOUSE**
Menacanthus stramineus
F: Menoponidae
Found worldwide as an
ectoparasite of chickens,
infestations of this pale,
flattened chewing louse can
cause feather loss and infection.

1/32–1/16 in
1–2 mm

HUMAN HEAD LOUSE
Pediculus humanus capitis
F: Pediculidae
This louse glues its eggs,
called "nits," to head hair.
Infestation outbreaks are
common among young
children. A related species
attacks chimpanzees.

GOAT LOUSE
Damalinia caprae
F: Trichodectidae
Found worldwide on
goats, this biting louse
can also survive for
a few days on sheep
but is not able to
breed on them.

BARKLICE AND BOOKLICE

Common on vegetation and in litter, insects of the order Psocoptera are small, squat, and soft-bodied. The head has threadlike antennae and bulging eyes. They eat microflora and some species are pests of stored products.

¼ in / 6 mm

PSOCOCERASTIS GIBBOSA
F: Psocidae
Native to Europe and parts of Asia, this relatively large bark louse can be found on a wide range of deciduous and coniferous trees.

¹⁄₂₄ in / 1.5 mm

LIPOSCELIS LIPARIUS
F: Liposcelididae
This very widespread species prefers dark, damp microhabitats and can be a pest in libraries and granaries if the humidity is too high.

THRIPS

Members of the order Thysanoptera are tiny insects, typically with two pairs of narrow, hair-fringed wings. They have large compound eyes and distinctive mouthparts for piercing and sucking.

¹⁄₃₂–¹⁄₂₄ in / 1–1.5 mm

FLOWER THRIP
Frankliniella sp.
F: Thripidae
Flower thrips are found all over the world and can be pests on crops such as peanuts, cotton, sweet potatoes, and coffee.

SNAKEFLIES

Snakeflies of the order Raphidioptera are woodland insects with a long prothorax, a broad head, and two pairs of wings. Adults and larvae eat aphids and other soft prey.

RAPHIDIA NOTATA
F: Raphidiidae
Found in deciduous or coniferous woodlands in Europe, this snakefly is usually associated with oak trees, where it eats aphids.

½–¾ in / 1.6–1.8 cm

ALDERFLIES AND DOBSONFLIES

Megaloptera have two pairs of wings, which are held rooflike over the body when they are at rest. The aquatic larvae, which are predatory and have abdominal gills, pupate on land in soil, moss, or rotting wood.

EASTERN DOBSONFLY
Corydalus cornutus
F: Corydalidae
Found in North America, the male of this species has very long mandibles used for combat and for gripping the female.

4 in / 10 cm

ALDERFLY
Sialis lutaria
F: Sialidae
The female of this widespread species lays a mass of up to 2,000 eggs on twigs or leaves near water.

½–¾ in / 1.4–1.8 cm

LACEWINGS AND RELATIVES

Insects of the order Neuroptera have conspicuous eyes and biting mouthparts. The pairs of net-veined wings are held rooflike over the body at rest. The larvae have sickle-shaped mouthparts, which form sharp, sucking tubes.

1¼ in / 3 cm

½ in / 1.4 cm

MANTISPA STYRIACA
F: Mantispidae
Like a miniature praying mantid, this, found in southern and central Europe, lives in lightly wooded areas, where it hunts small flies.

OWLFLY
Libelloides macaronius
F: Ascalaphidae
Capturing insect prey in midair, this species will only fly on warm sunny days. It can be found in central and southern Europe and parts of Asia.

GIANT LACEWING
Osmylus fulvicephalus
F: Osmylidae
This European species can be found among shady woodland vegetation by streams, where it eats small insects and pollen.

½ in / 1.5 cm

SPOON-WINGED LACEWING
Nemoptera sinuata
F: Nemopteridae
Common to areas of southeast Europe, this delicate species feeds on nectar and pollen at flowers in woodland and open grassland.

1½ in / 4 cm

³⁄₈–½ in / 1–1.2 cm

GREEN LACEWING
Chrysopa perla
F: Chrysopidae
This widespread European species has a characteristic bluish green tinge and black markings and is commonly found in deciduous woodland.

ANTLION
Palpares libelluloides
F: Myrmeleontidae
This large, day-flying, Mediterranean species with distinctively mottled wings can be found in rough grassland and warm scrubby habitats.

2–2¼ in / 5–5.5 cm

males have organs for clasping females

BEETLES

Members of the order Coleoptera—the largest insect order—range from minute to very large species. A distinguishing feature are the toughened front wings, called elytra, which meet down the body's midline and protect the larger membranous hind wings. Beetles occupy every aquatic and terrestrial habitat, where they are scavengers, herbivores, or predators.

VIOLET GROUND BEETLE
Carabus violaceus
F: Carabidae
A nocturnal hunter, this beetle is common in many habitats, including gardens. It is native to Europe and parts of Asia.

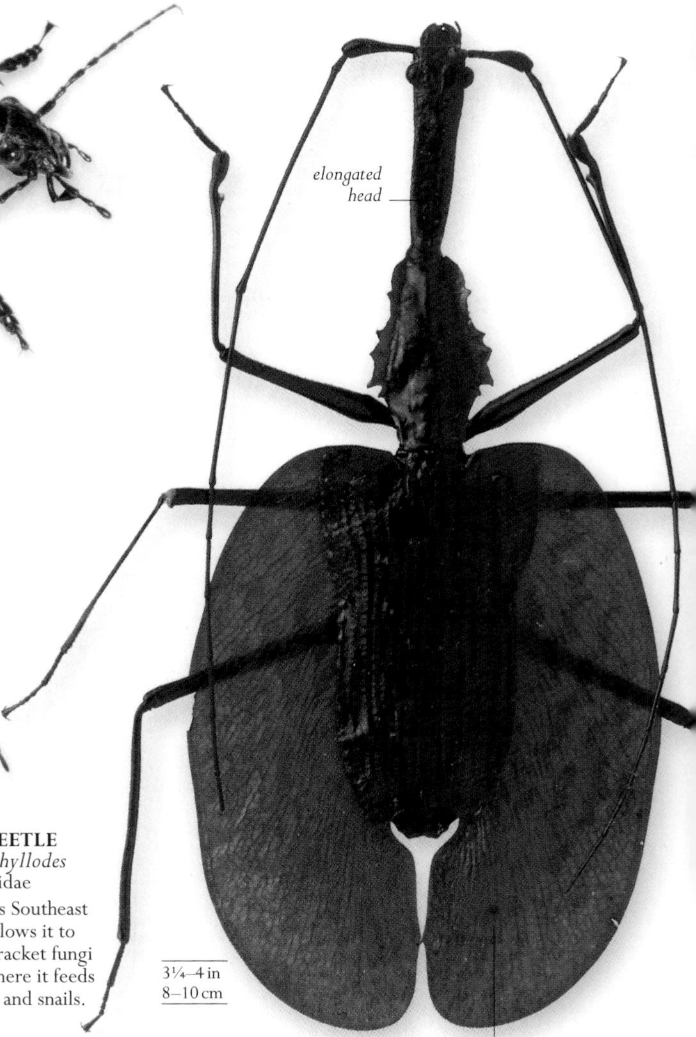

1–1⅖ in
2.8–3.4 cm

elongated head

⅛–³⁄₁₆ in
3–5 mm

1½ in
4 cm

⅜ in
1 cm

WOODWORM BEETLE
Anobium punctatum
F: Anobiidae
Adapted to breeding in timber in buildings and furniture, this species is now widespread and a serious pest.

JEWEL BEETLE
Chrysochroa chinensis
F: Buprestidae
This metallic species is native to India and southeast Asia, where its larvae burrow into the wood of deciduous trees.

SOLDIER BEETLE
Rhagonycha fulva
F: Cantharidae
A common sight on flowers in the summer, this European beetle can be found in meadows and woodland margins.

VIOLIN BEETLE
Mormolyce phyllodes
F: Carabidae
The shape of this Southeast Asian beetle allows it to squeeze under bracket fungi and tree bark, where it feeds on insect larvae and snails.

3¼–4 in
8–10 cm

wide, flat elytra

⁷⁄₃₂–⅜ in
7–10 mm

elytra with metallic sheen

³⁄₁₆–⁵⁄₁₆ in
5–8 mm

ANT BEETLE
Thanasimus formicarius
F: Cleridae
Associated with coniferous trees in Europe and northern Asia, the larvae and adults of this beetle prey on other beetle larvae.

⁵⁄₁₆–⅜ in
8–10 mm

1¼–1½ in
3–4 cm

LARDER BEETLE
Dermestes lardarius
F: Dermestidae
Found in Europe and parts of Asia, this beetle feeds on animal remains but also lives in buildings, where it eats stored produce.

¹⁄₁₆–1½ in
0.2–4 cm

GREAT DIVING BEETLE
Dytiscus marginalis
F: Dytiscidae
This large beetle lives in weedy ponds and lakes in Europe and northern Asia. It eats insects, frogs, newts, and small fish.

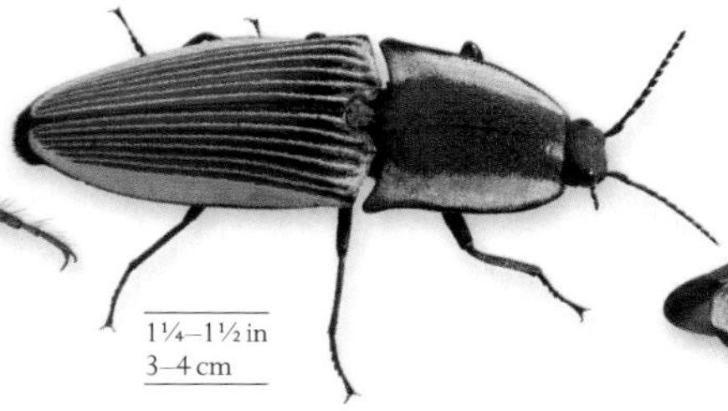

CLICK BEETLE
Chalcolepidius limbatus
F: Elateridae
This beetle is found in woodland and grassland in the warmer parts of South America. Its larvae are predators in rotten wood and soil.

WHIRLIGIG BEETLE
Gyrinus marinus
F: Gyrinidae
A common European species, this beetle is a surface-dweller on ponds and lakes. It uses its paddle-shaped legs to skim over the water.

1 in
2.5 cm

¼–⅜ in
6–10 mm

½–¾ in
1.5–2 cm

curved horns

⅜ in
1 cm

SCREECH BEETLE
Hygrobia hermanni
F: Hygrobiidae
This European beetle feeds on small invertebrate prey in slow-moving rivers and muddy ponds. It makes a squeaking noise if handled.

LAMPYRIS NOCTILUCA
F: Lampyridae
This beetle is found across Europe and Asia, where it prefers rough grassland. The wingless female—known as the glow worm—emits a greenish light to attract a mate.

FOUR-SPOTTED HISTER BEETLE
Hister quadrimaculatus
F: Histeridae
Widespread across Europe, this beetle is found in dung and sometimes carrion, where it feeds on small insects and their larvae.

MINOTAUR BEETLE
Typhoeus typhoeus
F: Geotrupidae
This species is found in sandy areas in western Europe and buries sheep and rabbit droppings, on which the larvae feed, in burrows.

POWDERPOST BEETLE
Lyctus opaculus
F: Lyctidae

This North American beetle breeds in dry wood and reduces it to a fine powder.

⅛–⁵⁄₃₂ in
3–4 mm

red parallel-sided elytra

⁵⁄₁₆ in
8 mm

yellow antennal tip

1 in
2.6 cm

SEXTON BEETLE
Nicrophorus investigator
F: Silphidae

This species is found all over the Northern Hemisphere in woodland and grassland. It buries the carcass of small animals and the female lays an egg on it, providing food for the larva.

2¼–6½ in
6–17 cm

HERCULES BEETLE
Dynastes hercules
F: Dynastidae

The largest species of rhinoceros beetle (genus *Dynastes*), it feeds on rotting fruit in rain forests in Central and South America. Its larvae breed in decaying wood.

2¼–4 in
5.5–10 cm

GOLIATH BEETLE
Goliathus cacicus
F: Cetoniidae

This species, the world's heaviest insect, lives in equatorial Africa, where the adults feed on ripe fruit or tree sap.

NET-WINGED BEETLE
Platycis minuta
F: Lycidae

This small beetle is associated with decaying wood in mature and ancient woodland across Eurasia.

massive mandibles

3 in
7.5 cm

STAG BEETLE
Lucanus cervus
F: Lucanidae

This beetle lives in woodland in southern and central Europe. Its larvae develop over 4–6 years, mainly in decaying oak stumps. Males use their enlarged mandibles in contests over females.

⁵⁄₃₂–³⁄₁₆ in
4–5 mm

POLLEN BEETLE
Glischrochilus hortensis
F: Nitidulidae

Often found feeding at fermenting sap flows or ripe fruit, this western European species is associated with decaying trees such as birch.

³⁄₅–1 in
1.6–2.6 cm

PHANAEUS DEMON
F: Scarabaeidae

Native to Central America, this metallic green beetle breeds in the dung of large herbivorous animals in grassland and pasture.

antennal club

golden iridescence

DEVIL'S COACH HORSE
Staphylinus olens
F: Staphylinidae

A European species found among woodland and garden leaf litter, this beetle raises its abdomen in a threatening display if disturbed.

³⁄₄–1¹⁄₁₀ in
2–2.8 cm

midline between elytra

EMUS HIRTUS
F: Staphylinidae

Native to southern and central Europe, this hairy rove beetle feeds on other insects attracted to cow and horse manure or carrion.

1¼ in
3 cm

hind leg claw

long, segmented antennae

GOLD BEETLE
Plusiotis resplendens
F: Scarabaeidae

Found in upland areas of Costa Rica and Panama, this beetle lives in wet forest or plantations. The larvae breed in rotten tree trunks.

¾ in
2 cm

»

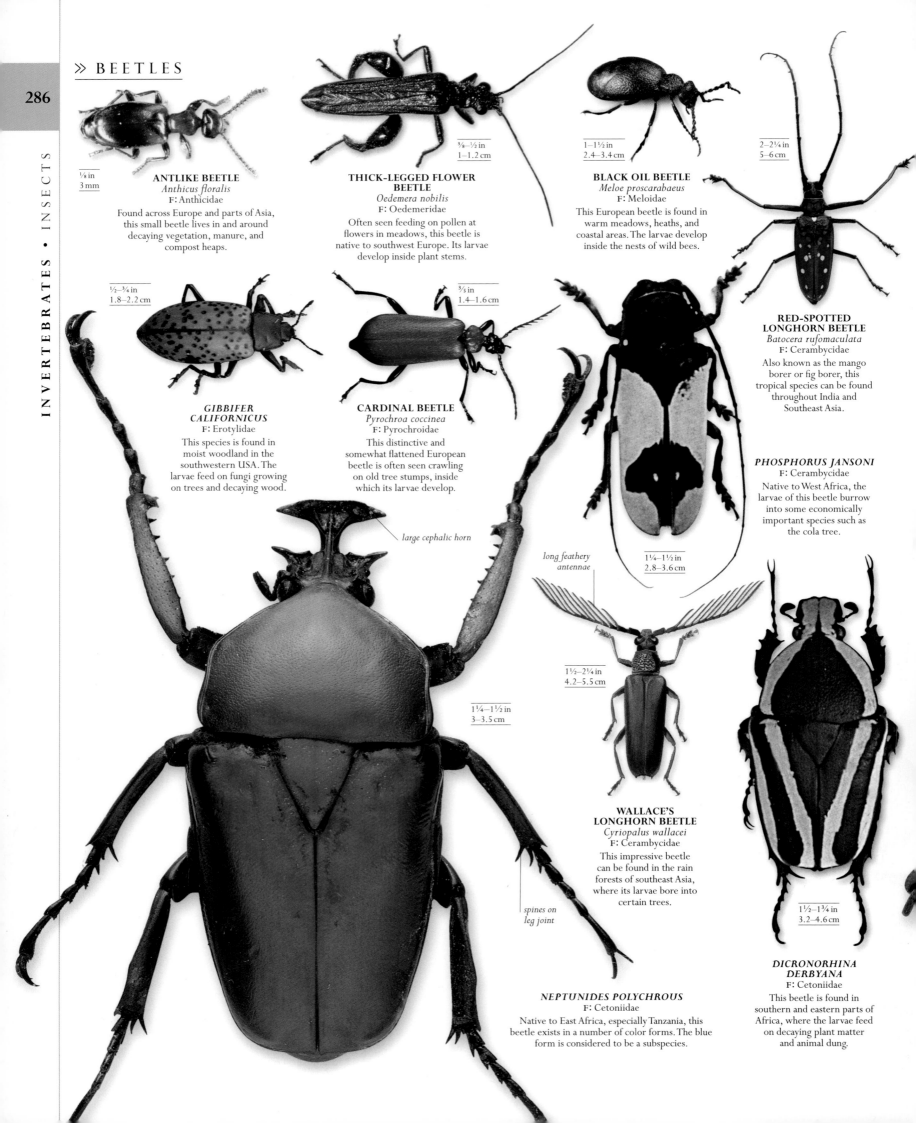

ANTLIKE BEETLE
Anthicus floralis
F: Anthicidae
Found across Europe and parts of Asia, this small beetle lives in and around decaying vegetation, manure, and compost heaps.

⅛ in
3 mm

THICK-LEGGED FLOWER BEETLE
Oedemera nobilis
F: Oedemeridae
Often seen feeding on pollen at flowers in meadows, this beetle is native to southwest Europe. Its larvae develop inside plant stems.

⅜–½ in
1–1.2 cm

BLACK OIL BEETLE
Meloe proscarabaeus
F: Meloidae
This European beetle is found in warm meadows, heaths, and coastal areas. The larvae develop inside the nests of wild bees.

1–1½ in
2.4–3.4 cm

2–2¼ in
5–6 cm

RED-SPOTTED LONGHORN BEETLE
Batocera rufomaculata
F: Cerambycidae
Also known as the mango borer or fig borer, this tropical species can be found throughout India and Southeast Asia.

GIBBIFER CALIFORNICUS
F: Erotylidae
This species is found in moist woodland in the southwestern USA. The larvae feed on fungi growing on trees and decaying wood.

½–¾ in
1.8–2.2 cm

CARDINAL BEETLE
Pyrochroa coccinea
F: Pyrochroidae
This distinctive and somewhat flattened European beetle is often seen crawling on old tree stumps, inside which its larvae develop.

⅗ in
1.4–1.6 cm

PHOSPHORUS JANSONI
F: Cerambycidae
Native to West Africa, the larvae of this beetle burrow into some economically important species such as the cola tree.

1¼–1½ in
2.8–3.6 cm

large cephalic horn

1¼–1½ in
3–3.5 cm

spines on leg joint

long feathery antennae

1½–2¼ in
4.2–5.5 cm

WALLACE'S LONGHORN BEETLE
Cyriopalus wallacei
F: Cerambycidae
This impressive beetle can be found in the rain forests of southeast Asia, where its larvae bore into certain trees.

NEPTUNIDES POLYCHROUS
F: Cetoniidae
Native to East Africa, especially Tanzania, this beetle exists in a number of color forms. The blue form is considered to be a subspecies.

1½–1¾ in
3.2–4.6 cm

DICRONORHINA DERBYANA
F: Cetoniidae
This beetle is found in southern and eastern parts of Africa, where the larvae feed on decaying plant matter and animal dung.

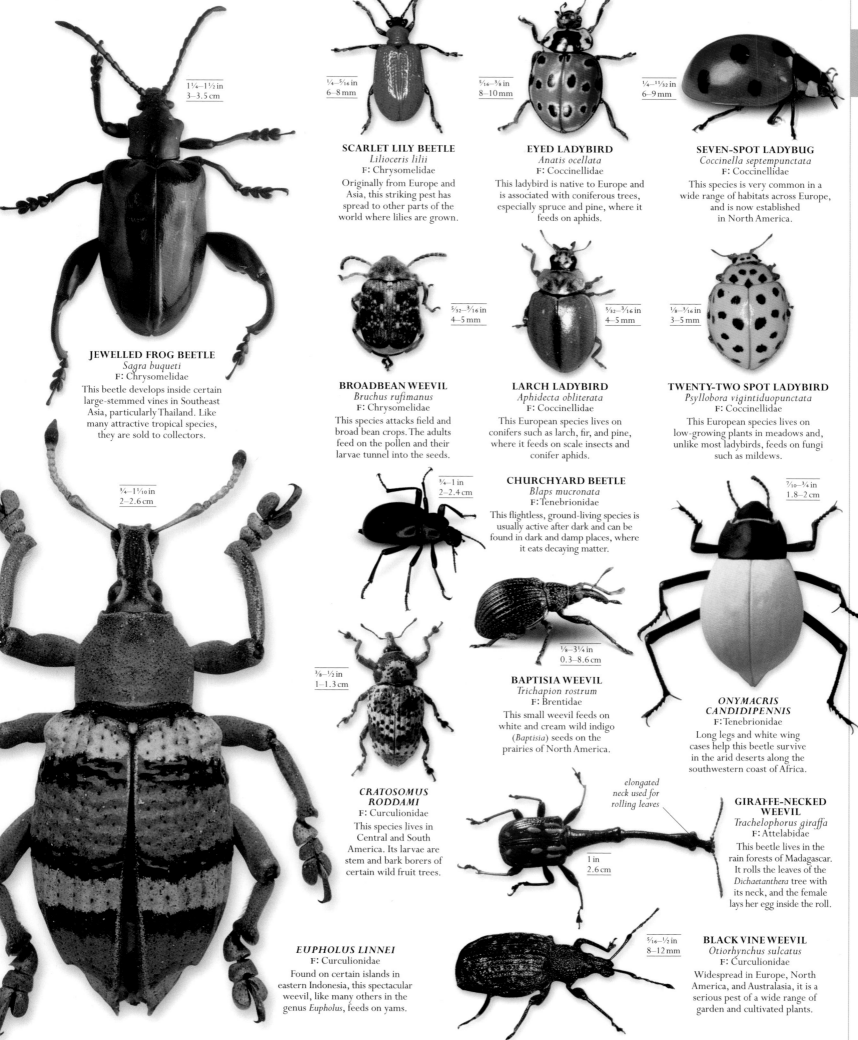

1¼–1½ in
3–3.5 cm

JEWELLED FROG BEETLE
Sagra buqueti
F: Chrysomelidae
This beetle develops inside certain
large-stemmed vines in Southeast
Asia, particularly Thailand. Like
many attractive tropical species,
they are sold to collectors.

¾–1 ¹⁄₁₀ in
2–2.6 cm

EUPHOLUS LINNEI
F: Curculionidae
Found on certain islands in
eastern Indonesia, this spectacular
weevil, like many others in the
genus *Eupholus*, feeds on yams.

¼–⁵⁄₁₆ in
6–8 mm

SCARLET LILY BEETLE
Lilioceris lilii
F: Chrysomelidae
Originally from Europe and
Asia, this striking pest has
spread to other parts of the
world where lilies are grown.

⁵⁄₃₂–³⁄₁₆ in
4–5 mm

BROADBEAN WEEVIL
Bruchus rufimanus
F: Chrysomelidae
This species attacks field and
broad bean crops. The adults
feed on the pollen and their
larvae tunnel into the seeds.

¾–1 in
2–2.4 cm

CHURCHYARD BEETLE
Blaps mucronata
F:Tenebrionidae
This flightless, ground-living species is
usually active after dark and can be
found in dark and damp places, where
it eats decaying matter.

⅜–½ in
1–1.3 cm

**CRATOSOMUS
RODDAMI**
F: Curculionidae
This species lives in
Central and South
America. Its larvae are
stem and bark borers of
certain wild fruit trees.

⅛–3¼ in
0.3–8.6 cm

BAPTISIA WEEVIL
Trichapion rostrum
F: Brentidae
This small weevil feeds on
white and cream wild indigo
(*Baptisia*) seeds on the
prairies of North America.

1 in
2.6 cm

*elongated
neck used for
rolling leaves*

**GIRAFFE-NECKED
WEEVIL**
Trachelophorus giraffa
F: Attelabidae
This beetle lives in the
rain forests of Madagascar.
It rolls the leaves of the
Dichaetanthera tree with
its neck, and the female
lays her egg inside the roll.

⁵⁄₁₆–³⁄₈ in
8–10 mm

EYED LADYBIRD
Anatis ocellata
F: Coccinellidae
This ladybird is native to Europe and
is associated with coniferous trees,
especially spruce and pine, where it
feeds on aphids.

⁵⁄₃₂–³⁄₁₆ in
4–5 mm

LARCH LADYBIRD
Aphidecta obliterata
F: Coccinellidae
This European species lives on
conifers such as larch, fir, and pine,
where it feeds on scale insects and
conifer aphids.

¼–¹¹⁄₃₂ in
6–9 mm

SEVEN-SPOT LADYBUG
Coccinella septempunctata
F: Coccinellidae
This species is very common in a
wide range of habitats across Europe,
and is now established
in North America.

⅛–³⁄₁₆ in
3–5 mm

TWENTY-TWO SPOT LADYBIRD
Psyllobora vigintiduopunctata
F: Coccinellidae
This European species lives on
low-growing plants in meadows and,
unlike most ladybirds, feeds on fungi
such as mildews.

⁷⁄₁₀–¾ in
1.8–2 cm

**ONYMACRIS
CANDIDIPENNIS**
F:Tenebrionidae
Long legs and white wing
cases help this beetle survive
in the arid deserts along the
southwestern coast of Africa.

⁵⁄₁₆–½ in
8–12 mm

BLACK VINE WEEVIL
Otiorhynchus sulcatus
F: Curculionidae
Widespread in Europe, North
America, and Australasia, it is a
serious pest of a wide range of
garden and cultivated plants.

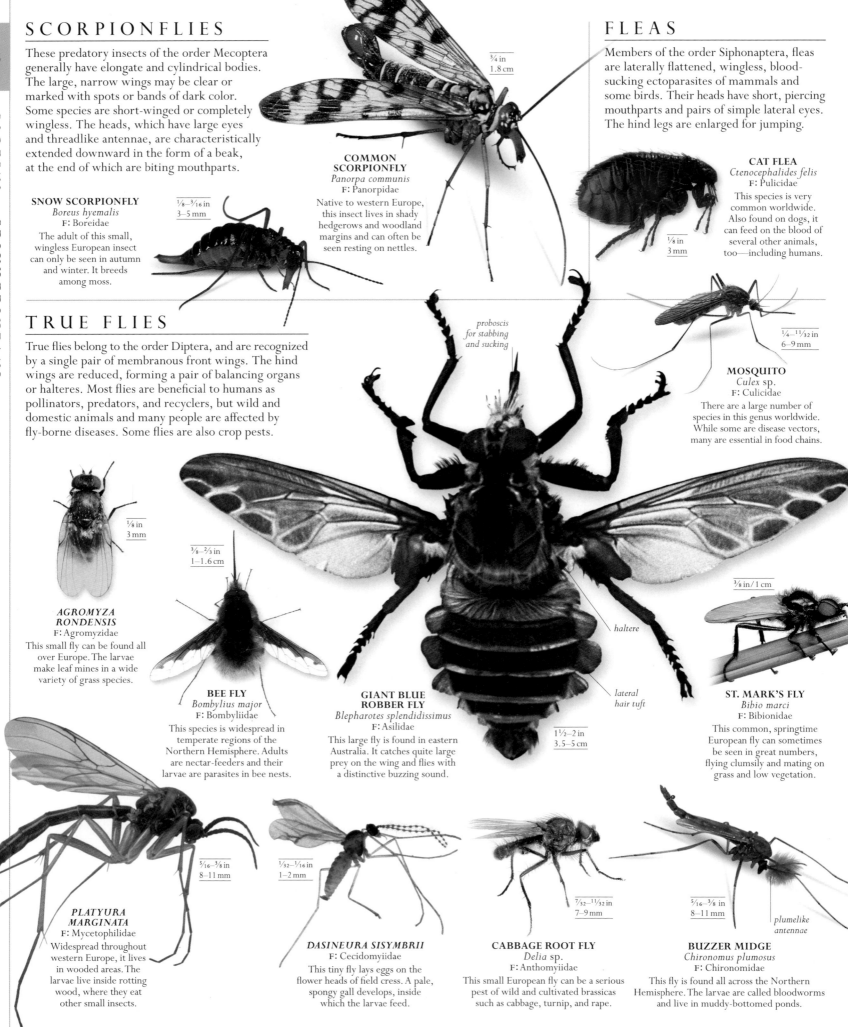

SCORPIONFLIES

These predatory insects of the order Mecoptera generally have elongate and cylindrical bodies. The large, narrow wings may be clear or marked with spots or bands of dark color. Some species are short-winged or completely wingless. The heads, which have large eyes and threadlike antennae, are characteristically extended downward in the form of a beak, at the end of which are biting mouthparts.

SNOW SCORPIONFLY
Boreus hyemalis
F: Boreidae
The adult of this small, wingless European insect can only be seen in autumn and winter. It breeds among moss.

⅛–³⁄₁₆ in
3–5 mm

COMMON SCORPIONFLY
Panorpa communis
F: Panorpidae
Native to western Europe, this insect lives in shady hedgerows and woodland margins and can often be seen resting on nettles.

¾ in
1.8 cm

FLEAS

Members of the order Siphonaptera, fleas are laterally flattened, wingless, blood-sucking ectoparasites of mammals and some birds. Their heads have short, piercing mouthparts and pairs of simple lateral eyes. The hind legs are enlarged for jumping.

CAT FLEA
Ctenocephalides felis
F: Pulicidae
This species is very common worldwide. Also found on dogs, it can feed on the blood of several other animals, too—including humans.

⅛ in
3 mm

TRUE FLIES

True flies belong to the order Diptera, and are recognized by a single pair of membranous front wings. The hind wings are reduced, forming a pair of balancing organs or halteres. Most flies are beneficial to humans as pollinators, predators, and recyclers, but wild and domestic animals and many people are affected by fly-borne diseases. Some flies are also crop pests.

proboscis for stabbing and sucking

MOSQUITO
Culex sp.
F: Culicidae
There are a large number of species in this genus worldwide. While some are disease vectors, many are essential in food chains.

¼–¹¹⁄₃₂ in
6–9 mm

AGROMYZA RONDENSIS
F: Agromyzidae
This small fly can be found all over Europe. The larvae make leaf mines in a wide variety of grass species.

⅛ in
3 mm

³⁄₈–²⁄₃ in
1–1.6 cm

BEE FLY
Bombylius major
F: Bombyliidae
This species is widespread in temperate regions of the Northern Hemisphere. Adults are nectar-feeders and their larvae are parasites in bee nests.

haltere

lateral hair tuft

GIANT BLUE ROBBER FLY
Blepharotes splendidissimus
F: Asilidae
This large fly is found in eastern Australia. It catches quite large prey on the wing and flies with a distinctive buzzing sound.

1½–2 in
3.5–5 cm

³⁄₈ in / 1 cm

ST. MARK'S FLY
Bibio marci
F: Bibionidae
This common, springtime European fly can sometimes be seen in great numbers, flying clumsily and mating on grass and low vegetation.

⁵⁄₁₆–³⁄₈ in
8–11 mm

PLATYURA MARGINATA
F: Mycetophilidae
Widespread throughout western Europe, it lives in wooded areas. The larvae live inside rotting wood, where they eat other small insects.

¹⁄₃₂–¹⁄₁₆ in
1–2 mm

DASINEURA SISYMBRII
F: Cecidomyiidae
This tiny fly lays eggs on the flower heads of field cress. A pale, spongy gall develops, inside which the larvae feed.

⁷⁄₃₂–¹¹⁄₃₂ in
7–9 mm

CABBAGE ROOT FLY
Delia sp.
F: Anthomyiidae
This small European fly can be a serious pest of wild and cultivated brassicas such as cabbage, turnip, and rape.

⁵⁄₁₆–³⁄₈ in
8–11 mm

plumelike antennae

BUZZER MIDGE
Chironomus plumosus
F: Chironomidae
This fly is found all across the Northern Hemisphere. The larvae are called bloodworms and live in muddy-bottomed ponds.

HOUSE FLY
Musca domestica
F: Muscidae
This species is found worldwide and is the most common fly around dwellings, where it transmits a number of disease-causing organisms to foodstuffs.

prominent
red eyes

MEROMYZA PRATORUM
F: Chloropidae
This fly occurs across the Northern Hemisphere, especially in sandy coastal regions. The larvae burrow in stems of marram grass and reed grass.

3/8 in
1 cm

1 in
2.5 cm

long legs can be
shed if caught

MARSH CRANE FLY
Tipula oleracea
F: Tipulidae
Originally from Europe, this crane fly is now present in North America and some upland parts of South America. It is often found by water.

5/16–3/8 in
8–10 mm

red-orange
wing base

1/4–7/32 in
6–7 mm

POECILOBOTHRUS NOBILITATUS
F: Dolichopodidae
This European species can be found in damp habitats near water. Males display in sunny patches by waving their wings.

5/32–1/4 in
4–6 mm

LESSER HOUSE FLY
Fannia canicularis
F: Fanniidae
Able to breed in almost any decaying, semiliquid matter, this species is particularly associated with human habitation.

5/32 in
4 mm

SIMULIUM ORNATUM
F: Simuliidae
This small fly lives in Europe and Asia but has been introduced elsewhere. The adults feed on animal blood and can transmit bovine onchocerciasis, a disease commonly known as river blindness.

1/8–3/16 in
3–5 mm

1/2–2/3 in
1.4–1.8 cm

MOTH FLY
Clogmia albipunctata
F: Psychodidae
This very widespread fly species resembles a tiny moth. The larvae breed in dark, wet places such as drains, tree holes, and sewage.

FLESH FLY
Sarcophaga carnaria
F: Sarcophagidae
Widely distributed through Europe and Asia, this fly feeds on nectar and liquids from rotting matter. The female lays live larvae on carrion.

males with longer
eyestalks win fights
for territory

SICUS FERRUGINEUS
F: Conopidae
This European fly lays eggs in the abdomen of certain bumble bees. The larvae develop as internal parasites, killing the host.

ROTHSCHILD'S ACHIAS
Achias rothschildi
F: Platystomatidae
Males of this Papua New Guinean fly have very elongate eyestalks. The head capsule is the biggest of any insect in the world.

1/2–2/3 in
1.5–1.8 cm

pale face with large
compound eyes

1/2 in
1.4 cm

1/16 in
2 mm

3/8–1/2 in
1–1.2 cm

FARMYARD MIDGE
Culicoides nubeculosus
F: Ceratopogonidae
This widespread European fly breeds in mud contaminated by dung or sewage. Adults suck the blood of horses and cattle.

BLUEBOTTLE
Calliphora vicina
F: Calliphoridae
Found in Europe and North America, this species is especially common in urban areas and breeds in dead pigeons and rodents.

»

YELLOW DUNG FLY
Scathophaga stercoraria
F: Scathophagidae
This very common fly is found
in many parts of the Northern Hemisphere,
where it breeds in cattle and horse dung. The
larvae feed on the dung, while the adults prey
on other insects attracted to the dung.

⁵⁄₁₆–³⁄₈ in
8–11 mm

*fleshy, toothed
proboscis for
attacking prey*

*yellow, bristly
body*

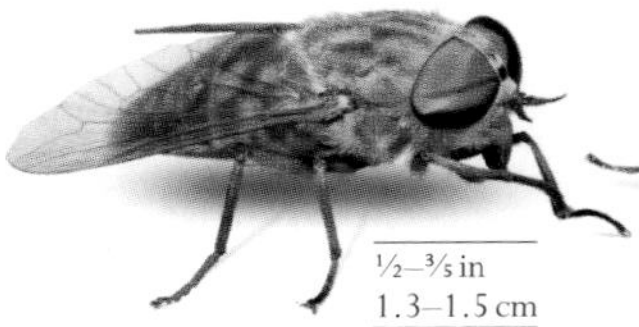

½–³⁄₅ in
1.3–1.5 cm

BANDED BROWN
HORSE FLY
Tabanus bromius
F: Tabanidae
Widespread in Europe and
the Middle East, this fly
mainly attacks horses but will
feed on some other animals,
including humans.

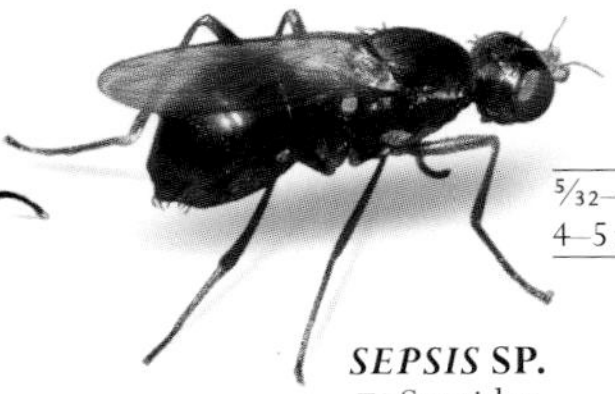

⁵⁄₃₂–³⁄₁₆ in
4–5 mm

SEPSIS SP.
F: Sepsidae
These flies are common and
widespread in a wide variety
of habitats. The larvae
develop in animal dung
and rotting matter.

⅛ in / 3 mm

COMMON FRUIT FLY
Drosophila melanogaster
F: Drosophilidae
A common laboratory
animal, this widespread
species has a distinctive dark
patch on the abdomen and
breeds in rotting fruit.

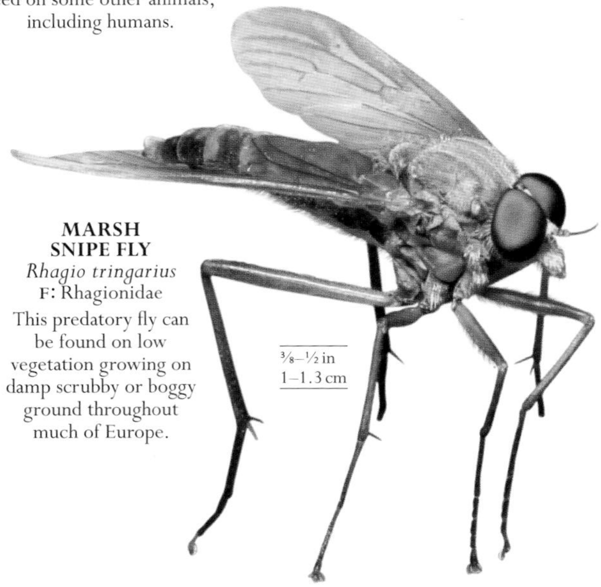

MARSH
SNIPE FLY
Rhagio tringarius
F: Rhagionidae
This predatory fly can
be found on low
vegetation growing on
damp scrubby or boggy
ground throughout
much of Europe.

³⁄₈–½ in
1–1.3 cm

⁵⁄₁₆ in
8 mm

FOREST FLY
Hippobosca equina
F: Hippoboscidae
Occurring mainly in wooded
areas in Europe and parts of
Asia, this bloodsucking fly
attacks horses, deer, and
sometimes cattle.

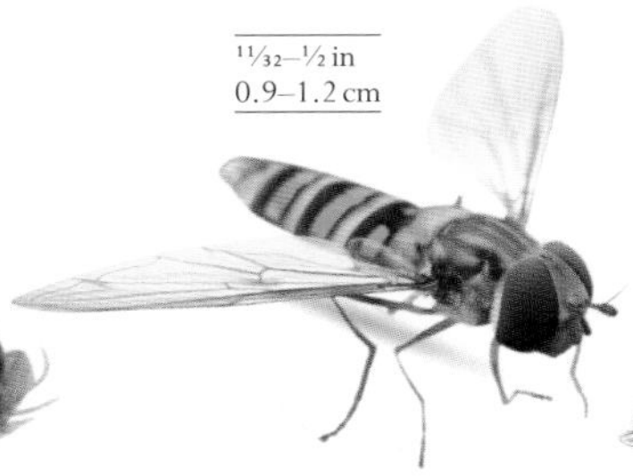

¹¹⁄₃₂–½ in
0.9–1.2 cm

³⁄₈–½ in
1–1.2 cm

²⁄₅–½ in
1.1–1.3 cm

MARMALADE
HOVER FLY
Episyrphus balteatus
F: Syrphidae
Very common in a variety
of habitats including gardens, this
European fly feeds on pollen and
nectar. The larvae eat aphids.

LEUCOZONA
LEUCORUM
F: Syrphidae
This species from Europe
can be found visiting flowers
in damp woodlands in the
spring and early summer. Its
larvae prey on aphids.

DRONE FLY
Eristalis tenax
F: Syrphidae
This European fly, which
has been introduced to
North America, is a mimic of
the honey bee. The larvae
breed in stagnant water.

⁵⁄₃₂ in
4 mm

½ in
1.2 cm

2¼ in
6 cm

PENICILLIDIA FULVIDA
F: Nycteribiidae
Widespread in sub-Saharan
Africa, this bloodsucking,
wingless, ectoparasitic fly can
be found on a wide range
of bat species.

SYRPHUS RIBESII
F: Syrphidae
The larvae of this nectar-feeding
fly are major predators of aphids
across Europe. Its coloration
mimics that of wasps and bees,
deterring predators.

GAUROMYDAS HEROS
F: Mydidae
This South American insect,
the world's largest fly, breeds in
the nests of leaf-cutter ants,
where the young are thought to
eat scarab beetle larvae.

RODENT BOT FLY
Cuterebra fontinella
F: Oestridae
This parasitic fly lives in North America. The larvae develop inside the bodies of rodents such as white-footed mice.

1 in
2.5 cm

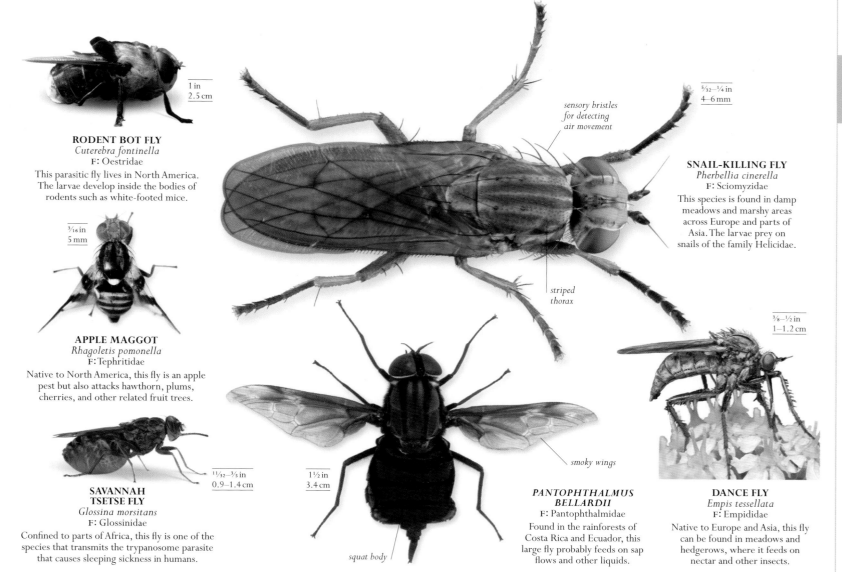

³/₁₆ in
5 mm

APPLE MAGGOT
Rhagoletis pomonella
F: Tephritidae
Native to North America, this fly is an apple pest but also attacks hawthorn, plums, cherries, and other related fruit trees.

sensory bristles for detecting air movement

⁵/₃₂–¹/₄ in
4–6 mm

SNAIL-KILLING FLY
Pherbellia cinerella
F: Sciomyzidae
This species is found in damp meadows and marshy areas across Europe and parts of Asia. The larvae prey on snails of the family Helicidae.

striped thorax

³/₈–¹/₂ in
1–1.2 cm

SAVANNAH TSETSE FLY
Glossina morsitans
F: Glossinidae
Confined to parts of Africa, this fly is one of the species that transmits the trypanosome parasite that causes sleeping sickness in humans.

¹¹/₃₂–³/₅ in
0.9–1.4 cm

1½ in
3.4 cm

smoky wings

squat body

PANTOPHTHALMUS BELLARDII
F: Pantophthalmidae
Found in the rainforests of Costa Rica and Ecuador, this large fly probably feeds on sap flows and other liquids.

DANCE FLY
Empis tessellata
F: Empididae
Native to Europe and Asia, this fly can be found in meadows and hedgerows, where it feeds on nectar and other insects.

CADDISFLIES

Closely related to the Lepidoptera, these slender, rather mothlike insects of the order Trichoptera are covered with hairs, not scales. The head has long, threadlike antennae and weakly developed mouthparts. The two pairs of wings are held tent-like over the body at rest. Caddisfly larvae are aquatic and often construct species-specific portable shelters out of small stones or plant fragments.

SALT AND PEPPER MICROCADDIS
Agraylea multipunctata
F: Hydroptilidae
This small caddisfly is widespread and common in North America. It has narrow wings and breeds in algae-rich ponds and lakes.

¹/₈–¹/₅ in
3–4.5 mm

GREAT RED SEDGE
Phryganea grandis
F: Phryganeidae
This European caddisfly breeds in weedy lakes and slow rivers. The larvae make a case from cut leaf sections arranged in a spiral.

1¼ in
3 cm

²/₅–¹/₂ in
1.1–1.3 cm

DARK-SPOTTED SEDGE
Philopotamus montanus
F: Philopotamidae
This European caddisfly has shortish antennae and breeds in fast-flowing rocky streams. The larvae make tubelike nets on the undersides of rocks.

³/₅ in
1.5 cm

³/₅–⁷/₁₀ in
1.6–1.7 cm

MARBLED SEDGE
Hydropsyche contubernalis
F: Hydropsychidae
This caddisfly flies in the evening and breeds in rivers and streams. The aquatic larvae weave a net to catch food.

MOTTLED SEDGE
Glyphotaelius pellucidus
F: Limnephilidae
This European caddisfly breeds in lakes and small ponds. The larvae make a case out of fragments of withered tree leaves.

MOTHS AND BUTTERFLIES

Members of the order Lepidoptera are covered with minute scales. They have large compound eyes and a mouthpart called a proboscis. Larvae, known as caterpillars, pupate, metamorphosing into their adult form. Most moths are nocturnal and rest with their wings open; the majority of butterflies are diurnal and rest with their wings closed.

8–10½ in
20–27 cm

hind wing tail

HERCULES MOTH
Coscinocera hercules
F: Saturniidae

Found in New Guinea and Australia, this moth is one of the biggest in the world. Only the male has the long hind wing tails.

2¾–4¼ in
7–11 cm

AMERICAN MOON MOTH
Actias luna
F: Saturniidae

This lime-green North American moth has distinctive long hind wing tails. The caterpillars feed on a range of deciduous trees.

4–6 in / 10–15 cm

POLYPHEMUS MOTH
Antheraea polyphemus
F: Saturniidae

Common and widespread in the USA and southern Canada, this moth has large eyespots on the wings, to startle predators.

2¼–3½ in
6–9 cm

GIANT LEOPARD MOTH
Hypercompe scribonia
F: Arctiidae

The distribution of this striking moth extends from southeast Canada as far south as Mexico. The caterpillars feed on a variety of plants.

2–3 in
5–7.5 cm

OAK EGGAR
Lasiocampa quercus
F: Lasiocampidae

Ranging from Europe to North Africa, the caterpillars of this moth feed on the foliage of bramble, oak, heather, and other plants.

2–3 in
5–7.5 cm

LAPPET MOTH
Gastropacha quercifolia
F: Lasiocampidae

Found in Europe and Asia, the specific name of this large moth refers to its resting appearance, which resembles a cluster of dead oak leaves.

2–3¼ in
5–8 cm

PINE-TREE LAPPET
Dendrolimus pini
F: Lasiocampidae

This moth is widespread in coniferous forests across Europe and Asia, where the caterpillars feed on pine, spruce, and fir.

white chevrons on forewing

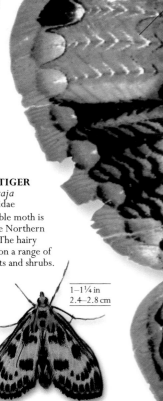

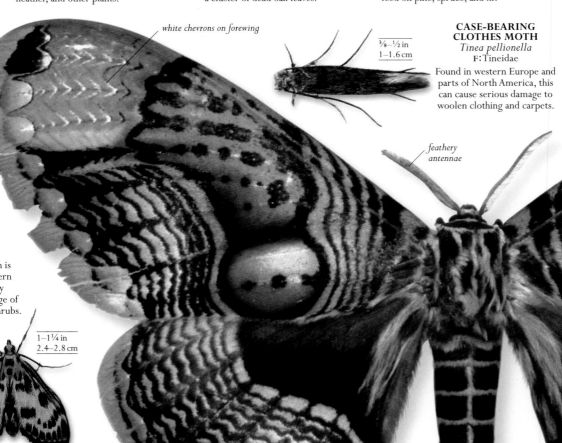

1½–1¾ in
3.5–4.5 cm

AUSTRALIAN MAGPIE MOTH
Nyctemera amica
F: Arctiidae

Widespread in Australia and also found in New Zealand, the caterpillars of this day-flying moth feed on plants such as groundsel and ragwort.

⅜–½ in
1–1.6 cm

CASE-BEARING CLOTHES MOTH
Tinea pellionella
F: Tineidae

Found in western Europe and parts of North America, this can cause serious damage to woolen clothing and carpets.

feathery antennae

2–3 in / 5–7.5 cm

GARDEN TIGER
Arctia caja
F: Arctiidae

This unmistakable moth is found across the Northern Hemisphere. The hairy caterpillars feed on a range of low-growing plants and shrubs.

1½–2 in / 4–5 cm

1–1¼ in
2.4–2.8 cm

SNOUT MOTH
Vitessa suradeva
F: Pyralidae

This moth occurs in India, parts of southeast Asia, and New Guinea. The caterpillars feed in a web on young leaves of poisonous shrubs.

SMALL MAGPIE
Eurrhypara hortulata
F: Pyralidae

Found in hedgerows and waste ground, the caterpillars of this very common European species feed on rolled nettle leaves.

abdomen

9½–12 in/24–31 cm

GIANT AGRIPPA
Thysania agrippina
F: Noctuidae
Found in Central and parts of
South America, this species has one
of the largest wingspans of any
moth in the world.

ILIA UNDERWING
Catocala ilia
F: Noctuidae
This widespread species
from North America has
distinctive red-banded hind
wings. The caterpillars feed
on the foliage of oak trees.

2¾–3¼ in
7–8 cm

1–1¼ in/2.5–3 cm

VAPOURER MOTH
Orgyia antiqua
F: Lymantriidae
Now found across the
Northern Hemisphere, the
female of this European
species has tiny wings and
is unable to fly.

2¼–4 in
6–10 cm

**SILVER-SPOTTED
GHOST MOTH**
Sthenopis argenteomaculatus
F: Hepialidae
This species occurs in southern
Canada and parts of the USA. The
caterpillars mainly feed inside the
roots of alder trees.

**ACACIA CARPENTER
MOTH**
Xyleutes eucalypti
F: Cossidae
The robust, white caterpillars
of this large and distinctive
Australian moth burrow into
the woody tissue of certain
species of the acacia tree.

5–8 in
13–20 cm

2½–3¾ in
6.5–9.5 cm

DIVA MOTH
Divana diva
F: Castniidae
Found in tropical forests in
South America, this
day-flying species is well
camouflaged at rest despite
brightly colored hind wings.

large eyespot to
scare predators

2–2¼ in
5–6 cm

LARGE EMERALD
Geometra papilionaria
F: Geometridae
This moth occurs across Europe
and the temperate parts of Asia.
The caterpillars feed mainly on
the foliage of birch trees.

1¼–1½ in
3–4 cm

ARGENT AND SABLE
Rheumaptera hastata
F: Geometridae
The common name for this strikingly patterned
moth from the Northern Hemisphere comes from
the heraldic terms for silvery-white and black.

1½–2 in
4–5 cm

CLARA'S SATIN MOTH
Thalaina clara
F: Geometridae
The caterpillars of this moth,
which occurs in eastern and
southeastern Australia and
northern Tasmania, feed on
acacia foliage.

wavy pattern
of black and
orange lines

4–6½ in
10–16 cm

**WALLICH'S
OWL MOTH**
Brahmaea wallichii
F: Brahmaeidae
This large moth occurs
from northern India to
China and Japan. The
caterpillars eat the foliage
of ash, privet, and lilac.

2¾–3 in
7–7.5 cm

**COPPERY
DYSPHANIA**
Dysphania cuprina
F: Geometridae
Widespread across southeast Asia, this brightly
colored moth flies during the day and is thought
to be unpalatable to birds.

SILK MOTH
Bombyx mori
F: Bombycidae
Originating in China, the
silk moth has been bred in
captivity, on a diet of white
mulberry leaves, for
thousands of years. Its
cocoon is used to make silk.

1½–2¼ in
4–6 cm

**WHITE PLUME-
MOTH**
Pterophorus pentadactyla
F: Pterophoridae
Common across Europe
on dry grassland, waste
ground, and gardens,
the caterpillars of this
distinctive species feed
on hedge bindweed.

1–1¼ in
2.5–3 cm

long
legs

2–2¼ in
5–6 cm

REGENT SKIPPER
Euschemon rafflesia
F: Hesperiidae
This brightly colored species is native
to tropical and subtropical forests in
eastern Australia, where
it can be seen feeding at flowers.

GUAVA SKIPPER
Phocides polybius
F: Hesperiidae
This species can be found from
southern Texas to as far south as
Argentina. The caterpillars feed
inside rolled-up guava leaves.

1¾–2½ in
4.5–6.2 cm

HORNET MOTH
Sesia apiformis
F: Sesiidae
To deter predators, the adults of this
moth look very similar to hornets
but they are harmless. The
caterpillars bore into the trunk and
roots of poplars and willows.

1¼–1¾ in
3–4.5 cm

BUFF-TIP
Phalera bucephala
F: Notodontidae
This moth is found in Europe
and eastward to Siberia. At rest
its folded wings wrap around its
body for camouflage, resembling
a broken-off twig.

1¾–2½ in
4.5–6.5 cm

2¼ in
5.5–6 cm

ELEPHANT HAWK MOTH
Deilephila elpenor
F: Sphingidae
This pretty, pinkish hawk moth is
widespread in the temperate parts
of Europe and Asia. The caterpillars
feed on bedstraw and willowherb.

2¾–4¼ in
7–11 cm

VERDANT SPHINX
Euchloron megaera
F: Sphingidae
This distinctive moth is widespread
throughout sub-Saharan Africa. The
caterpillars feed on the foliage of
creepers of the grape family.

**COMMON
MORPHO**
Morpho peleides
F: Nymphalidae
This butterfly is widespread
in tropical forests in Central
and South America. Adults
feed on the juices of
rotting fruit.

3¾–6 in
9.5–15 cm

*metallic blue color is
attractive to mates*

1–1½ in
2.5–3.8 cm

SIX-SPOT BURNET
Zygaena filipendulae
F: Zygaenidae
Brightly colored and distasteful
to birds, this day-flying moth can
be found in meadows and woodland
clearings across Europe.

LITTLE WOOD SATYR
Euptychia cymela
F: Nymphalidae
This woodland butterfly is found
from southern Canada to northern
Mexico. The caterpillars feed on
grasses in clearings close to water.

1¾–2 in
4.5–5 cm

3–4 in
7.5–10 cm

MONARCH BUTTERFLY
Danaus plexippus
F: Nymphalidae
A well-known migrant, this butterfly has
spread from the Americas to many other
parts of the world. The caterpillars feed
on the milkweed plant.

QUEEN CRACKER
Hamadryas arethusa
F: Nymphalidae
The common name for this
butterfly, found in forests
from Mexico to Bolivia,
refers to the clicking noise
it makes when it flies.

WHITE ADMIRAL
Ladoga camilla
F: Nymphalidae
This butterfly can be found
across temperate Europe and
Asia to Japan. The caterpillars
feed on the foliage of the
honeysuckle.

2–2¼ in
5–6 cm

hind wing tails

long antennae

2¼–2¾ in
6–7 cm

PURPLE EMPEROR
Apatura iris
F: Nymphalidae
Found in woodland throughout Europe
and Asia as far east as Japan, the male of
this butterfly is iridescent purple
whereas the female is dull brown.

2¾–3½ in
7–9 cm

INDIAN LEAF BUTTERFLY
Kallima inachus
F: Nymphalidae
With undersides that resemble a brown
leaf, this species is perfectly camouflaged
at rest with its wings folded. It occurs
from India to southern China.

3½–4¾ in
9–12 cm

SMALL WHITE
Pieris rapae
F: Pieridae

This butterfly can now
be found worldwide. The
caterpillars feed on wild and
cultivated cabbages and
mustards, and can be pests.

1½–2¼ in / 3.5–5.5 cm

pointed front wing

1½–2½ in
4–6.5 cm

CALIFORNIA DOG-FACE
Zerene eurydice
F: Pieridae

This butterfly is restricted to parts
of California and, sometimes,
western Arizona. It breeds on
a shrub called Napa False Indigo.

1½–2 in
4–5 cm

ORANGE TIP
Anthocharis cardamines
F: Pieridae

This butterfly can be found
in meadows across temperate
Europe and Asia as far as Japan.
The caterpillars feed on lady's
smock and garlic mustard.

2¾–3¾ in
7–9.5 cm

ORANGE-BARRED
SULPHUR
Phoebis philea
F: Pieridae

The distribution of this
butterfly extends from
southern Brazil to Central
America, southern USA, and
sporadically farther north. The
caterpillars feed on sennas.

3–3¾ in
7.5–9.5 cm

MADAGASCAN
SUNSET MOTH
Chrysiridia rhipheus
F: Uraniidae

This colorful, day-flying moth,
with iridescent wing scales, is endemic
to Madagascar. It feeds on certain
poisonous shrubs of the spurge family.

iridescent red
wing marking

dark spots on
hind wing

1½–1¾ in
4–4.5 cm

TIGER
PIERID
Dismorphia amphione
F: Pieridae

This colorful butterfly, which
mimics unpalatable butterfly species
for protection from predators, is
widespread and common from
Mexico to South America.

2¼–3¼ in
5.5–8 cm

SMALL POSTMAN
Heliconius erato
F: Nymphalidae

This butterfly is common along
forest edges and open ground from
Central America to southern Brazil.
Caterpillars eat passionflower foliage.

OWL BUTTERFLY
Caligo idomeneus
F: Nymphalidae

The underside of this large species, native to
South America, has prominent owl-like eyespots
to deter predators when the butterfly is at rest.

4¾–6 in
12–15 cm

UNDERSIDE

2¼–3 in
5.5–7.5 cm

BLACK-VEINED
WHITE
Aporia crataegi
F: Pieridae

Occurring across Europe, North
Africa, and Asia as far as Japan, this
distinctive butterfly breeds on
hawthorn and blackthorn.

2–2¾ in
5–7 cm

CLEOPATRA
Gonepteryx cleopatra
F: Pieridae

This species is common in countries
around the Mediterranean,
particularly in lightly wooded coastal
areas. The caterpillars feed
on species of buckthorn.

>>

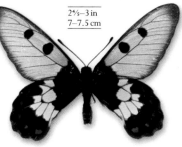

2⅖–3 in
7–7.5 cm

BIG GREASY BUTTERFLY
Cressida cressida
F: Papilionidae
This species is found in grassland
and drier forests in Australia and
Papua New Guinea, where its
pipevine foodplants grow.

2¼–3¼ in
6–8 cm

**ZEBRA
SWALLOWTAIL**
Eurytides marcellus
F: Papilionidae
Found in damp woodlands in eastern North
America, this butterfly has distinctive black and
white markings. The caterpillars feed on pawpaw.

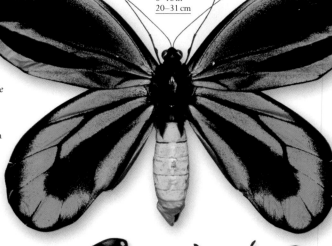

8–12 in
20–31 cm

**QUEEN
ALEXANDRA'S
BIRDWING**
Ornithoptera alexandrae
F: Papilionidae
Found only in
Southeast Papua New
Guinea east of the Owen
Stanley Ranges, this
endangered species,
the largest butterfly
in the world,
is now protected.

4¾–7 in
12–18 cm

RAJAH BROOKE'S BIRDWING
Troides brookiana
F: Papilionidae
This butterfly lives in the tropical forests of Borneo
and Malaysia, where the adults sip fruit juices and
nectar. The caterpillars feed on pipevines.

4¼–5 in
11–13 cm

CAIRNS BIRDWING
Ornithoptera priamus
F: Papilionidae
This large butterfly occurs from Papua New Guinea
and the Solomon Islands to tropical northern
Australia. The caterpillars eat pipevines.

3¼–4 in
8–10 cm

SWALLOWTAIL
Papilio machaon
F: Papilionidae
This species is found in wet meadow,
marshland, and other habitats across
the Northern Hemisphere.
The caterpillars feed on
various umbellifers.

1½–2 in
4–5 cm

**GREEN
DRAGONTAIL**
Lamproptera meges
F: Papilionidae
This unmistakable butterfly,
which hovers while it feeds at
flowers, can be found from
India to China and through
South and Southeastern Asia.

*short, robust
antenna*

APOLLO
Parnassius apollo
F: Papilionidae
This butterfly can be found in flower-rich meadows
in mountainous regions of Europe and Asia.
The caterpillars feed on stonecrop.

2¼–3½ in
6–9 cm

red spot

1¾–2 in
4.5–5 cm

**BLUE
TRIANGLE**
Graphium sarpedon
F: Papilionidae
Common and widespread
from India to China,
Papua New Guinea, and
Australia, this butterfly
feeds on nectar and
drinks from puddles.

*zigzag pattern
on hind wing*

2¾–3¼ in
7–8 cm

3¼–3½ in
8–9 cm

**SCARCE
SWALLOWTAIL**
Iphiclides podalirius
F: Papilionidae
Despite its name, this species is quite widespread
in Europe and across temperate Asia to China.
The caterpillars feed on blackthorn.

SPANISH FESTOON
Zerynthia rumina
F: Papilionidae
This species lives in scrub, meadows, and
rocky hillsides in southeast France, Spain,
Portugal, and parts of North Africa. The
caterpillars feed on species of birthwort.

3½–5½ in
9–14 cm

TIGER SWALLOWTAIL
Papilio glaucus
F: Papilionidae
This species is widespread in North America. The young caterpillars, which resemble bird droppings, feed on a variety of trees and shrubs.

SMALL COPPER
Lycaena phlaeas
F: Lycaenidae
Common in Europe, North Africa, and Asia as far as Japan, this species can also be found in North America. The caterpillars feed on sorrels and docks.

1–1¼ in
2.5–3 cm

1½–1¾ in
3.5–4.5 cm

BROWN HAIRSTREAK
Thecla betulae
F: Lycaenidae
This species can be found in hedgerows, scrubland, and woodland across Europe and temperate Asia. The caterpillars feed at night on blackthorn.

BLUE THAROPS
Menander menander
F: Lycaenidae
This fast-flying butterfly occurs in tropical forests from Panama to northern South America. Little is known about its life cycle or caterpillars.

1¼–1½ in
3–4 cm

1–1½ in / 2.5–3.5 cm

ADONIS BLUE
Lysandra bellargus
F: Lycaenidae
This European species can be found in chalky grassland, where the caterpillars feed on horseshoe vetch. The male is sky blue while the female is brown.

¾–1 in
2–2.5 cm

SONORAN BLUE
Philotes sonorensis
F: Lycaenidae
This rare butterfly is restricted to rocky washes and desert cliffs in California. The caterpillars feed on succulents known as stonecrop or "live-forever" plants.

1¼–1½ in
3–4 cm

DUKE OF BURGUNDY FRITILLARY
Hamearis lucina
F: Lycaenidae
This species ranges across central Europe as far as the Urals and prefers flower meadows, where its food plants, cowslip and primrose, grow.

SAWFLIES, WASPS, BEES, AND ANTS

Hymenoptera typically have two pairs of wings, joined in flight by tiny hooks. Apart from sawflies, they have a "wasp-waist" and females have an ovipositor that may sting. Many wasps are predatory or parasitic. Bees are vital pollinators and ants play a role in many ecosystems.

STEM SAWFLY
Cephus nigrinus
F: Cephidae
This slender, all-black sawfly is widespread in western Europe. The larvae bore downward inside the stems of wild and some cultivated grasses.

7/32–11/32 in
7–9 mm

7/32–11/32 in
7–9 mm

¾–9/10 in
1.8–2.2 cm

CIMBICID SAWFLY
Trichiosoma lucorum
F: Cimbicidae
This stout-bodied European sawfly is found in woodland, hedgerows, and scrubby areas. The larvae feed on birch and willow.

fine tufts on wing margin

TENTHREDO ARCUATA
F: Tenthredinidae
Widespread in meadows, this distinctive black-and-yellow-marked European sawfly lays its eggs on clovers.

PERGID SAWFLY
Unknown sp.
F: Pergidae
These herbivorous insects are found in Australia and South America. Many species attack eucalyptus trees, and young larvae feed gregariously.

¾ in
2 cm

slender antennae

1½ in
3.5–4 cm

large head

ROSE SAWFLY
Arge ochropus
F: Argidae
Recognizable by the black along the wing edge, this European sawfly visits flat-topped flowers to feed. The larvae feed on wild roses.

7/32–3/8 in
7–10 mm

LEAF ROLLING SAWFLY
Acantholyda erythrocephala
F: Pamphilidae
Originally found in Europe and Asia, this sawfly has spread to Canada. The larvae feed communally under silken webs or in rolled up foliage.

7/32–11/32 in
7–9 mm

HORNTAIL
Urocerus gigas
F: Siricidae
This impressive species can be found across the Northern Hemisphere. The female drills deep into pine trees to lay her eggs.

ovipositor

>>

1⁄32–1⁄8 in
1–3 mm

FIG WASP
Unknown sp.
F: Agaonidae
Fig wasps are found in tropical and subtropical regions. Many are essential pollinators and specific to certain species of fig.

1⁄8–3⁄8 in
3–10 mm

BRACONID WASP
Unknown sp.
F: Braconidae
Found worldwide, braconid wasps generally live as parasites on caterpillars and the larvae of beetles and flies. The females of some species have a long ovipositor.

large compound eye

bright metallic coloration

3⁄16–3⁄10 in
5–6.5 mm

OAK APPLE GALL WASP
Biorhiza pallida
F: Cynipidae
This wasp emerges from root galls on oak trees in Europe and Asia. It lays eggs in buds, inducing the growth of oak apple galls.

3⁄4–9⁄10 in
2–2.2 cm

GIANT WOOD WASP
Rhyssa sp.
F: Ichneumonidae
Found in pine forests across the Northern Hemisphere, these very large wasps drill into trees to lay their egg on the larvae of horntails.

1 2⁄5–1 1⁄2 in
3.6–4 cm

LISSONOTA SP.
F: Ichneumonidae
The genus *Lissonota* contains many very similar wasps. This species drills into a trunk to lay its eggs in a wood-boring moth larva.

7⁄10–3⁄4 in
1.8–2 cm

SPLENDID EMERALD WASP
Stilbum splendidum
F: Chrysididae
This large wasp from N. Australia lives as a parasite on the larvae of solitary mud-nesting wasps. They are known to visit flowers to feed.

5⁄32 in
4 mm

TORYMUS SP.
F: Torymidae
This wasp uses her long ovipositor to drill through gall tissue and lay eggs on the developing larvae inside.

1⁄8–5⁄32 in
3–4 mm

PTEROMALID WASP
Mesopolobus typographi
F: Pteromalidae
This European and Asian wasp is a hyperparasite; it is a parasite on another wasp that is itself a parasite on the larvae of bark beetles.

3⁄8–1⁄2 in
1–1.2 cm

CHALCID WASP
Chalcis sispes
F: Chalcididae
This wasp is found in Europe and parts of Asia, where it lives as a parasite on the aquatic larvae of large soldier flies.

EUROPEAN BEEWOLF
Philanthus triangulum
F: Sphecidae
Found in southern and central Europe and North Africa, it makes a nest in sandy areas stocked with honey bees as food for its larvae.

1⁄2–3⁄5 in
1.3–1.5 cm

hard, dimpled body surface gives protection from stings

2 3⁄4–3 1⁄4 in
7–8 cm

TARANTULA HAWK
Pepsis heros
F: Pompilidae
This large S. American wasp hunts for tarantulas, which it paralyzes and buries as food for its larvae.

smoky orange wing

1 3⁄4–2 1⁄4 in
4.5–5.5 cm

MAMMOTH WASP
Scolia procer
F: Scoliidae
Native to Borneo, Java, and Sumatra, this wasp paralyzes the larvae of scarab beetles and lays its eggs on them.

2⁄5–7⁄10 in
1.1–1.7 cm

VELVET ANT
Mutilla europaea
F: Mutillidae
This European species lives in sandy areas and rough grassland. The female is wingless, and the larvae eat the larvae of bumble bees.

TIPHIID WASP
Methocha ichneumonides
F: Tiphiidae

The wingless females of this European species live in sandy areas, where they are a parasite on the larvae of tiger beetles inside their burrows.

11/32–7/16 in
9–11 mm

HORNET
Vespa crabro
F: Vespidae

This large social wasp lives in Europe and Asia and has been introduced elsewhere. It prefers woodland, where it makes its nest in hollow trees.

1–1½ in
2.5–3.5 cm

9/10–1 in
2.3–2.5 cm

**BUFF-TAILED
BUMBLE BEE**
Bombus terrestris
F: Apidae

This social bee of central and southern Europe and North Africa has been introduced elsewhere. It is an important crop pollinator.

very hairy body

**COMMON
WASP**
Vespula vulgaris
F: Vespidae

Native to the Northern Hemisphere, this wasp feeds its young on insects and makes a paper nest from chewed wood fibers.

½–7/10 in
1.2–1.7 cm

**TAWNY
MINING BEE**
Andrena fulva
F: Andrenidae

Found in central Europe, this early spring bee makes a nest underground in grassy areas, raising a little mound of soil at the entrance.

3/8–½ in
1–1.2 cm

½ in
1.2 cm

**HONEY
BEE**
Apis mellifera
F: Apidae

The western honey bee, an important crop pollinator, is now found worldwide. In the wild it nests in hollow trees, but is widely kept in domestic hives.

½ in
1.2–1.4 cm

**ORCHID
BEE**
Euglossa asarophora
F: Apidae

Like all orchid bees, this species lives in colonies in the rain forests of South America. Males collect oils and resins from orchids to attract a mate.

2/5–½ in
1.1–1.3 cm

PLASTERER BEE
Colletes sp.
F: Colletidae

These solitary, ground-nesting bees are common across the Northern Hemisphere. They line their nests with a waterproof abdominal secretion.

**GREAT
CARPENTER
BEE**
Xylocopa latipes
F: Anthophoridae

Found all over southeast Asia, this large bee excavates nest cavities in tree branches or wooden beams and posts.

1¼–1½ in
3.3–3.6 cm

5/16–3/8 in
8–10 mm

WOOD ANT
Formica rufa
F: Formicidae

Common throughout Europe, this ant is an important predator of insects in forests. It squirts formic acid from the rear of its abdomen as a defense.

3/5 in
1.6 cm

**LEAF-CUTTER
ANT**
Atta sp.
F: Formicidae

Leaf-cutter ants live in Central and South America. They can make vast underground nests, where they cultivate a special fungus on chewed leaf fragments as food.

compound eye

band of pale hair

5/32–½ in
4–12 mm

WOOL CARDER BEE
Anthidium manicatum
F: Megachilidae

This European bee makes nests in existing holes in wood or masonry, which it lines with hairs that it collects from certain plants.

2/5 in
1 cm

½–3/5 in
1.3–1.5 cm

SWEAT BEE
Halictus quadricinctus
F: Halictidae

This solitary bee is found across southern Europe and the Mediterranean. It makes large clusters of cells underground in which to rear its young.

ARMY ANT
Eciton burchellii
F: Formicidae

This South American ant forms large colonies of up to two million individuals and attacks large prey.

RIBBON WORMS

Marine worms of the phylum Nemertea are voracious predators, seizing their prey with a proboscis and swallowing it whole or sucking its fluids.

Ribbon worms are soft, slimy animals, cylindrical in shape or somewhat flattened. Many move by gliding, but some can also swim. One species,*Lineus longissimus*—the bootlace worm—has been recorded growing to 100 ft (30 m) in length. Like others, it has a fragile body that is easily broken.

The proboscis of a ribbon worm is not part of its gut, but emerges from a sac on its head just above the mouth. In some species the proboscis has sharp spikes for gripping or piercing prey, and even venom for immobilizing it. These well-armed species prey on other invertebrates, such as crustaceans, bristle worms, and mollusks. Other ribbon worms feed on dead material.

PHYLUM	NEMERTEA
CLASSES	2
ORDERS	3
FAMILIES	41
SPECIES	About 1,150

BRYOZOANS

These minute animals form coral-like colonies, filter-feeding with microscopic tentacles and, in turn, providing food for grazing invertebrates.

Many bryozoans resemble coral, but are more advanced animals than cnidarians. A colony—genetically a single individual—is made up of thousands of tiny bodies called zooids, each consisting of a retractable fan of tentacles surrounding the mouth. Microscopic, beating hairs on the tentacles waft particles of food down to the mouth, which leads into a U-shaped gut. Waste exits via the anus in the body wall. Although a hand lens is needed to see individual zooids, colonies assume a variety of forms, depending on species. Some are encrustations on rock or seaweed, others grow into bristly thickets or fleshy lobes. Many are heavily calcified like stony corals; others are soft.

PHYLUM	BRYOZOA
CLASSES	3
ORDERS	4
FAMILIES	About 160
SPECIES	About 4,150

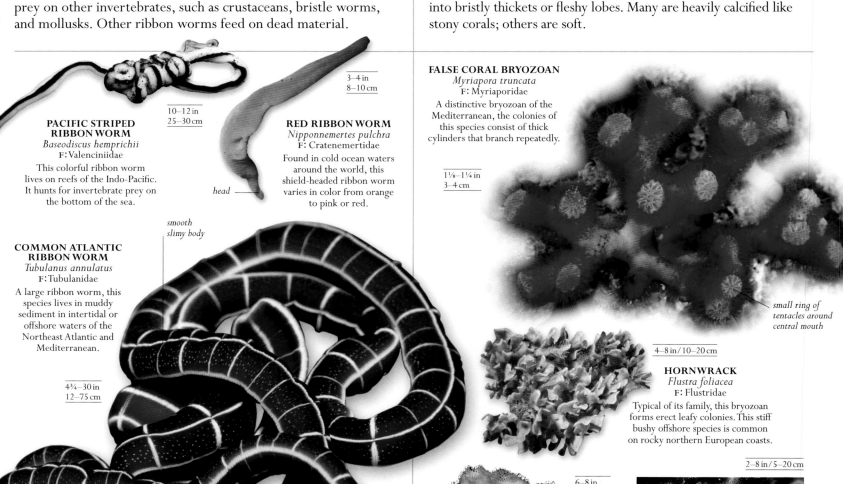

PACIFIC STRIPED RIBBON WORM
Baseodiscus hemprichii
F: Valenciniidae
This colorful ribbon worm lives on reefs of the Indo-Pacific. It hunts for invertebrate prey on the bottom of the sea.

10–12 in
25–30 cm

3–4 in
8–10 cm

head

RED RIBBON WORM
Nipponnemertes pulchra
F: Cratenemertidae
Found in cold ocean waters around the world, this shield-headed ribbon worm varies in color from orange to pink or red.

COMMON ATLANTIC RIBBON WORM
Tubulanus annulatus
F: Tubulanidae
A large ribbon worm, this species lives in muddy sediment in intertidal or offshore waters of the Northeast Atlantic and Mediterranean.

smooth slimy body

4¾–30 in
12–75 cm

FALSE CORAL BRYOZOAN
Myriapora truncata
F: Myriaporidae
A distinctive bryozoan of the Mediterranean, the colonies of this species consist of thick cylinders that branch repeatedly.

1⅛–1¼ in
3–4 cm

small ring of tentacles around central mouth

4–8 in/10–20 cm

HORNWRACK
Flustra foliacea
F: Flustridae
Typical of its family, this bryozoan forms erect leafy colonies. This stiff bushy offshore species is common on rocky northern European coasts.

2–8 in/5–20 cm

6–8 in
15–20 cm

PINK LACE BRYOZOAN
Iodictyum phoeniceum
F: Phidoloporidae
Lace bryozoans—sometimes inaccurately called "lace corals"—form tall, rigid colonies. This colorful species lives on coastlines around southern and eastern Australia.

LACY CRUST BRYOZOAN
Membranipora membranacea
F: Membraniporidae
The bryozoans known as sea mats grow in encrusting lacy sheets. This species forms fast-growing colonies on kelp fronds of the Northeast Atlantic.

LAMPSHELLS

Lampshells look very like bivalve mollusks, but feed instead with tentacles. They belong to a different, very ancient phylum.

The soft bodies of lampshells are enclosed within a two-valved shell, one valve larger than the other.

PHYLUM	BRACHIOPODA
CLASSES	2
ORDERS	5
FAMILIES	About 25
SPECIES	About 300

The shell is either attached to the seabed with a rubbery stalk or is directly cemented to rock. The shell valves are positioned top-and-bottom around the animal—not around the sides, as in bivalve mollusks. Like mollusks, lampshells have a fleshy mantle attached to the inner surface of the shell, which encloses a mantle cavity. In the mantle cavity there is a ring of tentacles with microscopic beating hairs that drive particles to a central mouth—just like the feeding structure in bryozoans.

The fossil record shows that lampshells were much more common and demonstrated much greater diversity in the warm, shallow seas of the Palaeozoic Era. They began to decline dramatically during the time of the dinosaurs—perhaps because bivalve mollusks were more successful.

¾–1¼ in
2–3 cm

EUROPEAN LAMPSHELL
Terebratulina retusa
F: Cancellothyrididae
Found from the Northeast Atlantic to the Mediterranean, this lampshell has pear-shaped valves and attaches to vertical rocks with a short stalk.

1¼–2¼ in
3–5.5 cm

hinged, two-valved shell

PACIFIC LAMPSHELL
Terebratalia transversa
F: Terebrataliidae
An abundant lampshell of the Northern Pacific, this short-stalked species has a variable smooth or ribbed shell.

⅜–½ in
1–1.5 cm

INARTICULATED LAMP SHELL
Novocrania anomala
F: Craniidae
A brachiopod of North Atlantic waters, this species cements its shell to rock, and superficially resembles a limpet.

MOLLUSKS »

Mollusks are a large, highly diverse group, ranging from blind filter-feeding bivalves cemented to rock, through voraciously grazing snails and slugs, to lively, intelligent octopus and squid.

A typical mollusk has a soft body carried on a large muscular foot, and a head that senses the world using eyes and tentacles. The internal organs (viscera) are contained in a visceral hump, which is covered by a fleshy mantle. This mantle overhangs the edge of the hump, creating a groove called the mantle cavity, which is used for breathing. In most species, this mantle also secretes substances used to make the shell. Most mollusks feed with a tongue called a radula. Coated in teeth made of chitin, the radula moves forward and backward through the mouth to rasp at food. Bivalves lack a radula. They feed instead by siphoning particles through their shells. Most species trap particles of food in the mucus on their gills.

PHYLUM	MOLLUSCA
CLASSES	7
ORDERS	46
FAMILIES	609
SPECIES	About 110,000

These limpets have grazed the rock around them bare, but they cannot reach the green algae growing on top of their own shells.

WITH AND WITHOUT SHELLS

A shell is not only a refuge from predators, it is also protection from drying out. Some snails even have a trapdoor called an operculum to seal the opening of the shell. The shell is hardened with minerals obtained from the diet and surrounding water. It is coated in tough protein; and smooth on the inside, to allow the body to slide in and out—some groups have a lining of mother-of-pearl. Many shell-less mollusks defend themselves with distasteful or poisonous chemicals, and advertise this with striking colors.

Most cephalopods—tentacled mollusks, such as octopus and squid—lack an obvious shell. Members of this group are entirely predatory and have a horny beak for chewing on flesh. Their muscular foot is modified into their characteristic tentacles, which are used for grasping and swimming.

GETTING OXYGEN

Most mollusks are aquatic and so they have gills for breathing, usually projecting into the mantle cavity and irrigated by a current of water. In most land snails and slugs the mantle cavity is filled with air and functions as a lung. Because many freshwater snails probably evolved from this group, they have a lung too, and must frequently surface to breathe.

APLACOPHORANS

More wormlike than mollusklike, the Aplacophora are small cylindrical burrowers that feed on detritus or other invertebrates in the sediment of deep oceans. They lack a shell, but their clearly molluskan features include a rasping, tonguelike radula. The mantle cavity is reduced to an opening at the rear of the body, into which the animal discharges waste from the gut.

⅛–3¼ in
0.3–8 cm

COMMON GLISTENWORM
Chaetoderma sp.
F: Chaetodermatidae
A typical member of its class, this wormlike mollusk of the North Atlantic burrows through muddy sediment of ocean depths.

BIVALVES

Highly specialized aquatic mollusks, bivalves are instantly recognizable, with hinged shells that open up to access food and oxygen-rich water.

Bivalves are identifed from their shells, which consist of two plates, called valves, joined together by a hinge. To protect against predators and the dangers of drying out, the shells clamp tightly shut by contracting powerful muscles, sealing most or all of the body inside. Seashore species are regularly exposed by receding tides.

Within the class there is a huge variation in size, ranging from the tiny fingernail clam at about ¼ in (6 mm) to the giant clam, which can grow to 4½ ft (1.4 m) across. Some bivalves attach to rocks and hard surfaces by a bundle of tough

threads, called a byssus. Others use a well-developed muscular foot for burrowing into muddy sediment. A few bivalves, such as scallops, swim freely, using a method of jet propulsion.

Water is pumped in and out of bivalve shells through tubes, called siphons. The water supplies oxygen and food, which are collected as the water passes over modified gills. Food particles caught in the sticky mucus around the gills is wafted toward the mouth by microscopic hairs.

BIVALVES AND PEOPLE
Mussels, clams, and oysters are important sources of food, and some oysters create fine quality pearls by coating a foreign object with layers of nacre. Bivalves are also useful indicators of water quality as many cannot survive high levels of pollution.

PHYLUM	MOLLUSCA
CLASS	BIVALVIA
ORDERS	10
FAMILIES	105
SPECIES	About 8,000

A queen scallop claps the two valves of its shell together to accelerate away from the unwelcome attention of a predatory starfish.

OYSTERS AND SCALLOPS

The Ostreoida feed on tiny food particles filtered from the seawater. Many oyster species live permanently submerged in coastal waters, fixed to the rock by the gland that produces byssus thread in other bivalves. Scallops, by contrast, can swim freely by clapping their valves open and shut.

4¾–6 in
12–15 cm

GREAT SCALLOP
Pecten maximus
F: Pectinidae
In addition to swimming freely, scallops have jet-propelled escape responses. This commercial species inhabits fine sands of European coasts.

spiny left valve

4–4¾ in
10–12 cm

CAT'S TONGUE OYSTER
Spondylus linguafelis
F: Spondylidae
One of a family of thorny oysters with a colorful mantle, the cat's tongue oyster of the Pacific is named for its spiny exterior.

4–4¾ in
10–12 cm

COCK'S COMB OYSTER
Lopha cristagalli
F: Ostreidae
Close relatives of scallops, many species of oysters are prized as food and for their pearls. This is an Indo-Pacific species.

3¼–4 in
8–10 cm

EDIBLE OYSTER
Ostrea edulis
F: Ostreidae
Formerly abundant in Europe, this commercially important species is now over-fished in some areas. The species is inedible during summer months when breeding.

ARC CLAMS AND RELATIVES

Shells of these bivalves are closed by two strong muscles, and hinge along a straight line bearing a continuous series of teeth. Like their relatives, the Arcoida have a reduced foot and large gills that help trap food particles.

2–2¾ in
5–7 cm

NOAH'S ARK
Arca noae
F: Arcidae
Arc shells are square-edged, thick-shelled bivalves. Noah's Ark is a byssus-attached intertidal species on rocky coasts of the eastern Atlantic.

2–2¼ in
5–6 cm

EUROPEAN BITTERSWEET
Glycymeris glycymeris
F: Glycymerididae
Dog shells are rounded relatives of arc shells. This Northeast Atlantic species, fished in Europe, has a sweet flavor but is tough when overcooked.

MARINE MUSSELS

The Mytiloida have distinctively shaped elongate, asymmetrical shells that attach to rocks by byssus threads. Only one of their two shell-closing muscles is well developed.

3¼–4 in
8–10 cm

COMMON MUSSEL
Mytilus edulis
F: Mytilidae
The most commercially important mussel in Europe, this long-lived species occurs in dense beds. It can tolerate the low salinity of estuarine waters.

FAN MUSSELS AND RELATIVES

Members of the Pterioida include long-hinged, winged oysters and T-shaped hammer-oysters as well as fan mussels. The group also contains commercially important marine pearl oysters.

10–16 in
25–40 cm

FLAG FAN MUSSEL
Atrina vexillum
F: Pinnidae
Fan mussels, such as this species from western European coastlines, have triangular shells attached by a byssus in soft sediment.

FRESHWATER MUSSELS

The Unionoida are the only exclusively freshwater order of bivalves. The tiny larvae use their valves to clamp onto fish, forming cysts on the gill fins and feeding on blood or mucus before dropping off as juvenile mussels.

3½–4 in
9–10 cm

SWAN MUSSEL
Anodonta sp.
F: Unionidae
Bitterling fish lay eggs in living freshwater mussels, such as this Eurasian species. The mollusk becomes a nursery for fish fry.

4–6 in
10–15 cm

FRESHWATER PEARL MUSSEL
Margaritifera margaritifera
F: Margaritiferidae
Well known for its good-quality pearls, this mussel lives buried in sand or gravel beneath fast-flowing rivers in Eurasia and N. America.

TRUE CLAMS AND BORING BIVALVES

Long-siphoned bivalves in the order Myoida typically burrow in mud or bore through wood or rock. In piddocks the front of the shell acts like a file to bore burrows in soft rock. Shipworms use their shell valves to drill through timber.

4¾–6 in / 12–15 cm

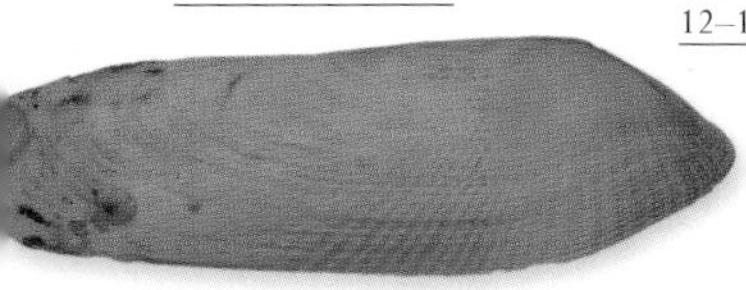

COMMON PIDDOCK
Pholas dactylus
F: Pholadidae
This common piddock of the northeastern Atlantic is phosphorescent and, like other species, lives in burrows bored in wood or clay.

4¾–6 in
12–15 cm

SOFT SHELL CLAM
Mya arenaria
F: Myidae
An edible clam with a thin shell, this North Atlantic species is especially abundant in muddy estuaries, where it burrows in soft sediment.

SHIPWORM
Teredo navalis
F: Teredinidae
This highly modified, widespread bivalve uses its ridged shell valves as a drill to bore deep, chalk-lined burrows in timber, damaging ships.

½–¾ in
1.5–2 cm

WATERING POTS AND RELATIVES

The order Pholadomyoida includes clams and tropical watering pots. Scarcely recognizable as bivalves, watering pots live encased in chalky tubes, and draw detritus and water into the tube through a perforated plate at the anterior end.

6–6½ in
15–17 cm

PHILIPPINE WATERING POT
Penicillus philippinensis
F: Clavagellidae
This bizarre Indo-Pacific watering pot belongs to a family named for their wide, perforated ends, fringed with tubules. It lives partially buried in sediment.

COCKLES AND RELATIVES

The largest order of bivalves, the Veneroida include a large range of marine animals. Most have short siphons that are often fused together. Some, notably the cockles, are agile, capable of burrowing or even leaping with their foot. Others are attached to rocks by a byssus.

1¼–1½ in
3–4 cm

ZEBRA MUSSEL
Dreissena polymorpha
F: Bivalvia
This freshwater mussel is anchored by byssus threads, but can detach and crawl on a slender foot. Native to eastern European waters, it has spread elsewhere.

COMMON EDIBLE COCKLE
Cerastoderma edule
F: Cardiidae
Sand-burrowing cockles have shells with radiating ribs. This northeastern Atlantic species, often found in vast numbers, is fished in northern Europe.

1½–2 in
4–5 cm

1–1¼ in
2.5–3 cm

concentric rings

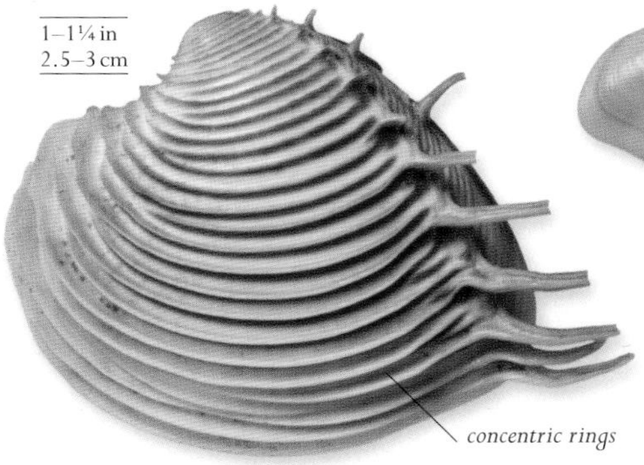

ROYAL COMB VENUS
Pitar dione
F: Veneridae
Named for the comb-like spines on its shell, this Venus clam occurs on tropical American coastlines.

2–3¼ in
5–8 cm

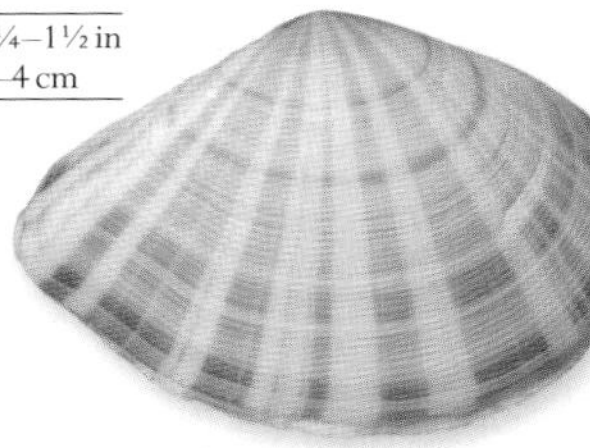

RED CALLISTA
Callista erycina
F: Veneridae
Many Venus clams are prized by collectors, including this Indo-Pacific species, also known as the sunset clam.

algae within mantle exposed to sunlight

RINGED DOSINIA
Dosinia anus
F: Veneridae
This Venus clam is found on the coastlines of New Zealand, and is one of many species harvested for human consumption.

2¼–3¼ in
6–8 cm

3¼–4½ ft / 1–1.4 m

GIANT CLAM
Tridacna gigas
F: Cardiidae
The world's largest bivalve, this long-lived and now endangered coral-dwelling animal lives on sandy beds in the Indo-Pacific.

¾–1½ in
2–4 cm

6–8 in
15–20 cm

SWORD RAZOR CLAM
Ensis siliqua
F: Pharidae
Razor shells, such as this northeast Atlantic species, live in burrows, feeding and respiring through their emergent siphon, but drop quickly using their muscular foot when disturbed.

CRADLE DONAX
Donax cuneatus
F: Donacidae
One of a family of triangular wedge shells that are fast burrowers in surf, this is a tropical Indo-Pacific species.

2–2¼ in / 5–6 cm

OBLONG TRAPEZIUM
Trapezium oblongum
F: Trapezidae
An Indo-Pacific bivalve, this animal lives attached to rocks by byssus threads, usually in crevices or beneath coral rubble.

2–2¾ in
5–7 cm

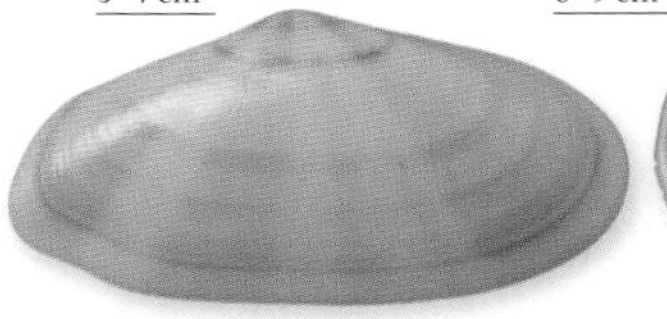

SUNRISE TELLIN
Tellina radiata
F: Tellinidae
The shell of this tellin from the Caribbean has a variable banding pattern and is often found on beaches.

3¼–3½ in
8–9 cm

WEST AFRICAN TELLIN
Tellina madagascariensis
F: Tellinidae
Tellins, such as this rose-colored tropical species, feed on particles in sediment, drawn in through their long, extendable siphons.

1¼–1½ in
3–4 cm

STRIPED TELLIN
Tellina virgata
F: Tellinidae
The shells of many tellins, such as this Indo-Pacific species, have decorative patterns.

large, scale-like flutes

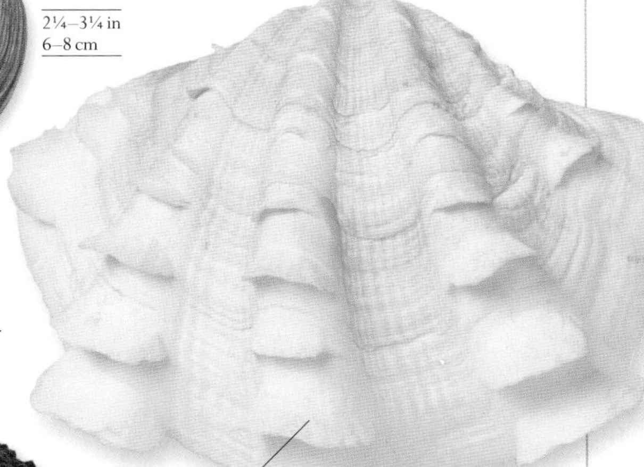

12–16 in
30–40 cm

FLUTED GIANT CLAM
Tridacna squamosa
F: Cardiidae
Like other Indo-Pacific giant clams, this species opens its shell during the day so that algae in its colored mantle can photosynthesize to make food.

GASTROPODS

Gastropods are by far the largest class of mollusks. The word gastropod means "stomach-foot," because the animal appears to crawl on its belly.

Most snails and slugs glide along a stream of slime on a single muscular foot, rasping at their food with a radula, a tonguelike emery board. Most gastropods use their radula to rasp at vegetation, algae, or the fine film of microbes that coat underwater rocks, but some are predatory. They usually have a distinct head with well-developed sensory tentacles. Snails have a coiled or helical shell into which they can withdraw; slugs lost their shells during the course of evolution. The class also includes a number of animals that deviate from this basic plan, such as swimming

sea slugs. Gastropods originated in the oceans and they still have their greatest diversity there; other species live in fresh water or on land.

TORSION

The juvenile gastropod undergoes a process called torsion: the whole of the body within the shell twists round 180 degrees, so that the respiratory mantle cavity comes to lie above the animal's head. This allows the vulnerable head to withdraw into the safety of the shell. In marine snails, such as periwinkles and limpets, this body form persists into adulthood. Because of this, marine snails are described as prosobranchs, meaning "forward gills". In sea slugs, however, the body twists back again, and they are called opisthobranchs, or "hind gills".

PHYLUM	MOLLUSCA
CLASS	GASTROPODA
ORDERS	21
FAMILIES	409
SPECIES	About 90,000

TRUE LIMPETS

Primitive algae-grazing limpets, the Patellogastropoda have a slightly coiled, conical shell. Their powerful muscles clamp down firmly on intertidal rocks, protecting them from predators, desiccation, and waves.

1¼–2 in
3–5 cm

COMMON LIMPET
Patella vulgata
F: Patellidae

This species grazes algae from rocks on shores of the Northeast Atlantic at high tide, then returns to a depression in the rock that matches the shape of its shell.

NERITES AND RELATIVES

A small but diverse group, Cycloneritimorpha are well known in the fossil record. They include marine, freshwater, and land-living forms, some with coiled shells, and a few like limpets. There are also species with trapdoors to their shells.

ZIGZAG NERITE
Neritina communis
F: Neritidae

½–¾ in / 1.2–2 cm

An inhabitant of Indo-Pacific mangroves, this is a highly variable species, with white, black, red, or yellow shells even in the same population.

¾–2 in
2–5 cm

BLEEDING TOOTH NERITE
Nerita peloronta
F: Neritidae

Named for the blood-red mark on its shell opening, this Caribbean intertidal mollusk can survive out of water for prolonged periods.

TOP SHELLS AND RELATIVES

The Vetigastropoda are marine gastropods that graze algae and microbes with a brushlike radula. Their shells range from those of keyhole limpets, with the barest suggestion of a spire, to the coiled globes and pyramids of top shells, which can seal themselves inside using an operculum, or trapdoor.

3¼–4¾ in
8-12 cm

GIANT TOP SHELL
Tectus niloticus
F: Turbinidae

Many top shells have a thick mother-of-pearl lining. This large Indo-Pacific species is exploited to make jewelry.

shell shaped like spinning top

banded pattern

LISTER'S KEYHOLE LIMPET
Diodora listeri
F: Fissurellidae

Keyhole limpets are close relatives of abalones and top shells. Like others, this western Atlantic species ejects oxygen-depleted water through its "keyhole."

¾–1 in
2–2.5 cm

CHECKERED TOP SHELL
Osilinus turbinatus
F: Trochidae

Top shells have "spinning-top" shells that are closed with a circular operculum. This is a Mediterranean species.

½–1¾ in
1.5–4.5 cm

2–2¾ in / 5–7 cm

SILVER MOUTH TURBAN SHELL
Turbo argyrostomus
F: Turbinidae

Turban shells are close relatives of top shells, but have a calcified operculum. This is an Indo-Pacific species.

8–12 in
20–30 cm

RED ABALONE
Haliotis rufescens
F: Haliotidae

Abalones have ear-shaped shells lined with thick mother-of-pearl and punctuated with holes for exhaling water. This Northeast Pacific kelp-grazing species is the largest.

TOWER SHELLS AND RELATIVES

These tall-spired snails typically live in muddy or sandy sediment. They feed on particles in the water that they circulate through their mantle cavity. The Cerithioidea are found in marine, freshwater, and estuarine habitats. They are slow moving and often group together in very large numbers.

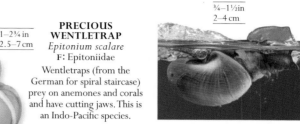

GREAT SCREW SHELL
Turritella terebra
F: Turritellidae

1–2¼ in
2.5–5.5 cm

6–17 cm
2⅜–6½ in

A filter feeder of muddy sediments, this Indo-Pacific snail belongs to a group variously known as tower shells or auger shells.

ROUGH CERITH
Rhinoclavis asper
F: Cerithiidae

Ceriths are snails often abundant in shallow-water tropical marine sediments. Like others, this Indo-Pacific species lays strings of eggs attached to solid material.

WEST INDIAN WORM SHELL
Vermicularia spirata
F: Turritellidae

Free-floating males of this Caribbean unwound screw shell attach to solid material, often embedding themselves in sponges. They then develop into larger sedentary females.

2.5–16 cm
1–6½ in

SUNBURST CARRIER
Stellaria solaris
F: Xenophoridae

spikes help attach objects

Carrier snails cement objects such as pebbles or shells of other animals to their shells for camouflage. This species inhabits Indo-Pacific waters.

2¾–5 in
6–13 cm

attached detritus

PERIWINKLES, WHELKS, AND RELATIVES

The largest and most diverse order of marine snails, the Caenogastropoda are divided into three groups. Ptenoglossids—wentletraps and violet snails—are free-floating or swimming specialist predators of cnidarians. Littorinids graze on algae, and include periwinkles, cowries, and conches. Whelks and their relatives are predators that project a long siphon out through a groove in the shell.

PRECIOUS WENTLETRAP
Epitonium scalare
F: Epitoniidae

1–2¾ in
2.5–7 cm

Wentletraps (from the German for spiral staircase) prey on anemones and corals and have cutting jaws. This is an Indo-Pacific species.

¾–1½ in
2–4 cm

LARGE VIOLET SNAIL
Janthina janthina
F: Janthinidae

Violet snails float on tropical oceans, preying on floating cnidarians. They secrete mucus to create rafts of bubbles that keep them buoyed up.

TIGER COWRIE
Cypraea tigris
F: Cypraeidae

4–6 in
10–15 cm

The fleshy lobes of a cowrie's mantle wrap right around the smooth shell when it crawls. This Indo-Pacific species preys on other invertebrates.

FOOL'S CAP
Capulus ungaricus
F: Capulidae

½–2¼ in
1.5–6 cm

This North Atlantic gastropod resembles unrelated limpets, attaching itself to stones or even shells of other mollusks, such as scallops.

ATLANTIC SLIPPER LIMPET
Crepidula fornicata
F: Calyptraeidae

¾–2 in
2–5 cm

Unrelated to true limpets, filter-feeding slipper limpets form mating towers, with smaller males on top changing sex to replace dead females beneath.

COMMON PELICAN'S FOOT
Aporrhais pespelecani
F: Aporrhaididae

Cousins of conches, pelican's foot snails are mud-dwelling detritus-eaters with extensions to their shells that look like webbed feet. This species inhabits the Mediterranean and the North Sea.

12–16½ in / 30–42 cm

COMMON PERIWINKLE
Littorina littorea
F: Littorinidae

¾–1¼ in
2–3 cm

Periwinkles, such as this European species, form a family of intertidal snails with a globular spired shell closed by an operculum.

PINK CONCH
Strombus gigas
F: Strombidae

6–12 in
15–31 cm

Conches are mostly tropical, medium to large marine grazing snails. *Strombus* conches, such as this giant West Atlantic species, have flare-lipped shells.

GIANT FROG SHELL
Tutufa bubo
F: Bursidae

4–12½ in
10–32 cm

Tropical marine frog shells are warty. This Indo-Pacific species eats bristleworms using a proboscis, first anesthetizing them with saliva.

— warty shell

canal for siphon

TRUMPET TRITON
Charonia tritonis
F: Ranellidae

4–20 in
10–50 cm

Tropical intertidal tritons and related frog and helmet shells are predators of other invertebrates. This Indo-Pacific species eats the aggressive crown of thorns starfish.

POLI'S NECKLACE SHELL
Euspira pulchella
F: Naticidae

³⁄₁₆–1¼ in
0.5–3 cm

This is a European member of a family of sand-burrowing predators of bivalves. Necklace shells are named for their necklacelike ribbons of spawn.

COMMON NORTHERN WHELK
Buccinum undatum
F: Buccinidae

3¼–4¼ in
8–11 cm

A large predator of other mollusks and tube worms, this North Atlantic snail also consumes carrion, and is a popular seafood.

PRICKLY PACIFIC DRUPE
Drupa ricinus
F: Muricidae

¾–1¼ in
2–3 cm

A member of the dog whelk family, this snail lives on coral reefs of the Indo-Pacific, where it preys on bristle worms.

DOG WHELK
Nucella lapillus
F: Muricidae

1¼–1½ in
3–4 cm

This North Atlantic snail belongs to a whelk family that catch prey by using the radula to drill holes in the shells of barnacles and mollusks, helped by secretions of enzymes.

TEXTILE CONE
Conus textile
F: Conidae

3½–6 in
9–15 cm

Cone shells use their radula to harpoon their prey and inject venom. Some, such as this Indo-Pacific species, can be harmful to humans.

TENT OLIVE
Oliva porphyria
F: Olividae

1⅛–5 in
3–13 cm

The largest of the olive snails, this species from the Pacific coasts of Mexico and South America has a colorful glossy shell.

BANDED TULIP
Cinctura lilium
F: Fasciolariidae

3–4½ in
7.5–11.5 cm

This is a Caribbean coral-dwelling snail related to *Buccinum* whelks. Sand-dwelling species with longer siphonal canals are called spindle shells.

IMPERIAL HARP
Harpa costata
F: Harpidae

2¼–4¼ in
6–11 cm

Harp snails are sand-living predators of crabs, trapping them with their broad foot and digesting them with saliva. This is an Indian Ocean species.

SUBULATE AUGER
Terebra subulata
F: Terebridae

2¾–8 in
7–20 cm

Typical of many augers, this Indo-Pacific species has a patterned shell. Augers burrow through surface layers of sand to prey on worms.

FRESHWATER GILLED SNAILS

The Archaeaenioglossa are the only order of gilled snails containing no marine species. Most live in freshwater, but some live on land. Apple snails have gills in the mantle cavity that can function as a lung, allowing them to survive periods of drought. All species have an operculum for closing their shells.

CHANNELED APPLE SNAIL
Pomacea canaliculata
F: Ampullariidae

4–6 in
10–15 cm

A typical apple snail, this tropical American species has been introduced elsewhere and has become an invasive pest.

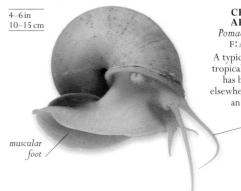

muscular foot

sensory tentacles

COMMON RIVER SNAIL
Viviparus viviparus
F: Viviparidae

A European freshwater snail with gills, this is a relative of the apple snails and, like them, has a trapdoor (operculum) to close its shell.

1¼–1½ in / 3–4 cm

SEA HARES

Sensory water-tasting head stalks, a common feature of sea slugs, are so big in the large sea hares that they resemble ears. The Anaspidea have a small internal shell and, like sea angels, can swim with their foot flaps.

foot flap

2¾–8 in
7–20 cm

SPOTTED SEA HARE
Aplysia punctata
F: Aplysiidae

Like other seaweed-eating sea hares, this European species gathers in large breeding groups. As a defense, it releases ink when disturbed by predators.

SWIMMING SEA SLUGS

The Gymnosomata are also called sea angels, because they have evolved flaps from their muscular foot and use them to fly through water. The related sea butterflies or Thecosomata also have flaps, but retain fragile shells.

winglike flaps
for swimming

COMMON SEA ANGEL
Clione limacina
F: Clionidae

1½–2 in / 4–5 cm

Sea angels swim freely with soft transparent "wings." This cold-water species preys on a related gastropod called the Arctic sea butterfly (*Limacina helicina*).

TRUE SEA SLUGS

The Nudibranchia are the largest group of sea slugs. Nudibranch means "naked gills" and refers to the fact that these animals' gills are exposed on their back, rather than contained in a mantle cavity. Many species have vibrant color patterns to advertise their toxicity.

4–5 in
10–13 cm

VARICOSE SEA SLUG
Phyllidia varicosa
F: Phyllidiidae

This Indo-Pacific sea slug is a common inhabitant of rocks, rubble, and sand in coastal waters, where it preys on sponges.

2–3¼ in
5–8 cm

**BLACK-MARGINED
SEA SLUG**
Glossodoris atromarginata
F: Chromodorididae

This common Indo-Pacific sea slug lives in shallow waters, feeding on sponges. It varies in color from off-white to pale yellow.

OPALESCENT SEA SLUG
Hermissenda crassicornis
F: Facelinidae

1½–2 in
4–5 cm

This intertidal northern Pacific species belongs to a group of sea slugs with body outgrowths that store the still potent stinging organs of their jellyfish prey.

¾–2 in
2–5 cm

ANNA'S SEA SLUG
Chromodoris annae
F: Chromodorididae

Like other *Chromodoris* sea slugs, this variable species from the western Pacific Ocean specializes in feeding on sponges.

external gills

rhinophores
(scent detectors)

2¾–3¼ in
7–8 cm

head end

ELEGANT SEA SLUG
Okenia elegans
F: Goniodorididae

Typical of its family, this sea slug feeds on sea squirts. It occurs in waters around Europe, including the Mediterranean.

gills obtain oxygen

4–4¾ in / 10–12 cm

VARIABLE NEON SEA SLUG
Nembrotha kubaryana
F: Polyceridae

This Indo-Pacific sea slug preys on sea squirts and incorporates their defensive chemicals into its mucus. Many of its relatives prey on bryozoans.

mantle

12–16 in / 30–40 cm

SPANISH DANCER
Hexabranchus sanguineus
F: Chromodorididae

A giant Indo-Pacific sea slug, this species gets its name from its free-swimming movements; its red mantle is said to resemble the ruffled skirt of a flamenco dancer.

FRESHWATER AIR-BREATHING SNAILS

Snails of the order Basommatophora have their mantle cavity developed as a lung; unlike most other marine gastropods, they must return to the water surface to breathe. Mainly herbivorous, they are common in alkaline or neutral, weed-choked, stagnant or slow-moving freshwater.

shell opening to left

coiled shell

GREAT POND SNAIL
Lymnaea stagnalis
F: Lymnaeidae

1–2 in
2.5–5 cm

Widespread throughout temperate parts of the Northern Hemisphere, this is a common pond snail in still or slow-moving freshwater.

EUROPEAN BLADDER SNAIL
Physella acuta
F: Physidae

⅜–⅝ in
1–1.6 cm

Most snails have shells that open to the right (when the opening faces the observer), but the freshwater physid family have left-opening shells.

COMMON RIVER LIMPET
Ancylus fluviatilis
F: Planorbidae

³⁄₁₆–⁵⁄₁₆ in
5–8 mm

Not a true limpet, but a limpetlike member of the ramshorn snail family, this species is common in fast-flowing European waters.

GREAT RAMSHORN SNAIL
Planorbarius corneus
F: Planorbidae

1¼–1½ in
3–3.5 cm

Ramshorn snails have flat-coiled shells, instead of the spire shells typical of snails. Most—like this Eurasian species—occur in still freshwater.

LAND AIR-BREATHING SNAILS AND SLUGS

The Stylommatophora are land snails and slugs that breathe air using a lung in their mantle cavity. They have eyes at the tips of their tentacles. Many are hermaphrodite and have both male and female reproductive organs. Their courtship behavior involves the exchange of spearlike darts before mating.

GIANT AFRICAN SNAIL
Achatina fulica
F: Achatinidae

6–9 in
15–22 cm

This giant land snail from East Africa has been introduced to many other warm parts of the world, becoming an invasive pest.

CUBAN LAND SNAIL
Polymita picta
F: Cepolidae

1¼–1½ in
3–3.5 cm

This snail is only found in the mountain forests of Cuba. Its shells, which exist in numerous colors, are prized by collectors.

4–8 in
10–20 cm

BROWN GARDEN SNAIL
Helix aspersa
F: Helicidae

1–1¾ in
2.5–4.5 cm

This variably colored species with a wrinkled shell is widespread in European woods, hedgerows, dunes, and gardens.

COMMON SHELLED SLUG
Testacella haliotidea
F: Testacellidae

3¼–4¾ in
8–12 cm

A typical member of a family of slugs with small external shells, this European species preys on earthworms.

shell

DOMED DISK SNAIL
Discus patulus
F: Discidae

¼ in / 7–8 mm

Discid snails make up a family characterized by certain primitive features and a flat, coiled shell. This is a North American woodland species.

EUROPEAN BLACK SLUG
Arion ater
F: Arionidae

Black specimens of this European slug predominate in the north, but other individuals are orange. Although it eats plants, it also clears garden waste.

PACIFIC BANANA SLUG
Ariolimax columbianus
F: Ariolimacidae

6–10 in / 15–25 cm

Named for its yellow color, the Pacific banana slug lives in wet coniferous forests on the western coast of North America.

1¼–2 in
3–5 cm

eye on tip of tentacle

ROMAN SNAIL
Helix pomatia
F: Helicidae

Widespread in central Europe on calcium-rich soils, this is the largest snail of the region and is bred locally for human consumption.

BROWN-LIPPED SNAIL
Cepaea nemoralis
F: Helicidae

¾–1¼ in
2–3 cm

A close relative of the Roman snail, this western European species has a highly variable shell color, and banding patterns for camouflage in different habitats.

ASHY-GREY SLUG
Limax cinereoniger
F: Limacidae

4–12 in
10–30 cm

Keel-back slugs have a small internal shell. This large species of European woodland exists in a number of color varieties.

CEPHALOPODS

Cephalopods are agile hunters in the mollusk phylum. Their sophisticated nervous system enables them to hunt down fast-moving prey.

Cephalopods include the most intelligent of all invertebrates. Many species signal their mood using pigment-containing structures in their skin called chromatophores. The class is organized by number of arms. Squid and cuttlefish have eight arms used for swimming, plus two longer retractable arms, called tentacles, with suckers for gripping prey. Octopus lack tentacles, but have suckers on all eight of their arms. The cephalopod mantle encloses a cavity, which contains the gills. To obtain oxygen, cephalopods take in water through the sides of the mantle. It passes over the gills and is then expelled through a short funnel. Cephalopods can move backward quickly by expelling water through the funnel, to propel themselves forwards. Squid, cuttlefish, and some octopus have fins on the sides of the mantle, used for open-water swimming. Most octopus spend their time on the sea floor.

Only the open-ocean nautilus has a coiled shell. In squid the shell is reduced to a pen-like structure that gives some internal support, in cuttlefish to a similar, calcified structure called a cuttlebone. Most species of octopus have lost their shell.

Cephalopods are fast-moving predators that use their arms to grasp prey, and a parrotlike beak to despatch it. Squid catch free-swimming prey, while cuttlefish and octopus take slower, bottom-crawling crustaceans, such as crabs.

PHYLUM	MOLLUSCA
CLASS	CEPHALOPODA
ORDERS	About 8
FAMILIES	About 45
SPECIES	About 750

A North Pacific giant octopus discharges black-pigmented ink in self-defense as it jets away from a predator.

CHAMBERED NAUTILUS
Nautilus pompilius
F: Nautilidae
6–9½ in
15–24 cm
This Indo-Pacific species is the largest and best-known living member of a small group of cephalopods otherwise known only as fossils.

2¼–2¾ in
6–7 cm

tentacles at front

FLAMBOYANT CUTTLEFISH
Metasepia pfefferi
F: Sepiidae
Unusual among cuttlefish for its habit of using its tentacles to walk on ocean floors, this Indo-Pacific cuttlefish has recently been shown to be poisonous.

BROADCLUB CUTTLEFISH
Sepia latimanus
F: Sepiidae
18–20 in
45–50 cm
This large cuttlefish is wide-ranging throughout the Indo-Pacific region. It is an abundant predator on coral reefs, feeding on prawns and shrimps.

COMMON CUTTLEFISH
Sepia officinalis
F: Sepiidae
16–20 in
40–50 cm
Like many other cuttlefish, this species migrates inshore to spawn on muddy sediments. It occurs off the coasts of Europe and South Africa.

AUSTRALIAN GIANT CUTTLEFISH
Sepia apama
F: Sepiidae
18–20 in/45–50 cm
The largest known cuttlefish, this animal inhabits waters off the southern coast of Australia, living in seagrass beds and rocky reefs.

WHIP-LASH SQUID
Mastigoteuthis sp.
F: Mastigoteuthidae
8–20 in
20–50 cm
Reddish whip-lash squid such as this species hover with broad fins in deep water, waiting for prey with its long tentacles extended.

prey-snatching tentacles

BERRY'S BOBTAIL SQUID
Euprymna berryi
F: Sepiolidae
¾–1¼ in
2–3 cm
Bobtail squid, such as this Indo-Pacific species, are small cuttlefish relatives. They have a pen, not cuttlebone, and the rounded body has lobed fins.

BIGFIN REEF SQUID
Sepioteuthis lessoniana
F: Loliginidae
10–14 in
25–35 cm
A pair of large fins makes this Indo-Pacific squid resemble a cuttlefish. It communicates by flashing light from special organs, which contain light-emitting bacteria.

COMMON SQUID
Loligo vulgaris
F: Loliginidae
12–18 in
30–45 cm
This commercially important squid is common in the northeast Atlantic and Mediterranean. Like related species, it has prominent side fins.

RAM'S HORN SQUID
Spirula spirula
F: Spirulidae
1½–1¾ in
3.5–4.5 cm
This small, deep-ocean cephalopod contains a spiral gas-filled shell that acts as a float; the animal rises toward the ocean surface at night.

»

COMMON OCTOPUS
Octopus vulgaris

Whether it is extricating a lobster from a lobster pot or hunting a scuttling crab, the common octopus is one of the most intelligent of all the invertebrates. This sea-living mollusk has excellent vision and eight grasping arms or tentacles which are also used for crawling. The octopus can change color in an instant, and it can squeeze through very narrow crevices. Its horny beak can bite into prey, whose scattered shell fragments frequently litter the sand around its lair. But in spite of its versatility, the common octopus is short-lived. Upon hatching, an infant octopus enters the oceanic plankton with half a million of its siblings, before descending to the seabed two months later. If it is not eaten, it takes little more than a year to mature and spawn its own offspring before dying. The common octopus is widespread in tropical and warm temperate oceanic waters, and may consist of several similar species.

SIZE Tentacle span 5–9¾ ft (1.5–3 m)
HABITAT Rocky coastal waters
DISTRIBUTION Widespread in tropical and warm temperate waters
DIET Crustaceans, shelled mollusks

fleshy mantle encloses a cavity that contains vital organs, including gills for breathing

> INTELLIGENT MOLLUSK

An octopus is an agile predator with an advanced nervous system. Two-thirds of its neurons are located in its arms, which operate with a remarkable degree of independence from its brain. Experiments show that it also has considerable problem-solving intelligence, and excellent long- and short-term memories.

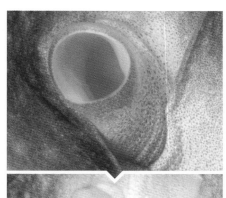

∧ FUNNEL, OPEN AND CLOSED

On one side of the mantle, just behind the head, is a funnel, which the octopus uses in three ways: to expel water after gills have extracted oxygen; to rapidly discharge water as jet propulsion for a quick getaway; and to squirt a confusing cloud of ink at the enemy before fleeing.

∨ UNDERSIDE

The octopus is a cephalopod, which literally means "head-footed." Its eight mobile arms radiate from the underside of the head, where the mouth can be seen at the center.

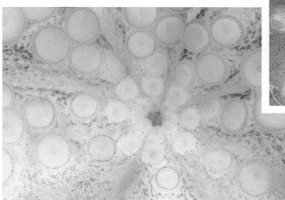

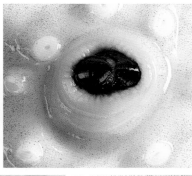

< BEAK

This octopus is a predator of crustaceans, among other prey. Its jaws are a parrotlike beak, which is strong enough to penetrate the tough carapace of a crab or lobster.

< SKIN

The skin contains special cells called chromatophores. These contain pigment that can change the octopus's color—to blend in with its surroundings or to signal its mood if angry or afraid.

∧ SUCKER

Each arm has two series of suckers. These give the octopus an excellent grip, allowing it to negotiate the sea bottom and coral reefs with ease. The suckers also have receptors that enable the octopus to taste what it is touching.

prominent eye with a horizontal, slit-shaped pupil

tough, warty skin can change color, texture, and shape for camouflage

long, muscular arms for moving over the sea bed and grasping objects

cuplike suckers

VAMPIRE SQUID
Vampyroteuthis infernalis
F: Vampyroteuthidae

This deep-sea cephalopod—with characteristics seemingly intermediate between squid and octopus—has fins projecting from its mantle and light-producing organs covering the body.

4–6 in
10–15 cm

DUMBO OCTOPUS
Grimpoteuthis plena
F: Grimpoteuthidae

Named for its earlike fins used in swimming, this cephalopod lives at depths of 10–13,000 ft (3–4,000 m), preying on other invertebrates.

8 in
20 cm

WINGED ARGONAUT
Argonauta hians
F: Argonautidae

Argonauts—also known as paper nautiluses—are octopus relatives. Females produce paper-thin egg cases that resemble shells. This species is distributed widely around the world.

1¼–2¼ in
3–6 cm

NORTH PACIFIC GIANT OCTOPUS
Enteroctopus dofleini
F: Octopodidae

Perhaps the largest of all octopuses, this animal is surprisingly short-lived. Females are diligent caretakers of their huge broods.

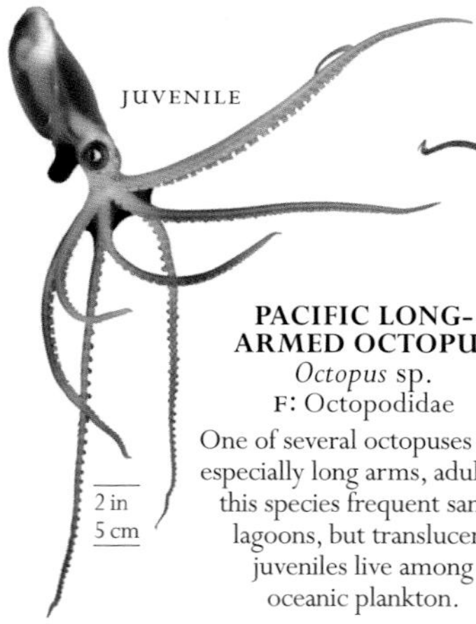

6½ ft
2 m

JUVENILE

PACIFIC LONG-ARMED OCTOPUS
Octopus sp.
F: Octopodidae

One of several octopuses with especially long arms, adults of this species frequent sandy lagoons, but translucent juveniles live among oceanic plankton.

2 in
5 cm

CARIBBEAN REEF OCTOPUS
Octopus briareus
F: Octopodidae

An octopus of the western Atlantic and Caribbean, this coral-dwelling species often traps its prey by spreading its webbed arms like a net.

39–59 in
1–1.5 m

COMMON OCTOPUS
Octopus vulgaris
F: Octopodidae

Globally widespread in tropical and warm temperate waters, this octopus typically has a warty body and two rows of suckers on its arms.

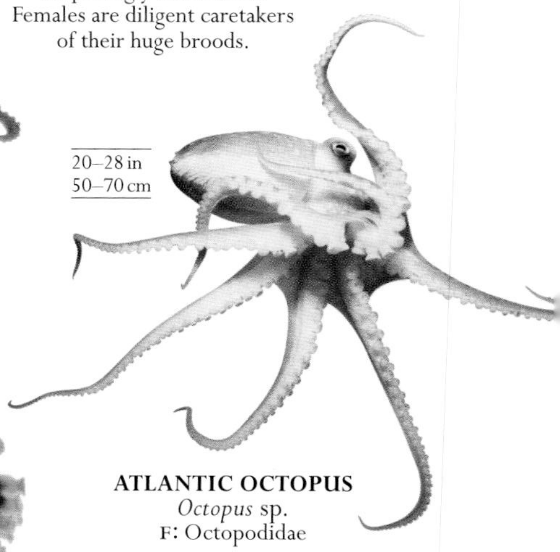

4½ ft
1.3 m

20–28 in
50–70 cm

ATLANTIC OCTOPUS
Octopus sp.
F: Octopodidae

DNA analysis has revealed a number of similar octopus species, including this one, that are related to the common octopus, revealing hidden biodiversity.

MIMIC OCTOPUS
Thaumoctopus mimicus
F: Octopodidae

Many octopuses can change color, but this Asian species can change shape too—even disguising itself to look like other marine animals such as sponges, corals, and jellyfish.

39 in/1 m

yellow background color

funnel

BLUE-RINGED OCTOPUS
Hapalochlaena lunulata
F: Octopodidae

This octopus preys on crustaceans and fish in the western Pacific, paralyzing them with venomous saliva that is also potentially fatal to humans.

6–8 in
15–20 cm

black and blue ring markings warn that this is a highly venomous species

CHITONS

Flattened, rock-hugging chitons are among the most primitive of mollusks. Most are grazers of algae and microbe films, and occur in coastal waters.

A chiton's shell, popularly known as a "coat-of-mail" shell, is made up of eight interlocking plates that are sufficiently flexible to allow the animal to bend when gliding over uneven rocks—or even to roll up when disturbed. Chitons have no eyes or tentacles, although the shell itself contains cells that enable them to react to light. The plates are fringed by the edge of the mantle, known as the girdle. The mantle forms a fleshy skirt around the chiton. This overhangs grooves running down each side, which channel water, delivering oxygen to gills that project into the grooves. The radula, or rasping tongue, is coated with minute teeth reinforced with iron and silica, so chitons can graze on the toughest encrusting algae.

PHYLUM	MOLLUSCA
CLASS	POLYPLACOPHORA
ORDERS	4
FAMILIES	10
SPECIES	About 850

girdle fringe

eight-plated shell

3¼ in/8 cm

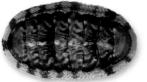

MARBLED CHITON
Chiton marmoratus
F: Chitonidae
Like other chitons, the shell plates of this Caribbean mollusk are composed entirely of a chalky mineral called aragonite.

1½–2 in/4–5 cm

GREEN CHITON
Chiton glaucus
F: Chitonidae
A variable-colored chiton, this species occurs along the coastlines of New Zealand and Tasmania and, like most chitons, is active at night.

¾–3¼ in/2–8 cm

**WEST INDIAN
FUZZY CHITON**
Acanthopleura granulata
F: Chitonidae
This spiky-fringed Caribbean chiton can tolerate exposure to sun, and can live high up in the intertidal zone.

1 in/2.5 cm

DECKED CHITON
Ischnochiton comptus
F: Ischnochitonidae
Ischnochiton chitons have a spiny or scaly girdle fringe. This is a common intertidal species of western Pacific shores.

**12–13 in
30–33 cm**

**GUMBOOT
CHITON**
Cryptochiton stelleri
F: Acanthochitonidae
In one chiton family the fleshy girdle overlaps the chain mail plates. In this North Pacific species, the largest chiton, it forms a leathery skin.

2–3¼ in/5–8 cm

**BRISTLED
CHITON**
Mopalia ciliata
F: Mopaliidae
Named for its prominent hairy girdle, this chiton of the Pacific coast of North America is sometimes found beneath long-moored boats.

1½–2¾ in/4–7 cm

**HAIRY
CHITON**
Chaetopleura papilio
F: Ischnochitonidae
A bristly-girdled species with brown-striped chainmail plates, the hairy chiton is found beneath rocks on South African coasts.

**½–2 in
4–5 cm**

LINED CHITON
Tonicella lineata
F: Ischnochitonidae
This brightly colored chiton occurs on North Pacific coasts, where it may be camouflaged when feeding on red encrusting algae.

TUSK SHELLS

Curious, mud-shoveling tusk shells live in marine sediments and have a tubular curved shell—open at both ends and unlike the shell of any other mollusk.

PHYLUM	MOLLUSCA
CLASS	SCAPHOPODA
ORDERS	2
FAMILIES	12
SPECIES	About 500

Tusk shells are common, but are rarely seen because they usually occur far offshore, with the wide base of the hollow tusk buried in mud. The eyeless head and foot reach deeper into the sediment, probing for food with their tentacle-like structures called captacula, and tasting the mud with chemical detectors. When food is found—small invertebrates or detritus—the tentacles bring it to the mouth, where it is ground up by the radula, a rasping tongue typical of mollusks. Tusk shells lack gills; the animal's fleshy mantle encloses a water-filled, oxygen-extracting tube that runs the length of the shell. When levels of oxygen run low, the tusk shell contracts its foot and squirts the stale water out through the top end of the shell. Freshwater enters by the same route.

**1¼–1½ in
3–4 cm**

EUROPEAN TUSK
Antalis dentalis
F: Dentaliidae
O: Dentaliida
This tusk shell of the northeast Atlantic is widespread in sandy areas of coastal waters. In places its empty shells are found in huge numbers.

**2–3¼ in
5–8 cm**

BEAUTIFUL TUSK
Pictodentalium formosum
F: Dentaliidae
O: Dentaliida
This colorful tropical tusk shell has been recorded in marine sediments in Japan, the Philippines, Australasia, and New Caledonia.

ECHINODERMS

Echinoderms include an astonishing array of sea creatures, ranging from filter-feeding feather stars to grazing urchins and predatory starfish.

This is the only large invertebrate phylum entirely restricted to saltwater. Echinoderms are sluggish, slow-moving inhabitants of the seafloor and most have a body form built around a five-part radial symmetry. Echinoderm means "spiny-skinned"—a reference to the hard, calcified structures, called ossicles, that make up the internal skeleton of these animals. In starfish these are spaced out within the soft tissue so that the animal is reasonably flexible, but in urchins they are fused together to form a solid internal shell. In sea cucumbers the ossicles are so minute and sparsely distributed—or even absent altogether— that the entire animal is soft-bodied.

HYDRAULIC FEET
Echinoderms are the only animals with a water transport system. Seawater is drawn into the center of the body through a perforated, sievelike plate, usually on the upper surface of the body.

The water circulates through tubes that run through the animal and is forced into tiny soft projections near the surface, called tube feet, that can move back and forth and stick to surfaces. In feather stars these tube feet point upward on the feathery arms, capturing particles of food, which are then moved toward the central mouth. In other echinoderms thousands of downward-pointing tube feet are used in locomotion—together they pull the animal over sediment and rock.

DEFENSE
The tough prickly skin of most echinoderms offers good protection from potential predators, but these animals defend themselves in other ways too. Many urchins are covered with formidable spines, too; some can inflict serious wounds in humans. Many have tiny pincerlike projections, sometimes venomous, that can remove cluttering debris as well as deter potential predators. Soft-bodied sea cucumbers rely instead on noxious chemicals to protect themselves—and are often brightly colored as a warning. As a last resort, some sea cucumbers can even spew out sticky tangles or eject their gut.

PHYLUM	ECHINODERMATA
CLASSES	5
ORDERS	31
FAMILIES	147
SPECIES	About 7,000

The tiny tube feet on which echinoderms move around the seabed are seen here on a crown of thorns starfish, *Acanthaster planci*.

FEATHER STARS

A feather star carries five basic filter-feeding arms, which in some species branch into a dense cluster. A mouth and anus point upward from the center of the star. Some members of the class Crinoidea crawl along on moving fingerlike roots. Others, known as sea lilies, are attached by a stalk.

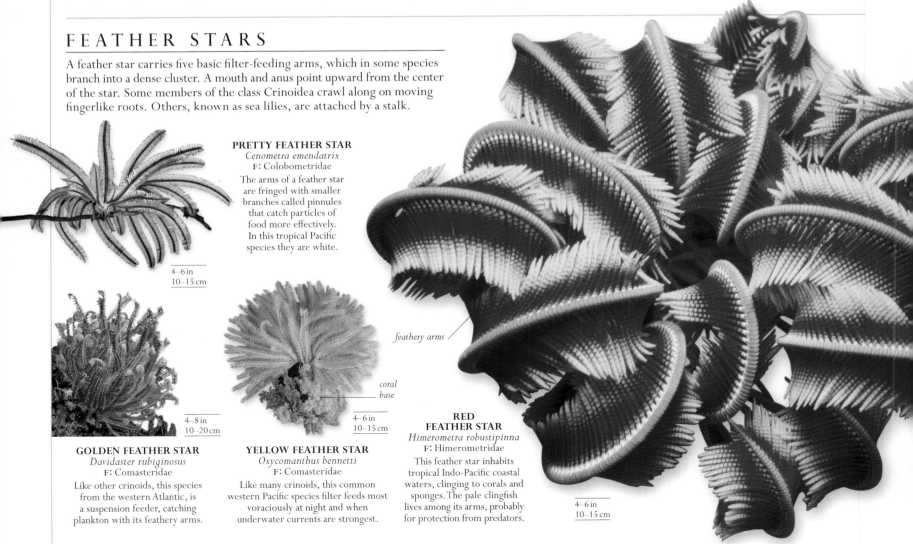

PRETTY FEATHER STAR
Cenometra emendatrix
F: Colobometridae
The arms of a feather star are fringed with smaller branches called pinnules that catch particles of food more effectively. In this tropical Pacific species they are white.

4–6 in
10–15 cm

feathery arms

coral base

GOLDEN FEATHER STAR
Davidaster rubiginosus
F: Comasteridae
Like other crinoids, this species from the western Atlantic, is a suspension feeder, catching plankton with its feathery arms.

4–8 in
10–20 cm

YELLOW FEATHER STAR
Oxycomanthus bennetti
F: Comasteridae
Like many crinoids, this common western Pacific species filter feeds most voraciously at night and when underwater currents are strongest.

4–6 in
10–15 cm

RED FEATHER STAR
Himerometra robustipinna
F: Himerometridae
This feather star inhabits tropical Indo-Pacific coastal waters, clinging to corals and sponges. The pale clingfish lives among its arms, probably for protection from predators.

4–6 in
10–15 cm

URCHINS AND RELATIVES

Sea urchins, which belong to the class Echinoidea, gave their name to the whole phylum. The hard ossicles of an urchin are plates that interlock to form a shell called a test. The animal's spines help in locomotion and defense. The mouth generally faces downward and the anus upward—with rows of tube feet running between the two.

4–4¼ in
10–11 cm

INDO-PACIFIC SAND DOLLAR
Echinodiscus auritus
F: Astriclypeidae

This Indo-Pacific urchin of coastal sands is a typical member of a group of burrowing sea dollars, named for their flattened appearance.

1½–2 in
4–5 cm

BANDED SEA URCHIN
Echinothrix calamaris
F: Diadematidae

The shorter spines of this Indo-Pacific reef-dwelling urchin are painfully venomous. Cardinal fish (Apogonidae) often hide among them for protection.

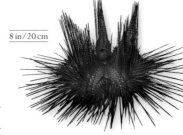

8 in / 20 cm

RED URCHIN
Astropyga radiata
F: Diadematidae

This Indo-Pacific lagoon species belongs to a family of tropical urchins with long hollow spines. It is often carried by a crab, *Dorippe frascone*.

3¼–4 in
8–10 cm

MATHA'S URCHIN
Echinometra mathaei
F: Echinometridae

This Indo-Pacific species belongs to a group of urchins characterized by thick spines. It lives in crevices on coral reefs.

6½–7 in
16–18 cm

EDIBLE URCHIN
Echinus esculentus
F: Echinidae

This large globular urchin common on northeast Atlantic coasts has a variable color. It grazes on algae, especially kelp.

FIRE URCHIN
Asthenosoma varium
F: Echinothuriidae

This somewhat flat Indo-Pacific species of sandy lagoons has a shell sufficiently flexible for the urchin to enter crevices. It can deliver a painful sting.

flexible test

8–10 in
20–25 cm

short venomous spines

3¼–4 in / 8–10 cm

PURPLE URCHIN
Strongylocentrotus purpuratus
F: Strongylocentrotidae

An inhabitant of the underwater kelp forests along the Pacific coast of North America, this urchin is often used in biomedical research.

SEA POTATO
Echinocardium cordatum
F: Loveniidae

Heart urchins, such as this globally widespread species, are detritus-feeding burrowers that lack the obvious radial symmetry of other urchins.

2¾–3¼ in
7–8 cm

SEA CUCUMBERS

Sea cucumbers are soft tubular animals, with a mouth surrounded by a ring of food-collecting tentacles at one end, and an anus at the other. Some Holothuroidea shovel or burrow in sediments using their tentacles; others use rows of tube feet to crawl along the seabed.

2–3¼ in
5–8 cm

YELLOW SEA CUCUMBER
Colochirus robustus
F: Cucumariidae

This echinoderm belongs to a family of thick-skinned sea cucumbers. Like this Indo-Pacific species, many are covered in knobbly projections.

6–7 in
15–18 cm

SEA APPLE CUCUMBER
Pseudocolochirus violaceus
F: Cucumariidae

Sea apples are brightly colored, highly poisonous, reef-dwelling sea cucumbers. This species has yellow or orange tube feet, but variable body color.

warning coloration

sticky mass ejected to repel predators

15–23½ in
38–60 cm

OCELLATED SEA CUCUMBER
Bohadschia argus
F: Holothuriidae

Distributed from the western Indian Ocean around Madagascar to the South Pacific, this large species is variable in color, but is always spotted.

10–12 in / 25–30 cm

EDIBLE SEA CUCUMBER
Holothuria edulis
F: Holothuriidae

Large sea cucumbers are especially abundant in tropical waters. This Indo-Pacific species is harvested and dried for human consumption.

23½ in
60 cm

VERMIFORM SEA CUCUMBER
Synapta maculata
F: Synaptidae

Typical of its family, this Indo-Pacific species is a soft-bodied, wormlike sea cucumber that burrows in soft sediment. It lacks tube feet.

spikes provide grip on ocean floor

14–16 in
35–40 cm

GIANT CALIFORNIA SEA CUCUMBER
Stichopus californicus
F: Stichopodidae

One of the largest sea cucumbers of North America's Pacific coast, this fleshy spiked species is fished as a local delicacy.

PRICKLY REDFISH
Thelenota ananas
F: Stichopodidae

This echinoderm belongs to a family of sea cucumbers covered in fleshy spikes. It occurs on sandy seabeds around Indo-Pacific coral reefs.

23½–30 in
60–75 cm

12 in
30 cm

DEEP SEA CUCUMBER
Kolga hyalina
F: Elpidiidae

Virtually worldwide in distribution at depths of up to 5,000 ft (1,500 m), this is one of many little-known sea cucumbers found on ocean floors.

BRITTLE STARS

These animals are named for the fact that their long, thin, sometimes branching, arms easily break off. There is one opening to the gut in the central disk—a downward-facing mouth. Some Ophiuroidea use their arms to trap food particles; others are predatory.

8–12 in
20–30 cm

SHORT-SPINED BRITTLE STAR
Ophioderma sp.
F: Ophiodermatidae
A brittle star of western Atlantic coastal waters, this species lives in submerged grass beds, where it preys on other invertebrates, such as shrimp.

long, flexible arms

COMMON BRITTLE STAR
Ophiothrix fragilis
F: Ophiotrichidae
This northeast Atlantic brittle star—often found in dense populations—has spiny arms, some of which it holds up to trap food particles.

4¾–6 in
12–15 cm

8–12 in
20–30 cm

BLACK BRITTLE STAR
Ophiocoma nigra
F: Ophiocomidae
This large brittle star of European waters can either filter feed on particles or scavenge on detritus. It can be common on rocky coasts with strong currents.

8–10 in
20–25 cm

GORGON'S HEAD BRITTLE STAR
Gorgonocephalus caputmedusae
F: Gorgonocephalidae
This species, named for its coiled, snakelike branching arms, is common around European coasts. Larger animals are found in places of strong current, where they can collect more food.

STARFISH

Like most other echinoderms, starfish crawl on the seabed using their rows of tube feet. These run along grooves in the starfish's arms. Most species of Asteroidea are five-armed, but a few have more than five arms, and some are virtually spherical in shape. Many prey on other slow-moving invertebrates; others scavenge or feed on detritus. Although their skins are studded with hard ossicles, starfish are sufficiently flexible to catch prey.

14–16 in
35–40 cm

PURPLE SUNSTAR
Solaster endeca
F: Solasteridae
Sunstars are large spiny starfish with many arms. Living on offshore mud in the cold waters of the northern hemisphere, this species has from 7 to 13 arms.

20–24 in
50–60 cm

SEVEN ARM STARFISH
Luidia ciliaris
F: Luidiidae
Unlike most other starfish, this large Atlantic species has seven arms. It has long tube feet and is fast moving in its pursuit of other echinoderms.

8–9½ in
20–24 cm

MOSAIC STARFISH
Plectaster decanus
F: Echinasteridae
Like a number of other members of its family, this striking species of southwest Pacific rocky coasts varies greatly in color.

WARTY STARFISH
Echinaster callosus
F: Echinasteridae
Belonging to a family of stiff-bodied starfishes with conical arms, this western Pacific species is unusual for its pink and white warts.

8–10 in
20–25 cm

tube feet on underside of arms

broad base to arms

coarse, spiny upper surface

32–39 in
80–100 cm

GIANT SUNFLOWER STARFISH
Pycnopodia helianthoides
F: Pycnopodiidae
One of the world's largest starfish, this many-armed species preys on mollusks and other echinoderms. It lives among offshore seaweed on the northeast Pacific coast.

4¾–5 in
10–12 cm

4–10 in
10–25 cm

16–20 in
40–50 cm

4–4¾ in
10–12 cm

flattened body

BLOOD HENRY STARFISH
Henricia oculata
F: Echinasteridae
This relative of the warty starfish lives in tidal pools and kelp forests of the northeast Atlantic. Its body secretes mucus to trap food particles.

OCHER STARFISH
Pisaster ochraceus
F: Asteriidae
This North American Pacific relative of the common Atlantic starfish—like others of its family—hunts invertebrates. It is an important predator of mussels.

COMMON STARFISH
Asterias rubens
F: Asteriidae
This common northeast Atlantic predator of other invertebrates is unusual among echinoderms in tolerating estuarine conditions. It often occurs in huge local populations.

ICON STARFISH
Iconaster longimanus
F: Goniasteridae
Many echinoderms develop from planktonic larvae, but this Indo-Pacific starfish lacks a larval stage. It produces large yolky eggs.

venomous spines

CROWN OF THORNS STARFISH
Acanthaster planci
F: Acanthasteridae

This giant Indo-Pacific starfish is a predator of coral polyps, threatening some reef ecosystems. It has 10–20 arms and sharp, mildly venomous spines that can cause injury to humans.

20–24 in / 50–60 cm

8–12 in
20–30 cm

BLUE STARFISH
Linckia laevigata
F: Ophidiasteridae

Snake-armed starfish have smooth, spineless surfaces. This Indo-Pacific species is usually blue, but a minority are purple or orange. It frequently lives in association with detritus-eating shrimp.

8–10 in
20–25 cm

4–4¾ in
10–12 cm

PURPLE STARFISH
Ophidiaster ophidianus
F: Ophidiasteridae

This snake-armed starfish occurs in warmer waters of the northeast Atlantic, on coarse seabeds around the Mediterranean and as far as West Africa.

NECKLACE STARFISH
Fromia monilis
F: Ophidiasteridae

Like many other members of the family, this snake-armed starfish is brightly colored to deter predators. It is an Indo-Pacific species, occurring as far west as the Red Sea.

10–12 in
25–30 cm

2¼–2¾ in
6–7 cm

KNOBBLY STARFISH
Pentaceraster cumingi
F: Oreasteridae

This large strikingly patterned cushion star occurs on sandy and rubble-strewn sea bottoms of tropical central and eastern Pacific.

CUMING'S PLATED STARFISH
Neoferdina cumingi
F: Ophidiasteridae

All snake-armed starfish have rows of granular plates on their upper surfaces, but these are especially conspicuous in plated sea stars, such as this Pacific species.

rows of red tubercles

10–12 in
25–30 cm

8–10½ in
20–27 cm

8–10 in
20–25 cm

RED GENERAL STARFISH
Protoreaster lincki
F: Oreasteridae

From the cushion star family, this Indian Ocean species grazes on algae as a juvenile, but preys on other invertebrates when adult.

GRANULATED STARFISH
Choriaster granulatus
F: Oreasteridae

A giant Indo-Pacific echinoderm of rubble slopes and coral reefs, this cushion star grazes on algae and detritus in shallow waters.

INDO-PACIFIC CUSHION STAR
Culcita novaeguineae
F: Oreasteridae

This coral-feeder of the Indo-Pacific belongs to a family of starfish that become shorter-armed as they get bigger, eventually maturing into dumpy cushion shapes.

RED CUSHION STAR
Porania pulvillus
F: Poraniidae

This distinctive smooth-topped starfish with short arms and fringe of spines occurs on European rocky coasts, often on kelp holdfasts.

6–8 in / 15–20 cm

groups of stiff spines on short, blunt arms

1½–2 in
4–5 cm

GOOSEFOOT STARFISH
Anseropoda placenta
F: Asterinidae

A thin flat starfish with ill-defined arms, the goosefoot starfish of the eastern Atlantic is a predator of bottom-dwelling crustaceans.

SMALL CUSHION STAR
Asterina gibbosa
F: Asterinidae

This short-armed variable-colored starfish, from the same family as the goosefoot, scavenges dead material on the rocky coasts of the northeast Atlantic.

4–4¾ in
10–12 cm

CHORDATES

Less than three percent of animal species are chordates, yet they include the largest, fastest, and most intelligent animals alive today. Most chordates have skeletons made of bone, but their defining feature is a rodlike structure called a notochord—the evolutionary forerunner of the spine.

The earliest known chordates were small, streamlined animals, with bodies just a few inches long. Living over 550 million years ago, they had no hard parts, except for a stiff but flexible cartilaginous notochord. The notochord ran the length of their bodies, creating a framework for their muscles to pull against. Today's chordates have all inherited this feature, and a small number keep it throughout life. However, in the vast majority of chordates—from fish and amphibians to reptiles, birds, and mammals—the notochord is present only in the early embryo. As the embryo develops, the notochord disappears, replaced by an internal skeleton made of cartilage or bone. These animals are known as vertebrates, from the column of vertebrae that makes up their spines.

Unlike shells or body cases, bony skeletons work on an amazing spectrum of scales. The smallest vertebrate, a freshwater fish called *Paedocypris progenetica*, is less than ⅜ in (1 cm) long and weighs several billion

times less than a blue whale, which is the largest chordate, and the biggest animal that has ever lived.

LIFESTYLES

Some of the simplest chordates, such as sea squirts, do not have skeletons, and spend their adult lives fastened in one place. But they are an exception. Chordates generally—and vertebrates in particular—are often fast-moving animals, with rapid reactions, well-developed nervous systems, and sizable brains. Birds and mammals use energy from food to keep their bodies at a constant optimum temperature.

Chordates vary in breeding techniques and in the way they raise their young. Apart from mammals, most vertebrates spawn or lay eggs, although species bearing live young exist in almost every vertebrate group apart from birds. The number of offspring produced is directly related to the amount of parental care. Some fish produce millions of eggs, and play no part in raising their young, whereas mammals and birds have much smaller families.

TUNICATES
The subphylum Tunicata contains about 3,000 species. Larval tunicates have a notochord, and resemble tadpoles. The adults are filter feeders, the most common type being sea squirts.

AMPHIBIANS
Numbering over 6,000 species, amphibians include frogs and toads, salamanders, newts, and caecilians. Most start their lives as aquatic larvae, but as adults they take up life on land.

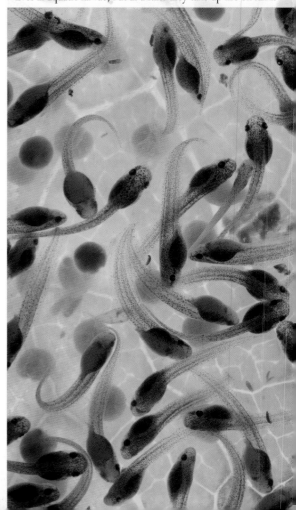

CHORDATE TREE

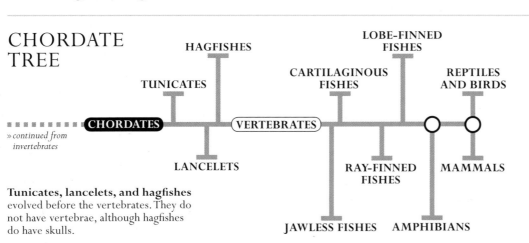

» *continued from invertebrates*

HAGFISHES

TUNICATES

LOBE-FINNED FISHES

CARTILAGINOUS FISHES

REPTILES AND BIRDS

CHORDATES — VERTEBRATES

LANCELETS

RAY-FINNED FISHES

MAMMALS

JAWLESS FISHES   AMPHIBIANS

Tunicates, lancelets, and hagfishes evolved before the vertebrates. They do not have vertebrae, although hagfishes do have skulls.

LANCELETS
Small marine animals with slender bodies and a notochord throughout life, lancelets live half buried in the seabed. There are 20 species in the subphylum Cephalochordata.

HAGFISHES
These eel-like, seabed scavengers belong to the subphylum Craniata, which includes all animals with skulls. Unlike other craniates, the 60 species of hagfish lack a backbone.

FISH
Fish were the earliest animals to evolve vertebrae. They are classified in several different classes, reflecting their varied evolutionary past. Over 30,000 living species are known.

REPTILES
With nearly 8,000 species, reptiles inhabit every continent on Earth, with the exception of Antarctica. Unlike birds and mammals, they are cold-blooded, and covered in scales.

BIRDS
Birds are the only living animals that have feathers. There are around 10,000 species of birds. All of these lay eggs, and many show highly developed parental care.

MAMMALS
Readily identified by their fur, mammals are the only chordates that raise their young on milk. There are about 5,400 known mammal species.

FISH

Fish are the most diverse chordates and can be found in all types of watery habitat—from tiny freshwater pools to the deep ocean. Almost without exception, they breathe using feathery gills to absorb oxygen from the water, and almost all swim using fins.

PHYLUM	CHORDATA
CLASSES	PETROMYZONTIDA
	CHONDRICHTHYES
	ACTINOPTERYGII
	SARCOPTERYGII
ORDERS	63
FAMILIES	538
SPECIES	31,254

Overlapping scales cover the body of a bicolor parrotfish, forming a flexible layer that protects against injury.

Predatory sharks lunge into a huge school of sardines, which swim ever closer together for protection.

The huge mouth of a male gold-specs jawfish provides an unlikely nest for its eggs. He will not feed during incubation.

Fish are not a natural, single group but are in fact four classes of vertebrate chordates, of which the familiar ray-finned (bony) fishes are by far the most populous. The majority of fishes are cold-blooded so their body temperature matches that of the surrounding water. A few top predators, such as white sharks, can maintain a flow of warm blood to the brain, eyes, and main muscles, which allows them to hunt actively even in very cold water. Most fishes are protected by scales or bony plates embedded in the skin. In fast swimmers these are lightweight, and provide streamlining as well as protection from abrasion and disease. Although a few fishes slither or shuffle over the seabed, most swim using fins. With the notable exception of rays, the tail fin usually provides the main propulsive force. Paired pectoral and pelvic fins (on the sides and underside respectively) provide stability and maneuverability, helped by up to three dorsal fins on the fish's upper side and one or two anal fins. Fishes avoid collisions—especially when in large schools—by using special sensory organs that detect the vibrations made as they or other animals move through the water. Most fishes have a row of these organs running along both sides of the body, called the lateral line. While they also use the senses of hearing, touch, sight, taste, and smell found in other chordates, only fishes have a lateral line system.

REPRODUCTIVE STRATEGIES

The four classes of fish have very different ways of reproducing. Most ray-finned fishes, and some lobe-finned fishes, have external fertilization and shed masses of eggs and sperm directly into the water to compensate for the huge numbers that are eaten or die before they can develop into juvenile fishes. In contrast, cartilaginous fishes (sharks, rays, and chimaeras) have internal fertilization and produce eggs or young at an advanced developmental stage. This requires a high input of energy—only a few young are produced at one time, but these have a good chance of survival. The eggs of lamprey (jawless fishes) hatch into larvae that live and feed for many months before undergoing metamorphosis into the adult form.

BRIGHT SPECTACLE >
The fabulous colors of the Banggai cardinalfish, seen here sheltering among giant anemones, attract aquarium owners.

JAWLESS FISHES

Unlike all other vertebrates, jawless fishes have no biting jaws, although they do have teeth. This ancient group has only a few living representatives.

Instead of jaws, lampreys have a circular sucker disk at the end of the snout with concentric rows of rasping teeth. A row of seven small, circular gill openings runs along each side of the body just behind the eyes. Their skin is smooth with no scales and they have either one or two separate dorsal fins on the back near the tail. Lampreys have a cranium—a skull without jaws—but it is made of cartilage. They lack a bony skeleton and have only a partially formed vertebral column. A flexible rod —the notochord—supports the body.

DIFFERING LIFECYCLES

Lampreys live in temperate coastal waters and freshwaters worldwide. All species breed in freshwater. The coastal species are anadromous fish—that is, like salmon, they swim up rivers into freshwater to spawn, after which they die. The eggs hatch into wormlike, burrowing larvae, known as ammocoetes. These live in mud and feed on detritus. After about three years the larvae of the anadromous species change into adults and swim out to sea, where they feed for several years. The freshwater species remain in rivers and lakes.

UNWELCOME PARASITES

Lampreys are notorious for feeding parasitically on larger fish when they are adult. Clinging on with its sucker mouth, a lamprey grinds its way through its victim's skin, eating both flesh and blood. They can be a nuisance to fishermen because they damage and kill fish in nets and fish farms. However, most lampreys are not totally reliant on sucking blood but also eat invertebrates. In freshwater there are some completely nonparasitic species. These reproduce up to six months after undergoing metamorphosis from the larval to the adult form but the adults do not feed. The sucker mouth is also useful when the fish are swimming upstream against the current—they use it to cling onto rocks so they can have a rest.

Three fossil lamprey species dating from the late Carboniferous period have been found in the USA, and one of these appears to be very similar to living lampreys.

PHYLUM	CHORDATA
CLASS	PETROMYZONTIDA
ORDERS	1
FAMILIES	3
SPECIES	38

DEBATE

ARE LAMPREYS AND HAGFISHES RELATED?

Soft mud in the ocean depths is home to slimy scavengers called hagfishes. These too are jawless, eel-like fishes with no bony skeleton. But unlike lampreys, they have an underslung, slitlike mouth, only vestigial eyes, and eggs develop directly into miniature adults. Traditionally hagfishes and lampreys were classed together as Cyclostomata, separate from jawed vertebrates; this was the "cyclostome hypothesis." But detailed morphological comparisons led lampreys to be allied with the jawed vertebrates. The latest molecular studies, however, are supporting the cyclostome hypothesis.

LAMPREYS

Jawless fishes in the order Petromyzontiformes have a suckerlike mouth armed with concentric arcs of small, sharp teeth. The gills open through a row of seven circular openings on each side behind the eyes. All lampreys spawn in temperate rivers and streams, but some live in coastal waters.

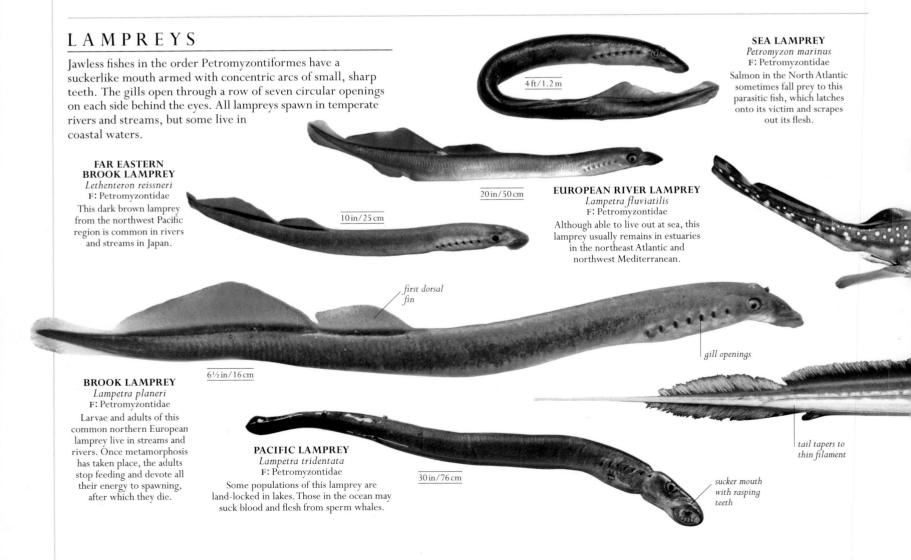

SEA LAMPREY
Petromyzon marinus
F: Petromyzontidae
Salmon in the North Atlantic sometimes fall prey to this parasitic fish, which latches onto its victim and scrapes out its flesh.

4 ft/1.2 m

FAR EASTERN BROOK LAMPREY
Lethenteron reissneri
F: Petromyzontidae
This dark brown lamprey from the northwest Pacific region is common in rivers and streams in Japan.

20 in/50 cm

10 in/25 cm

EUROPEAN RIVER LAMPREY
Lampetra fluviatilis
F: Petromyzontidae
Although able to live out at sea, this lamprey usually remains in estuaries in the northeast Atlantic and northwest Mediterranean.

first dorsal fin

gill openings

BROOK LAMPREY
Lampetra planeri
F: Petromyzontidae
Larvae and adults of this common northern European lamprey live in streams and rivers. Once metamorphosis has taken place, the adults stop feeding and devote all their energy to spawning, after which they die.

6½ in/16 cm

PACIFIC LAMPREY
Lampetra tridentata
F: Petromyzontidae
Some populations of this lamprey are land-locked in lakes. Those in the ocean may suck blood and flesh from sperm whales.

30 in/76 cm

tail tapers to thin filament

sucker mouth with rasping teeth

CARTILAGINOUS FISHES

These fishes have a skeleton of pliable cartilage rather than the hard bone found in most other vertebrate groups. Most are predators with acute senses.

Sharks, rays, and chimaeras are all cartilaginous fishes although chimaeras have distinct anatomical differences from the other two groups. In chimaeras, the upper jaw is fused to the braincase and cannot be moved independently. They also have teeth that grow continuously. In contrast, sharks and rays regularly lose their hard enamel-covered teeth but replace them from extra rows lying flat behind the active teeth—a characteristic that helps make sharks some of the most formidable predators on Earth. The skin of all three groups—sharks, rays, and chimaeras—is protected by toothlike scales called dermal denticles.

SEARCHING FOR PREY

The ocean is home to most cartilaginous fishes and while most rays live on the seabed, the larger predatory shark species roam in open water. The bullshark and over 100 other species can enter estuaries and swim up rivers and a few live entirely in rivers. Open-water cartilaginous fishes are continually on the move because, unlike bony fish, they do not have a gas-filled swim bladder to maintain neutral buoyancy and therefore sink if they stop swimming. The whale shark and other surface-living species have a large oily liver to prevent this. Predatory sharks are well known for their amazing ability to smell blood and home in on wounded fish and mammals. Cartilaginous fishes also have the ability to detect the weak electrical fields that surround living creatures and while not unique to them, this sense is developed to an extraordinary extent in this group.

REPRODUCTION

All cartilaginous fishes mate and have internal fertilization. Chimaeras and some sharks and rays lay eggs, each protected by a tough egg capsule and commonly known as a "mermaid's purse." However, most sharks and skates give birth to well-developed live young that are nourished in the uterus by egg yolk or by a placental connection to the mother. Unlike mammals, the young are independent from birth and neither parent takes any interest in their offspring.

PHYLUM	CHORDATA
CLASS	CHONDRICHTHYES
ORDERS	12
FAMILIES	51
SPECIES	1,171

A white shark, *Carcharodon carcharias*, shows the continually developing rows of teeth that make it such a fearsome predator.

CHIMAERAS

Also known as rabbitfishes because of their fused platelike teeth, the Chimaeriformes are a small order of cartilaginous fishes with about 34 species. A strong, venomous spine in front of the first of two dorsal fins provides protection in their deepwater habitat.

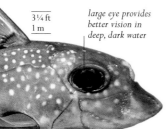

3¼ ft
1 m

large eye provides better vision in deep, dark water

SPOTTED RAT FISH
Hydrolagus colliei
F: Chimaeridae
Like most chimaeras, this northeast Pacific species uses its large pectoral fins to glide and flap along while searching for food.

4¼ ft
1.3 m

PACIFIC SPOOKFISH
Rhinochimaera pacifica
F: Rhinochimaeridae
The long, conical snout of this chimaera is covered in sensory pores that detect the electrical fields of its prey.

4¼ ft
1.3 m

ELEPHANT FISH CHIMAERA
Callorhinchus milii
F: Callorhinchidae
Using its long fleshy snout as a plow, this chimaera unearths shellfish from muddy seabeds in the southwest Pacific and Australia.

5 ft/1.5 m

pectoral fins flap for swimming

RAT FISH
Chimaera monstrosa
F: Chimaeridae
Typically living below 1,000 ft (300 m) in the Mediterranean and eastern Atlantic, rat fishes swim in small groups and search for seabed invertebrates.

SIX- AND SEVEN-GILL SHARKS

While most sharks have five pairs of gill slits, those of the order Hexanchiformes have either six or seven pairs. Also known as cow and frill sharks, they have a single dorsal fin on the back near the tail. The six known species live mainly in deep water.

18 ft
5.5 m

BLUNTNOSE SIXGILL SHARK
Hexanchus griseus
F: Hexanchidae
Rocky seamounts around the world are the haunt of this huge, green-eyed shark that weighs up to 1,325 lb (600 kg).

6½ ft
2 m

FRILLED SHARK
Chlamydoselachus anguineus
F: Chlamydoselachidae
Scattered records indicate this shark has a worldwide distribution. It has brilliant white teeth that may attract its prey of fish and squid.

DOGFISH SHARKS AND RELATIVES

A large and varied order, the Squaliformes contains at least 130 species and includes gulper, lantern, sleeper, rough, and kitefin sharks. These all have two dorsal (back) fins and no anal fin. All species so far studied bear live young.

5 ft
1.5 m

PIKED DOGFISH
Squalus acanthias
F: Squalidae

Occurring worldwide in temperate waters this once abundant shark is now endangered. Living to an age of 100, it grows and breeds very slowly.

LARGETOOTH COOKIECUTTER SHARK
Isistius plutodus
F: Dalatiidae

In the western Atlantic and northwest Pacific, dolphins and large fish suffer attacks from this shark, which bites and twists out plugs of flesh, hence its name.

22 in
56 cm

18 in
45 cm

VELVET BELLY LANTERN SHARK
Etmopterus spinax
F: Etmopteridae

The lantern shark lives in the deep eastern Atlantic. It has tiny light organs on its belly, which help in finding a mate.

8–14 ft
2.4–4.3 m

GREENLAND SHARK
Somniosus microcephalus
F: Somniosidae

One of only a few sharks that live in Arctic waters, the Greenland shark often scavenges for drowned land animals.

rough skin

spines in sail-
like dorsal fins

5 ft
1.5 m

ANGULAR ROUGHSHARK
Oxynotus centrina
F: Oxynotidae

This shark has two sail-like dorsal fins and rough skin. It lives in the eastern Atlantic at depths of at least 330 ft (100 m).

CARPET SHARKS

The 33 or so sharks of the Orectolobiformes have two dorsal fins, one anal fin, and sensory barbels hanging down from the nostrils. With the exception of the whale shark, they live quietly on the seabed, feeding on fish and invertebrates.

white spots

mouth at
end of snout

WHALE SHARK
Rhincodon typus
F: Rhincodontidae

This, the largest known fish, cruises tropical oceans feeding on plankton and tiny fish. Each shark has its own unique pattern of spots.

40–65 ft
12–20 m

3½ ft / 1.1 m

EPAULETTE CATSHARK
Hemiscyllium ocellatum
F: Hemiscylliidae

The boldly marked epaulette shark clambers among coral on its fins. This small shark has a long tail and lives in the South Pacific.

TASSELLED WOBBEGONG
Eucrossorhinus dasypogon
F: Orectolobidae

With its fringe of skin tassels, flattened body, and camouflage pattern, this southwest Pacific coral reef resident is hard to spot.

4 ft
1.2 m

"beard" of
branching tassels

NURSE SHARK
Ginglymostoma cirratum
F: Ginglymostomatidae

By day this shark lies hidden in rock crevices, emerging at night to hunt in the warm coastal waters of the Atlantic and eastern Pacific.

9¾ ft / 3 m

SAWSHARKS

Sawsharks have a flat head with gills on the sides and a rostrum, a long snout edged with teeth like a saw. Two long sensory barbels hang from the rostrum and help find buried food. Most of the nine species of Pristiophoriformes live in the tropics.

LONGNOSE SAWSHARK
Pristiophorus cirratus
F: Pristiophoridae
Sandy seabeds off southern Australia are home to this sawshark. It uses its rostrum to slash into shoals of fish.

4½ ft
1.4 m

BULLHEAD SHARKS

The Heterodontiformes are small bottom-living sharks with paddlelike pectoral fins. They have a blunt, sloping head, crushing teeth, and two dorsal (back) fins each preceded by a sharp spine.

5½ ft
1.7 m

spine

PORT JACKSON SHARK
Heterodontus portusjacksoni
F: Heterodontidae

Using its paddlelike front fins, this shark crawls over the seabed off southern Australia in search of sea urchins to eat.

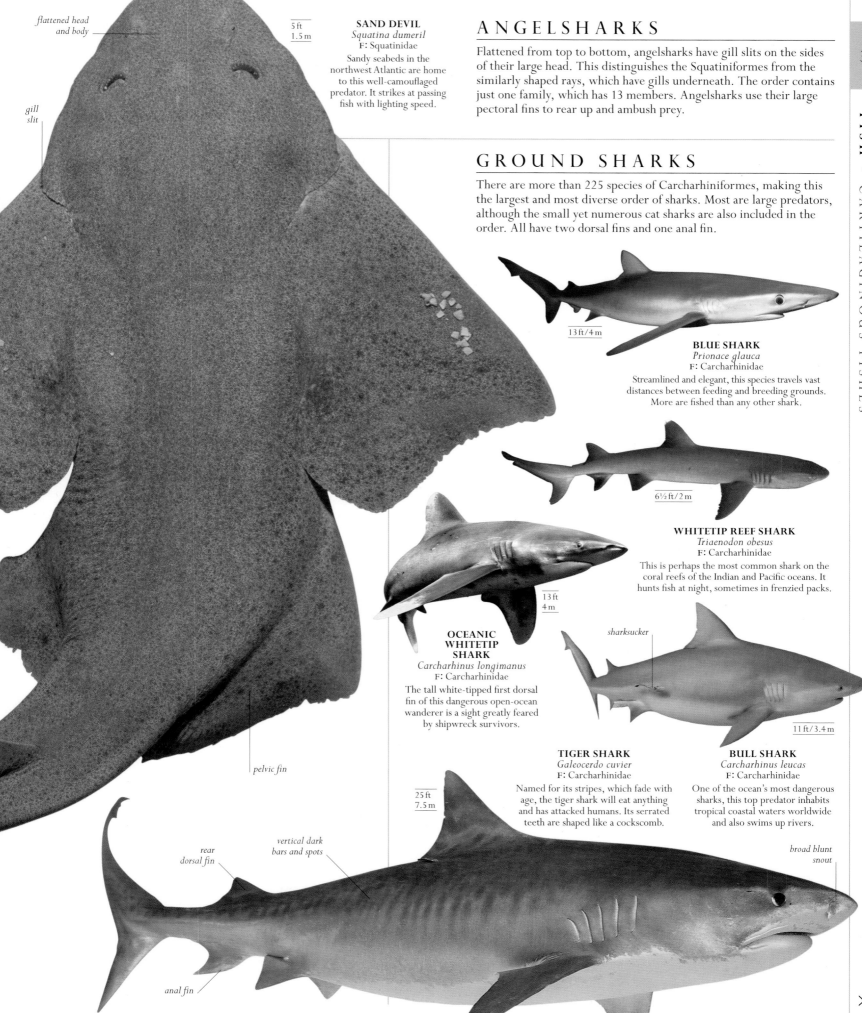

flattened head
and body

SAND DEVIL
Squatina dumeril
F: Squatinidae
Sandy seabeds in the
northwest Atlantic are home
to this well-camouflaged
predator. It strikes at passing
fish with lighting speed.

5 ft
1.5 m

gill
slit

ANGELSHARKS

Flattened from top to bottom, angelsharks have gill slits on the sides
of their large head. This distinguishes the Squatiniformes from the
similarly shaped rays, which have gills underneath. The order contains
just one family, which has 13 members. Angelsharks use their large
pectoral fins to rear up and ambush prey.

GROUND SHARKS

There are more than 225 species of Carcharhiniformes, making this
the largest and most diverse order of sharks. Most are large predators,
although the small yet numerous cat sharks are also included in the
order. All have two dorsal fins and one anal fin.

13 ft / 4 m

BLUE SHARK
Prionace glauca
F: Carcharhinidae
Streamlined and elegant, this species travels vast
distances between feeding and breeding grounds.
More are fished than any other shark.

6½ ft / 2 m

WHITETIP REEF SHARK
Triaenodon obesus
F: Carcharhinidae
This is perhaps the most common shark on the
coral reefs of the Indian and Pacific oceans. It
hunts fish at night, sometimes in frenzied packs.

13 ft
4 m

**OCEANIC
WHITETIP
SHARK**
Carcharhinus longimanus
F: Carcharhinidae
The tall white-tipped first dorsal
fin of this dangerous open-ocean
wanderer is a sight greatly feared
by shipwreck survivors.

sharksucker

11 ft / 3.4 m

TIGER SHARK
Galeocerdo cuvier
F: Carcharhinidae
Named for its stripes, which fade with
age, the tiger shark will eat anything
and has attacked humans. Its serrated
teeth are shaped like a cockscomb.

BULL SHARK
Carcharhinus leucas
F: Carcharhinidae
One of the ocean's most dangerous
sharks, this top predator inhabits
tropical coastal waters worldwide
and also swims up rivers.

25 ft
7.5 m

pelvic fin

rear
dorsal fin

vertical dark
bars and spots

broad blunt
snout

anal fin

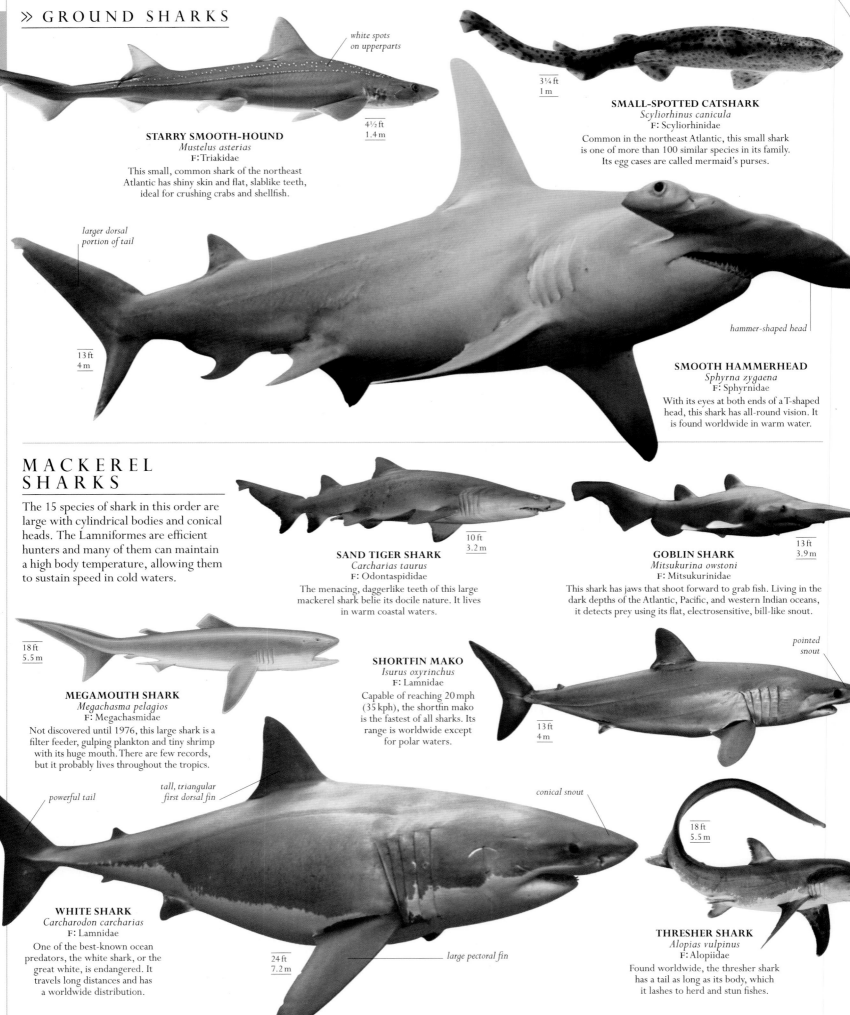

» GROUND SHARKS

white spots on upperparts

4½ ft
1.4 m

STARRY SMOOTH-HOUND
Mustelus asterias
F: Triakidae
This small, common shark of the northeast
Atlantic has shiny skin and flat, slablike teeth,
ideal for crushing crabs and shellfish.

3¼ ft
1 m

SMALL-SPOTTED CATSHARK
Scyliorhinus canicula
F: Scyliorhinidae
Common in the northeast Atlantic, this small shark
is one of more than 100 similar species in its family.
Its egg cases are called mermaid's purses.

larger dorsal
portion of tail

13 ft
4 m

hammer-shaped head

SMOOTH HAMMERHEAD
Sphyrna zygaena
F: Sphyrnidae
With its eyes at both ends of a T-shaped
head, this shark has all-round vision. It
is found worldwide in warm water.

MACKEREL SHARKS

The 15 species of shark in this order are
large with cylindrical bodies and conical
heads. The Lamniformes are efficient
hunters and many of them can maintain
a high body temperature, allowing them
to sustain speed in cold waters.

10 ft
3.2 m

SAND TIGER SHARK
Carcharias taurus
F: Odontaspididae
The menacing, daggerlike teeth of this large
mackerel shark belie its docile nature. It lives
in warm coastal waters.

13 ft
3.9 m

GOBLIN SHARK
Mitsukurina owstoni
F: Mitsukurinidae
This shark has jaws that shoot forward to grab fish. Living in the
dark depths of the Atlantic, Pacific, and western Indian oceans,
it detects prey using its flat, electrosensitive, bill-like snout.

18 ft
5.5 m

MEGAMOUTH SHARK
Megachasma pelagios
F: Megachasmidae
Not discovered until 1976, this large shark is a
filter feeder, gulping plankton and tiny shrimp
with its huge mouth. There are few records,
but it probably lives throughout the tropics.

SHORTFIN MAKO
Isurus oxyrinchus
F: Lamnidae
Capable of reaching 20 mph
(35 kph), the shortfin mako
is the fastest of all sharks. Its
range is worldwide except
for polar waters.

pointed
snout

13 ft
4 m

powerful tail

tall, triangular
first dorsal fin

conical snout

18 ft
5.5 m

WHITE SHARK
Carcharodon carcharias
F: Lamnidae
One of the best-known ocean
predators, the white shark, or the
great white, is endangered. It
travels long distances and has
a worldwide distribution.

24 ft
7.2 m

large pectoral fin

THRESHER SHARK
Alopias vulpinus
F: Alopiidae
Found worldwide, the thresher shark
has a tail as long as its body, which
it lashes to herd and stun fishes.

SKATES AND RAYS

In this order of fishes, a flat, disk-shaped body and winglike pectoral fins joined to the head are adaptations for bottom-living. However, some members of the order Rajiformes are free-swimming. A long, thin tail helps them keep their balance when swimming. The tail is sometimes armed with a sting.

tail as long as body

brownish upperparts

pectoral fin

4½ ft
1.4 m

35 in
90 cm

35 in
90 cm

9½ ft
2.9 m

COMMON STINGRAY
Dasyatis pastinaca
F: Dasyatidae
While most stingrays live in the tropics, the range of this fish is from the Mediterranean into northern Europe, where it lives inshore on sediment.

BLUE-SPOTTED RIBBONTAIL RAY
Taeniura lymma
F: Dasyatidae
Found on most coral reefs in the Indian Ocean and the western Pacific, this stingray has a venomous spine on its tail.

THORNBACK RAY
Raja clavata
F: Rajidae
This European ray is protected from predators by the row of distinctive curved spines with wide bases that runs down its back.

BLUE SKATE
Dipturus batis
F: Rajidae
The largest European ray, the blue skate preys upon midwater fish and seabed invertebrates. It is distinguished by its long, pointed snout and blue underparts.

13 ft
4 m

huge, pointed "wings"

20 in
50 cm

23 in
58 cm

11 ft
3.3 m

SPINY BUTTERFLY RAY
Gymnura altavela
F: Gymnuridae
This ray glides over the sea bed in the warmer parts of the Atlantic, searching for invertebrates and fishes.

OCELLATE RIVER STINGRAY
Potamotrygon motoro
F: Potamotrygonidae
Unlike most rays, this South American stingray lives in freshwater. It inhabits rivers and is a danger to fishermen, as its tail delivers a painful sting.

HALLER'S ROUND RAY
Urobatis halleri
F: Urotrygonidae
At home on shallow, sandy flats, the venomous spine on this stingray poses a threat to bathers. It is found in coastal waters from California to Panama.

SPOTTED EAGLE RAY
Aetobatus narinari
F: Myliobatidae
This ray is found in tropical waters worldwide. It swims by beating its pectoral fins like a bird's wings.

GIANT MANTA RAY
Manta birostris
F: Myliobatidae
A filter feeder, this huge, tropical ray funnels plankton into its mouth, using special mouth flaps called cephalic horns.

30 ft
9 m

30 in
75 cm

ATLANTIC GUITARFISH
Rhinobatos lentiginosus
F: Rhinobatidae
Unusual for rays, this species has small pectoral fins and therefore swims using its tail. It uses its snout to dig for mollusks and crabs.

pointed snout with rounded tip

cephalic horns to funnel plankton

SAWFISHES

These large modified rays have a tough, bladelike snout, armed with regular-sized teeth on both edges. Superficially, the Pristiformes resemble sawsharks, but like rays, they have gills on their underside.

SMALLTOOTH SAWFISH
Pristis pectinata
F: Pristidae
Born live with a flexible saw that helps prevent injury to its mother, this rare fish is found worldwide in warm coastal waters and estuaries.

25 ft
7.6 m

ELECTRIC RAYS

The wings of the Torpediniformes have special organs that produce enough electricity to stun prey and discourage predators. These rays have a circular disk-shaped body and a thick tail that ends in a fanlike tip.

MARBLED ELECTRIC RAY
Torpedo marmorata
F: Torpedinidae
When it pounces on a seabed fish, this electric ray can shock its prey with up to 200 volts. Electricity can be produced by even the newborn of this species.

3¼ ft
1 m

BLUE-SPOTTED RIBBONTAIL RAY
Taeniura lymma

Stingrays are notorious for the painful wounds they can inflict with their barbed tail, which can, in exceptional circumstances, be lethal. However, the tropical blue-spotted ribbontail ray, like other stingrays, uses its sting just to defend itself. Much of its time is spent resting motionless on sandy patches among coral, hidden under overhangs. Often only its blue-edged tail gives it away to divers—if disturbed, it will swim away, flapping its two winglike pectoral fins. The best time to see one is on a rising tide when the stingrays swim inshore to feed on invertebrates in shallow water.

SIZE 28–35 in (70–90 cm), including tail
HABITAT Sandy patches in coral reefs
DISTRIBUTION Indian Ocean, W. Pacific
DIET Mollusks, crabs, shrimps, worms

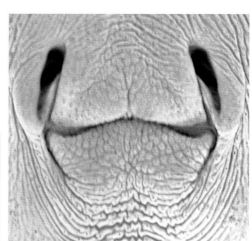

< MOUTH
The mouth is on the underside of the stingray, allowing it to extract mollusks and crabs hidden beneath the surface of the sand. Two plates within the mouth are used to crush the shells of its prey.

PELVIC FINS >
In this female the urogenital opening is visible between the pelvic fins on the underside. After mating, females produce up to seven live young following a few months' to a year's gestation.

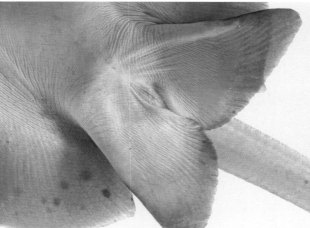

< GILL SLITS
After passing over the gills, water leaves the body through five pairs of gill slits on the underside.

∨ TOP HOLE
Oxygen-rich water is drawn in over the gills through two spiracles on top of the head behind the eyes. Their elevated position helps prevent sand from getting in.

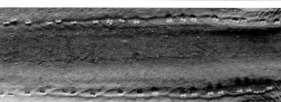

< BACK SPINES
This species has relatively smooth skin, but has two parallel rows of tiny spines running down its back, as well as other scattered spines.

spiracle behind eye

∧ ARMED TAIL
The tail is armed with one or two sharp, barbed spines that cause physical wounds and inject venom if the ray is attacked or stepped on.

barbed tail

pectoral fin

∧ HIDDEN SPOTS
Unlike most of its relatives,
the blue-spotted ribbontail
ray rarely buries itself in
sand, relying instead on its
camouflage colors. Although
bright, its blue spots break up
the ray's outline when seen
from above in the shifting
sunlight of a shallow coral reef.

mouth

< THE FLIP SIDE
The mouth, nostrils, and gill slits
are all on the underside. The skin
is plain white with no blue spots
because this side is usually hidden.

tail spine

∨ FRONT VIEW
The eyes are raised up on top of
the head so that the ray can watch
for predators and prey, even when
it is partially buried in sand.

eye

pelvic fin

RAY-FINNED FISHES

These fishes are bony, with a hard, calcified skeleton. Their fins are supported by a fan of jointed rods called rays, made of bone or cartilage.

Ray-finned fish are able to swim with far greater precision than cartilaginous fish. Using their highly mobile and versatile fins, they can execute maneuvers such as hovering, braking, and even swimming backward. The fins themselves can be delicate and flexible or strong and spiny and they often have important secondary uses, such as in defense, display, and camouflage.

With the exception of bottom-living species, most ray-finned fish have a buoyancy aid in the form of a gas-filled swim bladder. This allows them to maintain their position at a certain depth and to rise or fall by adjusting the pressure of the gas.

MYRIAD ADAPTATIONS

The great majority of fish are ray-finned and the group is hugely diverse—ranging from tiny gobies to the gigantic ocean sunfish. Species have evolved to inhabit every conceivable aquatic niche from tropical coral reefs to the waters beneath the ice-shelves of Antarctica and from the depths of the ocean to shallow desert pools. Herbivores, carnivores, and scavengers are all well represented in the group and its members exhibit many ingenious hunting and defense strategies, as well as cooperation between species.

SAFETY IN NUMBERS

Most ray-finned fish shed eggs and sperm into the water and fertilization is external. In some cases, fewer eggs are laid and parental care is provided. For example, jawfish and some cichlids protect their eggs and young in their mouths, while sticklebacks and many wrasses build nests of weed and debris. Some species protect their eggs with such vigor that even scuba divers are warned off.

Most species, however, lay eggs in vast numbers. The millions of floating eggs and fish larvae are an important food source for other aquatic creatures, but those that survive drift and disperse the species. Populations of fish reproducing in this way are less vulnerable to overfishing because numbers can recover when fishing stops, but even stocks of such prolific breeders as Atlantic cod will eventually succumb if intense levels of fishing are continued.

PHYLUM	CHORDATA
CLASS	ACTINOPTERYGII
ORDER	46
FAMILIES	482
SPECIES	30,033

A pair of bluecheek butterflyfishes *(Chaetodon semilarvatus)* swim in unison as they patrol their patch of coral reef.

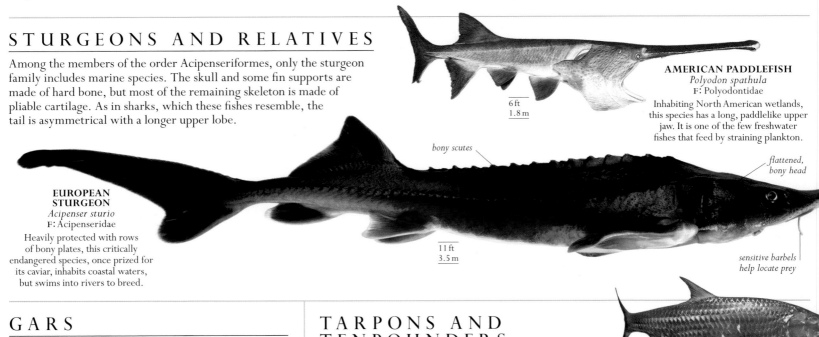

STURGEONS AND RELATIVES

Among the members of the order Acipenseriformes, only the sturgeon family includes marine species. The skull and some fin supports are made of hard bone, but most of the remaining skeleton is made of pliable cartilage. As in sharks, which these fishes resemble, the tail is asymmetrical with a longer upper lobe.

6 ft
1.8 m

AMERICAN PADDLEFISH
Polyodon spathula
F: Polyodontidae
Inhabiting North American wetlands, this species has a long, paddlelike upper jaw. It is one of the few freshwater fishes that feed by straining plankton.

flattened, bony head

bony scutes

EUROPEAN STURGEON
Acipenser sturio
F: Acipenseridae
Heavily protected with rows of bony plates, this critically endangered species, once prized for its caviar, inhabits coastal waters, but swims into rivers to breed.

11 ft
3.5 m

sensitive barbels help locate prey

GARS

Fishes in the order Lepisosteiformes are primitive, freshwater predators in North America, with long, cylindrical bodies, protected by heavy, close-fitting scales. Their long jaws have needlelike teeth.

6 ft
1.8 m

LONGNOSE GAR
Lepisosteus osseus
F: Lepisosteidae
A skilled predator, this long, thin fish hangs motionless in the water, hidden by vegetation, before thrusting forward to capture its prey.

TARPONS AND TENPOUNDERS

The order Elopiformes is small and its members are silvery with only one dorsal fin and a forked tail. They resemble giant herring. They have special throat bones (gular plates). Although these are marine fishes, some species also swim up estuaries and rivers.

LADYFISH
Elops saurus
F: Elopidae
The ladyfish moves in large schools, close to the western Atlantic shores. When alarmed, it skips over the water's surface.

3¼ ft
1 m

8¼ ft
2.5 m

TARPON
Megalops atlanticus
F: Megalopidae
Found along Atlantic coasts, the tarpon sometimes enters rivers. In stagnant water, it gulps air from the surface using its swim bladder as a primitive lung.

BONYTONGUES AND RELATIVES

As their name suggests, the fishes of the order Osteoglossiformes have many sharp teeth on the tongue and roof of the mouth, which help them seize and hold prey. These fishes live in freshwater, mainly in the tropics. The group has many fishes with unusual shapes.

dorsal fin set far back on body

15 ft
4.5 m

gray to green body

ARAPAIMA
Arapaima gigas
F: Arapaimidae
Weighing up to 440 lb (200 kg), this South American species is one of the largest freshwater fish. Its long fins help it lunge forward to catch prey.

ELEPHANTNOSE FISH
Gnathonemus petersii
F: Mormyridae
This African fish finds its way in its murky water habitat by generating weak electrical pulses. Its long lower jaw probes the mud for food.

9 in
23 cm

CLOWN KNIFEFISH
Chitala chitala
F: Notopteridae
This slender, humpbacked fish lives in wetlands of southeast Asia. In stagnant water, it gulps air and absorbs oxygen from its swim bladder.

34 in
87 cm

EELS

The Anguilliformes have long, thin, snakelike bodies with smooth skin, in which the scales are either absent or deeply embedded. The fins are often limited to one long fin runing down the back, around the tail and along the belly. Eels are found in marine and freshwater habitats.

long dorsal fin

5 ft
1.5 m

ZEBRA MORAY
Gymnomuraena zebra
F: Muraenidae
This tropical, boldly striped eel has dense pebblelike teeth to eat tough-shelled crabs, mollusks, and sea urchins.

23½ in
60 cm

large jaws

spotted, camouflaging skin

JEWEL MORAY EEL
Muraena lentiginosa
F: Muraenidae
Inhabiting coral reefs in the eastern Pacific, this eel rhythmically opens and closes its mouth for respiration.

SPOTTED GARDEN EEL
Heteroconger hassi
F: Congridae
Sandy patches near coral reefs house colonies or "gardens" of this eel. With their tails hidden in burrows, these eels sway like plants, retreating if disturbed.

16 in
40 cm

CONGER EEL
Conger conger
F: Congridae
This eel of the North Atlantic and Mediterranean finds an ideal home in shipwrecks. It hides in crevices during the day, emerging only at night to hunt other fishes.

thick body

RIBBON EEL
Rhinomuraena quaesita
F: Muraenidae
Juvenile ribbon eels are black with yellow fins and mature into bright blue males. These later change sex and become yellow females. They are found in the Indian and western Pacific oceans.

4¼ ft
1.3 m

9¾ ft
3 m

smooth skin

BANDED SNAKE EEL
Myrichthys colubrinus
F: Ophichthidae
Closely resembling a venomous sea snake, this harmless eel of the Indian and western Pacific oceans is left alone by predators. It searches in sand burrows for small fishes.

38 in
97 cm

EUROPEAN EEL
Anguilla anguilla
F: Anguillidae
Spending most of its life in freshwater, this endangered, snakelike eel migrates across the Atlantic to the Sargasso Sea to spawn and then die.

4¼ ft
1.3 m

SWALLOWERS AND GULPERS

Living in the deep ocean, these bizarre, eel-like fishes, belonging to the order Saccopharyngiformes, have no tail fin or pelvic fins, and no scales. They do not have ribs and their large jaws are modified for a wide gape. They are believed to spawn once and then die, like true eels.

3¼ ft
1 m

long, whiplike tail

huge, loosely hinged mouth

PELICAN GULPER EEL
Eurypharynx pelecanoides
F: Eurypharyngidae
This deep-sea, eel-like fish has huge jaws and an expandable stomach, which enable it to swallow prey almost as large as itself.

MILKFISH AND RELATIVES

With only two exceptions, including the milkfish itself, the Gonorynchiformes live in freshwater. They have a pair of pelvic fins that are set well back on the belly.

streamlined body

MILKFISH
Chanos chanos
F: Chanidae

This fast swimmer has a large forked tail. Feeding solely on plankton, it is farmed in southeast Asia.

6 ft
1.8 m

pelvic fins

20 in
50 cm

BEAKED SALMON
Gonorynchus greyi
F: Gonorynchidae

Native to Australia and New Zealand, this fish lives near the seabed and dives under sand if threatened.

SARDINES AND RELATIVES

This predominantly marine order includes many commercially important species. The Clupeiformes are silvery, with loose scales, one dorsal fin, a forked tail, and a keel-shaped belly. Most live in large schools and are preyed on by sharks, tuna, and other large fishes.

PERUVIAN ANCHOVETA
Engraulis ringens
F: Engraulidae

This tiny plankton-eater lives in enormous schools along the western coast of South America, where it is a major source of food for humans, pelicans, and larger fishes.

8 in
20 cm

18 in
45 cm

ALLIS SHAD
Alosa alosa
F: Clupeidae

In spring, adults of this species migrate from the sea into European rivers to spawn, sometimes swimming very long distances.

33 in
83 cm

ATLANTIC HERRING
Clupea harengus
F: Clupeidae

This silver-scaled fish lives in large schools, gathering plankton. Distinct local races exist in the northeast Atlantic and the North Sea.

CARP AND RELATIVES

This is one of the largest freshwater fish orders, with over 3,000 species worldwide. The Cypriniformes have the standard "fish shape," with a single dorsal fin. They typically have large scales. The teeth are in the throat instead of the jaws. Many are familiar aquarium fishes, including loaches, minnows, and carp.

red-edged dorsal fin

tigerlike black stripes

12 in
30 cm

deeply forked tail

CLOWN LOACH
Chromobotia macracanthus
F: Cobitidae

Native to wetlands in southeast Asia, this loach is a bottom-feeder. It has sharp spines by its eyes, which it uses to defend itself.

TIGER BARB
Puntius tetrazona
F: Cyprinidae

Native to the Indonesian islands of Sumatra and Borneo, this fish has been widely introduced elsewhere and is bred for the aquarium trade.

2¾ in
7 cm

BITTERLING
Rhodeus amarus
F: Cyprinidae

This European fish lays its eggs inside the mantle cavity of a mussel, where they grow and hatch before the fry swim away.

4¼ in
11 cm

5 ft
1.5 m

GRASS CARP
Ctenopharyngodon idella
F: Cyprinidae

A native of Asia, this species feeds on aquatic plants. For this reason, it has been introduced in Europe and the USA to keep drainage channels clear of weeds.

4 in
10 cm

GOLDFISH
Carassius auratus
F: Cyprinidae

Originally native to central Asia and China, the goldfish has been introduced all over the world and there are now many varieties.

large, silvery scales

4 ft
1.2 m

protrusible mouth

COMMON CARP
Cyprinus carpio
F: Cyprinidae

With its mouth that can be thrust out and sensory barbels, a carp finds food by grubbing through the bottom mud. Now introduced worldwide, it was originally from China and central Europe.

2¼ in
6 cm

ZEBRAFISH
Brachydanio rerio
F: Cyprinidae

This active little fish spawns frequently and is common in ponds and lakes in South Asia. It is bred in aquaria and laboratories.

CHARACINS AND RELATIVES

These freshwater fishes are mostly carnivores, with well-developed teeth. As well as a normal dorsal fin, most also have a small, fatty adipose fin near the tail. Of the 18 families in the order Characiformes, the piranhas are the most notorious predators.

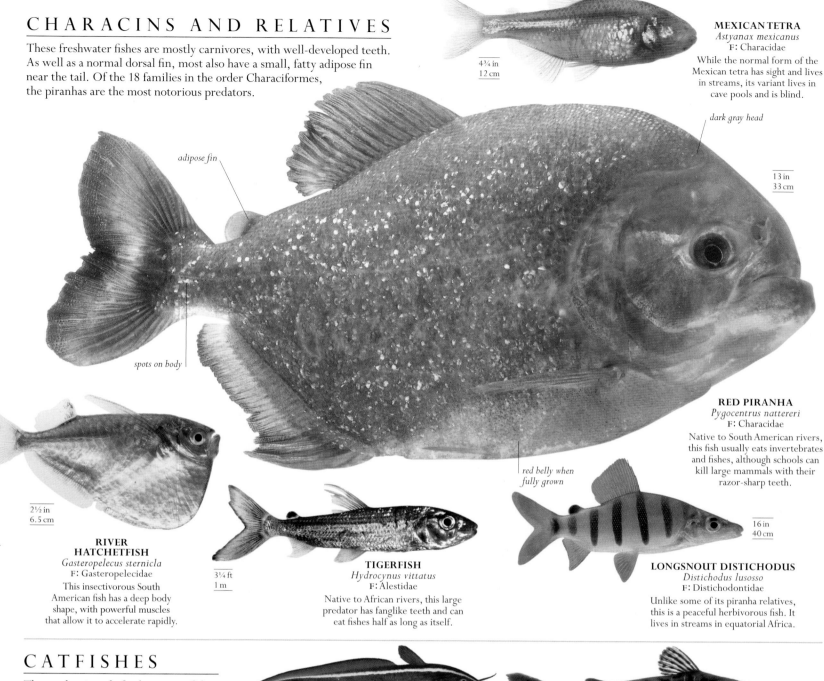

4¾ in
12 cm

MEXICAN TETRA
Astyanax mexicanus
F: Characidae
While the normal form of the Mexican tetra has sight and lives in streams, its variant lives in cave pools and is blind.

dark gray head

13 in
33 cm

adipose fin

spots on body

RED PIRANHA
Pygocentrus nattereri
F: Characidae
Native to South American rivers, this fish usually eats invertebrates and fishes, although schools can kill large mammals with their razor-sharp teeth.

red belly when fully grown

2½ in
6.5 cm

RIVER HATCHETFISH
Gasteropelecus sternicla
F: Gasteropelecidae
This insectivorous South American fish has a deep body shape, with powerful muscles that allow it to accelerate rapidly.

3¼ ft
1 m

TIGERFISH
Hydrocynus vittatus
F: Alestidae
Native to African rivers, this large predator has fanglike teeth and can eat fishes half as long as itself.

16 in
40 cm

LONGSNOUT DISTICHODUS
Distichodus lusosso
F: Distichodontidae
Unlike some of its piranha relatives, this is a peaceful herbivorous fish. It lives in streams in equatorial Africa.

CATFISHES

The predominantly freshwater catfishes have a long body and many mouth barbels. There is a sharp spine in front of the dorsal and pectoral fins. Most species in the order Siluriformes have a small adipose fin on the back near the tail.

12½ in
32 cm

STRIPED EEL CATFISH
Plotosus lineatus
F: Plotosidae
Juveniles of this tropical marine species form dense ball-shaped schools for protection. Solitary adults defend themselves with their venomous fin spines.

3¼ ft
1 m

TIGER SHOVELNOSE CATFISH
Pseudoplatystoma fasciatum
F: Pimelodidae
The long barbels of this South American fish help it find food as it hunts over riverbeds at night searching out small fishes.

16 ft
5 m

SHEATFISH
Silurus glanis
F: Siluridae
From wetlands in central Europe and Asia, this huge fish has been known to weigh over 660 lb (300 kg), but due to heavy fishing, none this large exists today.

long anal fin

spine visible through transparent body

BROWN BULLHEAD
Ameiurus nebulosus
F: Ictaluridae
This North American catfish has venomous fin spines that help ward off predators when it is guarding its nest.

6 in
15 cm

GLASS CATFISH
Kryptopterus bicirrhis
F: Siluridae
This native of southeast Asia often hangs motionless in the water; its transparent body makes it hard to see.

sensory mouth barbels help find food

20½ in
52 cm

SALMON AND RELATIVES

This order of marine and freshwater fishes also includes many anadromous members (moving from sea to fresh water to breed). Powerful predators, the Salmoniformes have a large tail, a single dorsal fin, and a much smaller adipose fin.

adipose fin

SOCKEYE SALMON
Oncorhynchus nerka
F: Salmonidae
From the North Pacific, this fish moves up rivers into Asian and North American lakes to spawn. At this time, both sexes turn red and the male's jaw develops a hook.

33 in
84 cm

hooked jaw on male

CISCO
Coregonus artedi
F: Salmonidae
Widespread in North American lakes and large rivers, this fish forms schools that feed on plankton and invertebrates.

22½ in
57 cm

ARCTIC CHAR
Salvelinus alpinus
F: Salmonidae
Clean, cold water is a must for this fish. Some individuals live in high-altitude lakes, while others migrate from sea to river.

3¼ ft
1 m

RAINBOW TROUT
Oncorhynchus mykiss
F: Salmonidae
Although native to North America, this fish has been introduced into freshwaters for food and sport throughout the world.

4 ft
1.2 m

PIKES AND RELATIVES

Found in cool freshwaters across the Northern Hemisphere, the Esociformes are fast and agile. The dorsal and anal fins are set far back, near the tail, to give these predatory fishes an instant forward thrust.

single dorsal fin set far back

MUD MINNOW
Umbra krameri
F: Umbridae
The European mud minnow is now rare, as the small ditches and canals of its home—the Danube and Dneister river systems—disappear.

6½ in
17 cm

distinctive markings

LIZARDFISHES AND RELATIVES

This diverse order of marine fishes inhabits shallow coastal waters as well as the deep sea. The Aulopiformes have large mouths, with many small teeth, and can catch large prey. They have a single dorsal fin and a much smaller adipose fin.

elongated pectoral fins

triangular head

16 in
40 cm

long pelvic fin used as prop

VARIEGATED LIZARDFISH
Synodus variegatus
F: Synodontidae
This resident of tropical reefs of the Indian and Pacific oceans perches on a coral head, remaining completely still, and then darts out to catch fish.

16 in
40 cm

BOMBAY DUCK
Harpadon nehereus
F: Synodontidae
During the monsoon season in the Indo-Pacific, schools of this small fish gather near river deltas to feed on material that is washed down.

FEELER FISH
Bathypterois longifilis
F: Ipnopidae
This globally distributed fish perches above the deep, muddy ocean floor, and uses its filament-like pectoral fins to catch food.

14½ in
37 cm

elongated pelvic fins

LANTERNFISHES AND RELATIVES

Lanternfishes are small, slim fishes with many photophores (light organs) that help them communicate in their deep, dark ocean habitat. Members of the order Myctophiformes, these species have large eyes and many migrate toward the surface at night to feed.

4¼ in
11 cm

SPOTTED LANTERNFISH
Myctophum punctatum
F: Myctophidae
Living in the depths of the Atlantic, this lanternfish uses its impressive array of light-emitting organs, or photophores, to signal to others.

DRAGONFISHES AND RELATIVES

A majority of these deep-sea fishes have photophores to help them hunt, hide, and find mates. Most members of the order Stomiiformes are fearsome-looking predators, with large teeth and sometimes a long chin barbel.

long, slender body

14 in
35 cm

9½ in
24 cm

NORTHERN STOPLIGHT LOOSEJAW
Malacosteus niger
F: Stomiidae
Found globally in temperate, tropical, and subtropical oceans, this fish uses a beam of red bioluminescent light, which its shrimp prey cannot see.

SLOANE'S VIPERFISH
Chauliodus sloani
F: Stomiidae
The Sloane's viperfish has long, transparent fangs that protrude when its mouth is closed. It uses photophores to emit light in the depths of tropical and subtropical oceans.

silvery body

large eyes

2¾ in
7 cm

PACIFIC HATCHETFISH
Argyropelecus affinis
F: Sternoptychidae
The silvery, thin body of this fish helps camouflage and hide it from predators. It is found in temperate, tropical, and subtropical waters.

duckbill-shaped snout

3¼ ft
1 m

CHAIN PICKEREL
Esox niger
F: Esocidae
When hunting, this North American pike makes delicate movements with its fins to hover motionlessly before striking with lightning speed.

KNIFEFISHES

Knifefishes have flattened, eel-like bodies, and a single long anal fin, which they use to move backward and forward. The electric eel is atypical, with a long, round body. The Gymnotiformes live in fresh water and can produce electrical impulses.

23½ in
60 cm

ELECTRIC EEL
Electrophorus electricus
F: Gymnotidae
This large South American fish can produce an electric shock of up to 600 volts—enough to kill other fish and stun a human.

8¼ ft
2.5 m

BANDED KNIFEFISH
Gymnotus carapo
F: Gymnotidae
Found in murky wetlands in Central and South America, this fish produces mild electric currents, which it uses to sense its surroundings.

SMELTS AND RELATIVES

Smelts resemble small, slim salmon and, like them, most have an adipose fin on the back near the tail. Some of the Osmeriformes have a distinctive smell. The European smelt smells like fresh cucumber.

4 in
10 cm

BARREL-EYE
Opisthoproctus soleatus
F: Opisthoproctidae
Living in semidarkness in oceans worldwide, this fish has tubular eyes that point upward. These help it make full use of the available light.

10 in
25 cm

CAPELIN
Mallotus villosus
F: Osmeridae
Found in cold Arctic and nearby waters, this fish forms large schools that are a crucial food source for many seabirds; its abundance or scarcity determines the breeding success of these birds.

OARFISHES AND RELATIVES

The 18 members of this marine order, called Lampriformes, are colourful fishes of the open ocean, the adults having crimson fins. In many species, the rays of the dorsal fin extend as long streamers. Most fishes in this order are ocean wanderers that are rarely seen.

crest of elongated dorsal fin rays

OARFISH
Regalecus glesne
F: Regalecidae
The longest bony fish in the world, this species has been the subject of many stories about sea serpents. It is found in tropical, subtropical, and temperate waters worldwide.

JUVENILE

36 ft
11 m

pectoral fin

6½ ft
2 m

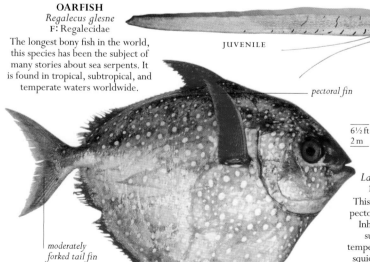

moderately forked tail fin

OPAH
Lampris guttatus
F: Lampridae
This fish beats its long pectoral fins like wings. Inhabiting tropical, subtropical, and temperate oceans, it eats squid and small fishes.

ANGLERFISHES AND RELATIVES

The 300 or so Lophiiformes include some of the most bizarre of all marine fishes. A modified fin ray on top of the head acts as a fishing lure (bioluminescent in deep-water species), which attracts prey toward the fishes' cavernous mouths.

8 in
20 cm

RED-LIPPED BATFISH
Ogcocephalus darwini
F: Ogcocephalidae
The oddly shaped batfish props itself up on its paired pectoral and pelvic fins and shuffles around in search of food.

COFFINFISH
Chaunax endeavouri
F: Chaunacidae
Coffinfishes lie on muddy seabeds in the southwest Pacific, waiting for small fish to stray within reach.

9 in
22 cm

ANGLER
Lophius piscatorius
F: Lophiidae
A fringe of seaweed-shaped flaps around its mouth helps disguise this northeast Atlantic angler. It can strike with lightning speed.

6½ ft
2 m

4½ in
11.5 cm

8 in
20 cm

FANFIN ANGLER
Caulophryne jordani
F: Caulophrynidae
In the dark ocean depths it is difficult to find a mate. Once successful, the tiny male fanfin angler latches onto the female permanently.

WARTY FROGFISH
Antennarius maculatus
F: Antennariidae
The well-camouflaged warty frogfish clambers over coral reefs using its limblike pectoral fins.

8 in
20 cm

SARGASSUMFISH
Histrio histrio
F: Antennariidae
While most frogfish live on the seabed, this one hides among rafts of floating *Sargassum* seaweed, which it closely resembles.

large pectoral fins used for scrambling

skin flaps for camouflage

COD AND RELATIVES

The Gadiformes include many important marine commercial species. Most have two or three soft dorsal fins on their back and many have a chin barbel. Grenadiers live in deep water and have a long thin tail.

6½ ft
2 m

chin barbel

ATLANTIC COD
Gadus morhua
F: Gadidae
Overfishing has reduced the average weight of Atlantic cod to 24 lb (11 kg) from a historical maximum weight of over 200 lb (90 kg).

3 ft/91 cm

ALASKA POLLOCK
Theragra chalcogramma
F: Gadidae
The lack of a chin barbel and a protruding lower jaw help distinguish this fish from cod. It lives in cold Arctic waters.

20 in/50 cm

colorful tail fin with rounded edge

SHORE ROCKLING
Gaidropsarus mediterraneus
F: Lotidae
Equipped with three sensory mouth barbels, this eel-like rockling searches out food in northeast Atlantic rock pools.

4 ft
1.2 m

BURBOT
Lota lota
F: Lotidae
Unlike most of its relatives, the burbot lives in freshwater. It is found in deep lakes and rivers across the Northern Hemisphere.

3¼ ft
1 m

PACIFIC GRENADIER
Coryphaenoides acrolepis
F: Macrouridae
A long scaly tail and bulbous head give this abundant deepwater cod relative its alternative name of Pacific rat-tail.

CUSK EELS

Most members of the order Ophidiiformes live in the ocean and are long, slim, eel-like fishes. They have thin pelvic fins and long dorsal and anal fins, which in many species join onto the tail fin.

PEARLFISH
8½ in / 21 cm
Carapus acus
F: Carapidae

The adult pearlfish finds refuge inside a sea cucumber, entering tail-first through the anus, exiting at night to feed.

GREY MULLETS

Grey mullets are silvery striped fishes with two widely separated dorsal fins; the first has sharp spines, the second has soft rays. Members of the Mugiliformes are distributed worldwide. They are vegetarian, feeding on fine algae and detritus.

30 in
75 cm

GOLDEN GREY MULLET
Liza aurata
F: Mugilidae

Harbors, estuaries, and coastal waters in the northeast Atlantic are all likely haunts of this mullet, which often lives in schools.

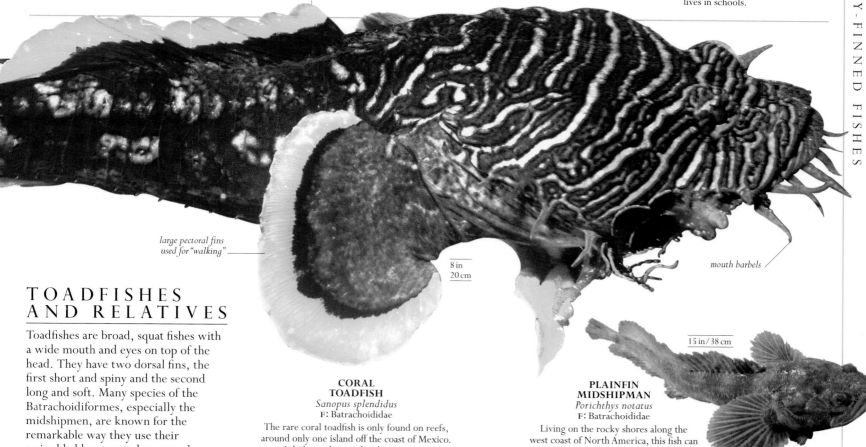

large pectoral fins used for "walking"

8 in
20 cm

mouth barbels

TOADFISHES AND RELATIVES

Toadfishes are broad, squat fishes with a wide mouth and eyes on top of the head. They have two dorsal fins, the first short and spiny and the second long and soft. Many species of the Batrachoidiformes, especially the midshipmen, are known for the remarkable way they use their swim bladders to produce sound.

CORAL TOADFISH
Sanopus splendidus
F: Batrachoididae

The rare coral toadfish is only found on reefs, around only one island off the coast of Mexico. It hides under coral and in crevices.

15 in / 38 cm

PLAINFIN MIDSHIPMAN
Porichthys notatus
F: Batrachoididae

Living on the rocky shores along the west coast of North America, this fish can breathe air when the tide goes out.

SILVERSIDES AND RELATIVES

These small, slim silvery fishes often live in large schools. There are over 300 Atheriniformes, occuring both in marine and freshwater habitats. Most have two dorsal fins, the first of which has flexible spines, and a single anal fin.

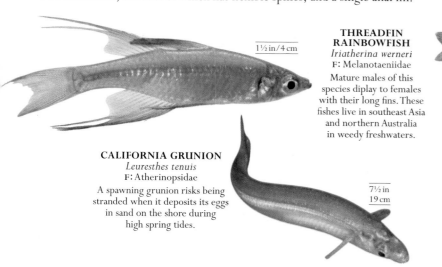

THREADFIN RAINBOWFISH
1½ in / 4 cm
Iriatherina werneri
F: Melanotaeniidae

Mature males of this species diplay to females with their long fins. These fishes live in southeast Asia and northern Australia in weedy freshwaters.

CALIFORNIA GRUNION
Leuresthes tenuis
F: Atherinopsidae

A spawning grunion risks being stranded when it deposits its eggs in sand on the shore during high spring tides.

7½ in
19 cm

NEEDLEFISHES AND RELATIVES

With long, thin, rodlike bodies and extended beaklike jaws, these silvery fishes are well camouflaged in the open ocean. Flying fishes—with their large, paired pectoral (side) and pelvic (belly) fins—also belong to this order, the Beloniformes.

3 ft / 93 cm

GARFISH
Belone belone
F: Belonidae

Remaining close to the ocean surface in the northeast Atlantic, the needlefish chases small fish, especially members of the herring family.

ATLANTIC FLYINGFISH
Cheilopogon heterurus
F: Exocoetidae

Pursued by predators or startled by boats, this fish escapes by breaching the surface and gliding away on outspread fins.

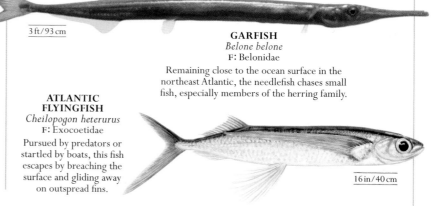

16 in / 40 cm

KILLIFISHES AND RELATIVES

Most members of the Cyprinodontiformes are small freshwater fishes with a single dorsal fin and a large tail. Of the ten families, the guppies are perhaps the best known because they produce live young and are popular aquarium fish.

2¾ in/7 cm

tail important in display

FOUR-EYED FISH
Anableps anableps
F: Anablepidae
The bulging eyes of this South American fish are divided so that it can see clearly both above and below water.

AMIET'S LYRETAIL
Fundulopanchax amieti
F: Nothobranchiidae
This small colorful fish lives in rain forest streams in Cameroon, Africa. There are many other similar species, known collectively as killifishes.

12½ in/32 cm

AMERICAN FLAGFISH
Jordanella floridae
F: Cyprinodontidae
Swamps and streams in Florida are home to this peaceful vegetarian. The male courts a female and guards their eggs.

3 in/7.5 cm

BLACK PEARLFISH
Austrolebias nigripinnis
F: Rivulidae
Subtropical rivers in South America are home to this species and all the other members of its colorful family.

2¾ in/7 cm

2 in/5 cm

SAILFIN MOLLY
Poecilia latipinna
F: Poeciliidae
The large dorsal fin of this North American fish is displayed in courtship. The female bears live young.

DORIES AND RELATIVES

The 42 species belonging to the order Zeiformes are all marine, deep-bodied, but thin fishes with long dorsal and anal fins. Dories have protrusible jaws—they can be shot out to capture prey. They feed on a wide range of small fishes that they stalk head on.

JOHN DORY
Zeus faber
F: Zeidae
Viewed face on, this ultrathin fish is hard to see and can stalk and strike its prey very effectively.

3 ft
90 cm

SQUIRRELFISHES AND RELATIVES

These marine fishes, in the order Beryciformes, are deep-bodied with large scales, a forked tail, and sharp fin spines. Most are nocturnal and are often a red color. The red end of the light spectrum is the first to be absorbed by water. This makes them appear black below a certain depth.

CROWN SQUIRRELFISH
Sargocentron diadema
F: Holocentridae
This is a typical squirrelfish and one of many similar species found in tropical waters. By day it hides in crevices.

6½ in/17 cm

PINEAPPLEFISH
Cleidopus gloriamaris
F: Monocentridae
Few predators would attack this spiny fish, which has an armor of thick scales. Its color warns that it is unpalatable.

9 in/22 cm

12 cm/4¾ in

EYELIGHT FISH
Photoblepharon palpebratum
F: Anomalopidae
At night this fish signals to others using a light organ under each eye. It has a black membrane that it uses to cover the light, creating a flashing effect.

COMMON FANGTOOTH
Anoplogaster cornuta
F: Anoplogastridae
The huge fangs of this deep-ocean predator make escape impossible for its prey, which it swallows whole.

7 in/18 cm

30 in
75 cm

ORANGE ROUGHY
Hoplostethus atlanticus
F: Trachichthyidae
This is one of the longest-lived and slowest-growing of any fish and can live to at least 150 years old.

protrusible jaw

STICKLEBACKS AND SEAMOTHS

Most sticklebacks inhabit still and slow-moving freshwater, but some live in the sea as do seamoths. The Gasterosteiformes have a long, thin, stiff body protected by bony scutes (plates) along the sides and sharp spines along the back.

spines

4¼ in
11 cm

bony scutes

males have red belly
when breeding

THREE-SPINED STICKLEBACK
Gasterosteus aculeatus
F: Gasterosteidae
This little fish is widespread in freshwater and shallow seas in the Northern Hemisphere. The male performs an elaborate courtship dance.

SEAMOTH
Eurypegasus draconis
F: Pegasidae
Unlike the closely related sticklebacks, the tropical marine seamoth is flat and has large winglike pectoral (side) fins.

2¾ in/7 cm

CLINGFISHES AND RELATIVES

Clingfishes are small, usually marine, bottom-dwelling fishes. Most of the Gobiesociformes possess a sucker formed from modified pelvic fins with which they cling to rocks. Their eyes are set high on the head and they have one dorsal fin.

dorsal fin

3¼ in/8 cm

CONNEMARA CLINGFISH
Lepadogaster candolii
F: Gobiesocidae
Living in rocky shallows in the northeast Atlantic, this little fish is exposed to strong waves, but can cling on tightly.

PIPEFISHES AND SEAHORSES

Seahorses and other similar fishes in the order Syngnathiformes are encased in a body armor of bony plates, making their bodies very stiff. The group includes both marine and freshwater species. Sea horses have a small mouth at the end of a tubular snout, which they use to feed on tiny planktonic crustaceans.

tiny pectoral
fin helps
maintain
position

6½ in
16 cm

ROBUST GHOST PIPEFISH
Solenostomus cyanopterus
F: Solenostomidae
Large pelvic fins enable the ghost pipefish to drift and swim slowly among weeds and seagrasses as it hunts minute invertebrates.

long,
tubular
snout

46 cm
18 in

WEEDY SEADRAGON
Phyllopteryx taeniolatus
F: Syngnathidae
This large, bizarrely shaped Australian seadragon hides among seaweeds on rocky reefs, camouflaged by its many leaflike skin flaps.

YELLOW SEAHORSE
Hippocampus kuda
F: Syngnathidae
As in all seahorses, the male of this species has a belly pouch in which he broods eggs laid there by the female.

12 in
30 cm

6 in
15 cm

RAZORFISH
Aeoliscus strigatus
F: Centriscidae
Hanging head down among urchin spines disguises the razorfish, which habitually swims in this position. This is an Indo-Pacific species.

RINGED PIPEFISH
Doryrhamphus dactyliophorus
F: Syngnathidae
The long thin body of this coral-reef dweller is typical of pipefishes. It hovers under and between corals and rocks.

7 in
18 cm

skin flaps
provide
camouflage
in seaweed

TRUMPETFISH
Aulostomus chinensis
F: Aulostomidae
Trumpetfishes often shadow moray eels, which hunt over coral reefs in order to snap up small fish that the eels have flushed out.

32 in/80 cm

SWAMPEELS AND RELATIVES

These tropical and subtropical freshwater fishes have an eel-like body and no pelvic fins. Other fins are often reduced as well. A few of the Synbranchiformes survive in brackish water such as mangrove swamps.

FIRE EEL
Mastacembelus erythrotaenia
F: Mastacembelidae

3¼ ft
1 m

Living in flooded lowland plains and slow rivers in southeast Asia, this edible spiny eel eats insect larvae and worms.

MARBLED SWAMPEEL
Synbranchus marmoratus
F: Synbranchidae

Able to breathe air if necessary, this almost finless fish lives in small bodies of water in Central and South America.

5 ft
1.5 m

FLATFISHES

Pleuronectiformes start life as normal juveniles with a left and a right side but as they grow the body becomes flattened and they lie on the seabed. The underneath eye migrates round to join the other on the upper side.

EUROPEAN PLAICE
Pleuronectes platessa
F: Pleuronectidae

This important North Atlantic commercial flatfish lies camouflaged on the sea bed with its right side uppermost. It emerges at night to feed.

3¼ ft
1m

ATLANTIC HALIBUT
Hippoglossus hippoglossus
F: Pleuronectidae

One of the largest flatfish, the Atlantic halibut lies on its left side with both eyes on the upward-facing right side.

8¼ ft
2.5 m

COMMON SOLE
Solea solea
F: Soleidae

Although sole can live to 30 years old, most do not survive this long because they are a valuable commercial fish.

28 in
70 cm

right eye has moved to top of the fish

TURBOT
Psetta maxima
F: Scophthalmidae

The ability to alter its color to match the seabed allows turbot to escape predators' attention. It lives in the North Atlantic

3¼ ft
1 m

TRIGGERFISHES, PUFFERFISHES, AND RELATIVES

This diverse group of marine and freshwater fishes includes the huge ocean sunfish and the poisonous pufferfish. Instead of normal teeth, the Tetraodontiformes have fused tooth plates or just a few large teeth. Their scales are modified to form protective spines or plates.

20 in
50 cm

CLOWN TRIGGERFISH
Balistoides conspicillum
F: Balistidae

This brightly colored coral-reef fish can wedge itself in a crevice by erecting its dorsal spines, which lock in position.

large tail

10 in
25 cm

SPOTTED BOXFISH
Ostracion meleagris
F: Ostraciidae

Encased in a rigid box of fused bony plates, and with poisonous skin, this Indo-Pacific reef fish is left well alone by predators.

males have violet blue sides

top is the left side of the fish

6 in
15 cm

WHITE-SPOTTED PUFFER
Arothron hispidus
F:Tetraodontidae

The neurotoxin contained within the skin and organs of this fish can easily kill a human and deters natural predators.

20 in / 50 cm

LONG-SPINE PORCUPINEFISH
Diodon holocanthus
F: Diodontidae

Found in all tropical seas, this spiny fish can suck in water to inflate its body and turn itself into a prickly ball—a very effective deterrent against predators.

SCORPIONFISHES AND RELATIVES

Mostly bottom-living and marine, fishes in this large order have a large spiny head with a unique bony strut across the cheek. Most Scorpaeniformes have sharp, sometimes venomous spines in their dorsal fins and many are camouflage experts.

MIMIC FILEFISH
Paraluteres prionurus
F: Monacanthidae
The mimic filefish is avoided by predators because it closely resembles the saddled pufferfish, which has toxic flesh.

4¼ in
11 cm

flat back forms top of bony protective box

SMALLSCALED SCORPIONFISH
Scorpaena porcus
F: Scorpaenidae
Well concealed by skin flaps on the head and its ability to change color, the scorpionfish is hard to spot.

14½ in
37 cm

JUVENILE

TUB GURNARD
Chelidonichthys lucerna
F: Triglidae
Walking over the seabed on three mobile rays on each of its pectoral fins, the tub gurnard can probe for hidden invertebrates.

30 in
75 cm

15 in
38 cm

RED LIONFISH
Pterois volitans
F: Scorpaenidae
The striped colors warn predators that this coral-reef fish has venomous dorsal fin spines. Adults are less white and may have white spots on their sides.

8½ in
21 cm

23½ in
60 cm

LUMPSUCKER
Cyclopterus lumpus
F: Cyclopteridae
A strong sucker on the belly allows this rotund North Atlantic fish to cling onto wave-battered rocks and guard its eggs.

BIG BAIKAL OILFISH
Comephorus baikalensis
F: Comephoridae
About a quarter of the body of this fish is oil, which provides it with buoyancy. It is endemic to Lake Baikal, Russia.

small mouth has stout teeth to tear off sponges

venomous dorsal fin spines

large, upturned mouth

OCEAN SUNFISH
Mola mola
F: Molidae
The ocean sunfish is the heaviest of all bony fish and can weigh 5,070 lb (2,300 kg). It feeds on jellyfish.

10 in
25 cm

LONGSPINED BULLHEAD
Taurulus bubalis
F: Cottidae
The color of this shallow coastal fish varies widely, according to its background. For example, fish living among red seaweeds are red.

STONEFISH
Synanceia verrucosa
F: Synanceiidae
It is extremely difficult to spot this well-camouflaged tropical reef fish. A sting from its venomous spines can be fatal to humans.

16 in
40 cm

20 in
50 cm

13 ft
4 m

7 in / 18 cm

BULLHEAD
Cottus gobio
F: Cottidae
The bullhead lives among stones and vegetation in freshwater streams and rivers throughout much of Europe. The male guards the eggs.

FLYING GURNARD
Dactylopterus volitans
F: Dactylopteridae
This species uses its huge fanlike pectoral fins to "fly" through the water. It "takes off" from the seabed if disturbed.

RED LIONFISH
Pterois volitans

The red lionfish is a night hunter, patrolling tropical coral and rocky reefs in the western Pacific in search of small fish and crustaceans. It corrals its prey against the reef by spreading out its wide pectoral fins—one on each side of its body—before snapping up the prey with lightning speed. Sometimes it stalks its prey out in the open, relying on stealth and a final quick rush, similar to a lion hunting on an African plain. Protected by venomous spines, it will often face a diver or potential predator head-on if approached. One male may collect a small harem of females and will charge other males that stray too close. When ready to spawn, the male displays to the female, circling around her before they swim up toward the surface to shed eggs and sperm into the water. After a few days the eggs hatch into planktonic larvae that drift in the plankton for about a month before settling on the seabed.

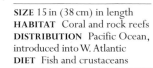

SIZE 15 in (38 cm) in length
HABITAT Coral and rock reefs
DISTRIBUTION Pacific Ocean, introduced into W. Atlantic
DIET Fish and crustaceans

> VIBRANT WARNING
The contrasting striped pattern of the red lionfish acts as a warning to predators that the fish is venomous and they should stay away. On land, stinging wasps use a similar method to avoid being eaten.

eye is disguised by a dark stripe that confuses potential predators

head tentacle

< MENACING PREDATOR
Dusk is a prime hunting time for this fish. It has large eyes to help it see in the dim light, an acute sense of smell, and fleshy "whiskers" around its mouth.

∧ VARIED STRIPES
The striped pattern differs among individuals, and becomes much less obvious in breeding males, which can be very dark.

< PECTORAL FINS
The soft rays of the pectoral fins are joined together for part of their length by a fine membrane, which is marked with circular splotches of color.

fleshy "whiskers" may help disguise the fish's large, open mouth when approaching prey

VENOMOUS SPINES >
Sharp spines in this fish's dorsal, anal, and pelvic fins can be used to inject venom that causes extreme pain in humans, but is rarely fatal. The spines are purely defensive.

< TAIL UP
Lionfish normally hang slightly head down and tail up, ready to pounce on passing prey. The tail helps the fish maintain this position, rather than to swim fast.

dorsal fin made up
of several spines

wide rays of pectoral
fins spread out fully
when the lionfish
ambushes prey

tail

anal fin

PERCH AND RELATIVES

With 156 families and nearly 10,000 species, the Perciformes form the largest and most diverse vertebrate order. At first sight, there seems little to connect some of these fishes, but they all have a similar anatomy. Most have both spines and soft rays in their dorsal and anal fins. The pelvic fins are set well forward, close to the pectoral fins.

23½ in
60 cm

LONGFIN SPADEFISH
Platax teira
F: Ephippidae

Living on coral reefs in the Indo-Pacific, this flattened species patrols in small groups, feeding on algae and invertebrates.

6½ in
16 cm

FOURSPOT BUTTERFLYFISH
Chaetodon quadrimaculatus
F: Chaetodontidae

Many different species of butterflyfishes are found on coral reefs around the world. This fish lives in the western Pacific.

35 in
90 cm

BLUE-SPOTTED SEABREAM
Pagrus caeruleostictus
F: Sparidae

Found in the eastern Atlantic, this is a typical seabream with a steep head profile, forked tail, and long dorsal fin.

4¾ in
12 cm

OCHRE-STRIPED CARDINALFISH
Apogon compressus
F: Apogonidae

This coral-reef resident of the western Pacific is a small, nocturnal species with two dorsal fins and large eyes. The male incubates the eggs in his mouth.

16 in
40 cm

RED MULLET
Mullus surmuletus
F: Mullidae

Common in the Mediterranean and northeast Atlantic, this species detects buried prey with its mobile barbels. It is a relative of the tropical goatfish.

32 in
80 cm

RED BANDFISH
Cepola macrophthalma
F: Cepolidae

Living in vertical mud burrows in the northeast Atlantic, the red bandfish feeds on passing plankton.

GREEN SUNFISH
Lepomis cyanellus
F: Centrarchidae

Well known in North America, this large species is one of the most common fishes in lakes and rivers.

12 in
30 cm

12 in
30 cm

BANDED ARCHERFISH
Toxotes jaculatrix
F: Toxotidae

Living mainly in brackish mangrove estuaries in southeast Asia, Australia, and the western Pacific, it catches insects from overhanging branches by shooting a jet of water from its mouth.

28 in
72 cm

juvenile has brown body with black-edged white spots

HARLEQUIN SWEETLIPS
Plectorhinchus chaetodonoides
F: Haemulidae

The adult harlequin sweetlips is cream with black spots. The color and movement of the juvenile mimics a toxic flatworm.

4 ft
1.2 m

BLUEFISH
Pomatomus saltatrix
F: Pomatomidae

This widespread, voracious, and aggressive predator roams tropical and subtropical oceans in schools that herd and attack smaller fish.

4 in
10 cm

YELLOWHEAD JAWFISH
Opistognathus aurifrons
F: Opistognathidae

After the female of this Caribbean fish has laid her eggs, the male broods them in his mouth.

20 in
51 cm

EUROPEAN PERCH
Perca fluviatilis
F: Percidae

Native to Europe and Asia, this widespread, predatory freshwater fish has been introduced as a sport fish for anglers in Australia and elsewhere, where it has become a pest.

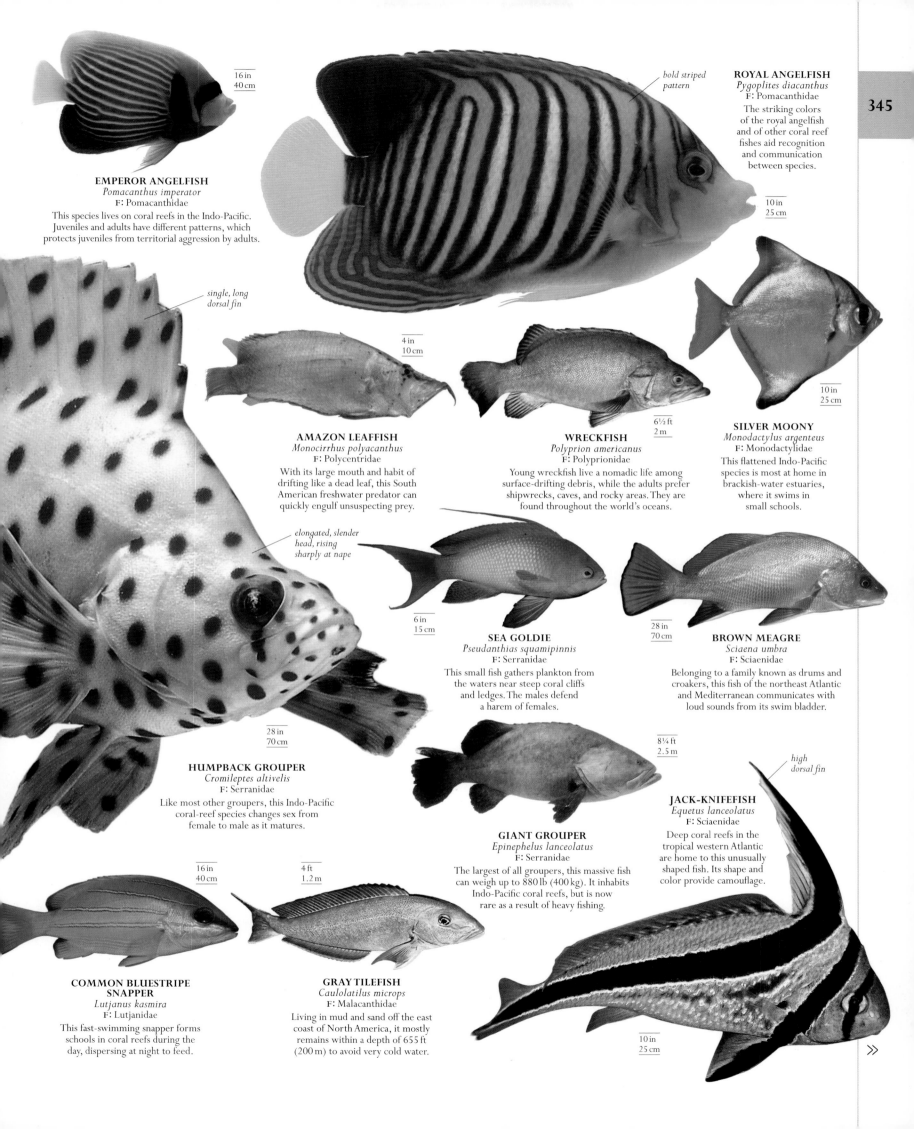

EMPEROR ANGELFISH
Pomacanthus imperator
F: Pomacanthidae

This species lives on coral reefs in the Indo-Pacific. Juveniles and adults have different patterns, which protects juveniles from territorial aggression by adults.

16 in
40 cm

bold striped pattern

ROYAL ANGELFISH
Pygoplites diacanthus
F: Pomacanthidae

The striking colors of the royal angelfish and of other coral reef fishes aid recognition and communication between species.

10 in
25 cm

single, long dorsal fin

4 in
10 cm

AMAZON LEAFFISH
Monocirrhus polyacanthus
F: Polycentridae

With its large mouth and habit of drifting like a dead leaf, this South American freshwater predator can quickly engulf unsuspecting prey.

WRECKFISH
Polyprion americanus
F: Polyprionidae

Young wreckfish live a nomadic life among surface-drifting debris, while the adults prefer shipwrecks, caves, and rocky areas. They are found throughout the world's oceans.

6½ ft
2 m

SILVER MOONY
Monodactylus argenteus
F: Monodactylidae

This flattened Indo-Pacific species is most at home in brackish-water estuaries, where it swims in small schools.

10 in
25 cm

elongated, slender head, rising sharply at nape

6 in
15 cm

SEA GOLDIE
Pseudanthias squamipinnis
F: Serranidae

This small fish gathers plankton from the waters near steep coral cliffs and ledges. The males defend a harem of females.

28 in
70 cm

BROWN MEAGRE
Sciaena umbra
F: Sciaenidae

Belonging to a family known as drums and croakers, this fish of the northeast Atlantic and Mediterranean communicates with loud sounds from its swim bladder.

28 in
70 cm

HUMPBACK GROUPER
Cromileptes altivelis
F: Serranidae

Like most other groupers, this Indo-Pacific coral-reef species changes sex from female to male as it matures.

8¼ ft
2.5 m

high dorsal fin

JACK-KNIFEFISH
Equetus lanceolatus
F: Sciaenidae

Deep coral reefs in the tropical western Atlantic are home to this unusually shaped fish. Its shape and color provide camouflage.

GIANT GROUPER
Epinephelus lanceolatus
F: Serranidae

The largest of all groupers, this massive fish can weigh up to 880 lb (400 kg). It inhabits Indo-Pacific coral reefs, but is now rare as a result of heavy fishing.

16 in
40 cm

4 ft
1.2 m

COMMON BLUESTRIPE SNAPPER
Lutjanus kasmira
F: Lutjanidae

This fast-swimming snapper forms schools in coral reefs during the day, dispersing at night to feed.

GRAY TILEFISH
Caulolatilus microps
F: Malacanthidae

Living in mud and sand off the east coast of North America, it mostly remains within a depth of 655 ft (200 m) to avoid very cold water.

10 in
25 cm

»

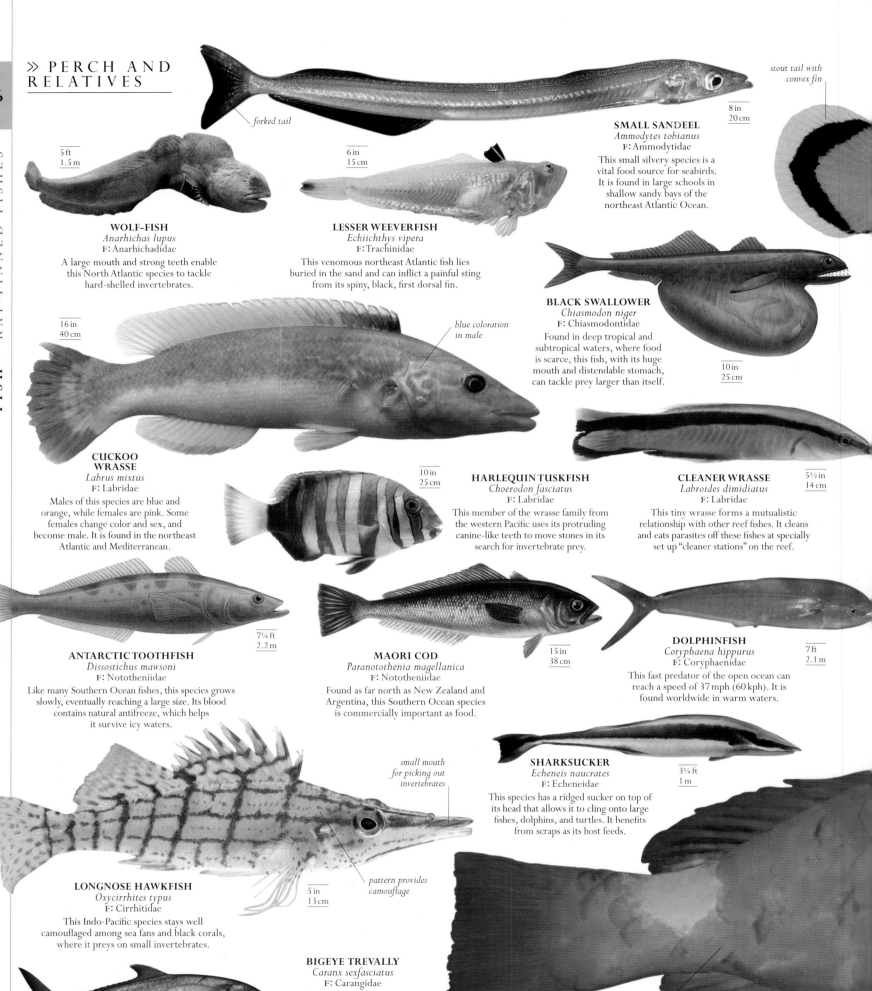

forked tail

stout tail with convex fin

8 in
20 cm

SMALL SANDEEL
Ammodytes tobianus
F: Ammodytidae

This small silvery species is a
vital food source for seabirds.
It is found in large schools in
shallow sandy bays of the
northeast Atlantic Ocean.

5 ft
1.5 m

WOLF-FISH
Anarhichas lupus
F: Anarhichadiidae

A large mouth and strong teeth enable
this North Atlantic species to tackle
hard-shelled invertebrates.

6 in
15 cm

LESSER WEEVERFISH
Echiichthys vipera
F: Trachinidae

This venomous northeast Atlantic fish lies
buried in the sand and can inflict a painful sting
from its spiny, black, first dorsal fin.

BLACK SWALLOWER
Chiasmodon niger
F: Chiasmodontidae

Found in deep tropical and
subtropical waters, where food
is scarce, this fish, with its huge
mouth and distendable stomach,
can tackle prey larger than itself.

10 in
25 cm

16 in
40 cm

*blue coloration
in male*

**CUCKOO
WRASSE**
Labrus mixtus
F: Labridae

Males of this species are blue and
orange, while females are pink. Some
females change color and sex, and
become male. It is found in the northeast
Atlantic and Mediterranean.

10 in
25 cm

HARLEQUIN TUSKFISH
Choerodon fasciatus
F: Labridae

This member of the wrasse family from
the western Pacific uses its protruding
canine-like teeth to move stones in its
search for invertebrate prey.

CLEANER WRASSE
Labroides dimidiatus
F: Labridae

This tiny wrasse forms a mutualistic
relationship with other reef fishes. It cleans
and eats parasites off these fishes at specially
set up "cleaner stations" on the reef.

5½ in
14 cm

ANTARCTIC TOOTHFISH
Dissostichus mawsoni
F: Nototheniidae

Like many Southern Ocean fishes, this species grows
slowly, eventually reaching a large size. Its blood
contains natural antifreeze, which helps
it survive icy waters.

7¼ ft
2.2 m

MAORI COD
Paranototothenia magellanica
F: Nototheniidae

Found as far north as New Zealand and
Argentina, this Southern Ocean species
is commercially important as food.

15 in
38 cm

DOLPHINFISH
Coryphaena hippurus
F: Coryphaenidae

This fast predator of the open ocean can
reach a speed of 37 mph (60 kph). It is
found worldwide in warm waters.

7 ft
2.1 m

*small mouth
for picking out
invertebrates*

SHARKSUCKER
Echeneis naucrates
F: Echeneidae

This species has a ridged sucker on top of
its head that allows it to cling onto large
fishes, dolphins, and turtles. It benefits
from scraps as its host feeds.

3¼ ft
1 m

LONGNOSE HAWKFISH
Oxycirrhites typus
F: Cirrhitidae

This Indo-Pacific species stays well
camouflaged among sea fans and black corals,
where it preys on small invertebrates.

5 in
13 cm

*pattern provides
camouflage*

BIGEYE TREVALLY
Caranx sexfasciatus
F: Carangidae

Built for speed, this predator
hunts other fishes at night over
reefs in the Indian and Pacific
oceans. Juveniles live inshore
and may enter estuaries.

4 ft
1.2 m

reddish coloration overall

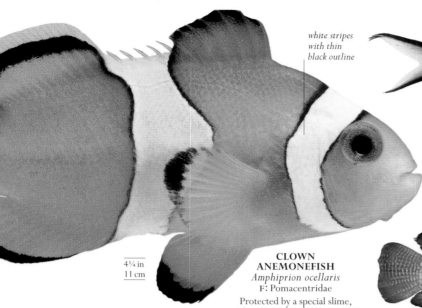

white stripes with thin black outline

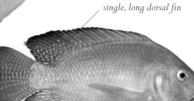

FISH · RAY-FINNED FISHES

BLUE CHROMIS
Chromis cyanea
F: Pomacentridae
6 in / 15 cm

Found in the tropical western Atlantic, this is one of the most common fishes on coral reefs. It lays its eggs through an orange breeding tube.

SERGEANT MAJOR
Abudefduf saxatilis
F: Pomacentridae
9 in / 23 cm

The most common member of the damselfish family, this brightly striped, small fish can be easily spotted on coral reefs in the Atlantic Ocean.

CLOWN ANEMONEFISH
Amphiprion ocellaris
F: Pomacentridae
4¼ in / 11 cm

Protected by a special slime, this colorful fish of the tropical western Pacific finds a safe home among the stinging tentacles of giant sea anemones.

single, long dorsal fin

TILAPIA
Oreochromis niloticus eduardianus
F: Cichlidae
19½ in / 49 cm

This common subspecies of African lakes is a locally important source of food. The female broods about 2,000 eggs in her mouth.

CHIPOKAE
Melanochromis chipokae
F: Cichlidae
JUVENILE / 5 in / 13 cm

This species is found only around the rocky shores of Lake Malawi in Africa. Other fishes in this family are endemic to other lakes.

ORCHID DOTTYBACK
Pseudochromis fridmani
F: Pseudochromidae
2½ in / 6.5 cm

One of the most brightly colored species of all coral-reef fishes, this Red Sea denizen hides beneath overhangs in steep areas.

COMMON STARGAZER
Kathetostoma laeve
F: Uranoscopidae
30 in / 75 cm

Found around southern Australia, the common stargazer buries itself in sand with just its eyes and mouth exposed.

CONVICT CICHLID
Amatitlania nigrofasciata
F: Cichlidae
4 in / 10 cm

Native to C. American rivers and streams, this cichlid has been introduced elsewhere and is a potential pest competing for food and space.

FRESHWATER ANGELFISH
Pterophyllum scalare
F: Cichlidae
6 in / 15 cm

This unusual cichlid, with its disklike and laterally compressed body, is native to South American swamps. Both parents guard the eggs and fry.

elongated pelvic fins

BLACKFIN ICEFISH
Chaenocephalus aceratus
F: Channichthyidae
28 in / 72 cm

A natural antifreeze in the blood of this Southern Ocean fish enables it to survive at temperatures as low as 28.4°F (-2°C).

GUNNEL
Pholis gunnellus
F: Pholidae
10 in / 25 cm

Found in rocky pools in the North Atlantic, this eel-like fish with no scales can be very slippery, sliding away easily from predators.

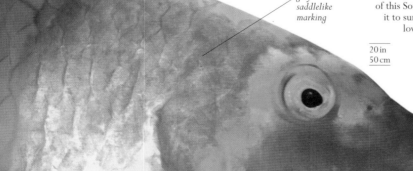
grayish saddlelike marking

MEDITERRANEAN PARROTFISH
Sparisoma cretense
F: Scaridae
20 in / 50 cm

This Mediterranean species is the only parrotfish in which the female is more colorful than the male. Most other parrotfish species are tropical.

beaklike teeth

GREEN HUMPHEAD PARROTFISH
Bolbometopon muricatum
F: Scaridae
4¼ ft / 1.3 m

hump helps break up coral

A tough beaklike mouth allows this Indo-Pacific coral reef resident to crunch up living coral, expelling the remains as sand.

SPOTTED CTENOPOMA
Ctenopoma acutirostre
F: Anabantidae

This tropical freshwater fish is found in the Congo basin of Africa. It often stalks its prey with its head held down.

4¾ in
12 cm

BUTTERFLY BLENNY
Blennius ocellaris
F: Blenniidae

Like many of its relatives, this fish of the northeast Atlantic lives on the sea bed and guards its eggs, which are often laid in an empty shell.

8 in
20 cm

FIREFISH
Nemateleotris magnifica
F: Microdesmidae

Hovering above its coral reef burrow, this fish gathers plankton from the water and darts back into safety if it senses danger.

3½ in
9 cm

elongated first ray on dorsal fin

broad tail fin

SIAMESE FIGHTING FISH
Betta splendens
F: Osphronemidae

The original range of this Asian freshwater fish is unclear, but it has been bred for centuries for its fighting ability, particularly strong in males.

2½ in
6.5 cm

POWDERBLUE SURGEONFISH
Acanthurus leucosternon
F: Acanthuridae

This Indian Ocean fish has a sharp bladelike structure hidden on either side of its tail base, with which it can slash its opponent if attacked.

9 in
23 cm

MANDARINFISH
Synchiropus splendidus
F: Callionymidae

Native to the Pacific, this is one of the most colorful of all tropical reef fish. Its vivid colors warn predators of its foul taste.

2¼ in
6 cm

huge, sail-like dorsal fin

ATLANTIC SAILFISH
Istiophorus albicans
F: Istiophoridae

This ocean predator uses its upper jaw, which extends as a long spear, to slash into schools of fishes and stun them.

10 ft
3.2 m

spear-shaped snout

torpedo-shaped body

projecting lower jaw

GREAT BARRACUDA
Sphyraena barracuda
F: Sphyraenidae

Found globally in tropical and subtropical waters, this solitary predator stalks its prey and then accelerates rapidly as it strikes.

6½ ft
2 m

23½ in
60 cm

15 ft
4.5 m

ATLANTIC MACKEREL
Scomber scombrus
F: Scombridae

Living in large schools in the North Atlantic, this fish feeds voraciously on small fishes and plankton. Its streamlined body makes it a fast swimmer.

BUMBLEBEE FISH
Brachygobius doriae
F: Gobiidae

Found in southeast Asia, this bottom-dwelling goby is tolerant of brackish water and lives in estuaries and mangroves.

1½ in
4 cm

3½ in
9 cm

SAND GOBY
Pomatoschistus minutus
F: Gobiidae

This fish is common in sandy inshore areas of the northeast Atlantic. During breeding, males develop a spot at the rear of the first dorsal fin.

NORTHERN BLUEFIN TUNA
Thunnus thynnus
F: Scombridae

Found globally, this tuna is one of the world's most valuable commercial fishes. A fast, dynamic predator, it roams widely, hunting small fishes.

long tail

10 in
25 cm

high-set, bulbous eyes

ATLANTIC MUDSKIPPER
Periophthalmus barbarus
F: Gobiidae

As long as it stays moist, this mudskipper can stay out of water for hours by absorbing oxygen through its skin.

LOBE-FINNED FISHES

Considered to be the ancestors of land vertebrates, lobe-finned fishes have fins that resemble primitive limbs, with a fleshy base preceding the fin membrane.

Like the ray-finned fishes, these are bony fishes with a hard skeleton but their fins have a different structure. The fin membrane is supported on a muscular lobe that projects out from the body and is strong enough to let some of these fish shuffle along on their paired pectoral and pelvic fins. Bones and cartilage inside the lobes provide attachment for muscles. There are many fossil groups of lobe-finned fishes but living representatives are confined to the marine coelacanths and the freshwater lungfish.

A LIVING FOSSIL

Coelacanths are nocturnal and secretive. The first modern specimen was discovered in 1938—until then only fossil species over 65 million years old were known. The historic find belonged to a species living in deep rocky areas in the western Indian Ocean.

A second species was found in Indonesian waters in 1998. The formation of the backbone from the notochord is incomplete and the tail fin has a characteristic extra middle lobe. Their scales are heavy bony plates and they are not long-distance swimmers. Unlike the egg-laying lungfish, coelacanths bear live young because the eggs hatch internally. Gestation may be as long as three years, longer than any other vertebrate, and this makes their survival precarious if individuals continue to be caught in the nets of deep-sea trawlers.

BREATHING OUT OF WATER

Although most of their fossil ancestors lived in the ocean, modern lungfishes are limited to freshwater habitats in South America, Africa, and Australia. All can to some extent breathe air via a connection to the swim bladder—useful when pools dry out seasonally. Some species can survive buried in mud for many months and would die if kept permanently submerged in water, while others still rely mainly on gills for breathing. Their shape and the fact that the larvae of some species have external gills led early zoologists to believe that lungfishes were amphibians.

PHYLUM	CHORDATA
CLASS	SARCOPTERYGII
ORDER	3
FAMILIES	4
SPECIES	8

DEBATE
FISH ON LAND

While it is generally accepted that land-living vertebrates evolved from a fishlike ancestor in the sea, finding that ancestor is more difficult. Recent work suggests that coelacanths and lungfishes (lobe-finned fishes) are more closely related to tetrapods (four-limbed vertebrates such as mammals) than to other fish groups such as sharks. Modern classifications group lobe-finned fishes with tetrapods in the same group, the Sarcopterygii. So coelacanths are not the direct ancestors of tetrapods but a side branch. In 2002 a fossil lobe-finned fish, *Styloichthys*, was found in China. It appears to show close links between lungfishes and tetrapods.

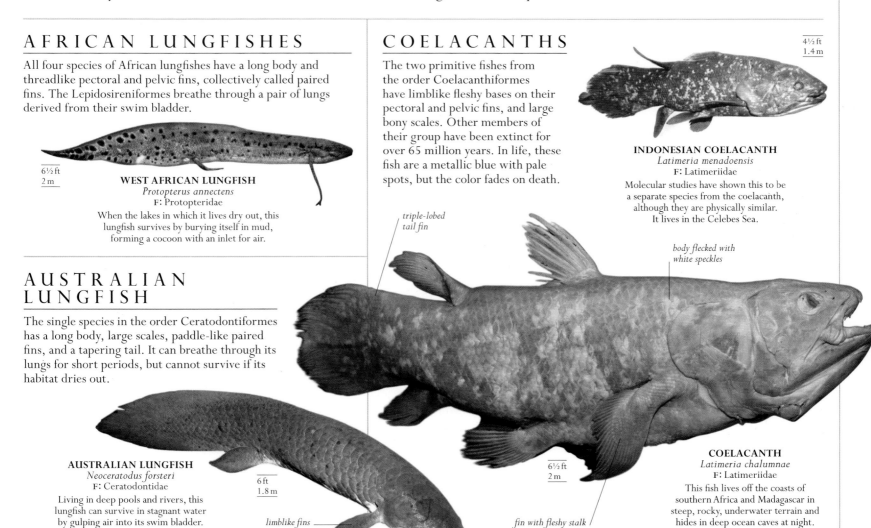

AFRICAN LUNGFISHES

All four species of African lungfishes have a long body and threadlike pectoral and pelvic fins, collectively called paired fins. The Lepidosireniformes breathe through a pair of lungs derived from their swim bladder.

6½ ft
2 m

WEST AFRICAN LUNGFISH
Protopterus annectens
F: Protopteridae
When the lakes in which it lives dry out, this lungfish survives by burying itself in mud, forming a cocoon with an inlet for air.

AUSTRALIAN LUNGFISH

The single species in the order Ceratodontiformes has a long body, large scales, paddle-like paired fins, and a tapering tail. It can breathe through its lungs for short periods, but cannot survive if its habitat dries out.

AUSTRALIAN LUNGFISH
Neoceratodus forsteri
F: Ceratodontidae
Living in deep pools and rivers, this lungfish can survive in stagnant water by gulping air into its swim bladder.

6 ft
1.8 m

limblike fins

COELACANTHS

The two primitive fishes from the order Coelacanthiformes have limblike fleshy bases on their pectoral and pelvic fins, and large bony scales. Other members of their group have been extinct for over 65 million years. In life, these fish are a metallic blue with pale spots, but the color fades on death.

triple-lobed tail fin

4½ ft
1.4 m

INDONESIAN COELACANTH
Latimeria menadoensis
F: Latimeriidae
Molecular studies have shown this to be a separate species from the coelacanth, although they are physically similar. It lives in the Celebes Sea.

body flecked with white speckles

6½ ft
2 m

fin with fleshy stalk

COELACANTH
Latimeria chalumnae
F: Latimeriidae
This fish lives off the coasts of southern Africa and Madagascar in steep, rocky, underwater terrain and hides in deep ocean caves at night.

AMPHIBIANS

Amphibians are cold-blooded vertebrates that thrive in freshwater habitats. Some spend their whole lives in water; others just require water to breed. On land, they must find damp places because their skin is permeable and does not protect them from drying out.

PHYLUM	CHORDATA
CLASS	AMPHIBIA
ORDERS	3
FAMILIES	54
SPECIES	About 6,670

Large numbers of male and female Costa Rican golden toads gathered in a pool to mate. The species is now extinct.

A female of the North American Jefferson's salamander attaches a mass of eggs to a submerged twig in early spring.

In this species of poison frog, *Ranitomeya reticulata*, parents carry tadpoles from one small pool to another.

DEBATE
OUR BIGGEST CHALLENGE?

A third of the world's amphibians face extinction in the near future and this poses a huge conservation challenge. This danger of extinction is largely due to the destruction and pollution of freshwater habitats, but amphibians are also threatened by the global spread of the disease chytridiomycosis, caused by a fungus that invades their soft skin.

The three orders of living amphibians are thought to share a common ancestor, but their origins are still uncertain. There is a huge gap in the fossil record between the first land animals to evolve from fish—the tetrapods, some 375 million years ago—and a froglike creature that lived 230 million years ago.

Amphibians have a unique and complex life cycle and occupy very different ecological niches at different stages of their lives. Their eggs hatch into larvae – called tadpoles in frogs and toads—that live in water, often in very high densities, and feed on algae and other plant material. As larvae, they grow rapidly and then undergo a complete change of form—metamorphosis—to become land-living adults. As adults, all amphibians are carnivores, most feeding on insects and other small invertebrates. They typically lead secretive and solitary lives, except when they return to ponds and streams to breed. Amphibians thus require two very different habitats during their lives: water and land. During metamorphosis, they undergo a wide variety of anatomical and physiological changes—from aquatic creatures that swim with tails and breathe through gills, to terrestrial animals that move with four limbs and breathe through lungs.

DIVERSE PARENTAL CARE

Some amphibians produce huge numbers of eggs that are left to fend for themselves, so that only a very few will survive. Many others have evolved various forms of parental care. This is generally associated with the production of far fewer offspring, so that reproductive success is achieved by caring for a manageable number of young, rather than by producing as many eggs as possible. Parental care takes many forms, including defending eggs or larvae against predators, feeding tadpoles with unfertilized eggs, and carrying tadpoles from one place to another. In some species, such as the midwife toads and some poison frogs, parental care is the duty of the father. In many salamanders and caecilians, the mother is the sole protector of the young. In a few frog species, mother and father establish a durable pair bond and share parental duties.

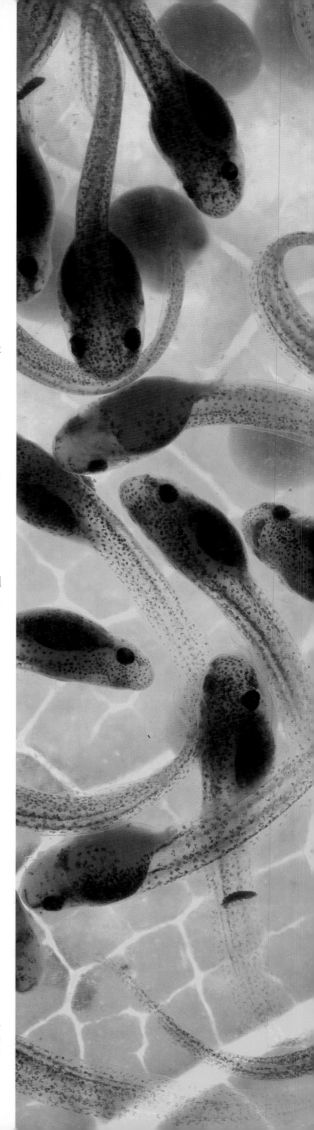

READY TO HATCH >
Tadpoles of the Tanzanian Mitchell's reed frog wriggle within their egg membranes, just prior to hatching.

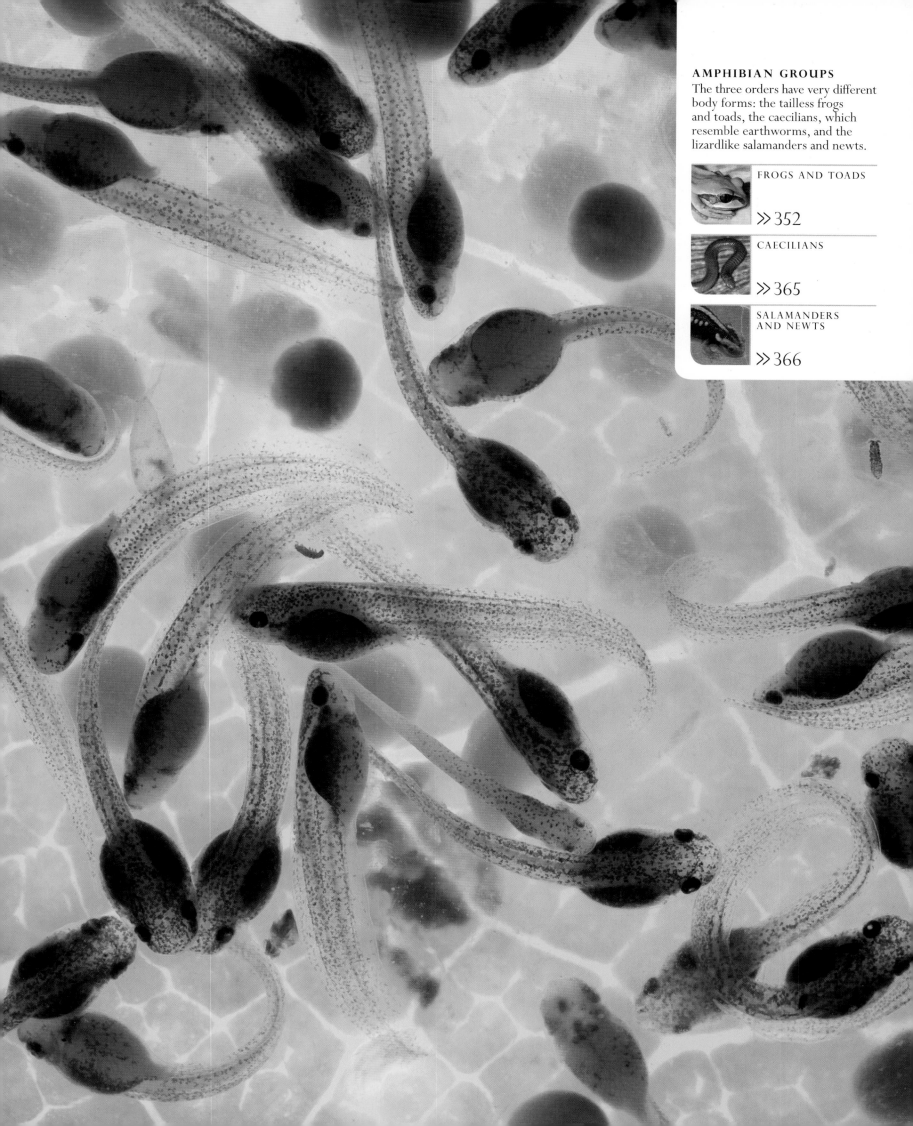

With powerful hind limbs tucked under its body, a wide mouth, and protruding eyes, the typical frog has a unique and unmistakable body shape.

The name of the order of frogs and toads, Anura means "animals without a tail." As adults, all other amphibians have tails, but in the anurans, the tail gradually disappears during metamorphosis from the larval to the adult stage of the life cycle. The larvae, known as tadpoles, feed chiefly on plant material and have a spherical body that contains the long, coiled gut required by such a diet. The adults, in contrast, are totally carnivorous, feeding on a wide range of insects and other invertebrates, with larger species also taking small reptiles and mammals, as well as other frogs.

INGENIOUS ADAPTATIONS

Frogs and toads catch their prey by ambush, most spectacularly by jumping. In many frogs and toads, the hind limbs are modified for jumping, being much longer than the forelimbs and very muscular. Jumping is also an effective method of escaping from predators, of which frogs and toads

have many. Not all anurans jump, however. Many have hind limbs adapted for other kinds of locomotion, including swimming, burrowing, climbing, and, in a few species, gliding through the air. Most frogs and toads live in damp habitats, close to the pools and streams in which they breed, but there are several species adapted to life in very arid areas. The greatest diversity of anurans is found in the tropics, particularly in rain forests. Many are active by day, while others are nocturnal. Some species are cleverly camouflaged; in contrast, others are brightly colored, advertising that they are poisonous or unpleasant to taste.

COURTSHIP AND BREEDING

Anurans differ from other amphibians in having a voice and very good hearing. Males of most species call to attract females, making sounds characteristic of their species. In all but a very few species, fertilization is external, the male shedding sperm onto the eggs as they emerge from the female's body. To do this, the male clasps the female from above—a position known as amplexus. The duration of amplexus varies between species—from a few minutes to several days.

PHYLUM	CHORDATA
CLASS	AMPHIBIA
ORDER	ANURA
FAMILIES	38
SPECIES	5891

MIDWIFE TOADS

Males of the Alytidae, a small family of terrestrial frogs, call at night to attract females. When mating, the male attaches fertilized eggs to his back. He carries them until they are ready to hatch, when he releases the tadpoles into water. Occasionally, a male may carry egg strings from more than one female.

SQUEAKERS AND RELATIVES

The Arthroleptidae are a large, diverse family of frogs occurring across sub-Saharan Africa—in forest, woodland, and grassland, some at high altitudes. Its members range from tiny "squeakers," named for their high-pitched call, to large tree frogs.

¾–1¼ in
2–3 cm

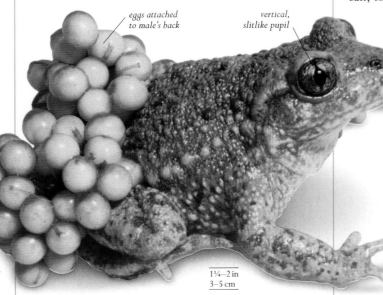

eggs attached to male's back

vertical, slitlike pupil

MIDWIFE TOAD
Alytes obstetricans
Found across much of mainland Europe; the midwife toad has a plump body and powerful forelimbs adapted for digging. It hides in a burrow by day.

1¼–2 in
3–5 cm

1¼–1½ in
3–4 cm

RIO BENITO LONG-FINGERED FROG
Cardioglossa gracilis
An inhabitant of lowland forest, this frog breeds in streams. Males call from nearby slopes.

very long third finger

WEST AFRICAN SCREECHING FROG
Arthroleptis poecilonotus
The female of this small species lays large eggs in soil cavities. The male is noted for his loud call.

1–1½ in
2.5–4 cm

1½–2¼ in
4–5.5 cm

WEST CAMEROON FOREST TREE FROG
Leptopelis nordequatorialis
This large tree frog lives in montane grasslands of West Africa. Males call to females near water when breeding, and eggs are deposited in ponds or marshes.

AFRICAN TREE FROG
Leptopelis modestus
This species is found near streams in forests of West and Central Africa. Females are larger than males.

GLASS FROGS

Found in Central and South America, frogs of the family Centrolenidae are known as "glass frogs" because many species have transparent skin on the underside, through which their internal organs are visible.

silver eyes with black reticulations

LIMON GIANT GLASS FROG
Sachatamia ilex
This arboreal frog lives in wet vegetation near streams. It has dark green bones, visible through its skin.

¾–1¼ in
2–3 cm

yellowish green feet

1–1½ in
2.5–3.5 cm

FLEISCHMANN'S GLASS FROG
Hyalinobatrachium fleischmanni
Males of this species are territorial, using calls to defend their territory and to attract females. The females lay eggs on leaves over water.

WHITE-SPOTTED COCHRAN FROG
Sachatamia albomaculata
This species is found in wet lowland forests, and breeds near streams. Males call to females from low vegetation nearby.

¾–1¼ in
2–3 cm

EMERALD GLASS FROG
Espadarana prosoblepon
Males of this arboreal frog are fiercely territorial, defending their space by calling, and occasionally fighting rivals while hanging upside down.

¾–1¼ in
2–3 cm

CERATOPHRYIDAE

These South American horned frogs have very large heads and wide mouths that enable them to eat animals nearly as large as themselves. They are "sit-and-wait" predators, remaining still and well-camouflaged before their prey comes within range.

"horn" above eye

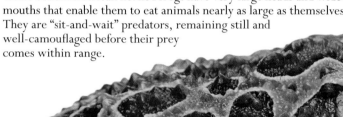

ORNATE HORNED FROG
Ceratophrys ornata
A voracious predator living in grasslands in Argentina; this species breeds after heavy rain, in temporary pools and ditches.

wide mouth

3½–5 in
8–13 cm

CRANWELL'S HORNED FROG
Ceratophrys cranwelli
Spending much of its life underground, this large frog emerges after heavy rain to mate and lay its eggs in pools.

1½–4 in
4–10 cm

3½–5½ in
9–14 cm

BUDGETT'S FROG
Lepidobatrachus laevis
This frog has a flattened body, a wide mouth, and fangs. It spends dry periods in a cocoon below ground, emerging to breed after rain.

PAINTED FROGS

Confined to Europe and northwest Africa, the Discoglossidae emerge at night from burrows, which they build by digging head first. In some species, the female calls in response to a male's call.

2¼–2¾ in
6–7 cm

PAINTED FROG
Discoglossus pictus
The painted frog gets its name from its bright coloration. A female mates with several males, producing up to 1,000 eggs.

ROBBER FROGS

Found in North, Central, and South America, the Craugastoridae produce eggs that develop directly into small adults, without a tadpole stage. Eggs may be deposited on the ground or in vegetation. Many species show parental care of the eggs.

¾–2 in
2–5 cm

1–2¼ in
2.5–5.5 cm

FITZINGER'S ROBBER FROG
Craugastor fitzingeri
In this forest species, the male calls to the larger female from an elevated perch. Eggs are laid in the ground and are guarded by the female.

1¼–2¾ in
3–7 cm

BROAD-HEADED RAIN FROG
Craugastor megacephalus
This Central American frog hides in a burrow by day and emerges at night. It lays its eggs in leaf litter.

ISLA BONITA ROBBER FROG
Craugastor crassidigitus
Native to the humid forests of Central America, this ground-living frog is also found in coffee plantations and pasture.

TRUE TOADS

Distributed worldwide, the family Bufonidae is large and diverse. Its members are characterized by shortened forelimbs, hind limbs that are used for walking or hopping, dry warty skin, and parotoid glands behind the eyes. However, this family also includes the more slender and long-limbed harlequin toads of South America.

horizontal pupil

green blotches on back

warty skin

3½–4¾ in
9–12 cm

RAUCOUS TOAD
Amietophrynus rangeri
Common in southern Africa, this thickset toad breeds in dams and ponds. Males produce rasping, ducklike calls to attract females.

2–4½ in
5–11.5 cm

2–4 in
5–10 cm

MALAYAN TREE TOAD
Pedostibes hosii
Found in eastern Asia, this toad is unusual in living a largely arboreal life. It has adhesive pads on its toes that enable it to climb trees.

GREEN TOAD
Pseudepidalea viridis
A native of sandy habitats, this colorful toad is found in Europe and western Asia. It emerges from its burrow in spring to breed in ponds.

2–4 in
5–10 cm

NATTERJACK
Epidalea calamita
Compared with other true toads, this species has short legs and runs like a mouse. Occurring across Europe, it breeds from spring to summer.

CERATOBATRACHIDAE

Found in Southeast Asia, China, and several Pacific islands, frogs of this family produce large eggs that hatch directly into small frogs. In many species, the tips of the fingers and toes are enlarged.

FIJI GROUND FROG
Platymantis vitianus
The populations of this species on several of the Fijian Islands have been wiped out following the introduction of mongooses.

1–4¼ in
2.5–11 cm

hornlike projection above eye

flat, triangular head

2–3¼ in
5–8 cm

SOLOMON ISLANDS HORNED FROG
Ceratobatrachus guentheri
This species has a pointed snout and hornlike projections above its eyes. It hides among dead leaves.

BOMBINATORIDAE

These small aquatic toads are found in Europe and Asia. They have flattened bodies and many are brightly colored. Fire-bellied toads are active by day, but the dull-colored barbourulas from the Philippines and Borneo are nocturnal.

prominent eyes

bright red underside

bright green coloration

1¼–2 in
3–5 cm

ORIENTAL FIRE-BELLIED TOAD
Bombina orientalis
Found in China and Korea, this small, squat frog can produce a toxic skin secretion. If attacked, it displays its bright belly colors.

RAIN FROGS

Members of the family Brevicipitidae are found in eastern and southern Africa. During mating, the much smaller male is glued onto a female's back by a special skin secretion.

1¼–2 in
3–5 cm

DESERT RAIN FROG
Breviceps macrops
Living away from standing water, this burrowing frog lives and breeds among Namibian sand dunes, occasionally moistened by sea fog.

AMERICAN TOAD
Anaxyrus americanus
Found in eastern
North America, this toad
is quite variable in color.
Breeding occurs in ponds
where males produce
long trilling calls.

2–3½ in
5–9 cm

large parotoid glands

tubular, warty legs

TRUANDO TOAD
Rhaebo haematiticus
Living among leaf litter in forests
of Central and South America, this
broad-headed toad lays its eggs in
long strings in rocky pools.

1½–3¼ in
4–8 cm

VARIABLE HARLEQUIN TOAD
Atelopus varius
This aggressive species from Panama
and Costa Rica has vivid and variable
coloration. It lives near streams and
is active during the day.

1–2¼ in
2.5–6 cm

GUYANAN STUBFOOT TOAD
Atelopus barbotini
This small toad from Guyana has a
flattened body. It breeds throughout
the year in forest streams.

1–1½ in
2.5–4 cm

PANAMANIAN GOLDEN TOAD
Atelopus zeteki
This brightly colored frog from Panama
breeds in pools after heavy rain. It may
now be extinct in the wild.

2–4 in
5–10 cm

EUROPEAN COMMON TOAD
Bufo bufo
Found throughout Europe and North
Africa, this species breeds in spring,
with males outnumbering the much
larger females by about three to one.

3¼–8 in
8–20 cm

GREEN CLIMBING FROG
Incilius coniferus
Native to Central and South
America, this nocturnal toad
is often found climbing
among vegetation.

2¼–3¾ in
5.5–9.5 cm

parotoid glands
secrete toxin

olive-brown, warty skin

CANE TOAD
Rhinella marina
One of the world's
largest toads, this
American species was
introduced in Australia,
where it is now a serious
threat to native wildlife.

4–9½ in
10–24 cm

CYCLORAMPHIDAE

This family, found in South America, includes
Darwin's frog and a number of species that
have coloration and hornlike projections
that enable them to mimic the
appearance of dead leaves.

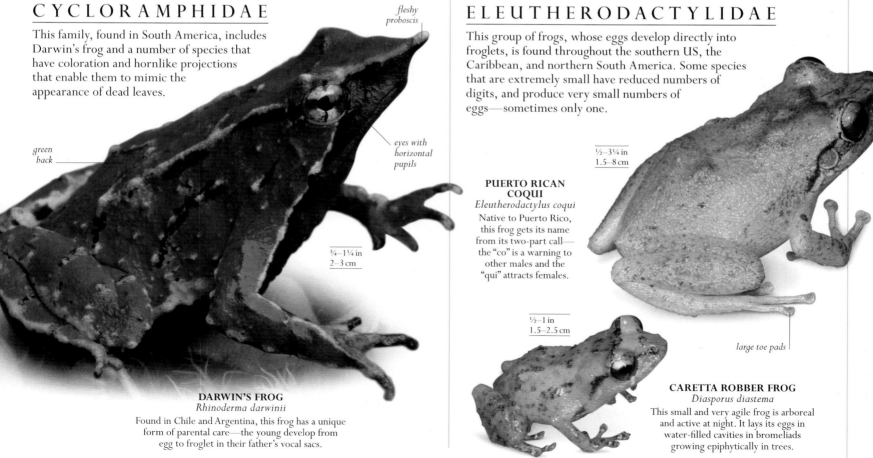

green back

fleshy proboscis

eyes with horizontal pupils

¾–1¼ in
2–3 cm

DARWIN'S FROG
Rhinoderma darwinii
Found in Chile and Argentina, this frog has a unique
form of parental care—the young develop from
egg to froglet in their father's vocal sacs.

ELEUTHERODACTYLIDAE

This group of frogs, whose eggs develop directly into
froglets, is found throughout the southern US, the
Caribbean, and northern South America. Some species
that are extremely small have reduced numbers of
digits, and produce very small numbers of
eggs—sometimes only one.

½–3¼ in
1.5–8 cm

**PUERTO RICAN
COQUI**
Eleutherodactylus coqui
Native to Puerto Rico,
this frog gets its name
from its two-part call—
the "co" is a warning to
other males and the
"qui" attracts females.

½–1 in
1.5–2.5 cm

large toe pads

CARETTA ROBBER FROG
Diasporus diastema
This small and very agile frog is arboreal
and active at night. It lays its eggs in
water-filled cavities in bromeliads
growing epiphytically in trees.

CANE TOAD
Rhinella marina

One of the largest toads in the world, the cane toad is a hardy creature with a huge appetite. Also known as the marine toad, it is mainly a resident of dry environments, scrub, and savanna. It commonly lives around human settlements and is often seen under streetlights, waiting for insects to fall. The female is larger than the male, and the largest females can lay more than 20,000 eggs in a single clutch. Males attract females with a slow, low-pitched trill. They have few enemies because at all stages of their life cycle they are distasteful or toxic to potential predators. In Australia they have become a major pest—they are poisonous to native and domestic animals, injurious to humans, and breed so prolifically as to be out of control.

SIZE 4–9½ in (10–24 cm)
HABITAT Nonforested habitats
DISTRIBUTION Central and South America; introduced to Australia and elsewhere
DIET Terrestrial invertebrates

NIGHT HUNTER >
Protected by their poisonous skin and unafraid of predators, cane toads emerge at night from their daytime hiding places to hop about in search of prey.

dark belly markings

short legs have powerful muscles for jumping

∧ PALE BELLY
The cane toad's belly and throat are relatively smooth, and mostly pale in color. Toads have permeable skin and must hide by day to conserve water.

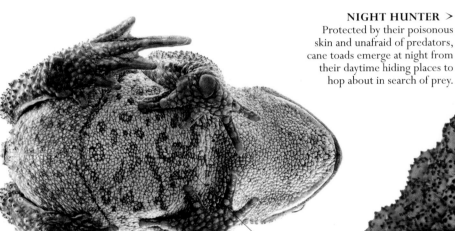

DEBATE
PEST CONTROL

The cane toad gets its name from Australia, where it was introduced into Queensland in 1935 to control insect pests on sugar cane farms. The toads flourished in Australia, feasting on native fauna and building up much denser populations than in their native habitat. Still spreading at an alarming rate, they now occur throughout eastern and northern Australia, and are likely to migrate further. Scientists are working on methods to limit their numbers and contain their territorial expansion.

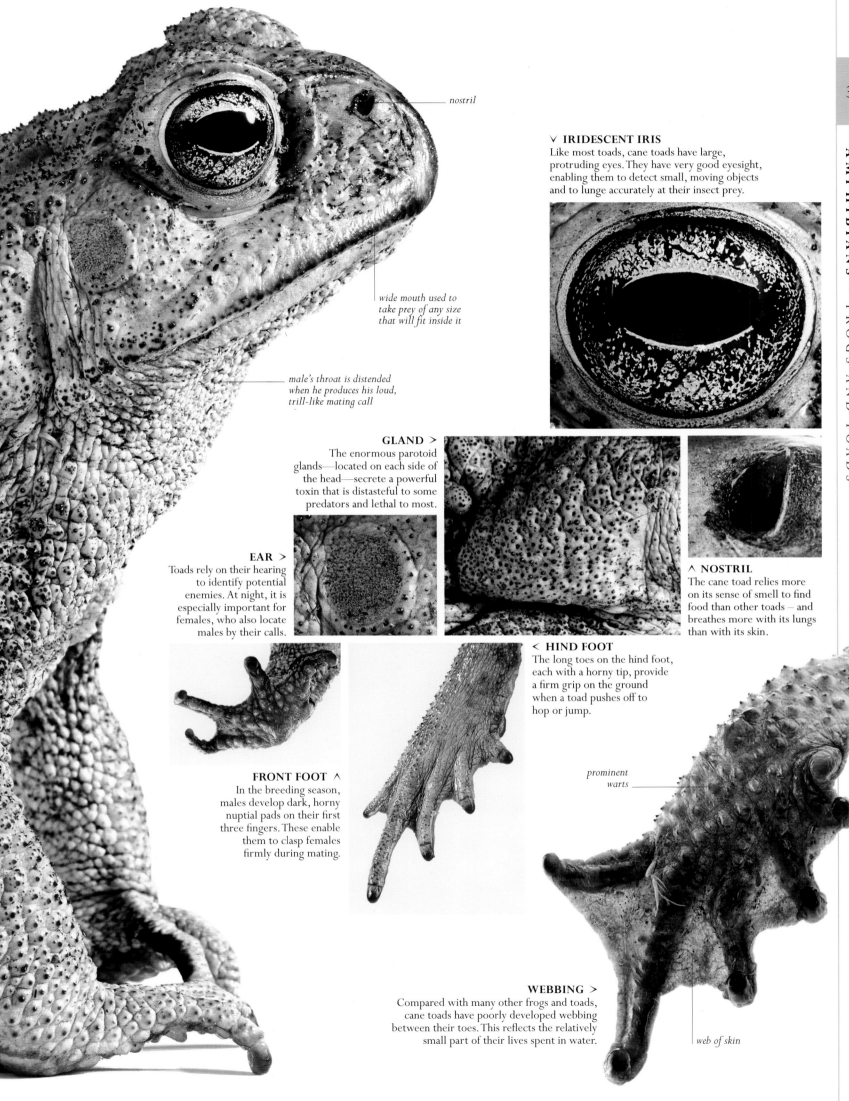

nostril

*wide mouth used to
take prey of any size
that will fit inside it*

*male's throat is distended
when he produces his loud,
trill-like mating call*

∨ IRIDESCENT IRIS
Like most toads, cane toads have large,
protruding eyes. They have very good eyesight,
enabling them to detect small, moving objects
and to lunge accurately at their insect prey.

GLAND >
The enormous parotoid
glands—located on each side of
the head—secrete a powerful
toxin that is distasteful to some
predators and lethal to most.

EAR >
Toads rely on their hearing
to identify potential
enemies. At night, it is
especially important for
females, who also locate
males by their calls.

∧ NOSTRIL
The cane toad relies more
on its sense of smell to find
food than other toads – and
breathes more with its lungs
than with its skin.

< HIND FOOT
The long toes on the hind foot,
each with a horny tip, provide
a firm grip on the ground
when a toad pushes off to
hop or jump.

FRONT FOOT ∧
In the breeding season,
males develop dark, horny
nuptial pads on their first
three fingers. These enable
them to clasp females
firmly during mating.

*prominent
warts*

WEBBING >
Compared with many other frogs and toads,
cane toads have poorly developed webbing
between their toes. This reflects the relatively
small part of their lives spent in water.

web of skin

CRYPTIC FOREST FROGS

Found in South and Central America, this small family of frogs, the Aromobatidae, is closely related to poison-dart frogs, but does not produce toxic skin secretions. Most are cryptically colored.

white stripe on dark brown back

1–1½ in
2.5–3.5 cm

BRILLIANT-THIGHED POISON FROG
Allobates femoralis
The male of this South American species guards the eggs laid by the female in leaf nests. Later, he carries the tadpoles to water on his back.

MARSUPIAL FROGS

Found in South and Central America, the Hemiphractidae carry their eggs on their back, where they hatch directly into froglets. Some also carry the eggs in a pouch, hence the name "marsupial frogs."

2½–3¼ in
6.5–8 cm

HORNED MARSUPIAL FROG
Gastrotheca cornuta
The female of this canopy-dwelling Central and South American species lays large eggs that develop in a pouch on her back.

1¾–2½ in
4.5–6.5 cm

SUMACO HORNED TREEFROG
Hemiphractus proboscideus
This species is found in Colombia, Ecuador, and Peru. The female carries her eggs on her back, but does not have a pouch.

POISON-DART FROGS

Called poison or poison-dart frogs, the Dendrobatidae are noted for their bright coloration. This warns predators that the skin of these frogs contains powerful toxins, which are derived from their insect food. These frogs are found in the forests of Central and South America. They are active by day.

¾ in
2 cm

long, slender legs

MIMIC POISON FROG
Ranitomeya imitator
Found in Peru, this frog is variable in color and is similar in appearance to at least three other frog species.

¾–1 in
2–2.5 cm

1¼–1¾ in
3–4.5 cm

LOVELY POISON-DART FROG
Phyllobates lugubris
This poisonous frog is found in leaf litter of lowland forests from Nicaragua to Panama. The male cares for the eggs and tadpoles.

GOLDEN POISON FROG
Phyllobates terribilis
Possibly the most toxic of all poison-dart frogs, this terrestrial species occurs in lowland forests of Colombia.

⅜–¾ in
1–2 cm

1½–1¾ in
3.5–4.5 cm

RAIN FORESTROCKET FROG
Silverstoneia flotator
This frog is found in Costa Rica and Panama. Its tadpoles have an upward-tilting mouth, which helps them feed from the water surface.

THREE-STRIPED POISON FROG
Ameerega trivittata
This South American frog calls by day, especially after rain. It lays its eggs in leaf litter and is common near human settlements.

REED AND SEDGE FROGS

Also called African tree frogs, this large family, the Hyperoliidae, includes many agile climbers that gather in trees, bushes, or reeds near water to find mates and lay eggs. Some of the species are brightly colored, with marked differences between the sexes.

red patch on leg

prominent eyes

FOULASSI BANANA FROG
Afrixalus paradorsalis
Found in West Africa, this frog lays its eggs in a folded leaf above water. The male attracts females with a clicklike call.

1–1½ in
2.5–3.5 cm

2¼–2½ in
5.5–6.5 cm

RED-LEGGED KASSINA
Kassina maculata
This is an aquatic East African frog with adhesive discs on its toes. Its eggs, laid on submerged vegetation, hatch into large tadpoles.

GREEN AND BLACK POISON-DART FROG
Dendrobates auratus
The males of this species fight to secure a territory. The male defends the eggs and, when they hatch, carries tadpoles to small pools in tree holes.

1–2¼ in
2.5–6 cm

1¼–1½ in
3–4 cm

YELLOW-HEADED POISON FROG
Dendrobates leucomelas
Found in wet forests of northern South America, this frog derives its skin toxin from the ants on which it feeds.

rounded snout

bright blue skin

1¼–1¾ in
3–4.5 cm

long forelimbs

DYEING POISON FROG
Dendrobates tinctorius
This South American species is quite variable in color. Both sexes protect the eggs and are very aggressive while defending their territory.

bright red body

toes with adhesive pads

¾–1 in
2–2.5 cm

STRAWBERRY POISON FROG
Oophaga pumilio
The female of this species cares for the young by carrying tadpoles to water-filled tree holes and feeding them with unfertilized eggs.

1¼–1½ in
3–4 cm

SPLASHBACK POISON FROG
Adelphobates galactonotus
Living in leaf litter in Brazilian forests, this frog lays eggs on the ground and carries the tadpoles to water.

½–¾ in
1.5–2 cm

RIO MADEIRA POISON FROG
Adelphobates quinquevittatus
Found in Brazil and Peru, this tiny frog carries its tadpoles to water-filled holes, where the female feeds them with unfertilized eggs.

¾ in
2 cm

GRANULAR POISON FROG
Oophaga granulifera
This species is found in Costa Rica and Panama. Females care for the young by feeding them unfertilized eggs.

¾–1 in
2–2.5 cm

BRAZIL-NUT POISON FROG
Adelphobates castaneoticus
This species is found in Brazil. The males of this frog place tadpoles individually into small, water-filled tree holes. The tadpoles are voracious feeders.

SMITH'S REED FROG
Hyperolius tuberilinguis
This agile frog is noted for its loud call. In the mating season, thousands of males may gather around a pond to produce a deafening chorus.

1¼–1¾ in
3–4.5 cm

large adhesive disc on toes

¾–1½ in
2–3.5 cm

large eyes

BOLIFAMBA REED FROG
Hyperolius bolifambae
This small West African frog occurs in bushland, and breeds in pools. The male's call is a high-pitched buzz.

AUSTRALIAN GROUND FROGS

The Limnodynastidae, found in Australia and New Guinea, include many terrestrial and burrowing frog species. Two recently extinct species uniquely brooded their eggs in the stomach.

1¼–2¼ in
3–6 cm

BROWN-STRIPED MARSH FROG
Limnodynastes peronii
This Australian frog survives dry periods by burying itself in the ground. It emerges to breed after heavy rain, and deposits its eggs in floating foam nests.

TREE FROGS

The Hylidae are a large family widespread worldwide. They are especially well represented in the New World. These frogs have long, slender limbs and adhesive discs on their fingers and toes. Most are arboreal and nocturnal. Many gather to breed in noisy choruses.

brown upperparts with dark patches

¾–1¼ in
2–3 cm

SPRING PEEPER
Pseudacris crucifer
Found in moist woodland in the eastern US and Canada, the spring peeper's distinctive high-pitched call indicates that spring is underway.

2–2¾ in
5–7 cm

PARADOX FROG
Pseudis paradoxa
This aquatic frog is so named because its tadpoles are four times longer than the adult. It is found in South America and Trinidad.

2–3½ in
5–9 cm

SPLENDID LEAF FROG
Cruziohyla calcarifer
This frog occurs in Central and northern South America and lives high up in trees. It glides from one tree to another, using its extended webbed feet as parachutes.

1–1½ in
2.5–4 cm

RUFOUS-EYED STREAM FROG
Duellmanohyla rufioculis
Found in forests of Costa Rica, this frog breeds in fast-flowing streams. Its tadpoles have modified mouths with which they attach themselves to rocks.

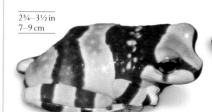

2¾–3½ in
7–9 cm

MISSION GOLDEN-EYED TREEFROG
Trachycephalus resinifictrix
Found high in the canopy of South American forests, this frog lays its eggs in water-filled tree holes. The tadpoles develop into froglets.

2¼–3 in
5.5–7.5 cm

ROSENBERG'S TREEFROG
Hypsiboas rosenbergi
Males of this Central and South American frog dig pools in damp ground in which the eggs are laid. They defend them against rivals in fights that can be fatal.

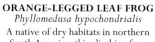

1½–2 in
4–5 cm

ORANGE-LEGGED LEAF FROG
Phyllomedusa hypochondrialis
A native of dry habitats in northern South America, this climbing frog reduces water loss by rubbing a waxy secretion over its skin.

adhesive discs on toes

2¼–4¼ in
5.5–11 cm

COPE'S BROWN TREEFROG
Ecnomiohyla miliaria
This large Central American tree frog has fringes of skin on the limbs that may enable it to glide from one tree to another.

1½–2¼ in
3.5–5.5 cm

BOULENGER'S SNOUTED TREE FROG
Scinax boulengeri
Found in Central America and Colombia, this frog breeds in temporary pools created after rain. Males return night after night to the same calling perch to attract mates.

1–4 in
2.5–10 cm

CUBAN TREEFROG
Osteopilus septentrionalis
Native to Cuba, the Cayman Islands, and the Bahamas, this tree frog has been introduced to Florida, where it preys on native frogs, reducing their population.

1¼–2 in
3–5 cm

EUROPEAN TREEFROG
Hyla arborea
Males of this European tree frog gather in spring, calling in loud choruses to attract females. Mating pairs descend to lay eggs in nearby ponds.

large, prominent red eyes

pale underparts

1½–2¾ in
4–7 cm

RED-EYED TREE FROG
Agalychnis callidryas
An excellent climber, this tree frog mates in trees overhanging water. It lays its eggs on leaves and, upon hatching, the tadpoles fall into the water.

1¼–2 in
3–5 cm

LEMUR FROG
Agalychnis lemur
This Central American tree frog is nocturnal, sleeping on the underside of leaves during the day. It lays its eggs on leaves over water.

2¾–4 in
7–10 cm

MANAUS SLENDER-LEGGED TREEFROG
Osteocephalus taurinus
This arboreal frog lives in South American forests. It mates after rain, and lays its eggs on the surface of pools.

SMALL-HEADED TREE FROG
Dendropsophus microcephalus
Found in Central and South America and Trinidad, this frog breeds in pools. It is pale yellow by day, and red-brown by night.

¾–1¼ in
2–3 cm

horizontal pupils

2–4 in
5–10 cm

AUSTRALIAN GREEN TREE FROG
Litoria caerulea
Found in northeast Australia and New Guinea, this agile climber is often found close to human habitations.

NEW ZEALAND FROGS

This family, the Leiopelmatidae, is composed of four species of frogs that are all confined to New Zealand. They are characterized by extra vertebrae and, unlike most frogs, they kick their legs alternately when swimming. They live in humid forests and are nocturnal.

1–1½ in
2.5–3.5 cm

COROMANDEL NEW ZEALAND FROG
Leiopelma archeyi
Found only in New Zealand's North Island, this terrestrial frog lays its eggs under logs. It is critically endangered, as a result of habitat loss and disease.

LEIUPERIDAE

This family is composed of a small group of South and Central American frogs. These frogs are active at night and mostly dwell on the ground. The best-known species of the family Leiuperidae is the Túngara frog.

warty skin

1¼–1½ in
3–4 cm

TÚNGARA FROG
Engystomops pustulosus
During mating, the female of this Central American species produces a secretion that the male whips into a floating foam nest in which the eggs are laid.

MANTELLAS

Found only on the islands of Madagascar and Mayotte, the Mantellidae are active by day. Many are brightly colored, warning predators of powerful toxins in their skin. Most species are threatened by the loss of habitat and by the international pet trade.

SPINY-HEADED TREEFROG
Anotheca spinosa
Native to Mexico and Central America, this frog lives in bromeliads and banana plants, laying its eggs in water-filled cavities.

2¼–3¼ in
6–8 cm

1½–3¼ in
4–8 cm

MASKED TREEFROG
Smilisca phaeota
Active at night in humid forests in Central and South America it lays its eggs in small pools.

WHITE-LIPPED BRIGHT-EYED FROG
Boophis albilabris
This large arboreal frog, found only in Madagascar, lives near the streams in which it breeds. It has fully webbed hind feet.

1¾–3¼ in
4.5–8 cm

2–2¼ in
5–6 cm

ELEGANT MADAGASCAN FROG
Spinomantis elegans
An inhabitant of rocky outcrops, this frog is found at high altitudes, even above the tree line. It breeds in streams.

bony projection

2–3 in
5–7.5 cm

YUCATECAN SHOVEL-HEADED TREEFROG
Triprion petasatus
An inhabitant of lowland forests in Mexico and Central America, this frog retreats into tree holes, using the bony projection on its head to seal the opening.

rough, moist skin

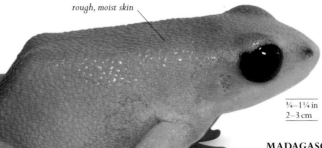

¾–1¼ in
2–3 cm

MADAGASCAN GOLDEN MANTELLA
Mantella aurantiaca
The bright color of this tiny frog from Madagascar rain forests warns potential predators that its skin secretes a powerful toxin.

¾–1 in
2–2.5 cm

PAINTED MANTELLA
Mantella madagascariensis
This Madagascan frog breeds in forest streams. It is under threat due to loss of habitat. The male's call consists of short chirps.

NARROW-MOUTHED FROGS

Frogs of the large and diverse family Microhylidae are found in the Americas, Asia, Australia, and Africa. Most are ground dwelling and some live in burrows. Most have stout hind legs, short snouts, and plump, often teardrop-shaped, bodies.

2–3 in
5–7.5 cm

PAINTED TOAD
Kaloula pulchra
Widespread in Asia, this species has adapted well to human settlements. It protects itself with a noxious, sticky skin secretion.

1¼–2¼ in
3–6 cm

3¼–4¾ in
8–12 cm

TOMATO FROG
Dyscophus antongilii
Native to Madagascar, this frog stays buried in soil by day, emerging to feed at night. A sticky skin secretion protects it against predators.

BLACK-SPOTTED NARROW-MOUTHED FROG
Kalophrynus pleurostigma
This frog from the Philippines protects itself by producing a sticky secretion. It breeds in small pools after rain.

¾–1½ in
2–3.5 cm

¾–1 in
2–2.5 cm

EASTERN NARROW-MOUTHED TOAD
Gastrophryne carolinensis
Found in the southeastern US, this burrowing toad breeds in water bodies of all sizes. The male's call sounds like a bleating lamb.

GIANT STUMP-TOED FROG
Stumpffia grandis
This small terrestrial frog is found in leaf litter in high-altitude forests in Madagascar.

MEGOPHRYIDAE

Found across Asia, this is a small family of frogs whose body shape and color patterns enable them to be camouflaged among leaves. They walk rather than jump, and most are ground living.

hornlike projections on eyelids

cryptic coloration with black markings

ASIAN HORNED FROG
Megophrys nasuta
The "horns" and coloration of this frog enable it to conceal itself among dead leaves, while it waits for prey.

2¾–5½ in
7–14 cm

PARSLEY FROGS

Made up of only three species, the family Pelodytidae is confined to Europe and the Caucasus. Named for the green markings on their skin, parsley frogs breed after rain, laying their eggs in broad strips.

COMMON PARSLEY FROG
Pelodytes punctatus
When climbing smooth, vertical surfaces, this European frog uses its underside as a sucker. Both sexes call during breeding.

1¼–2 in
3–5 cm

TONGUELESS FROGS

These aquatic frogs are well adapted to life in water: they have flattened bodies, fully webbed hind feet, and eyes that protrude upward, enabling them to see above the surface. As their common name suggests, the Pipidae lack tongues. They feed on a wide range of prey and scavenge on dead animals.

eggs on female's back

muscular hind legs

1–1¾ in
2.5–4.5 cm

DWARF SURINAM TOAD
Pipa parva
In this wholly aquatic species found in Venezuela and Colombia, the eggs develop on the female's back.

claws used for tearing food

1¼–2 in
3–5 cm

FRASER'S CLAWED FROG
Xenopus fraseri
Found in West and central Africa, this wholly aquatic frog thrives in human-altered habitats and is harvested by people for food.

SPADEFOOT TOADS

A small family found in Eurasia and North Africa, the Pelobatidae are characterized by horny projections on their hind feet. They use them to burrow into the ground, where they wait for rain.

COMMON SPADEFOOT TOAD
Pelobates fuscus
Found in Europe and Asia, this species is variable in color. It has a plump body, which it inflates when attacked.

1½–3¼ in
4–8 cm

DICROGLOSSIDAE

This diverse family of frogs is found across Africa, Asia, and several Pacific islands. Most are ground living, but are found close to water. Many lay their eggs in water and have free-living tadpoles.

½–¾ in
1.5–2 cm

MARTEN'S PUDDLE FROG
Occidozyga martensii
Found in China and Southeast Asia, this small frog occurs in pools near forest streams and rivers.

1½–2½ in
4–6.5 cm

INDIAN BULLFROG
Hoplobatrachus tigerinus
This large, voracious feeder from southern Asia breeds during the monsoon. The male has a particularly loud call.

2½–6½ in
6.5–17 cm

RAJAMALLY WART FROG
Fejervarya kirtisinghei
Found only in Sri Lanka, this frog lives in leaf litter near streams and thrives in plantations and gardens.

1–1¾ in
2.5–4.5 cm

COMMON SKITTERING FROG
Euphlyctis cyanophlyctis
Widespread in southern Asia, this aquatic frog is noted for its ability to skitter over the surface of water.

TRUE FROGS

Known as true frogs, this large family is found in most parts of the world. Most of the Ranidae have powerful hind limbs that enable them to jump athletically on land and to swim powerfully in water. They typically breed in early spring, with many laying their eggs communally.

powerful hind legs for swimming
long, webbed toes

GOLIATH FROG
Conraua goliath
The world's largest frog, this West African species has an aquatic lifestyle; its powerful legs and webbed feet make it a strong swimmer.
4–16 in
10–40 cm

EDIBLE FROG
Pelophylax esculenta
This frog is a hybrid between the widespread European *Pelophylax lessonae* and other more local species. It lives in and close to water.
3¼–4¾ in
8–12 cm

WOOD FROG
Lithobates sylvaticus
The only American frog found north of the Arctic Circle, it breeds in early spring in temporary pools free of fish.
1½–3¼ in
3.5–8 cm

PICKEREL FROG
Lithobates palustris
Found over much of North America, this frog breeds in spring, with females laying their eggs in clumps containing two to three thousand eggs.
2¼–2¾ in
6–7 cm

black spots on green to brown body
white vocal sac
2–4 in
5–10 cm
male has thick forelimbs

EUROPEAN COMMON FROG
Rana temporaria
Also known as the grass frog, this species lives mostly on land, migrating to ponds in spring to breed. It lays eggs in clumps.

AMERICAN BULLFROG
Lithobates catesbeianus
This voracious predator has tadpoles that may take four years to develop, reaching a large size. It is the largest North American frog.
3½–8 in
9–20 cm

PHRYNOBATRACHIDAE

This family of small, terrestrial or semiaquatic frogs is confined to sub-Saharan Africa. Most breed throughout the year, laying their eggs in water. They reach maturity in five months.

warty skin

½–¾ in
1.5–2 cm

GOLDEN PUDDLE FROG
Phrynobatrachus auritus
This frog is so named because it breeds in very small pools. A ground-dwelling species, it is found in central African rain forests.

ORNATE FROGS AND GRASS FROGS

Found in open country in Africa, Madagascar, and the Seychelles, the Ptychadenidae include many brightly colored frogs. Streamlined bodies and strong hind legs make them prodigious jumpers.

1¾–2¾ in
4.5–7 cm

MASCARENE RIDGED FROG
Ptychadena mascareniensis
Common on agricultural land, this frog has long legs and a pointed snout. It breeds in puddles, wheel ruts, and ditches.

AFRO-ASIAN TREE FROGS

The Rhacophoridae range across Africa and much of Asia, and are mostly arboreal frogs. The family includes flying frogs that glide from tree to tree. Many lay their eggs in foam nests, where the eggs and tadpoles stay protected against predators.

1½–2¼ in
4–6 cm

shiny green coloration

3½–4 in
9–10 cm

1¾–2¼ in
4.5–6 cm

SOUTHERN WHIPPING FROG
Polypedates longinasus
Endangered by the loss of much of its habitat, this arboreal frog occurs in Sri Lanka's remaining patches of rain forest.

AFRICAN FOAM-NEST TREE FROG
Chiromantis rufescens
Found in the forests of West and Central Africa, this frog lays its eggs in a foam nest attached to a branch overhanging water.

2¾–3½ in
7–9 cm

MOSSY FROG
Theloderma corticale
The warty skin and green color of this Vietnamese frog camouflages it against moss. It curls into a ball when threatened.

WALLACE'S FLYING FROG
Rhacophorus nigropalmatus
This arboreal species from the rain forests of Southeast Asia has webbed feet that enable it to glide between trees.

long, fully webbed fore- and hind feet

PYXICEPHALIDAE

Inhabiting sub-Saharan Africa, this family varies in size from huge bullfrogs to typical pond frogs and tiny moss frogs. Most lay eggs in water, although some small species lay eggs on land. All eggs hatch into tadpoles.

olive-green body with dark markings

3¼–9 in
8–23 cm

very wide mouth

AFRICAN BULLFROG
Pyxicephalus adspersus
Males of this large African species aggressively defend their eggs and tadpoles. They dig channels in soil to help the tadpoles reach open water.

powerful limbs

MEXICAN BURROWING TOAD

The only member of the family Rhinophrynidae, the burrowing toad specializes in digging in soil and eating ants. It has a long, thin tongue that it sticks out of its narrow mouth.

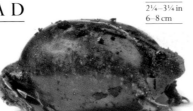

2¼–3¼ in
6–8 cm

MEXICAN BURROWING TOAD
Rhinophrynus dorsalis
This unusually shaped toad spends most of its life burrowing underground, emerging only after rain to breed in temporary pools.

AMERICAN SPADEFOOT TOADS

This family consists of toads that live on dry land and stay inactive underground for long periods. The Scaphiopodidae emerge after rain to breed in temporary pools. These may evaporate quickly, so the tadpoles develop very rapidly.

2¼–3½ in
5.5–9 cm

1½–2¼ in
4–6 cm

PLAINS SPADEFOOT
Spea bombifrons
Found in arid plains of North America, this frog burrows by day. It gathers to breed in large numbers after heavy rain.

COUCH'S SPADEFOOT
Scaphiopus couchii
This North American frog lives in arid areas, spending much of its time underground. It emerges at night to feed and, after heavy rain, to breed.

mottled greenish brown skin

STRABOMANTIDAE

Many small frogs native to South America and the Caribbean make up this family. They are all direct developers: there is no tadpole stage and the egg hatches straight into a miniature adult.

LIMON ROBBER FROG
Pristimantis cerasinus
Hiding in leaf litter by day, but arboreal by night, this small frog is found in humid lowland forests of Central America.

½–1½ in
1.5–3.5 cm

¾–1½ in
2–4 cm

CHIRIQUI ROBBER FROG
Pristimantis cruentus
This small, terrestrial frog, found in Central and South America, lays its eggs in crevices on tree trunks.

½–1 in
1.5–2.5 cm

PYGMY RAINFROG
Pristimantis ridens
Found in the forests of Central and South America, this tiny nocturnal frog thrives in gardens and lays its eggs in leaf litter.

CAECILIANS

Caecilians are long-bodied, limbless amphibians with little or no tail. Ring-shaped folds (annuli) in the skin give them a segmented appearance.

All caecilians live in the tropics. They vary in length from 4¾ in (12 cm) to 5¼ ft (1.6 m). Most live underground, burrowing in soft soil, using their pointed, bony head as a shovel. They emerge at night, especially after rain, to feed on earthworms, termites, and other insects. Others live in water and resemble eels, rarely moving onto the land. These have a fin on the tail. With only rudimentary eyes, all rely on smell to find food and mates. A pair of retractable tentacles between the eyes and the nostrils transmit chemical signals to the nose.

In all caecilians, the eggs are fertilized internally. Some species lay eggs, but in others the eggs are retained inside the female's body. The young emerge either as gilled larvae or as small adults.

PHYLUM	CHORDATA
CLASS	AMPHIBIA
ORDER	GYMNOPHIONA
FAMILIES	6
SPECIES	186

20 in
50 cm

PURPLE CAECILIAN
Gymnopis multiplicata
This terrestrial caecilian from Central America lives in a wide range of habitats. The eggs hatch inside the female.

ICHTHYOPHIIDAE

Found in Asia, these caecilians lay eggs in soil near water. Females remain with their clutches, defending them until the larvae have made their way to open water.

13 in
33 cm

KOH TAO CAECILIAN
Ichthyophis kohtaoensis
Found in a variety of habitats in Southeast Asia, this caecilian lays its eggs on land, but its larvae live in water.

yellow stripe along body

CAECILIIDAE

Most species in this family are burrowers. Found in most tropical regions of the world, they vary greatly in length, some growing to over 5 ft (1.5 m). In some the eggs hatch into larvae; in others the larvae develop inside the female.

26 in
65 cm

CONGO CAECILIAN
Herpele squalostoma
From West and Central Africa, this caecilian lives underground and is found near streams and rivers in lowland forests.

9 in
22 cm

SLENDER CAECILIAN
Dermophis parviceps
Living underground in the wet forests of Central and South America, this slender caecilian emerges to forage at night.

SALAMANDERS AND NEWTS

Unlike their fellow amphibians, the frogs, salamanders and newts normally have slender, lizardlike bodies, long tails, and four legs similar in size.

Newts and salamanders, also known as urodeles, are generally found in damp habitats and are largely confined to the northern hemisphere. They are numerous in the Americas, ranging from Canada to northern South America. They vary considerably in size, from species over 3¼ ft (1 m) in length, to tiny creatures about ¾ in (2 cm) long.

AMPHIBIOUS LIFESTYLES
Some species, notably the newts, spend part of their life in water, part on land. Some salamander species live their entire lives in water, while others are wholly terrestrial. Most have smooth, moist skin through which they breathe to a greater or lesser extent.

The salamanders of one family, the plethodontids, have no lungs and breathe entirely through their skin and the roof of their mouths. Urodeles have relatively small heads, compared with frogs and toads; they also have smaller eyes, smell being the most important sense used in finding food and in social interactions. Most species, especially the terrestrial ones, are nocturnal, hiding under a log or rock during the day.

REPRODUCTION
In the majority of species, the eggs are fertilized inside the female. Males, however, do not have a penis but package sperm in capsules called spermatophores which are passed to the female during mating. In many species sperm transfer is preceded by elaborate courtship in which the male induces the female to cooperate with him. She may, of course, reject his advances. In many newts, males develop dorsal crests and bright colors in the breeding season.

Many species lay their eggs in water. When the larvae hatch, they have long, slender bodies, deep, finlike tails, and large, feathery external gills. The larvae are carnivorous, feeding on tiny water creatures. The exceptions to this rule are the wholly terrestrial species of salamander. These lay their eggs on land and the larval stage is completed within the egg, which hatches to produce a miniature adult.

PHYLUM	CHORDATA
CLASS	AMPHIBIA
ORDER	CAUDATA
FAMILIES	10
SPECIES	585

A male alpine newt sniffs a female before courting her. Odor helps to identify the gender and species of potential partners.

SIRENS

The wholly aquatic salamanders of the family Sirenidae are found in the southern US and Mexico. They retain larval features in the adult stage and resemble eels. They have external gills and no hind limbs.

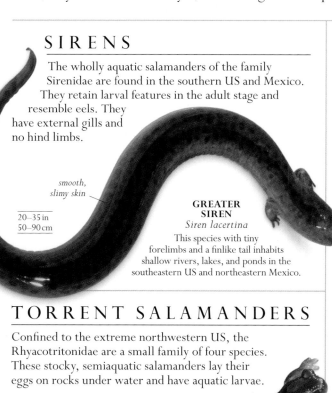

smooth, slimy skin

20–35 in
50–90 cm

GREATER SIREN
Siren lacertina
This species with tiny forelimbs and a finlike tail inhabits shallow rivers, lakes, and ponds in the southeastern US and northeastern Mexico.

TORRENT SALAMANDERS

Confined to the extreme northwestern US, the Rhyacotritonidae are a small family of four species. These stocky, semiaquatic salamanders lay their eggs on rocks under water and have aquatic larvae.

COLUMBIA TORRENT SALAMANDER
Rhyacotriton kezeri
Confined to the forests of Oregon and Washington, this salamander lays its eggs in springs. Intensive logging has caused its numbers to decline considerably.

3–4½ in
7.5–11.5 cm

NEWTS AND EUROPEAN SALAMANDERS

The Salamandridae are small- and medium-sized salamanders and newts found in North America, Europe, North Africa, and Asia. Eggs are fertilized inside the female, with sperm being transferred from the male in a spermatophore (sperm capsule) during an elaborate courtship.

4¾–8 in
12–20 cm

CALIFORNIA NEWT
Taricha torosa
This nocturnal newt enters ponds in spring to mate and lay eggs. It secretes a lethal nerve toxin to discourage predators.

orange poison glands

4¾–7 in
12–18 cm

CROCODILE NEWT
Tylototriton verrucosus
Found in central Asia, this robust newt breeds in ponds after the monsoon. Its orange markings signal distasteful secretions.

broad head with rounded jaws

flattened tail

6–12 in
15–30 cm

SHARP-RIBBED SALAMANDER
Pleurodeles waltl
This large Spanish and Moroccan species has a unique mode of defense. When grasped, it pushes out the sharp tips of its ribs through its skin.

cylindrical tail

blue markings on tail

2¼–4¾ in
6–12 cm

orange underside

ALPINE NEWT
Ichthyosaura alpestris
Found across much of
northern Europe, this newt
breeds in early spring. The
female wraps eggs individually
in the leaves of water plants.

2¾–4 in
7–10 cm

7–11 in
18–28 cm

SPECTACLED SALAMANDER
Salamandrina terdigitata
This secretive salamander is found only
in Italy and lives in streams in hilly areas.
It has a long, slender, and flattened body.

5–6½ in
13–17 cm

SPOTLESS STOUT NEWT
Pachytriton labiatus
Found in mountain streams in China,
this newt is equipped with a large tail for
swimming. It attaches its eggs to rocks.

large,
prominent eyes

4–5½ in
10–14 cm

**SARDINIAN BROOK
SALAMANDER**
Euproctus platycephalus
Found only in Sardinia, this slender
salamander lives in streams, laying its
eggs under rocks. It is endangered
due to degradation of its habitat.

FIRE SALAMANDER
Salamandra salamandra
This European salamander spends most
of its time on land, entering water only
to lay eggs. The glands on its head can
spray a toxic secretion at enemies.

large glands
secrete toxin

3½–5 in
9–13 cm

2½–5½ in
6.5–14 cm

EASTERN NEWT
Notophthalmus viridescens
This pond-breeding newt is from eastern
North America. Juveniles or "efts" are terrestrial,
bright red in color, and very toxic.

LORESTAN NEWT
Neurergus kaiseri
Found only in Iran, this species lives
in streams. It is critically endangered
due to loss of its habitat and its
popularity as a pet.

orange and
black limbs

3½–4¾ in
9–12 cm

2¾–4 in
7–10 cm

JAPANESE NEWT
Cynops pyrrhogaster
This newt spends much of its life in water. Its
brightly colored belly warns potential predators
that the glands in its skin secrete a toxin.

SMOOTH NEWT
Lissotriton vulgaris
Common across Europe and western
Asia, this small amphibian breeds in
ponds. The male mates with the female
after elaborate courtship.

GREAT CRESTED NEWT
Triturus cristatus
Found across Europe and central Asia, males of this large,
pond-breeding species develop a spectacular dorsal crest
in spring, and display it vigorously to potential mates.

4–7 in
10–18 cm

MARBLED NEWT
Triturus marmoratus
Found in France and Spain, this newt lives
in woodland, meadows, and hedgerows.
It enters ponds to breed in spring.

4–5½ in
10–14 cm

LUNGLESS SALAMANDERS

Containing over 390 species, the Plethodontidae are the largest salamander family. Members of this family have no lungs, but breathe through their mouth and skin. Except for six European species, all live in North, Central, and South America, inhabiting a wide variety of habitats and feeding mainly on small invertebrates.

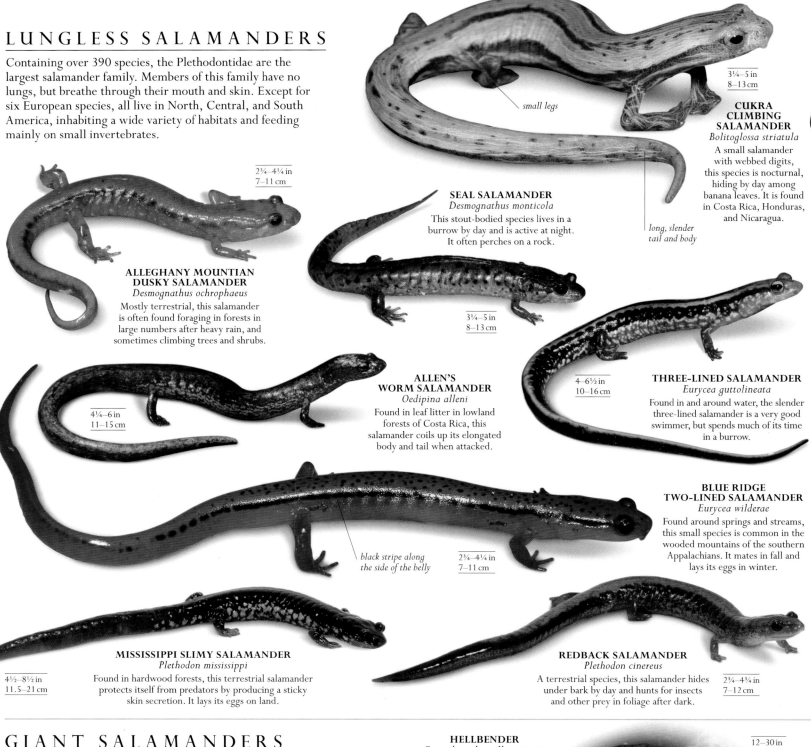

small legs

2¾–4¼ in
7–11 cm

3¼–5 in
8–13 cm

CUKRA CLIMBING SALAMANDER
Bolitoglossa striatula
A small salamander with webbed digits, this species is nocturnal, hiding by day among banana leaves. It is found in Costa Rica, Honduras, and Nicaragua.

long, slender tail and body

ALLEGHANY MOUNTIAN DUSKY SALAMANDER
Desmognathus ochrophaeus
Mostly terrestrial, this salamander is often found foraging in forests in large numbers after heavy rain, and sometimes climbing trees and shrubs.

SEAL SALAMANDER
Desmognathus monticola
This stout-bodied species lives in a burrow by day and is active at night. It often perches on a rock.

3¼–5 in
8–13 cm

4–6½ in
10–16 cm

THREE-LINED SALAMANDER
Eurycea guttolineata
Found in and around water, the slender three-lined salamander is a very good swimmer, but spends much of its time in a burrow.

4¼–6 in
11–15 cm

ALLEN'S WORM SALAMANDER
Oedipina alleni
Found in leaf litter in lowland forests of Costa Rica, this salamander coils up its elongated body and tail when attacked.

BLUE RIDGE TWO-LINED SALAMANDER
Eurycea wilderae
Found around springs and streams, this small species is common in the wooded mountains of the southern Appalachians. It mates in fall and lays its eggs in winter.

black stripe along the side of the belly

2¾–4¼ in
7–11 cm

4½–8½ in
11.5–21 cm

MISSISSIPPI SLIMY SALAMANDER
Plethodon mississippi
Found in hardwood forests, this terrestrial salamander protects itself from predators by producing a sticky skin secretion. It lays its eggs on land.

REDBACK SALAMANDER
Plethodon cinereus
A terrestrial species, this salamander hides under bark by day and hunts for insects and other prey in foliage after dark.

2¾–4¾ in
7–12 cm

GIANT SALAMANDERS

The Cryptobranchidae are three large, wholly aquatic species—one each from North America, Japan, and China. They feed on a wide variety of prey, from worms to small mammals. The family includes the world's largest salamander—the Chinese giant salamander, which is about 6 ft (1.8 m) long.

HELLBENDER
Cryptobranchus alleganiensis
This North American species has a flattened head that helps it dig a burrow under rocks, where the males guard the eggs. It has very wrinkled skin.

12–30 in
30–75 cm

flattened body

3¼–4½ ft
1–1.4 m

JAPANESE GIANT SALAMANDER
Andrias japonicus
This aquatic species is threatened by degradation of its habitat. Some males, called "den masters," defend burrows in which females lay eggs.

splayed legs

SPRING SALAMANDER
Gyrinophilus porphyriticus

4¾–7½ in
12–19 cm

An agile inhabitant of mountain streams and springs, this colorful salamander is most commonly found hiding under a log or rock.

2¾–4¾ in
7–12 cm

ITALIAN CAVE SALAMANDER
Speleomantes italicus

Found near streams and springs in the mountains of northern Italy, this salamander lives in caves and rock crevices.

3–6 in
7.5–15.5 cm

ENSATINA SALAMANDER
Ensatina eschscholtzii

This smooth-skinned, North American salamander has a plump tail, which is narrow at the base. It waves its tail at enemies in a defensive posture.

2–3½ in
5–9 cm

FOUR-TOED SALAMANDER
Hemidactylium scutatum

A terrestrial salamander, the adults of this species live in mosses, but the larvae grow in water. There is a marked narrowing at the base of its tail.

ASIATIC SALAMANDERS

Around 50 small- to medium-sized species form the Hynobiidae. They are confined to Asia, with some found in mountain streams. They lay their eggs in ponds or streams, and their larvae have external gills. Some species have claws to grasp rocks.

4–6½ in
10–16 cm

OITA SALAMANDER
Hynobius dunni

Females of this endangered Japanese species produce eggs in sacs, and males then compete to fertilize them externally.

MOLE SALAMANDERS

These large creatures mostly live in burrows and emerge to forage by night. The Ambystomatidae contain 37 species, all found in North America. Some species, notably the Mexican axolotl, are aquatic as adults and retain larval features, such as external gills. The four large and aggressive species of the genus *Dicamptodon*, found in western North America, were formerly classified as a separate family.

AXOLOTL
Ambystoma mexicanum

4–12 in
10–30 cm

Adult axolotls never leave the water and resemble very large salamander larvae, with their deep tails and feathery external gills.

broad head with small eyes

7–10 in
18–25 cm

yellow or white markings on brown or black skin

TIGER SALAMANDER
Ambystoma tigrinum

A thick-set salamander found throughout much of North America, this species migrates in spring to ponds to mate and lay eggs.

3½–4¼ in
9–11 cm

MARBLED SALAMANDER
Ambystoma opacum

This short-tailed, stocky salamander breeds in fall and lays its eggs in dried-up ponds that fill with rain in winter.

massive head

CALIFORNIA GIANT SALAMANDER
Dicamptodon ensatus

This large, nocturnal salamander is threatened by loss and degradation of its forest habitat. This species has an aquatic larval stage.

marbled coloration

6½–12 in
17–30 cm

MUDPUPPIES AND RELATIVE

Five of the six species in the Proteidae are found in North America; the sixth one—the cave-living olm—lives in Europe. Members of this family retain the larval form into their adult life, with long, slender bodies, external gills, and small eyes.

8–20 in
20–50 cm

MUDPUPPY
Necturus maculosus

A voracious predator, the mudpuppy, or waterdog, feeds on a variety of invertebrates, fishes, and amphibians. The female aggressively defends her developing eggs.

8–12 in
20–30 cm

EUROPEAN OLM
Proteus anguinus

The wholly aquatic olm lives in dark, flooded caves in Slovenia and Montenegro. It is blind and has white, pink, or gray skin.

AMPHIUMAS

Occurring in eastern North America, the Amphiumidae are a family of three large, wholly aquatic salamanders with eel-like bodies and tiny limbs. Females guard the eggs in a nest on land. Amphiumas can survive drought by burrowing into mud and forming a cocoon. They feed on worms, mollusks, fish, snakes, and small amphibians.

THREE-TOED AMPHIUMA
Amphiuma tridactylum

16–43 in
40–110 cm

This large salamander, with slimy skin and a long tail, can give a painful bite. Males breed every year, while females breed in alternate years.

REPTILES

Reptiles are a sophisticated, diverse, and successful group of ectothermic (cold-blooded) vertebrates. Although commonly associated with hot, dry environments, they are found in a wide range of habitats and climates around the world.

PHYLUM	CHORDATA
CLASS	REPTILIA
ORDERS	4
FAMILIES	60
SPECIES	About 7,700

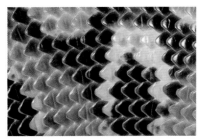

Reptile scales are a protective layer of skin with overlapping patches thickened with keratin and sometimes bone.

The shelled egg allows reptiles to reproduce out of water. The outer layer protects the embryo from dehydration.

Crocodiles can walk with their body raised well off the ground, whereas turtles manage only a sprawling gait.

Reptiles modify their behavior to regulate body temperature. They bask in the morning sun to absorb heat energy.

The first reptiles evolved from amphibians more than 295 million years ago. They are the ancestors not only of modern-day reptiles but also of mammals and birds. During the Mesozoic Era, reptiles such as dinosaurs, aquatic ichthyosaurs and plesiosaurs, and aerial pterosaurs dominated Earth. Reptile groups that still exist today evolved in this period and survived the mass extinction event that killed off the dinosaurs 65 million years ago.

All reptiles share certain characteristics, such as scaly skin and behavioral thermoregulation—using outside sources to maintain a constant body temperature, by basking in the sun, for example. But the various groups exhibit obvious differences. Turtles and tortoises are protected by a unique, heavily armored shell. Lizards, crocodilians, and tuataras have four limbs and long tails. Many squamates, or scaled reptiles—lizards, snakes, and amphisbaenians—have evolved limbless forms.

Hot desert ecosystems are typically dominated by lizards and snakes, but reptiles of all types are found in all habitats in the tropics and subtropics. Fewer species are found in cooler temperate climates. Within each ecosystem reptiles are important both as predators and as prey. Most are carnivores, feeding on a wide range of other animals. A few large lizards and tortoises are exclusively herbivores, but many species are opportunistic omnivores.

BEHAVIOR AND SURVIVAL

Some species are solitary whereas others are highly social. Although sluggish and inactive when cool, once their preferred body temperature is reached by basking or pressing against a warm rock, they can be very active.

Reproductive behavior can be complex, with males actively defending territories and courting females. Although some squamates give birth to live young, in most cases the male reptile fertilizes the female, who then lays eggs in an underground nest. All reptiles are independent and self-feeding from birth, although some crocodilians give parental care for two or more years.

Reptiles are exploited by humans for their skin or as food. This, along with habitat loss, pollution, and climate change, threatens the survival of many species.

SEA TURTLE IN ITS ELEMENT >
The green turtle may appear primitive but, with efficient flippers and a flattened shell, is well adapted to its aquatic life.

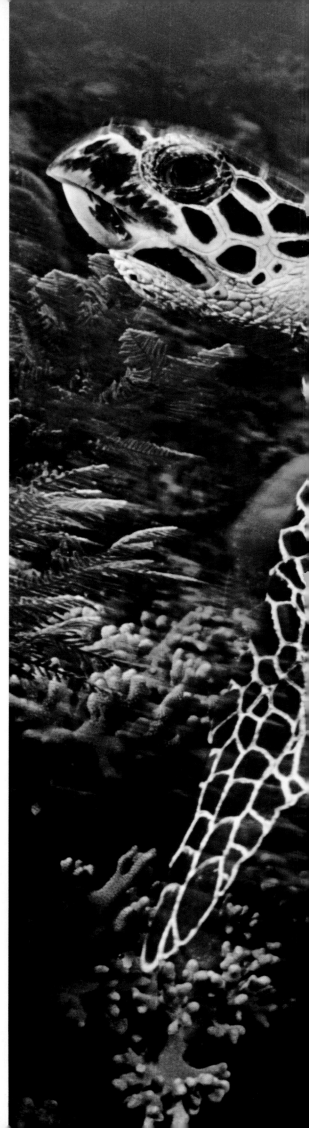

TURTLES AND TORTOISES

With their bony shell, stout limbs, and toothless, beaklike mouth, turtles and tortoises have changed little from species that lived 200 million years ago.

The order includes sea turtles, freshwater turtles, and terrestrial tortoises. Some extinct turtles were gigantic, but most modern ones are more modestly sized. The exceptions are the large sea turtles and a few terrestrial tortoises inhabiting isolated islands.

PROTECTION AND LOCOMOTION

A turtle's shell is formed of numerous fused bones, overlaid by scutes. The domed upperpart is known as the carapace and the lower part, the plastron. Scutes are horny plates and new ones form beneath the old ones each year. The degree of protection provided by the shell varies between species and is much reduced in some aquatic turtles. Not all species can retract the head into the shell—some tuck their head along one shoulder under the edge of the shell.

Tortoises are slow animals, but some sea turtles can reach speeds of 19 mph (30 kph) when swimming, as a result of the modification of their forelimbs into flippers. Although they need to breathe air, many turtles are tolerant of low oxygen levels and can remain submerged for hours. Turtles have a slow metabolism and are generally long lived.

Some species are carnivores, others herbivores, but most turtles are omnivorous. Animal prey is either slow moving or ambushed as the turtle lies hidden. Food is broken up by the "beak," a sharp, keratinous jaw covering.

NESTING AND BREEDING

Turtles do not defend territories, but they can have extensive home ranges and may develop social hierarchies. Otherwise turtles often congregate on river banks or lakesides to bask or nest.

Both in water and on land, males of some species engage in elaborate courtship of females before copulation. Fertilization is internal and the females lay shelled eggs, like those of other reptiles and birds. These are spherical or elongated, with rigid or flexible shells. Clutches are laid in nests dug up by the female in terrestrial sites. Nearly all marine species come to land only to nest. In many species, but not all, the gender of the hatchlings depends on incubation temperature.

PHYLUM	CHORDATA
CLASS	REPTILIA
ORDER	TESTUDINES
FAMILIES	14
SPECIES	300

A green sea turtle hatchling enters the sea. Many of its breeding beaches are protected, but it remains endangered.

AUSTRO-AMERICAN SIDE-NECKED TURTLES

From South America and Australasia, the Chelidae include carnivorous and omnivorous species. Their characteristically long neck cannot be retracted, so it is turned sideways under the edge of the shell. They lay elongated eggs with leathery shells.

13½ in
34 cm

MACQUARIE TURTLE
Emydura macquarii
Widely distributed through the Murray River basin of Australia, this species eats amphibians, fishes, and algae. Males are smaller than females.

REIMANN'S SNAKE-NECKED TURTLE
Chelodina reimanni
This turtle from New Guinea eats crustaceans and mollusks. When threatened, it tucks its large head under the side of its carapace.

30 in
75 cm

scalloped carapace with keel (central ridge)

COMMON SNAKE-NECKED TURTLE
Chelodina longicollis
A shy freshwater turtle of Australia, this species has a long neck, which enables it to raise its head out of water and capture prey.

10 in
25 cm

20 in
50 cm

MATAMATA
Chelus fimbriatus
This South American turtle uses its unusual appearance as camouflage when ambushing prey, which it sucks into its mouth.

long snout

AFRICAN SIDE-NECKED TURTLES

Most of the Pelomedusidae are carnivorous and occupy freshwater habitats. When threatened, they can hide their head and neck beneath the edge of the shell. Found throughout Africa and Madagascar, they survive dry conditions by burying themselves in mud.

brown carapace

8 in
20 cm

scales on head resemble helmet

AFRICAN HELMETED TURTLE
Pelomedusa subrufa
Widespread in sub-Saharan Africa, this carnivorous turtle is highly social and often hunts in packs to bring down large prey.

BIG-HEADED TURTLE

The sole member of the Platysternidae, the big-headed turtle is an endangered species from forest streams in South and Southeast Asia. It forages in shallow streams, bottom-walking rather than swimming.

7 in
18 cm

BIG-HEADED TURTLE
Platysternon megacephalum
This small, carnivorous turtle has a flattened body, a very large head with powerful jaws, and a long tail.

AMERICAN SIDE-NECKED RIVER TURTLES

Closely related to African side-necked turtles, most species of the family Podocnemididae are found in tropical South America, except one in Madagascar. These are herbivorous turtles found in various freshwater habitats. They cannot retract their necks into their shells.

RED-HEADED AMAZON RIVER TURTLE
Podocnemis erythrocephala
This species is found in swamps of the Rio Negro region of the Amazon basin in South America.

12½ in
32 cm

SNAPPING TURTLES

Native to North and Central America, these large, aquatic turtles are noted for their aggression. The Chelydridae have rough shells and a powerful head, with heavy, crushing jaws. They are effective predators, ambushing a variety of animals, but they also eat plants.

22 in
55 cm

COMMON SNAPPING TURTLE
Chelydra serpentina
A robust turtle, it often lies in wait for prey half-buried in mud. It occupies freshwater habitats from eastern North America as far south as Ecuador.

massive head with strong jaws and sharp, pointed beak

tongue with wormlike lure

heavily built carapace with three rows of conical scutes

32 in
80 cm

ALLIGATOR SNAPPING TURTLE
Macrochelys temminckii
One of the largest freshwater turtles in the world, this North American species has a worm-shaped lure on its tongue to attract prey as it lies in wait.

SOFTSHELL TURTLES

These aquatic predators inhabit freshwater habitats in North America, Africa, and southern Asia. The shells of the Trionychidae are flattened and covered with leathery skin rather than keratinous scutes. They range in size from 10 in (25 cm) to over 3¼ ft (1 m) in shell length.

grayish green carapace with raised ridges

EASTERN SPINY SOFTSHELL
Apalone spinifera
A native of eastern North America, this species eats mainly insects and aquatic invertebrates.

22 in
55 cm

INDIAN FLAP-SHELLED TURTLE
Lissemys punctata
This Indian turtle has a rear flap on either side of the plastron that protects the hind limbs when they are withdrawn into the body.

10½ in
27 cm

14 in
35 cm

CHINESE SOFT-SHELLED TURTLE
Pelodiscus sinensis
Hunted for its meat, this turtle of eastern Asia is rare in its native habitat, but tens of thousands are bred annually on farms.

PIG-NOSED TURTLE

The only species in the family Carettochelyidae, this turtle is an omnivore. Its carapace lacks hard scutes, but is still rigid. Its snout is adapted for breathing air while submerged.

PIG-NOSED RIVER TURTLE
Carettochelys insculpta
This nocturnal turtle is found in New Guinea and northern Australia. Like sea turtles, it has forelimbs modified as flippers for aquatic flight.

clawed flippers

28 in
70 cm

AMERICAN MUD AND MUSK TURTLES

These turtles from the New World emit a strong scent when threatened. The Kinosternidae tend to walk along the bottom of lakes and rivers, rather than swim, and are opportunistic omnivores. They lay elongated, hard-shelled eggs.

5 in
13 cm

MISSISSIPPI MUD TURTLE
Kinosternon subrubrum
An omnivore, this freshwater turtle feeds on the bottom of slow, shallow watercourses in the southeastern US.

COMMON MUSK TURTLE
Sternotherus odoratus
This freshwater turtle from eastern North America is an omnivore. When threatened, it not only exudes a nauseating musk, but also bites.

5 in
13 cm

LEATHERBACK SEA TURTLE

There is only one species in the Dermochelyidae. This turtle is capable of maintaining an elevated body temperature, which allows it to swim in cold waters. The shell has no scutes; the leathery skin covers a layer of insulating oily tissue.

leathery carapace with seven keels

5 ft
1.5 m

LEATHERBACK SEA TURTLE
Dermochelys coriacea
Feeding mainly on jellyfish, this oceanic sea turtle is the world's largest turtle. It has a worldwide distribution that includes subarctic waters.

clawless flippers

SEA TURTLES

Sea turtles occur throughout the world's oceans, mainly in coastal waters. They are highly adapted to the marine environment, with streamlined bodies and broad, paddlelike limbs. The Cheloniidae come to land to nest on beaches. Most species in this family are endangered.

vertebral keel in younger turtles

4 ft
1.2 m

large head with prominent eyes and beak

LOGGERHEAD SEA TURTLE
Caretta caretta
This carnivorous species occurs in coastal waters worldwide but undertakes long migrations between feeding and nesting grounds.

pale coloration on underside

30 in
75 cm

3¼ ft
1 m

almond-shaped eyes

4¼ ft
1.3 m

GREEN SEA TURTLE
Chelonia mydas
This is the only sea turtle known to bask on land. It is found in temperate and tropical oceans worldwide and is a strict herbivore.

flattened carapace with large scutes

OLIVE RIDLEY SEA TURTLE
Lepidochelys olivacea
A species of mainly tropical, shallow coastal waters, this turtle eats a wide range of invertebrates and algae.

HAWKSBILL SEA TURTLE
Eretmochelys imbricata
This turtle uses its horny jaws to forage for mollusks and other prey. It is found in tropical oceans throughout the world.

POND TURTLES

These turtles range from being fully aquatic to fully terrestrial. Most of them are found in North America, but one species in this family comes from Europe. Diet varies between species, although many are herbivores. The Emydidae are often brightly colored and intricately marked.

row of spines on carapace

FALSE MAP TURTLE
Graptemys pseudogeographica
This North American turtle occurs in freshwater habitats with abundant vegetation. Females are almost twice as large as males.

10½ in
27 cm

pattern of yellow lines on neck and head

strong, clawed front legs

11 in
28 cm

RED-EARED SLIDER
Trachemys scripta elegans
This North American species is mainly herbivorous and is common in the pet trade. Its ability to colonize new habitats has allowed it to become widely established in Europe and Asia.

10½ in
27 cm

YELLOW-BELLIED SLIDER
Trachemys scripta scripta
Named for its habit of sliding into water when disturbed, this turtle of the southern US is a diurnal omnivore.

9 in
23 cm

SALTWATER TERRAPIN
Malaclemys terrapin
This is a diurnal species of brackish waters in eastern North America. It has powerful jaws adapted for eating crustaceans and mollusks.

distinctively patterned shell

10 in
25 cm

PAINTED TURTLE
Chrysemys picta
Widely distributed in North America, this small, freshwater turtle is active during summer. During winter, it lies torpid under water.

8 in
20 cm

CAROLINA BOX TURTLE
Terrapene carolina
Males of this North American species have evolved highly curved claws that help grip the female's domed shell while mating.

5½ in
14 cm

ORNATE BOX TURTLE
Terrapene ornata
This is an omnivorous, terrestrial turtle of central North America. It digs burrows to escape extreme heat or cold.

strong claws used for burrowing

5 in
13 cm

SPOTTED TURTLE
Clemmys guttata
Identifiable by its spots, this small turtle feeds on aquatic invertebrates and plants in the marshes of eastern North America.

10 in
26 cm

CHICKEN TURTLE
Deirochelys reticularia
This shy turtle of swamps in eastern North America extends its long neck to strike at crayfish and other prey.

8½ in
21 cm

EUROPEAN POND TURTLE
Emys orbicularis
Widespread in Europe, this is a highly aquatic turtle. It basks on logs or rocks, but rapidly dives into the water if disturbed.

5 in
13 cm

WOOD TURTLE
Glyptemys insculpta
This turtle occupies the damp forests in northeastern North America. Unusually, males and females perform an elegant dance together during courtship.

olive skin with irregular yellow spots

10 in
26 cm

BLANDING'S TURTLE
Emydoidea blandingii
Found mainly across the Great Lakes region of North America, this turtle is an omnivore and especially skilled at capturing crayfish.

15 in
38 cm

FLORIDA REDBELLY TURTLE
Pseudemys nelsoni
Restricted to Florida, this turtle lives in lakes and slow-moving streams. During courtship, the male caresses the female's head with his forelimbs.

16 in
40 cm

RED-BELLIED TURTLE
Pseudemys rubriventris
Restricted to the northeastern US, this omnivorous, diurnal species prefers large, deep bodies of water. Females are larger than males.

ALDABRA GIANT TORTOISE
Aldabrachelys gigantea

The last remaining species of giant tortoise on islands in the Indian Ocean, the Aldabra giant tortoise can weigh over 660 lb (300 kg). Although resident on three islets of the Aldabra atoll, over 90 percent of the tortoises live on Grande-Terre—the largest islet by far— despite the scarcity of water and inadequate supply of vegetation. Poor conditions there have inhibited the growth of individual tortoises, and many do not reach sexual maturity. However, they are highly social, more so than the larger tortoises on other islets. Males are larger than females, but courtship is a gentle affair. Their eggs are buried underground, and hatch in the rainy season. The whole population is vulnerable to natural disasters and rising sea levels.

SIZE 4 ft (1.2 m)
HABITAT Grassy areas
DISTRIBUTION Aldabra, Indian Ocean
DIET Vegetation

< HORNY BEAK
The tortoise's large mouth contains no teeth. It cuts vegetation with its sharp, horny beak, takes it into the mouth with its tongue, and swallows it whole.

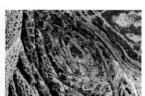

< EAR
Tortoises have no external ear flap; so the eardrum is situated in a hollow.

^ LEATHERY VISAGE
The skin is tough, leathery, and— depending on islet—either gray or brown. The skin is creased into folds around the neck. If it senses danger, the tortoise pulls its head back inside its shell.

< EYE
The eye is relatively large and has a well-developed eyelid. Tortoises can see in color, particularly in the red and yellow parts of the spectrum; this probably helps them find colorful fruit.

horny scales cover the front and hind limbs

hind limbs are elephantine, with strong claws

< DIGGING CLAWS
Tortoises have sturdy, elephant-like hind limbs, with five claws on each foot. Females have larger claws than males, which they use when excavating a nesting site.

^ FORELEG
The front limbs are cylindrical in shape, and longer than the hind limbs. This allows the tortoise to raise its body well above the ground as it walks. The legs are covered in large, tough scales.

< TAIL
The giant tortoise has a short tail, and can tuck it to one side under the back of the carapace. Males have longer tails than females.

scutes on the
upper shell show
growth rings

MADE TO LAST ∨
Compared to giant tortoises from the
Galapagos, this species has a more
rounded head, a pointed snout, and
a smiling mouth. It can live for
more than a hundred years.

POND AND RIVER TURTLES

Found both in the Old and New Worlds, the Geoemydidae are freshwater and terrestrial turtles, ranging in adult shell length from 5½ to 20 in (14–50 cm). They vary in their dietary preferences from being herbivorous to carnivorous. Many exhibit sexual dimorphism, with females being larger than males.

TORTOISES

Found in the Americas, Africa, and southern Eurasia, members of the family Testudinidae can reach great sizes. They have solid domed shells into which they can retract their heads. Fully terrestrial, tortoises are characterized by elephantine limbs. They lay hard-shelled eggs.

12 in
30 cm

DESERT TORTOISE
Gopherus agassizii
This tortoise lives in small burrows in deserts of southwestern North America. Mainly vegetarian, it sometimes preys on animals.

9 in
23 cm

BROWN LAND TURTLE
Rhinoclemmys annulata
A herbivorous tortoise, this species lives in tropical forests of Central America. It is mainly active in the morning and after rains.

13 in
33 cm

ELONGATED TORTOISE
Indotestudo elongata
Found in tropical Southeast Asia, this tortoise eats fruit and carrion. It hides in moist leaf litter when the weather is dry.

12 in
30 cm

GOLDEN COIN TURTLE
Cuora trifasciata
This is a carnivorous turtle from southern China. Its name derives from its value in the illegal wildlife market. Its use in traditional medicine threatens its survival.

16 in
40 cm

SERRATED HINGE-BACK TORTOISE
Kinixys erosa
This is an omnivorous species found in swamps in tropical West Africa. Older tortoises develop hinges in the rear of their carapaces.

pronounced spinal keel

6½ in
17 cm

yellow stripes behind eyes

YELLOW-MARGINATED BOX TURTLE
Cuora flavomarginata
An omnivore of China's rice paddies, this species avoids deep water and spends hours basking on land.

28 in
70 cm

RED-FOOTED TORTOISE
Chelonoidis carbonaria
Although known to eat carrion, this tortoise is mainly vegetarian. It is found in a variety of habitats of northeastern South America.

16 in
40 cm

RADIATED TORTOISE
Astrochelys radiata
Restricted to southern Madagascar, this endangered species eats mainly plants and is active during the early part of the day.

5 in
13 cm

5 in
13 cm

ASIAN LEAF TURTLE
Cyclemys dentata
This omnivorous turtle is found in slow-moving waters in Southeast Asia. It defends itself by emitting foul-smelling liquid.

BLACK-BREASTED LEAF TURTLE
Geoemyda spengleri
An inhabitant of wooded mountains in southern China, this turtle eats small invertebrates and fruit. Its carapace is rectangular, keeled, and spiky.

highly domed scutes

INDIAN STARRED TORTOISE
Geochelone elegans
Found in dry areas of India and Sri Lanka, this herbivorous species mates and breeds in the monsoon season.

15 in
38 cm

PANCAKE TORTOISE
Malacochersus tornieri
Found in East Africa, this omnivorous species inhabits rocky areas. Its flat shape allows it to hide in cracks.

7 in
18 cm

growth rings on the carapace of juveniles

4 ft
1.2 m

flattened scute

ALDABRA GIANT TORTOISE
Aldabrachelys gigantea
Restricted to the Aldabra atoll in the Indian Ocean, this large herbivore is able to drink through its nostrils.

large forelimbs

GALAPAGOS TORTOISE
Chelonoidis elephantopus
One of the world's largest tortoises, this mainly vegetarian species has 11 subspecies, from different islands in the Galapagos archipelago.

7½ in
19 cm

HERMANN'S TORTOISE
Testudo hermanni
This herbivorous species inhabits dry forests of coastal Italy and southern France. It hibernates during the cold winter months.

jaw with sharp cutting edge for eating vegetation

4 ft
1.2 m

STEPPES TORTOISE
Testudo horsfieldii
A herbivore from the dry deserts and steppes of Central Asia, this species escapes the daytime heat by sheltering in burrows.

11 in
28 cm

TUATARAS

Superficially like a lizard, New Zealand's tuatara belongs to a far more ancient order of reptiles. Its closest relatives became extinct 100 million years ago.

The tuatara has many anatomical features that set it apart from lizards. Most striking are its wedge-shaped "teeth"—actually serrations of the jawbone. The upper jaw has a double row that fits over a single row on the lower jaw.

Tuataras are very long lived, but are vulnerable to introduced ground predators. Inhabiting coastal forests, they are active at low body temperatures and emerge from their burrows at night to hunt for invertebrates, and birds' eggs and chicks. Males are territorial and nesting is communal. Eggs take up to four years to form and are incubated for 11 to 16 months before hatching. Incubation temperature determines the sex of the hatchlings.

PHYLUM	CHORDATA
CLASS	REPTILIA
ORDER	RHYNCHOCEPHALIA
FAMILIES	1
SPECIES	2

TUATARA

Often portrayed as "living fossils," the Sphenodontidae are modern representatives of reptiles that coexisted with dinosaurs. These primitive species are found only on offshore islands of New Zealand.

stout tail

TUATARA
Sphenodon punctatus
This reptile has clawed limbs to dig burrows. It can shed its tail to escape predators. The male uses its dorsal crest for display.

23½ in
60 cm

powerful clawed limbs

LIZARDS

Typically, lizards have four legs and a long, thin tail, but there are also many legless species. They have scaly skin and firm jaw articulation.

All lizards are ectothermic, obtaining heat energy from the environment. Although seen as predominately tropical or desert animals, they have a global distribution; lizards are found from beyond the Arctic Circle in Europe to the tip of South America. Highly adaptable, they occupy a wide range of terrestrial habitats with many being arboreal or rock dwelling. Many legless species are adapted for burrowing, a few lizards are effective gliders from trees, and others are semiaquatic, including a marine species on the Galapagos Islands.

STRATEGIES FOR SURVIVAL
Lizards range in length from about ½ in (1.5 cm) to 9¾ ft (3 m) in the case of the Komodo dragon, but the majority are between 4 and 12 in (10–30 cm). Although most are carnivores, around two percent of species are primarily herbivorous. Many lizards are prey for other carnivores. Their

defense mechanisms depend on agility, camouflage, and bluff. Many species are able to shed their tail to distract the attention of predators as they escape. The tail grows back. In some families of lizard, the skin can change color. This ability can be used for camouflage or for sexual or social signaling. Unusually, in many lizards the pineal gland on the top of the head acts as a light-sensitive "third eye."

VARIED LIFESTYLES
Although some species are solitary, many lizards have complex social structures, with the males maintaining their territories through visual signals. Many species lay eggs in underground nests, while others retain the eggs in the oviduct until hatching. There are also truly viviparous species, in which the mother provides nutrition via a placenta. Males have paired hemipenes— sexual organs used for internal fertilization— although some species are parthenogenetic, with females reproducing without the participation of a male. Some lizards tend their eggs during incubation but very few exhibit any maternal care of the hatchlings.

PHYLUM	CHORDATA
CLASS	REPTILIA
ORDER	SQUAMATA
FAMILIES (LIZARDS)	27
SPECIES	4,560

DEBATE

SQUAMATA: AN ALL-INCLUSIVE ORDER

Traditionally, lizards and snakes were considered two distinct groups, but modern genetic research makes it clear that they are not. Primitive squamates (scaled reptiles) were lizardlike animals that arose in the mid-Jurassic. Available evidence suggests that snakes evolved from lizards during the mid-Cretaceous. By this time various families of lizards had branched off on their own evolutionary course. Some lizard families are thus closer to snakes than they are to other lizards. Of living lizards, the monitors are thought to be those most closely related to snakes.

CHAMELEONS

Restricted to the Old World, the members of the family Chamaeleonidae have long limbs with grasping feet and a prehensile tail for gripping branches—an adaptation for life in trees. Their eyes are capable of moving independently in any direction to locate insects or small vertebrates. They catch their prey with a long, sticky tongue that shoots out from the mouth. Their ability to change color is for display and camouflage. Many species are threatened with extinction.

3¼ in
8 cm

BEARDED PYGMY-CHAMELEON
Rieppeleon brevicaudatus
This unusual, small chameleon is from East Africa. Drab coloration and patterning along its body allow it to resemble a dead leaf.

PARSON'S CHAMELEON
Calumma parsonii
The largest chameleon in the world, this species is restricted to Madagascar. It hunts invertebrates in the canopy of montane forests.

28 in
70 cm

prehensile tail

12 in
30 cm

green skin coloration sometimes changes to brown

12 in
30 cm

MEDITERRANEAN CHAMELEON
Chamaeleo chamaeleon
This chameleon spends its time in bushes searching for insects. It is found in North Africa and around the Mediterranean.

JACKSON'S CHAMELEON
Chamaeleo jacksonii
This diurnal, arboreal species is found in East Africa. The male is identifiable by the three horns on its snout, which are for display.

dorsal spines

grasping, four-toed feet

20 in
51 cm

22 in
56 cm

GIANT SPINY CHAMELEON
Furcifer verrucosus
This large chameleon is a native of the humid, coastal region of Madagascar. A shy species, it relies on camouflage to ambush its insect prey.

PANTHER CHAMELEON
Furcifer pardalis
Found only in dry forests in Madagascar, this lizard hunts insects in trees by stealth. Males are highly territorial.

23½ in
60 cm

VEILED CHAMELEON
Chamaeleo calyptratus
This species is from the southern coast of the Arabian Peninsula. Males have a large casque on the head, while females have a smaller one.

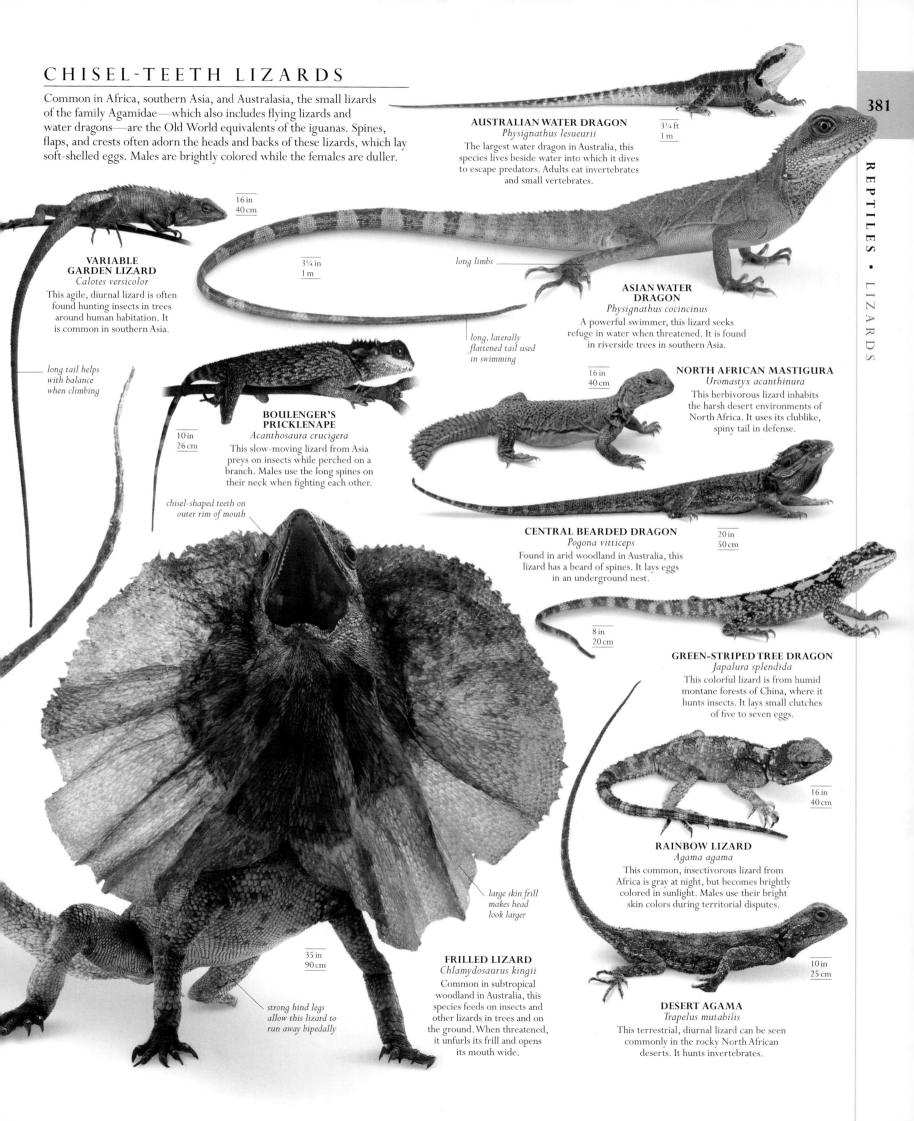

CHISEL-TEETH LIZARDS

Common in Africa, southern Asia, and Australasia, the small lizards of the family Agamidae—which also includes flying lizards and water dragons—are the Old World equivalents of the iguanas. Spines, flaps, and crests often adorn the heads and backs of these lizards, which lay soft-shelled eggs. Males are brightly colored while the females are duller.

AUSTRALIAN WATER DRAGON
Physignathus lesueurii

The largest water dragon in Australia, this species lives beside water into which it dives to escape predators. Adults eat invertebrates and small vertebrates.

3¼ ft
1 m

16 in
40 cm

VARIABLE
GARDEN LIZARD
Calotes versicolor

This agile, diurnal lizard is often found hunting insects in trees around human habitation. It is common in southern Asia.

3¼ in
1 m

long limbs

ASIAN WATER
DRAGON
Physignathus cocincinus

A powerful swimmer, this lizard seeks refuge in water when threatened. It is found in riverside trees in southern Asia.

long tail helps with balance when climbing

long, laterally flattened tail used in swimming

16 in
40 cm

NORTH AFRICAN MASTIGURA
Uromastyx acanthinura

This herbivorous lizard inhabits the harsh desert environments of North Africa. It uses its clublike, spiny tail in defense.

10 in
26 cm

BOULENGER'S
PRICKLENAPE
Acanthosaura crucigera

This slow-moving lizard from Asia preys on insects while perched on a branch. Males use the long spines on their neck when fighting each other.

chisel-shaped teeth on outer rim of mouth

CENTRAL BEARDED DRAGON
Pogona vitticeps

Found in arid woodland in Australia, this lizard has a beard of spines. It lays eggs in an underground nest.

20 in
50 cm

8 in
20 cm

GREEN-STRIPED TREE DRAGON
Japalura splendida

This colorful lizard is from humid montane forests of China, where it hunts insects. It lays small clutches of five to seven eggs.

large skin frill makes head look larger

RAINBOW LIZARD
Agama agama

This common, insectivorous lizard from Africa is gray at night, but becomes brightly colored in sunlight. Males use their bright skin colors during territorial disputes.

16 in
40 cm

35 in
90 cm

strong hind legs allow this lizard to run away bipedally

FRILLED LIZARD
Chlamydosaurus kingii

Common in subtropical woodland in Australia, this species feeds on insects and other lizards in trees and on the ground. When threatened, it unfurls its frill and opens its mouth wide.

DESERT AGAMA
Trapelus mutabilis

This terrestrial, diurnal lizard can be seen commonly in the rocky North African deserts. It hunts invertebrates.

10 in
25 cm

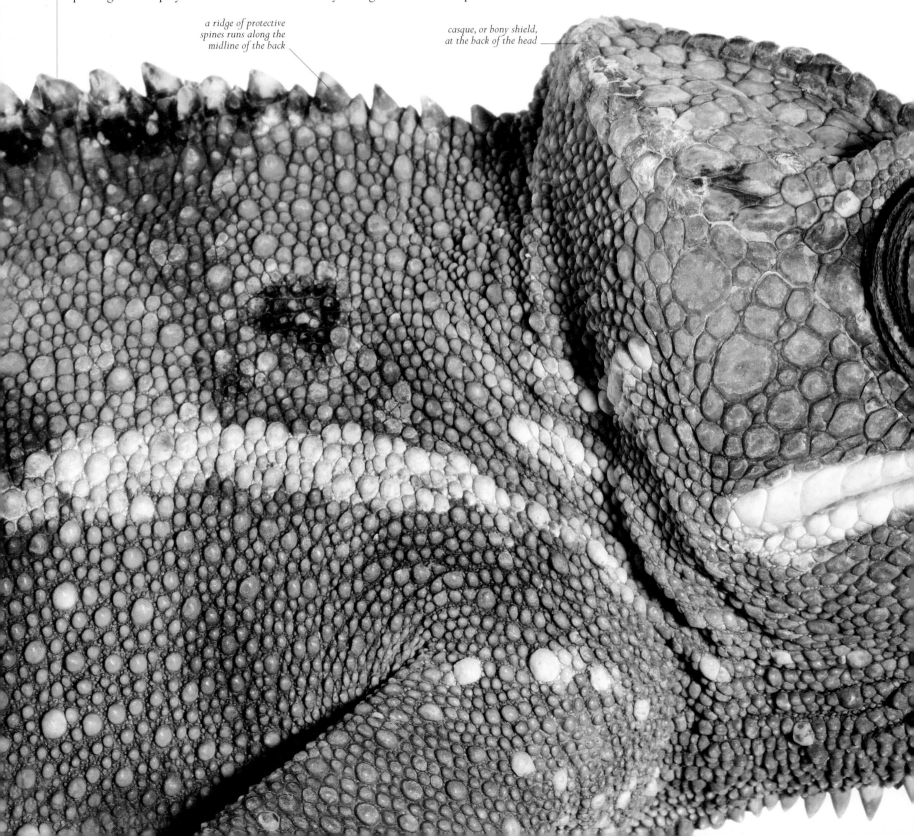

PANTHER CHAMELEON
Furcifer pardalis

This large chameleon is endemic to Madagascar, off the east coast of southern Africa, and has also been introduced to Mauritius and Réunion. This species lives in trees in humid scrubland, and its feet are so well adapted for grasping branches that it is difficult for the lizard to walk across a flat surface. Active during the day, it moves slowly through the branches to hunt its insect food by stealth. When it spots its next meal, the chameleon focuses on its victim with both eyes, then shoots out its long tongue to grasp the insect and pull it back into its large mouth. The chameleon's uncanny ability to change color is an indicator of mood and social status only, and is not used for camouflage. When faced with a rival, it rapidly inflates its body and changes color, putting on a display of dominance that is usually enough to decide a dispute.

SIZE 16–22 in (40–56 cm)
HABITAT Trees in humid scrub
DISTRIBUTION Madagascar
DIET Arthropods, crustaceans

a ridge of protective
spines runs along the
midline of the back

casque, or bony shield,
at the back of the head

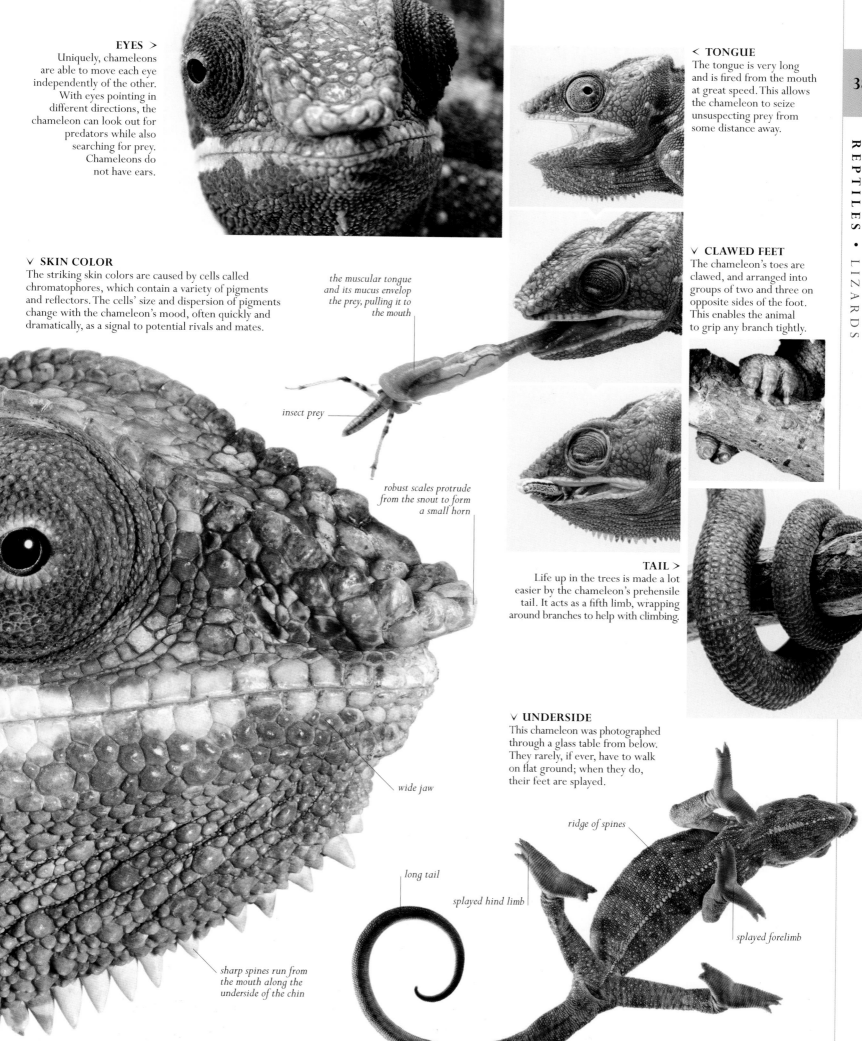

EYES >
Uniquely, chameleons are able to move each eye independently of the other. With eyes pointing in different directions, the chameleon can look out for predators while also searching for prey. Chameleons do not have ears.

< TONGUE
The tongue is very long and is fired from the mouth at great speed. This allows the chameleon to seize unsuspecting prey from some distance away.

∨ SKIN COLOR
The striking skin colors are caused by cells called chromatophores, which contain a variety of pigments and reflectors. The cells' size and dispersion of pigments change with the chameleon's mood, often quickly and dramatically, as a signal to potential rivals and mates.

the muscular tongue and its mucus envelop the prey, pulling it to the mouth

insect prey

robust scales protrude from the snout to form a small horn

∨ CLAWED FEET
The chameleon's toes are clawed, and arranged into groups of two and three on opposite sides of the foot. This enables the animal to grip any branch tightly.

TAIL >
Life up in the trees is made a lot easier by the chameleon's prehensile tail. It acts as a fifth limb, wrapping around branches to help with climbing.

wide jaw

∨ UNDERSIDE
This chameleon was photographed through a glass table from below. They rarely, if ever, have to walk on flat ground; when they do, their feet are splayed.

ridge of spines

long tail

splayed hind limb

splayed forelimb

sharp spines run from the mouth along the underside of the chin

GECKOS

Widely distributed in the tropics and subtropics, the Gekkonidae number over 1,000 species in 100 genera. They are noted for being very vocal and for their ability to climb smooth surfaces. Many lay hard-shelled eggs, while others lay soft-shelled eggs. Some species are viviparous.

WESTERN BANDED GECKO
Coleonyx variegatus
This terrestrial gecko hunts invertebrates in the deserts of the western US. Unlike many geckos, it has moveable eyelids.

4¾ in
12 cm

KUHL'S FLYING GECKO
Ptychozoon kuhli
This gecko's webbed toes and skin flaps help break its fall as it parachutes from trees in the rain forests of Southeast Asia.

8 in
20 cm

COMMON LEOPARD GECKO
Eublepharis macularius
Originally from southern central Asia, this gecko is a popular pet, which has been bred for different markings and colors.

8½ in
21 cm

MOORISH GECKO
Tarentola mauritanica
Common around the Mediterranean, this agile gecko preys on insects. It lives on rock faces but often enters houses.

6 in
15 cm

8 in
20 cm

WONDER GECKO
Teratoscincus scincus
To defend itself, this Central Asian species waves its tail slowly, causing a row of large scales on the tail to rub together, making a hissing sound.

bright coloration

MADAGASCAR DAY GECKO
Phelsuma madagascariensis
This colorful diurnal species is typically found on trees. Like many geckos, it lays hard-shelled eggs that stick to a branch.

adhesive pads on toes

10 in
25 cm

TOKAY
Gekko gecko
Named after its harsh "to-kay" call, this large gecko from Southeast Asia often lives in houses.

10 in
25 cm

AFRICAN FAT-TAILED GECKO
Hemitheconyx caudicinctus
Common in the western Sahara, this species has a tail that stores fat. It lacks the adhesive foot pads of many other geckos.

16 in
40 cm

4 in
10 cm

13½ in
34 cm

RING-TAILED GECKO
Cyrtodactylus louisiadensis
Found in New Guinea, this large, nocturnal gecko feeds aggressively on invertebrates and small frogs found on the ground.

6 in
15 cm

ASIAN SPINY GECKO
Hemidactylus brookii
This gecko lives in close proximity to humans. Native to northern India, it now also occurs in Hong Kong, Shanghai, and the Philippines.

MEDITERRANEAN GECKO
Hemidactylus turcicus
Often betrayed by its mewing cry, this small European gecko is common in houses, where it hunts insects attracted to lights.

IGUANAS

Largely confined to the Americas, the Iguanidae number nearly 30 species in 8 genera. They are diverse in color. Most are diurnal, carnivorous predators, although larger species are herbivorous. They all lay eggs.

2¼–5 ft
0.7–1.5 m

MARINE IGUANA
Amblyrhynchus cristatus
Native to the Galapagos Islands, this aquatic lizard is well adapted to feeding on underwater algae. Nasal glands help remove salt.

dorsal crest

GREEN IGUANA
Iguana iguana
Widespread in Central and South America, this large iguana is herbivorous. Males defend their territory by vigorously nodding their heads to display dominance.

6½ ft
2 m

BLACK IGUANA
Ctenosaura similis
This gregarious species from Central America lives in colonies with a dominant male. Although usually herbivorous, it sometimes eats small lizards.

35 in
90 cm

long, whiplike tail

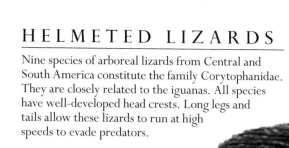

HELMETED LIZARDS

Nine species of arboreal lizards from Central and South America constitute the family Corytophanidae. They are closely related to the iguanas. All species have well-developed head crests. Long legs and tails allow these lizards to run at high speeds to evade predators.

sail-like dorsal crest

bright orange iris

HELMETED IGUANA
Corytophanes cristatus
This Central American iguana lives on a diet of large arthropods. By feeding infrequently, it minimizes the time it spends in the open or is conspicuous to predators.

hard, conelike casque at rear of head

GREEN BASILISK
Basiliscus plumifrons
Native to Central American rain forests, this species inhabits riverbanks. The crests on its head, back, and tail are supported by bony spines.

26 in
65 cm

long legs

EASTERN CASQUE-HEADED IGUANA
Laemanctus longipes
This large, insectivorous species lives in small groups of one male and two to three females. It is native to Central America.

slender green feet and legs

13½ in
34 cm

long tail aids balance when climbing or running

28 in
70 cm

NORTH AMERICAN SPINY LIZARDS

Restricted to North and Central America, the Phrynosomatidae are a diverse family of lizards that prefer arid environments and hunt insects. They are generally small, dull colored, and spiny. Most lay eggs, but those living at high altitudes are viviparous.

COLORADO DESERT FRINGE-TOED LIZARD
Uma notata
Closable nostrils and ear flaps, overlapping jaws, and interlocking eyelids help protect this desert-living species as it burrows in sand.

3¼ in
8 cm

GREEN SPINY IGUANA
Sceloporus malachiticus
This diurnal, arboreal lizard from Central America has stiff, keeled scales, giving it a spiny appearance.

8 in
20 cm

DESERT HORNED LIZARD
Phrynosoma platyrhinos
This lizard of North American deserts primarily eats ants. Its flattened body is an adaptation to maximize heat absorption while basking in the sun.

6 in
15 cm

ANOLES

Most anoles come from around the Caribbean. A diverse group, the Polychrotidae are typically small, arboreal, and insectivorous. Although often green or brown, they change skin color according to mood and environment. Both sexes aggressively defend their territories.

KNIGHT ANOLE
Anolis equestris
The largest anole, this species is restricted to Cuba. Adhesive pads on its toes enable it to scale smooth walls.

20 in
50 cm

CAROLINA ANOLE
Anolis carolinensis
Males of this species show their dominance by bobbing their heads and flaring their brightly colored throat fan.

8 in
20 cm

FLAP-FOOTED LIZARDS

All lizards of this group have elongated bodies, no front limbs, and much-reduced hind limbs. Restricted to Australasia, the 36 species in the family Pygopodidae hunt insects by burrowing or on the surface. They are related to geckos and lay soft-shelled eggs.

4¾ in
12 cm

FRASER'S DELMA
Delma fraseri
This insectivorous, Australian lizard inhabits spinifex grassland. It is well adapted to moving through the stiff blades of grass.

8½ in
21 cm

COMMON SCALY FOOT
Pygopus lepidopodus
Snakelike in appearance, this lizard is widespread in Australia. It is a diurnal predator of insects and tunnel-inhabiting spiders.

23½ in
60 cm

BURTON'S SNAKE-LIZARD
Lialis burtonis
Native to Australia, this lizard has an elongated, wedge-shaped snout that it uses to grasp skinks, which it consumes whole.

SKINKS

Distributed worldwide, the Scincidae are a diverse group of 1,400 species. While many are active diurnal predators, several others are nocturnal, limbless, burrowing lizards. They use chemical as well as visual communication. Although typically oviparous, many species are viviparous.

bright coloration on flanks

14 in
35 cm

FIRE SKINK
Lepidothyris fernandi
This insectivorous skink is native to humid, forested areas of West Africa. Its attractive coloration makes it a popular lizard to keep in captivity.

weak legs

EMERALD SKINK
Lamprolepis smaragdina
Native to islands of the western Pacific, this arboreal lizard preys on insects on bare tree trunks.

10 in
25 cm

8½ in
21 cm

FIVE-LINED SKINK
Plestiodon fasciatus
This North American skink coils around its eggs to protect them during incubation. It prefers woodland and feeds on ground-living insects.

PERCIVAL'S LANCE SKINK
Acontias percivali
This legless African skink burrows through leaf litter to hunt its invertebrate prey. It gives birth to up to three live young.

12 in
30 cm

WALL AND SAND LIZARDS

Lizards of the family Lacertidae are found in a wide variety of habitats throughout the Old World. They are active predators and have complex social systems, with males defending their territory. Almost all species lay eggs. These lizards typically have large heads.

LARGE PSAMMODROMUS
Psammodromus algirus
Found in the western Mediterranean region, this small lizard inhabits dense, bushy areas. During breeding, males develop a red patch on their throat.

3 in
7.5 cm

OCELLATED LIZARD
Timon lepidus
The largest European member of the family, this lizard lives in dry, bushy places. It eats insects, eggs, and small mammals.

8 in
20 cm

VIVIPAROUS LIZARD
Zootoca vivipara
Found throughout Europe and up to 9,900 ft (3,000 m) in the Alps, this ground-dwelling lizard lives in a variety of habitats. It gives birth to live young.

6 in
15 cm

FRINGE-TOED LIZARD
Acanthodactylus erythrurus
Found in the Iberian Peninsula and North Africa, this species has fringes of spiny scales on its toes that enable it to cross over loose sand.

3½ in
9 cm

GRAN CANARIA GIANT LIZARD
Gallotia stehlini
This large species is restricted to the shrubland of Grand Canary Island. It is diurnal and herbivorous.

32 in
80 cm

3 in
7.5 cm

ITALIAN WALL LIZARD
Podarcis siculus
A ground-living species, found around the northern Mediterranean, this lizard inhabits grassy places and often lives in close proximity to humans.

WHIPTAILS AND RACERUNNERS

These fast-running American lizards occupy a range of habitats. Smaller species of the Teiidae are insectivorous but larger ones are carnivorous. All 120 species are oviparous, although many whiptails are all-female and reproduce parthenogenetically, laying viable, fertile eggs without mating.

18 in
45 cm

AMAZON RACERUNNER
Ameiva ameiva
Powerful jaws allow this South American species to feed on small vertebrates and insects that it catches in open terrestrial habitats.

long tail used in defense

RED TEGU
Tupinambis rufescens
This large species of the arid regions in central South America is an active predator and scavenger, but is also known to eat plants.

4 ft
1.2 m

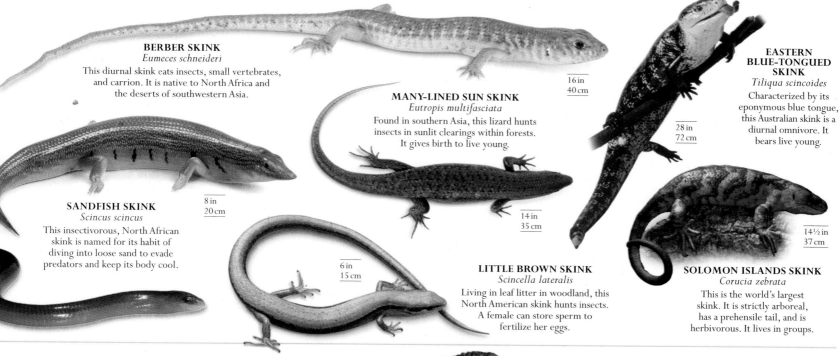

BERBER SKINK
Eumeces schneideri
This diurnal skink eats insects, small vertebrates, and carrion. It is native to North Africa and the deserts of southwestern Asia.

16 in
40 cm

MANY-LINED SUN SKINK
Eutropis multifasciata
Found in southern Asia, this lizard hunts insects in sunlit clearings within forests. It gives birth to live young.

14 in
35 cm

EASTERN BLUE-TONGUED SKINK
Tiliqua scincoides
Characterized by its eponymous blue tongue, this Australian skink is a diurnal omnivore. It bears live young.

28 in
72 cm

SANDFISH SKINK
Scincus scincus
This insectivorous, North African skink is named for its habit of diving into loose sand to evade predators and keep its body cool.

8 in
20 cm

6 in
15 cm

LITTLE BROWN SKINK
Scincella lateralis
Living in leaf litter in woodland, this North American skink hunts insects. A female can store sperm to fertilize her eggs.

SOLOMON ISLANDS SKINK
Corucia zebrata
This is the world's largest skink. It is strictly arboreal, has a prehensile tail, and is herbivorous. It lives in groups.

14½ in
37 cm

PLATED LIZARDS

There are 32 species of plated lizards. All are from sub-Saharan Africa and lay eggs. They have cylindrical bodies with well-developed legs for hunting insects in rocky areas and the savanna. Solitary in nature, the Gerrhosauridae are often aggressive to other members of their species.

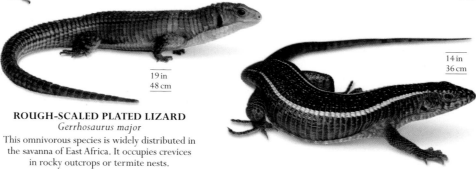

19 in
48 cm

ROUGH-SCALED PLATED LIZARD
Gerrhosaurus major
This omnivorous species is widely distributed in the savanna of East Africa. It occupies crevices in rocky outcrops or termite nests.

14 in
36 cm

MADAGASCAN GIRDLED LIZARD
Zonosaurus madagascariensis
This insectivorous lizard is native to Madagascar. It prefers to be solitary, and feeds on the ground in open, dry habitats.

GIRDLED LIZARDS

Limited to southern and eastern Africa, the Cordylidae are named for the rings of spiny scales that encircle their tail. Their flattened body has limited space for live young or eggs. In viviparous species, this restricts litter size; oviparous species lay only two eggs.

CAPE GIRDLED LIZARD
Cordylus cordylus
Endemic to southern Africa, this lizard lives in dense colonies. Adults are aggressive, and form social hierarchies under a dominant male.

8½ in
21 cm

spiny scales

MICROTEIID LIZARDS

There are 165 species of the Gymnophthalmidae in tropical South America. Generally small, they are characterized by large scales on their back. These are diurnal and secretive insectivorous lizards and their dull coloration provides camouflage in leaf litter. Most species lay eggs.

BROMELIAD LIZARD
Anadia ocellata
An arboreal species found in Central America, the bromeliad lizard hunts insects and takes refuge in foliage.

3¼ in
8 cm

muscular body

glossy head scales

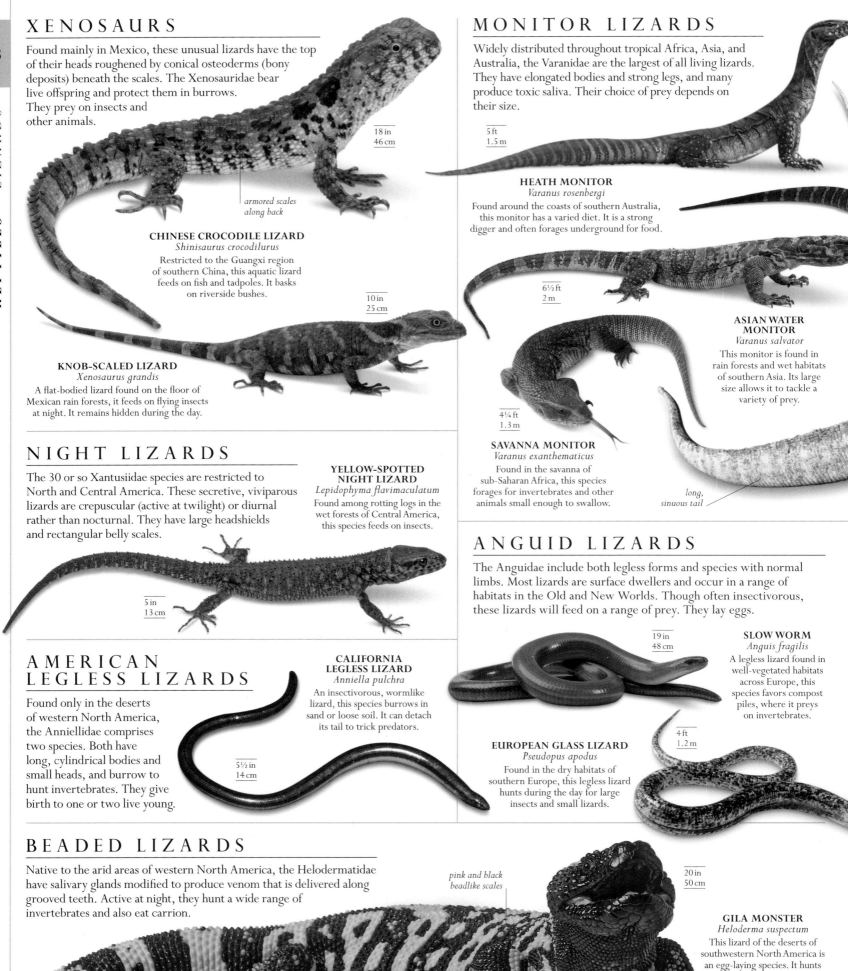

XENOSAURS

Found mainly in Mexico, these unusual lizards have the top of their heads roughened by conical osteoderms (bony deposits) beneath the scales. The Xenosauridae bear live offspring and protect them in burrows. They prey on insects and other animals.

18 in
46 cm

armored scales along back

CHINESE CROCODILE LIZARD
Shinisaurus crocodilurus
Restricted to the Guangxi region of southern China, this aquatic lizard feeds on fish and tadpoles. It basks on riverside bushes.

10 in
25 cm

KNOB-SCALED LIZARD
Xenosaurus grandis
A flat-bodied lizard found on the floor of Mexican rain forests, it feeds on flying insects at night. It remains hidden during the day.

NIGHT LIZARDS

The 30 or so Xantusiidae species are restricted to North and Central America. These secretive, viviparous lizards are crepuscular (active at twilight) or diurnal rather than nocturnal. They have large headshields and rectangular belly scales.

YELLOW-SPOTTED NIGHT LIZARD
Lepidophyma flavimaculatum
Found among rotting logs in the wet forests of Central America, this species feeds on insects.

5 in
13 cm

AMERICAN LEGLESS LIZARDS

Found only in the deserts of western North America, the Anniellidae comprises two species. Both have long, cylindrical bodies and small heads, and burrow to hunt invertebrates. They give birth to one or two live young.

CALIFORNIA LEGLESS LIZARD
Anniella pulchra
An insectivorous, wormlike lizard, this species burrows in sand or loose soil. It can detach its tail to trick predators.

5½ in
14 cm

MONITOR LIZARDS

Widely distributed throughout tropical Africa, Asia, and Australia, the Varanidae are the largest of all living lizards. They have elongated bodies and strong legs, and many produce toxic saliva. Their choice of prey depends on their size.

5 ft
1.5 m

HEATH MONITOR
Varanus rosenbergi
Found around the coasts of southern Australia, this monitor has a varied diet. It is a strong digger and often forages underground for food.

6½ ft
2 m

ASIAN WATER MONITOR
Varanus salvator
This monitor is found in rain forests and wet habitats of southern Asia. Its large size allows it to tackle a variety of prey.

4¼ ft
1.3 m

SAVANNA MONITOR
Varanus exanthematicus
Found in the savanna of sub-Saharan Africa, this species forages for invertebrates and other animals small enough to swallow.

long, sinuous tail

ANGUID LIZARDS

The Anguidae include both legless forms and species with normal limbs. Most lizards are surface dwellers and occur in a range of habitats in the Old and New Worlds. Though often insectivorous, these lizards will feed on a range of prey. They lay eggs.

19 in
48 cm

SLOW WORM
Anguis fragilis
A legless lizard found in well-vegetated habitats across Europe, this species favors compost piles, where it preys on invertebrates.

EUROPEAN GLASS LIZARD
Pseudopus apodus
Found in the dry habitats of southern Europe, this legless lizard hunts during the day for large insects and small lizards.

4 ft
1.2 m

BEADED LIZARDS

Native to the arid areas of western North America, the Helodermatidae have salivary glands modified to produce venom that is delivered along grooved teeth. Active at night, they hunt a wide range of invertebrates and also eat carrion.

pink and black beadlike scales

20 in
50 cm

GILA MONSTER
Heloderma suspectum
This lizard of the deserts of southwestern North America is an egg-laying species. It hunts insects and ground-dwelling vertebrates, including small mammals.

PERENTIE
Varanus giganteus
The largest lizard in Australia, this shy species inhabits arid regions. It has a varied diet and mildly venomous saliva.

spotted pattern of scales

8¼ ft
2.5 m

6½ ft
2 m

strong claws for digging out prey

NILE MONITOR
Varanus niloticus
The second largest reptile in Africa, this monitor preys on various vertebrates and mollusks, and also scavenges on carrion.

4½ ft
1.4 m

YELLOW-SPOTTED MONITOR
Varanus panoptes
This species from Australia and southern New Guinea is rarely found far from a permanent source of water. It preys on other reptiles.

4½ ft
1.4 m

GOULD'S GOANNA
Varanus gouldii
This monitor actively forages for any animal smaller than itself. It is found in open woodlands and grasslands throughout Australia.

10 ft
3.1 m

skin folds on neck

KOMODO DRAGON
Varanus komodoensis
The world's largest lizard, this species is restricted to a few Indonesian islands. It is a formidable predator of large mammals.

well-developed, strong limbs

grayish brown scaly skin

AMPHISBAENIANS

They may resemble legless lizards, but the burrowing reptiles in the suborder Amphisbaenia are quite distinct in their anatomy and behavior.

The amphisbaenians, or worm-lizards, are reptiles wholly adapted to a subterranean lifestyle, spending almost their entire lives underground. Most species have lost all trace of their limbs and have elongated bodies with smooth scales. They hunt for soil invertebrates by means of scent and sound, killing them in their crushing jaws. Burrowing involves the worm-lizard propelling itself forward by contracting and extending its body while the specially adapted head pushes through the soil. The eyes are protected by tough, translucent skin. Fertilization is internal, with some species laying eggs, others bearing live young. They are found in Florida, South America, Africa, the Middle East, and southern Europe.

PHYLUM	CHORDATA
CLASS	REPTILIA
ORDER	SQUAMATA
FAMILIES (AMPHISBAENIANS)	4
SPECIES	165

WORM-LIZARDS

All the Amphisbaenidae are burrowing reptiles with specially adapted heads for digging. They are found in sub-Saharan Africa and South America. They are formidable predators of soil invertebrates, hunting by scent and sound.

EUROPEAN WORM-LIZARD
Blanus cinereus
Rarely seen above ground, this burrowing reptile from Spain and Morocco forages in leaf litter for invertebrates, particularly ants.

12 in
30 cm

black and white markings

heavily ossified skull

18 in
45 cm

SPECKLED WORM-LIZARD
Amphisbaena fuliginosa
This worm-lizard hunts invertebrates while burrowing through leaf litter in the rain forests of South America. It surfaces during heavy rains.

6½ in
17 cm

LANG'S ROUND-HEADED WORM-LIZARD
Chirindia langi
Found in southern Africa, this species burrows through sandy soil to forage for termites. If caught, it can shed its tail in a bid to escape.

SNAKES

Snakes are predators with elongated bodies and skins of overlapping scales. In a number of species modified teeth, or fangs, are used to deliver venom.

Although most numerous in tropical regions, snakes have adapted to live at colder latitudes and at altitude and are found on all continents except Antarctica. They are usually terrestrial with many living in trees, but some species burrow, some are semiaquatic, and others live entirely in the sea. The smallest are tiny, threadlike creatures, but the largest can reach 33 ft (10 m) in length. Most are between 1 and 6½ ft (30 cm–2 m).

A UNIQUE ARRAY OF SENSES

Some snakes have external vestiges of hind limbs, but they move by muscular contractions creating traction between the scales on their undersides and the surface beneath. A large number of flexibly linked vertebrae allows bending and coiling in any direction. The scaly skin is regularly shed to allow growth. Many snakes rely on just one elongated lung for breathing, while the gut is a simple tube with a large, muscular stomach.

Instead of eyelids that can be opened and closed, each eye is covered by a clear, protective scale. Eyesight can be good, but varies according to lifestyle. Lacking external ears, snakes detect sounds as vibrations from the ground as well as in the air. Their key sense is smell, not through the nostrils, but from airborne chemicals collected by the forked tongue. Some species can also detect the body heat of mammals and birds.

KILLING TO LIVE

All snakes are carnivorous. Their sharp teeth curve backward to grasp and hold prey. Whether they eat their victim alive or kill it by injection of venom or by constriction, they consume their prey whole. Their uniquely elastic jaws allow them to swallow animals much larger than a snake's own head. Snakes defend themselves by camouflage, warning coloration, or mimicry. When threatened, they may strike out and bite.

Species in cold places tend to bear live offspring, whereas those in warmer climates lay eggs. Fertilization is internal. The male deposits sperm by means of one of his paired sexual organs, known as hemipenes.

PHYLUM	CHORDATA
CLASS	REPTILIA
ORDER	SQUAMATA
FAMILIES (SNAKES)	18
SPECIES	Over 2,700

If threatened, the Mojave rattlesnake uses its characteristic rattle, bares its fangs, and coils up, ready to strike at the aggressor.

BURROWING VIPERS

Members of this family of snakes from Africa and southwest Asia are all venomous. Despite their common name they are not closely related to the Viperidae—the position of the fangs differs from that of the true vipers. The burrowing vipers, the Atractaspididae, hunt small mammals and reptiles underground. Their relatively small size, smooth scales, and cylindrical bodies are all adaptations for their burrowing way of life. Most species in the family lay eggs.

BOAS

The majority of Boidae are found in Central and South America, but a few species of boas are also present in Madagascar and New Guinea. These snakes are found from trees to marshes, where they prey on vertebrates, killing them by constriction. This family includes the anaconda, the largest living snake, and other species of varying sizes. Most give birth to live young.

jaws that dislocate to swallow large prey

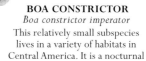

BOA CONSTRICTOR
Boa constrictor imperator
This relatively small subspecies lives in a variety of habitats in Central America. It is a nocturnal hunter of small mammals.

small, smooth granular scales

30 in
75 cm

MOLE VIPER
Atractaspis fallax
This poisonous snake from East Africa has large front fangs. It hunts other burrowing vertebrates underground.

saddlelike dark markings

6 ft
1.8 m

DUMERIL'S BOA
Acrantophis dumerili
Found in the humid forests of Madagascar, this snake becomes torpid when sheltering in burrows during the dry, cool months.

32 in
80 cm

RUBBER BOA
Charina bottae
Preferring cool, humid conditions, this burrowing species is found at high altitudes and as far north as British Columbia.

EAST AFRICAN SAND BOA
Gongylophis colubrinus
This African snake, with a stout body and short tail, waits in a burrow with only its head exposed to catch unsuspecting prey.

small, smooth scales
assist in burrowing

35 in
90 cm

3½ ft
1.1 m

head with
small eyes

headlike tail
used in defense

CALABAR GROUND BOA
Calabaria reinhardtii
Native to West Africa, the Calabar ground boa hunts in burrows for small mammals. It is the only egg-laying boa.

mottled brown
coloration for
camouflage

long,
muscular
body

8¼ ft
2.5 m

RAINBOW BOA
Epicrates cenchria
Microscopic ridges on the scales of this South American snake refract light to give it an iridescent sheen. It hunts for mammals in forests.

6½ ft
2 m

smooth, glossy scales

3¼ ft
1 m

**NEW GUINEA
GROUND BOA**
Candoia aspera
A sharply angled snout helps to identify this New Guinea snake. It is a slow-moving terrestrial hunter of small vertebrates.

**COOK'S
TREE BOA**
Corallus cookii
Endemic to St. Vincent Island in the Caribbean, this rare snake hunts in trees for birds and mammals under the cover of darkness.

5 ft
1.5 m

33 ft
10 m

ROSY BOA
Lichanura trivirgata
This is a slow-moving snake from the deserts of the western US. It ambushes mammals and kills them by constriction.

dull yellow color
with prominent
black spots

GREEN ANACONDA
Eunectes murinus
This aquatic South American species is the largest of the New World snakes. It hides in water and kills its prey by constriction.

3¼ ft
1 m

thick, muscular body
used to constrict prey

COMMON BOA
Boa constrictor

The common boa is a large terrestrial snake from the tropics of Central and South America. Although they are often found in woodland and scrub, boas easily adapt to a variety of habitats. Lying motionless on the ground, the snake ambushes its mammalian prey—grasping its victim with its jaws, it coils around, suffocating and killing by slow constriction before swallowing the body whole. The left and right sides of the boa's lower jaw separate in front, and can open wide to take astonishingly large prey. Usually a solitary species, males will actively seek out a partner at breeding times, attracted by the scent emitted by the female. Larger than males, female boas give birth to 30–50 live offspring, each measuring about 12 in (30 cm) in length. This species is regularly seen in captivity but its propensity to bite makes it an unpredictable pet.

SIZE Up to 8¼ ft (2.5 m)
HABITAT Open woodland and scrubland
DISTRIBUTION Central and South America
DIET Mammals, birds, reptiles

∧ COLOR VARIANTS
There are numerous subspecies of common boa, which are defined on the basis of size and color patterns. Color also varies considerably between localities.

> CAMOUFLAGE
Pigmentation of some scales forms patches of color that disrupt the outline of the snake. This helps camouflage it when hunting prey or evading predators.

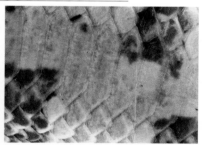

∧ BELLY SCALES
The robust belly scales grip most surfaces, allowing the common boa to pull itself along and climb trees. It periodically sheds its skin.

> SIZABLE FOE
The common boa hunts by sight and smell, so the eyes and nostrils are prominent on the triangular head. Feeding can be infrequent— a small prey item may last a snake two to three weeks. A large snake that kills a deer will not need to feed for at least six months.

distinctive, saddlelike markings

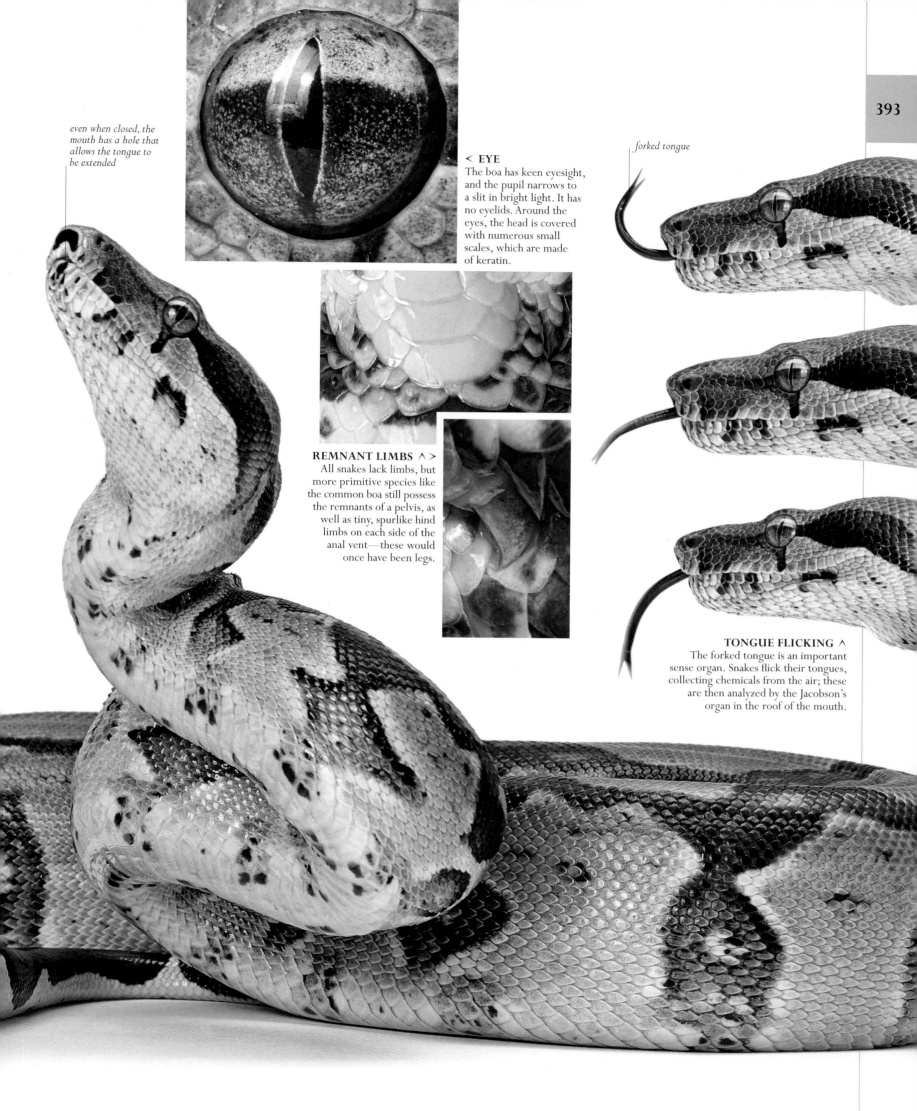

even when closed, the mouth has a hole that allows the tongue to be extended

forked tongue

< EYE
The boa has keen eyesight, and the pupil narrows to a slit in bright light. It has no eyelids. Around the eyes, the head is covered with numerous small scales, which are made of keratin.

REMNANT LIMBS ∧ >
All snakes lack limbs, but more primitive species like the common boa still possess the remnants of a pelvis, as well as tiny, spurlike hind limbs on each side of the anal vent—these would once have been legs.

TONGUE FLICKING ∧
The forked tongue is an important sense organ. Snakes flick their tongues, collecting chemicals from the air; these are then analyzed by the Jacobson's organ in the roof of the mouth.

COLUBRIDS

With over 1,600 species worldwide, colubrids form the largest family of snakes. The members of the family Colubridae inhabit a wide range of habitats, from deserts to wetlands. They have a varied diet. Many species lay eggs.

WESTERN LYRE SNAKE
Trimorphodon biscutatus
This secretive, nocturnal species of western North America is found in rocky habitats. It hunts bats and other small mammals, as well as lizards.

banded coloration

4 ft
1.2 m

COMMON GARTER SNAKE
Thamnophis sirtalis
This North American species is active in the daytime, hunting vertebrates in a range of habitats. In Manitoba, after hibernation it congregates in large numbers to mate.

4¼ ft
1.3 m

large eye

6 ft
1.8 m

GIANT MALAGASY HOGNOSE SNAKE
Leioheterodon madagascariensis
This large, diurnal snake, with an upturned snout, hunts lizards and amphibians in grasslands and forests in Madagascar.

RUTHVEN'S KINGSNAKE
Lampropeltis ruthveni
An egg-laying species, this colubrid is from the Mexican plateau, where it hunts rodents and lizards in dry woodland.

35 in
90 cm

GRAY EARTHSNAKE
Geophis brachycephalus
This small, Central American snake is terrestrial and nocturnal. It feeds mainly on earthworms and soft insect larvae.

18 in
46 cm

3½ ft
1.1 m

LONG-NOSED SNAKE
Rhinocheilus lecontei
Characterized by a pointed snout, this shy, nocturnal, burrowing species is from dry grasslands of North America, where it hunts lizards.

3¼ ft
1 m

MOLE SNAKE
Pseudaspis cana
This harmless species lives in underground burrows, where it feeds on moles and other small mammals after constricting them. It is found in southern Africa.

7 ft
2.1 m

CALIFORNIA MOUNTAIN KINGSNAKE
Lampropeltis zonata
This secretive snake prefers high-altitude wooded habitats. It is typically nocturnal, but is diurnal when nights are cold.

4 ft
1.2 m

AFRICAN EGG-EATING SNAKE
Dasypeltis scabra
This African snake has a highly specialized diet of eggs on which it feasts during the bird-nesting season. It then fasts for the rest of the year.

blotched, brown coloration

long, powerful body

9¼ ft
2.8 m

EASTERN PINE SNAKE
Pituophis melanoleucus
When threatened, this large, powerful snake of woodlands in North America ejects foul-smelling waste from its rear opening (cloaca).

GREEN HIGHLAND RACER
Drymobius chloroticus
This fast, agile snake is from the rain forests of Central America. It is usually found near water, where it feeds on frogs.

3¼ ft
1 m

long, slender body

WESTERN INDIGO SNAKE
Drymarchon corais
This is one of the longest snakes in North America. It often shares a burrow with the gopher tortoise.

4½ ft
1.4 m

9¾ ft
3 m

4 ft
1.2 m

BANDED FLYING SNAKE
Chrysopelea pelias
Despite its name, this southern Asian snake does not actively fly. Instead, it glides down from high branches by making its underside concave.

NORTHERN WATER SNAKE
Nerodia sipedon
This aquatic, viviparous species from eastern North America is active throughout the day and the night. It feeds on amphibians and fishes.

short snout

26 in
65 cm

6 ft
1.8 m

MOELLENDORFF'S RAT SNAKE
Orthriophis moellendorffi
This species is from dry limestone regions of China and Vietnam. It has an elongated snout and a relatively long tail.

7 ft
2.1 m

MUD SNAKE
Farancia abacura
This North American snake preys on aquatic salamanders, grasping them with strongly curved teeth. Females remain coiled around their eggs until they hatch.

FALSE CORAL SNAKE
Erythrolamprus mimus
The bright coloration of the harmless false coral snake from South America is similar to that of highly poisonous coral snakes.

striking red, white, and black bands

REDBACK COFFEE SNAKE
Ninia sebae
This harmless snake from Central America has an ability to extend its neck as a form of intimidation.

16 in
40 cm

3¼ ft
99 cm

NORTHERN CAT-EYED SNAKE
Leptodeira septentrionalis
This nocturnal, arboreal species from Central America has large eyes to help it hunt for vertebrates and eggs of tree frogs.

4¼ ft
1.3 m

BROWN BLUNT-HEADED VINE SNAKE
Imantodes cenchoa
The large eyes of this slender snake help it hunt lizards in the dark. It is found in tropical American rain forests.

3¼ ft
1 m

BROWN TREESNAKE
Boiga irregularis
Originally from Australia and New Guinea, this snake was accidentally introduced on the island of Guam in the western Pacific, where it has decimated the native fauna.

9¾ ft
3 m

FALSE WATER COBRA
Hydrodynastes gigas
Native to South American rain forests, this is a semiaquatic snake. Like cobras, it can flatten its neck to appear more intimidating.

GREEN VINE SNAKE
Oxybelis fulgidus
Long and delicate, this arboreal snake from Central and South American rain forests holds its prey in the air until its venom immobilizes the victims.

6½ ft
2 m

3½ ft
1.1 m

characteristic yellow collar

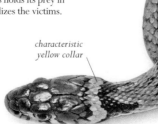

CALICO SNAKE
Oxyrhopus petola
This species feeds on lizards and other small vertebrates. It is a terrestrial and diurnal inhabitant of South American rain forests.

4 ft
1.2 m

GRASS SNAKE
Natrix natrix
Widely distributed throughout Europe, this species often feigns death when threatened. At home in water, it regularly feeds on amphibians.

olive-gray body color

» COLUBRIDS

5¼ ft / 1.6 m

RED-TAILED GREEN RATSNAKE
Gonyosoma oxycephalum
This fast-moving snake hunts for birds and mammals in trees. It is found in the rain forests of Southeast Asia.

7¾ ft / 2.4 m

long, slender body

5¼ ft / 1.6 m

4½ ft / 1.4 m

COMMON TRINKET SNAKE
Coelognathus helena
This Indian species enlarges its neck and rises up to intimidate its enemies. It hunts mammals, usually at night.

ROUGH GREENSNAKE
Opheodrys aestivus
Native to woodlands in southeastern North America, this arboreal snake is a diurnal predator of insects. It is oviparous.

23½ in / 60 cm

SMOOTH SNAKE
Coronella austriaca
This secretive European snake, found in meadows, kills its prey by constriction. Females produce eggs that hatch within their bodies.

3¼ ft / 1 m

4½ ft / 1.4 m

BALKAN RACER
Hierophis gemonensis
Restricted to the Balkans, this snake is found in dry scrubland and olive groves. It hunts during the day, preying on lizards.

DAHL'S WHIPSNAKE
Platyceps najadum
From the Mediterranean region, this snake is found in dry, stony habitats. It hunts small lizards and grasshoppers during the day.

SOUTHERN WATER SNAKE
Nerodia fasciata
Living in wetlands, this colubrid preys on amphibians and fishes. It is found in the southern US.

BROWN-HEADED CENTIPEDE SNAKE
Tantilla melanocephala ruficeps
This burrowing snake is found in the tropical forests of Central America. It is a mainly diurnal insect eater.

8 in / 20 cm

brown markings on muscular body

head with prominent eyes

6 ft / 1.8 m

pale underside

6 ft / 1.8 m

RED CORN SNAKE
Pantherophis guttatus
This snake is common in southeastern North America, but is rarely seen. It hunts small mammals in woodland.

6½ ft / 2 m

YELLOW RATSNAKE
Spilotes pullatus
A predator of small vertebrates, this large snake is rarely found far from water. It is widely distributed in South and Central America.

DIADEM SNAKE
Spalaerosophis diadema cliffordi
Native to deserts in North Africa, this subspecies is diurnal during cooler months, but turns nocturnal during summer months.

32 in / 80 cm

WESTERN HOGNOSED SNAKE
Heterodon nasicus
This snake of the North American prairies uses its specialized, enlarged teeth to puncture the lungs of toads, making them easier to swallow.

6½ ft / 2 m

MONTPELLIER SNAKE
Malpolon monspessulanus
Inhabiting dry scrub and rocky hillsides around the Mediterranean, this long, slender snake hunts small vertebrates during the day.

ASIAN PIPE SNAKES

Found in Sri Lanka and Southeast Asia, these small, burrowing snakes of the Cylindrophiidae have cylindrical bodies and smooth, shiny scales. They live in wet habitats, sheltering in tunnels, but emerge at night to hunt other snakes and eels. They give birth to live young.

CEYLONESE PIPESNAKE
Cylindrophis maculatus
Endemic to Sri Lanka, this harmless burrowing snake eats invertebrates. It uses its flattened tail to mimic the movements of the venomous cobra.

26 in / 65 cm

SEA SNAKES

These highly venomous snakes from the tropical coastal waters of the Indian and Pacific oceans belong to the subfamily Hydrophiinae, part of the larger Elapidae. They are well adapted for swimming, with a paddlelike tail. All but a few breed in water, bearing live young. They eat eels and other fish.

YELLOW-LIPPED SEAKRAIT
Laticauda colubrina
This snake hunts fish at night in tropical Indo-Pacific seas. It can move on land as well.

4½ ft / 1.4 m

COBRAS AND RELATIVES

Widely distributed in the tropics, the venomous snakes of the Elapidae have short, permanently erect fangs at the front of the mouth. Varied in body forms, they occupy a wide variety of habitats. Some species lay eggs, while others bear live young.

small, slender head

32 in
80 cm

CENTRAL AMERICAN CORAL SNAKE
Micrurus nigrocinctus
This venomous snake of Central America forages in leaf litter of tropical forests. Its bright coloration serves as a warning to potential aggressors.

4 ft
1.2 m

YELLOW-FACED WHIP SNAKE
Demansia psammophis
Widely distributed across Australia, this slender snake is an active diurnal hunter of lizards. It prefers dry, open habitats.

30 in
75 cm

EASTERN SHIELD-NOSE SNAKE
Aspidelaps scutatus fulafulus
This nocturnal predator of small lizards and mammals is found in southern African savanna. It burrows in sandy soil.

26 in
65 cm

ROSEN'S SNAKE
Suta fasciata
Found in arid areas of Western Australia, this venomous snake hunts for lizards.

spread hood acts as warning

20 in
50 cm

14 in
35 cm

28 in
70 cm

RINGED BROWN SNAKE
Pseudonaja modesta
Found in dry, rocky areas in Australia, this venomous snake feeds on small skinks. Its survival is threatened by the loss of its habitat.

SOUTHERN DESERT BANDED SNAKE
Simoselaps bertholdi
Widespread in Western Australia, this small snake burrows in search of its lizard prey. The banding may confuse predators.

DESERT DEATH ADDER
Acanthophis pyrrhus
This predator of the western deserts of Australia lures its prey of small lizards and mammals by wiggling its tail. It then ambushes them.

olive coloration on long, thin body

6½ ft
2 m

7¾ ft
2.4 m

16 ft
5 m

MONOCLED COBRA
Naja kaouthia
Common in Southeast Asia, this large snake hunts rats and other snakes in woodland and paddy fields, often close to human habitation.

EGYPTIAN COBRA
Naja haje
This large cobra hunts small vertebrates in the deserts of North and Central Africa. When threatened, it spreads its hood and raises its head.

KING COBRA
Ophiophagus hannah
A forest species of tropical Asia, the massive king cobra primarily preys on other snakes. Unusually for snakes, its nest of eggs is defended by both sexes.

smooth scales on brown body

30 in
75 cm

RED SPITTING COBRA
Naja pallida
This African species not only raises its hood when threatened, but also sprays streams of venom into the face of its aggressor.

PYTHONS

Members of the family Pythonidae can be found across Africa, Asia, and Australia. Pythons detect warm-blooded prey with their facial pits. They use their mouth to grasp prey but kill by constriction. Some species coil around their eggs to help incubate them using the heat generated by their shivering.

5 ft
1.5 m

GREEN TREEPYTHON
Morelia viridis
Found in the tropical forests of Australasia, this arboreal snake spends its time looped over a branch, ready to ambush lizards and small mammals.

bright green color aids in camouflage

BLOOD PYTHON
Python curtus
This relatively short python of Southeast Asian rain forests coils around its clutch of eggs to incubate and protect them.

6 ft
1.8 m

9¾ ft
3 m

BLACK-HEADED PYTHON
Aspidites melanocephalus
Endemic to Australia, this species is found in a variety of habitats. It preys upon other snakes and reptiles.

BURMESE PYTHON
Python molurus
Rare in its native homeland, this Asian snake is becoming superabundant in the Florida Everglades. Its broad diet of birds, mammals, and reptiles is a threat to many local animals.

23 ft
7 m

BLINDSNAKES

The Typhlopidae have eyes covered with scales, rendering them effectively blind. Found in leaf litter of tropical forests, these small, burrowing snakes mainly eat soil invertebrates and have teeth only on their upper jaw. Most of them lay eggs.

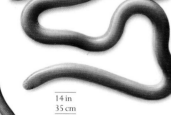

14 in
35 cm

30 in
75 cm

BLACKISH BLINDSNAKE
Austrotyphlops nigrescens
Tough scales allow this burrowing snake from eastern Australia to resist attack from ants while eating their eggs and larvae.

EURASIAN BLINDSNAKE
Typhlops vermicularis
This wormlike blind snake from Europe inhabits dry, open areas. It burrows to feed on invertebrates, mainly ant larvae.

VIPERS

The Viperidae are characterized by thick bodies, keeled scales, and a triangular head. Long, tubular, hinged fangs at the front of the mouth efficiently inject venom into their vertebrate prey. Pit vipers have sensory pits between their eyes and nostrils that detect heat. Most vipers bear live young.

5 ft
1.5 m

FER-DE-LANCE
Bothrops atrox
Found in the forests of tropical America, this venomous snake is characterized by a sharply pointed head. It hunts birds and mammals by night.

34 in
85 cm

23½ in
60 cm

THREAD SNAKES

These small, slender snakes are rarely seen because of their habit of burrowing to hunt invertebrates. Belonging to the Leptotyphlopidae, they are found in the tropics of the Americas, Africa, and southwest Asia.

12 in
30 cm

SENEGAL BLINDSNAKE
Myriopholis rouxestevae
This snake inhabits the soil of tropical forests in western Africa. First studied in 2004, it is believed to eat invertebrates.

DESERT HORNED VIPER
Cerastes cerastes
Occurring in the deserts of North Africa and the Sinai Peninsula, this snake lies buried in sand to ambush small mammals and lizards.

HOGNOSE PIT VIPER
Porthidium nasutum
This diurnal hunter of Central America prefers humid, open forests. Pit vipers have heat-detecting sensory pits between their eyes and nostrils and bear live young.

blotched markings

powerful, muscular body useful for constricting prey

netlike pattern

SPOTTED PYTHON
Antaresia maculosa
This python is found on the rocky hillsides of northern Australia. It feeds on bats by catching them at the entrance of their roosting caves.

4½ ft
1.4 m

33 ft
10 m

RETICULATED PYTHON
Python reticulatus
One of the longest snakes in the world, this species inhabits Asian rain forests. It kills large mammals by constriction.

SUNBEAM SNAKES

Characterized by iridescent scales, the Xenopeltidae include only two species, both restricted to Southeast Asia. They live in burrows in forest habitats and hunt amphibians, other reptiles, and small mammals. They reproduce by laying eggs.

4¼ ft
1.3 m

SUNBEAM SNAKE
Xenopeltis unicolor
A flat head helps this snake burrow into decaying vegetation. It emerges at dusk to feed on amphibians and small mammals.

4¼ ft
1.3 m

COPPERHEAD
Agkistrodon contortrix
Banded markings help this snake hide in leaf litter in rocky woodland. It is found in eastern North America.

23½ in
60 cm

ASP VIPER
Vipera aspis
This European viper prefers warm, dry habitats, where it feeds on small mammals. It bears up to 20 live young.

35 in
90 cm

COMMON ADDER
Vipera berus
This diurnal species hunts small mammals and lizards. It is widely distributed across a variety of habitats in Eurasia.

4 ft
1.2 m

short, dark-edged blotches

PRAIRIE RATTLESNAKE
Crotalus viridis
This snake from the midwestern US hunts mammals around dawn and dusk. It hides in crevices by day.

GABOON VIPER
Bitis gabonica
Occurring in tropical rain forests in Africa, this large, heavily built snake preys on mammals by ambushing them.

6½ ft
2 m

7 ft
2.1 m

WESTERN DIAMOND-BACKED RATTLESNAKE
Crotalus atrox
A nocturnal hunter of mammals, this large snake is widespread in arid areas of western North America.

large head

PUFF ADDER
Bitis arietans
Most active at night, this venomous snake usually lies still to ambush its vertebrate prey. It is found in rocky grassland in Africa.

6 ft
1.8 m

3¼ ft
1 m

MALAYAN PIT VIPER
Calloselasma rhodostoma
Occurring in Southeast Asia, this species forages by night in open areas adjacent to forests for rodents and lizards.

CROCODILES AND ALLIGATORS

Crocodilians are large, carnivorous, aquatic reptiles. Their armored skin and powerful jaws make them formidable predators, yet they are social animals and caring parents.

All crocodilians—crocodiles, alligators, and gharials—have a similar body plan: an elongated snout armed with numerous, sharp, unspecialized teeth, a streamlined body, a long, muscular tail, and skin armored with bony plates. The saltwater crocodile is the largest living reptile.

SWIMMING AND FEEDING

Crocodilians are found in tropical regions worldwide, inhabiting a variety of freshwater and marine habitats. The eyes, ears, and nostrils are on top of the head, which enables the animal to be almost completely submerged while hunting. The powerful tail is used for swimming and strong legs allow crocodilians to move easily on land with the body raised well above the ground.

Although generalized carnivores, feeding on fish, reptiles, birds, and mammals, crocodilians exhibit sophisticated feeding behaviors to locate and ambush specific prey, such as migrating mammals and fish. Small prey is swallowed whole but large prey is first drowned before the crocodilian rolls over while grasping the carcass to twist off pieces of meat. Digestion is aided by stones in the stomach and highly acidic gastric secretions.

SOCIAL BEHAVIOR AND BREEDING

Behaviorally, crocodilians have more in common with birds, their closest living relatives, than with other reptiles. Adults form loose groupings—particularly at good feeding sites—and use a wide range of vocalizations and body language to interact with each other.

In the breeding season, dominant males control territories and actively court females. Fertilization is internal and clutches of rigid-shelled eggs are laid in a nest constructed and guarded by the female. The sex of individuals is determined by the incubation temperature. Hatchlings call to stimulate the female to excavate the nest and carry them in her mouth to water. Mortality is high in young crocodilians but once they are longer than 3¼ft (1 m), they have few natural enemies.

PHYLUM	CHORDATA
CLASS	REPTILIA
ORDER	CROCODYLIA
FAMILIES	3
SPECIES	23

Exploiting a fast-flowing stream, these Yacare caimans are lying with their mouths open, waiting for fish to come within reach.

GHARIALS

The endangered Gavialidae are restricted to India. The slender snout has rows of numerous sharp teeth that help to catch fish. Males develop a wartlike growth on the tip of the snout.

olive-green body

23 ft
7 m

GHARIAL
Gavialis gangeticus
Among the largest of the Crocodylia, this Asian species has a fearsome array of teeth, which help it catch fish, but it is not known to attack humans.

CROCODILES

Relatively unspecialized in lifestyle, habitat, and diet, the Crocodylidae are most easily recognized by the exposure of the fourth tooth of the lower jaw, when the jaws are closed. These tropical reptiles occupy many habitats around rivers and coasts.

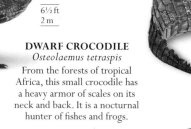

6½ ft
2 m

DWARF CROCODILE
Osteolaemus tetraspis
From the forests of tropical Africa, this small crocodile has a heavy armor of scales on its neck and back. It is a nocturnal hunter of fishes and frogs.

CUBAN CROCODILE
Crocodylus rhombifer
Endemic to Cuba, this medium-sized crocodile inhabits swamps, and hunts fish and small mammals. It lays its eggs in holes in the ground.

11 ft
3.5 m

13 ft
4 m

SIAMESE CROCODILE
Crocodylus siamensis
Restricted to Southeast Asia, this large crocodile is critically endangered in the wild. Found in freshwater marshes, it feeds on a variety of prey.

powerful tail propels crocodile through water

ALLIGATORS AND CAIMANS

Members of the family Alligatoridae feed on fish, birds, and mammals that share their aquatic habitat. These reptiles are distributed throughout freshwater swamps and rivers of the tropical and subtropical Americas. The only species living outside the Americas is the rare Chinese alligator.

coloration darkens with age

AMERICAN ALLIGATOR
Alligator mississippiensis
Conservation efforts have now made this North American species quite common. It preys on birds, small mammals, and turtles.

strong legs allow smooth movement on land

16 ft
5 m

8¼ ft
2.5 m

SPECTACLED CAIMAN
Caiman crocodilus
This caiman feeds on a wide range of prey. Widespread in Central and South America, it appears to be the only crocodilian that readily occupies man-made aquatic habitats.

CHINESE ALLIGATOR
Alligator sinensis
This endangered species of the Yangtze valley of China lies dormant in burrows during the cold winter months.

6½ ft
2 m

rounded head with broad snout

9¾ ft
3 m

BROAD-SNOUTED CAIMAN
Caiman latirostris
A mound-nesting species found across much of central South America, this caiman is characterized by a broad snout. It hunts mammals and birds.

mottled pattern

bony armor

5½ ft
1.7 m

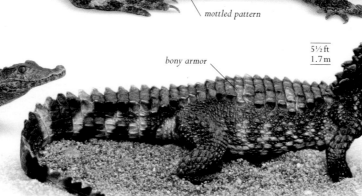

CUVIER'S DWARF CAIMAN
Paleosuchus palpebrosus
This South American species is the smallest of the New World members of this order. It has a doglike skull and bony armored skin.

5 ft
1.5 m

CUVIER'S DWARF CAIMAN
Paleosuchus trigonatus
This small crocodile of South American rain forests is semiterrestrial. It builds its nest against a termite mound, which keeps the eggs warm.

greenish brown body, often covered in algae

23 ft
7 m

SALTWATER CROCODILE
Crocodylus porosus
This largest of all living reptiles is common in the Indo-Pacific and readily crosses the open ocean. Its diet is wide-ranging and unspecialized.

eyes set high on head

16 ft
5 m

NILE CROCODILE
Crocodylus niloticus
Widely distributed in Africa, this large crocodile typically lives in freshwater but has been seen around coasts as well. Its diet varies with age, with adults taking larger prey.

CUBAN CROCODILE
Crocodylus rhombifer

This strikingly colored crocodile is native to Cuba.
It is medium-sized, robust, and armored with bony scales. More terrestrial than
other crocodilian species, it raises its belly off the ground and adopts a "high walk"
posture when moving on land. Its preferred prey are turtles, which are crushed by
the strong teeth at the rear of its mouth. Because of hunting pressure and habitat loss,
the few remaining wild Cuban crocodiles were captured in the 1960s and released
into Zapata Swamp, a reserve in the south of the island. Although protected, the
population remains small, and hybridization with American crocodiles on
the reserve threatens the purity of this species.

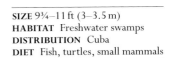

SIZE 9¾–11 ft (3–3.5 m)
HABITAT Freshwater swamps
DISTRIBUTION Cuba
DIET Fish, turtles, small mammals

*large scales armored
with bony deposits*

*bony projections at
back of the head
resemble horns*

*external ear flap,
which is closed while the
crocodile is underwater*

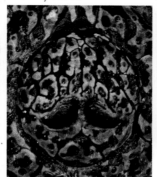

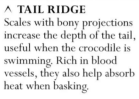

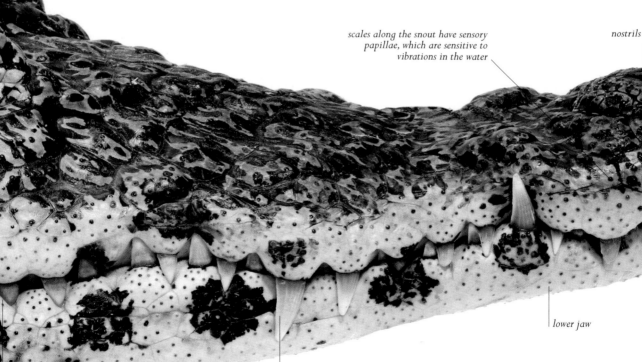

MUSCULAR LIMBS ∨ >

This crocodile can race over short distances, powered by its strong rear legs. Because they are used for walking rather than swimming, the five toes on each foot are not webbed.

∨ NOSTRILS

Paired nostrils are located on the nasal disk, an elevated pad of tissue at the end of the snout. Underwater, the nostrils can be closed by valves.

∧ FORMIDABLE JAWS

Crocodiles cannot chew their food, but they use their powerful jaws to grasp their prey. Sensitive taste buds on the tongue allow the crocodile to reject unpleasant tasting objects. Evaporation from the mouth cools the body.

∧ TAIL RIDGE

Scales with bony projections increase the depth of the tail, useful when the crocodile is swimming. Rich in blood vessels, they also help absorb heat when basking.

< BELLY SCALES

The scales on the crocodile's underside are small and uniform in size and pattern. These scales are prized by the leather trade.

scales along the snout have sensory papillae, which are sensitive to vibrations in the water

nostrils

lower jaw

sharp front teeth

blunt back teeth are more robust than the front teeth

∧ STEALTH ATTACK

Sharp eyes and sensitive ears are located on the top of the head, so that prey can be detected while the body is submerged. This enables the crocodile to take its victims by surprise. The long lower jaw can be closed with tremendous force to grasp struggling prey securely.

BIRDS

Birds have busy, active lives. Many are spectacularly beautiful; others sing complex—even musical—songs. They have the intelligence and parental devotion of a mammal combined with characteristics more reminiscent of a reptilian ancestor.

PHYLUM	CHORDATA
CLASS	AVES
ORDERS	29
FAMILIES	196
SPECIES	10,117

Asymmetrical flight feathers, with a narrow outer blade and broader inner blade, give birds better control in the air.

Many chicks are altricial—born in a helpless state. Blind and naked, they require prolonged parental care.

Elaborate nests are built by the male weaver bird to impress prospective mates, testament to the brain power of birds.

DEBATE

ARE BIRDS DINOSAURS?

Traditionally birds are in a class separate from reptiles, including dinosaurs. But today scientists hold that all descendants of a common ancestor should be classified together to reflect evolutionary relationships, a method known as cladistics. This makes birds as much part of the dinosaur group as *Tyrannosaurus*.

Birds are the only living animals with feathers. They are warm-blooded vertebrates that stand on two legs and their forelimbs are considerably modified as wings. In addition to facilitating flight, feathers insulate the body, so many birds stay active in ice-cold conditions, just like furry mammals. Feathers can also be colorful, showing that birds rely on flaunting visual cues to communicate and find a mate. Birds evolved from a group of two-legged, carnivorous dinosaurs that included *Tyrannosaurus*. It is possible—even likely—that these ancestors were already feathered.

TAKING TO THE AIR

No one knows for sure how—or why—the first birds took to the air, but the fact that they did had effects that would be forever imprinted upon the bird's body. Their wristbones became fused together as arms evolved into wings. The breastbone evolved a huge keel for the attachment of larger muscles for powering the flapping wings. Already their dinosaur ancestors had lightweight, air-filled bones. But birds coupled this with a strong cardiovascular system powered by a mammal-like four-chambered heart and a rapid metabolism to generate lots of energy. A system of air sacs—linked to the bone chambers—expanded their breathing capacity to flush stale air out of lungs more efficiently than in mammals.

Many reptilian features still remain. The bare parts of a bird's legs and toes have horny reptilian scales. Birds excrete a semisolid waste product called uric acid (not a solution of urea, as in mammals). Waste from the kidneys and gut mix and exit via a common reptilian opening called a cloaca. But a bird's brain is better developed than that of a reptile, because it is supported by a warm-blooded body and fast metabolism. This makes birds not only skilled at flying, but remarkably sophisticated in the ways they acquire food or raise a family. Young birds hatch from reptilian hard-shelled eggs, but they are reared by intelligent parents who will invest much time and energy into raising them to maturity. Sixty million years of evolution mean that a bird is no longer just a reptile with feathers.

THE PEACOCK'S DAZZLING DISPLAY >
The feathers of many birds are vibrantly colored, used as striking signals in courtship and other social interactions.

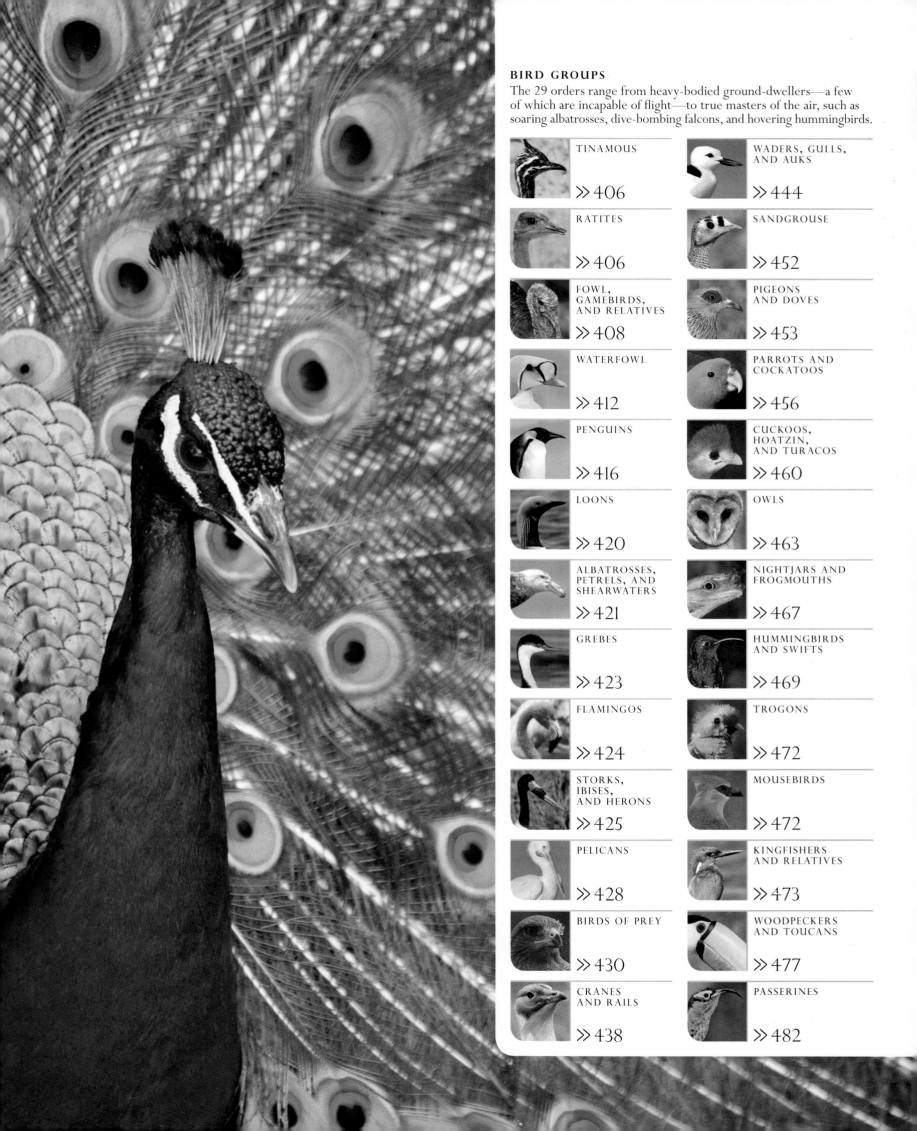

BIRD GROUPS

The 29 orders range from heavy-bodied ground-dwellers—a few of which are incapable of flight—to true masters of the air, such as soaring albatrosses, dive-bombing falcons, and hovering hummingbirds.

TINAMOUS

The ground-dwelling tinamous from Central and South America resemble Old World partridges, but they are most closely related to ratites.

Tinamous, the sole family in the order Tinamiformes, are small, ground-dwelling birds with rounded bodies and short legs. All have very short tails, which gives them a dumpy appearance, and some species are crested.

PHYLUM	CHORDATA
CLASS	AVES
ORDER	TINAMIFORMES
FAMILIES	1
SPECIES	47

Unlike the ratites, tinamous have a ridged breastbone for the attachment of flight muscles, a feature that is seen in all other birds. Their wings are much better developed than those of the ratites, and they can fly, although only over short distances. Usually they are reluctant to take to the air and prefer instead to run away from predators. Tinamous have comparatively small a heart and lungs, which perhaps explains why these birds become exhausted so easily.

Some groups of tinamous inhabit woodlands and forest, while others are found in open grassland. All feed on seeds, fruit, insects, and sometimes small vertebrates. The cryptic plumage patterns of most species makes them difficult to spot in the wild, and they are more likely to be recognized by their distinctive calls.

BEAUTIFUL EGGS
Male tinamous mate with many different females, and it is the males that tend the eggs and young. The nests are made in leaf litter on the ground. Tinamou eggs—brightly colored turquoise, red, or purple—have a porcelain-like gloss.

DEBATE
ORIGINS OF TINAMOUS

Traditionally, tinamous are grouped separately from the flightless ratites. But the two groups share a pattern of skull bones not found in other birds, suggesting that they have a common ancestor. Scientists are not certain whether this ancestor was flightless or able to fly; the latter is more likely.

16 in
40 cm

ELEGANT CRESTED TINAMOU
Eudromia elegans
F: Tinamidae
This bird of high-altitude shrubland occurs from southern Chile to Argentina. Unlike other tinamou species, it often lives in flocks.

12–14 in
31–35 cm

ORNATE TINAMOU
Nothoprocta ornata
F: Tinamidae
An Andean tinamou, this species is found on the high montane grasslands of western South America from Peru to northern Argentina.

RATITES

A group that includes the largest living birds, the ratites are flightless and strong-footed. Some flock in open habitats; others live alone in forests.

Ratites evolved in the Southern Hemisphere and there is evidence to suggest they have flying ancestors. They lack some of the features associated with flight, but retain others.

PHYLUM	CHORDATA
CLASS	AVES
ORDERS	4
FAMILIES	5
SPECIES	15

Ratites do not have the bony ridge on their breastbone that anchors powerful flight muscles in other birds, but they still have wings, and the part of the brain that controls flight is well developed.

The ostriches, which make up the order Struthioniformes, are found in dry open areas in Africa. They are powerful-limbed birds that can run at high speed on flat ground. Somewhat similar in appearance, though less massively built, are the rheas—the order Rheiformes—from the grasslands of South America. Like most ratites, both these groups of birds have a reduced number of toes: the rhea has three toes, the ostrich only two. Both have large wings that they use in displays or to assist their balance when sprinting.

AUSTRALASIAN SPECIES
The Australasian ratites include the cassowaries and emus of the order Casuariiformes and the kiwis, the Apterygiformes. All are forest birds, except the emu, which inhabits scrubland and grassland. The tiny wings of these ratites are invisible under their plumage, which resembles shaggy hair rather than feathers. The nocturnal kiwis, familiar as the national symbol of New Zealand, are the only ratites that do not have a reduced number of toes.

The male ostrich scrapes out a shallow nest in the ground, in which his harem of females lay up to 50 eggs.

20–26 in
50–65 cm

NORTH ISLAND KIWI
Apteryx mantelli
F: Apterygidae
One of several New Zealand kiwis, this nocturnal ratite detects buried invertebrate prey by smell, using nostrils at its probing bill tip.

furlike plumage

26–28 in
65–70 cm

TOKOEKA
Apteryx australis
F: Apterygidae
This is the South Island cousin of *A. mantelli*. Lighter colored than its relative, it has been recognized as a separate species after DNA analysis.

coarse, shaggy plumage

helmetlike casque

red wattles

5–6 ft
1.5–1.8 cm

SOUTHERN CASSOWARY
Casurius casuarius
F: Casuariidae
Cassowaries are fruit-eating rain forest birds. With a range that extends through New Guinea and northern Australia, this is the most widespread species of its family.

brown plumage

5½–7 ft
1.7–2.1 cm

EMU
Dromaius novaehollandiae
F: Dromaiidae
Australia's largest living bird is found across the continent in open grassy areas. Like the related cassowary, its chicks are striped for camouflage.

almost bare neck

black body plumage

3–3¼ ft
92–100 cm

LESSER RHEA
Pterocnemia pennata
F: Rheidae
Smaller than the greater rhea, its northern cousin, this ratite inhabits the southern Andes and Patagonia, where it lives in small flocks.

4¼–4½ ft
127–140 cm

GREATER RHEA
Rhea americana
F: Rheidae
This central South American ratite has a breeding system that does not involve pair-bonding. Males tend a large nest of eggs produced by many females.

white primary feathers

♂

5½–8¾ ft
1.7–2.7 cm

large eye

OSTRICH
Struthio camelus
F: Struthionidae
The largest bird in the world, the ostrich lives in savanna and semidesert areas of Africa. Breeding males mate with several females.

scaly legs

26–28 in
65–70 cm

GREAT SPOTTED KIWI
Apteryx haastii
F: Apterygidae
One of two species of kiwis with gray mottled plumage, this bird is confined to mountain areas of western South Island.

♀

featherless thighs

gray neck

grayish brown plumage

5 ½–8¾ ft
1.7–2.7 cm

SOMALI OSTRICH
Struthio camelus molydophanes
F: Struthionidae
Separated from other populations of ostrich by the Rift Valley, this East African gray-necked form is sometimes regarded as a separate species.

14–18 in
35–45 cm

LITTLE SPOTTED KIWI
Apteryx owenii
F: Apterygidae
Introduced mammals have driven this small kiwi to the edge of extinction. It survives on predator-free satellite islands around New Zealand.

two-toed feet

FOWL, GAMEBIRDS, AND RELATIVES

Strong footed gamebirds are adapted to a wide range of habitats. They are primarily ground-dwellers and feed mostly on plants.

Although they may be good fliers, the majority of gamebirds, the order Galliformes, take to the air only when threatened. None, except the common and Japanese quails, sustains a long migratory flight. Most gamebirds spend their lives on the ground. Only the members of the curassow family of tropical America are habitual tree-dwellers; these birds make their nests in trees, too. Most primitive of the gamebird group are the big-footed megapodes of the Indo-Pacific forests. They bury their eggs in heaps of compost or in volcanic sand and rely on heat generated by decomposition or volcanic action to incubate them. All other gamebirds take a more active role in parenthood, although only the female raises the brood. The precocious offspring can run and feed themselves soon after hatching.

Male gamebirds often have flamboyant plumage, which they use in courtship displays to attract a harem of females. Some species are armed with leg spurs to battle with competitors.

DOMESTICATED SPECIES

The relative ease with which many gamebird species, including the turkey, can be kept in captivity has made them economically important in many countries. Domesticated chickens around the world are all descendants of the red junglefowl of Southeast Asia.

PHYLUM	CHORDATA
CLASS	AVES
ORDER	GALLIFORMES
FAMILIES	5
SPECIES	290

DEBATE
WHAT IS A PARTRIDGE?

Traditional categories of birds may not reflect biological relationships. Many gamebirds are called partridges, but the European grey partridge differs subtly from others in that, for example, males are more distinctively patterned than females—perhaps making it a closer ally of pheasants.

22 in
55 cm

28 in
70 cm

BRUSH TURKEY
Alectura lathami
F: Megapodiidae
Males of this eastern Australian megapode control the temperature of their incubation mounds by adding or removing rotting vegetation. More females hatch at high temperatures.

MALEO
Macrocephalon maleo
F: Megapodiidae
This megapode (large-footed bird) from Sulawesi Island in Indonesia lays its eggs in sand and relies on heat from the sun or underground volcanic activity for incubation.

black markings down the front of the throat

31–36 in
78–92 cm

GREAT CURASSOW
Crax rubra
F: Cracidae
Ranging from Mexico to Ecuador, this large curassow has a curly crest like other curassows. Males are black; females are brown.

23½ in
60 cm

33 in
84 cm

MALLEEFOWL
Leipoa ocellata
F: Megapodiidae
Like many megapodes, this southern Australian species makes a compost mound to lay its eggs. It feeds primarily on various seeds, but is thought to be omnivorous.

brown, white, and black wings

BARE-FACED CURASSOW
Crax fasciolata
F: Cracidae
This South American curassow has a sparsely feathered face. Unlike other related species, it has no fleshy wattles or bill knob.

gray-brown
upperparts

olive-brown
breast

19–21 in
48–53 cm

18 in
46 cm

27 in
69 cm

23–26 in
59–65 cm

white-tipped tail

PLAIN CHACHALACA
Ortalis vetula
F: Cracidae

Ranging from Texas to Costa Rica, this northernmost member of the American curassow family is the only one found in the USA.

GREY-HEADED CHACHALACA
Ortalis cinereiceps
F: Cracidae

Named for their calls, chachalacas are brown relatives of curassows. This species lives in scrubby habitat from Honduras to Colombia.

BLUE-THROATED PIPING GUAN
Pipile cumanensis
F: Cracidae

This South American bird has glossy black plumage typical of tree-dwelling piping guans. They are named for their shrill breeding calls.

HIGHLAND GUAN
Penelopina nigra
F: Cracidae

This Central American bird is more ground dwelling than other guans. It is perhaps the only member of its family that nests on the ground.

22 in
55 cm

26–30 in
66–76 cm

26–30 in
67–75 cm

SPIX'S GUAN
Penelope jacquacu
F: Cracidae

This S. American species belongs to a group of brown guans that resemble chachalacas, but have shorter legs. It nests in trees and has a loud mating call.

BEARDED GUAN
Penelope barbata
F: Cracidae

Named for its streaked neck, the bearded guan, like related species in this genus, is a rain forest bird. It is found in Ecuador and Peru.

30 in
76 cm

red throat
wattle

CAUCA GUAN
Penelope perspicax
F: Cracidae

A bronze-sheened relative of the spix's guan, this bird is found only in the Cauca Valley region in the western and northern parts of Colombia.

bare
bluish head

24–28 in
61–71 cm

27–30 in
68–75 cm

DUSKY-LEGGED GUAN
Penelope obscura
F: Cracidae

The only brown guan with dark rather than reddish legs, this species occurs in central South America, from Brazil to northern Argentina.

CHESTNUT-BELLIED GUAN
Penelope ochrogaster
F: Cracidae

The throat wattle that is typical of brown guans is especially well developed in this species from south-central Brazil.

cape of
long feathers

26–31 cm
10–12 in

9½–10½ in
24–27 cm

10 in
25 cm

VULTURINE GUINEAFOWL
Acryllium vulturinum
F: Numididae

Guineafowl are African gamebirds that live in groups, but are monogamous. They all have bare heads. This is the largest guineafowl species and is found in eastern Africa.

MOUNTAIN QUAIL
Oreortyx pictus
F: Odontophoridae

This native of the Rocky Mountains, like other American quails, is a ground-nesting, monogamous species.

CALIFORNIAN QUAIL
Callipepla californica
F: Odontophoridae

Found from Oregon to California, this species is characterized by its forward-drooping crest, comprised of a cluster of six feathers.

GAMBEL'S QUAIL
Callipepla gambelii
F: Odontophoridae

This relative of the Californian quail has a longer crest. It occurs in the desert in the south of California and the ranges of the two do not overlap.

12½–14 in
32–35 cm

CHUKAR PARTRIDGE
Alectoris chukar
F: Phasianidae
With a range extending across central
Eurasia, this is the most widespread of the
red-legged partridges—typically birds of dry
country, with striking black markings.

black stripes on flanks

15 in
38 cm

ARABIAN PARTRIDGE
Alectoris melanocephala
F: Phasianidae
This large red-legged partridge occurs
in semi-desert regions of the Arabian
Peninsula and Yemen. Its preferred
habitat is juniper woodland.

8½ in
21 cm

**CHESTNUT-BELLIED
HILL PARTRIDGE**
Arborophila javanica
F: Phasianidae
Found in Southeast Asian rain
forests, hill partridges are small,
cryptically colored gamebirds
with stumpy tails. Many have
distinctive head markings.

11½–12½ in
29–32 cm

GRAY PARTRIDGE
Perdix perdix
F: Phasianidae
This is the most widespread of a
small genus of dark-bellied Eurasian
gamebirds that are more related to
pheasants than other partridges.

11–12 in
28–30 cm

GREY FRANCOLIN
Francolinus pondicerianus
F: Phasianidae
As with related junglefowl,
males of this southern Asian
francolin have leg spurs that
they use in fighting.

10–15 in
25–38 cm

RED-NECKED FRANCOLIN
Francolinus afer
F: Phasianidae
A ground-nesting bird of forest
and grassland, this species is
widespread through Africa,
from the Democratic Republic
of Congo to the Cape.

10 in
26 cm

ROULROUL
Rollulus rouloul
F: Phasianidae
Related to hill partridges, the
Southeast Asian roulroul has a more
colorful plumage. Males are dark blue
and have a reddish crest; females are
green and lack the crest.

12 in
31 cm

**CHINESE BAMBOO
PARTRIDGE**
Bambusicola thoracicus
F: Phasianidae
There are two species of bamboo
partridge, birds of eastern Asia
that are related to francolins
and junglefowl. This bird is
native to China.

6½–7 in
6–18 cm

COMMON QUAIL
Coturnix coturnix
F: Phasianidae
Unlike other gamebirds, this
tiny species of western Eurasian
grasslands and semi-desert is
migratory in the northern
parts of its range.

15½–16 in
39–40 cm

*black
throat*

SPRUCE GROUSE
Canachites canadensis
F: Phasianidae
Like other woodland
grouse, this North American
species can digest the needles
of conifer and spruce trees,
rejected by most animals.

16–20 in
40–50 cm

SOOTY GROUSE
Dendragapus obscurus
F: Phasianidae
One of several North
American grouse that have
an inflatable neck sac, this
dark-colored species
inhabits the pine forests
of the Pacific coast.

16–18½ in
41–47 cm

**SHARP-TAILED
GROUSE**
Tympanuchus phasianellus
F: Phasianidae
This N. American grouse
is a more widespread, more
northerly cousin of prairie
chickens. Males have purple
sacs for display.

17 in
43 cm

**GREATER
PRAIRIE CHICKEN**
Tympanuchus cupido
F: Phasianidae
This species is from the central
USA. As with other species of
prairie chicken, males engage
in group courtship displays by
inflating their colored neck sacs.

15–16 in
38–41 cm

**LESSER PRAIRIE
CHICKEN**
Tympanuchus pallidicinctus
F: Phasianidae
Found in southern N. America,
this small species of prairie
chicken has a colored neck sac
and drumming call, typical
of *Tympanuchus* grouse.

23½–34 in
60–87 cm

13½–14 in
34–36 cm

RED GROUSE
Lagopus lagopus scotica
F: Phasianidae
Unlike most
ptarmigans, this British
race of the circumpolar
willow grouse does not
turn white in winter.

*pale gray
tail feathers*

KALIJ PHEASANT
Lophura leucomelanos
F: Phasianidae
This pheasant occurs in forests
from the Himalayas to Myanmar.
There is also an introduced
population in the
Hawaiian Islands.

411

15–16 in
38–41 cm

WESTERN CAPERCAILLIE
Tetrao urogallus
F: Phasianidae
Found in western Eurasian
coniferous forests, this is the
largest member of the grouse
family. Males display to females
with rattling, popping calls.

ROCK PTARMIGAN
Lagopus muta
F: Phasianidae
One of three related grouse
that turn white in winter, the
rock ptarmigan is a tundra
and mountain bird with a
circumpolar distribution.

23½–32 in
60–80 cm

SIAMESE FIREBACK
Lophura diardi
F: Phasianidae
Like many pheasants, this S.E. Asian
bird has bare, red facial skin. As with
other species of *Lophura*, this feature
is common to both sexes.

22–30 in
55–75 cm

SATYR TRAGOPAN
Tragopan satyra
F: Phasianidae
This is a Himalayan species
of *Tragopan*, a genus of Asian
tree-nesting pheasants. Male
tragopans have expandable
fleshy lappets and horns that
are used for display.

21–35 in
53–89 cm

23½–28 in
60–70 cm

23½–47 in
60–120 cm

COMMON PHEASANT
Phasianus colchicus
F: Phasianidae
Native to the woodlands of
central and eastern Eurasia, this
species has been introduced
into western Europe, where it
is now common in farmland.

*white-tipped tail
fanned in display*

LADY AMHERST'S PHEASANT
Chrysolophus amherstiae
F: Phasianidae
Like many other pheasants, only males of
this species have spectacular plumage, and
only females care for the young. This
bird is from China and Myanmar.

16–20 in
40–50 cm

PALAWAN
PEACOCK-PHEASANT
Polyplectron napoleonis
F: Phasianidae
Male peacock-pheasants
lack the long plumes of their
peafowl relatives, but display
tail feathers marked with
iridescent eyespots.

bald head

2½–7¼ ft
0.8–2.2 m

INDIAN PEAFOWL
Pavo cristatus
F: Phasianidae
This tropical Asian species occurs
in India and Sri Lanka. Males court
females by raising their long plumes,
which are just above the tail.

16–31 in
41–78 cm

3½–4 ft
1.1–1.2 m

RED JUNGLEFOWL
Gallus gallus
F: Phasianidae
Unlike related francolins, males of this
Asian ancestor of domestic chickens mate
with many females. They have combs and
wattles, which are absent in females.

WILD TURKEY
Meleagris gallopavo
F: Phasianidae
Turkeys are large, wattled
gamebirds native to N. America.
This species from the southern
USA is the ancestor of
domesticated turkeys.

WATERFOWL

Species in this group are web-footed birds adapted for surface swimming. The majority feed on plants, although some take small aquatic animals.

Most waterfowl have short legs set well back on the body. These birds, the Anseriformes, swim by propelling themselves with webbed feet, and all have plumage kept waterproofed with oil taken from a gland near the tail. South American screamers and magpie-goose from Australasia have partially webbed feet, spending much of their time on land or wading in marshes. These birds belong to two ancient families. All other waterfowl are classified in a single family, the most primitive group being the whistling ducks.

SWANS, GEESE, AND DUCKS

In swans and geese, the sexes have similar plumage patterns. These long-necked, long-winged birds occur mostly outside the tropics. Many northern species breed close to the Arctic and migrate southwards to overwinter. Ducks are generally smaller and have shorter necks than others in this group. Unlike geese

and swans, male ducks are usually more brightly plumaged than the females, especially when breeding. Both sexes of many duck species have a vividly colored wing-patch called a speculum.

The typical "duckbills" of waterfowl have internal plates that have evolved to strain aquatic food drawn in by a muscular tongue. All species retain this feature, even though some have adopted different feeding methods. Geese graze on grasslands, but swans are more aquatic and use their long necks to plunge their heads deep in water. Many ducks dabble – up-ending to feed on or just below the water surface. Others, such as pochards, dive under water to gather food. The saw-billed ducks have narrow bills with serrated edges to catch fish. One group of ducks, including eiders, scoters, and mergansers, includes accomplished sea divers.

NESTING

Most waterfowl are monogamous, some pairing for life. They nest largely on the ground, but a few nest in trees. The marine ducks come inland to breed. All waterfowl produce downy chicks that can walk and swim soon after hatching.

PHYLUM	CHORDATA
CLASS	AVES
ORDER	ANSERIFORMES
FAMILIES	3
SPECIES	174

Flying in formation reduces air drag, enabling birds like these snow geese to conserve energy on long migrations.

21–22 in
53–56 cm

RED-BREASTED GOOSE
Branta ruficollis
F: Anatidae
This is the most brightly colored member of the *Branta* group of dark geese. It breeds in northwestern Siberia, where it nests close to raptors, possibly for protection from foxes.

striking red, black, and white plumage

20–43 in
50–110 cm

CANADA GOOSE
Branta canadensis
F: Anatidae
The largest of the *Branta* geese, this species is native to N. America, but has been introduced to northern Europe.

23–28 in
58–71 cm

BARNACLE GOOSE
Branta leucopsis
F: Anatidae
This goose breeds on Arctic tundra in Greenland and Russia, where it avoids predators by nesting on cliff tops.

22–28 in
56–71 cm

HAWAIIAN GOOSE
Branta sandvicensis
F: Anatidae
Confined to the Hawaiian Islands, this goose has reduced foot-webbing and strong claws, and is adapted to climbing on to rocky lava flows.

32–37 in
81–94 cm

CHINESE GOOSE
Anser cygnoides
F: Anatidae
Widely domesticated, this goose is threatened in the wild in its native C. Asia, where it occurs on the steppes.

23½–30 in
60–75 cm

PINK-FOOTED GOOSE
Anser brachyrhynchus
F: Anatidae
This small, grey goose breeds on rocky outcrops of tundra in Greenland and Iceland, and overwinters in western Europe.

BAR-HEADED GOOSE
Anser indicus
F: Anatidae
This bird is adapted to the thin air of montane C. Asia. Migrating high over the Himalayas, it overwinters in India and Myanmar.

28–30 in
71–76 cm

pale gray body

30–35 in
76–89 cm

GREYLAG GOOSE
Anser anser
F: Anatidae
Widespread across the grasslands and wetlands in Eurasia, this typical grey *Anser* goose is the wild ancestor of domesticated geese.

EMPEROR GOOSE
Anser canagicus
F: Anatidae

26–35 in
66–89 cm

finely barred, grey body

Found in northeastern Siberia and Alaska, this goose grazes on coastal grass and seaweed. It is less gregarious than most other geese.

ASHY-HEADED GOOSE
Chloephaga poliocephala
F: Anatidae

20–23½ in
50–60 cm

S. American *Chloephaga* geese are perhaps more related to ducks than to other geese. This species occurs in Chile and Argentina. Like others in its genus, it may perch in trees.

EGYPTIAN GOOSE
Alopochen aegyptiaca
F: Anatidae

28–29 in
71–73 cm

Belonging to a group of southern hemisphere geese that are related to ducks, the Egyptian goose is widespread through Africa.

PLUMED WHISTLING DUCK
Dendrocygna eytoni
F: Anatidae

15½–17½ in
39–44 cm

Distinct from geese and true ducks, whistling ducks are named for their unique call. This is an Australian species.

MAGPIE-GOOSE
Anseranas semipalmata
F: Anseranatidae

28–35 in
70–90 cm

From Australian wetlands, this species has long legs and partially webbed feet. It is not closely related to any other waterfowl.

CAPE BARREN GOOSE
Cereopsis novaehollandiae
F: Anatidae

30–39 in
75–100 cm

This distinctive goose is restricted to southern Australia and offshore islands, where it gathers in small flocks and grazes on grassland.

23½–30 in
60–75 cm

BLUE-WINGED GOOSE
Cyanochen cyanoptera
F: Anatidae

This goose has a thick plumage as an adaptation to the cool highlands of its native Eritrea and Ethiopia.

3–4 ft
0.9–1.2 m

COSCOROBA SWAN
Coscoroba coscoroba
F: Anatidae

Resembling a goose, this is the smallest of swans. It is confined to swamps in the southern parts of Chile and Argentina.

blackish plumage

BLACK SWAN
Cygnus atratus
F: Anatidae

red bill

3½–4½ ft
1.1–1.4 m

This sooty-black swan with white wing tips may nest in huge colonies. Native to Australia and Tasmania, it has been introduced to New Zealand, Europe, and North America.

TRUMPETER SWAN
Cygnus buccinator
F: Anatidae

Related to other vocal swans, such as the Eurasian whooper swan (*C. cygnus*), this North American species has a loud, honking call.

straight neck

5–6 ft
1.5–1.8 m

all-white body

4¼–5¼ ft
1.3–1.6 m

MUTE SWAN
Cygnus olor
F: Anatidae

This species breeds in Europe and central Asia. Like other swans, it grazes on underwater vegetation by submerging its head.

3½–4 ft
1–1.2 m

BLACK-NECKED SWAN
Cygnus melancoryphus
F: Anatidae

Found in southern South America, this species spends more time on water than other swans and nests on floating vegetation.

WHITE-BACKED DUCK
Thalassornis leuconotos
F: Anatidae

15–16 in
38–40 cm

This bird from Africa and Madagascar is related to whistling ducks, but spends more time on water, nesting on islands of vegetation.

AFRICAN PYGMY-GOOSE
Nettapus auritus
F: Anatidae

12–13 in
30–33 cm

Like other pygmy-geese, this African species nests in tree hollows. It is usually found in wetlands with water lilies, on which it feeds.

ORINOCO GOOSE
Neochen jubata
F: Anatidae

24–26 in
61–66 cm

A South American relative of ducks, this goose occurs on tropical wet savanna and forest edges along rivers.

SOUTHERN SCREAMER
Chauna torquata
F: Anhimidae

33–37 in
83–95 cm

Screamers are large, bulky birds that live in South American marshlands. Like other screamers, this species has bony wing spurs that may be used in fighting.

cream, black,
and green
head pattern

18–22 in
45–56 cm

15½–17 in
39–43 cm

17–22 in
43–56 cm

AMERICAN WIGEON
Anas americana
F: Anatidae

This species feeds by dabbling on
the surface of shallow waters and
occasionally up-ends. Huge flocks
winter in the Caribbean after
breeding in North America.

BAIKAL TEAL
Anas formosa
F: Anatidae

This distinctive duck
breeds in cold, open forest
in Siberia on the edge of
the tundra, and migrates to
Southeast Asia in the winter.

NORTHERN SHOVELER
Anas clypeata
F: Anatidae

Both sexes of dabbling (surface-feeding)
ducks typically have a colored wing-patch
called a speculum. This widespread
wetland species of the northern
hemisphere has a green speculum.

22–26 in
55–65 cm

15–20 in
38–51 cm

20–26 in
50–65 cm

20–26 in
50–65 cm

♀

MALLARD
Anas platyrhynchos
F: Anatidae

Widespread in the northern hemisphere, the
surface-feeding mallard can interbreed with
related species, perhaps indicating that this
group has only recently evolved.

♂

13–16 in
33–40 cm

orange cheek
plumes

INDIAN RUNNER
Anas platyrhynchos
F: Anatidae

A domestic descendant
of the wild mallard, this
long-necked breed originated
in the Malay Peninsula and
India in the 19th century.

WHITE-CHEEKED
PINTAIL
Anas bahamensis
F: Anatidae

This saltwater dabbling duck
is found in S. American estuaries
and mangrove swamps. Unlike
temperate northern pintails, the
sexes look alike.

DOMESTIC DUCK
Anas platyrhynchos
F: Anatidae

Most domesticated ducks have
descended from the mallard.
They are raised for their meat,
eggs, or down, or kept for
ornamental purposes.

BUFFLEHEAD
Bucephala albeola
F: Anatidae

The smallest sea duck in North America,
the bufflehead nests in tree-holes
and sometimes uses those
vacated by woodpeckers.

15–20 in
38–51 cm

17–20 in
43–51 cm

HARLEQUIN DUCK
Histrionicus histrionicus
F: Anatidae

A highly buoyant sea duck, this
species rides choppy
waters and nests beside fast-flowing streams in
eastern North America, Iceland,
and western Russia.

WOOD DUCK
Aix sponsa
F: Anatidae

The newly hatched chicks of this North
American tree-perching duck jump down
from their nests in high tree-holes to reach
the water below.

MANDARIN DUCK
Aix galericulata
F: Anatidae

Wrongly thought to be
monogamous, this tree-nesting
duck is a symbol of love in its
native northeastern Asia. It has
been introduced in Europe
and California.

16–20 in
41–51 cm

♂

TORRENT DUCK
Merganetta armata
F: Anatidae

A powerful swimmer, this
high-altitude, South American duck
lives in fast-flowing rivers of the
Andes and nests beneath
riverside rocks.

♀

RINGED
TEAL
Callonetta leucophrys
F: Anatidae

Like other tropical ducks,
this South American teal is
not migratory and retains
its plumage colors
throughout the year.

white flank
patch

17–18 in
43–46 cm

14–15 in
35–38 cm

SURF SCOTER
Melanitta perspicillata
F: Anatidae
The North American surf scoter, like other scoter species, breeds near fresh water and winters at sea. The male's body is entirely black.

18–22 in
46–55 cm

HOODED MERGANSER
Lophodytes cucullatus
F: Anatidae
Found in North America, this bird has a saw-edged bill for catching fish. It dives with powerful kicks of the feet.

16½–20 in
42–50 cm

SPECTACLED DUCK
Speculanas specularis
F: Anatidae
This duck is found along the rivers of S. America. It is locally called "dog-duck" after the barking call of the female.

18–22 in
46–54 cm

white cheek patch

PINK-EARED DUCK
Malacorhynchus membranaceus
F: Anatidae
The pink head spot of this widespread Australian species is less distinctive than its zebra pattern. It consumes plankton, strained from water by beak flaps.

14–18 in
36–45 cm

RUDDY DUCK
Oxyura jamaicensis
F: Anatidae
This North American duck, now introduced in Europe, has a stiff tail which it uses as a rudder when diving.

14–17 in
35–43 cm

TUFTED DUCK
Aythya fuligula
F: Anatidae
A Eurasian member of the pochard group, the tufted duck feeds mostly, though not entirely, on invertebrates in contrast to its largely vegetarian relatives.

16–18½ in
40–47 cm

CANVASBACK
Aythya valisineria
F: Anatidae
This North American bird is the largest species of pochards, a group of ducks with typically stocky bodies and large heads.

19–24 in
48–61 cm

LONG-TAILED DUCK
Clangula hyemalis
F: Anatidae
Unlike most other Arctic sea ducks, the long-tailed duck breeds in saltwater as well as freshwater habitats. Males have a distinctive long tail.

15–23 in
38–58 cm

SMEW
Mergellus albellus
F: Anatidae
A member of the group of cavity-nesting mergansers, this species is the only small white duck found in northern Eurasia.

14–17½ in
35–44 cm

COMMON SHELDUCK
Tadorna tadorna
F: Anatidae
This goose-like duck is largely a coastal resident in Europe, but in Asia it migrates southwards from inland regions during winter. It nests in burrows.

24–25 in
61–63 cm

RED-BREASTED MERGANSER
Mergus serrator
F: Anatidae
Widespread across the northern hemisphere, the red-breasted merganser breeds on coasts and spends more time at sea than other mergansers.

20½–23 in
52–58 cm

orange head lobe

KING EIDER
Somateria spectabilis
F: Anatidae
This bird breeds along Arctic tundra coastlines. Its large body size may help it to dive deeply for invertebrate prey.

17–25 in
43–63 cm

rose blush on breast

ROSYBILL
Netta peposaca
F: Anatidae
This S. American duck is related to the diving pochards, but spends more time feeding at the water surface. Only the males have a red bill.

22 in
55–56 cm

CRESTED DUCK
Lophonetta specularioides
F: Anatidae
This Andean species is a possible relic of a South American lineage that was ancestral to more widespread dabbling ducks, such as the mallard.

20–24 in
51–61 cm

STELLER'S EIDER
Polysticta stelleri
F: Anatidae
Like other related sea ducks of Arctic and subarctic regions, this eider overwinters farther south in huge flocks, sometimes numbering 20,000 birds.

17–19 in
43–48 cm

PENGUINS

With their two-toned plumage, erect posture, and waddling walk, penguins are instantly recognizable as the classic symbol of the southern oceans.

Residents of the coastal regions of the Southern Hemisphere, all penguins are adapted to life in cold water. Most species live around islands encircling Antarctica, although some are found on southern coastlines of South America, Africa, and Australasia.

These flightless birds, order Sphenisciformes, probably descended from a common ancestor shared with the albatrosses. They may also be distant cousins of the loons of the Northern Hemisphere.

SPECIAL ADAPTATIONS

Penguins have legs set far back toward the tail, a feature that provides excellent propulsion in water and is seen also in birds such as loons and grebes. On land penguins walk upright, but with webbed feet placed flat on the ground their gait is ungainly. In common with other flightless birds, penguins have reduced wings, but these are modified in such a way that they can be used as flippers. In effect, penguins "fly" underwater.

A penguin's short, densely packed feathers have downy bases that trap warm air, and a layer of fat beneath the skin provides further insulation. The feather tips are greasy and waterproof, lubricated by oil produced from a large gland on the rump. A complex system of bloodflow through the legs and feet ensures that the penguin's body is not chilled by standing on snow or ice. The largest penguin species incubate their eggs on their feet. All penguins have counter-shaded plumage (dark above and pale below), which in the sea provides camouflage against ocean-going predators, such as leopard seals.

FORAGING AND NESTING

Penguins can dive for fishes, shrimps, and krill more than 200 times each day. When they have eggs to incubate and young to tend, the parents take turns foraging. In the case of emperor penguins, one of the few species to breed on the Antarctic, the males incubate the eggs alone throughout the polar winter while the females remain feeding at sea. Most penguins are colonial breeders and often return to the same nest site each season.

PHYLUM	CHORDATA
CLASS	AVES
ORDER	SPHENISCIFORMES
FAMILIES	1
SPECIES	18

KING PENGUIN
Aptenodytes patagonicus
F: Spheniscidae
Resembling the emperor penguin, this sub-Antarctic species has an orange-yellow neck and breast markings, and incubates a single egg on its feet.

3–3¼ ft
90–100 cm

EMPEROR PENGUIN
Aptenodytes forsteri
F: Spheniscidae
The largest penguin, this species breeds in colonies on Antarctic ice. Males incubate eggs during the bitter polar winter.

18–23 in
45–58 cm

yellow plume

white feathers contrast with black head and wings

ROCKHOPPER PENGUIN
Eudyptes chrysocome
F: Spheniscidae
The smallest of the sub-Antarctic crested penguins, this bird gets its name from its habit of clambering over rocks and boulders.

3½–4 ft
1.1–1.2 m

28 in
70 cm

14–16 in
35–40 cm

22–23½ in
55–60 cm

LITTLE PENGUIN
Eudyptula minor
F: Spheniscidae
A burrow-nester, this bird is the smallest of all penguins. It occurs along the coasts of southern Australia and New Zealand.

FIORDLAND PENGUIN
Eudyptes pachyrhynchus
F: Spheniscidae
Nesting in the cool coastal forests of southern New Zealand, this species has hairlike crest feathers and a red bill typical of *Eudyptes* penguins.

MACARONI PENGUIN
Eudyptes chrysolophus
F: Spheniscidae
This bird inhabits islands in the far southern Atlantic and Indian oceans but is the only crested penguin to breed on the Antarctic Peninsula.

stubby bill

white eye ring

28—32 in
71—80 cm

18—30 in
46—75 cm

CHINSTRAP PENGUIN
Pygoscelis antarcticus
F: Spheniscidae
The chinstrap penguin dives for krill and fish. It breeds on Antarctic coasts and islands of the South Atlantic.

thin black streak across face

26—28 in
67—72 cm

GENTOO PENGUIN
Pygoscelis papua
F: Spheniscidae
This penguin breeds on the Antarctic Peninsula and Southern Ocean islands. Its nest is a simple cluster of sticks, stones, and feathers.

30 in
75 cm

blue-black upperparts

ADELIE PENGUIN
Pygoscelis adeliae
F: Spheniscidae
One of three "brush-tailed" *Pygoscelis* penguins from Antarctica and adjacent islands, the Adelie penguin has a simple countershading that is typical of the genus.

YELLOW-EYED PENGUIN
Megadyptes antipodes
F: Spheniscidae
A cousin of *Eudyptes* crested penguins, this rare New Zealand bird nests in scrub, but not in dense colonies like the other *Eudyptes* penguins.

GALAPAGOS PENGUIN
Spheniscus mendiculus
F: Spheniscidae
This is the only penguin to breed in tropical waters cooled by the Humboldt Current, which flows along the western coast of South America. It nests in rock crevices.

black face

black breast band

19—20 in
48—51 cm

HUMBOLDT PENGUIN
Spheniscus humboldti
F: Spheniscidae
Found along the Pacific coast of southern South America, this species belongs to a group of burrow-nesters characterized by bold banded patterns from flanks to thighs.

27—28 in
68—70 cm

26—28 in
65—70 cm

MAGELLANIC PENGUIN
Spheniscus magellanicus
F: Spheniscidae
Closely related to the Humboldt penguin, this banded penguin lives in colonies around the southern tip of South America and the Falkland Islands.

24—30 in
61—76 cm

JACKASS PENGUIN
Spheniscus demersus
F: Spheniscidae
Named for its donkeylike braying call, this bird is the only penguin to breed in Africa, in colonies on southwestern coasts.

KING PENGUIN
Aptenodytes patagonicus

The king penguin is the second largest penguin species. Only its close cousin, the emperor penguin, is bigger. Unlike the emperor, the king inhabits sub-Antarctic islands. It hunts fish, ignoring the krill taken by many of its rivals, and dives to extraordinary depths to get them—sometimes more than 660 ft (200 m). It produces just one egg at a time and takes more than a year to raise the single young. Because adults cannot breed annually, the enormous breeding colonies—containing juveniles of different ages—maintain a permanent presence on favored islands of the southern oceans.

SIZE 37–39 in (94–100 cm)
HABITAT Flat coastal plains and waters around sub-Antarctic islands
DISTRIBUTION Islands of S. Atlantic and S. Indian Oceans
DIET Mostly lanternfish, occasionally squid

black upper mandible

penguins drink seawater and excrete excess salt as brine through the nostrils

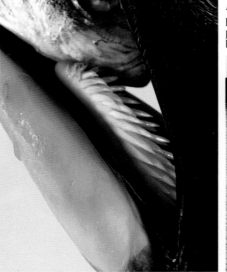

< SPINY TONGUE
A penguin's tongue is muscular and spiny: projections on the tongue surface called papillae have evolved into backward-facing barbs that help grip fish caught in dives.

∨ KEEN EYE
Penguins hunt by sight and have good underwater vision. Bioluminescent lanternfish, caught on nighttime dives, predominate in the diet of this species.

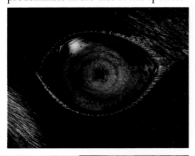

∧ WING PROPULSION
Penguins are flightless birds, but they are propelled by feet and wings during dives. Their flipperlike wings effectively make them "fly" under water.

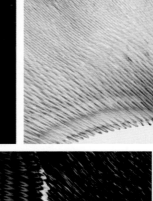

< DENSE FEATHERS
The feathers, arranged in an outer oily waterproof layer and inner downy insulating layers, are adapted for diving in cold water.

∧ SCALY SKIN
Scales on the legs and feet are reminiscent of the reptilian roots of all birds; the dark skin may help spread heat to eggs and young.

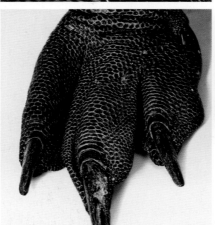

∧ EGG PROTECTOR
The single egg is incubated above the feet and beneath a fold of warm skin called a brood pouch. Once the chick has hatched, it uses the brood pouch for shelter.

WEBBED FOOT >
The kicking action of webbed feet helps propel the bird forward under water—and on land, when "tobogganing" over snow on its belly.

< FIRM TAIL
The short tail consists of stiffened feathers, used as a rudder under water. In smaller species, it is used as a prop on land.

∨ **COUNTERSHADED DIVER**
Below its distinctive yellow markings, the
king penguin has the tuxedo-style plumage
pattern typical of penguins: white beneath
and dark above. When diving, this pattern
offers camouflage from aquatic predators.
Seen from below, the pale belly is harder
to see against the sunlit surface, and when
seen from above, the bird's dark back blends
in with the dark of the water below.

*yellow markings are
due to pigments called
carotenoids, which are
absent from some
penguin species*

*yellow stripe
on lower
mandible*

*yellow
breast*

LOONS

These web-footed, fish-eating birds of Arctic waters have rear-set legs that limit their movement on land but provide excellent thrust for swimming.

Loons are the only family in the order Gaviiformes. Their name may originate from their haunting "lunatic" wailing cries uttered during the breeding season, or possibly from their ungainly motion when out of water. A loon's legs are set so far back on the body that the bird can only manage an awkward shuffle on land. In water, loons swim and dive with ease, possibly leading to their other name of diver. Their streamlined bodies and spearlike bills closely resemble those of penguins, which have similar habits—it is possible, although not certain, that these birds may share a common ancestor.

Although their pointed wings are relatively small for their body size, loons are fast fliers. To get themselves airborne, the larger species have to patter across open water; only the red-throated loon is able to take off from land. All species fly southward to overwinter.

SHARED PARENTING

Territorial male loons choose nest sites in vegetation on the shores of clear Arctic lakes, where both sexes incubate and tend the young. Young loon chicks ride on the backs of their parents, but they can swim and even dive soon after hatching. After breeding, loons lose their striking head and neck patterns, and their duller body plumage makes it more difficult to distinguish the different species.

PHYLUM	CHORDATA
CLASS	AVES
ORDER	GAVIIFORMES
FAMILIES	1
SPECIES	5

Loons fly with their outstretched heads held slightly lower than the body, which gives them a hunchbacked appearance.

COMMON LOON
Gavia immer
F: Gaviidae
One of the largest loons, this bird breeds on lakes in subarctic regions of North America and Iceland. It overwinters on coasts farther south, including those around Britain.

27–36 in
69–91 cm

striped neck patch

black head and neck

30–36 in
76–91 cm

WHITE-BILLED LOON
Gavia adamsii
F: Gaviidae
A large species of Arctic waters, this white-billed bird can be distinguished from other loons by its yellowish white bill.

grayish head and neck

21–27 in
53–69 cm

23–29 in
58–74 cm

23–29 in
58–73 cm

BLACK-THROATED LOON
Gavia arctica
F: Gaviidae
Largely confined to Eurasia when breeding, this species sometimes reaches Alaska, but winters farther south, including the Pacific coast of North America.

RED-THROATED LOON
Gavia stellata
F: Gaviidae
The smallest loon, this bird breeds on small tundra pools in circumpolar regions and migrates south to Europe, China, and the southeastern USA to overwinter.

PACIFIC LOON
Gavia pacifica
F: Gaviidae
This species has similar striped plumage to that of the black-throated loon. Both species have white throats when not breeding.

white spots in summer

ALBATROSSES, PETRELS, AND SHEARWATERS

Long-winged albatrosses and their relatives spend much of their lives in the air, traveling great distances as they scan the surface of the oceans for fish.

These birds, the order Procellariiformes, are master aeronauts, rarely returning to land except to breed. Albatrosses, petrels, and shearwaters—called "tubenoses" on account of the tubular nasal protuberances on their bills—are globally widespread, ocean-going birds, but show the greatest diversity in the Southern Hemisphere.

Unusual for birds, they locate their sparsely scattered sea prey by smell. All but the smallest diving species in this group have elongated wings, and almost all have their webbed feet positioned so far back that some cannot easily walk. These birds deter predators by regurgitating a noxious oil that they produce in their stomachs and sometimes forcefully eject. This oil is also nutritious enough to feed their chicks.

SLOW REPRODUCTION
Tubenoses form pair bonds, sometimes for life, which in the larger species can be several decades. Many species breed in colonies on remote islands, often returning to the same site year after year. Smaller species nest in cavities and burrows. These birds have low breeding rates, but the parents invest heavily in rearing their young. Typically, tubenoses make only one breeding attempt per season and produce a single egg. Despite a lengthy incubation period, chicks are helpless upon hatching and mature very slowly.

PHYLUM	CHORDATA
CLASS	AVES
ORDER	PROCELLARIIFORMES
FAMILIES	4
SPECIES	133

Long-lived wandering albatrosses are monogamous. They reinforce their pair-bonding with an elaborate courtship dance.

WANDERING ALBATROSS
Diomedea exulans
F: Diomedeidae
The largest of the giant Southern Ocean *Diomedea* albatrosses, this bird forms a life-long pair bond and produces a single young every two years.

dark wings turn white with age

3½–4½ ft
1.1–1.4 m

mostly white body

pale pink bill

LAYSAN ALBATROSS
Phoebastria immutabilis
F: Diomedeidae
Albatrosses from the northern Pacific include those that breed in the tropics, like this small species that nests on islands—including Hawaii—when breeding.

30–32 in
77–80 cm

27–29 in
68–74 cm

BLACK-FOOTED ALBATROSS
Phoebastria nigripes
F: Diomedeidae
Like other northern Pacific albatrosses, this small, dark species frequently interrupts its gliding flight with bursts of wing-flapping.

black "brow"

18–20 in
45–50 cm

NORTHERN FULMAR
Fulmarus glacialis
F: Procellariidae
This gull-like petrel is common throughout the Northern Hemisphere. It nests on cliffs and can eject a foul-smelling stomach oil to repel predators.

white body, gray in parts

BLACK-BROWED ALBATROSS
Thalassarche melanophrys
F: Diomedeidae
This is one of several dark-backed albatrosses called mollymawks, from the Southern Hemisphere. It nests in dense colonies, producing one egg each year.

32–37 in
80–95 cm

sooty brown head

12 in / 30 cm

AUDUBON'S SHEARWATER
Puffinus lherminieri
F: Procellariidae

This small bird breeds on tropical oceanic islands. Different populations are considered by some to be separate species.

19 in / 48 cm

PINK-FOOTED SHEARWATER
Puffinus creatopus
F: Procellariidae

A variable species existing in dark and light forms, this shearwater nests on islands off Chile. It migrates to the eastern Pacific during summer.

BULLER'S SHEARWATER
Puffinus bulleri
F: Procellariidae

This shearwater nests on islands off northern New Zealand but roams across the Pacific outside the breeding season.

18–18½ in / 45–47 cm

6½–8 in / 17–20 cm

18–22 in / 45–56 cm

CORY'S SHEARWATER
Calonectris diomedea
F: Procellariidae

A large shearwater that glides with bowed wings, this species breeds in the Mediterranean region and overwinters in the Atlantic.

ANTARCTIC PRION
Pachyptila desolata
F: Procellariidae

Prions are small, gray Southern Ocean petrels that skim the sea to strain plankton with their flattened bills.

12 in / 31 cm

JOUANIN'S PETREL
Bulweria fallax
F: Procellariidae

A tropical petrel of the northwestern Indian Ocean, this bird has a weaving flight. It is closely related to the gliding shearwaters.

dark zigzag pattern

14–16 in / 36–41 cm

16 in / 41 cm

17 in / 43 cm

ANTARCTIC PETREL
Thalassoica antarctica
F: Procellariidae

A bird of sub-Antarctic waters, this large species dives for fish and squid. It breeds on islands around the Antarctic.

SNOW PETREL
Pagodroma nivea
F: Procellariidae

One of the few birds that breed on the Antarctic, this petrel breeds farther south than any other bird, even wandering to the South Pole.

BLACK-CAPPED PETREL
Pterodroma hasitata
F: Procellariidae

Like many of the small, fast species known as gadfly petrels, this one has a tropical distribution. It breeds on islands of the West Indies.

scattered dark feathers on white body

massive yellowish bill

15½–16 in / 39–40 cm

CAPE PETREL
Daption capense
F: Procellariidae

One of several members of this family, mostly circumpolar in the southern hemisphere, the cape petrel breeds on islands around Antarctica and winters farther north.

SOUTHERN GIANT PETREL
Macronectes giganteus
F: Procellariidae

This carrion-eating bird breeds in the South Atlantic. Unlike many other petrels, it has strong enough legs to walk well on land.

34–39 in / 86–99 cm

BAND-RUMPED STORM PETREL
Oceanodroma castro
F: Hydrobatidae

Typical of Northern Hemisphere storm petrels, this tiny bird has a white-banded rump and forked tail. It is seen in both the Atlantic and Pacific.

7½–8½ in / 19–21 cm

GREBES

Birds of ponds and lakes, grebes characteristically swim low in the water and dive propeled by their feet. They feed on small aquatic animals.

Like many other diving birds, grebes have legs that are set far back on the body, making them clumsy on land but agile in water. The toes of their feet are lobed, providing good power strokes during diving, but minimizing drag in the water between strokes. The feet may be used for steering, too; in other diving birds, the tail is used for steering.

A grebe's tail is nothing more than a tuft of feathers. This serves more as a social signal than as a rudder, and is often cocked to expose white feathers beneath. The plumage is dense and well waterproofed by an oil gland. Uniquely in birds, the gland produces a secretion that is 50 per cent paraffin. Grebes have small wings and many species are often reluctant to fly, although northern species are migratory, traveling from inland habitats to the coast to overwinter.

Traditionally, grebes, the order Podicipediformes, are classified as close relatives of loons, penguins, and albatrosses. However, new research suggests that they may be related to flamingos.

RITUAL AND REPRODUCTION
In the breeding season, some grebes perform elaborate courtship rituals. They nest on floating mats of vegetation in freshwater habitats. The young are fully mobile upon hatching and can swim, but seek refuge on a parent's back for the first few weeks.

PHYLUM	CHORDATA
CLASS	AVES
ORDER	PODICIPEDIFORMES
FAMILIES	1
SPECIES	22

The courtship ritual of the great crested grebe culminates with both birds rising up from the water clutching a clump of weed.

9–11½ in
23–29 cm

LITTLE GREBE
Tachybaptus ruficollis
F: Podicipedidae
Found across the Old World, this bird is the most widespread of the small, dumpy grebes known as dabchicks. It has reddish neck when breeding.

9½–14 in
24–36 cm

WHITE-TUFTED GREBE
Rollandia rolland
F: Podicipedidae
A native of southern South America, this is a bird of open lakes with abundant water weeds. A related short-winged Andean species is flightless.

12–15 in
30–38 cm

PIED-BILLED GREBE
Podilymbus podiceps
F: Podicipedidae
This American bird is more stocky and has a stubbier bill than other grebes. Northern birds overwinter in the Caribbean, but tropical populations are sedentary.

16–20 in
40–50 cm

11–13½ in
28–34 cm

gray flanks

RED-NECKED GREBE
Podiceps grisegena
F: Podicipedidae
This species breeds in Eurasia and North America and overwinters farther south in coastal waters. Like other grebes, it migrates at night.

BLACK-NECKED GREBE
Podiceps nigricollis
F: Podicipedidae
Like other *Podiceps* grebes, this sharp-billed diver has a colorful head plumage when breeding. It is found across the Northern Hemisphere.

black head

18–20 in
46–51 cm

10–11½ in
25–29 cm

SILVERY GREBE
Podiceps occipitalis
F: Podicipedidae
This South American grebe is a colonial breeder, congregating on alkaline or salty lakes. It is found from the Andes to the Falkland Islands.

22–30 in
55–75 cm

WESTERN GREBE
Aechmophorus occidentalis
F: Podicipedidae
One of two similar western North American grebes, this species has a range from Canada to Mexico; northern populations winter off the Pacific coast.

GREAT CRESTED GREBE
Podiceps cristatus
F: Podicipedidae
This Old World grebe is known for its courtship display. Like other grebes, both sexes are colorful.

dark gray back

white throat, breast, and belly

FLAMINGOS

These remarkable birds live on salty lagoons and alkaline lakes. Once classified with storks, flamingos are now thought to be related to grebes.

Extreme gregariousness defines flamingo life. These birds of the order Phoenicopteriformes congregate in huge flocks, sometimes numbering hundreds of thousands. These vast groups are so tightly packed that individual birds cannot easily take flight and must initially walk or run if disturbed. But the open habitat that flamingos favor, coupled with the vigilance of many birds, ensures that predators are easily seen.

The large flocks are necessary to stimulate breeding, and flamingo courtship involves group displays. Pairs build mud nests and

territory is simply determined by how far a bird's neck will stretch from the nest. A few days after hatching, flamingo chicks gather together in large crèches. The parent birds feed their young with a liquid food called crop milk.

FILTER FEEDERS

Flamingos have a unique style of filter-feeding using a specially adapted bill. Holding its head upside-down, a flamingo strains planktonic algae and shrimps from the water through hairlike structures lining the bill. Pigments absorbed from the food give flamingos their characteristic pink color. There are few competitors for food—the organisms are taken from otherwise barren inland waters that are either very salty or caustic.

PHYLUM	CHORDATA
CLASS	AVES
ORDER	PHOENICOPTERIFORMES
FAMILIES	1
SPECIES	6

Some of the biggest flocks of flamingos are seen in the African Rift Valley, where lesser flamingos gather to feed.

CHILEAN FLAMINGO
Phoenicopterus chilensis
F: Phoenicopteridae
The most widespread flamingo in South America, the Chilean flamingo has a range from Peru to Tierra del Fuego, and is distinguished by gray legs with pink knees.

pinkish white plumage

pink knee

slender gray leg

3¼–4¼ ft
1–1.3 m

black-tipped bill

CARIBBEAN FLAMINGO
Phoenicopterus ruber ruber
F: Phoenicopteridae
This Caribbean species differs from the greater flamingo in being somewhat smaller and having a pinker plumage.

extremely long neck

pale bill

bright red wing feathers

4–4½ ft
1.2–1.4 m

3½–5 ft
1.1–1.5 m

GREATER FLAMINGO
Phoenicopterus ruber roseus
F: Phoenicopteridae
With a range spanning Africa, southern Europe, and Central Asia, this is the largest and most widespread of the flamingos.

3¼–3½ ft
1–1.1 m

32–39 in
80–100 cm

ANDEAN FLAMINGO
Phoenicoparrus andinus
F: Phoenicopteridae
One of two flamingo species restricted to the higher altitudes of the Andes, this distinctively yellow-legged flamingo may wander nomadically between lakes for food.

LESSER FLAMINGO
Phoeniconaias minor
F: Phoenicopteridae
The smallest flamingo, this species from Africa and southern Asia occurs in huge numbers on highly alkaline lakes.

STORKS, IBISES, AND HERONS

Most species in this group are wetland birds with long legs for walking through marshy or grassy habitats. They have long bills for snatching prey.

These birds, the Ciconiiformes, are hunters, predominantly of fishes and amphibians, but also sometimes of small mammals and insects. The majority frequent the margins of rivers and freshwater lakes and ponds, but some species of the stork family prefer drier habitats, such as pastureland.

Bill shape is one of the features that distinguish members of this order. The bills of herons and storks are usually straight and sharp-tipped but those of ibises and spoonbills are highly modified. Ibises have thin, curved bills that they use to probe in mud and soft earth; spoonbills sweep the distinctive, flattened tips of their bills in shallow water, snapping them shut when they detect prey. Herons and bitterns have modified vertebrae that allow the neck to form an S-shape, which is important for lightning-fast spearing of prey. The same skeletal modification enables these species to retract their necks in flight, unlike storks and ibises, which fly with their necks outstretched.

COLONIAL NESTING
Many birds of this group are gregarious when breeding, and their nesting colonies may include several species. Among the exceptions are the largely secretive and solitary bitterns. All species produce helpless young that must be reared in the nest for some weeks.

CLASS	CHORDATA
AVES	AVES
ORDER	CICONIIFORMES
FAMILIES	3
SPECIES	121

Storks typically nest in trees, but in western Europe white storks make good use of level platforms high on buildings.

red and black bill

4½–5 ft
1.4–1.5 m

black and white plumage

30–36 in
75–91 cm

WOOLLY NECKED STORK
Ciconia episcopus
F: Ciconiidae
The most widespread tropical stork, this bird occurs in both Africa and Asia. It prefers wetland habitats but may stray into pasture.

3¼–4 ft
1–1.2 m

black plumage

3–4 ft
0.9–1.2 m

WOOD STORK
Mycteria americana
F: Ciconiidae
This North American bird belongs to a group of storks with ibislike bills. It submerges its open bill in shallow water, snapping it shut on moving prey.

EUROPEAN WHITE STORK
Ciconia ciconia
F: Ciconiidae
Three *Ciconia* species breed outside the tropics. This European bird uses thermal updrafts on migration routes over land to winter in Africa.

gap in bill

SADDLE-BILL STORK
Ephippiorhynchus senegalensis
F: Ciconiidae
Related to the jabiru, this African stork has a slightly upturned bill with a yellow "saddle." It occurs in solitary pairs, even when nesting.

long, heavy bill

4–5 ft
1.2–1.5 m

32–37 in
81–94 cm

MARABOU
Leptoptilos crumeniferus
F: Ciconiidae
Like other *Leptoptilos* storks, this African species is bareheaded, which enables it to scavenge in carrion without soiling its feathers. It flies with its head retracted.

4–4½ ft
1.2–1.4 m

JABIRU
Jabiru mycteria
F: Ciconiidae
A large American stork, the jabiru is the tallest flying bird in South America. It expands its featherless neck sac when excited.

AFRICAN OPENBILL
Anastomus lamelligerus
F: Ciconiidae
Small, tropical wetland storks, openbills use their distinctive bill to catch and manipulate mollusks before eating them. This species occurs in mainland Africa and Madagascar.

greenish black cap

gray-green back

16–22 in
40–55 cm

yellow legs and feet

28–32 in
70–80 cm

10½–15 in
27–38 cm

GREEN HERON
Butorides virescens
F: Ardeidae

A small heron of North American wetlands, the green heron may sometimes bait fishes with food at the water's edge.

23½–30 in
60–75 cm

AMERICAN BITTERN
Botaurus lentiginosus
F: Ardeidae

This is a typical solitary bittern. Its cryptic coloration and still, vertical posture help camouflage it in reed beds.

EURASIAN BITTERN
Botaurus stellaris
F: Ardeidae

Like other bitterns, this species is more frequently heard than seen. It has a loud, booming call.

LITTLE BITTERN
Ixobrychus minutus
F: Ardeidae

Among the smallest of the heron-bittern family, this skulking Old World species clambers around reeds — sometimes freezing upright in typical bittern manner.

long, narrow crest

32–39 in
80–100 cm

35–39 in
90–98 cm

32–39 in
80–100 cm

black tinge to white foreneck

GREAT EGRET
Ardea alba
F: Ardeidae

Not a true egret but a large white heron, this marsh bird is widespread throughout much of the world.

GREY HERON
Ardea cinerea
F: Ardeidae

Familiar across Eurasia and Africa, the grey heron, like other larger herons, breeds in colonies. It builds stick nests in trees.

WHITE-NECKED HERON
Ardea pacifica
F: Ardeidae

This large heron of wet regions in Australia and New Guinea hunts for insects and small vertebrates in marshes and grassland.

22–28 in
55–70 cm

19–21 in
48–53 cm

CATTLE EGRET
Bubulcus ibis
F: Ardeidae

Actually a small white heron, this globally widespread species forages on grassland and often follows livestock to catch disturbed prey.

23–25 in
58–63 cm

YELLOW-CROWNED NIGHT HERON
Nyctanassa violacea
F: Ardeidae

This American night heron is sedentary in tropical regions, but in the colder parts of its range, it migrates southward during winter.

22–26 in
55–65 cm

22–22½ in
55–57 cm

23½–28 in
60–70 cm

22–26 in
55–65 cm

LITTLE BLUE HERON
Egretta caerulea
F: Ardeidae

This American species belongs to the egret group. The purplish head and neck turn gray-blue outside the breeding season.

18–20 in
45–50 cm

LITTLE EGRET
Egretta garzetta
F: Ardeidae

Found from Europe to Australia, this species has a distinct black bill and legs, and the more widespread race has yellow feet.

WHITE-FACED HERON
Egretta novaehollandiae
F: Ardeidae

A member of the egret group from Indonesia, Australia, and New Zealand, this species has a varied diet that includes insects and frogs.

TRICOLORED HERON
Egretta tricolor
F: Ardeidae

This is an inhabitant of American swamps, where, like related species, it uses its daggerlike bill to spear small animals.

WESTERN REEF HERON
Egretta gularis
F: Ardeidae

A coastal species from Africa and India, this heron has white and dark gray color forms, as well as intermediate forms.

BOAT-BILLED HERON
Cochlearius cochlearius
F: Ardeidae

This heron has a broad bill, adapted for scooping and stabbing animal prey. It inhabits mangrove swamps of Central and South America.

BLACK-CROWNED NIGHT HERON
Nycticorax nycticorax
F: Ardeidae
Night herons have good night vision. Found in most warm regions except Australia, this is the most widespread species.

23–26 in
58–65 cm

AUSTRALIAN IBIS
Threskiornis molucca
F: Threskiornithidae
This is a common species throughout Australia. It often invades urban areas, where it is sometimes considered a pest.

27–30 in
69–76 cm

bare head and neck

black wing plumes

26–30 in
65–75 cm

16½–18 in
42–45 cm

INDIAN POND HERON
Ardeola grayii
F: Ardeidae
Common in southern Asia, this heron stalks aquatic prey, but it may also flush fishes by flying low over water.

black-tipped bill

27–32 in
68–82 cm

23–30 in
59–76 cm

SACRED IBIS
Threskiornis aethiopicus
F: Threskiornithidae
This is a common bird of the wetlands and grasslands of Africa and Madagascar. It has been introduced into America and Europe.

STRAW-NECKED IBIS
Threskiornis spinicollis
F: Threskiornithidae
So called because of the strawlike feathers at the base of its neck, this nomadic ibis occurs in New Guinea and Australia.

neck plumes when breeding

gray body

HADADA IBIS
Bostrychia hagedash
F: Threskiornithidae
A common African species found in grasslands, forests, parks, and gardens, the hadada ibis is named after its distinctive flight call.

30–35 in
76–89 cm

30 in
75–77 cm

22–26 in
55–65 cm

REDDISH EGRET
Egretta rufescens
F: Ardeidae
This American species has white and reddish gray color forms. When fishing, it sometimes holds out its wings to reduce sun glare.

GLOSSY IBIS
Plegadis falcinellus
F: Threskiornithidae
Found throughout warm parts of the world, this is the most widespread ibis species. Its tree-nesting colonies may be accompanied by herons.

BLACK-FACED IBIS
Theristicus melanopis
F: Threskiornithidae
Found in South America, this species occurs in temperate grassland habitat from the Andes to Patagonia.

pinkish red patch on wing

pink breast tuft when breeding

black wing tip

gray legs

22–24 in
56–61 cm

35–36 in
90–92 cm

32–35 in
80–90 cm

28–34 in
71–86 cm

SCARLET IBIS
Eudocimus ruber
F: Threskiornithidae
The national bird of Trinidad, the scarlet ibis occurs in tropical America and acquires its scarlet pigment from eating crustaceans.

AFRICAN SPOONBILL
Platalea alba
F: Threskiornithidae
The only spoonbill confined to wetlands of Africa, this species is characterized by its red face and legs.

EURASIAN SPOONBILL
Platalea leucorodia
F: Threskiornithidae
This species breeds in Eurasia and overwinters in Africa. Unlike the African spoonbill, it does not nest with herons and storks.

ROSEATE SPOONBILL
Ajaia ajaja
F: Threskiornithidae
Like other spoonbills, this distinctive American species feeds on small aquatic animals by swinging its bill from side to side in water.

PELICANS AND RELATIVES

Long-winged, web-footed, fish-eating birds are typical of this order. Most hunt either by plunge-diving or by skimming the water surface.

The common characteristics of this group include reduced nostrils, a particularly important feature in plunge-diving cormorants, shags, gannets, and boobies. Birds of this order, Pelecaniformes, also share the same foot structure, with webbing that spans all four forward-pointing toes. Many of these birds incubate their eggs by holding them on their feet. Frigatebirds, which have reduced webbing between their toes, use a featherless breast area called a brood patch, rather than feet, to keep their eggs warm. Remarkably, cormorants, shags, and anhingas

lack good waterproofing, and after diving spread their wings to dry. Frigatebirds are also poorly waterproofed, they avoid landing in water and skim prey from the surface.

Although awkward and weak-footed on land, frigatebirds and the plunge-diving tropicbirds are superb fliers. Frigatebirds can stay aloft night and day, traveling long distances to feed.

THROAT POUCHES

Almost all these birds have throat pouches, which serve a variety of purposes. Pelicans use their capacious, highly flexible pouch to scoop large catches of fish. Cormorants and anhingas stretch their pouches in courtship display, something male frigatebirds take to an extreme with their hugely inflated bright red throat sacs.

PHYLUM	CHORDATA
CLASS	AVES
ORDER	PELECANIFORMES
FAMILIES	8
SPECIES	67

This pelican chick reaches deep into its parent's throat to feed on regurgitated, partly digested fish.

4–5 ft
1.2–1.5 m

gray plumage

large bill

SHOEBILL
Balaeniceps rex
F: Balaenicipitidae
Restricted to swamps from Sudan to Zambia, this wading bird uses its huge bill to scoop vertebrate prey from muddy waters.

25–30 in
64–75 cm

BROWN BOOBY
Sula leucogaster
F: Sulidae
Like other boobies, this widespread species—the only one with predominantly dark plumage—nests in colonies on tropical coasts, often on islands.

31–36 in
80–92 cm

MASKED BOOBY
Sula dactylatra
F: Sulidae
This is the largest of the tropical boobies, with a black and white plumage and long, yellow bill. It most closely resembles the related cold-water gannets.

32 in
81 cm

BLUE-FOOTED BOOBY
Sula nebouxii
F: Sulidae
Found on rocky coasts from California to Peru and the Galapagos Islands, this booby flaunts its blue feet as part of its courtship display.

yellow tinge to back of head

3–3¼ ft
90–100 cm

white upperparts

30–32 in
76–80 cm

WHITE-TAILED TROPICBIRD
Phaethon lepturus
F: Phaethontidae
Tropicbirds are streamer-tailed, tern-like seabirds with feet so feeble that they use their bellies to shuffle on land. This species frequents most tropical coastlines.

3¼–3½ ft
1–1.1 m

22 in
56 cm

3–3½ ft
0.9–1.1 m

MAGNIFICENT FRIGATEBIRD
Fregata magnificens
F: Fregatidae
Infrequent feeding opportunities and long hours in flight cause frigatebirds, such as this American species, to have a low breeding rate and the longest period of parental care of any bird.

HAMMERKOP
Scopus umbretta
F: Scopidae
This African wetland bird builds huge, heavy-walled nests using sticks and mud to protect its young, who are often left for long periods.

NORTHERN GANNET
Morus bassanus
F: Sulidae
The three similar species of cold-water gannets breed in large colonies on rocky coasts. This one is a native of the North Atlantic.

RED-BILLED TROPICBIRD
Phaethon aethereus
F: Phaethontidae
Found from the eastern Pacific to the Atlantic, this typical, hole-nesting tropicbird rears a single, slow-growing chick, an adaptation to scattered ocean food resources.

ANHINGA
Anhinga anhinga
F: Anhingidae
This American species has a long, cormorant-like neck and a straight bill for spearing fish. These birds are also called snake birds or darters.

snakelike neck

spearlike bill

30–37 in
75–95 cm

28–35 in
70–90 cm

28 in
71 cm

32–39 in
80–100 cm

DOUBLE-CRESTED CORMORANT
Phalacrocorax auritus
F: Phalacrocoracidae
A typical, medium-sized cormorant, this American species is named for its two white head tufts that develop during breeding.

RED-FACED SHAG
Phalacrocorax urile
F: Phalacrocoracidae
A deep-diving seabird found from Japan to the Bering Sea, this species belongs to a group of North Pacific cormorants that are highly adapted to the marine environment.

GREAT CORMORANT
Phalacrocorax carbo
F: Phalacrocoracidae
This cormorant is widespread in temperate and tropical regions. It usually hunts in shallow waters, but can dive up to 100 ft (30 m).

20–22 in
50–55 cm

26–32 in
65–80 cm

3–3¼ ft
90–100 cm

30 in
75–76 cm

orange throat pouch

LITTLE PIED CORMORANT
Phalacrocorax melanoleucos
F: Phalacrocoracidae
This is an Australasian member of a group of primitive, short-billed "micro-cormorants" that are largely associated with freshwater or estuarine habitats.

EUROPEAN SHAG
Phalacrocorax aristotelis
F: Phalacrocoracidae
The European shag develops a frontal crest when breeding. It is a common seabird of northeast Atlantic rocky coasts, where it nests on ledges.

FLIGHTLESS CORMORANT
Phalacrocorax harrisi
F: Phalacrocoracidae
The only living flightless cormorant, this bird inhabits the Galapagos Islands. It evolved from an American group that includes the double-crested cormorant.

RED-LEGGED CORMORANT
Phalacrocorax gaimardi
F: Phalacrocoracidae
An unusually patterned coastal cormorant of temperate South America, this species resembles certain sub-Antarctic shags, but apparently has no close relatives.

18–22 in
45–55 cm

4¼–5 ft
1.3–1.5 m

all-white plumage

PYGMY CORMORANT
Phalacrocorax pygmeus
F: Phalacrocoracidae
The smallest cormorant, this species frequents reed bed habitats on fresh or brackish water from central Europe to Central Asia.

SPOT-BILLED PELICAN
Pelecanus philippensis
F: Pelecanidae
A species from southern Asia, the spot-billed pelican, like most other pelicans, fishes by scooping prey while swimming on the water surface.

red-tipped bill

3¼–4½ ft
1–1.4 m

BROWN PELICAN
Pelecanus occidentalis
F: Pelecanidae
This gray and brown pelican from the southern USA to South America is a coastal-breeding species and, unlike other pelicans, plunge-dives for fish.

4¼–5¼ ft
1.3–1.6 m

AMERICAN WHITE PELICAN
Pelecanus erythrorhynchos
F: Pelecanidae
This pelican breeds on inland lakes of N. America and overwinters on the coasts. In the breeding season, it develops a flat "horn" on its bill.

BIRDS OF PREY

The largest and most important group of day-flying hunters, almost all the birds in this order are exclusively meat-eaters. In some habitats they are the top predators.

Even the smallest members of the order Falconiformes, such as the tiny pygmy falcon, are killers, capable of dispatching birds as big as themselves. Others such as the harpy eagle and Philippine eagle of tropical rain forests, are so powerful they can tackle and kill large monkeys or even small deer.

Birds of prey typically have keen vision and hunt by sight. Only one species, the turkey vulture, detects food sources by smell. Birds of this order have strong clawed feet with opposable hind talons for gripping food and a sharply hooked bill for tearing flesh. Feeding behavior varies from group to group. Large eagles can kill their prey with their talons. Falcons typically deliver a killing bite to the neck of their victim, severing the spinal cord.

The birds of this order are strong winged and some species fly very high in the sky. The largest birds of prey, including eagles, buteos, and vultures, conserve energy by soaring.

SCAVENGERS

The Old World vultures and the American condors follow a scavenging lifestyle. These huge birds have weaker bills for tearing carrion. The bare-headed American vultures may be more closely related to storks than to other birds of prey, but the issue is unresolved.

PHYLUM	CHORDATA
CLASS	AVES
ORDER	FALCONIFORMES
FAMILIES	3
SPECIES	319

25–32 in
64–81 cm

22–26 in
56–66 cm

white streaks on massive wings

TURKEY-VULTURE
Cathartes aura
F: Cathartidae
Unusual for birds, this widespread American vulture locates rotting carcasses by smell. It often nests in dark recesses, such as under large rocks or stumps.

BLACK VULTURE
Coragyps atratus
F: Cathartidae
More gregarious than the related turkey-vulture, the black vulture is an opportunistic scavenger found from the central USA to Chile.

26–32 in
67–81 cm

9–12 in
23–30 cm

contrasting black and white wings

ANDEAN CONDOR
Vultur gryphus
F: Cathartidae
South America's largest flying land bird, this species soars on updraughts in the Andes, locating carrion by sight or by following other scavengers, such as turkey-vultures.

3¼–4½ ft
1–1.4 m

KING VULTURE
Sarcoramphus papa
F: Cathartidae
This large bird soars high above forests of tropical America looking for carrion. It is distinguished by its colorful head and fleshy bill wattle.

BAT FALCON
Falco rufigularis
F: Falconidae
This swift-flying American bird hunts birds, bats, and large insects at twilight. It is found from Mexico to Argentina.

AMERICAN KESTREL
Falco sparverius
F: Falconidae
Kestrels are small falcons that habitually hover when hunting. This species is distributed throughout the Americas, including the Caribbean islands.

8–12 in
20–31 cm

MERLIN
Falco columbarius
F: Falconidae
A nimble predator with a dashing flight, this falcon captures birds in midair over hills and moorland across the Northern Hemisphere.

9½–13 in
24–33 cm

AMUR FALCON
Falco amurensis
F: Falconidae
Unusual for falcons, this raptor habitually gathers in flocks; it breeds in marshy woodland across Siberia and China and migrates to southern Africa to overwinter.

10–12 in
26–30 cm

PEREGRINE FALCON
Falco peregrinus
F: Falconidae
As the fastest bird of prey, this falcon dives steeply on prey in midair. It is found throughout the world in open country, including tundra and semidesert.

dark "mustache"

slate-gray upperparts

13½–23 in
34–58 cm

yellow feet

COMMON KESTREL
Falco tinnunculus
F: Falconidae
Like other kestrels, this species of open country across Eurasia and Asia uses updrafts to maintain its hovering as it scans the ground for prey.

12½–15½ in
32–39 cm

AFRICAN PYGMY FALCON
Polihierax semitorquatus
F: Falconidae
This African bird dives for insects and lizards on ground. It breeds in the nests of weaver birds and may raise a brood cooperatively with others.

7–8½ in
18–21 cm

white neck ruff

fleshy comb

MOUNTAIN CARACARA
Phalcoboenus megalopterus
F: Falconidae
Relatives of falcons, caracaras are more sluggish and longer legged. Like other caracaras, this species of the high Andes is a scavenger, but it also hunts small animals.

19–21 in
48–53 cm

STRIATED CARACARA
Phalcoboenus australis
F: Falconidae
The tendency of this fearless caracara to attack newborn lambs has led to persecution in its native Falkland Islands.

21–24 in
53–62 cm

head plumes

YELLOW-HEADED CARACARA
Milvago chimachima
F: Falconidae
A hawklike scavenger that also eats oil palm fruit, this bird of southern South America frequents savanna and forest edges.

16–18 in
40–46 cm

black crown and crest

yellowish red facial skin

elongated central tail feathers

CRESTED CARACARA
Caracara cheriway
F: Falconidae
This common species of caracara is found in open country, from the southern USA to northern South America. It nests in trees or on the ground.

19½–23 in
49–58 cm

OSPREY
Pandion haliaetus
F: Accipitridae
Found almost worldwide, the fish-eating osprey plunges for its prey and has a reversible outer toe for a better grip on its slippery catch.

20½–26 in
52–66 cm

SECRETARY BIRD
Sagittarius serpentarius
F: Accipitridae
One of the few raptors to hunt on ground, this long-legged bird of African savanna chases small animals, often stamping to disable them.

4¼–5 ft
1.3–1.5 m

long legs

white head and neck

chestnut upperparts

17–20 in
43–51 cm

red-brown tail

BRAHMINY KITE
Haliastur indus
F: Accipitridae

With a range from India to Australasia, this riverside and coastal scavenger also hunts for live prey such as fishes and small mammals.

20–25 in
50–64 cm

SWALLOW-TAILED KITE
Elanoides forficatus
F: Accipitridae

This insect-eating bird of prey is a graceful and agile flier. It breeds in the southeastern USA and Central America, and winters in South America.

12½–15 in
32–38 cm

WHITE-TAILED KITE
Elanus leucurus
F: Accipitridae

Typical of its genus, this sharp-browed kite habitually hovers when hunting. It occurs from the USA to South America, outside the Amazon basin.

20½–23½ in
52–60 cm

EUROPEAN HONEY BUZZARD
Pernis apivorus
F: Accipitridae

Belonging to a group of tropical raptors that eat bee and wasp larvae, this species breeds in Eurasia and overwinters in Africa.

pale head

GOLDEN EAGLE
Aquila chrysaetos
F: Accipitridae

A graceful soaring bird, this large, long-tailed eagle occurs in open country across the Northern Hemisphere. It frequents forests in some regions.

23½–39 in
60–100 cm

feathered legs

24–30 in
61–75 cm

CHANGEABLE HAWK-EAGLE
Spizaetus cirrhatus
F: Accipitridae

Often crested, Asian hawk eagles are forest raptors. This variable species has dark and pale forms and is found from the Himalayas to Indonesia.

28–33 in
70–83 cm

EASTERN IMPERIAL EAGLE
Aquila heliaca
F: Accipitridae

True eagles, including those of the *Aquila* genus, such as this Eurasian species, are described as "booted" because of their fully feathered legs.

28–38 in
71–96 cm

BALD EAGLE
Haliaeetus leucocephalus
F: Accipitridae

This North American sea eagle and USA national symbol, captures or scavenges fish, sometimes hunting cooperatively. It breeds in woodland near water.

28–35 in
70–90 cm

WHITE-BELLIED SEA EAGLE
Haliaeetus leucogaster
F: Accipitridae

With a distribution from India to Australasia along lakes and rivers, like other large eagles, this fishing raptor builds huge stick nests.

22–28 in
55–72 cm

BONELLI'S EAGLE
Hieraaetus fasciatus
F: Accipitridae

A woodland- and mountain-dwelling species, this long-winged buzzardlike eagle has a range that extends from southern Eurasia to North Africa.

22–26 in
55–65 cm

AFRICAN HAWK EAGLE
Hieraaetus spilogaster
F: Accipitridae

A small raptor from sub-Saharan Africa, this hawk eagle hunts in wooded savanna and hilly country.

dark brown flight feathers

AFRICAN WHITE-BACKED VULTURE
Gyps africanus
F: Accipitridae

One of the most common vultures of the sub-Saharan savanna, this species gathers in large numbers near carcasses. These birds are seen in towns and villages.

34–38 in
85–97 cm

3¼–4¼ ft
1–1.3 m

BEARDED VULTURE
Gypaetus barbatus
F: Accipitridae

Native to the mountains of Africa and Eurasia, this diamond-tailed, solitary vulture subsists largely on marrow obtained by dropping bones on rocks to split them open.

LAPPET-FACED VULTURE
Torgos tracheliotus
F: Accipitridae

Like the related *Gyps* vultures, this carrion-eater of arid Africa has a long neck and a bare head to prevent carcasses from soiling its plumage.

3¼–4 ft
1–1.2 m

35–39 in
90–98 cm

bulbous bill

EURASIAN GRIFFON
Gyps fulvus
F: Accipitridae

This vulture occurs in mountainous regions of southwestern Eurasia and northeastern Africa. It breeds and roosts among rocks and on ledges.

neck ruff turns white with age

3–3½ ft
0.9–1.1 m

RÜPPELL'S VULTURE
Gyps rueppelli
F: Accipitridae

An African relative of the Eurasian griffon, this darker species inhabits arid areas. It has been recorded as flying at higher altitudes than any other bird.

28–34 in
72–85 cm

23½–28 in
60–70 cm

23½ in
60 cm

PALM-NUT VULTURE
Gypohierax angolensis
F: Accipitridae

Unusual for a vulture, this African species has a largely vegetarian diet, consisting of oil palm fruit. However, it also feeds on fishes and carrion.

WHITE-HEADED VULTURE
Trigonoceps occipitalis
F: Accipitridae

Found in northern, eastern, and southern Africa, this vulture is often seen in pairs, and is usually outnumbered by other vulture species when feeding on carcasses.

EGYPTIAN VULTURE
Neophron percnopterus
F: Accipitridae

A relative of the palm-nut vulture, this bird of southern Eurasia and Africa routinely uses rocks to crack open ostrich eggs.

dark brown to white coloration

15–17 in
38–43 cm

20–26 in
50–65 cm

20–22½ in
51–57 cm

WHITE-EYED BUZZARD
Butastur teesa
F: Accipitridae

This small buzzard-hawk from southern Asia is more terrestrial than related species, hunting small animals and insects on the ground.

LONG-LEGGED BUZZARD
Buteo rufinus
F: Accipitridae

This buzzard of semideserts and mountains breeds in central Europe and Central Asia, with some populations migrating to North Africa in winter.

EURASIAN BUZZARD
Buteo buteo
F: Accipitridae

A common raptor, this species has light and dark color forms. Its northern populations overwinter in tropical Africa and Asia.

>>

RÜPPELL'S VULTURE
Gyps rueppellii

One of the iconic scavengers of the African plains, from Senegal east to Sudan and Tanzania, Rüppell's vulture is able to fly at great heights in its search for food because its blood is specially adapted to capture oxygen in the thin air. This vulture patrols dry mountainous terrain, leaving its cliff-top roost sites in early morning to rise on updrafts instead of thermals. It uses acute vision to find carcasses and will wait patiently—days, if necessary—for predators to leave a kill. Like most vultures, it eats soft, rotting flesh and offal. However, its longer neck allows it to reach deeper into corpses than many of its competitors, gorging its fill until, with some difficulty, it returns to the skies.

SIZE 34–38 in (85–97 cm)
HABITAT Gorges in dry, open country
DISTRIBUTION North and East Africa
DIET Carrion

> THIRD EYELID
A typical feature of birds, this membrane cleans the surface of the eye, and may help protect against flying debris during frenzied feeding.

∨ WHITE COLLAR
White fluffy feathers encircling the base of the neck form a ruff. This may be discolored with dust and blood from carcasses.

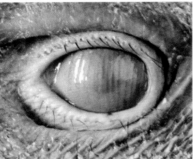

< SCALLOPING
This vulture's dark wing feathers are broadly tipped with lighter marking, giving the bird a scalloped appearance at a distance.

nostril

< PLUMAGE
Beneath the patterned feathers that mark the contours of the body are fluffy down feathers that trap body heat—vital at high altitude.

hooked bill

rows of covert feathers smooth passage of air flow

longer, stiffer flight feathers provide thrust and lift for flying

< LEG
Although the bird's thigh is feathered, its strong legs are bare and so remain relatively clean when the bird is feeding on carcasses.

> A HEAD FOR GORE
Although the sparse down of the vulture's head and neck is frequently blood-stained, a fully feathered head would get clogged with sticky debris as the bird reached deep inside the carcasses of large mammals to feed. The vulture's hooked bill is used to tear semi rotten flesh, and is long enough to probe carcasses.

∧ WING
Long, broad wings help the vulture soar or glide, saving energy. Takeoff can be a struggle after a heavy meal.

< FOOT
Because vultures use their feet for walking rather than killing, they lack the big talons that are typical of predatory birds of prey.

pinkish gray skin of head and neck has a light coating of down

18–21 in
46–53 cm

18–22 in
45–56 cm

17–24 in
43–61 cm

19–22 in
48–56 cm

18–20 in
46–51 cm

RED-SHOULDERED HAWK
Buteo lineatus
F: Accipitridae

Among the most vocal of American hawks, this species from eastern and southern North America is found in woodland near water.

SWAINSON'S HAWK
Buteo swainsoni
F: Accipitridae

Huge flocks of the Swainson's hawk migrate from Canada to Argentina, a distance that— among raptors—is exceeded only by the peregrine.

BLACK-COLLARED HAWK
Busarellus nigricollis
F: Accipitridae

A relative of the snail kite, this species occurs in Central and South American wetlands, where it fishes by dropping feet first into floating vegetation.

RED-BACKED HAWK
Buteo polyosoma
F: Accipitridae

A South American hawk of mountain slopes, sometimes above the treeline, this species habitually hovers. It has gray and brown color phases.

RED-TAILED HAWK
Buteo jamaicensis
F: Accipitridae

The most common hawk in North America, it has an aerial courtship display in which pairs often clasp feet and spiral downward.

13½–14½ in
34–37 cm

21½–24 in
54–61 cm

23½–26 in
60–66 cm

narrow, strongly
hooked bill

MISSISSIPPI KITE
Ictinia mississippiensis
F: Accipitridae

Related to hawks and buzzards, this North American raptor inhabits open woodland and is a colonial nester. It overwinters in South America.

SAVANNA HAWK
Buteogallus meridionalis
F: Accipitridae

This South American hawk hunts on the ground, sometimes following forest and grass fires to catch small animals that have been disturbed.

14–16 in
36–40 cm

18–23 in
46–59 cm

22–23½ in
55–60 cm

deeply
forked tail

HARRIS'S HAWK
Parabuteo unicinctus
F: Accipitridae

Found from California into S. America, this co-operative hunter works in small groups to flush out prey.

BLACK KITE
Milvus migrans
F: Accipitridae

Found in open country across Eurasia, Africa, and Australasia, the black kite has a varied diet that includes live fish and small mammals, carrion, and sometimes human refuse.

RED KITE
Milvus milvus
F: Accipitridae

Like other *Milvus* kites, this species from Europe and the Middle East is slightly weak footed, but accomplished at soaring. It frequently feeds on carrion.

red feet

18–22 in
46–56 cm

AFRICAN HARRIER-HAWK
Polyboroides typus
F: Accipitridae

This raptor from sub-Saharan Africa eats oil-palm fruit and hunts small vertebrates. Its flexible "double-jointed" legs help it take prey from tree-holes.

DARK CHANTING GOSHAWK
Melierax metabates
F: Accipitridae

An African hawk of dry, open country, this bird resembles a harrier when in flight. It has a musical piping call.

SNAIL KITE
Rostrhamus sociabilis
F: Accipitridae

This bird from Florida and Central and South American marshes has a strongly curved bill, adapted for feeding on aquatic snails.

WHITE HAWK
Leucopternis albicollis
F: Accipitridae

This forest hawk of Central and South America hunts for reptiles, especially snakes. It has been described as lethargic and easily approached.

23½–26 in
60–66 cm

17–22 in
43–56 cm

black-tipped wings

white underparts

NORTHERN HARRIER
Circus cyaneus
F: Accipitridae
Harriers are characterized by a narrow tail; narrow, pointed wings; and long legs. This species is widespread in the Northern Hemisphere.

17½–20½ in
44–52 cm

17–18½ in
43–47 cm

MONTAGU'S HARRIER
Circus pygargus
F: Accipitridae
Eurasian harriers, such as the Montagu's harrier, migrate to Africa and southern Asia. This species inhabits grassland and reed beds.

19–22 in
48–56 cm

WESTERN MARSH HARRIER
Circus aeruginosus
F: Accipitridae
The males of this species are brown, like the females, with gray on their wings and tail. Males of other harrier species are all gray.

12–14½ in
30–37 cm

LIZARD BUZZARD
Kaupifalco monogrammicus
F: Accipitridae
A native of the African savanna, this bird mostly preys on large insects, such as grasshoppers, but also eats small vertebrates.

10–14 in
25–35 cm

SHIKRA
Accipiter badius
F: Accipitridae
A typical *Accipiter* hawk with a long tail and short wings, the Old World shikra has a dashing flight for catching small animals, including birds.

19–24 in
48–62 cm

NORTHERN GOSHAWK
Accipiter gentilis
F: Accipitridae
This large hawk from North America and Eurasia is capable of crashing through dense forest to catch squirrels and grouse.

11–16 in
28–40 cm

EURASIAN SPARROWHAWK
Accipiter nisus
F: Accipitridae
One of nearly 50 species of *Accipiter* hawks, this hunter of small birds is found in woodland habitats from Europe to Japan.

red facial skin

BLACK-BREASTED SNAKE EAGLE
Circaetus pectoralis
F: Accipitridae
A bird of African grasslands, this eagle feeds on lizards and small mammals as well as snakes.

25–27 in
63–68 cm

BROWN SNAKE EAGLE
Circaetus cinereus
F: Accipitridae
Snake eagles typically watch for prey from exposed perches. This African species may hover before plunging to kill snakes on the ground.

28–30 in
71–76 cm

long, broad wings

24–26 in
62–67 cm

SHORT-TOED EAGLE
Circaetus gallicus
F: Accipitridae
A Eurasian member of a group of snake-eating eagles, this species occurs on rocky hillsides and coastal plains.

22–30 in
55–75 cm

CRESTED SERPENT EAGLE
Spilornis cheela
F: Accipitridae
Belonging to a group of Asian snake eagles, this species inhabits a range from India to the Philippines, and is often found near freshwater.

22–28 in
55–70 cm

BATELEUR
Terathopius ecaudatus
F: Accipitridae
This African savanna bird is the only snake eagle that regularly eats carrion. *Bateleur* means "acrobat" in French, a reference to this bird's acrobatic flight.

red feet

CRANES AND RAILS

From graceful dancing cranes to small skulking rails, this order contains a wide variety of ground-dwelling birds of both dry and wetland habitats.

Mostly long-legged, long-billed birds, members of the order Gruiformes are behaviourally very diverse. It is possible that many of the groups traditionally included here may belong in another order. The sunbittern and kagu are almost certainly not allied to cranes and rails, but their biological relationships are still uncertain.

Cranes and their relatives spend most of their time on the ground and their foot structure is adapted accordingly: without the need to perch, their hind toe is reduced or absent. The most aquatic members of the order,

such as the finfoots and coots, have lobed, not webbed, feet. Nearly three-quarters of species in the group belong to the rail family. Many rails live in wetlands and have flat-sided bodies to enable them to move easily through thick reed beds.

The much larger bustards and cranes live in more open country, where they can show off their elaborate courtship displays. Bustards prefer dry, sometimes semidesert habitats.

FLIGHTLESS SPECIES

Many species in this group have colonized oceanic islands, where a few have evolved into flightless forms in habitats that were originally predator-free. Rats and other introduced animals now pose a threat to such species.

PHYLUM	CHORDATA
CLASS	AVES
ORDER	GRUIFORMES
FAMILIES	11
SPECIES	228

Red-crowned cranes, like other crane species, engage in elaborate dances to attract mates or reinforce pair bonding.

GREAT BUSTARD
Otis tarda
F: Otidae
Male bustards are bigger than females, especially so in this species of Eurasian steppes. Adult birds take six years to mature and develop plumes.

rufous breast band

28–43 in
70–110 cm

16–18 in
40–45 cm

LITTLE BUSTARD
Tetrax tetrax
F: Otidae
This diminutive bustard breeds in open country across Eurasia and migrates southward in winter. In flight, it resembles a shelduck.

22–26 in
55–65 cm

HOUBARA BUSTARD
Chlamydotis undulata
F: Otidae
A native of arid country, this bird is found in open plains and barren desert areas of northern Africa and the Canary Islands.

orange-brown wings

mottled black and white markings

21 in
53 cm

RED-CRESTED BUSTARD
Lophotis ruficrista
F: Otidae
Like other bustards, this southern African species has a spectacular courtship display. Males perform tumbling aerial flights, and pairs call in duet.

3¼–4½ ft
1–1.4 m

KORI BUSTARD
Ardeotis kori
F: Otidae
Among the heaviest flying birds, weighing up to 42 lb (19 kg), this bustard occurs across eastern and southern Africa. It feeds on small vertebrates, carrion, and seeds.

AUSTRALIAN BUSTARD
Ardeotis australis
F: Otidae
This bustard inhabits grasslands and open woodland of Australia and southern New Guinea. Males have a throat sac, which they inflate when displaying.

2½–5 ft
0.8–1.5 m

RED-LEGGED SERIEMA
Cariama cristata
F: Cariamidae
Related to the extinct, predatory, giant terror birds, seriemas are natives of the grasslands of South America. This species has a sicklelike claw to dismember small prey.

30–35 in
75–90 cm

long crest

off-white plumage

KAGU
Rhynochetos jubatus
F: Rhynochetidae
Confined to the forests of New Caledonia in the southwestern Pacific, the kagu lacks strong muscles necessary for flight. Instead, it uses its wings to glide and for display.

22 in
55 cm

white stripes

SUNBITTERN
Eurypyga helias
F: Eurypygidae
This heronlike predator inhabits the humid forests of Central and South America. It flashes its colored wing spots for display or to startle intruders.

17–19 in
43–48 cm

mottled plumage

YELLOW-LEGGED BUTTONQUAIL
Turnix tanki
F: Turnicidae
Like other buttonquails, this east Asian species inhabits tropical grasslands. The females are more colorful, and compete for the drab males, who take care of the young.

6 in
15 cm

olive-brown wings and body

LIMPKIN
Aramus guarauna
F: Aramidae
Mostly nocturnal, this small cousin of cranes inhabits tropical American wetlands and uses its tweezerlike bill to extract snails from shells.

26–28 in
65–70 cm

long, gray legs

BUFF-BANDED RAIL
Gallirallus philippensis
F: Rallidae
Unlike some of the other Indo-Pacific *Gallirallus* species that are flightless or nearly so, this rail has managed to disperse to many oceanic islands, from the Philippines to New Zealand.

11–13 in
28–33 cm

4–6 in
10–15 cm

BLACK RAIL
Laterallus jamaicensis
F: Rallidae
Belonging to a group of mostly tropical American ruddy-patterned crakes, this species breeds in North America, where it is the smallest rail.

YELLOW RAIL
Coturnicops noveboracensis
F: Rallidae
One of several barred-backed rails, this tiny, secretive North American species is most easily detected by its nocturnal clicking call.

5–7 in
13–18 cm

BLACK CRAKE
Amaurornis flavirostra
F: Rallidae
This African crake is widespread south of the Sahara. Unlike many of its secretive relatives, it is often seen out in the open.

7½–9 in
19–23 cm

CORNCRAKE
Crex crex
F: Rallidae
This shy grassland crake is best identified by its rasping "crex crex" call. It breeds in Eurasia and overwinters in Africa.

9–12 in
22–30 cm

WATER RAIL
Rallus aquaticus
F: Rallidae
Rallus rails are long-billed marsh birds with a narrow body for passing through reed beds. Like other members of the genus, this Eurasian bird is rarely seen away from dense cover.

9–11 in
23–28 cm

KING RAIL
Rallus elegans
F: Rallidae
This species is found in eastern North America, Mexico, and Cuba. It forages in concealed locations for insects, spiders, shrimps, and snails.

15–19 in
38–48 cm

grayish brown upperparts

dull flank stripes

CLAPPER RAIL
Rallus longirostris
F: Rallidae
Unlike many other rails, this tropical American species prefers salt marshes and mangrove swamps. It has a distinctive clacking call.

12½–16 in
32–41 cm

VIRGINIA RAIL
Rallus limicola
F: Rallidae
This long-distance migrant has a range from North America to northern South America. It is a secretive bird that is difficult to spot.

8–10½ in
20–27 cm

red bill with yellow tip

8½–10½ in
21–27 cm

purplish blue underparts

7–9 in
18–22 cm

WHITE-BROWED CRAKE
Porzana cinerea
F: Rallidae
Shorter bills and distinctive calls distinguish crakes from rails. This gray-fronted bird is a typical *Porzana* crake. Its range extends from the Malay Peninsula to Polynesia.

RUDDY-BREASTED CRAKE
Porzana fusca
F: Rallidae
This crake occurs in wetlands of eastern Asia, but is also seen in mangroves or drier habitats. It has distinctive chestnut-colored underparts.

8–10 in
20–25 cm

9–9½ in
22–24 cm

green wings

12–14 in
30–36 cm

yellow legs

SORA RAIL
Porzana carolina
F: Rallidae
This crake is the most common member of the rail family in North America. It breeds in shallow wetland and winters in the Caribbean.

SPOTTED CRAKE
Porzana porzana
F: Rallidae
This shy bird hides in thick vegetation, and is detected by its whipcracklike call. It breeds in marshes across Eurasia and feeds on small aquatic animals.

PURPLE GALLINULE
Porphyrio martinica
F: Rallidae
Occurring in tropical American marshes, this species belongs to a group of rails called swamphens. It is characterized by blue-purple plumage and a blue forehead shield.

12½–14 in
32–35 cm

COMMON MOORHEN
Gallinula chloropus
F: Rallidae
Moorhens are noisy birds with dark plumage and jerky movements. One of several *Gallinula* species, this one has an almost global distribution.

RED-KNOBBED COOT
Fulica cristata
F: Rallidae
Coots are black rails with a white forehead shield and lobed toes. Occurring in Africa, Madagascar, and southern Spain, this species has two red knobs on its forehead.

15–16½ in
38–42 cm

AMERICAN COOT
Fulica americana
F: Rallidae
Found from North America to northern South America, this bird, unlike true rails, can swim. It forages in shallow water and feeds on land.

15½–16 in
39–40 cm

red wattle

19–22 in
48–56 cm

3½ ft
1.1 m

GREY CROWNED CRANE
Balearica regulorum
F: Gruidae
African crowned cranes are the only cranes that can grip branches, enabling them to roost in trees. This is the most southerly species.

AFRICAN FINFOOT
Podica senegalensis
F: Heliornithidae
Streamlined swimming cousins of rails, finfoots are named for their web-lobed toes. This species occurs in wetlands in sub-Saharan Africa.

20–25 in
50–63 cm

SUNGREBE
Heliornis fulica
F: Heliornithidae
This species is a tropical American finfoot. Like all finfoots, it is a secretive bird of slow-moving water, where it feeds on small animals.

10–13 in
26–33 cm

GREY-WINGED TRUMPETER
Psophia crepitans
F: Psophiidae
Named for their loud calls, trumpeters are gregarious Amazonian ground birds that are weak fliers. This is a black-bodied species, which—like other trumpeters—has a "hunchback" appearance.

gray-white plumage

bare red head

3½–4 ft
1–1.2 m

BROLGA
Grus rubicunda
F: Gruidae
This Australian crane has a bare red head and black dewlap (flap under the chin). It performs a spectacular prancing courtship display.

BLUE CRANE
Anthropoides paradiseus
F: Gruidae
The long wing plumes of this African crane resemble a tail. When not breeding, it is nomadic, frequenting lakeside, grassland, and farmland.

3¼–3½ ft
1–1.1 m

slate-gray plumage

3½–4 ft
1.1–1.2 m

4½–5 ft
1.4–1.5 m

RED-CROWNED CRANE
Grus japonensis
F: Gruidae
Declining in numbers, this bird breeds in Siberia and overwinters in Korea and China. The heaviest of all cranes, like related species it has a bare, red crown patch.

black wing feathers

3½–4 ft
1.1–1.2 m

SANDHILL CRANE
Grus canadensis
F: Gruidae
A North American crane, this species also reaches westward into Siberia. It migrates south in family groups, as far as Mexico.

COMMON CRANE
Grus grus
F: Gruidae
Frequenting marsh, heathland, and tundra, this bird breeds in Eurasia and migrates to northern Africa and southern Asia, often in a V-formation.

juveniles lose their buff face feathers as they mature, leaving a patch of white cheek skin

red throat sac

∧ **CROWN AND COLOUR**
This species sports a bristly golden crown, a neat black forehead, and bare white cheeks, edged with a patch of red, which is broader in East African birds. Both males and females have red throat sacs that can be inflated with air, then rapidly deflated to give a booming call.

neck feathers lack interlocking barbs, giving a loose, "hairy" appearance

GREY CROWNED CRANE
Balearica regulorum

Grey crowned cranes belong to the Gruidae family of birds, famous for their choreography; for these birds, dance is an important part of life. On the open savanna they display by jumping, flapping their wings, and bowing their heads—sometimes, it seems, just to relieve aggression or reinforce pair-bonding—but mainly to court their mates by flaunting their elaborate head ornamentation. They lack the long, coiled windpipe of the longer-billed crane species, so instead of calling like a bugle they honk like a goose. They also boom during courtship by puffing air from an inflated red throat sac. During the breeding season pairs retire to wetter habitats, where thicker vegetation conceals their nest—a circular platform made from grasses and sedges. Here, the chicks are hidden from predators, allowing the parents to roost, unlike other cranes, perched high in trees.

SIZE 3½ ft (1.1 m)
HABITAT Open country
DISTRIBUTION East and South Africa
DIET Grasses, seeds, invertebrates, small vertebrates

black head feathers give the appearance of a bulging forehead

nostril

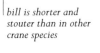

bill is shorter and stouter than in other crane species

THIRD EYELID >
Also known as a nictitating membrane, from the Latin *nictare*, meaning "to blink," this translucent eyelid—the same as in other birds—moves across the eye to clean its surface.

GOLDEN PLUMES >
When the wings are folded, long, golden feathers on the upper wings—just above the flight feathers—hang down over the side of the body.

∨ RUFFLED NECK
Long, tapered contour feathers give the crowned crane a shaggy appearance around the upper part of the body and the lower part of the neck. Most of the bird's plumage is gray.

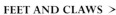

FEET AND CLAWS >
Crowned cranes have long hind toes, setting them apart from other crane species. This enables them to perch in trees—perhaps a vestige of tree-dwelling ancestors.

∧ LONG LEGS
Although long legs help in dancing displays, and are good for wading, crowned cranes have shorter legs than those of other crane species.

white wing covert feathers

black primary flight feathers

brown secondary flight feathers

∧ WING
When crowned cranes fly overhead, their white underwing patches are clearly visible. Although they have strong wings, these tropical birds—unlike other cranes—do not sustain long-distance migrations.

WADERS, GULLS, AND AUKS

Largely birds of coastal areas, waders are diverse in form and habit. Many are adapted for feeding in mud and water, with long legs and probing bills.

The birds collectively known as waders, or Charadriiformes, fall into three main groups. Of these, two groups are made up of shorebirds. Plovers and allied species are mostly short-legged, short-billed birds that feed on small invertebrates found close to the surface of the ground. Some birds in this group, such as lapwings, favor drier habitats inland. Other species in the plover group are more adapted to wetland. Stilts and avocets sweep shallow water with their needlelike bills, while oystercatchers use their long stout bills for opening mollusk shells. In the group consisting of sandpipers, snipe, and allies, birds also have long bills for probing deep mud. Some of their relatives, such as shanks and curlews, have long legs, too, so they can wade into deeper water to feed.

OCEANIC BIRDS

The final group in this order is made up of gulls, terns, skuas, and auks—web-footed birds that are the most ocean-going of the order. They may spend much of their lives at sea and some travel immense distances on migration. Gulls are opportunistic predators that are also seen foraging far inland. Auks, distributed around the Arctic, are specialized for diving for ocean-swimming prey. Their black and white coloration gives them a superficial resemblance to penguins, but they are not close relatives.

PHYLUM	CHORDATA
CLASS	AVES
ORDER	CHARADRIIFORMES
FAMILIES	19
SPECIES	379

DEBATE

ORIGIN OF A GULL

New species appear when populations diverge too much to interbreed with others. In this way it was thought that European herring gulls diverged from Asian ancestors that spread eastward to encircle the Arctic. Recent evidence suggests herring gulls descended from gulls isolated in the North Atlantic.

EURASIAN STONE CURLEW
Burhinus oedicnemus
F: Burhinidae
Named for their shrill curlewlike calls, the mostly nocturnal stone curlews are related to plovers. This widespread Eurasian species occurs in dry mudflats inland.

16–17½ in
40–44 cm

GREAT STONE CURLEW
Esacus recurvirostris
F: Burhinidae
Large *Esacus* stone curlews have chisel-like bills for hunting prey, such as crabs, near water. This is a southern Asian species.

19½–22 in
49–55 cm

SNOWY SHEATHBILL
Chionis albus
F: Chionidae
This is one of two white sheathbills, Antarctic cousins of plovers, which scavenge and consume carrion and the food and young of other birds.

13½–16 in
34–41 cm

8–9 in
20–22 cm

MAGELLANIC PLOVER
Pluvianellus socialis
F: Chionidae
This unusual South American wader—the only one that regurgitates food for its chicks—is more closely related to sheathbills than to plovers.

IBISBILL
Ibidorhyncha struthersii
F: Ibidorhynchidae
The only plover relative with a long, downcurved bill, this bird plucks invertebrates from stony riverbeds in mountains of Central Asia.

15–16 in
38–41 cm

dark brown to black body

long, red bill

AMERICAN BLACK OYSTERCATCHER
Haematopus bachmani
F: Haematopodidae
Black oystercatcher species generally have more restricted ranges than pied (having two or more colors) species. This species is confined to the west coast of North America.

16½–18½ in
42–47 cm

pink legs

EURASIAN OYSTERCATCHER
Haematopus ostralegus
F: Haematopodidae
The most widespread oystercatcher, this pied species breeds in northern Eurasia. Like others, it uses its long bill to open bivalve mollusks.

16–18 in
40–45 cm

CRAB PLOVER
Dromas ardeola
F: Dromadidae
The crab plover may be more closely related to gulls than to true plovers. Found around Indian Ocean coastlines, it uses its stout bill for eating crabs.

13–16 in
33–40 cm

RED-NECKED AVOCET
Recurvirostra novaehollandiae
F: Recurvirostridae
A nomadic bird of Australian wetlands, this distinctively marked wader is a social species that feeds in large flocks.

16–18 in
40–46 cm

PIED AVOCET
Recurvirostra avosetta
F: Recurvirostridae
This Eurasian wader is typical of avocets: relatives of stilts that take small aquatic animals by sweeping their upturned bills through water.

16½–18 in
42–45 cm

needlelike bill

BANDED STILT
Cladorhynchus leucocephalus
F: Recurvirostridae
Stilts hunt small swimming invertebrates. This Australian species gathers in large flocks on salt lakes to feed on brine shrimp.

14–16 in
36–45 cm

black and white body

upturned bill

8 in
20 cm

WRYBILL
Anarhynchus frontalis
F: Charadriidae
Related to lapwings, this wader from New Zealand is the only bird with a bill bent sideways, which it uses to reach invertebrates under stones.

BLACK-WINGED STILT
Himantopus himantopus
F: Recurvirostridae
This highly variable stilt is nearly global in distribution. There are white-necked and black-necked forms that may be separate species.

13–14 in
33–36 cm

INLAND DOTTEREL
Peltohyas australis
F: Charadriidae
This sand-colored relative of lapwings is found in arid parts of Australia, often far from water.

7½–9 in
19–23 cm

SPUR-WINGED PLOVER
Vanellus spinosus
F: Charadriidae
Occurring in wetlands across Africa and the Middle East, this bird is one of several lapwings, such as masked and southern, with a spur on each wing.

10–10½ in
25–27 cm

white crown

10–11½ in
26–29 cm

10–12 in
25–30 cm

14–15 in
35–38 cm

11–12 in
28–31 cm

black underparts when breeding

BLACK-BELLIED PLOVER
Pluvialis squatarola
F: Charadriidae
The only member of the golden plover group with gray, not yellow, speckled upperparts, this species breeds around the Arctic coastal tundra.

EURASIAN GOLDEN PLOVER
Pluvialis apricaria
F: Charadriidae
Golden plovers are more closely related to stilts and oystercatchers than they are to other plovers. This species develops a black belly when breeding.

blue-gray feet and legs

MASKED LAPWING
Vanellus miles
F: Charadriidae
Many lapwings have yellow, fleshy wattles. These facial adornments are especially prominent on this species from Australasia.

NORTHERN LAPWING
Vanellus vanellus
F: Charadriidae
Lapwings are named for the lapping sound of their flight. This Eurasian species has a distinctive, pointed crest. It is also called a peewit for its unique call.

7–8 in
18–20 cm

9–10½ in
23–27 cm

8–9 in
20–22 cm

COMMON RINGED PLOVER
Charadrius hiaticula
F: Charadriidae
One of the most widespread ring-necked plovers, this species breeds around the Arctic and overwinters in Africa and southwestern Asia.

KILLDEER
Charadrius vociferus
F: Charadriidae
This long-tailed, grassland plover migrates between North and South America, but has resident populations in Peru and Chile.

EURASIAN DOTTEREL
Charadrius morinellus
F: Charadriidae
Females of this tundra-breeding bird are more brightly colored than males, but like many other plovers both become duller in winter.

»

11–12 in
28–31 cm

9–12 in
23–31 cm

long tail
when breeding

golden
neck patch

8–10½ in
20–27 cm

12–23 in
31–58 cm

6½–9 in
17–23 cm

AFRICAN JACANA
Actophilornis africanus
F: Jacanidae
Typical of the jacana family, this African wetland bird has large feet that enable it to walk on floating vegetation.

BRONZE-WINGED JACANA
Metopidius indicus
F: Jacanidae
This species is widespread across India and Southeast Asia. Males have a flattened "forearm" bone that helps them lift chicks with their wings.

COMB-CRESTED JACANA
Irediparra gallinacea
F: Jacanidae
Found from Asia to Australia, this species is named for its fleshy head wattle. As with other jacanas, males incubate the eggs and care for the young.

PHEASANT-TAILED JACANA
Hydrophasianus chirurgus
F: Jacanidae
This south Asian spur-winged jacana is the only species that changes plumage—losing its long tail—outside the breeding season.

WATTLED JACANA
Jacana jacana
F: Jacanidae
The dominant female of this South American species breeds with many males, perhaps to compensate for crocodile predation of eggs.

6–7½ in
15–19 cm

9–10 in
23–25 cm

9–10 in
23–25 cm

6½–8 in
16–20 cm

GREATER PAINTED-SNIPE
Rostratula benghalensis
F: Rostratulidae
Related to the jacanas, the short-legged painted-snipes share their female-dominant mating system. This bird is from Old World tropical wetlands.

9–10 in
23–26 cm

PLAINS WANDERER
Pedionomus torquatus
F: Pedionomidae
This quail-like bird of Australian grassland—the only member of its family—is actually related to the wetland jacanas.

SHORT-BILLED DOWITCHER
Limnodromus griseus
F: Scolopacidae
Dowitchers are related to snipes and are reddish in their breeding plumage. The short-billed dowitcher is confined to the Americas.

RED KNOT
Calidris canutus
F: Scolopacidae
Like many migratory waders, the Arctic-breeding red knot molts its distinctive summer plumage to more drab colors in winter.

DUNLIN
Calidris alpina
F: Scolopacidae
Globally common, the dunlin is typical of waders that breed around the Arctic and flock to the warmer south in the winter.

6½–7½ in
17–19 cm

COMMON SNIPE
Gallinago gallinago
F: Scolopacidae
Typical of waders, chicks of this globally widespread species are active soon after hatching. Unlike other waders, however, snipes feed their chicks.

10–10½ in
25–27 cm

JACK SNIPE
Lymnocryptes minimus
F: Scolopacidae
Birds of the snipe-woodcock group have long bills, short legs, and are camouflaged by their plumage. This Old World bird is the smallest in this group.

long, slightly
upturned bill

8–8½ in
20–21 cm

SANDERLING
Calidris alba
F: Scolopacidae
This sandpiper breeds within the Arctic Circle, farther north than other waders. In winter, it flocks on sandy shores farther south.

rufous (reddish)
underparts
during breeding

black and white
upperparts

HUDSONIAN GODWIT
Limosa haemastica
F: Scolopacidae
This is one of two American godwits, and is named after its breeding habitat, which includes the shores of Hudson Bay.

14½–16½ in
37–42 cm

16–17½ in
40–44 cm

7–7½ in
18–19 cm

BLACK-TAILED GODWIT
Limosa limosa
F: Scolopacidae
Typical of godwits, this Old World species has a slightly upturned bill. It ranges far inland to forage on grassland, heath, and pasture.

RED-NECKED PHALAROPE
Phalaropus lobatus
F: Scolopacidae
This species has a breeding range that includes much of the Arctic region. Females are brightly colored and attract males with displays during breeding; males care for offspring.

LONG-BILLED CURLEW
Numenius americanus
F: Scolopacidae

A typical curlew, this American species is a large wader with a long, downcurved bill for probing deeply in mud for invertebrate prey.

18–26 in
45–66 cm

WHIMBREL
Numenius phaeopus
F: Scolopacidae

The smallest curlew, it has a distinctive trilling call and breeds in circumpolar regions. Some individuals winter as far away as Australia.

16–16½ in
40–42 cm

BUFF-BREASTED SANDPIPER
Tryngites subruficollis
F: Scolopacidae

This wader breeds in the tundra of North America and extreme eastern Siberia. It winters in the grasslands of South America.

7–8 in
18–20 cm

large neck ruff in breeding male

COMMON REDSHANK
Tringa totanus
F: Scolopacidae

Most waders breed near fresh water, but the Old World common redshank—named after its colored legs—may sometimes breed on salt marshes.

10½–11½ in
27–29 cm

LESSER YELLOWLEGS
Tringa flavipes
F: Scolopacidae

Yellowlegs belong to the shank group. This species breeds in forests in Alaska and Canada, and overwinters in the Caribbean.

9–10 in
23–25 cm

WANDERING TATTLER
Tringa incana
F: Scolopacidae

This shank species breeds in Alaska, during which its breast is barred. It winters farther south along the Pacific coastline of America.

10–12 in
26–30 cm

AMERICAN WOODCOCK
Scolopax minor
F: Scolopacidae

Like other woodcocks and snipes, this American bird is superbly camouflaged and its eyes are placed high, enabling all-round vision to spot predators.

10–11 in
26–28 cm

RUFF
Philomachus pugnax
F: Scolopacidae

Males of this Old World marsh and meadow species transform during breeding: they lose their gray winter plumage and take on a rufous and black coloration, with a showy neck ruff.

8–12 in
20–30 cm

SPOTTED SANDPIPER
Actitis macularius
F: Scolopacidae

Like the related common sandpiper (*A. hypoleucos*) from Eurasia, this American sandpiper has a short bill for pecking food from dry ground.

7–8 in
18–20 cm

black-spotted chest and belly in summer

reddish legs

RUDDY TURNSTONE
Arenaria interpres
F: Scolopacidae

This bird, which breeds across the Northern Hemisphere, is named for the way it flips objects to search for prey.

9–9½ in
22–24 cm

SPOON-BILLED SANDPIPER
Eurynorhynchus pygmeus
F: Scolopacidae

Like spoonbills, this wader from eastern Asia uses its distinctive spoon-shaped bill to sweep shallow water for invertebrate prey.

5½–6½ in
14–16 cm

WHITE-BELLIED SEEDSNIPE
Attagis malouinus
F: Thinocoridae

Four species of short-billed, herbivorous seedsnipes live in the open habitats of South America. This species is confined to the southern tip.

10½–11½ in
27–29 cm

CREAM-COLORED COURSER
Cursorius cursor
F: Glareolidae

7½–8½ in
19–21 cm

Often nocturnal, coursers are inconspicuous, long-legged ground birds, similar to plovers. This is a migratory species from Asia and Africa.

AUSTRALIAN PRATINCOLE
Stiltia isabella
F: Glareolidae

7½–9½ in
19–24 cm

This pratincole from Australia and Indonesia usually stays close to freshwater, but has special glands that enable it to drink saltwater too.

BLACK-WINGED PRATINCOLE
Glareola nordmanni
F: Glareolidae

9½–11 in
24–28 cm

Like other pratincoles, this species is migratory. It breeds in eastern Europe and Central Asia and overwinters in Africa.

black-bordered creamy throat

HEUGLIN'S COURSER
Rhinoptilus cinctus
F: Glareolidae

10½–11 in
27–28 cm

Most coursers are restricted to arid deserts and scrub, but this African species also ventures into woodland habitats.

LITTLE PRATINCOLE
Glareola lactea
F: Glareolidae

6½–7½ in
17–19 cm

This small species is from southern Asia. Like its relatives, the little pratincole feeds on insects in flight. It has a forked tail.

COLLARED PRATINCOLE
Glareola pratincola
F: Glareolidae

9–10 in
23–26 cm

Like most pratincoles, this species from southern Europe and Africa gathers in large, noisy flocks on open wetlands.

forked tail

SWALLOW-TAILED GULL
Creagrus furcatus
F: Laridae

20–23½ in
50–60 cm

The only nocturnal gull, this species breeds on the Galapagos Islands and overwinters in South America. It feeds on fishes and squid.

SOOTY GULL
Larus hemprichii
F: Laridae

16½–18 in
42–45 cm

As with many warm-region gulls, this species from Asia and Africa has dark plumage, which may have evolved as an adaptation to strong sunshine.

HERRING GULL
Larus argentatus
F: Laridae

20½–23½ in
52–60 cm

Related to the lesser black-backed gull (*L. fuscus*), this species is often seen in coastal towns in eastern North America and in Europe.

COMMON BLACK-HEADED GULL
Larus ridibundus
F: Laridae

13½–14½ in
34–37 cm

Like other members of the hooded gull group, this dark-headed gull becomes white headed in winter. It is abundant in the Northern Hemisphere.

LAUGHING GULL
Larus atricilla
F: Laridae

14–16 in
36–41 cm

An American hooded gull, this species is named for its laughing call. It is a colonial breeder on coastal estuaries and salt marshes.

HEERMANN'S GULL
Larus heermanni
F: Laridae

This dark gull from North America is actually related to northern white-headed gulls. It often forages with brown pelicans and steals their food.

grey body

GLAUCOUS GULL
Larus hyperboreus
F: Laridae

18–21 in
46–53 cm

black-tipped red bill

A large, Arctic-breeding species, this coastal gull is much paler than most related white-headed gulls found in the Northern Hemisphere.

RING-BILLED GULL
Larus delawarensis
F: Laridae

24–27 in
62–68 cm

Breeding in North America and overwintering in the Caribbean, this species has a dark bill ring. It often forages on agricultural land.

WHITE-EYED GULL
Larus leucophthalmus
F: Laridae

18–20 in
46–51 cm

15½–17 in
39–43 cm

This relative of the sooty gull is confined to the Red Sea region, where it is threatened by oil spills.

20–26 in
50–67 cm

PACIFIC GULL
Larus pacificus
F: Laridae

Belonging to a group of southern gulls
with banded tails, this large-billed
Australian species cracks open
shellfish by dropping them on rocks.

11–12 in
28–30 cm

BONAPARTE'S GULL
Larus philadelphia
F: Laridae

A North American hoodedf gull,
this species breeds in wet
coniferous forests in Canada and
winters along Caribbean coasts.

22–26 in
55–66 cm

WESTERN GULL
Larus occidentalis
F: Laridae

This large gull is found on the
Pacific coast of North America.
It nests in colonies, usually
offshore on islands or rocks.

large
white head

16–16½ in
40–42 cm

MEW GULL
Larus canus
F: Laridae

This Northern Hemisphere
species, whose name derives from
its distinctive call, is also called
the common gull. It breeds inland
on moors, as well as on the coast.

18–18½ in
45–47 cm

GREY GULL
Larus modestus
F: Laridae

Confined to Peru and
Chile, this species breeds
in the Atacama Desert,
one of the driest regions
of the world.

slaty black back
and wings

25–31 in
64–78 cm

GREAT BLACK-BACKED GULL
Larus marinus
F: Laridae

The world's largest gull, this North Atlantic
coastal bird is an aggressive predator and
preys on other seabirds and their chicks.

pink feet

16–18 in
40–45 cm

SILVER GULL
Larus novaehollandiae
F: Laridae

Despite its different
appearance, this Australian
gull is related to the
common black-headed
gull. It visits human
refuse sites to feed.

white-edged eye

tricolored bill

GREAT BLACK-HEADED GULL
Larus ichthyaetus
F: Laridae

One member of an Asian group
of black-headed gulls, this species
breeds in Russia and overwinters on
Mediterranean and Indian ocean coasts.

22½–24 in
57–61 cm

narrow black collar

15–16 in
38–40 cm

wedge-shaped
tail

yellow legs

16½–17½ in
42–44 cm

DOLPHIN GULL
Leucophaeus scoresbii
F: Laridae

This gull is famously
aggressive to other birds. It
is confined to the southern
tip of South America and
the Falkland Islands.

16–17 in
40–43 cm

IVORY GULL
Pagophila eburnea
F: Laridae

This Arctic gull is rarely
found far from pack ice.
It follows polar bears to
scavenge at their kills.

ROSS'S GULL
Rhodostethia rosea
F: Laridae

A distinctive pink gull, Ross's gull
breeds in swampy and wooded
Arctic tundra, and winters
along coasts and at sea.

10½–12½ in
27–32 cm

SABINE'S GULL
Xema sabini
F: Laridae

A relative of the ivory gull, this bird is an
Arctic breeder, but it migrates long distances
to overwinter in South America and Africa.

14–16 in
35–40 cm

RED-LEGGED KITTIWAKE
Rissa brevirostris
F: Laridae

This kittiwake breeds only on islands in
the Bering Sea in the North Pacific and
spends winter at sea farther south.

15–16 in
38–40 cm

BLACK-LEGGED KITTIWAKE
Rissa tridactyla
F: Laridae

The world's most abundant species of gull,
this colonial cliff-nesting bird is found
in the North Atlantic and Pacific oceans.

11–13 in
28–33 cm

WHITE TERN
Gygis alba
F: Laridae
This small all-white tern from the islands of the tropical Atlantic and Indian oceans is notable for laying its eggs on bare branches in trees.

16–16½ in
40–42 cm

white cheek stripe

INCA TERN
Larosterna inca
F: Laridae
A distinctively patterned bird, the inca tern is found in Peru and Chile, where it breeds on rocky coastlines.

bright red legs

14–14½ in
35–37 cm

LESSER CRESTED TERN
Sterna bengalensis
F: Laridae
In this relative of the greater crested tern, the bill changes from yellow to orange during breeding.

9–9½ in
22–24 cm

BLACK TERN
Chlidonias niger
F: Laridae
This small freshwater tern breeds in the wetlands of the Northern Hemisphere and winters in South America and Africa.

13–14 in
33–35 cm

ARCTIC TERN
Sterna paradisaea
F: Laridae
This tern migrates the greatest distance of any animal: from its Arctic breeding grounds to Antarctica. It feeds on fishes and crustaceans.

black cap

18½–21½ in
47–54 cm

9–9½ in
22–24 cm

LITTLE TERN
Sterna albifrons
F: Laridae
The Old World little tern belongs to a group of coastal species with a white "blaze" above the eye.

black markings on tips of underwings

CASPIAN TERN
Sterna caspia
F: Laridae
The largest tern species, this bird is found in most continents. It is a colonial ground-nester, like most sea terns.

13–14 in
33–36 cm

SOOTY TERN
Sterna fuscata
F: Laridae
This white-bridled sea tern breeds on tropical islands. It is also known as the wideawake tern because of its noisy colonies.

long, black legs

13–15 in
33–38 cm

SAUNDERS'S TERN
Sterna saundersi
F: Laridae
A small tern from the Red Sea and Indian Ocean, Saunders's tern used to be considered a race of the little tern.

9–9½ in
23–24 cm

12–12½ in
30–32 cm

BRIDLED TERN
Sterna anaethetus
F: Laridae
This tern of tropical and subtropical regions has a white forehead, or bridle, similar to little and sooty terns. It spends much time at sea.

ROSEATE TERN
Sterna dougallii
F: Laridae
Although widespread around the world, the roseate tern is not common anywhere, and breeds in a number of isolated colonies.

12½–13½ in
32–34 cm

WHITE-CHEEKED TERN
Sterna repressa
F: Laridae
This species from the Red Sea and Indian Ocean can be identified by its plumage, which is darker than that of other gray terns.

18–19½ in
46–49 cm

GREAT CRESTED TERN
Sterna bergii
F: Laridae
This Old World tern belongs to a group of related crested terns, characterized by a black nape tuft.

12–12½ in
30–32 cm

BLACK-NAPED TERN
Sterna sumatrana
F: Laridae
Found on the Indian and Pacific oceans, the black-naped tern nests in small colonies, usually apart from other terns.

16–20 in
40–50 cm

BLACK SKIMMER
Rynchops niger
F: Laridae

Skimmers are the only birds with projecting lower beak mandibles, used to skim water for fish. This species is found in North and South America.

16–18 in
40–45 cm

BROWN NODDY
Anous stolidus
F: Laridae

Noddies are dark or white, tropical members of the tern family. The largest species, the brown noddy, is widespread around the world.

stout, hooked bill

20½–21 in
52–54 cm

brownish gray body

SOUTH POLAR SKUA
Stercorarius maccormicki
F: Stercorariidae

This large bird, with a reputation for attacking other seabirds, is one of the few waders to breed on the Antarctic coast.

18–20 in
46–51 cm

POMARINE JAEGER
Stercorarius pomarinus
F: Stercorariidae

Jaegers (also called skuas in Europe) are aggressive gull-like birds. This Arctic species will kill and eat other seabirds and even attack humans that threaten its nest.

19–21 in
48–53 cm

LONG-TAILED JAEGER
Stercorarius longicaudus
F: Stercorariidae

Like other species, this smallest member of the jaeger family is migratory. It breeds in circumpolar regions and winters farther south.

16–18 in
41–46 cm

PARASITIC JAEGER
Stercorarius parasiticus
F: Stercorariidae

The most common jaeger in the Arctic, like most of its relatives it mobs other seabirds to steal their prey.

9½–10 in
24–25 cm

MARBLED MURRELET
Brachyramphus marmoratus
F: Alcidae

This small American auk nests in trees in coniferous woods. Fledged chicks leave the nest at night and make for the sea.

14½–15½ in
37–39 cm

RAZORBILL
Alca torda
F: Alcidae

The North Atlantic razorbill has a flattened, white-barred bill. As with other auks, its pointed eggs cannot roll off cliff-top breeding sites.

9½–10½ in
24–27 cm

CRESTED AUKLET
Aethia cristatella
F: Alcidae

Like other auklets, this North Pacific species feeds on planktonic crustaceans. Pairs anoint themselves with secretions from their backs during courtship.

6½–7½ in
17–19 cm

DOVEKIE
Alle alle
F: Alcidae

This tiny auk breeds on Arctic islands and winters at sea farther south. It feeds on tiny fishes and crustaceans.

dark brown to black head

12–14 in
30–36 cm

PIGEON GUILLEMOT
Cepphus columba
F: Alcidae

This North Pacific auk, fully adapted to cold conditions, cannot migrate south through warmer waters, just as Southern Hemisphere penguins cannot move northward.

BLACK GUILLEMOT
Cepphus grylle
F: Alcidae

This species of northern North American and Eurasian coasts breeds in sparser colonies than other auks and winters mainly inshore.

12–12½ in
30–32 cm

11–11½ in
28–29 cm

RHINOCEROS AUKLET
Cerorhinca monocerata
F: Alcidae

This North Pacific relative of puffins shares their burrow-nesting habits, and is named for the hornlike bill projection in breeding adults.

white wing patch

red feet

10–11½ in
26–29 cm

ATLANTIC PUFFIN
Fratercula arctica
F: Alcidae

A small, North Atlantic member of the auk family, this species—like other puffins—nests in colonies, in burrows, usually below turf.

13½–14 in
34–36 cm

TUFTED PUFFIN
Fratercula cirrhata
F: Alcidae

Like its Atlantic relative, this larger Pacific puffin catches small fish and can hold many at a time, crosswise, in its bill.

15–16 in
38–41 cm

COMMON MURRE
Uria aalge
F: Alcidae

A typical diving member of the auk family, this bird breeds on North Atlantic and Pacific coasts, and winters at sea.

SANDGROUSE

Sand-colored plumage camouflages these birds in their desert habitats, where they are well adapted for living in an extremely dry environment.

With their rounded bodies and short legs, sandgrouse might be mistaken for partridges until they take flight, darting quickly and acrobatically in the air. These birds, the Pteroclidiformes, are found in arid areas of Asia, Africa, Madagascar, and southern Europe. They are not related to the subarctic grouse species but are more closely related to pigeons.

Sandgrouse have long pointed wings and all species have cryptic plumage—mottled on the back, but often boldly marked with brown or white stripes or blotches on the head and underparts. They

are social birds that gather in flocks to make early-morning, or sometimes evening, forays to drinking holes, often traveling considerable distances. They feed exclusively on seeds.

WATER CARRIERS

Sandgrouse breed during the rainy season to take advantage of seed harvests. Their nest is nothing more than shallow depression on the ground. Both parents incubate and care for their young. Remarkably, the males supply water to their chicks, which are normally raised far from water sources. The male's belly feathers soak up and retain moisture when the bird visits a drinking hole. Back at the nest, the chicks drink from the parent's soaked feathers.

PHYLUM	CHORDATA
CLASS	AVES
ORDER	PTEROCLIDIFORMES
FAMILIES	1
SPECIES	16

Namaqua sandgrouse drink at a watering hole. Like all sandgrouse, they gather in large flocks to confuse predators.

PALLAS'S SANDGROUSE
Syrrhaptes paradoxus
F: Pteroclididae
One of two Central Asian sandgrouse, this large species has feathered toes, a long tail, and a long flight feather on each wing.

barred buff plumage

12–16 in
30–41 cm

long, pointed tail

CHESTNUT-BELLIED SANDGROUSE
Pterocles exustus
F: Pteroclididae
This bird of open desert gathers in huge flocks, in a range that extends from Senegal to Kenya, and east as far as India.

12–13 in
31–33 cm

CROWNED SANDGROUSE
Pterocles coronatus
F: Pteroclididae
This yellow-throated sandgrouse is found from the stony desert of the Sahara to Pakistan. It can tolerate high temperatures and even brackish water.

10½–12 in
27–30 cm

DOUBLE-BANDED SANDGROUSE
Pterocles bicinctus
F: Pteroclididae
A bird of southern African savanna and open woodland, this is one of several sandgrouse species in which the males have distinctive belly stripes.

10–11 in
25–28 cm

black and white bands on forecrown (male only)

white-barred wings

black breast band

LICHTENSTEIN'S SANDGROUSE
Pterocles lichtensteinii
F: Pteroclididae
This small bird is less gregarious than other species of sandgrouse. It inhabits bushy scrubland, and semidesert regions from North and East Africa to Pakistan.

9½–10 in
24–26 cm

PIGEONS AND DOVES

This is a highly successful group of seed-eating and fruit-eating birds that are distributed almost worldwide outside the coldest regions.

After the parrots, pigeons and doves form the largest group of vegetarian tree-perching birds: the order Columbiformes. However, whereas parrots have heavy hooked bills for cracking large nuts, pigeons have less robust bills for feeding on smaller seeds and grain. Some tropical groups, such as the Indo-Pacific fruit doves, are specialized fruit-eaters of the rain forest canopy.

Most birds in this order are short-legged and a few spend almost all of their time on the ground. Pigeons and doves are unusual among birds in being able to suck up water when drinking without tilting back their head, owing to a pumping action in their food pipe. This allows them to drink continuously, which is a particular benefit in arid environments. They can store food in their bill, and they feed their chicks on a secretion from the bill that is similar to mammalian milk.

THREATS AND EXTINCTIONS

The success of many pigeon species is due to high rates of reproduction. However, some species are threatened by humans and others have already been lost. The 17th-century demise of the dodo, a vulnerable flightless species, was the result of human impact. And the passenger pigeon, once one of North America's most common birds, was hunted to extinction in the 1900s.

PHYLUM	CHORDATA
CLASS	AVES
ORDER	COLUMBIFORMES
FAMILIES	2
SPECIES	321

DEBATE
A MODIFIED PIGEON?

Mid-19th-century scientists placed the extinct flightless dodo from Mauritius in the pigeon order, and recent analysis indicates that it was related to the Nicobar pigeon. So the dodo was really a modified pigeon, with Indo-Pacific ancestors that might already have been flightless.

EUROPEAN TURTLE DOVE
Streptopelia turtur
F: Columbidae
With a range from Africa to Eurasia, turtle doves are small, slim relatives of *Columba* pigeons. Many, such as this western species, have distinctive neck markings.

black and white striped neck patch

10–11 in
26–28 cm

10–10½ in
25–27 cm

red patch around eyes

white-speckled wings

LAUGHING DOVE
Streptopelia senegalensis
F: Columbidae
Common in villages and oases in Africa and southern Asia, this turtle dove is named after its distinctive chuckling song.

15–17 in
38–43 cm

BROWN CUCKOO-DOVE
Macropygia amboinensis
F: Columbidae
Long-tailed *Macropygia* doves are cuckoolike birds of Indo-Pacific rain forests. Like others, this species from the Moluccas (Indonesia), New Guinea, and Australia has many races.

NAMAQUA DOVE
Oena capensis
F: Columbidae
This long-tailed, ground-feeding species belongs to a group of small African wood doves. Its range extends into Madagascar and Saudi Arabia.

10–11 in
26–28 cm

13–15 in
33–38 cm

SPECKLED PIGEON
Columba guinea
F: Columbidae
This large African pigeon is common in open country south of the Sahara, and it often gathers around towns and villages.

12½ in
32 cm

15–17 in
38–43 cm

12–14 in
31–35 cm

12–14 in
31–35 cm

PINK PIGEON
Nesoenus mayeri
F: Columbidae
Possibly related to turtle doves, this rare bird is restricted to Mauritius. Its once threatened population recovered due to a successful captive-breeding program.

WOODPIGEON
Columba palumbus
F: Columbidae
This large western Eurasian pigeon is commonly found in woods and farmland, and often enters parks and gardens.

ROCK PIGEON
Columba livia
F: Columbidae
True wild pigeons inhabit cliffs and hedges. In its natural environment, this species lives in mountainous regions of Europe and Asia.

DOMESTIC PIGEON
Columba livia
F: Columbidae
Domesticated and feral descendants of the rock pigeon are found in urban areas worldwide. These birds have variable plumage patterns.

»

13–16 in
33–40 cm

6½–9 in
17–23 cm

WOMPOO FRUIT DOVE
Ptilinopus magnificus
F: Columbidae
Fruit doves are multicolored relatives
of imperial pigeons (*Ducula* sp.).
This large species from New Guinea
and Australia lives in the
rain forest canopy.

INCA DOVE
Columbina inca
F: Columbidae
From arid regions of the southern
USA and Central America, this
species belongs to a group of
mostly drab-colored tropical
American ground-doves.

yellow markings
on wings

8 in
20 cm

NICOBAR PIGEON
Caloenas nicobarica
F: Columbidae
Possibly related to the extinct
Mauritian dodo, this pigeon
inhabits coastal regions and
island forests from Malaysia
to New Guinea.

DIAMOND DOVE
Geopelia cuneata
F: Columbidae
A small, nomadic bird from
the dry interiors of Australia,
this dove occurs in groups,
flocking in large numbers
at watering holes.

11½–22 in
29–55 cm

10–12 in
25–31 cm

**WHITE-TIPPED
DOVE**
Leptotila verreauxi
F: Columbidae
A tropical relative of
the mourning dove, this
species is widespread in
Central and South
America, and reaches as
far north as Texas.

dark green tail

14 in
35 cm

9–13½ in
23–34 cm

12 in
30 cm

dark pink head
and breast

**MINDANAO
BLEEDING-HEART PIGEON**
Gallicolumba criniger
F: Columbidae
The five species of Philippine
bleeding-heart pigeons are named
for their blood-red breast spots.
This bird inhabits the southern
islands of the archipelago.

**SULAWESI
GROUND DOVE**
Gallicolumba tristigmata
F: Columbidae
From the forests of Sulawesi,
Indonesia, this terrestrial bird
is related to Philippine bleeding-
heart pigeons, and perhaps to
various doves of arid Australia.

**MOURNING
DOVE**
Zenaida macroura
F: Columbidae
Named for its mournful call,
this long-tailed dove occurs
in open country throughout
North and Central America,
and the Caribbean.

emerald-
green wings
and back

18 in
45 cm

15½–17½ in
39–44 cm

16–18 in
40–46 cm

**PIED
IMPERIAL PIGEON**
Ducula bicolor
F: Columbidae
Imperial pigeons are large,
fruit-eating rain forest birds.
This species from southeast
Asia and Australasia has a white
plumage that is often stained
due to its diet.

9–11 in
23–28 cm

EMERALD DOVE
Chalcophaps indica
F: Columbidae
An iridescent green bird, this
ground-feeding species occurs in
rain forests from India to islands
of the southwest Pacific. It has a
diet of fruits and seeds.

TOPKNOT PIGEON
Lopholaimus antarcticus
F: Columbidae
A large hawklike pigeon
from eastern Australia, this
species is named for the
double crest of feathers on
its forecrown and crown.

GREEN IMPERIAL PIGEON
Ducula aenea
F: Columbidae
Found in rain forest canopy from
India to southeast Asia, this
large bird has a deep, booming
call and its diet consists
primarily of fruits.

VICTORIA CROWNED PIGEON
Goura victoria
F: Columbidae
Crowned pigeons are the largest of the pigeons. This bird from northern New Guinea is distinguished from the southern crowned pigeon by its white-tipped crest.

29–30 in
74–75 cm

side-flattened tail

14–15 in
36–38 cm

WONGA PIGEON
Leucosarcia melanoleuca
F: Columbidae
Confined to eastern Australia, this distinctively patterned pigeon inhabits woodland and scrub from southern Queensland to Victoria.

18–20 in
45–50 cm

PHEASANT PIGEON
Otidiphaps nobilis
F: Columbidae
Recent research suggests this New Guinean ground pigeon belongs to a group that includes crowned pigeons and perhaps even the extinct dodo.

fanlike crown

SOUTHERN CROWNED PIGEON
Goura scheepmakeri
F: Columbidae
This species inhabits the forests of the southern part of New Guinea. It has bluish gray upperparts, a maroon chest, and a lacy crest.

bluish gray plumage

maroon chest

30 in
75 cm

KEY WEST QUAIL-DOVE
Geotrygon chrysia
F: Columbidae
Quail-doves are birds of tropical American forests. This iridescent species occurs in the Caribbean, including the Bahamas.

10½–12 in
27–31 cm

10–11 in
25–28 cm

AFRICAN GREEN PIGEON
Treron calvus
F: Columbidae
More than 20 species of *Treron* pigeons occur in the tropical regions of Africa and Asia. This species is widespread south of the Sahara.

13–14 in
33–36 cm

COMMON BRONZEWING
Phaps chalcoptera
F: Columbidae
Bronzewings are fast-flying Australian pigeons that feed on the ground. They have iridescent wing patches, which are particularly extensive in this widespread woodland species.

grayish body

8–9 in
20–22 cm

white wing patch

iridescent wing patch

12–14 in
31–35 cm

SPINIFEX PIGEON
Geophaps plumifera
F: Columbidae
This Australian bronzewing pigeon is found in arid, rocky habitats with an abundance of spinifex grass, in which it nests.

CRESTED PIGEON
Ocyphaps lophotes
F: Columbidae
One of several species of Australian bronzewing pigeons, this species is widespread in open country throughout the continent.

PARROTS AND COCKATOOS

Most parrots are birds of tropical forests, although a few favour open habitats. This order, the Psittaciformes, is diverse and often colorful.

The most recognizable feature of a parrot is its decurved bill. Both mandibles of the bill articulate with its skull: the upper mandible can move up, just as the lower can move down. This allows the bird not only to crack open hard seeds and nuts, but also to use its bill for gripping when clambering through trees. The legs are strong and the feet, with two forward-facing and two backward-facing toes, can grasp and manipulate foodstuffs. Bright colors are common, often in both sexes, with green plumage predominating. This color is the result of feather structure,

which scatters light on yellow pigment. The Australasian cockatoos—parrots with distinctive erectile crests—lack this feather texture and so lack green and blue colors.

Australasia's high parrot diversity, which includes the brush-tongued, nectar-feeding lorikeets, suggests that the group originated in this part of the world. The most primitive parrots, the kea and the nocturnal, flightless kakapo, are both found in New Zealand.

SOCIAL SPECIES

Parrots are sociable, often living in large flocks, and almost all species form strong pair bonds. Their engaging personalities make them popular as pets, but the international cagebird trade has driven many species to the edge of extinction.

PHYLUM	CHORDATA
CLASS	AVES
ORDER	PSITTACIFORMES
FAMILIES	1
SPECIES	375

Green-winged macaws socialize at a clay-lick, which provides sodium in an area where the mineral is very limited.

KEA
Nestor notabilis
F: Psittacidae
An opportunistic alpine omnivore, the kea eats carrion and live petrel chicks. It belongs to an ancient group of parrots in New Zealand that includes the kakapo.

19 in
48 cm

short tail

barred green plumage

23½ in
60 cm

14 in
36 cm

KAKAPO
Strigops habroptila
F: Psittacidae
The only flightless parrot, this large, nocturnal bird occurs in the small islands off New Zealand. Males make booming calls to attract females.

GALAH
Eolophus roseicapilla
F: Psittacidae
The only cockatoo with a deep pink neck and underparts, this small relative of the white cockatoos is widespread throughout Australia in areas with scattered trees.

5–6 in
13–15 cm

♀

4¾–6 in
12–15 cm

♂

19½ in
49 cm

20–24 in
50–61 cm

VERNAL HANGING PARROT
Loriculus vernalis
F: Psittacidae
Hanging parrots are Indo-Pacific birds, even though recent genetic evidence suggests affinities with African lovebirds. This species is found from India to Thailand.

BLUE-CROWNED HANGING PARROT
Loriculus galgulus
F: Psittacidae
Hanging parrots, unusual among birds, sleep upside-down. Like other such parrots, this small forest bird from Southeast Asia has a short tail and primarily green plumage.

SULPHUR-CRESTED COCKATOO
Cacatua galerita
F: Psittacidae
White cockatoos are screeching parrots with a range that extends from Indonesia to the Pacific Rim. This species occurs in New Guinea and Australia.

RED-TAILED BLACK COCKATOO
Calyptorhynchus banksii
F: Psittacidae
One of several Australian black cockatoos with colored tail panels, this glossy species has a typical wailing call and ponderous wing beats.

COCKATIEL
Nymphicus hollandicus
F: Psittacidae

A bird of Australia's dry interior, this species resembles a parakeet, but its DNA shows that it is a miniature cockatoo.

♀ ♂

12½ in
32 cm

CHATTERING LORY
Lorius garrulus
F: Psittacidae

One of a group of green-winged red lories from New Guinea and surrounding islands, this species is from the Moluccas.

12 in
30 cm

DUSKY LORY
Pseudeos fuscata
F: Psittacidae

The patchy brown coloration of this species is unique among lories. This bird occurs in New Guinea and nearby islands.

10 in
25 cm

BLACK-WINGED LORY
Eos cyanogenia
F: Psittacidae

The *Eos* lories are vividly red and violet Indonesian birds. This is the easternmost species, restricted to the area around Geelvink Bay, New Guinea.

12 in
30 cm

bright scarlet head

blue band across back and chest

♀

♂

OLIVE-HEADED LORIKEET
Trichoglossus euteles
F: Psittacidae

Found on the island of Timor, this long-tailed relative of the rainbow lorikeet has bright green plumage, typical of many parrots.

9½ in
24 cm

VARIED LORIKEET
Psitteuteles versicolor
F: Psittacidae

A small lorikeet from northern Australian woodlands, this species nests in eucalyptus tree cavities, like many other parrots in its habitat.

7 in
18 cm

RAINBOW LORIKEET
Trichoglossus haematodus
F: Psittacidae

A variable species, this bird is found in many habitats with nectar-giving blossom. Its range includes most of Australasia and the islands of the southwest Pacific.

pale green streak on green back

17 in
43 cm

purple rump

10–12 in
25–30 cm

mainly green body

ECLECTUS PARROT
Eclectus roratus
F: Psittacidae

Males and females of this parrot are so different they were originally classified as separate species. They are found in tropical Australasian rain forests.

13–15½ in
33–39 cm

AUSTRALIAN KING PARROT
Alisterus scapularis
F: Psittacidae

King parrots are rain forest birds from tropical Australasia and represent an evolutionary link to Asian parakeets. This species occurs in eastern Australia.

BUDGERIGAR
Melopsittacus undulatus
F: Psittacidae

A nomadic Australian bird, this small parrot from arid regions flocks to watering holes. Despite being a seed-eater, it is related to nectar-feeding lorikeets.

7 in
18 cm

RED-FRONTED PARAKEET
Cyanoramphus novaezelandiae
F: Psittacidae

Small parakeets with colored foreheads have diversified in the southwest Pacific, and include the only parakeets—like this species—from New Zealand.

10½ in
27 cm

AUSTRALIAN RINGNECK
Barnardius zonarius
F: Psittacidae

Named for its yellow neck-ring, this Australian bird has black- and green-headed forms. It is widespread in woodland.

13½–15 in
34–38 cm

≫

black mask

14 in
36 cm

♂

8 in
20 cm

♀

18½ in
47 cm

yellow belly

12 in
30 cm

TURQUOISE PARROT
Neophema pulchella
F: Psittacidae
A member of a group of small,
green grass parrots from
Australia, the turquoise parrot
is an open woodland bird from
the southeast of the continent.

EASTERN ROSELLA
Platycercus eximius
F: Psittacidae
Rosellas are broad-tailed parrots
with a range from Australia to
nearby Pacific islands. A typical
rosella, this variable eastern
Australian bird has white cheeks.

RED-FAN PARROT
Deroptyus accipitrinus
F: Psittacidae
Possibly related more closely
to macaws than to other
short-tailed parrots, this
South American parrot raises
a red neck ruff when excited.

**ROSE-RINGED
PARAKEET**
Psittacula krameri
F: Psittacidae
This is the most widespread Asian
parakeet that reaches westwards into
northern Africa. This species has also
been introduced in Europe.

MASKED SHINING PARROT
Prosopeia personata
F: Psittacidae
One of the three species of
shining parrots restricted to Fiji,
this parrot is declining in numbers
due to loss of its native forest.

*blue plumage
and upperparts*

15–16½ in
38–42 cm

16–18½ in
40–47 cm

♀   ♂

*gray-black
undertail*

white face

13 in
33 cm

16 in
40 cm

SUPERB PARROT
Polytelis swainsonii
F: Psittacidae
This southeastern Australian bird
belongs to a group of long-tailed
Australian parrots and breeds in
red gum forests. It is rapidly
declining in number.

PRINCESS PARROT
Polytelis alexandrae
F: Psittacidae
A nomad from central
Australia, this bird follows
watercourses and gathers
near spinifex grass. It often
breeds in small colonies
in eucalyptus trees.

*yellow
undertail*

*pale gray
underparts*

6 in
15 cm

6½–7 in
17–18 cm

14–14½ in
35–37 cm

GREY PARROT
Psittacus erithacus
F: Psittacidae
A native of African rain forests,
this highly intelligent species is an
accomplished mimic. This has led to
its exploitation in the bird trade.

red tail

**YELLOW-COLLARED
LOVEBIRD**
Agapornis personatus
F: Psittacidae
Lovebirds are small African parrots
that live in small flocks. Unlike
most parrots, this Tanzanian
species builds a nest—a domed
structure in a tree cavity.

**ROSY-FACED
LOVEBIRD**
Agapornis roseicollis
F: Psittacidae
This social lovebird from dry
woodland and semidesert
habitats of southwestern
Africa habitually gathers
near watering holes.

BROWN-NECKED PARROT
Poicephalus robustus
F: Psittacidae
The largest of a group of mainly green-gray
parrots from Africa, this bird occurs in forests
from Gambia to the Cape of Good Hope.

white facial patch with lines of black feathers

34 in
85 cm

BLUE-AND-YELLOW MACAW
Ara ararauna
F: Psittacidae
Macaws are large, long-tailed parrots with sparsely feathered facial patches. One of two blue and yellow species, this bird comes from northern South America.

powerful bill

BLUE-FRONTED PARROT
Amazona aestiva
F: Psittacidae
A member of a large group of mostly green parrots, this species occurs in open forest habitats of central-eastern South America.

15 in
38 cm

ST VINCENT PARROT
Amazona guildingii
F: Psittacidae
Several Caribbean *Amazona* parrots are faced with extinction. This large species is being conserved by an ongoing breeding program in its native island of St. Vincent.

16 in
40 cm

22–23½ in
55–60 cm

RED-FRONTED MACAW
Ara rubrogenys
F: Psittacidae
This small macaw is restricted to the arid scrub of central Bolivia. Its small population is threatened by habitat loss and wildlife trafficking.

bright yellow feathers on blue wings

31–35 in
79–89 cm

SCARLET MACAW
Ara macao
F: Psittacidae
Like other macaws, this noisy, flocking bird has a heavy bill for breaking nuts and palm fruit. Its range extends from southern Mexico to central Brazil.

long, red tail

BLUE-HEADED PARROT
Pionus menstruus
F: Psittacidae
A relative of the *Amazona* parrots, this small parrot is common in lowland forests from Costa Rica to Bolivia.

24–28 in
9½–11 in

4¾–5½ in
12–14 cm

PACIFIC PARROTLET
Forpus coelestis
F: Psittacidae
Parrotlets are tiny green American parrots; only New Guinean pygmy parrots are smaller. This species occurs in western Ecuador and Peru.

3¼ ft
1 m

HYACINTH MACAW
Anodorhynchus hyacinthinus
F: Psittacidae
Unlike other macaws, this giant Brazilian parrot has its facial patch reduced to an eye-ring—similar to that of the related *Aratinga* parakeets.

8–10 in
20–25 cm

12 in
30 cm

JANDAYA PARAKEET
Aratinga jandaya
F: Psittacidae
Aratinga is a genus of predominantly green "mini-macaws" found in northeastern Brazil. Some have extensive splashes of golden coloration.

white eye-ring

10 in
25 cm

red belly

17½–18 in
44–46 cm

BURROWING PARAKEET
Cyanoliseus patagonus
F: Psittacidae
A Patagonian relative of macaws, this colonial breeder nests in tunnels in earth banks. It forms monogamous pair bonds, unlike most birds.

YELLOW-CHEVRONED PARAKEET
Brotogeris chiriri
F: Psittacidae
Native to central South America, escaped cagebirds of this species have established feral populations in warmer parts of the USA.

11½ in
29 cm

MONK PARAKEET
Myiopsitta monachus
F: Psittacidae
This colonial breeder from temperate South America is the only parrot to build a stick nest, often a large communal structure.

MAROON-BELLIED PARAKEET
Pyrrhura frontalis
F: Psittacidae
Relatives of *Aratinga*, most *Pyrrhura* parakeets have prominent splashes of maroon or red, such as this species from eastern South America.

CUCKOOS, HOATZIN, AND TURACOS

This order includes cuckoos, mostly softly feathered birds drably colored in browns and grays, as well as the enigmatic hoatzin and the striking colored turacos.

All members of the Cuculiformes have feet with two toes pointing forward and two backward. The most primitive cuckoos are anis, heavy-bodied American birds that forage on or near the ground—a lifestyle taken to the extreme by the sprinting roadrunners. In spite of their reputation, not all cuckoos lay their eggs in other birds' nests. Almost all these ground cuckoos build nests and raise their own young, as do their Old World counterparts, such as the tropical coucals and couas. Cuckoos that lay eggs in nests of other bird species are called "brood parasites." Remarkably, this habit has evolved independently at least twice in Old World cuckoos. The chicks of crested cuckoos (*Clamator* sp.) outgrow their host nest so quickly that host chicks die from starvation. Another line of cuckoos, which includes the Eurasian common cuckoo, evolved chicks with more murderous intent: they actively eject host eggs and chicks.

DISTANT RELATIVES

Turacos, thought to be distantly related to the cuckoos, are fruit-eaters confined to Africa. They share several features with the cuckoos, including foot structure. Even more uncertainly related is the hoatzin, an enigmatic South American bird that has an entirely vegetarian diet of leaves.

PHYLUM	CHORDATA
CLASS	AVES
ORDER	CUCULIFORMES
FAMILIES	3
SPECIES	170

DEBATE

HOATZIN: BIRD ENIGMA

Some features of the hoatzin, such as strong feet, are those of gamebirds. But the hoatzin also shares features with cuckoos, a relationship that appeared to be confirmed by early DNA studies. Later research has thrown doubt on this—leaving the hoatzin in a group of its own.

HOATZIN
Opisthocomus hoazin
F: Opisthocomidae
Of uncertain taxonomic affiliation, the hoatzin is a plant-eater of South American riverine forests. The claws on chicks' wings enable them to clamber through branches.

spiky crest

24–26 in
61–66 cm

long neck

ROSS'S TURACO
Musophaga rossae
F: Musophagidae
The second-largest turaco, this species is distinguished by its upright, red crest. It is an eastern African counterpart of the violet turaco.

20–21½ in
51–54 cm

18–20 in
45–50 cm

VIOLET TURACO
Musophaga violacea
F: Musophagidae
One of two glossy, violet-colored turacos, this species ranges through forests of western and central Africa.

long tail

fanlike crest

28–30 in
70–75 cm

yellow bill with red tip

18½–20 in
47–50 cm

19 in
48 cm

blue body

BARE-FACED GO-AWAY BIRD
Corythaixoides personatus
F: Musophagidae
Like other go-away birds, this eastern African species is found in savanna woodland. It raises its crest when excited.

20 in
50 cm

GREY GO-AWAY BIRD
Corythaixoides concolor
F: Musophagidae
Named after its call, "kay-waaay," this southern African species is a typical go-away bird with a pointed, shaggy crest.

long, broad tail

GREAT BLUE TURACO
Corythaeola cristata
F: Musophagidae
This fan-crested turaco from western and central Africa is the largest and most distinctive member of the turaco family.

EASTERN GREY PLANTAIN-EATER
Crinifer zonurus
F: Musophagidae
Like the related go-away birds, plantain-eaters are drably colored. Despite their name, they favor figs not plantains. This species is from eastern Africa.

RUSPOLI'S TURACO
Tauraco ruspolii
F: Musophagidae

A distinctively white-crested green turaco, this species has one of the most restricted ranges of any turaco, being confined to forests in southern Ethiopia.

16 in
40 cm

RED-CRESTED TURACO
Tauraco erythrolophus
F: Musophagidae

This Angolan bird is one of a few turaco species with a red crest. The red color is attributed to a pigment possibly unique to turacos.

16–17 in
40–43 cm

red crest

white face

17 in
43 cm

HARTLAUB'S TURACO
Tauraco hartlaubi
F: Musophagidae

A blue-crested member of the green turaco group, this species is found in highland forest in eastern Africa.

16–17 in
40–43 cm

18–18½ in
45–47 cm

GREEN TURACO
Tauraco persa
F: Musophagidae

With a range from Senegal to Angola, this is the most widespread of a group of green turacos with crimson wing-feathers visible in flight.

KNYSNA TURACO
Tauraco corythaix
F: Musophagidae

A South African relative of the green turaco, this species shares the white eye-stripes, but differs in having a white-tipped crest.

bright crimson flight feathers

JACOBIN CUCKOO
Clamator jacobinus
F: Cuculidae

Clamator cuckoos are large crested Old World cuckoos. Like others, this tropical species from Africa and Asia eats hairy caterpillars that are avoided by competitors.

13½ in
34 cm

GREAT SPOTTED CUCKOO
Clamator glandarius
F: Cuculidae

With a range from Europe to Africa, this cuckoo lays eggs in magpie nests, but, like other *Clamator* species, does not evict the host's young.

14–16 in
35–40 cm

KLAAS'S CUCKOO
Chrysococcyx klaas
F: Cuculidae

This small African cuckoo is a close relative of the dideric cuckoo, differing in being greener and lacking white wing spots.

6½–7 in
16–18 cm

DIDERIC CUCKOO
Chrysococcyx caprius
F: Cuculidae

An African member of a group of small bronze-sheened tropical cuckoos, this species lays its eggs in weaver bird nests.

6½–7½ in
17–19 cm

FAN-TAILED CUCKOO
Cacomantis flabelliformis
F: Cuculidae

This brown-breasted cuckoo of Australasian woodlands is one of very few cuckoos that are found in the Pacific—and the only one in Fiji.

9½–11 in
24–28 cm

BRUSH CUCKOO
Cacomantis variolosus
F: Cuculidae

Related to Asian brown-breasted cuckoos, this plain-colored species breeds in southern Asia. The montane populations migrate to warm lowlands in winter.

9 in
23 cm

thick, black bars on belly

11–13½ in
28–34 cm

HIMALAYAN CUCKOO
Cuculus saturatus
F: Cuculidae

This species from Asia and Australia is typical of a group of Old World barred cuckoos that resemble sparrowhawks—which may help them flush hosts from nests.

12–13½ in
32–34 cm

GREY-BELLIED CUCKOO
Cacomantis passerinus
F: Cuculidae

Found from the Malay Peninsula to Australia, this bird is one of many Indo-Pacific, brown-breasted species that are closely related to *Cuculus* cuckoos.

9½ in
24 cm

PALLID CUCKOO
Cuculus pallidus
F: Cuculidae

Many Old World cuckoos have barred plumage, but in some, such as this Australian species, the barring occurs only in immature birds.

12–13 in
30–33 cm

long wings

COMMON CUCKOO
Cuculus canorus
F: Cuculidae

This widespread Eurasian species overwinters in Africa and southern Asia. It has the well-known "cooc-coo" call that gives cuckoos their name.

» CUCKOOS, HOATZIN, AND TURACOS

10–12½ in
26–32 cm

15 in
38 cm

24 in
62 cm

13½ in
34 cm

GUIRA CUCKOO
Guira guira
F: Cuculidae
This shaggy-looking
bird belongs to the ani
group of S. American
cuckoos. It occurs in
noisy flocks and nests
communally in trees.

GIANT COUA
Coua gigas
F: Cuculidae
Couas are Madagascan ground-cuckoos
with blue facial skin and long eyelashes.
This bird lives in dry coastal forests.

**YELLOW-BILLED
CUCKOO**
Coccyzus americanus
F: Cuculidae
One of several American tree
cuckoos with brown plumage
and white-spotted tails, this
species migrates between
North and South America.

**BLACK-BELLIED
CUCKOO**
Piaya melanogaster
F: Cuculidae
Like other American tree
cuckoos, this species builds
its own nest and cares for
its young. It is found from
Colombia to Bolivia.

19–20½ in
48–52 cm

**GREATER
COUCAL**
Centropus sinensis
F: Cuculidae
Like other coucals, this
southern Asian species
is a strong-legged bird
with spurred feet and
long hind claws.

*rufous
head crest*

11 in
28 cm

14 in
36 cm

*streaked, reddish
brown wings*

23½–32 in
60–80 cm

PHEASANT-CUCKOO
Dromococcyx phasianellus
F: Cuculidae
This American ground
cuckoo inhabits the
forest floor in tropical
South America, where it
lays its eggs in the nests of
smaller passerines.

PHEASANT-COUCAL
Centropus phasianinus
F: Cuculidae
Parental care in coucals is provided mostly
by the males. This Australasian species has
a black body when breeding, and builds
a cup-shaped nest in grass.

GROOVE-BILLED ANI
Crotophaga sulcirostris
F: Cuculidae
A heavy-billed American cuckoo
occurring from California to
Argentina, the communal-
nesting ani is a clumsy flier
but runs well.

*white
cheek
stripe*

long tail

**PAVONINE
CUCKOO**
Dromococcyx pavoninus
F: Cuculidae
A smaller relative of
the pheasant-cuckoo, this
tropical S. American bird
is a ground predator.
It feeds on invertebrates.

long, graduated tail

*gray-brown
upperparts with
white spots*

13 in
33 cm

♀

**CORAL-BILLED
GROUND CUCKOO**
Carpococcyx renauldi
F: Cuculidae
One of three species
of Asian ground cuckoos
related to Madagascan
couas, this bird inhabits rain
forests of Southeast Asia.

26–27 in
65–68 cm

**GREATER
ROADRUNNER**
Geococcyx californianus
F: Cuculidae
This fast-running predator is an
American ground cuckoo. The roadrunner
inhabits deserts and nests in cacti in the
southern USA and Mexico.

22 in
56 cm

15½–18 in
39–46 cm

COMMON KOEL
Eudynamys scolopaceus
F: Cuculidae
Unlike most cuckoos, this tropical
brood parasite, found from Asia to
Australasia, is a fruit-eater. Males are
black and females are gray-brown.

OWLS

With acute senses, powerful weaponry, and noiseless flight, owls are superbly adapted as nighttime hunters. Only a few species are active by day.

Although not related to birds of prey, owls, which belong to the order Strigiformes, are similarly equipped with hooked bills and strong talons. Most species have cryptic plumage for camouflage when roosting during the day. Barn owls are distinct in having heart-shaped facial disks; owls with round facial disks are more typical, although they are a family of great diversity, ranging from hawklike owlets to massive eagle owls.

SIGHT AND HEARING

All owls have large, forward-facing eyes that are efficient at letting in light and allow the birds to see well in dim conditions. Their binocular vision is good for judging distance during attacks, but they are fixed in their sockets. To track moving prey, an owl must turn its whole head. The bird can do this through a wide range of movement by means of a long flexible neck concealed beneath the plumage. Owls also have highly sensitive hearing,

helped by the facial disk, which serves to guide sounds to the large ear openings. A downward-pointing bill minimizes interference with sound. An owl can determine the direction of prey with great accuracy from sounds. Most of the nocturnal species have one ear slightly higher than the other, an arrangement that helps them pick up sounds in a vertical direction, too.

SILENT HUNTERS

Hunting owls usually swoop down on their prey and extend their feet to make the grab. The approach is almost silent on large rounded wings that require minimal flapping. A soft fringe of serrated flight feathers breaks up air turbulence, muffling the sound of wing beats. Some day-flying species lack these fringed feathers. Most owls prey on small mammals such as mice and voles, but some of the smaller species take large insects, and a few owls are specialized fish-eaters. An owl tears larger carcasses apart with its hooked bill, but swallows smaller prey whole. The indigestible parts of a meal, such as bones and fur, are compressed in part of the stomach and later regurgitated in the form of a pellet.

PHYLUM	CHORDATA
CLASS	AVES
ORDER	STRIGIFORMES
FAMILIES	2
SPECIES	202

DEBATE
SCREECH OWLS

There are over 60 species of owls in the genus *Otus*. In their native forests, these small, cryptically patterned birds are most reliably identified by their call. Although individual species' calls differ in frequency and duration of notes, *Otus* owls fall into two well-defined groups: scops owls, from the Old World, which call with slow notes; and American screech owls, which have faster, piercing trills. The distinction is enough for some ornithologists to advocate placing the screech owls in their own genus, *Megascops*. Recent research shows that this division may be underpinned by differences in DNA.

BARN OWL
Tyto alba
F: Tytonidae
Found virtually worldwide outside deserts and polar regions, this is the most widespread species of owl. It is famous for its blood-curdling screech.

heart-shaped facial disks

golden upperparts

10–18 in
25–45 cm

10–17 in
26–43 cm

ASHY-FACED OWL
Tyto glaucops
F: Tytonidae
Confined to the Caribbean island of Hispaniola, this owl of dry forest is threatened by competition from the stronger barn owl.

7½–8 in
19–20 cm

EURASIAN SCOPS OWL
Otus scops
F: Strigidae
This small, agile owl from western Eurasia is a typical tufted scops owl of the Old World. It nests in cavities of trees or buildings.

rufous body

9–9½ in
22–24 cm

RAIN FOREST SCOPS OWL
Otus rutilus
F: Strigidae
Restricted to Madagascar, this owl is common in wooded areas. Most are gray; the rarer rufous (reddish) birds are confined to rain forests.

7½–10 in
19–25 cm

WESTERN SCREECH OWL
Otus kennicottii
F: Strigidae
A common owl of western North America, this species favors wooded areas near rivers but will also enter parkland and towns.

6½–10 in
16–25 cm

EASTERN SCREECH OWL
Otus asio
F: Strigidae
Widespread throughout eastern North America, this owl occurs in gray or rufous (reddish) color varieties, with more rufous birds farther east in its range.

»

concentric dark rings on facial disk

26–28 in
65–70 cm

18½–21 in
47–53 cm

BROWN WOOD OWL
Strix leptogrammica
F: Strigidae

This owl occurs in lowland tropical forests throughout India and southeast Asia. Difficult to see, this wood owl can be recognized by its distinctive call.

facial disk

23½–24 in
60–62 cm

pale gray plumage with brown markings

14½–15½ in
37–39 cm

TAWNY OWL
Strix aluco
F: Strigidae

Occurring throughout Eurasia, the tawny owl frequents farmland, towns, and gardens. It belongs to a group of related wood owls.

17–20 in
43–50 cm

GREAT GRAY OWL
Strix nebulosa
F: Strigidae

A large owl of the circumpolar regions, this species occurs in coniferous forest. It sometimes hunts during the day and preys on large rodents and birds.

BARRED OWL
Strix varia
F: Strigidae

A large wood owl native to eastern North America, this aggressive species is spreading westward and displacing the smaller spotted owl.

URAL OWL
Strix uralensis
F: Strigidae

A relative of the great gray owl, this north Eurasian species frequents coniferous and broadleaf forests, and may also enter towns.

18½–19 in
47–48 cm

SPOTTED OWL
Strix occidentalis
F: Strigidae

Native to western North America, this wood owl occurs in mature coniferous forest, where it hunts flying squirrels and similar-sized prey.

18–20 in
45–50 cm

RUFESCENT SCREECH OWL
Otus ingens
F: Strigidae

This little-known bird from wet montane forests in northern South America is one of the larger screech owls, but has smaller ears.

8½–10 in
21–25 cm

10–11 in
25–28 cm

heavily barred underparts

DESERT EAGLE-OWL
Bubo ascalaphus
F: Strigidae

A bird of the Sahara Desert, this species is a smaller, paler, and longer-legged relative of the Eurasian eagle-owl.

23½–30 in
60–75 cm

EURASIAN EAGLE-OWL
Bubo bubo
F: Strigidae

One of the largest species of owls, this widespread eagle-owl occurs throughout Eurasia, and hunts animals as large as deer.

TROPICAL SCREECH OWL
Otus choliba
F: Strigidae

The most widespread tropical American screech owl, this species occurs in a range from Costa Rica to Argentina. It exists in brown and gray forms.

8½–9 in
22–23 cm

BLACK-CAPPED SCREECH OWL
Otus atricapilla
F: Strigidae

Screech owls have diversified in forests throughout the Americas. Many have conspicuous ear tufts. This species is confined to central and southern Brazil.

GREAT HORNED OWL
Bubo virginianus
F: Strigidae

The most widespread American owl, this eagle-owl is found from Alaska to Argentina, in habitats as diverse as forest and desert.

18–27 in
46–68 cm

26–30 in
66–75 cm

VERREAUX'S EAGLE-OWL
Bubo lacteus
F: Strigidae

This is the largest African owl. It is widespread south of the Sahara Desert, where it feeds on small game animals.

ear tuft

white
facial disk

14 in
36 cm

14–18 in
36–45 cm

ELF OWL
Micrathene whitneyi
F: Strigidae
Like other insect-eaters,
this tiny Mexican desert owl
does not need to fly silently,
so it lacks the fringed,
sound-muffling wing
feathers of larger species.

5–6 in
13–15 cm

brilliant
yellow eyes

9–9½ in
22–24 cm

NORTHERN HAWK-OWL
Surnia ulula
F: Strigidae
This relative of pygmy owls
is found in subarctic forests.
With its small head, long tail,
and daytime activity, it is the
most hawk-like owl.

broad
ear tufts

**SOUTHERN
WHITE-FACED OWL**
Ptilopsis granti
F: Strigidae
Like many other owl
species, this bird of
sub-Saharan Africa
breeds in the stick nests
of other birds.

STRIPED OWL
Pseudoscops clamator
F: Strigidae
A South American
eared owl, this species
lives in open, marshy
habitats and nests in
ground vegetation
or low tree holes.

18 in
46 cm

SNOWY OWL
Nyctea scandiaca
F: Strigidae
An Arctic species of eagle-owl,
this large, ground-nesting predator
breeds on open tundra, where it
feeds on lemmings and ptarmigan.

52–71 cm
20½–28 in

SPECTACLED OWL
Pulsatrix perspicillata
F: Strigidae
This is a distinctive species
from Central and South American
forests. Its fledglings are white with
a contrasting black face.

5–6 in
13–15 cm

**LEAST
PYGMY OWL**
*Glaucidium
minutissimum*
F: Strigidae
Like other related small
owls, this woodland
bird of Paraguay and
southeastern Brazil is
active during the day
and night.

**NORTHERN
PYGMY OWL**
Glaucidium gnoma
F: Strigidae
Pygmy owls often fearlessly
tackle prey bigger than
themselves. This little
predator from western
North America has been
seen attacking grouse.

6–6½ in
15–17 cm

6½–7 in
17–18 cm

**FERRUGINOUS
PYGMY OWL**
Glaucidium brasilianum
F: Strigidae
Typical of pygmy owls,
this American species has
eyespots on the back of the
head to confuse predators. It
flicks its tail when excited.

CUBAN PYGMY OWL
Glaucidium siju
F: Strigidae
Many owls nest in decayed
tree cavities, but pygmy
owls habitually choose old
woodpecker holes. This
species is restricted to Cuba.

6–7 in
15–18 cm

long claws

18–18½ in
46–47 cm

BUFFY FISH OWL
Ketupa ketupu
F: Strigidae
This southeast Asian relative of
eagle-owls catches fish and
other aquatic prey with the
help of its long-clawed feet.

7–8½ in
18–21 cm

13½–17 in
34–43 cm

yellow
eye

LONG-EARED OWL
Asio otus
F: Strigidae
Found in forests and heathland
through much of the Northern
Hemisphere, this owl can
flatten its long ear tufts,
rendering them invisible.

8½–11 in
21–28 cm

TENGMALM'S OWL
Aegolius funereus
F: Strigidae
This owl of northern circumpolar
forests is adept at locating small
mammals under snow. It hunts
during daylight.

NORTHERN
SAW-WHET OWL
Aegolius acadicus
F: Strigidae
A close relative of
the Tengmalm's owl,
this species is restricted to
North America, where it
primarily hunts small rodents.

SHORT-EARED
OWL
Asio flammeus
F: Strigidae
This owl is found in open
country in a range that
includes the Americas, Eurasia,
and North Africa, and is even
found on many Pacific islands.

12–14½ in
31–37 cm

pale brown,
heart-shaped face

LITTLE OWL
Athene noctua
F: Strigidae
Widespread in Eurasia
and North Africa, this
partly diurnal predator of
invertebrates and small
vertebrates occurs in open
woodland, farmland,
and semideserts.

BURROWING OWL
Athene cunicularia
F: Strigidae
A long-legged owl, this
partly diurnal species
inhabits grassland and
deserts across the Americas,
nesting in burrows of prairie
dogs or other animals.

yellow
eye

8½–10½ in
21–27 cm

7½–10 in
19–25 cm

heavily streaked
upperparts

15 in
38 cm

BLACK-AND-WHITE OWL
Ciccaba nigrolineata
F: Strigidae
This barred owl of dense
tropical forests from Mexico
to Ecuador is a close
relative of the *Strix*
wood owls.

barred
underparts

15–17 in
38–43 cm

BARKING OWL
Ninox connivens
F: Strigidae
Found from the Moluccas,
Indonesia, to Australia,
this woodland hawk-owl
is named for its doglike
barking call.

12–14 in
30–35 cm

MOREPORK
Ninox novaeseelandiae
F: Strigidae
Typical of Australasian
hawk-owls, this bird has large,
yellow eyes. Unlike most owls,
the male of the species
is larger than the female.

25–26 in
63–65 cm

PEL'S FISHING OWL
Scotopelia peli
F: Strigidae
The largest African fishing owl,
this bird frequents riverside forest,
where it hunts from low perches in
slow-running waters. It eats mainly
fishes, but also crabs and frogs.

NIGHTJARS AND FROGMOUTHS

All nightjars and their relatives are nocturnal. Most are insect-eaters, snatching their prey on the wing using an exceptionally wide bill gape.

Voracious predators by night, concealed by cryptic plumage during the day, nightjars have similarities to owls, and some ornithologists have thought the two groups to be related. Recent studies, however, link nightjars with swifts and hummingbirds. The connection is supported by features such as the weakness of their feet and the ability of some species to enter a torpid state.

Nightjars belong to the order Caprimulgiformes. Typically, they have big heads and long wings for quick aerial maneuvers. Their short bills open amazingly wide to catch flying insects. The gape is

particularly wide in the aptly named frogmouths from Australasia, which are predators of small vertebrates. Of the group, only the oilbird is vegetarian: a fruit-eater that roosts by day in caves.

All these birds are masters of camouflage. Nightjars blend perfectly with forest litter when on the ground; in trees they perch lengthways on a branch to minimize detection. Frogmouths and potoos freeze when disturbed, to mimic tree stumps, closing their eyes to complete the effect.

MINIMAL NESTS
All in this group are minimalist nesters. Nightjars simply lay their eggs in leaf litter on the ground. For oilbirds, a nest is a pile of droppings on a cave ledge, while potoos use nothing more than a depression on a branch.

PHYLUM	CHORDATA
CLASS	AVES
ORDER	CAPRIMULGIFORMES
FAMILIES	5
SPECIES	125

A great potoo opens its enormous gape, displaying a key feature of a group of birds that can scoop up flying insects on the wing.

OILBIRD
Steatornis caripensis
F: Steatornithidae
A fruit-eating relative of nightjars, this cave-nesting species from northern South America is the only nocturnal bird that navigates by echolocation.

16–19 in
41–48 cm

orange-yellow eye

long bristles at base of bill

9–9½ in
22–24 cm

12½–18 in
32–46 cm

TAWNY FROGMOUTH
Podargus strigoides
F: Podargidae
This nocturnal Australian bird hunts from perches. Like other frogmouths, it roosts motionless during the day, resembling a broken branch.

COMMON NIGHTHAWK
Chordeiles minor
F: Caprimulgidae
Nighthawks lack the bill bristles of nightjars, and this absence perhaps helps in their aerial pursuit of insects. This insect-eater migrates between North and South America.

erect, stumplike posture

14–16 in
36–41 cm

gray-brown, barklike plumage

8 in
20 cm

OCELLATED POORWILL
Nyctiphrynus ocellatus
F: Caprimulgidae
Poorwills are small American nightjars, so called because of their mournful, monotonous calls. This dark-colored species occurs in tropical forests.

gray mottled plumage

COMMON POTOO
Nyctibius griseus
F: Nyctibiidae
Potoos are nocturnal insectivores of Central and South America. This species has a haunting call. Its plumage provides excellent camouflage against tree bark.

7½–8½ in
19–21 cm

COMMON POORWILL
Phalaenoptilus nuttallii
F: Caprimulgidae
From arid parts of the USA and Mexico, this small nightjar is one of the few birds that enters a hibernation-like torpor in winter.

9½–11 in
24–28 cm

COMMON PAURAQUE
Nyctidromus albicollis
F: Caprimulgidae
This Central and South American nighthawk frequents scrubby habitats. It is often seen resting on dirt roads at night.

»

rufous collar

EUROPEAN NIGHTJAR
Caprimulgus europaeus
F: Caprimulgidae
Like many nightjars of northern temperate regions, this species is migratory. Found in western Eurasian heath and woodland, it overwinters in Africa.

10–11 in
26–28 cm

SPOT-TAILED NIGHTJAR
Caprimulgus maculicaudus
F: Caprimulgidae
Found from Mexico to Paraguay, this species may be more closely related to tropical American nighthawks than Old World nightjars.

8 in
20 cm

LARGE-TAILED NIGHTJAR
Caprimulgus macrurus
F: Caprimulgidae
Among the most widespread of the Old World nightjars, this species ranges from Pakistan to Australia and New Guinea.

10–10½ in
25–27 cm

PLAIN NIGHTJAR
Caprimulgus inornatus
F: Caprimulgidae
This African bird occurs from Mauritania to Saudi Arabia, and migrates south to Liberia, Congo, and Tanzania.

9 in
22 cm

9 in
23 cm

DUSKY NIGHTJAR
Caprimulgus saturatus
F: Caprimulgidae
Perhaps genetically closer to American poorwills than Old World nightjars, this species occurs in montane forests of Costa Rica and Panama.

LONG-TAILED NIGHTJAR
Caprimulgus climacurus
F: Caprimulgidae
The long tail of some nightjars may be used in a courtship display. This African species has a range from Senegal to Ethiopia.

10–14 in
25–35 cm

LADDER-TAILED NIGHTJAR
Hydropsalis climacocerca
F: Caprimulgidae
This Amazonian species belongs to a group of S. American nightjars, in which males have long, forked tails with white markings. These give the bird its name and are possibly used to attract females.

9–11 in
23–28 cm

MADAGASCAR NIGHTJAR
Caprimulgus madagascariensis
F: Caprimulgidae
This nightjar from Madagascar and Aldabra Island (Seychelles) has a call that resembles the sound of a marble bouncing on a hard floor.

8½ in
21 cm

9 in
22 cm

mottled plumage for camouflage

SHORT-TAILED NIGHTHAWK
Lurocalis semitorquatus
F: Caprimulgidae
One of several tropical American nighthawks, this bird has a short tail and long wings, which give it a bat-like appearance on its insect-chasing flights.

LONG-TRAINED NIGHTJAR
Macropsalis creagra
F: Caprimulgidae
As with many nightjars, males have elaborate tails. This species, limited to eastern South America, evolved from tropical American nighthawks.

elongated tail feather

13½–30 in
34–76 cm

STANDARD-WINGED NIGHTJAR
Macrodipteryx longipennis
F: Caprimulgidae
This African species eats insects, including moths and beetles. Breeding males have flaglike flight feathers, longer than the body, that they raise in courtship displays.

8½–9 in
21–23 cm

flaglike flight feathers

HUMMINGBIRDS AND SWIFTS

The dashing flight of swifts—fastest of all birds—and the audible hovering of hummingbirds characterize the master aeronauts united in this order.

Both swifts and hummingbirds, members of the order Apodiformes, have tiny feet that are useful only for perching, but in compensation the birds have exceptional maneuverability on the wing. Swifts skim through the skies to catch flying insects.

Hummingbirds have the wing control to fly backward, some of them whirring at over 70 wing beats per second—sustained by high metabolism fueled by energy-rich nectar. Insects and spiders provide hummingbirds with protein to feed to their chicks. Hummingbirds use cobwebs to glue together their thimble-sized nests, while swifts use saliva. The swifts' range is almost worldwide, reaching oceanic islands. Hummingbirds are found only in the Americas.

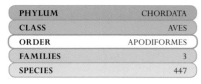

PHYLUM	CHORDATA
CLASS	AVES
ORDER	APODIFORMES
FAMILIES	3
SPECIES	447

COMMON SWIFT
Apus apus
F: Apodidae
This Eurasian swift nests in holes in cliffs or buildings, making it a common bird in towns. It overwinters in Africa.

6½ in
16–17 cm

CHIMNEY SWIFT
Chaetura pelagica
F: Apodidae
Originally nesting in caves and tree-holes, this eastern North American species now nests primarily in chimneys in urban centers. It winters in South America.

6–7 in
15–18 cm

4¾–6 in
12–15 cm

8–9 in
20–22 cm

ALPINE SWIFT
Tachymarptis melba
F: Apodidae
Distinctively white-bellied, with a dark undertail, the Alpine swift occurs in southern Eurasia, Africa, and Madagascar. It feeds on large insects.

WHITE-THROATED SWIFT
Aeronautes saxatalis
F: Apodidae
Found in canyons and mountains of western North America and Central America, this bird gathers in flocks in the evenings to roost.

SCALE-THROATED HERMIT
Phaethornis eurynome
F: Trochilidae
Hermits are dull-colored hummingbirds with long, curved bills for probing *Heliconia* flowers. This species is from eastern South America.

deeply forked tail

4¾ in
12 cm

decurved bill

WHITE-TIPPED SICKLEBILL
Eutoxeres aquila
F: Trochilidae
This bird has a range that extends from Costa Rica to Peru. It uses its decurved bill to feed from similarly curved *Heliconia* flowers.

5 in
13 cm

4¼ in
11 cm

violet "ear" patch

metallic blue throat

violet-blue underparts

4 in
10 cm

5½–6 in
14–15 cm

GREEN-FRONTED LANCEBILL
Doryfera ludovicae
F: Trochilidae
This inhabitant of the forests of the Andes takes nectar from five species of epiphytes (plants that grow on other plants), one of which is a mistletoe.

WHITE-VENTED VIOLET-EAR
Colibri serrirostris
F: Trochilidae
Most of the violet-eared hummingbirds occur in highland forests, but this species is native to the South American savanna.

VIOLET SABREWING
Campylopterus hemileucurus
F: Trochilidae
Sabrewings are named for the thickened shafts of their flight feathers. This Central American bird is the largest hummingbird outside South America.

RUBY TOPAZ
Chrysolampis mosquitus
F: Trochilidae

This South American, open-country species has a ruby-red crown and topaz-yellow throat, but in poor light the bird can appear all black.

brown-black plumage

3½ in
9 cm

orange-red tail

HOODED VISORBEARER
Augastes lumachella
F: Trochilidae

Visorbearers, with their green facial masks, are hummingbirds of arid habitats, such as highland savanna. This species occurs in eastern Brazil.

3¼ in
8 cm

BRAZILIAN RUBY
Clytolaema rubricauda
F: Trochilidae

Named for the male's rufous tail, this hummingbird is found in southeastern Brazil, but may be closely related to species found in the Andes.

2¾ in
7 cm

RUBY-THROATED HUMMINGBIRD
Archilochus colubris
F: Trochilidae

This tiny bee hummingbird is the only hummingbird breeding in the eastern USA. It flies nonstop to the Gulf of Mexico to overwinter.

2¾–3½ in
7–9 cm

LONG-TAILED SYLPH
Aglaiocercus kingi
F: Trochilidae

Belonging to a diverse group of Andean hummingbirds called coquettes, this bird inhabits woodland borders and gardens. Males have long, deeply forked tails.

4–7 in
10–18 cm

BUFF-TAILED CORONET
Boissonneaua flavescens
F: Trochilidae

Found in Colombia, Venezuela, and Ecuador, this relative of Andean hummingbirds forages near flowering trees from mid-level to forest canopy, often with other birds.

4¼ in
11 cm

BUFF-BELLIED HUMMINGBIRD
Amazilia yucatanensis
F: Trochilidae

Belonging to a large group of "emerald" hummingbirds found in Central America, this bird is a Mexican species of open woodland.

4–4¼ in
10–11 cm

BLUE-CHINNED SAPPHIRE
Chlorostilbon notatus
F: Trochilidae

One of many "emerald" hummingbirds with metallic green plumage, this forest and farmland bird is from northern South America.

3½ in
9 cm

dark ear patch

short, straight bill

BLUE-THROATED HUMMINGBIRD
Lampornis clemenciae
F: Trochilidae

This large Mexican species belongs to a group of hummingbirds called mountain gems that originated in Central America. The northerly populations are migratory.

4¾ in
12 cm

4 in
10 cm

COLLARED INCA
Coeligena torquata
F: Trochilidae

Incas form a group of forest hummingbirds native to the Andes. The collared inca is one of the most widespread of the group, found from Colombia to Bolivia.

4¼ in
11 cm

SPECKLED HUMMINGBIRD
Adelomyia melanogenys
F: Trochilidae

A common hummingbird from the Andes, this species has duller plumage than many of its relatives in the coquette group. The sexes are similar.

3½ in
9 cm

ANNA'S HUMMINGBIRD
Calypte anna
F: Trochilidae

A bee hummingbird of western North America, this species overwinters farther north than other hummingbirds. The bird feeds on nectar from a wide range of flowers.

WHITE-EARED HUMMINGBIRD
Basilinna leucotis
F: Trochilidae

This "emerald" hummingbird frequents pine and oak montane forests, often near streams, from southern Arizona to Nicaragua.

3½–4 in
9–10 cm

ANDEAN HILLSTAR
Oreotrochilus estella
F: Trochilidae

An Andean coquette hummingbird, this species lives higher than other species of the group, and feeds in air so thin that it cannot hover.

5 in
13 cm

4 in
10 cm

3½ in
9 cm

straight, black bill

FORK-TAILED WOODNYMPH
Thalurania furcata
F: Trochilidae

A South American relative of the Central American "emerald" hummingbirds, this species occurs in lowland forests as far south as northern Argentina.

BROAD-BILLED HUMMINGBIRD
Cynanthus latirostris
F: Trochilidae

Males of this Mexican scrub hummingbird fly back and forth in a pendulum display to attract females. The species belongs to the "emerald" group.

LUCIFER HUMMINGBIRD
Calothorax lucifer
F: Trochilidae

This bee hummingbird has a range from the southern USA to Mexico. It lives in semideserts, especially those with *Agave* plants.

purple throat patch

iridescent green throat

4 in
10 cm

4 in
10 cm

9–10 in
23–26 cm

SWORD-BILLED HUMMINGBIRD
Ensifera ensifera
F: Trochilidae

The only bird with a bill longer than its body, this member of the Andean inca group favors the trumpet-shaped flowers of passion vines.

BEE HUMMINGBIRD
Mellisuga helenae
F: Trochilidae

This tiny hummingbird is confined to Cuba, but is related to migratory North American hummingbirds. Males of the species are the smallest of all birds.

2–2¼ in
5–6 cm

4¾ in
12 cm

RACKET-TAILED PUFFLEG
Ocreatus underwoodii
F: Trochilidae

Belonging to the inca group, this bee-sized hummingbird of Andean forests is attracted to brush blossoms, such as those of legumes.

white spot near the eye

2¾–3½ in
7–9 cm

rufous upperparts

RUFOUS HUMMINGBIRD
Selasphorus rufus
F: Trochilidae

Sustaining the longest migration of comparably small birds, this aggressively territorial bee hummingbird takes a route from Alaska to Mexico.

3½ in
9 cm

STRIPE-BREASTED STARTHROAT
Heliomaster squamosus
F: Trochilidae

Male starthroats have a vividly colored band around their throats. Although starthroats are related to Central American mountain gems, this species is from eastern Brazil.

4¾ in
12 cm

3½ in
9 cm

WHITE-NECKED JACOBIN
Florisuga mellivora
F: Trochilidae

Genetic research suggests that this large, canopy-dwelling bird of tropical American lowlands belongs to a small lineage independent of other groups in its family.

GORGETED SUNANGEL
Heliangelus strophianus
F: Trochilidae

Sunangels belong to the coquette group of Andean hummingbirds. This species is found from Colombia to Ecuador and lives in wet forest thickets.

3½ in
9 cm

CALLIOPE HUMMINGBIRD
Stellula calliope
F: Trochilidae

One of the migratory bee hummingbirds, this bird breeds in open forests of western North America and winters in the Mexican semidesert.

PLOVERCREST
Stephanoxis lalandi
F: Trochilidae

This distinctive bird may be related to larger sabrewings of Central America, but is found in montane forests of eastern South America.

TROGONS

Gaudily colored birds of tropical forests, the fruit-eating trogons have wide bills, delicate plumage, and a foot structure unique to this group.

Crow-sized birds belonging to the order Trogoniformes are found in tropical America, Africa, and Asia. The males have brighter plumage than the females. Long-tailed and short-winged, trogons are capable fliers, though reluctant to break cover. The first and second toes of a trogon's foot face backward whereas the third and fourth toes face forward, an arrangement seen in no other birds. These unusual feet are so weak that a trogon can barely manage more than a shuffle when it is perched.

Trogons use their wide bills for tackling large fruit and taking invertebrates such as caterpillars. They also use their bills for excavating nesting holes in rotten wood and termite mounds.

PHYLUM	CHORDATA
CLASS	AVES
ORDER	TROGONIFORMES
FAMILIES	1
SPECIES	40

dark violet-blue crown

iridescent green-blue plumage

14–39 in 35–100 cm

long tail (only in male)

RESPLENDENT QUETZAL
Pharomachrus mocinno
F: Trogonidae
Quetzals are iridescent American trogons. The male's long tail in this Central American species is almost half its overall length.

12–14 in 31–36 cm

RED-HEADED TROGON
Harpactes erythrocephalus
F: Trogonidae
An Asian trogon found from the Himalayas to Sumatra, this elusive bird, like many of its cousins, perches motionless for long periods.

10½–12½ in 27–32 cm

ORANGE-BREASTED TROGON
Harpactes oreskios
F: Trogonidae
The color pattern of this bird, dull upperparts, and bright underparts, is typical of many Asian trogons. This species is a native of southeast Asia.

10–11 in 26–28 cm

CUBAN TROGON
Priotelus temnurus
F: Trogonidae
One of two species of *Priotelus* trogons confined to the Caribbean, this is the national bird of Cuba.

11–12 in 28–30 cm

ELEGANT TROGON
Trogon elegans
F: Trogonidae
The only trogon with a range that extends into the USA, this bird occurs in montane forests from southern Arizona to Central America.

10–10½ in 25–27 cm

MASKED TROGON
Trogon personatus
F: Trogonidae
Males of this species from South American montane forests have plumage that resembles that of the elegant trogon.

jagged-tipped tail

MOUSEBIRDS

Named for their scurrying, rodentlike behavior, the mousebirds are a small group of drab-colored, long-tailed birds confined to sub-Saharan Africa.

PHYLUM	CHORDATA
CLASS	AVES
ORDER	COLIIFORMES
FAMILIES	1
SPECIES	6

Mousebirds, the sole family of the order Coliiformes, are softly feathered in shades of brown or gray and have erectile crests and long tails. Gregarious and agile, they resemble parakeets in behavior. These tree-dwelling birds build twiggy, cup-shaped nests. The chicks hatch at an advanced stage of development and quickly master flight. Ornithologists think that mousebirds are the remnant of a more diverse group that, in prehistoric times, ranged beyond Africa—fossil mousebirds have been found in Europe. Their relationship to other birds is uncertain: mousebirds could be allied with trogons, kingfishers, or woodpeckers.

SPECKLED MOUSEBIRD
Colius striatus
F: Coliidae
The largest and one of the most widespread of mousebirds, this species occurs from Nigeria to South Africa in savanna and open woodland.

12–14 in 30–35 cm

BLUE-NAPED MOUSEBIRD
Urocolius macrourus
F: Coliidae
Urocolius mousebirds are stronger fliers and less "mouselike" than *Colius* mousebirds. This scrubland dweller is found from Senegal to Tanzania.

13–14 in 33–35 cm

KINGFISHERS AND RELATIVES

These birds are generally sit-and-wait predators of comparatively large prey. Some, such as hornbills, eat fruit too. Most dig cavity nests in banks or trees.

This order, the Coraciiformes, has a worldwide distribution. Many species are brightly plumaged, and all have the same foot structure: they perch with three toes pointing forward, and their two outer toes are fused at the base. The three main groups of kingfishers—river, tree, and water—follow diverse lifestyles. Many kingfisher species hunt land animals, including lizards, rodents, and insects, as well as fish. Large heads and strong neck muscles help these birds dive efficiently, and their long bills are sharp edged to grip slippery prey. Other members of this

order, the motmots of Central and South America and the Old World rollers, are all land hunters. Kingfisher relatives have a wide variety of bill shapes. In the aerial-hunting bee-eaters, the long bill holds stinging insects away from the head, and is used like forceps to squeeze out insect venom. Hoopoes probe the ground for grubs with their curved bills. Hornbills use their huge bills to take fruit and catch mammals, reptiles, and even other birds.

SEALED NESTS
Some male hornbills partially seal up their nesting cavities when the female is incubating. Using their bill as a spatula, they spread mud over the entrance, leaving a slit through which the temporarily imprisoned female can be fed.

PHYLUM	CHORDATA
CLASS	AVES
ORDER	CORACIIFORMES
FAMILIES	10
SPECIES	218

Most hornbills, such as this rhinoceros hornbill, have a hollow projection (casque) above the bill that helps to resonate calls.

LILAC-BREASTED ROLLER
Coracias caudatus
F: Coraciidae
Like other *Coracias* rollers, this African bird is a predator of ground animals, swooping to take lizards, rodents, and large invertebrates.

12½–14 in
32–36 cm

RUFOUS-CROWNED ROLLER
Coracias naevius
F: Coraciidae
Somewhat dull colored compared with other species, this large roller inhabits arid areas across sub-Saharan Africa.

14–16 in
36–41 cm

11–12 in
28–30 cm

BLUE-BELLIED ROLLER
Coracias cyanogaster
F: Coraciidae
This central African roller, like many of its relatives, nests in tree cavities. The species is common throughout its extensive range.

14–15 in
36–38 cm

11½–12½in
29–32 cm

EUROPEAN ROLLER
Coracias garrulus
F: Coraciidae
The most widespread of the *Coracias* rollers, named for their somersaulting courtship displays, this western Eurasian species overwinters in Africa.

black stripe
through eye

white throat

DOLLARBIRD
Eurystomus orientalis
F: Coraciidae
Named for its pale, coin-shaped underwing markings, this roller occurs in woodland from the Himalayas to Australia.

RACQUET-TAILED ROLLER
Coracias spatulatus
F: Coraciidae
This distinctive fork-tailed roller from East Africa has outer tail feathers that have projecting "flag-tipped" vanes.

PITTA-LIKE GROUND ROLLER
Atelornis pittoides
F: Brachypteraciidae
As the name suggests, ground rollers are mostly terrestrial. This rain forest species is the most brightly colored of the family.

10 in
26 cm

long, brown tail with dark barring

18½–20½in
47–52 cm

10½–12 in
27–30 cm

yellow bill

BROAD-BILLED ROLLER
Eurystomus glaucurus
F: Coraciidae
Typical of its genus, this African species is longer winged and more agile in pursuit of flying prey than *Coracias* rollers.

LONG-TAILED GROUND ROLLER
Uratelornis chimaera
F: Brachypteraciidae
Like other ground rollers, this species from the dry forests of southwest Madagascar nests in holes excavated in the ground.

10½–12 in
27–30 cm

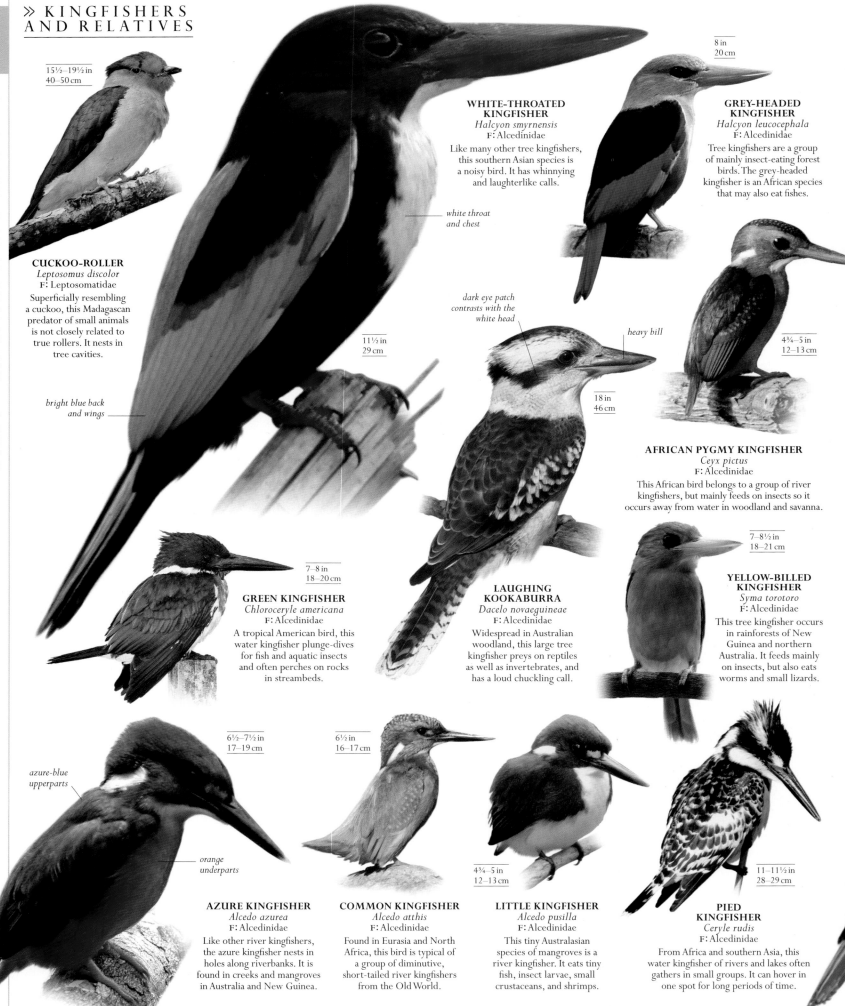

15½–19½ in
40–50 cm

CUCKOO-ROLLER
Leptosomus discolor
F: Leptosomatidae
Superficially resembling
a cuckoo, this Madagascan
predator of small animals
is not closely related to
true rollers. It nests in
tree cavities.

bright blue back
and wings

WHITE-THROATED KINGFISHER
Halcyon smyrnensis
F: Alcedinidae
Like many other tree kingfishers,
this southern Asian species is
a noisy bird. It has whinnying
and laughterlike calls.

white throat
and chest

8 in
20 cm

GREY-HEADED KINGFISHER
Halcyon leucocephala
F: Alcedinidae
Tree kingfishers are a group
of mainly insect-eating forest
birds. The grey-headed
kingfisher is an African species
that may also eat fishes.

dark eye patch
contrasts with the
white head

heavy bill

4¾–5 in
12–13 cm

AFRICAN PYGMY KINGFISHER
Ceyx pictus
F: Alcedinidae
This African bird belongs to a group of river
kingfishers, but mainly feeds on insects so it
occurs away from water in woodland and savanna.

11½ in
29 cm

18 in
46 cm

7–8 in
18–20 cm

GREEN KINGFISHER
Chloroceryle americana
F: Alcedinidae
A tropical American bird, this
water kingfisher plunge-dives
for fish and aquatic insects
and often perches on rocks
in streambeds.

LAUGHING KOOKABURRA
Dacelo novaeguineae
F: Alcedinidae
Widespread in Australian
woodland, this large tree
kingfisher preys on reptiles
as well as invertebrates, and
has a loud chuckling call.

7–8½ in
18–21 cm

YELLOW-BILLED KINGFISHER
Syma torotoro
F: Alcedinidae
This tree kingfisher occurs
in rainforests of New
Guinea and northern
Australia. It feeds mainly
on insects, but also eats
worms and small lizards.

azure-blue
upperparts

6½–7½ in
17–19 cm

6½ in
16–17 cm

orange
underparts

AZURE KINGFISHER
Alcedo azurea
F: Alcedinidae
Like other river kingfishers,
the azure kingfisher nests in
holes along riverbanks. It is
found in creeks and mangroves
in Australia and New Guinea.

COMMON KINGFISHER
Alcedo atthis
F: Alcedinidae
Found in Eurasia and North
Africa, this bird is typical of
a group of diminutive,
short-tailed river kingfishers
from the Old World.

4¾–5 in
12–13 cm

LITTLE KINGFISHER
Alcedo pusilla
F: Alcedinidae
This tiny Australasian
species of mangroves is a
river kingfisher. It eats tiny
fish, insect larvae, small
crustaceans, and shrimps.

11–11½ in
28–29 cm

PIED KINGFISHER
Ceryle rudis
F: Alcedinidae
From Africa and southern Asia, this
water kingfisher of rivers and lakes often
gathers in small groups. It can hover in
one spot for long periods of time.

BELTED KINGFISHER
Megaceryle alcyon
F: Alcedinidae
A water kingfisher from
North America, the belted
kingfisher is a fishing specialist
that often hovers over water
before diving for prey.

11–14 in
28–35 cm

single blue chest band

**BUFF-BREASTED
PARADISE KINGFISHER**
Tanysiptera sylvia
F: Alcedinidae
Many tree kingfishers, such as this
species from New Guinea and
northern Australia, nest in holes
in tree-top termite mounds.

12–14 in
30–35 cm

white central tail feathers

**BROWN-WINGED
KINGFISHER**
Pelargopsis amauroptera
F: Alcedinidae
A tree kingfisher of
mangrove forests, this species
occupies a range from India
to the Malay Peninsula.

14½ in
37 cm

STORK-BILLED KINGFISHER
Pelargopsis capensis
F: Alcedinidae
Occurring in southern Asia, this bird is found
near water in forests, especially outside the
range of the related brown-winged kingfisher.

14½–16 in
37–41 cm

**COLLARED
KINGFISHER**
Todiramphus chloris
F: Alcedinidae
Prefering mangroves,
this tree kingfisher is
widespread in coastal
areas across southern
Asia to the Pacific.

10–11 in
25–28 cm

black eye-stripe

**WHITE-FRONTED
BEE-EATER**
Merops bullockoides
F: Meropidae
This African bee-eater breeds
in large colonies, which have
a complex social system with
nonbreeding birds assisting
in raising the young.

8½–9½ in
22–24 cm

GREEN BEE-EATER
Merops orientalis
F: Meropidae
Found in dry, open habitats
in Africa and southern Asia,
the green bee-eater nests in
looser colonies than other
species of bee-eater.

8½–10 in
22–25 cm

**WHITE-THROATED
BEE-EATER**
Merops albicollis
F: Meropidae
Occurring in Africa south
of the Sahara, like other
bee-eaters, the white-throated
bee-eater nests in tunnels
in sand banks.

8–12½ in
20–32 cm

long, slightly curved bill

EUROPEAN BEE-EATER
Merops apiaster
F: Meropidae
Like other bee-eaters, this bird
from southwestern Eurasia and
Africa catches insects in flight.
Before eating bees, it removes
the sting from its prey.

10–11½ in
25–29 cm

**RAINBOW
BEE-EATER**
Merops ornatus
F: Meropidae
The only bee-eater in Australia,
southernmost populations of
this species migrate northward
to overwinter in northern
Australia and Indonesia.

9 in
23 cm

turquoise-blue crown

**BLUE-CROWNED
MOTMOT**
Momotus momota
F: Momotidae
Like the related bee-eaters and
many kingfishers, this typical
brightly colored motmot from
tropical America nests in
tunnels in riverbanks.

RUFOUS MOTMOT
Baryphthengus martii
F: Momotidae
Like other motmots,
this stout-billed bird from
C. and northern S. America
sallies for large insects
and small vertebrates.

18 in
46 cm

**TURQUOISE-BROWED
MOTMOT**
Eumomota superciliosa
F: Momotidae
As with other motmots,
the racket-tipped tail of this
Central American motmot is
formed by weakened feather
barbs that break away.

13 in
33 cm

racketlike tail tip

16 in
41 cm

JAMAICAN TODY
Todus todus
F: Todidae
Todies are tiny, green
kingfisher-like birds from the
Caribbean, with flattened bills
for catching insects in flight.
This species is found only in
the woodlands of Jamaica.

4½ in
11 cm

» KINGFISHERS AND RELATIVES

RED-BILLED HORNBILL
Tockus erythrorhynchus
F: Bucerotidae
Tockus hornbills are small African predators, most with red or yellow bills. This species inhabits savanna and open woodland from Senegal to Namibia.

reddish bill

gray, white, and black plumage

16½–18 in
42–45 cm

white underparts

12–14 in
30–36 cm

GREEN WOOD HOOPOE
Phoeniculus purpureus
F: Phoeniculidae
This African wood hoopoe climbs trees like a woodpecker, but has a hoopoelike bill that it uses to probe rotting wood for invertebrates.

black-tipped crest, raised when landing

HOOPOE
Upupa epops
F: Upupidae
This species nests in tree-holes in Africa and Eurasia. It has a flickering, butterfly-like flight and a strong bill to dig for ground-level invertebrates.

black and white bars on wings

10–12½ in
25–32 cm

30–32 in
75–80 cm

23–26 in
58–65 cm

TRUMPETER HORNBILL
Bycanistes bucinator
F: Bucerotidae
This hornbill inhabits the forests of eastern to southern Africa. It is similar to the related silvery-cheeked hornbill, but has red facial skin.

SILVERY-CHEEKED HORNBILL
Bycanistes brevis
F: Bucerotidae
This fruit-eating pied hornbill is a forest species from eastern Africa. As with other hornbills, the male has a larger bill than the female, with a more prominent casque.

3¼ ft
1 m

blue throat pouch (only in female)

dark plumage

28 in
70 cm

28 in
70 cm

ORIENTAL PIED HORNBILL
Anthracoceros albirostris
F: Bucerotidae
The most widespread of the Asian pied hornbills, this bird occurs from the Himalayas to Bali (Indonesia), in cultivated land as well as forest.

MALABAR PIED HORNBILL
Anthracoceros coronatus
F: Bucerotidae
Like most other Asian hornbills, this pied species from India and Sri Lanka is omnivorous, and eats a great deal of fruit.

NORTHERN GROUND HORNBILL
Bucorvus abyssinicus
F: Bucorvidae
One of two predatory ground hornbills of African grasslands, this species from Senegal to Kenya tolerates drier conditions than its southern counterpart.

WOODPECKERS AND TOUCANS

Mostly tree-dwelling cavity nesters, birds of this order, the Piciformes, have a common foot structure. More than half of the species are woodpeckers.

The woodpeckers are almost worldwide in distribution. These birds cling to trunks with zygodactyl feet typical of the order (two toes pointing forward, two backward) and, propped by their stiffened tail, drum through the wood with their strong bills. They extract prey with a long barbed tongue. Other birds in the group are largely confined to the tropics. Honeyguides feed on the wax from raided bee nests. Jacamars hunt large insects, while puffbirds consume flies. Barbets have serrated bills for eating fruits; the tropical American barbets are cousins of the toucans. Honeyguides lay eggs in other birds' nests, but others in this group nest in tree-holes, burrows, or dug-out termite mounds.

PHYLUM	CHORDATA
CLASS	AVES
ORDER	PICIFORMES
FAMILIES	5
SPECIES	411

21–23½ in
53–60 cm

WHITE-THROATED TOUCAN
Ramphastos tucanus
F: Ramphastidae
Like other toucans, this species from northern South America breeds in tree cavities, often making use of abandoned woodpecker nests.

17 in
43 cm

RED-BREASTED TOUCAN
Ramphastos dicolorus
F: Ramphastidae
Among the smallest of the *Ramphastos* toucans, this species from eastern South America is the only one with extensive red underparts.

22–23½ in
55–60 cm

CUVIER'S TOUCAN
Ramphastos tucanus cuvieri
F: Ramphastidae
This Amazonian bird is often regarded as a subspecies of the white-throated toucan, from which it differs by having a darker bill.

19 in
48 cm

CHANNEL-BILLED TOUCAN
Ramphastos vitellinus
F: Ramphastidae
Ramphastos toucans are black with a white or yellow chest. In this variable South American species, the chest color depends on race.

pale orange around eye

22–26 in
55–65 cm

12–14 in
30–35 cm

EMERALD TOUCANET
Aulacorhynchus prasinus
F: Ramphastidae
The most widespread of a group of small green toucans, this species occurs from Mexico to Bolivia. The species consists of several races.

14 in
35 cm

♀

SPOT-BILLED TOUCANET
Selenidera maculirostris
F: Ramphastidae
This south Brazilian species is one of several dichromatic toucanets, which are the only toucans with different-colored sexes. The females are marked with brown.

TOCO TOUCAN
Ramphastos toco
F: Ramphastidae
This bird from northern South America is the largest of the toucans. Unlike other forest species, it also inhabits more open woodland habitats.

massive, black-tipped orange bill

SAFFRON TOUCANET
Baillonius bailloni
F: Ramphastidae
Found in southeastern Brazil, this toucan has a yellow-olive plumage. It is related to the more brightly colored aracaris.

14–16 in
35–40 cm

16 in
41 cm

COLLARED ARACARI
Pteroglossus torquatus
F: Ramphastidae
With the most northerly range of the aracaris, this species occurs in humid forests from southern Mexico to northern South America.

14½ in
37 cm

blue eye-skin

CHESTNUT-EARED ARACARI
Pteroglossus castanotis
F: Ramphastidae
Aracaris are long-tailed, gregarious toucans. Most have a red rump and prominent bands on the belly. This species is from northwestern South America.

ornate black bill with yellow bands

≫

BLACK-BILLED BARBET
Lybius guifsobalito
F: Ramphastidae
Many African barbets occur in more
open country than the forest-dwelling
species of the Americas and Asia. This
species occurs in East Africa.

7½ in
19 cm

6½ in
17 cm

9 in
23 cm

10 in
26 cm

BLACK-SPOTTED BARBET
Capito niger
F: Ramphastidae
This barbet from northern
South America is rarely seen, but
its froglike call is often heard.

RED-HEADED BARBET
Eubucco bourcierii
F: Ramphastidae
This species is found from
Costa Rica to Peru. It is unusual
among barbets in being
largely silent.

BEARDED BARBET
Lybius dubius
F: Ramphastidae
Barbets have sensory bristles at the base of
their bill, which are much enlarged in this
species from western and central Africa.

red forehead

green
upperparts

BLUE-THROATED BARBET
Megalaima asiatica
F: Ramphastidae
Asian barbets often forage
with other fruit-eaters in the forest
canopy. This species occurs from
the Himalayas to Thailand.

9 in
23 cm

6½ in
17 cm

8 in
20 cm

large, yellowish
bill

11 in
28 cm

12½–13 in
32–33 cm

**COPPERSMITH
BARBET**
Megalaima haemacephala
F: Ramphastidae
This is a widespread barbet
found in forest edges and scrub.
Its incessant "tonk-tonk"
hammering call is a common
sound in southern Asia.

**CRIMSON-FRONTED
BARBET**
Megalaima rubricapillus
F: Ramphastidae
A small Asian barbet, this
species is restricted to Sri
Lanka and southwestern
India. It is a common species
that is seen in towns.

**BROWN-HEADED
BARBET**
Megalaima zeylanica
F: Ramphastidae
A typical Asian barbet,
this species from the
Himalayas, India, and Sri
Lanka is a fruit-eater and
especially favors figs.

**RED-FRONTED
TINKERBIRD**
Pogoniulus pusillus
F: Ramphastidae
Restricted to riverside forests of
the East African coast, this species
feeds on insects and fruits,
including mistletoe berries.

tail narrows
to point

4–4¼ in
10–11 cm

GREAT BARBET
Megalaima virens
F: Ramphastidae
The largest Asian barbet,
this species is a noisy bird
of highland forests from
the eastern Himalayas
to Thailand.

red vent

4–4¼ in
10–11 cm

4¼ in
11 cm

**YELLOW-RUMPED
TINKERBIRD**
Pogoniulus bilineatus
F: Ramphastidae
Tinkerbirds are small,
black and white African barbets
with repetitive calls that continue
throughout the day. This species is
widespread south of the Sahara.

**YELLOW-FRONTED
TINKERBIRD**
Pogoniulus chrysoconus
F: Ramphastidae
A more widespread relative of the
red-fronted tinkerbird, this species
occurs in dry, open woodland and
savanna in sub-Saharan Africa.

beardlike bristles

RED-FRONTED BARBET
Tricholaema diademata
F: Ramphastidae

An East African barbet related to the spot-flanked barbet, this species occurs in drier habitats. Like other barbets, it nests in tree-holes.

9 in
22 cm

SPOT-FLANKED BARBET
Tricholaema lacrymosa
F: Ramphastidae

A barbet of damp woodlands from central to eastern Africa, the spot-flanked barbet feeds mainly on figs and berries.

9 in
22 cm

white-spotted upperparts

6–6½ in
15–16 cm

D'ARNAUD'S BARBET
Trachyphonus darnaudii
F: Ramphastidae

Trachyphonus barbets from Africa live in open country and spend much time on the ground. D'Arnaud's barbet occurs throughout East Africa.

TOUCAN-BARBET
Semnornis ramphastinus
F: Ramphastidae

A bird of the rain forests of Colombia and Ecuador, this species is intermediate between barbets and toucans. It is exclusively a fruit-eater.

8 in
20 cm

striking head pattern

9 in
23 cm

RED-AND-YELLOW BARBET
Trachyphonus erythrocephalus
F: Ramphastidae

A typical African ground barbet, this bird eats insects, fruits, seeds, and even small lizards. It often burrows into termite mounds to nest.

FIRE-TUFTED BARBET
Psilopogon pyrolophus
F: Ramphastidae

The only Asian barbet with a graduated tail, this Southeast Asian species also has tufts of facial bristles and a cicada-like call.

11 in
28 cm

4¾–5 in
12–13 cm

6½ in
17 cm

GREEN-BACKED HONEYBIRD
Prodotiscus zambesiae
F: Indicatoridae

Honeybirds eat insects, fruits, and even beeswax. This African species lays eggs in the nests of the white-eye, a small woodland bird, and its young kill host nestlings.

NORTHERN WRYNECK
Jynx torquilla
F: Picidae

This Eurasian woodland ant-eater has a weaker bill than true woodpeckers and is named for its ability to twist its neck.

BAR-BREASTED PICULET
Picumnus aurifrons
F: Picidae

Piculets are tiny nuthatchlike members of the woodpecker family. They use their short bills for taking insects from decayed wood. This species is from central South America.

4 in
10 cm

4 in
10 cm

GOLDEN-SPANGLED PICULET
Picumnus exilis
F: Picidae

This South American piculet, like other piculets, lacks stiff, supporting tail feathers of bigger woodpeckers, so it spends less time on vertical trunks.

4 in
10 cm

OCHRACEOUS PICULET
Picumnus limae
F: Picidae

Confined to eastern Brazil, the ochraceous piculet—like other piculets—reuses woodpecker nesting holes as its bill is too small to excavate its own nest.

4 in
10 cm

OCHRE-COLLARED PICULET
Picumnus temminckii
F: Picidae

Confined to the forests of eastern Paraguay, southeastern Brazil, and northeastern Argentina, this piculet is well camouflaged by its pattern and coloration.

SPOTTED PICULET
Picumnus pygmaeus
F: Picidae

The spotted piculet occurs only in the tropical forests of northeastern Brazil, where it is relatively common.

**GROUND
WOODPECKER**
Geocolaptes olivaceus
F: Picidae
Unusual in living on the
ground, this South African
ant-eating woodpecker
inhabits barren, rocky areas
and tunnels into earth banks
to make a nesting chamber.

12 in
30 cm

7 in
18 cm

NUBIAN WOODPECKER
Campethera nubica
F: Picidae
This relative of the European
green woodpecker occurs in dry
parts of northeastern Africa,
where it is often seen in pairs.

7–9 in
18–22 cm

**YELLOW-BELLIED
SAPSUCKER**
Sphyrapicus varius
F: Picidae
Sapsuckers make holes in
trees to drink sap. This
fork-tailed species breeds
in North America and
migrates to the Caribbean.

11 in
28 cm

**COMMON
FLAME-BACKED
WOODPECKER**
Dinopium javanense
F: Picidae
A tropical woodpecker,
this species occurs in a
variety of woodland
types—including
mangrove—and ranges
from India eastward
to Borneo and Java.

**HEART-SPOTTED
WOODPECKER**
Hemicircus canente
F: Picidae
One of two related small,
crested woodpeckers
from southeast Asia, this
species is named for the
black, heart-shaped
marks on its back.

6–6½ in
15–17 cm

16–19½ in
40–49 cm

red cap

PILEATED WOODPECKER
Dryocopus pileatus
F: Picidae
This is the largest North American
woodpecker. Unlike the Eurasian black
woodpecker, American members of
the genus are crested.

18–22½ in
45–57 cm

9 in
23 cm

**GOLDEN-GREEN
WOODPECKER**
Piculus chrysochloros
F: Picidae
Typical of a group of
green-backed woodpeckers
from tropical America,
this bird often follows
mixed-species flocks,
gleaning bark surfaces.

male has
red face streak

BLACK WOODPECKER
Dryocopus martius
F: Picidae
This large woodpecker
from woodlands across
northern Eurasia belongs
to a small group of
predominantly black
species with red crowns.

white patch
on wing

7½–9 in
19–23 cm

white
underparts

9½ in
24 cm

**RED-BELLIED
WOODPECKER**
Melanerpes carolinus
F: Picidae
Like other *Melanerpes*
woodpeckers, this common
North American species stores
food in crevices. Despite its name,
the belly only has a tinge of red.

12–13 in
31–33 cm

**GREEN
WOODPECKER**
Picus viridis
F: Picidae
This European member
of a group of Old
World, green-backed
woodpeckers is also
called the yaffle after its
distinct yelping call.

**RED-HEADED
WOODPECKER**
Melanerpes erythrocephalus
F: Picidae
A distinctive woodpecker
from North America, this is
an aggressive species that
destroys nests and eggs of
other birds in its territory.

7½ in
19 cm

**YELLOW-FRONTED
WOODPECKER**
Melanerpes flavifrons
F: Picidae
Most *Melanerpes* woodpeckers
have partly barred plumage.
They may also have striking
colors, as does this South
American species.

12 in
31 cm

**ROBUST
WOODPECKER**
Campephilus robustus
F: Picidae
The *Campephilus* woodpeckers
are black and white birds
with red heads. The robust
woodpecker is restricted
to eastern South America.

HAIRY WOODPECKER
Picoides villosus
F: Picidae
A North American relative of
the three-toed woodpecker,
this species becomes more
abundant in response to
the availability of its prey—
the larvae of bark beetle.

11–12 in
28–31 cm

*red patch
(in male)*

*dark bars
on wings*

7–10 in
18–26 cm

9 in
22–23 cm

*red nape
(in male)*

8–9 in
20–22 cm

**MIDDLE-SPOTTED
WOODPECKER**
Dendrocopos medius
F: Picidae
A pied woodpecker confined
to Europe and southwestern
Asia, this species drums less
than the related great
spotted woodpecker.

*red patch
beneath tail*

NORTHERN FLICKER
Colaptes auratus
F: Picidae
American flickers are
named for the way their
colored underwings flash
in flight. This species has
red- and yellow-shafted
forms that hybridize.

**THREE-TOED
WOODPECKER**
Picoides tridactylus
F: Picidae
This northern Eurasian and
American bird is the most
northerly woodpecker.
Most woodpeckers have
four toes.

8½–9 in
21–22 cm

**GREAT SPOTTED
WOODPECKER**
Dendrocopos major
F: Picidae
A widespread member of a Eurasian
group of pied woodpeckers, this bird
is common in forests and gardens
from Europe to southeast Asia.

8 in
20 cm

**CHESTNUT
JACAMAR**
*Galbalcyrhynchus
purusianus*
F: Galbulidae
Jacamars specialize in
catching large insects, such as
butterflies, from the air. This is
one of two chestnut-colored
species from South America.

7 in
18 cm

*iridescent green
upperparts*

11 in
28 cm

red underparts

**RUFOUS-TAILED
JACAMAR**
Galbula ruficauda
F: Galbulidae
This Central and South American
species is typical of its genus,
characterized by iridescent
green coloration above and
rufous below.

9 in
23 cm

**THREE-TOED
JACAMAR**
Jacamaralcyon tridactyla
F: Galbulidae
The most dull-colored jacamar,
this dry forest species nests in
earthbanks and has two toes
pointing front and one back.

**GREAT
JACAMAR**
Jacamerops aureus
F: Galbulidae
This is the largest species
of jacamar. It occurs from
Costa Rica to Bolivia and
feeds mainly on insects,
supplemented by small lizards.

5½ in
14 cm

RUSTY-BREASTED NUNLET
Nonnula rubecula
F: Bucconidae
Nunlets are small, drab-colored
members of the puffbird family.
This bird is a typical South American
species of vine-bordered forest.

**SWALLOW-WINGED
PUFFBIRD**
Chelidoptera tenebrosa
F: Bucconidae
Described as martinlike
when perched and batlike in
flight, this bird of northern
South America sallies for flying
insects along rivers.

6 in
15 cm

8–8½ in
20–22 cm

11 in
28 cm

7½ in
19 cm

8–9 in
20 cm

**BLACK-FRONTED
NUNBIRD**
Monasa nigrifrons
F: Bucconidae
Nunbirds are black-bodied relatives
of puffbirds. This noisy South
American species often forages
beneath monkey troops, feeding on
small animals they have disturbed.

**WHITE-WHISKERED
PUFFBIRD**
Malacoptila panamensis
F: Bucconidae
Malacoptila puffbirds are brown-
colored birds. The white-whiskered
puffbird of Central America and
northwestern South America has
been described as rather tame.

**BLACK-BREASTED
PUFFBIRD**
Notharchus pectoralis
F: Bucconidae
This black and white
puffbird of northwestern
South America may follow
army ant swarms to catch
insects fleeing the ants.

WHITE-EARED PUFFBIRD
Nystalus chacuru
F: Bucconidae
Like other puffbirds, this species
from central South America is a
big-headed, "puff-bodied" bird
with a heavy bill for catching
small animals.

PASSERINES

This huge order comprises nearly 60 percent of all bird species. They are known collectively as perching birds, for their specialized feet.

Like many birds, passerines have four-toed feet with three toes pointing forward and one back. As a passerine lands, a muscle automatically tightens tendons running through the legs so that the toes lock around a perch, securing the bird even in sleep.

These birds, of the order Passeriformes, are found in almost all land habitats worldwide, from dense rain forest to arid desert and even freezing Arctic tundra. They range in size from the tiny American flycatchers, no bigger than hummingbirds, to the powerfully built Eurasian ravens.

DIVERSITY

The order includes birds adapted to a variety of feeding habits. The insect-eaters have bills like needles for probing foliage or with a wide gape for catching prey in flight. Other passerines have stubby bills for cracking open seeds, while some have long, curved bills for extracting nectar.

Passerines combine a high metabolic rate with a comparatively large brain size, giving some the resilience to tolerate cold and a few the intelligence to master the use of simple tools. They produce helpless, naked young, which are reared in nests ranging from simple cups to elaborate mud-chambers or hanging purses of woven grass. A few passerines are brood parasites, laying their eggs in the nests of other species.

SONGBIRDS

Passerines are split into two groups, largely on the basis of the structure of their vocal box. The first group, the Suboscines, includes about a fifth of all passerines. It occurs in the Old World tropics but has greatest diversity in South America; it includes broadbills, pittas, antbirds, and tyrant flycatchers. The second group, the Oscines, includes all remaining passerine families, popularly known as songbirds. Birds often communicate by voice, but the structure of their vocal box enables many passerines to deliver complex songs that are important in courtship and defending territory. Many songs are so distinctive that they can be used to identify species.

PHYLUM	CHORDATA
CLASS	AVES
ORDER	PASSERIFORMES
FAMILIES	96
SPECIES	5,962

BROADBILLS

The family Eurylaimidae consists of forest-dwelling birds from tropical Africa and Asia. They mostly use their wide bills for catching insects in trees, but members of one group of Asian green broadbills are fruit-eaters.

6½–7 in
17–18 cm

GREEN BROADBILL
Calyptomena viridis
One of the three southeast Asian green broadbills, this bird is a specialist fruit-eater and builds globular hanging nests.

blue bill with yellow at base

10 in
25 cm

BLACK-AND-RED BROADBILL
Cymbirhynchus macrorhynchos
This distinctively colored southeast Asian broadbill frequents forests near water. It builds a pouchlike nest suspended from the tips of branches.

6 in
15 cm

BLACK-AND-YELLOW BROADBILL
Eurylaimus ochromalus
An Asian insect-eating broadbill, this species forages in the middle and upper levels of rain forests from Myanmar to Borneo and Sumatra.

ASITIES

The brush-tipped tongue of these Madagascan birds, the Philepittidae, suggests that they may have evolved from nectar-feeding ancestors: one genus is now fruit-eating; the other resembles the unrelated sunbirds that feed on nectar.

thin, decurved bill

3½ in
9 cm

COMMON SUNBIRD-ASITY
Neodrepanis coruscans
One of two long-billed birds from eastern Madagascar, this species evolved the nectar-feeding habit of true sunbirds, but is unrelated to them.

PITTAS

Birds of this Old World family forage for insects on the forest floor in tropical regions. The Pittidae have rounded bodies and short bills, and many species are brilliantly plumaged. Both adults incubate the eggs.

8 in
20 cm

BLUE-WINGED PITTA
Pitta moluccensis
Found from southern China to Borneo and Sumatra, the blue-winged pitta inhabits dense forest when breeding, but overwinters in coastal scrub.

7½ in
19 cm

INDIAN PITTA
Pitta brachyura
Like its relatives, this pitta from the southern Himalayas, India, and Sri Lanka builds domed nests on or near the ground.

MANAKINS

These birds, the Pipridae, are related to cotingas. The brightly colored males gather in groups called leks to display to females. For many species, courtship involves elaborate dancing movements. Females build the nest and rear offspring alone.

red cap

6 in
15 cm

5½–6 in
14–15 cm

ARARIPE MANAKIN
Antilophia bokermanni
Described as recently as 1998, this critically endangered bird is found only in the Araripe highlands of northeastern Brazil.

BLUE MANAKIN
Chiroxiphia caudata
One of the most colorful birds of the rain forest of southern Brazil, the blue manakin has a whining, catlike call.

4¼–5 in
11–13 cm

3½–4 in
9–10 cm

PIN-TAILED MANAKIN
Ilicura militaris
Males of this manakin from southeastern Brazil have elongated central tail feathers and display to females with raised rump feathers.

STRIPED MANAKIN
Machaeropterus regulus
This is an inconspicuous species of northern South America. Males of the species may make insectlike buzzing sounds during their display.

3½ in
9 cm

GOLDEN-HEADED MANAKIN
Pipra erythrocephala
The male of this South American species displays by jumping and whirring its wings—and sliding to one side or backward.

COTINGAS AND RELATIVES

A diverse group of tropical American passerines, the Cotingidae consists of fruit-eating or insect-eating forest birds. Males are brightly colored and are highly vocal when courting females; some species display in trees and others on the ground.

10½–11 in
27–28 cm

BARE-THROATED BELLBIRD
Procnias nudicollis
Bellbirds are named for their metallic calls. Like other bellbirds, males of this eastern South American species have white plumage.

5 in
13 cm

bright yellow throat

FIERY-THROATED FRUIT-EATER
Pipreola chlorolepidota
Fruit-eaters are stocky green birds, males of which are usually black headed and red or yellow throated. This is a small Andean species.

glossy black body

purple-red patch

11–12 in
28–30 cm

bare red skin around eyes

9 in
22 cm

PURPLE-THROATED FRUITCROW
Querula purpurata
Males of this distinctive Amazonian cotinga have a prominent, iridescent purplish red throat patch. It can pluck fruits while hovering in midair.

BLACK-TAILED TITYRA
Tityra cayana
This canopy bird belongs to a group of big-headed fruit-eaters. It nests in tree-holes, often those abandoned by woodpeckers.

brilliant red plumage

large crest hides the bill

11–12½ in
28–32 cm

ANDEAN COCK-OF-THE-ROCK
Rupicola peruvianus
The colorful crested males of this large Andean cotinga display in groups to females. Females nest among rocks and rear young alone.

TYRANT FLYCATCHERS AND RELATIVES

Widespread throughout the Americas, these birds of the family Tyrannidae account for a third of all passerines in many South American bird communities. These birds are insectivorous and typically perch and wait for prey or glean it from foliage.

6½–8½ in
17–21 cm

6 in
15 cm

gray head

9 in
22 cm

brown tail
with pale
cinnamon edge

**GREAT CRESTED
FLYCATCHER**
Myiarchus crinitus
A widespread tyrant flycatcher, this large, migratory species, like its relatives, catches insects on the wing. It often hovers when foraging.

EASTERN WOOD PEWEE
Contopus virens
This bird, with its characteristic "pewee" call, hunts by sallying—taking off from perches to catch insects in midair. It breeds in eastern North America.

**TROPICAL
KINGBIRD**
Tyrannus melancholicus
A large, sallying tyrant flycatcher, this aggressively territorial bird breeds in open habitats from southern North America to South America.

6 in
15 cm

dark brown
upperparts

red
underparts

4 in
10 cm

7½ in
19 cm

6½ in
17 cm

**VERMILION
FLYCATCHER**
Pyrocephalus rubinus
A species of open country, this bird forages near the ground. Males are vivid red, whereas females are mainly gray and white.

COMMON TODY-FLYCATCHER
Todirostrum cinereum
This tiny Central and South American tyrant flycatcher is typical of a group that forages by upward striking, but prefers more open habitat than its relatives.

CLIFF FLYCATCHER
Hirundinea ferruginea
This flycatcher from northern and central South America is like a swallow in its aerial foraging, and perches on rocky outcrops.

BLACK PHOEBE
Sayornis nigricans
This tail-wagging, tyrant flycatcher forages close to the ground in tropical regions, often near water. It dives into ponds to capture minnows.

TYPICAL ANTBIRDS

From tropical American forests, these heavy-billed birds of the family Thamnophilidae hunt insects near the ground, with some following army ants to feed on insects fleeing the ants. Some are long-clawed for gripping vertical stems.

7 in
18 cm

**WHITE-BEARDED
ANTSHRIKE**
Biatus nigropectus
This little-known, uncommon species is restricted to southeastern Brazil, where it feeds on insects in bamboo forests. It is threatened by deforestation.

ANTPITTAS AND ANT-THRUSHES

Short-tailed antpittas spend more time on the ground than the tree-dwelling ant-thrushes. Both are insectivores of South American forests and belong to the family Formicariidae.

7 in
18 cm

MOUSTACHED ANTPITTA
Grallaria alleni
A rare species found in isolated localities in Colombia and Ecuador, the moustached antpitta inhabits undergrowth in humid montane forest.

AUSTRALASIAN WRENS

The Maluridae are small, cock-tailed insect-eaters that resemble wrens of the Northern Hemisphere, but are more closely related to nectar-feeding honeyeaters. Male fairy-wrens are patterned in blue and black. Emu-wrens and grasswrens are browner birds of grassy habitats.

STRIATED GRASSWREN
Amytornis striatus
Like most other grasswrens, this central Australian species is found in spinifex grass, where family groups dart beneath bushes.

6–7 in
15–18 cm

TAPACULOS AND CRESCENT-CHESTS

Strong-legged and weak-flying, these South American birds, the Rhinocryptidae, are among the most ground-adapted of South American passerines. Some have long hind claws to scratch for food in soil and leaf litter.

5½–6 in
14–15 cm

**COLLARED
CRESCENT-CHEST**
Melanopareia torquata
Crescent-chests have longer tails than tapaculos and may belong to a separate family. This Brazilian species occurs in arid habitats.

GNATEATERS

Dumpy, short-tailed, long-legged insect-eaters of forest undergrowth, gnateaters of the family Conophagidae are secretive birds that forage near the ground using tyrantlike sallying or gleaning techniques. They are related to antbirds.

RUFOUS GNATEATER
Conopophaga lineata
More abundant than other gnateater species, this bird from eastern South America often moves in mixed-species flocks and uses degraded habitats.

5 in
13 cm

**VARIEGATED
FAIRY-WREN**
Malurus lamberti
This most widespread Australian fairy-wren, like others, builds a domed nest. Young may stay near by to help raise the next brood.

6 in
15 cm

**SPLENDID
FAIRY-WREN**
Malurus splendens
Fairy-wrens form strong pair bonds, but partners may mate with other individuals. It is found mostly in southern Australia.

5½ in
14 cm

white throat

bright yellow underparts

9 in
22 cm

GREAT KISKADEE
Pitangus sulphuratus
Widespread through tropical America and named for its call, this species is a typical sallying tyrant flycatcher, but also forages near the ground.

BOWERBIRDS AND CATBIRDS

The Australasian family Ptilonorhynchidae consists predominantly of fruit-eaters. Male bowerbirds—often brightly colored—build structures called bowers to attract females. They mate with many females and play no part in raising the young.

9 in
23 cm

GREEN CATBIRD
Ailuroedus crassirostris
Male catbirds—named for their catlike call—attract mates by laying leaves on the ground. This species is found in New Guinea and eastern Australia.

olive upperparts

9–10 in
23–25 cm

GOLDEN BOWERBIRD
Prionodura newtoniana
The male of this small species from northern Australia attracts a mate by building a tower of sticks up to 9¾ ft (3 m) high.

LYREBIRDS

The Menuridae are large Australian birds that eat ground insects. They have complex vocal organs and excel at mimicking forest sounds. Males court females on display mounds by fanning their long tails and plumes.

32–38 in
80–96 cm

SUPERB LYREBIRD
Menura novaehollandiae
The most common lyrebird, this species is found in forests of southeastern Australia and Tasmania. Its lyre-shaped outer tail feathers are adorned with notches.

AUSTRALASIAN TREECREEPERS

The birds of the family Climacteridae have evolved to resemble the northern treecreepers, but are unrelated and, unlike them, do not use their tail for support when climbing up trees.

6½–7 in
16–18 cm

BROWN TREECREEPER
Climacteris picumnus
This common eastern Australian species has a distinct dark-backed form in the north of the range and a brown-backed form in the south.

OVENBIRDS AND RELATIVES

These American birds, the Furnariidae, are adept at hunting hidden invertebrates and are known for the immense diversity of their nest architecture. This includes stick nests, tunnel nests, and those that resemble clay ovens.

7½–8 in
19–20 cm

WHITE-EYED FOLIAGE-GLEANER
Automolus leucophthalmus
A South American bird with distinctive white irises, like many insect-eaters, this species forages in mixed flocks to flush out prey.

7–8 in
18–20 cm

RUFOUS HORNERO
Furnarius rufus
Widespread in central and southern South America, this species, typical of the hornero group, builds an ovenlike mud nest.

HONEYEATERS

The birds of the family Meliphagidae are from Australia and the southwest Pacific islands. They feed on nectar with a long, brush-tipped tongue and are important pollinators of plants in that region. Other nectar-feeders, such as sunbirds, have evolved similar features.

olive-green wings

10–12 in
25–30 cm

4–4¼ in
10–11 cm

SCARLET HONEYEATER
Myzomela sanguinolenta
This bird is an eastern Australian member of a group of long-billed, flower-visiting honeyeaters that often carry daubs of pollen on their foreheads.

BLUE-FACED HONEYEATER
Entomyzon cyanotis
A large, noisy honeyeater from Australia and New Guinea, this species is more insectivorous than many others, but it also feeds on fruits.

7½–8½ in
19–21 cm

LEWIN'S HONEYEATER
Meliphaga lewinii
A member of a short-billed group of honeyeaters, this species from eastern Australia eats insects, fruits, and berries.

WOODCREEPERS

The tropical American birds of the family Dendrocolaptidae are specialists at climbing up tree trunks. They have stiffened tails for support and strongly clawed front toes for gripping the bark.

5–6½ in
13–16 cm

EASTERN SPINEBILL
Acanthorhynchus tenuirostris
Spinebills are an ancient group of heathland honeyeaters and the most specialized nectar-feeders of the family. This species is from eastern Australia.

11½–12½ in
29–32 cm

TUI
Prosthemadera novaeseelandiae
Although confined to New Zealand, this species is related to short-billed honeyeaters of Australia. It has an extraordinary vocal range.

6½–7½ in
16–19 cm

NEW HOLLAND HONEYEATER
Phylidonyris novaehollandiae
Like other honeyeaters, this white-whiskered species from southern Australia and Tasmania also takes honeydew—the waste sugar solution produced by certain sap-sucking insects.

7½ in
19 cm

SCALLOPED WOODCREEPER
Lepidocolaptes falcinellus
This typical buff and brown woodcreeper is found only in the forests of southeastern South America.

THORNBILLS AND RELATIVES

The Acanthizidae is a small family of warbler- and wrenlike, insect-eating birds from Australia and adjacent islands. It also includes Australia's smallest bird—the weebill. These birds have short wings and tails, and drab-colored, longish legs.

dark facial mask between white stripes

4¼–5½ in
11–14 cm

4¼ in
11 cm

BUFF-RUMPED THORNBILL
Acanthiza reguloides
Thornbills are mostly gray, brown, or yellow. This eastern species has a freckled forehead, like many of the group.

WHITE-BROWED SCRUBWREN
Sericornis frontalis
Scrubwrens are birds of Australasian thickets. Mostly brown, some have white head markings, such as this species, which is widespread across Australia and Tasmania.

PARDALOTES

Dumpy birds from Australia, the Pardalotidae have stubby bills for gleaning sap-sucking scale insects from trees. Brightly colored, they nest in deep tunnels in earth banks.

3¼–4 in
8–10 cm

SPOTTED PARDALOTE
Pardalotus punctatus
Three of the four species of pardalotes have white spots. This highly active bird is from dry forests in southern and eastern Australia.

WOODSWALLOWS

From Southeast Asia, New Guinea, and Australia, the members of the family Artamidae are unique among passerines for their powder down feathers that give them soft plumage. They catch insects in flight and are among the few small passerines that can soar.

MASKED WOODSWALLOW
Artamus personatus
This dark-faced, thick-billed woodswallow occurs in drier parts of inland Australia and is highly nomadic. Like other woodswallows, it often gathers in large flocks.

7½ in
19 cm

BUTCHERBIRDS AND RELATIVES

The Australasian family Cracticidae consists of highly vocal, intelligent, omnivorous passerines. It includes predatory butcherbirds, crow-like currawongs, and mostly ground-dwelling Australian magpies. Related to woodswallows, these birds build untidy nests of twigs.

AUSTRALIAN MAGPIE
Gymnorhina tibicen
A widespread Australian species with a highly variable black and white plumage, this bird has a varied and melodious song and is a capable mimic.

13½–17½ in
34–44 cm

IORAS

These rainforest birds, the Aegithinidae, are usually active in the high canopy. Their green or yellow coloration camouflages them in foliage, where they glean for insects. Males can engage in elaborate courtship displays.

6 in
15 cm

COMMON IORA
Aegithina tiphia
The smallest and most widespread iora, it is found across tropical Asia, from India to Borneo, sometimes in disturbed habitats. This bird builds a cup-shaped nest.

CROWS AND JAYS

The family Corvidae, occurring worldwide, includes some of the biggest passerines. They are intelligent, opportunistic birds that have complex social organization and strong pair-bonding. Crows have demonstrated tool use, play behavior, and perhaps even self-awareness.

13–15½ in
33–39 cm

EURASIAN JACKDAW
Corvus monedula
A small crow of western Eurasia and North Africa, this bird nests in cavities on crags, and is found on coastal cliffs and in urban areas.

22–27 in
56–69 cm

COMMON RAVEN
Corvus corax
Found in a wide range of open-country habitats across the Northern Hemisphere, this is the most widespread crow and the largest passerine.

PIED CROW
Corvus albus
A relative of the raven, this heavy-billed bird of open country may be the most common member of the crow family in Africa and Madagascar.

18–20 in
46–50 cm

very long tail

BLUE JAY
Cyanocitta cristata
This colorful jay of North America lives in tightly bonded family groups. It is fond of acorns and disperses them, which helps distribute oak trees.

10–12 in
25–30 cm

COMMON MAGPIE
Pica pica
This Eurasian bird is common in habitats ranging from open woodland to semidesert. It is related more closely to *Corvus* crows than Asian magpies.

18 in
46 cm

18–19 in
45–48 cm

ROOK
Corvus frugilegus
Routinely flocking throughout the year in Eurasia, this bare-faced crow nests in colonies in trees in open countryside.

18½–20½ in
47–52 cm

CARRION CROW
Corvus corone
This common Eurasian, largely solitary species has a broad diet. It feeds on small animals and vegetable matter, as well as carrion.

DRONGOS

Black, long-tailed birds of Old World tropics, the Dicruridae sally for insects, bolting out suddenly to catch their prey. They are aggressive birds, and sometimes attack larger species to defend their nests.

10 in
26 cm

CRESTED DRONGO
Dicrurus forficatus
Like other drongos, this Madagascan species has a long, forked tail and red eyes. It has a distinctive tuft of feathers at the base of its bill.

SHRIKES

The family Laniidae contains predators of open country. Many species store their prey (insects and small vertebrates) by impaling them on thorns. Most occur in Africa and Eurasia; two species live in North America.

6½–7 in
17–18 cm

RED-BACKED SHRIKE
Lanius collurio
This bird breeds from Europe to Siberia and overwinters in Africa. Like other *Lanius* shrikes, it has a musical call.

BUSH-SHRIKES AND RELATIVES

The family Malaconotidae is entirely confined to Africa. They are found mostly in scrubby, open woodland, and have hooked bills for catching large insects.

8 in
20 cm

WHITE HELMET-SHRIKE
Prionops plumatus
Widespread across sub-Saharan Africa, the white helmet-shrike often gathers in small groups. This bird has a wide range of different calls.

9 in
23 cm

CRIMSON-BREASTED GONOLEK
Laniarius atrococcineus
Members of the African bush-shrike family, gonoleks have red and black plumage. This species occurs in southern Africa.

---

orange-red bill

26 in
67 cm

RED-BILLED BLUE MAGPIE
Urocissa erythrorhyncha
A forest bird from the Himalayas to eastern Asia, this species robs chicks from nests and plucks at carcasses.

GREEN JAY
Cyanocorax yncas
This bird feeds on fruits and seeds. South American populations of the green jay differ sufficiently from those of Central America to be considered as separate species.

11½ in
29 cm

AZURE-WINGED MAGPIE
Cyanopica cyanus
A social species, this bird breeds in colonies. Two separate populations (in Portugal and in eastern Asia) of this gregarious woodland bird may constitute separate species.

12–14 in
31–35 cm

13½ in
34 cm

EURASIAN JAY
Garrulus glandarius
More closely related to Old World crows than American jays, this colorful woodland bird habitually hoards acorns in autumn.

ORIOLES

Old World passerines related to shrikes and crows, the birds of the family Oriolidae live in forest canopies and eat insects and fruits. Many species have striking yellow and black plumage. The females are usually greener than males.

9½ in
24 cm

10½–11½ in
27–29 cm

EURASIAN GOLDEN ORIOLE
Oriolus oriolus
Breeding in woodlands in western and central Eurasia, this oriole migrates southwards to overwinter in Africa.

AUSTRALASIAN FIGBIRD
Sphecotheres vieilloti
Stout-billed relatives of orioles from Australasia, figbirds are gregarious fruit-eaters. This species occurs in northern and eastern Australia.

VANGAS AND RELATIVES

Related to helmet-shrikes of Africa, the members of the family Vangidae are predatory passerines from Madagascar. They feed on invertebrates, reptiles, and frogs, and have a range of bill shapes—chisel-, sickle-, and dagger-shaped—for different prey and feeding techniques.

8 in
20 cm

RUFOUS VANGA
Schetba rufa
Common in the forests of Madagascar, this bird resembles the shrike, but is not closely related to the shrike family.

WATTLE-EYES AND RELATIVES

The Platysteiridae, wattle-eyes and their relatives, are a family of insect-eating birds from Africa. They have flat, hooked bills with bristles at the base. Like flycatchers, they snatch their prey suddenly.

BROWN-THROATED WATTLE-EYE
Platysteira cyanea
Wattle-eyes get their name from the red skin around their eyes. This common species occurs in woodland throughout sub-Saharan Africa.

5 in
13 cm

VIREOS

Although superficially resembling American warblers but with somewhat thicker bills, the members of the family Vireonidae are more closely related to crows and to Old World orioles and shrikes. They take insects by gleaning or fly catching, and also eat some fruits.

BLACK-CAPPED VIREO
Vireo atricapilla
This species breeds in N. America and migrates to Mexico. Unlike other vireos, the sexes differ. The male is black-capped, while the female is gray-capped.

4¼ in
11 cm

4¾–5 in
12–13 cm

RED-EYED VIREO
Vireo olivaceus
North American populations of this very vocal vireo migrate to South America, where they join resident races of the same species.

TYPICAL TITS

These small, acrobatic, usually hole-nesting birds of the family Paridae occur in wooded habitats across North America, Eurasia, and Africa. They frequently hang upside-down to glean insects from foliage, and manipulate seeds and nuts to crack them open.

5½–6½ in
14–16 cm

4¾–5½ in
12–14 cm

VARIED TIT
Parus varius
This species is found in forest habitats, including conifers and bamboo forests, in northeastern Asia, Japan, and Taiwan.

4¾–6 in
12–15 cm

BLACK-CAPPED CHICKADEE
Parus atricapillus
A typically inquisitive and acrobatic tit, this is a common North American bird. Like other tits, it hoards seeds for later use.

TUFTED TITMOUSE
Parus bicolor
Like other tits, this eastern North American species supplements its insect diet with seeds, holding them firmly to smash them with its bill.

5½ in
14 cm

GREAT TIT
Parus major
Widespread across Eurasia, this tit occurs in habitats from forest to heathland, and has a diverse range of vocalizations.

4½–4¾ in
11–12 cm

BLUE TIT
Parus caeruleus
This common bird of broadleaf woodlands in Europe, Turkey, and North Africa is a frequent visitor to bird feeders in gardens.

PENDULINE TITS

These small, needle-billed birds from the family Remizidae occur in Africa and Eurasia, with one species in America. Most use cobwebs and other soft material to build flask-shaped nests that hang from branches, often over water.

4¼ in
11 cm

3½–4¼ in
9–11 cm

PENDULINE TIT
Remiza pendulinus
The only member of the family to have a wide range across Eurasia, this species occurs on marshes with trees, where it builds its pendulous nest.

VERDIN
Auriparus flaviceps
Unlike most other penduline tits, the verdin makes a spherical nest. It is found in desert scrub of the southern USA and Mexico.

BIRDS-OF-PARADISE

Mainly found in the rain forests of New Guinea, members of the family Paradisaeidae are mostly fruit-eating birds. The males display their bright, gaudy plumes in elaborate courtship displays, and spend most of their energies on this mating ritual, leaving the females to rear the young alone.

12½ in
32 cm

LESSER BIRD-OF-PARADISE
Paradisaea minor
This species occurs across northern and western New Guinea. Males use their long, yellow flank plumes and distinctive cape for courtship displays.

yellow flank plumes

AUSTRALIAN ROBINS

The Petroicidae are chunky, round-headed insect-eaters. Unrelated to the robins of Europe or America, they are found from Australasia to the islands of the southwest Pacific. Some exhibit co-operative breeding, in which young birds help their parents to raise a new brood.

5 in
13 cm

6 in
15 cm

JACKY WINTER
Microeca fascinans
This common robin uses its broad bill for catching flies. It is widespread in woodland throughout Australia and New Guinea.

EASTERN YELLOW ROBIN
Eopsaltria australis
Common in the woodlands and gardens of eastern Australia, this bird sallies from low perches for invertebrates, which it snatches from the ground.

LONG-TAILED TITS

Birds of the family Aegithalidae are small, restless insect-eaters that build dome-shaped nests woven with cobwebs and lined with feathers. Most species are found in Eurasia, but one, the bushtit, occurs in North America.

5½ in
14 cm

LONG-TAILED TIT
Aegithalos caudatus
The most widespread of the family, this woodland bird inhabits a range from northern to central Eurasia, gathering in restless flocks when not breeding.

WAXWINGS

From the family Bombycillidae, these berry-eating birds are named for the waxy, red-tipped shafts on their wings. Three species occur in the cool northern forests of North America and Eurasia.

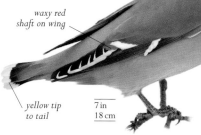

waxy red shaft on wing

yellow tip to tail

7 in
18 cm

BOHEMIAN WAXWING
Bombycilla garrulus
This sleek bird is pinkish brown, with chestnut undertail feathers. It breeds in northern taiga forest and is attracted to berried shrubs during its migration south.

SILKY FLYCATCHERS

The family Ptilogonatidae from Central America has just four species. These birds are named for their soft plumage (similar to that of related waxwings) and their feeding habits.

PHAINOPEPLA
Phainopepla nitens
This bird of the southern USA and Mexico nests colonially in woodland, but is territorial when breeding in deserts.

7–8½ in
18–21 cm

MONARCHS AND RELATIVES

Generally long-tailed birds, members of the Monarchidae have broad bills for flycatching. Most species occur in Old World tropical forests. With the exception of the magpie-lark, they are tree-dwellers that build cup-shaped nests decorated with lichen.

black head

10–12 in
26–30 cm

rufous upperparts

AFRICAN PARADISE-FLYCATCHER
Terpsiphone viridis
This bird has variable color forms, but all males have long tail streamers. This species occurs in savanna, south of the Sahara.

long tail feather

6½–15 in
17–38 cm

MAGPIE-LARK
Grallina cyanoleuca
Unlike other members of the monarch family, this Australian bird spends much time on the ground and builds a large mud nest.

AUSTRALIAN MUDNESTERS

The family Corcoracidae contains two species of social birds that feed on the ground and build large, cup-shaped nests of grass. These are held together with mud and are made in trees on horizontal branches.

APOSTLEBIRD
Struthidea cinerea
This ground-living bird associates in flocks of 6 to 20 individuals. It occurs in woodlands of northern and eastern Australia.

11½–12½ in
29–32 cm

LARKS

Brown in color, with a melodious call, these birds from the family Alaudidae inhabit arid, open habitats. Most occur in Africa, with one species in North America. Most have long hind claws that provide the stability they need to spend so much time on the ground.

7–8 in
18–20 cm

GREATER HOOPOE-LARK
Alaemon alaudipes
This long-legged lark has a curved bill and occurs in arid habitats of North Africa and the Middle East, where it habitually runs on the ground.

HORNED LARK
Eremophila alpestris
This lark breeds on Arctic North American and Eurasian tundra. It overwinters on coastlines farther south.

5½–6½ in
14–17 cm

7–7½ in
18–19 cm

EURASIAN SKYLARK
Alauda arvensis
Common across Eurasia, from the British Isles to Japan, this open-country bird is notable for its musical aerial song.

BULBULS

Found across the warmer parts of Eurasia and Africa, most bulbuls of the family Pycnonotidae are gregarious, noisy, fruit-eating birds. The soft plumage of many species is drably colored, marked by red or yellow feathers beneath the tail.

red cheek patch

9–10 in
23–25 cm

BLACK BULBUL
Hypsipetes leucocephalus
Common in forests and gardens in India, China, and Thailand, this species exists in dark-headed and white-headed races.

8 in
20 cm

RED-WHISKERED BULBUL
Pycnonotus jocosus
A common Asian bulbul found from India to the Malay Peninsula, this is an opportunistic woodland species that is also found near villages.

SWALLOWS AND MARTINS

From the family Hirundinidae, these swiftlike birds have long wings and forked tails. Their short, flattened bills and wide gapes help them take insects on the wing. They build mud nests or use tree-holes or tunnels in banks.

8 in
20 cm

GREATER STRIPED SWALLOW
Cecropis cucullata
This African grassland swallow breeds toward the south of the continent and migrates northward to overwinter.

4¾–5½ in
12–14 cm

BANK SWALLOW
Riparia riparia
Like other swallows, this species migrates south to overwinter in the tropics. It nests in colonies in riverside banks across the Northern Hemisphere.

4¾–6 in
12–15 cm

TREE SWALLOW
Tachycineta bicolor
This North American swallow of wooded swamps supplements its insect diet with berries. This enables it to breed farther north than other swallows.

6–7½ in
15–19 cm

BARN SWALLOW
Hirundo rustica
Occurring worldwide, this is the most widespread species of swallow. It was originally a cave-nester, but now uses buildings as well.

BABBLERS AND RELATIVES

Generally more gregarious, noisier, and less migratory than warblers, babblers, from the family Timaliidae, are widespread through warmer parts of the Old World, and have evolved into a variety of thrushlike or warblerlike forms. Some species are brightly colored.

4¾ in
12 cm

VINOUS-THROATED PARROTBILL
Paradoxornis webbianus
Despite its stumpy seed-cracking bill, this long-tailed Asian parrotbill may be related to the insectivorous Old World warblers. This species occurs in China.

5½ in
14 cm

9 in
23 cm

RED-TAILED MINLA
Minla ignotincta
This small babbler is similar to a tit. It is a noisy inhabitant of montane forest canopies in Nepal, China, and Myanmar.

6 in
15 cm

WRENTIT
Chamaea fasciata
A dull-colored bird with a cocked tail, the wrentit is the only American member of the babbler family, which may be related to parrotbills.

13 in
33 cm

floppy tail

WHITE-EARED SIBIA
Heterophasia auricularis
Sibias are nectar-feeding babblers. This species is confined to Taiwan, where its distinctive call is often heard in mountain forests.

6½ in
16–17 cm

5 in
13 cm

rufous collar

GREATER NECKLACED LAUGHING-THRUSH
Garrulax pectoralis
Laughing thrushes are large forest babblers with laughing calls. They often move in mixed flocks. This species inhabits the Himalayas and southeast Asia.

CHESTNUT-CAPPED BABBLER
Timalia pileata
A bird of low thickets in southeastern Asia, this babbler is often found near water, along with flycatchers and other babblers.

WHITE-NAPED YUHINA
Yuhina bakeri
Like the closely related white-eyes, crested yuhinas are adapted to take nectar. The white-naped yuhina is an eastern Himalayan species.

GNATCATCHERS

These small American insect-eaters from the family Polioptilidae are related to wrens, but are more warblerlike in appearance. Some gnatcatchers cock their tails when foraging, a behavior shared with several members of the wren family.

4¾ in
12 cm

BLUE-GRAY GNATCATCHER
Polioptila caerulea
This North American gnatcatcher may flick its white-edged tail to flush insects. Unlike related gnatcatcher species, males do not have dark head markings.

OLD WORLD WARBLERS AND RELATIVES

The family Sylviidae includes a number of thin-billed, insectivorous birds. Many Old World warblers lack bright colors and are distinguished by their call rather than their appearance. Recent DNA analysis shows that African, reed, and "grassbird" warblers belong to separate families.

5½ in
14 cm

7½–9 in
19–23 cm

4¾–5 in
12–13 cm

SUBALPINE WARBLER
Sylvia cantillans
Like many *Sylvia* warblers, this species breeds in scrubby Mediterranean habitats and winters in Africa.

BLACKCAP
Sylvia atricapilla
Male *Sylvia* warblers are typically patterned with patches of black or brown. In this widespread Eurasian species, females have a brown cap.

CAPE GRASSBIRD
Sphenoeacus afer
A bird of South African shrubland, it belongs to an ancient African group that evolved separately from other Old World warblers.

TREECREEPERS

From the family Certhiidae, these are small insect-eating birds of the Northern Hemisphere. They forage on vertical tree trunks using their tail as a prop, habitually climbing one tree and flying to the bottom of the next.

EURASIAN TREECREEPER
Certhia familiaris
This is the most widespread *Certhia* treecreeper. It is found across Eurasia, from Britain to Japan, in broadleaf and coniferous forests.

5 in
13 cm

ICTERINE WARBLER
Hippolais icterina
This Eurasian woodland species has a more musical call than related reed warblers. It migrates to southern Africa in winter.

5–6 in
13–15 cm

lemon yellow underparts

5 in
13 cm

7–9½ in
18–24 cm

SEDGE WARBLER
Acrocephalus schoenobaenus
This is one of many species of reed warblers (inhabitants of wetland areas) that breeds in Eurasia and winters in Africa.

BROWN SONGLARK
Cincloramphus cruralis
This "grassbird" warbler from Australia is a nomad of open habitats. Like larks, it sallies skyward from exposed perches.

dark red
wing patch

silver-gray
ear patch

7 in
18 cm

**SILVER-EARED
MESIA**
Leiothrix argentauris
A skulking, montane forest
bird, this southeast Asian
species belongs to a group
of "song babblers," which
includes minlas, sibias,
and laughing thrushes.

BEARDED TIT
Panurus biarmicus
This Eurasian reed-bed specialist,
possibly related to larks, eats insects
in summer and hardens its stomach
in winter to digest reed seeds.

6½ in
16–17 cm

GOLDCRESTS

Among the smallest of passerines, these color-crested birds of
the family Regulidae are found in cool northern forests. They
have a high metabolic rate, which forces them to feed constantly
when awake. They glean tiny, soft-bodied invertebrates
from foliage with their needlelike bills.

3½ in
9 cm

GOLDCREST
Regulus regulus
All kinglets are adapted to
coniferous forests. The Eurasian
goldcrest has especially grooved
feet and enlarged toe-pads for
clinging to needlelike leaves.

4½ in
11 cm

RUBY-CROWNED KINGLET
Regulus calendula
The red head patch of this North American
species is visible when it raises its crown
feathers—a trait shared by all kinglets.

NUTHATCHES

From the family Sittidae, nuthatches
are more acrobatic than the related
treecreepers, being able to move
down trunks as well as up. They
feed on insects and seeds,
sometimes hoarding them
in crevices for harsher times.

WALLCREEPER
Tichodroma muraria
This central
Eurasian mountain
bird gleans insects
from rocks with
its pointed bill.

6½ in
16–17 cm

5½ in
14 cm

4¼ in
11 cm

black stripe
across eye

**EURASIAN
NUTHATCH**
Sitta europaea
A widespread species, this
forest bird can—like other
nuthatches—crack open
nuts by wedging them
in cracks in tree bark.

**RED-BREASTED
NUTHATCH**
Sitta canadensis
This North American
species is similar in body
pattern to the Eurasian
species, but males are
more vividly colored.

orange-brown
underparts

WHITE-EYES

Most of the Zosteropidae are characterized by
a ring of white feathers around the eye. The
birds of this uniform family, closely related
to babblers, have brush-tipped tongues
and are nectar-feeding specialists.

BROAD-RINGED WHITE-EYE
Zosterops poliogastrus
This open woodland white-eye is
restricted to isolated mountain ranges in
Ethiopia, Kenya, and Tanzania. Distinct
races are found in different regions.

4¼ in
11 cm

FAIRY-BLUEBIRDS

Two species of fairy-bluebirds, from the family Irenidae,
are found in Southeast Asia, where they feed on
fruits—especially figs—in forest canopies.
Only males have a vivid blue color;
females are dull green.

bright blue
upper parts

**ASIAN
FAIRY-BLUEBIRD**
Irena puella
The most widespread fairy-
bluebird, found from India to
Indonesia, this species often feeds
with other fruit-eaters, such as
hornbills and pigeons.

10 in
25 cm

WRENS

With the exception of the winter
wren, the family Troglodytidae is
confined to the Americas. Most
species are highly vocal but visually
inconspicuous, short-winged birds
that forage on insects in undergrowth;
some even sleep on the ground.

4 in
10 cm

WINTER WREN
Troglodytes troglodytes
The only wren whose range
includes Eurasia, this bird is
found across the Northern
Hemisphere. Over 40 local
races have been described.

CACTUS WREN
Campylorhynchus brunneicapillus
The largest wren, this species
occurs in deserts of California and
Mexico, where it forages in
flocks on the ground.

5½ in
14 cm

7–9 in
18–23 cm

BEWICK'S WREN
Thryomanes bewickii
This long-tailed wren inhabits
dry, open, wooded habitats in
California and Mexico. It has a
variety of songs in its repertoire.

MOCKINGBIRDS AND RELATIVES

Birds of the family Mimidae occur throughout much of the Americas, and the Caribbean and Galapagos Islands. Usually gray or brown, these strong-legged birds are highly vocal; some are accomplished mimics.

**10½ in
27 cm**

**8½–9½ in
21–24 cm**

GRAY CATBIRD
Dumetella carolinensis
Named for its catlike mewing call, this N. American bird forages on the ground. It overwinters in Central America and the Caribbean.

CURVE-BILLED THRASHER
Toxostoma curvirostre
A bird of arid, scrubby habitats in the southern USA and Mexico, this species uses its long bill to probe the soil for invertebrates.

pale gray upperparts

long tail

**8½–10 in
21–26 cm**

NORTHERN MOCKINGBIRD
Mimus polyglottos
This North American bird is famous for its extraordinary repertoire of varied songs, which it delivers day and night.

STARLINGS AND MYNAS

From the family Sturnidae, these are mostly gregarious, noisy birds, many with a glossy plumage that has a metallic sheen. This family is divided into mynas and relatives from southern Asia and the Pacific, and true starlings from Africa and Eurasia.

BALI MYNA
Leucopsar rothschildi
Found only in the rain forests of Bali, Indonesia, this striking bird is endangered because of habitat destruction and the bird trade.

**10 in
25 cm**

**7½–9 in
19–22 cm**

YELLOW-BILLED OXPECKER
Buphagus africanus
Common in savanna across sub-Saharan Africa, this bird habitually perches on large mammals, eating parasites but also pecking at wounds.

EMERALD STARLING
Lamprotornis iris
This glossy West African species feeds mainly on fruits, especially figs, but also takes ants.

HILL MYNA
Gracula religiosa
Found in forests of tropical Asia, this species is a popular cage bird. It is notable for its melodious call and its mimicking ability.

**10½–12 in
27–31 cm**

**20 in
50 cm**

WHITE-NECKED MYNA
Streptocitta albicollis
This long-tailed, magpielike myna is restricted to rain forests in Sulawesi, Indonesia, and adjacent islands, where it usually associates in pairs.

**7–7½ in
18–19 cm**

OLD WORLD FLYCATCHERS AND CHATS

Related to thrushes, the family Muscicapidae is divided into two groups: true flycatchers, with broad bills for snatching flying insects; and chats, which include robins, nightingales, and wheatears. Some species are brightly colored, but most of these small birds have predominantly gray or brown plumage.

**7 in
18 cm**

**5 in
13 cm**

rufous base of tail

**5½ in
14 cm**

EUROPEAN ROBIN
Erithacus rubecula
Related to chats, this bird of hedgerows and woodland is found in western Eurasia and North Africa. In Britain, it is a common garden visitor.

**6–6½ in
15–16 cm**

NORTHERN WHEATEAR
Oenanthe oenanthe
Wheatears are white-rumped chats of open country. This is the most widespread species across Eurasia. It winters in Africa.

BLUE-AND-WHITE FLYCATCHER
Cyanoptila cyanomelaena
Belonging to a large group of vivid blue flycatchers from tropical Asia, this species from eastern Asia forages high in forests.

COMMON STONECHAT
Saxicola torquatus
Typical of chats, this is a small, upright, perching insectivore with a harsh call. It is common in grassland in Eurasia and Africa.

**7½–8½ in
19–21 cm**

MOCKING CLIFFCHAT
Thamnolaea cinnamomeiventris
Belonging to a group of dark-colored African chats, this bird occurs in bushy, rocky habitats and can become tame near villages.

**5½ in
14 cm**

COMMON REDSTART
Phoenicurus phoenicurus
Named for their rufous tails, redstarts are chatlike birds found mainly in Asia. This species occurs from western to central Eurasia and migrates to East Africa.

HILDEBRANDT'S STARLING
Lamprotornis hildebrandti
This East African glossy starling occurs in wooded savanna, where it preys on large ground insects, often flocking with other starling species.

7 in
18 cm

bronze neck patch

glossy blue body

12 in
30 cm

brown wings

9 in
22 cm

shiny black plumage with white spots

SPLENDID GLOSSY STARLING
Lamprotornis splendidus
Widespread in sub-Saharan Africa, this species belongs to a group of African starlings characterized by a metallic sheen on their plumage.

EUROPEAN STARLING
Sturnus vulgaris
Native to Eurasia but introduced to North America, this common starling roosts communally, in groups that gather after mass aerial maneuvers.

red spot on blue throat

WHITE-TAILED ROBIN
Myiomela leucura
A bird of riverine forests from the Himalayas to Indochina, it usually keeps close to the ground unless disturbed.

5½ in
14 cm

18 cm
7 in

BLUETHROAT
Luscinia svecica
One of the 10 races of bluethroats breeds in the far north, from Sweden to Alaska, and migrates to Africa, the Middle East, and southeast Asia.

6½ in
17 cm

COMMON NIGHTINGALE
Luscinia megarhynchos
This brown bird of western to central Eurasian thickets is known for its loud, rich song, which it sings night or day.

5 in
13 cm

PIED FLYCATCHER
Ficedula hypoleuca
This large genus of mainly Asian flycatchers is closely related to chats. This woodland species ranges from Europe to Siberia.

THRUSHES AND RELATIVES

Belonging to the family Turdidae, most thrushes are woodland birds that forage on the ground for invertebrate prey, such as earthworms, snails, and insects. They are found worldwide, but the majority of species occur in the Old World. Many have melodious songs.

5–5½ in
13–14 cm

WHITE-BROWED SHORTWING
Brachypteryx montana
Perhaps more closely related to flycatchers than thrushes, shortwings are running birds found in Asian forests. This species occurs from the Himalayas to Java, Indonesia.

9 in
22 cm

ORANGE-HEADED THRUSH
Zoothera citrina
One of many tropical Old World *Zoothera* thrushes, this species is found in forests from the Himalayas to Bali, Indonesia.

EASTERN BLUEBIRD
Sialia sialis
A bird of eastern North America, this is a common species of open woodland and fields. It sometimes nests in old woodpecker holes.

6½–8½ in
16–21 cm

black band on orange breast

faint spots on breast

VARIED THRUSH
Ixoreus naevius
Found in mature coniferous forests of western North America, this bird overwinters in parks and gardens. Like other thrushes, it forages in ground litter.

7½–10 in
19–26 cm

8–9 in
20–23 cm

SONG THRUSH
Turdus philomelos
Found from Europe to Siberia, this woodland and garden bird habitually uses a hard surface as an "anvil" to smash open snail shells.

8–11 in
20–28 cm

AMERICAN ROBIN
Turdus migratorius
Unrelated to the European robin, this North American thrush sometimes gathers in massive winter roosts thousands of birds.

EURASIAN BLACKBIRD
Turdus merula
A common long-tailed thrush of woodland habitats ranging from Europe and North Africa to India, the highly territorial blackbird often visits gardens.

9½–11½ in
24–29 cm

9–10½ in
22–27 cm

FIELDFARE
Turdus pilaris
This bird breeds in northern Eurasia and winters farther south, where it flocks together in fields.

LEAFBIRDS

The fruit-eating birds of the family Chloropseidae occur in forests in southeastern Asia. They use their brush-tipped tongues to take nectar, with which they supplement their diet. Male leafbirds are characteristically green, with blue or black throats.

8 in
20 cm

ORANGE-BELLIED LEAFBIRD
Chloropsis hardwickei
This melodious bird inhabits the forest canopy at high altitudes, in a range from the Himalayas to the Malay Peninsula.

WHYDAHS AND RELATIVES

African whydah birds and related indigobirds of the family Viduidae are cuckoolike brood parasites of waxbills. The mouthparts and begging behavior of the chicks usually resemble those of the host chicks, thus fooling the host parents.

4¾–15 in
12–38 cm

EASTERN PARADISE WHYDAH
Vidua paradisaea
Typical of whydahs, breeding males of this East African species have extraordinarily long tail feathers, which they use in display flights.

FLOWERPECKERS

These dumpy birds of the family Dicaeidae are related to the sunbirds from tropical Asia and Australasia. Largely fruit-eating, like sunbirds, they take nectar from flowers. The flowerpeckers differ in having shorter bills.

MISTLETOEBIRD
Dicaeum hirundinaceum
This Australian flowerpecker has a small gut to process mistletoe berries rapidly, and plays an important role in dispersing seeds of this parasitic plant.

4–4¼ in
10–11 cm

WAXBILLS AND RELATIVES

Consisting of small, highly gregarious, often brightly colored seed-eaters, the family Estrildidae occurs in tropical Africa, Asia, and Australia. Many birds live in grassland or open woodland and build domelike nests. Both parents share parental duties.

5½ in
14 cm

PURPLE GRENADIER
Uraeginthus ianthinogaster
An East African bird of dry woodland, the purple grenadier belongs to a genus of predominantly blue waxbills.

4 in
10 cm

GREEN-BACK TWINSPOT
Mandingoa nitidula
Found in western to southern Africa, this thicket-dwelling twinspot has white-spotted underparts and is more secretive than other waxbills.

COMMON WAXBILL
Estrilda astrild
Abundant throughout Africa, this tiny waxbill, like related species, is a restless, flocking bird of open land, feeding on grass seeds.

blackish bill

4¾ in
12 cm

black and white "scaly" belly

SCALY-BREASTED MUNIA
Lonchura punctulata
This bird is common in southern Asian scrub. Male and female are alike in appearance.

4 in
10 cm

ZEBRA FINCH
Taeniopygia guttata
Native to the drier parts of Australia, the zebra finch is a popular cagebird around the world.

JAVA SPARROW
Lonchura oryzivora
This vulnerable species from Java and Bali frequents farmland and feeds on cereal crops. It has been hunted as a rice pest and for the pet trade.

6½ in
16 cm

4¾ in
12 cm

♀

♂

GREEN-WINGED PYTILIA
Pytilia melba
Male *Pytilia* waxbills have red splashes on their wings. This African species is host to the paradise whydah (*Vidua paradisaea*), a brood parasite.

4¾ in
12 cm

RED-THROATED PARROTFINCH
Erythrura psittacea
Parrotfinches are waxbills with mostly green bodies from southeastern Asia to the Pacific Ocean. This species occurs in grasslands of the island of New Caledonia.

OLD WORLD SPARROWS

These stubby-billed, seed-eating birds, family Passeridae, are found across Africa and Eurasia. In addition to the true sparrows, the group includes snowfinches, mountain-dwelling birds that occur from the Pyrenees to Tibet.

HOUSE SPARROW
Passer domesticus
Originally native to Eurasia and North Africa, the bird has adapted to human settlements around the world.

6 in
15 cm

DIPPERS

Belonging to the family Cinclidae, dippers are the only passerines that can dive and swim underwater. They have adaptations for their aquatic lifestyle, such as well-oiled, waterproof feathers and oxygen-storing blood.

7 in
18 cm

WHITE-THROATED DIPPER
Cinclus cinclus
Widespread through temperate Eurasia, the white-throated dipper breeds near fast-flowing streams but may move to slower-flowing rivers in winter.

CUT-THROAT
Amadina fasciata
Named for the male's red neck patch, this is a common bird of dry African woodlands, and is often found near human habitations.

multicolored body

purple chest

5½ in
14 cm

GOULDIAN FINCH
Erythrura gouldiae
A brilliantly colored relative of parrotfinches, this endangered bird of northern Australia is nomadic. Males can have red or black faces.

PIPITS AND WAGTAILS

Present on every continent, the birds of the family Motacillidae inhabit open country, where they feed on insects. Most wagtails have longer tails and are more brightly colored than more somber pipits, and some are associated with water.

6 in
15 cm

RED-THROATED PIPIT
Anthus cervinus
This pipit breeds in Arctic tundra, during which it develops a colored throat: rufous in males and pink in females.

5½–6½ in
14–17 cm

BUFF-BELLIED PIPIT
Anthus rubescens
A typical ground-running pipit, this species breeds in Arctic tundra and winters in fields and on coasts further south.

olive back

6 in
15 cm

GOLDEN PIPIT
Tmetothylacus tenellus
A bird of open bush and grassland, the golden pipit is confined to East Africa, from Sudan to Tanzania.

6½ in
16–17 cm

YELLOW WAGTAIL
Motacilla flava
This widespread Eurasian species overwinters in Africa, India, and Australia. There are many races, including many with gray or black head patterns.

yellow underparts

6½–8 in
17–20 cm

WHITE WAGTAIL
Motacilla alba
A typical wagtail, this species is widespread throughout Eurasia— often on farmland or in towns.

SUNBIRDS

The small, fast-moving, nectar-feeding Old World tropical birds of the family Nectarinidae are similar to American hummingbirds, with their long, decurved bills and long tongues. Males are usually brightly colored with a metallic sheen. These birds are fiercely territorial.

long, decurved bill

4 in
10 cm

6 in
15 cm

brilliant scarlet breast

7 in
18 cm

STREAKY-BREASTED SPIDERHUNTER
Arachnothera affinis
Spiderhunters are drab, long-billed members of the sunbird family. Like other sunbirds, this Southeast Asian species eats invertebrates as well as nectar.

SCARLET-CHESTED SUNBIRD
Chalcomitra senegalensis
Common throughout much of sub-Saharan Africa, this large sunbird occupies a variety of wooded habitats.

PURPLE SUNBIRD
Cinnyris asiaticus
Like other sunbirds, this species from southern Asia feeds its young mostly on insects. The male loses its brilliant plumage after breeding.

WEAVERS

Gregarious seed-eaters, the Ploceidae build elaborate nests. Males usually take sole responsibility for this duty; females use the nest as the basis for choosing mates. Most species are African, with a few found in southern Asia.

6 in
15 cm

CHESTNUT WEAVER
Ploceus rubiginosus
The genus *Ploceus* includes the largest number of species of weavers. This one occurs in East Africa.

4¼–5 in
11–13 cm

RED-BILLED QUELEA
Quelea quelea
Widely considered the world's most abundant bird, this African species gathers in giant flocks that can cause serious damage to crops.

6–16 in
15–40 cm

RED-COLLARED WIDOWBIRD
Euplectes ardens
Breeding males of widowbirds are black and some fan enlarged tails during display flights. This species is widespread in sub-Saharan Africa.

4–4¼ in
10–11 cm

YELLOW-CROWNED BISHOP
Euplectes afer
This African bird is related to widowbirds. Breeding males are vibrantly patterned; females and nonbreeding males are red or black.

ACCENTORS

Mostly ground-dwelling passerines, the Prunellidae have narrow bills and are found in Eurasia. Most species are adapted to high altitudes, but they move to lower altitudes in winter to supplement their insectivorous diet with seeds.

DUNNOCK
Prunella modularis
Unlike other accentors, this is a lowland bird, and it does not usually gather in flocks. It is widespread in temperate Eurasia.

6 in
15 cm

4¾ in
12 cm

FINCHES AND RELATIVES

Finches of the family Fringillidae have diversified in Eurasia, Africa, and the tropical regions of the Americas. From thin-billed nectar-feeders to heavy-billed, seed-cracking grosbeaks and hawfinches, these birds have evolved to cope with a wide range of foods.

4¾ in
12 cm

4¾–5 in
12–13 cm

6 in
15 cm

6½ in
17 cm

4¾ in
12 cm

6–6½ in
15–17 cm

large, white
wing patch

EURASIAN GOLDFINCH
Carduelis carduelis
Goldfinches have thinly pointed bills for taking seeds from the seedheads of tall plants, such as thistles. This species is widespread in Eurasia.

AMERICAN GOLDFINCH
Carduelis tristis
Bright yellow goldfinches and siskins have diversified in the Americas, especially in South America. This migratory species is from North America.

CHAFFINCH
Fringilla coelebs
The most common finch in Europe, the chaffinch is also found throughout northern Asia. It often forages with other finch species in winter.

RED CROSSBILL
Loxia curvirostra
Crossbills use their uniquely crossed mandibles for extracting seeds from cones. This species is widespread in coniferous forests across the Northern Hemisphere.

YELLOW-FRONTED CANARY
Serinus mozambicus
This bird belongs to a group of mostly yellow-colored African serins and canaries. It is common south of the Sahara.

GRAY-CROWNED ROSY FINCH
Leucosticte tephrocotis
A member of a group of mountain finches related to bullfinches, this North American species is an inhabitant of high, rocky landscapes.

AMERICAN BLACKBIRDS AND RELATIVES

These American species of the family Icteridae superficially resemble the common blackbird, to which they are not related. As a group, the birds are more closely related to the finches. They have strong bills, which have a powerful gaping action for prying apart tough food.

8½–10 in
21–26 cm

6½–9½ in
17–24 cm

7½–10 in
19–26 cm

11–13½ in
28–34 cm

COMMON GRACKLE
Quiscalus quiscula
An opportunistic North American species, this bird raids refuse dumps and corn crops. Its internally keeled bill can saw open acorns.

YELLOW-HEADED BLACKBIRD
Xanthocephalus xanthocephalus
This marshland bird of western North America breeds in colonies and sites its nests over water, perhaps to deter predators.

RED-WINGED BLACKBIRD
Agelaius phoeniceus
A North American bird of marshy habitats, this colonial nester often breeds in the company of the larger, more dominant, yellow-headed blackbird.

EASTERN MEADOWLARK
Sturnella magna
A species from eastern North America, this open-country bird builds a nest on the ground, covering it with a grass roof.

7½–9 in
19–22 cm

7–8 in
18–20 cm

black head

6–8 in
15–20 cm

14½–18 in
37–46 cm

orange
underparts

BROWN-HEADED COWBIRD
Molothrus ater
This North American bird produces many eggs, laying them in the nests of other passerines. Its young are raised by many different host species.

BALTIMORE ORIOLE
Icterus galbula
Feeding on insects and caterpillars in spring and summer, this species changes its diet in winter to include nectar and berries.

BOBOLINK
Dolichonyx oryzivorus
This ground-nesting North American bird gets its name from its bubbling flight song. It migrates to central South America.

CRESTED OROPENDOLA
Psarocolius decumanus
Like other tropical American oropendolas, this colonial species weaves long nests that hang from tips of branches in open forest.

yellow
forehead

EVENING GROSBEAK
Hesperiphona vespertina
Grosbeaks of the finch family have
heavy seed-crushing bills, a feature they
share with the unrelated grosbeaks
of the cardinal-grosbeak family.
This is a North American species.

8 in
20 cm

HOUSE FINCH
Carpodacus mexicanus
The North American house
finch is a species of rosefinch,
which are are especially diverse
in temperate Asia. The males
are red or pale colored.

5½ in
14 cm

black crown
and chin

pale gray back

5½ in
14 cm

6–6½ in
15–16 cm

9 in
22 cm

JAPANESE GROSBEAK
Eophona personata
A strikingly patterned bird,
the Japanese grosbeak of cold
northern forests breeds in
Siberia and northern Japan and
winters in southern China.

8 in
20 cm

PINE GROSBEAK
Pinicola enucleator
The pine grosbeak occurs
in coniferous forests in the
Northern Hemisphere. The bill
of this grosbeak resembles that
of the closely related bullfinches.

IIWI
Vestiaria coccinea
This long-billed nectar-eater is
found only in Hawaii. It is a bird
of montane forests, the largest
populations occurring at
higher elevations.

EURASIAN BULLFINCH
Pyrrhula pyrrhula
Bullfinches have short, thick bills
and thick-set heads. This is the most
widespread of the bullfinches and
is found in woodlands across
temperate Eurasia.

AMERICAN WARBLERS

Unrelated to warblers of the Old World, these
finch-related insect-eaters of the family Parulidae
are found throughout North and South America.
The tropical species are sedentary, but
temperate ones are migratory.
The males molt their bright
colors in winter.

yellow body

4¼–5½ in
11–14 cm

**COMMON
YELLOWTHROAT**
Geothlypis trichas
A bird of wet habitats,
this is a migratory species
like other North American
warblers. It winters in
California and Mexico.

dull brown
feet and legs

4¾–5 in
12–13 cm

YELLOW WARBLER
Dendroica petechia
Widespread from
North America to the
Caribbean, there is a
large number of local
races of yellow warblers.

5½ in
14 cm

BAY-BREASTED WARBLER
Dendroica castanea
This bird breeds in spruce forests
in eastern North America. Its
numbers fluctuate according to
the abundance of its prey, the
spruce budworm.

4¼–5 in
11–13 cm

black and white
streaks overall

4½–5½ in
11–14 cm

AMERICAN REDSTART
Setophaga ruticilla
An active forager, this North American
warbler habitually flashes its orange and
black wings and tail to flush out insects.
It also catches prey in the air.

YELLOW-BREASTED CHAT
Icteria virens
This North American bird is large
for a warbler. It sings at night as
well as during the day, and mimics
the calls of other birds.

7 in
18 cm

BLACK-AND-WHITE WARBLER
Mniotilta varia
This North American warbler
moves up and down tree trunks like
a nuthatch, and has long hind claws
for clinging to bark.

6 in
15 cm

**NORTHERN
WATERTHRUSH**
Seiurus noveboracensis
A large, tail-bobbing ground
warbler of North America,
this bird forages in leaf litter
of wet woodlands and nests
in thickets near water.

**PROTHONOTARY
WARBLER**
Protonotaria citrea
Unusual for American
warblers, this species of
densely wooded swamps nests
in tree cavities—sometimes
using old woodpecker holes.

5 in
13 cm

5 in
13 cm

HOODED WARBLER
Wilsonia citrina
Like the American redstart, this
bird from broadleaf woodland of
the eastern USA feeds on insects
by catching flies while in flight.

4¾ in
12 cm

GOLDEN-WINGED WARBLER
Vermivora chrysoptera
A warbler of eastern North America,
this bird breeds in scrubby, open
habitats and farmland, and benefits
from deforestation.

BUNTINGS AND RELATIVES

True buntings are conical-billed, mostly ground-feeding seedeaters of the Old World. New World members of this family, the Emberizidae, are collectively known as American sparrows, but they are unrelated to the sparrows of Africa and Eurasia.

red eye

9 in
22 cm

white underparts

6½ in
17 cm

7 in
18 cm

SPOTTED TOWHEE
Pipilo maculatus
Towhees are long-tailed American sparrows. The spotted towhee is a rufous-sided bird found in thicket habitats across North America.

BLACK-HEADED BUNTING
Emberiza melanocephala
An inhabitant of scrub and olive groves, the black-headed bunting breeds in the Middle East and winters in India.

LARK-BUNTING
Calamospiza melanocorys
A bird of the North American prairies, the lark-bunting is a ground-nesting member of the American sparrow group.

long, dark, rounded tail

7½ in
19 cm

6 in
15 cm

5½ in
14 cm

7½ in
19 cm

YELLOW-THIGHED FINCH
Pselliophorus tibialis
A noisy, tropical member of the American sparrow group, this species is found only in the montane rain forests of Costa Rica and Panama.

LARGE GROUND FINCH
Geospiza magnirostris
This seed-eater is a native of the Galapagos Islands. The species feeds less from the ground than other ground finches.

BLACK-MASKED FINCH
Coryphaspiza melanotis
A ground-feeding finch of grassland habitats in central South America, this species has uncertain affinities, but may belong to the tanager family.

RED-CRESTED CARDINAL
Paroaria coronata
Birds of the *Paroaria* genus are found in South America. This is a common species in open woodland.

TANAGERS AND RELATIVES

An American family, the Thraupidae consist of gaudily colored tanagers, found in the forests of tropical South America. These finch-related birds have evolved to exploit a range of food sources, including fruits, insects, seeds, and nectar.

blue mantle

4¼ in
11 cm

green wing

5½ in
14 cm

7–7½ in
18–19 cm

7 in
18 cm

GREEN HONEYCREEPER
Chlorophanes spiza
This stocky species lives in forest canopy and uses its stout, decurved bill to eat fruits. It is often seen in mixed flocks of tanagers.

SCARLET TANAGER
Piranga olivacea
In most species of *Piranga* tanager, the males turn red during breeding. This migratory species is from North America.

WESTERN TANAGER
Piranga ludoviciana
This bird breeds in western North America and overwinters in Central America. It is the only *Piranga* tanager with a predominantly yellow breeding male.

BLUE-NAPED CHLOROPHONIA
Chlorophonia cyanea
The fruit-eating chlorophonias are mainly green. This species is widespread in forests across South America. Unlike other tanagers, it makes a dome-shaped nest.

7 in
18 cm

6 in
15 cm

BLUE-WINGED MOUNTAIN TANAGER
Anisognathus somptuosus
Most mountain tanagers of northern South America are colored blue and yellow, including this inhabitant of montane rain forests.

GOLDEN-COLLARED TANAGER
Iridosornis jelskii
A member of a group of Andean forest tanagers with a yellow head streak, this species occurs in Peru and Bolivia.

brown and gray head

4¼ in
11 cm

6 in
15 cm

6–6½ in
15–16 cm

dark gray body

long, white-edged tail

VARIABLE SEEDEATER
Sporophila corvina
Belonging to a group of tropical American, stubby-billed seedeaters that may be allied to tanagers, this bird has a variable plumage pattern.

LAPLAND LONGSPUR
Calcarius lapponicus
Relatives of buntings, longspurs have been named for their long hind claw. This species has a circumpolar distribution during the breeding season.

DARK-EYED JUNCO
Junco hyemalis
Juncos are gray-brown, ground-flocking North American sparrows. This species, locally called the snowbird, is a common winter visitor of garden bird feeders.

white belly with brown streaks

6½–7½ in
17–19 cm

5–5½ in
13–14 cm

6½–7½ in
17–19 cm

5½–6½ in
14–16 cm

SONG SPARROW
Melospiza melodia
A common North American bird from Alaska to Mexico, there are a large number of races. It is named for its melodious song.

WHITE-CROWNED SPARROW
Zonotrichia leucophrys
This North American bird is usually found in low vegetation or on the ground. It has distinctive black and white head markings.

CHIPPING SPARROW
Spizella passerina
A rufous-capped bird common in open woodland across North America, the chipping sparrow has a distinctive "chipping" trill.

FOX SPARROW
Passerella iliaca
Widespread throughout North America, this large sparrow has a red-streaked back and chest. It usually forages in low vegetation.

PURPLE-THROATED EUPHONIA
Euphonia chlorotica
Euphonias are small birds that mainly eat fruits, especially mistletoe berries. This species is from northern South America.

5 in
13 cm

4 in
10 cm

breeding male has blue underparts and head

RED-LEGGED HONEYCREEPER
Cyanerpes cyaneus
Tropical American honeycreepers feed on nectar. The most widespread species, this bird has a decurved bill. Like females, males develop a dull green "eclipse" plumage after breeding.

5 in
13 cm

BLUE-NECKED TANAGER
Tangara cyanicollis
This open forest bird from northern South America belongs to a group of tanager species with particularly colorful, iridescent plumage.

CARDINALS AND RELATIVES

These mostly chunky-billed birds are seedeaters like buntings, and many have bright colors like tanagers. Related to both, the Cardinalidae belong to a large group of American passerines that are derived from finches.

7–8½ in
18–21 cm

8½–9 in
21–23 cm

ROSE-BREASTED GROSBEAK
Pheucticus ludovicianus
This migratory American cardinal grosbeak has a typically heavy bill for feeding on large seeds and on insects, such as beetles.

NORTHERN CARDINAL
Cardinalis cardinalis
The males of this resident of the eastern USA and Mexico are red, due to chemicals called carotenoids that are acquired from their diet.

green back

5½ in
14 cm

5 in
13 cm

PAINTED BUNTING
Passerina ciris
This species breeds in the southern USA and winters in Central America and the Caribbean. Only males have the striking tricolored pattern.

INDIGO BUNTING
Passerina cyanea
The most wide ranging of the blue buntings, this species migrates between Canada and South America. The males lose their blue plumage in winter.

MAMMALS

The mammals are very successful animals. They are able to occupy almost every terrestrial habitat, and some visit the deep oceans between breaths of air. However, this has not always been the case; at 200 million years old, mammals are a relatively recent group.

PHYLUM	CHORDATA
CLASS	MAMMALIA
ORDERS	29
FAMILIES	153
SPECIES	About 5,500

A single jawbone articulated directly to the skull gives mammals, such as this Tasmanian devil, a powerful bite.

Baleen is made of a protein called keratin, and baleen whales use it to strain sea water, trapping food in their mouths.

By sucking milk from their mother, young warthogs obtain all the nourishment they need during the first weeks of life.

DEBATE
EQUALLY WELL ADAPTED

Marsupial features, including slightly reduced body temperature, are sometimes described as primitive. This implies that marsupials, whose young are born in an embryonic state and develop in a pouch, are less advanced than placental mammals, which bear conventional young. In fact, both lineages evolved at about the same time.

Mammals owe their success to a unique combination of adaptations that allowed them to replace reptiles as the dominant form of animal life on Earth. Like reptiles, mammals breathe air, but unlike their scaly ancestors, they are warm-blooded, so they are able to burn fuel (food) in order to maintain a constant, warm body temperature. This creates conditions in which the internal chemical processes that sustain life can happen most efficiently, without relying directly on heat from the sun. Uniquely, mammals have hair, which reduces heat loss and allows an animal to be active in cold climates and at night. Because fur can be molted and regrown, it is seasonally adjustable too.

VARIATIONS ON A THEME

The basic mammal skeleton is strong, with upright limbs supporting the body from below. This arrangement allows land-dwelling mammals to walk, run, and leap, but the basic structure is highly adaptable, and has been modified for swimming, as in seals and cetaceans; for flying, as in bats; and for climbing or swinging through the trees, as in some primates. The skulls of mammals have powerful jaws, with a single lower jawbone articulating directly with the skull, and a diverse range of teeth adapted to a huge variety of food types. Certain other bones, which in reptiles contribute to the lower jaw, are put to a new use by mammals—they have became the three tiny bones of the inner ear, allowing for greatly enhanced hearing. The mammalian skull also serves to protect the brain, which is larger in mammals than in other groups, hinting at increased processing power. With a big brain comes intelligence, and an unrivaled ability to learn, to remember, and to engage in complex behavior.

These abilities take time to fine-tune, and young mammals do so during the extended period of parental care that begins while they are being fed on milk produced by their mother's mammary glands. These unique organs evolved from sebaceous glands, which originally provided secretions to condition the skin and perhaps to prevent eggs from drying out, and it is for them that the entire group is named.

FUR AS CAMOUFLAGE >
The markings on the fur of this jaguar break up its outline, making it more difficult for the animals it hunts to see it.

MAMMAL GROUPS

There are three groups of mammals—the egg-laying monotremes, the pouched mammals (or marsupials), and the placental mammals. The latter group is the most diverse, and is here divided into further groups.

EGG-LAYING MAMMALS

There are just five species in this group of mammals, which are known as monotremes. They all have specialized snouts and lay eggs.

The monotremes, which are the duck-billed platypus and the long-nosed and short-nosed echidnas, form the order Monotremata. They are found in various habitats in New Guinea, Australia, and Tasmania.

Platypuses and echidnas lay soft-shelled eggs that hatch after around ten days' incubation. The young feed on milk secreted from the female's mammary glands; monotremes have no teats. Infant echidnas live in their mother's pouch until their spines appear and then, like young platypuses, stay in a burrow for several months.

Monotremes have specialized snouts for finding and eating prey. The platypus, which is partly aquatic, has a flattened, duck like bill covered with sensory receptors that allow the animal to locate invertebrates under water, even in murky conditions. The long, cylindrical snout and long tongue of the land-dwelling echidna are ideal for probing the nests of ants and termites, and for catching worms. Neither the platypus nor the echidnas have teeth; instead they have grinding surfaces or horny spines on their tongues.

WHAT'S IN A NAME
The word "monotreme" means "single hole" and makes reference to the cloaca, a single posterior body opening into which the digestive, urinary, and reproductive tracts empty.

PHYLUM	CHORDATA
CLASS	MAMMALIA
ORDER	MONOTREMATA
FAMILIES	2
SPECIES	5

The large, webbed feet of the duck-billed platypus are used for propulsion. The hind feet act as rudders.

DUCK-BILLED PLATYPUS

The sole member of the family Ornithorhynchidae, the duck-billed platypus is well adapted to a semi-aquatic lifestyle. It has a streamlined body, waterproof fur, webbed feet, and a flattened tail. Males have a venomous horny spur on each hind foot.

short, dense body fur

small eyes

sensitive, ducklike bill

16–23½ in
40–60 cm

venomous spur of male

PLATYPUS
Ornithorhynchus anatinus
Found in rivers and streams in eastern Australia and Tasmania, this rare species uses its soft bill, which is covered in electroreceptors, to hunt for invertebrates.

ECHIDNAS

The family Tachyglossidae is made up of the long- and short-beaked echidnas. Their rounded bodies are covered with fur and spines, and the long snouts are ideal for seeking out insects, ants, and worms.

sharp spines for defence

23½–39 in
60–100 cm

12–18 in
30–45 cm

SHORT-BEAKED ECHIDNA
Tachyglossus aculeatus
Widespread in Australia, Tasmania, and New Guinea, the short-beaked echidna lays a single egg into a pouch on its abdomen.

EASTERN LONG-BEAKED ECHIDNA
Zaglossus bartoni
The largest of the monotremes, this species lives in the forested highlands of eastern New Guinea.

POUCHED MAMMALS

These mammals bear live young that are born in a very immature state, and typically complete their growth in the mother's abdominal pouch.

Marsupials inhabit a vast array of habitats from desert and dry scrub to tropical rain forest. Most of them are either terrestrial or arboreal, but several glide, one is aquatic, and there are two marsupial moles that live underground. The diets of marsupials are equally wide-ranging: among the different species are carnivores, insectivores, herbivores, and omnivores. There are even marsupials that feed on nectar and pollen. Marsupials also differ markedly in size, from the tiny planigales, which are among the smallest mammals in the world, weighing less than about ⅓ oz (4.5 g), to the red kangaroo, males of which can weigh over 200 lb (90 kg).

EARLY DEVELOPMENT
Young marsupials are born blind and furless. They make their way through the mother's fur and latch onto a nipple and suckle. In around half of marsupial species, the nipples are located in a protective pouch. Some marsupials have only one offspring at a time, but others may have up to a dozen or more. The phase during which young are in the pouch is equivalent to the gestation period of placental mammals.

Kangaroos and some other marsupials are able to stop the development of an embryo before it implants in the uterus if they already have an infant occupying the pouch. The pregnancy continues when the pouch is vacated.

THE SEVEN ORDERS
Pouched mammals, known as marsupials, are now divided into seven major groups. These are: the American opossums, order Didelphimorphia; the shrew opossums, order Paucituberculata; monito del monte, the sole member of the order Microbiotheria; the Australasian carnivorous marsupials, order Dasyuromorphia; bandicoots, order Peramelemorphia; the marsupial moles, order Notoryctemorphia; and the Australasian marsupials of Diprotodontia, the largest order of pouched mammals, which includes the koala, wombats, possums, wallabies, and kangaroos.

PHYLUM	CHORDATA
CLASS	MAMMALIA
ORDER	7
FAMILIES	18
SPECIES	289

DEBATE
NOT SO PRIMITIVE?

Pouched mammals were once considered primitive because their young are born at an early stage without prolonged nourishment by a placenta inside the mother. Scientists now know that this reproductive strategy is an adaptation to their environment and that pouched mammals are just as advanced as placental mammals. Kangaroos can support two dependent juveniles of different ages, and a fertile egg that can develop almost immediately should anything happen to the other two. This strategy allows pouched mammals to recover quickly when harsh environmental conditions abate.

SHREW OPOSSUMS

Fewer incisor teeth distinguish the members of the family Caenolestidae from other American marsupials. All six species occur in the Andes in western South America.

3½–5½ in
9–14 cm

SHREW OPOSSUM
Caenolestes obscurus
Living at high altitudes in Colombia, Ecuador, and Venezuela, this shrew opossum uses its large lower incisors to kill prey.

MONITO DEL MONTE

The sole member of the family Microbiotheriidae, the monito del monte, or colocolo, is well adapted to its cold habitat, by having dense fur on its body and ears and by hibernating during winter.

3¼–5 in
8–13 cm

MONITO DEL MONTE
Dromiciops gliroides
Inhabiting cool bamboo forest and temperate rainforest in Chile and Argentina, this species has dense fur to limit heat loss.

MARSUPIAL MOLES

Two species of Australian marsupial moles form the family Notoryctidae. Their short limbs, large claws, and horned nose-shield help them burrow. They lack external ears and have nonfunctional eyes.

5–5¾ in
13–14.5 cm

SOUTHERN MARSUPIAL MOLE
Notoryctes typhlops
This marsupial mole is specialized for burrowing in the sandy desert and spinifex grassland of central Australia.

OPOSSUMS

The New World opossums of the family Didelphidae have pointed muzzles with sensitive whiskers, and naked ears. Many species have a grasping, prehensile tail that helps when climbing. Some opossums lack pouches.

VIRGINIA OPOSSUM
Didelphis virginiana
The largest American marsupial, this species lives in grasslands and temperate and tropical forests in the USA, Mexico, and Central America.

13–20 in
33–50 cm

7–11½ in
18–29 cm

BROWN-EARED WOOLLY OPOSSUM
Caluromys lanatus
Also known as the western woolly opossum, this arboreal, solitary opossum inhabits moist forests in western and central South America.

6½–11 in
16–28 cm

BARE-TAILED WOOLLY OPOSSUM
Caluromys philander
This woolly opossum's long, prehensile tail helps it move through the canopy of moist rain forests across eastern and central South America.

»

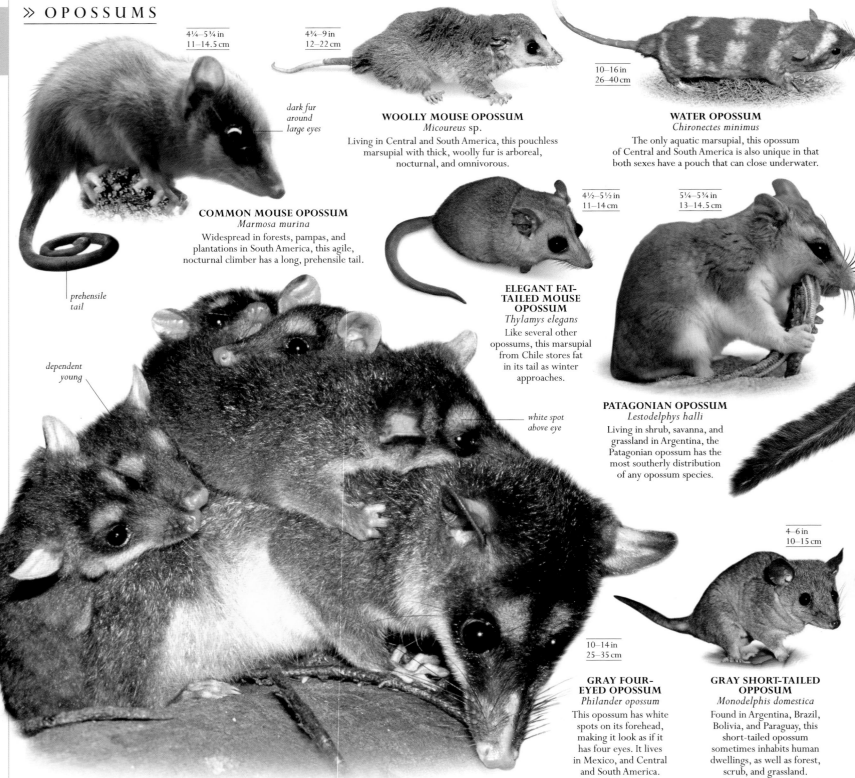

4¼–5¾ in
11–14.5 cm

4¾–9 in
12–22 cm

10–16 in
26–40 cm

dark fur
around
large eyes

WOOLLY MOUSE OPOSSUM
Micoureus sp.
Living in Central and South America, this pouchless
marsupial with thick, woolly fur is arboreal,
nocturnal, and omnivorous.

WATER OPOSSUM
Chironectes minimus
The only aquatic marsupial, this opossum
of Central and South America is also unique in that
both sexes have a pouch that can close underwater.

COMMON MOUSE OPOSSUM
Marmosa murina
Widespread in forests, pampas, and
plantations in South America, this agile,
nocturnal climber has a long, prehensile tail.

prehensile
tail

4½–5½ in
11–14 cm

5¼–5¾ in
13–14.5 cm

ELEGANT FAT-TAILED MOUSE OPOSSUM
Thylamys elegans
Like several other
opossums, this marsupial
from Chile stores fat
in its tail as winter
approaches.

dependent
young

PATAGONIAN OPOSSUM
Lestodelphys halli
Living in shrub, savanna, and
grassland in Argentina, the
Patagonian opossum has the
most southerly distribution
of any opossum species.

white spot
above eye

4–6 in
10–15 cm

10–14 in
25–35 cm

GRAY FOUR-EYED OPOSSUM
Philander opossum
This opossum has white
spots on its forehead,
making it look as if it
has four eyes. It lives
in Mexico, and Central
and South America.

GRAY SHORT-TAILED OPPOSSUM
Monodelphis domestica
Found in Argentina, Brazil,
Bolivia, and Paraguay, this
short-tailed opossum
sometimes inhabits human
dwellings, as well as forest,
scrub, and grassland.

NUMBAT

The numbat is the only member of the family Myrmecobiidae.
It has a distinctive striped coat, strong claws for digging,
and a very long tongue for extracting
termites from their nests.

NUMBAT
Myrmecobius fasciatus
Found only in eucalyptus
forest and woodland in
Western Australia, this
diurnal marsupial is a
specialized termite-eater.

8–11 in
20–28 cm

BILBIES

The family Thylacomyidae has only one
species, since the lesser bilby has been declared
extinct. Living in arid habitats, this nocturnal
marsupial does not need to drink water because
it absorbs moisture from its food.

12–22 in
30–55 cm

GREATER BILBY
Macrotis lagotis
A burrowing species of the desert of
central Australia, this marsupial has silky fur,
a tricolored tail, and long, rabbitlike ears.

QUOLLS, DUNNARTS, AND RELATIVES

The family Dasyuridae is made up of more than 70 large and small carnivorous marsupials with strong jaws and sharp canine teeth. These animals also have sharp claws, except on their big toe.

white band on rump and chest

fat store in base of tail

TASMANIAN DEVIL
Sarcophilus harrisii
The largest carnivorous marsupial in the world, this nocturnal hunter lives in a variety of habitats across Tasmania.

20½–32 in
52–80 cm

3¾–4¼ in
9.5–10.5 cm

FAT-TAILED FALSE ANTECHINUS
Pseudantechinus macdonnellensis
This nocturnal insect-eater stores fat in the base of its tapered tail. It inhabits arid rocky habitats of central and Western Australia.

5½–10 in
14–25 cm

BROWN ANTECHINUS
Antechinus stuartii
Endemic to forests in eastern Australia, the brown antechinus is unusual because all males die within a month of mating.

THREE-STRIPED DASYURE
Myoictis melas
This marsupial's coloration helps it to remain camouflaged on the rainforest floor in the islands of Indonesia and New Guinea.

6½–10 in
17–25 cm

4¾–8 in
12–20 cm

10–16 in
26–40 cm

WESTERN QUOLL
Dasyurus geoffroii
Also called the chuditch, this nocturnal hunter from southwest Australia is mainly ground-dwelling, although it can climb trees.

RED-TAILED PHASCOGALE
Phascogale calura
Possessing a brush of black hair on its red-based tail, this carnivorous marsupial inhabits woodland in southwest Australia.

3½–4¾ in
9–12 cm

MULGARA
Dasycercus cristicauda
This carnivore from west and central Australia inhabits arid and semi-arid habitats, such as desert, heath, and grassland. It stores fat in its tail.

2¾–4 in
7–10 cm

2¼–2½ in
5.5–6.5 cm

2–3 in
5–7.5 cm

2¼–3½ in
6–9 cm

KULTARR
Antechinomys laniger
This fast, agile marsupial uses its large hind feet to bound across woodland, grassland, and semi-desert habitats in southern and central Australia.

NARROW-NOSED PLANIGALE
Planigale tenuirostris
This flat-headed, nocturnal, rodentlike marsupial lives in low shrub and dry grassland in southeast Australia.

INLAND NINGAUI
Ningaui ridei
A shrewlike marsupial with a pointed snout, this nocturnal predator hunts insects in arid spinifex grassland in central Australia.

FAT-TAILED DUNNART
Sminthopsis crassicaudata
Living in open grassland in southern Australia, this small, nocturnal marsupial stores fat reserves in its tail.

BANDICOOTS

These omnivorous marsupials are found across Australia and New Guinea. Characterized by fused toes on their hind feet and more than two well-developed lower incisor teeth, the Peramelidae also have short, coarse, or spiny hair.

8–20 in
20–50 cm

11–14 in
28–36 cm

SPINY BANDICOOT
Echymipera kalubu
This forest-dwelling, nocturnal insect-eater of New Guinea has a conical snout, spiny coat, and a hairless tail.

coarse, yellowish brown fur

12–16½ in
31–42 cm

10½–14 in
27–35 cm

LONG-NOSED BANDICOOT
Perameles nasuta
Found in rain forest and woodland in eastern coastal Australia, this nocturnal bandicoot digs for its insect food.

EASTERN BARRED BANDICOOT
Perameles gunnii
Named for the creamy bands of fur on its flanks, this bandicoot lives in grassland and grassy woodland in Australia and Tasmania.

SOUTHERN BROWN BANDICOOT
Isoodon obesulus
This short-nosed bandicoot lives in the shrubby heathland in southern Australia and several islands, including Kangaroo Island and Tasmania.

KOALA

The sole member of the family Phascolarctidae, the koala is adept at climbing because of its powerful forearms, opposable fingers and toes, and sharp, curved claws. It sleeps for 20 hours each day because its diet of eucalyptus leaves is low in nutrients.

large, white, rounded ears

dense fur

26–32 in
65–82 cm

KOALA
Phascolarctos cinereus
Feeding almost exclusively on eucalyptus leaves, the koala lives in forest and woodland in eastern Australia. It is solitary and nocturnal.

long, curved claws

PYGMY POSSUMS

These small, nocturnal marsupials with prehensile tails are omnivorous, feeding on insects, fruits, nectar, and pollen. Four species of the family Burramyidae are endemic to Australia; the fifth lives in Australia and New Guinea.

4¼ in
10.5 cm

LONG-TAILED PYGMY POSSUM
Cercartetus caudatus
This arboreal marsupial is found in temperate rain forest in New Guinea and northeast Queensland.

4–5 in
10–13 cm

dull, grayish brown upperparts

MOUNTAIN PYGMY POSSUM
Burramys parvus
A ground-dweller of rocky, upland habitats in Australia, the mountain pygmy possum hibernates under snow for several months during winter.

RINGTAILS AND RELATIVES

This family consists of ringtail possums and the lemurlike greater glider. The Pseudocheiridae are all arboreal and are specialized leaf-eaters. They have an enlarged pouch at the start of the large intestine for fermentation of cellulose from their diet.

COMMON RINGTAIL
Pseudocheirus peregrinus
This agile marsupial lives in a wide range of habitats in eastern Australia and Tasmania. It has become a pest in New Zealand.

12–14 in
30–35 cm

WOMBATS

Stocky marsupials with short tails and limbs, wombats have large forepaws and long claws for burrowing. The family Vombatidae feed on coarse grass, which they grind using strong jaws, and digest with an elongated gut.

COMMON WOMBAT
Vombatus ursinus
Found in forest, heathland, and coastal scrub in southeast Australia, this species is capable of digging tunnels up to 655 ft (200 m) long.

28–47 in
70–120 cm

silky fur mottled brown and gray

SOUTHERN HAIRY-NOSED WOMBAT
Lasiorhinus latifrons
An inhabitant of central southern Australia, this wombat lives colonially in warrens but feeds alone.

30–37 in
77–95 cm

12–16 in
31–40 cm

BRUSHY-TAILED RINGTAIL
Hemibelideus lemuroides
Found only in a small area of rain forest in northeast Queensland, Australia, this species of ringtail is nocturnal.

CUSCUS AND RELATIVES

The family Phalangeridae includes cuscuses, brushtail possums, and their relatives. Most members are arboreal, with an opposable thumb on the hind limbs and a prehensile tail. In cuscuses, part or all of the tail is naked, whereas in brushtail possums the tail has fur.

14–26 in
35–65 cm

SULAWESI BEAR CUSCUS
Ailurops ursinus
The largest of the cuscuses, this species inhabits the canopy of temperate rain forest in Sulawesi and several other Indonesian islands.

large eyes assist sight at night

COMMON SPOTTED CUSCUS
Spilocuscus maculatus
This cuscus is sexually dimorphic (males and females look different)—only the male has spots. It lives in the rain forests of New Guinea and northeastern Australia.

24 in
61 cm

16–20 in
40–50 cm

strong, curved claws

furry tail

MOUNTAIN BRUSHTAIL POSSUM
Trichosurus cunninghami
The mountain brushtail possum inhabits dense, wet forests in southeast Australia, typically living at altitudes above 1,000 ft (300 m).

13–23½ in
33–60 cm

16 in
40 cm

SCALY-TAILED POSSUM
Wyulda squamicaudata
Found only in the Kimberley in northwest Australia, this nocturnal possum is solitary and bears just one young at a time.

CUSCUS
Phalanger sp.
Cuscuses of this genus live on New Guinea and nearby islands. They live at different altitudes to avoid competition with one another.

GLIDING AND STRIPED POSSUMS

The Petauridae include striped possums and gliders, except the greater glider (left). Gliders have a thin, furred membrane between their fore and hind limbs. Striped possums emit strong smells and have an elongated fourth finger to probe for wood-boring beetles.

grizzled, gray fur

14–19 in
35–48 cm

GREATER GLIDER
Petauroides volans
The largest gliding marsupial, this species can travel distances of over 330 ft (100 m) between trees. It lives in eastern Australia.

11–15 in
28–38 cm

GREEN RINGTAIL
Pseudochirops archeri
Named for its thick, greenish fur, this solitary possum is found only in rain forest in the far north of Queensland, Australia.

9½–11 in
24–28 cm

gray body with dark stripe along its back

6–6½ in
15–17 cm

LEADBEATER'S POSSUM
Gymnobelideus leadbeateri
Found in moist, high-altitude forests in Victoria, Australia, this possum feeds on insects, and sap and gum from trees.

DAINTREE RIVER RINGTAIL POSSUM
Pseudochirulus cinereus
Resembling a lemur, this ringtail possum lives in the montane tropical rain forest in the Daintree River area in northeast Queensland, Australia.

prehensile tail

STRIPED POSSUM
Dactylopsila trivirgata
Skunklike in appearance and odor, this possum is a nocturnal tree-dweller from northeastern Queensland, Australia, and New Guinea.

6–8½ in
15–21 cm

club-shaped tail

14 in
35 cm

thick tail

SUGAR GLIDER
Petaurus breviceps
Partial to the sweet sap of eucalyptus trees, the sugar glider is native to northeast Australia, New Guinea, and neighboring islands.

HONEY POSSUM

This diminutive species is the only member of the family Tarsipedidae. It has fewer teeth than other possums and a long, brush-covered tongue for probing flowers.

HONEY POSSUM
Tarsipes rostratus
Inhabiting heathland and woodland in southwest Australia, this possum is a specialized feeder on nectar and pollen.

2½–3½ in
6.5–9 cm

long, pointed snout

PYGMY GLIDER

The family Acrobatidae contains just two species—the pygmy or feather-tailed glider and the feather-tailed possum. Both have tails fringed with rows of stiff hairs.

PYGMY GLIDER
Acrobates pygmaeus
This species is the smallest gliding marsupial. It lives in eastern Australian forests, where it feeds on nectar.

2½–3¼ in
6.5–8 cm

KANGAROOS
AND RELATIVES

The kangaroos and wallabies of the family Macropodidae are medium to large animals, with long hind legs for jumping. The fourth and fifth toes of the hind feet are strong for load-bearing, while the second and third are reduced. The first toe has been lost in evolution.

2½–4½ ft
0.8–1.4 m

prominent ears

COMMON WALLAROO
Macropus robustus
Widespread throughout most of mainland Australia, the common or hill wallaroo often seeks shade in rocky outcrops.

23–41 in
59–105 cm

AGILE WALLABY
Macropus agilis
Unusual for wallabies, this species occurs in both Australia and New Guinea, where it inhabits grassland and open woodland.

26–36 in
66–92 cm

red-brown upperparts

18–21 in
45–53 cm

PARMA WALLABY
Macropus parma
Native to the mountains of the Great Dividing Range in Australia, the parma wallaby inhabits a range of forest habitats.

3¼–5¼ ft
1–1.6 m

RED KANGAROO
Macropus rufus
The largest extant marsupial, this kangaroo is widespread in Australia, where it lives in savanna grassland and desert.

joey in pouch

3–4½ ft
0.9–1.4 m

3–4½ ft
0.9–1.4 m

EASTERN GREY
KANGAROO
Macropus giganteus
Widespread in eastern Australia, this species lives in dry woodland, scrub, and shrub. A subspecies is found in Tasmania.

WESTERN GREY KANGAROO
Macropus fuliginosus
The only kangaroo that does not reproduce using the method of delayed implantation of the embryo in the uterus, this species occurs in southern Australia, including Kangaroo Island.

RED-NECKED
WALLABY
Macropus rufogriseus
This species lives in coastal forest and scrub in southeastern Australia, including the islands of Tasmania and the Bass Strait.

long tail used for support at rest, for balance when moving

POTOROOS

This family contains potoroos, bettongs, and rat kangaroos, small marsupials with many similarities to the larger Macropodidae. However, unlike them, the Potoroidae have a serrated premolar for tearing vegetation.

RAT-KANGAROO

The musky rat-kangaroo is the sole member of the family Hypsiprymnodontidae. It is relatively primitive; its hind foot has a first toe, which has been lost in other kangaroo species.

LONG-NOSED POTOROO
Potorous tridactylus

13½–15 in
34–38 cm

Inhabiting heath and forest in southeast Australia, this potoroo has strong, curved claws that help it to dig for underground fungi.

12–15 in
30–38 cm

BRUSH-TAILED BETTONG
Bettongia penicillata

This bettong lives in forest and grassland in southwest Australia. It has a prehensile tail that it uses for moving nesting material.

6–11 in
15–28 cm

MUSKY RAT-KANGAROO
Hypsiprymnodon moschatus

This diurnal species inhabits tropical rain forest in northern Queensland, Australia, where it forages for fallen fruits, seeds, and fungi.

28 in
70 cm

NORTHERN NAIL-TAILED WALLABY
Onychogalea unguifera

Also called the sandy nail-tailed wallaby, this medium-sized kangaroo occurs across northern Australia.

17–28 in
43–71 cm

BRIDLED NAIL-TAIL WALLABY
Onychogalea fraenata

This nocturnal wallaby was once thought to be extinct; the only remaining wild population is in a small area of Queensland, Australia.

RED-NECKED PADEMELON
Thylogale thetis

A forest wallaby from eastern Australia, this species ventures to the forest edge at night to feed on grass, leaves, and shoots.

muscular thighs

11½–25 in
29–63 cm

long, narrow sole of foot

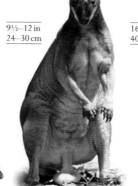

9½–12 in
24–30 cm

16–21½ in
40–54 cm

coarse, dark fur

26–34 in
66–85 cm

SWAMP WALLABY
Wallabia bicolor

Darker in color than other wallabies, the swamp wallaby lives in tropical and temperate forests, and swampland in eastern Australia.

12–15½ in
31–39 cm

RUFOUS HARE-WALLABY
Lagorchestes hirsutus

Previously inhabiting the Australian mainland, the mala or rufous hare-wallaby now exists in the wild on only two islands in Western Australia.

BROWN DORCOPSIS
Dorcopsis muelleri

This forest-dwelling kangaroo is endemic to the low-altitude rain forests of western New Guinea, and three offshore islands.

QUOKKA
Setonix brachyurus

Rare in mainland Australia, this small marsupial is found on Rottnest and Bald Islands, off the southwestern coast.

20–32 in
51–81 cm

DORIA'S TREE KANGAROO
Dendrolagus dorianus

The heaviest of the tree-dwelling marsupials, Doria's tree kangaroo frequently descends to the ground. It lives in montane forest in New Guinea.

reddish brown upperparts

22–30 in
55–77 cm

GOODFELLOW'S TREE-KANGAROO
Dendrolagus goodfellowi

Also called the ornate tree-kangaroo, this species is native to the mountainous rain forests of New Guinea, where it feeds on leaves and fruit.

nonprehensile tail

19–26 in
48–65 cm

LUMHOLTZ'S TREE-KANGAROO
Dendrolagus lumholtzi

The smallest of the tree-kangaroos, this marsupial is found in the rain forests of north Queensland, Australia.

20–23½ in
50–60 cm

BRUSH-TAILED ROCK WALLABY
Petrogale penicillata

This resident of southeast Australia has roughened, padded hind feet that help it to grip when jumping in rocky habitats.

⌄ BIG BOYS
Male red kangaroos are much larger than females, sometimes weighing twice as much. Only male red kangaroos are red; the females, known as blue fliers, have bluish gray fur all over. The males fight for access to mates, their battles taking the form of boxing matches.

velvety nose

glands in the chest produce scent, which male kangaroos rub on bushes to assert their dominance

kangaroos lick their wrists when hot—this cools the blood close to the skin

elastic tendon stores energy when the leg is flexed, and releases it to power the next hop

RED KANGAROO
Macropus rufus

Kangaroos evolved in Australia, filling the niche occupied elsewhere by grazing animals such as antelope. In common with those species, kangaroos have a big stomach to take in large quantities of grass. Living out in the open is risky for all herbivores, and kangaroos share with antelope a number of adaptations for avoiding predators. They live in herds (called mobs), have sharp senses, are tall enough to scan their surroundings for danger, and possess a wonderful turn of speed. The red is the largest and swiftest kangaroo, able to hop at more than 30 miles (50 km) an hour. Only the female has a pouch, in which she carries her young—known as a joey—for the first seven months of its life. Red kangaroos are well adapted to drought—they are able to eat scrubby saltbush, which is toxic to other animals.

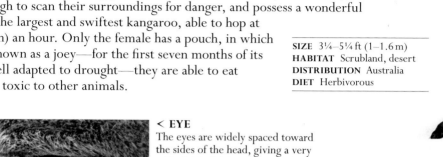

SIZE 3¼–5¼ ft (1–1.6 m)
HABITAT Scrubland, desert
DISTRIBUTION Australia
DIET Herbivorous

< EYE
The eyes are widely spaced toward the sides of the head, giving a very wide field of vision. This aids the kangaroo's awareness of predators.

> EARS
The ears are large and sensitive, and can be swiveled in different directions to focus on sounds that might mean danger is approaching.

∨ STRONG FIGHTER
The male kangaroo has a broad, muscular chest. The forelimbs are used to punch rivals in fights, but the most forceful blows come from double kicks of the hind feet.

tail acts as a counterbalance when hopping, and as a supportive fifth limb when standing still

∧ CLAWS
The feet have surprisingly sharp claws. These provide grip when hopping, but also serve as weapons, and as combs for grooming.

< ∨ HIND FOOT
The scientific name *Macropus* literally means "big foot." Each hind foot has four toes: an outer, weight-bearing pair, and an inner pair used for grooming.

∧ HOPPING
Red kangaroos hop on their enormous hind legs with apparently effortless grace. They are able to cover up to 30 ft (9 m) in a single bound.

SENGIS

Formerly known as elephant-shrews, sengis form a single family. They are superficially similar to the true shrews, although they are not related.

Distinguished by an elongated, flexible snout, sengis are small, with long legs and tail. They move either on all fours or, especially if they need to move fast, by hopping. Sengis live in pairs within a defined area, although there is little interaction between the partners. Males and females even sometimes make separate nests, each defending the jointly held territory against intruders of its own sex. Active mostly by day, many sengis maintain a network of cleared paths around their territory, which they patrol for prey and use as quick escape routes if threatened.

Found only in Africa, sengis occupy a range of habitats, from forests and savanna to extremely arid deserts. Some take shelter in rock crevices or nests of dried leaves, while the larger species dig out shallow burrows.

Sengis are primarily insectivorous, although some species take spiders and worms. They use their sensitive noses to root on the ground and under leaf litter for food, extending their long tongues to scoop up their prey.

ACTIVE YOUNG

Depending on species, a pair of sengis may breed several times during the year. Litters are small, numbering between one and three young, which are well developed at birth and rapidly become active.

PHYLUM	CHORDATA
CLASS	MAMMALIA
ORDER	MACROSCELIDEA
FAMILIES	1
SPECIES	15

SENGIS

Found only in Africa, the Macroscelididae occur in a wide range of habitats, from deserts to mountains to forests. Largely insectivorous, sengis (or elephant-shrews) use their elongated, movable snout to search for prey, which they flick into the mouth with the help of their tongue.

hind legs longer than forelegs

3½–4¼ in
9–11 cm

BUSHVELD SENGI
Elephantulus intufi
Widespread in arid shrubland in southern Africa, this sengi occupies large territories, which it defends vigorously against rivals of the same sex.

long furless tail

4¾–5½ in
12–14 cm

WESTERN ROCK SENGI
Elephantulus rupestris
Inhabiting rocky shrubland in southwest Africa, this species is active by day, leaving the shade of a rock crevice to foray for food.

4¼–5 in
11–12.5 cm

NORTH AFRICAN SENGI
Elephantulus rozeti
The only North African elephant-shrew, this desert species slaps its very long tail and drums its feet when alarmed.

3½–5½ in
9–14 cm

RUFOUS SENGI
Elephantulus rufescens
Found in southern and eastern Africa, this sengi has long, sandy reddish fur and a tail as long as its head and body.

DUSKY-FOOTED SENGI
Elephantulus fuscipes
From hot, dry grassland in central Africa, this poorly known species may belong to a different genus of sengis.

4–4¾ in
10–12 cm

6½–8 in
16–20 cm

FOUR-TOED SENGI
Petrodromus tetradactylus
Found in a range of moist habitats, this is one of the most widespread elephant-shrews. It occurs from central to South Africa.

9–12½ in
23–32 cm

BLACK AND RUFOUS SENGI
Rhynchocyon petersi
This giant elephant-shrew inhabits the coastal forests of East Africa. Its orange foreparts graduate through deep red to black on the rump.

4–4¾ in
10–12 cm

ROUND-EARED SENGI
Macroscelides proboscideus
A medium-sized sengi, this southern African species occupies some of the world's driest habitats. Strictly monogamous, males strongly guard their mates.

rounded ears

TENRECS AND GOLDEN MOLES

Two families make up the order known as Afrosoricida. The golden moles are exclusively burrowers, but tenrecs have evolved into a range of niches.

Most species in this order are native to mainland Africa, with tenrecs additionally found on Madagascar. Golden moles are all very similar in appearance, with a cylindrical body and other anatomical adaptations to suit their burrowing lifestyle. In contrast, tenrecs display diverse characteristics that reflect the wide range of habitats that they occupy. Various species of tenrecs are found in tropical forests, where they may be terrestrial or semiarboreal; others are aquatic and otterlike, living in streams and rivers; and there are also burrowing tenrecs. Until recently, tenrecs and golden moles were classified with true moles, shrews, and hedgehogs because of their largely insectivorous diet and often similar appearance. It is now known that the order Afrosoricida evolved independently and is most closely related to the elephants, hyraxes, and sirenians.

UNUSUAL FEATURES

Tenrecs and golden moles have features once considered primitive but now recognized as adaptations to harsh environments. Such features include a low metabolic rate and body temperature. These mammals also have the ability to enter a state of torpor—for up to three days—to save energy in cold conditions, and possess highly efficient kidneys, which reduce the need to drink water.

PHYLUM	CHORDATA
CLASS	MAMMALIA
ORDER	AFROSORICIDA
FAMILIES	2
SPECIES	About 57

A Grant's golden mole feasts on a locust. At night it forages for food on the surface, feeding mainly on termites.

GOLDEN MOLES

These moles resemble both true moles (Talpidae) from Europe, Asia, and North America, and marsupial moles (Notoryctemorphia) from Australia, as all have similar burrowing habits. The Chrysochloridae from southern Africa have short legs with powerful digging claws, dense fur that repels moisture, and toughened skin, especially on the head. They have nonfunctional eyes covered with skin, and lack external ears.

4–5 in
10–13 cm

JULIANA'S GOLDEN MOLE
Neamblysomus julianae
Unique to the dry highlands of South Africa, typically in sandy soils, this species frequents well-irrigated gardens within its range.

soft, dense, glossy fur

webbed toes kick soil backward

4–4¼ in
10–11 cm

CAPE GOLDEN MOLE
Chrysochloris asiatica
Although secretive, this is a common species in parts of South Africa. It uses the enlarged second toe on its front feet for digging.

eyes covered with thick layer of skin

leathery pad protects nostrils and aids burrowing

3–3¼ in
7.5–8.5 cm

HOTTENTOT GOLDEN MOLE
Amblysomus hottentotus
This golden mole inhabits tunnel systems up to 655 ft (200 m) long, using the large second and third toes on its front feet for digging.

4½–5¾ in
11.5–14.5 cm

GRANT'S DESERT GOLDEN MOLE
Eremitalpa granti
Inhabiting the southwest African coastal dunes, one of the world's driest habitats, this species "swims" through sand rather than building tunnels.

TENRECS

From Africa and Madagascar, tenrecs have diverse body forms, resembling the unrelated shrews, mice, hedgehogs, and otters. They range in weight from ³/₁₆ oz (5 g) to more than 2¼ lb (1 kg). Largely nocturnal and with poor eyesight, these insectivores use their sensitive whiskers to locate food.

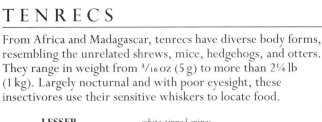

10–15½ in
26–39 cm

LOWLAND STREAKED TENREC
Hemicentetes semispinosus
A tail-less tenrec, the coarse, spiny coat of this species is distinctively two-tone—black, with yellow stripes and crown bristles.

5½–7½in
14–19cm

COMMON TENREC
Tenrec ecaudatus
A large terrestrial tenrec, with red-brown fur and spines along the body, the females have up to 29 teats, more than any other mammal.

LESSER HEDGEHOG TENREC
Echinops telfairi
Its hairs modified into spines, this species bears a remarkable resemblance to a true hedgehog, sharing its defense behavior of rolling into a ball.

white-tipped spines on back

4–6 in
10–15 cm

moderately large ears

4–4¾ in
10–12 cm

RICE TENREC
Oryzorictes sp.
A burrower with well-developed forelimbs, long claws, and small eyes and ears, this tenrec can become very numerous in marshes and rice paddies.

6–8½ in
15–21 cm

GREATER HEDGEHOG TENREC
Setifer setosus
Widespread throughout Madagascar, even in urban habitats, this species has a diverse diet, from insects, earthworms, and carrion to fruit.

sharp claws

AARDVARK

The only species in the Orycteropodidae, the aardvark is adapted for digging. It has an arched back, thick skin, long ears, and a tubular snout.

The aardvark's toes (four on the front feet and five on the hind) each have a flattened nail used to dig burrows and excavate insect nests, which it detects through its keen sense of smell.

It uses its long, thin tongue to catch huge numbers of insects. Its teeth are very unusual, and are one of the main reasons for it being classified in its own order. A young aardvark is born with incisors and canines at the front of the jaw, but when these fall out they are not replaced. The rear teeth lack enamel and grow through life; they are columnar and rootless, and unlike those of any other mammal.

PHYLUM	CHORDATA
CLASS	MAMMALIA
ORDER	TUBULIDENTATA
FAMILIES	1
SPECIES	1

AARDVARK

Inhabiting savanna and bushland in sub-Saharan Africa, aardvarks are solitary and nocturnal. Powerful front legs with flattened claws are used to dig into the nests of ants and termites for food.

AARDVARK
Orycteropus afer
The yellowish skin of the aardvark is often stained red by soil, in which it burrows and forages. It also has a sparse covering of bristly hairs.

3¼–4¼ ft
1–1.3 m

hair lighter colored on body

long, blunt, shovel-shaped claws

DUGONG AND MANATEES

A small order of fully aquatic herbivores, sirenians inhabit a range of tropical habitats, from swamps and rivers to marine wetlands and coastal waters.

Also known as sea cows, sirenians are supremely adapted to an aquatic way of life. Their front legs are paddlelike and adapted for steering; the hind legs are not visible, being restricted to two small remnant bones floating in the muscle. The body is therefore streamlined, and the flattened tail provides propulsion. A layer of blubber below its skin provides insulation, but to help counteract its natural buoyancy, the bones are dense and the lungs and diaphragm extend for the full length of the spine. The result is a hydrodynamic body, albeit rather slow-moving, capable of fine adjustments to shift its position in the water. All four species are considered to be at risk of extinction.

PHYLUM	CHORDATA
CLASS	MAMMALIA
ORDER	SIRENIA
FAMILIES	2
SPECIES	4

HYRAXES

Furry and rotund, modern hyraxes belong to the only extant family in the order Hyracoidea. Hyraxes were once the primary terrestrial herbivores in the Old World. Now just four species remain.

Widely represented in the fossil record of Asia, Africa, and Europe, some extinct species of hyrax were over three feet in height.

Unlike most browsing and grazing animals, hyraxes use their cheek teeth (rather than incisors) to slice vegetation, before chewing it. With a relatively indigestible diet, they have a complex stomach in which bacteria break down tough plant matter.

With a number of primitive mammalian features, such as poor control of their internal body temperature, hyraxes show several affinities with elephants: their upper incisor teeth are often extended into short tusks, their soles have sensitive pads, and they have high brain function.

PHYLUM	CHORDATA
CLASS	MAMMALIA
ORDER	HYRACOIDEA
FAMILIES	1
SPECIES	4

DUGONG

There is only one species in the Dugongidae. Slow-moving, marine herbivores, dugongs graze on submerged seagrass beds, using the downturned muscular snout to uproot food.

8¼–9¾ ft
2.5–3 m

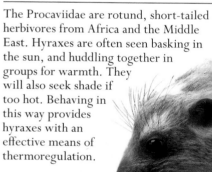

DUGONG
Dugong dugon
From the Indo-Pacific region, especially around Australia, the dugong's cylindrical body has no dorsal fin or hind limbs. The tail has flukes.

MANATEES

With shorter snouts than dugongs, the Trichechidae also differ in the shape of their tail, which is paddle-shaped rather than fluked. Rather sluggish, they spend much of the day sleeping under water, surfacing every 20 minutes to breathe.

AMAZONIAN MANATEE
Trichechus inunguis
A freshwater manatee from the Amazon Basin, this species usually has a characteristic white chest patch.

9¾–15 ft
3–4.6 m

FLORIDA MANATEE
Trichechus manatus latirostris
The largest living sirenian, the Florida manatee inhabits both freshwater and coastal waters of the southeastern USA. It feeds on aquatic plants.

6½–9¼ ft
2–2.8 m

9¾–13 ft
3–4 m

ANTILLEAN MANATEE
Trichechus manatus manatus
Ranging from Mexico to Brazil, this manatee is smaller and more likely to be found farther from the shore than the Florida subspecies.

HYRAXES

The Procaviidae are rotund, short-tailed herbivores from Africa and the Middle East. Hyraxes are often seen basking in the sun, and huddling together in groups for warmth. They will also seek shade if too hot. Behaving in this way provides hyraxes with an effective means of thermoregulation.

12½–23½ in
32–60 cm

long, silky fur

short snout

WESTERN TREE HYRAX
Dendrohyrax dorsalis
A dark, semiarboreal hyrax from West and central Africa, this species has a distinctive white patch on its lower back and rump.

12–15 in
30–38 cm

12–28 in
30–70 cm

BUSH HYRAX
Heterohyrax brucei
There are 25 subspecies of bush hyrax in rocky habitats across Africa. It feeds on grass and fruit, and also catches and eats small vertebrates.

18–22 in
45–55 cm

SOUTHERN TREE HYRAX
Dendrohyrax arboreus
An able climber from southern Africa, usually inhabiting hollow trees, this hyrax is known for its shrieking territorial calls at night.

ROCK HYRAX
Procavia capensis
From Africa and the Middle East, this colonial hyrax basks for long periods. Moist, rubbery foot pads provide grip on smooth rocky surfaces.

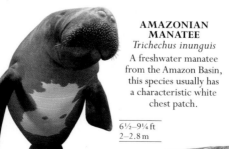

ELEPHANTS

Largest of all land mammals, elephants are distinguished by their colossal bulk, long flexible trunk, large ears, and curving ivory tusks.

The only members of the Proboscidea, elephants inhabit grassland and forest in tropical Africa and Asia. Sometimes weighing more than five tons, elephants have a modified skeleton to bear their weight. The limb bones are extremely stout, and the toes are splayed around a pad of connective tissue. This large frame needs vast amounts of food and elephants eat for up to 16 hours a day, consuming up to 550 lb (250 kg) of vegetation.

EARS, NOSE, AND MOUTH

The elephant's muscular trunk, a fusion of the nose and upper lip, is remarkably versatile and is able to grasp morsels of food. It can also draw in water and serve as a snorkel when swimming. Sensory organs in the trunk pick up scents and vibrations, aiding communication. An elephant's large ears are also used for communication and when spread out are a sign of aggression. Constant flapping of the ears helps an elephant lose heat. Tusks are incisor teeth that grow continually. They are used to dig for roots and salt, clear pathways, and mark territories.

SOCIAL STRUCTURES

Male and female elephants have different social behavior. Females form herds of related cows and their calves, led by an elderly matriarch. Bull elephants form short-term associations, but will fight with any intruding rival during the breeding season.

PHYLUM	CHORDATA
CLASS	MAMMALIA
ORDER	PROBOSCIDEA
FAMILIES	1
SPECIES	3

DEBATE

TWO SPECIES OR ONE?

Most zoologists recognize two species of African elephant, the savanna elephant and the smaller forest elephant. Physical differences between the species are reflected in their DNA, but even so, the two types may interbreed where their ranges overlap. Because of this, many conservation bodies still regard all African elephants a single species.

ELEPHANTS

The extinct mammoths and the living elephants form a family—the Elephantidae—of large herbivores, with a long trunk, large ears, thick skin, and tusks. The trunk is used in feeding, drinking, bathing, and social interactions. They make many sounds, from trumpeting to subsonic rumblings, enabling them to communicate over long distances.

6½–12 ft
2–3.6 m

large ears

9¾–12 ft
3–3.6 m

long, curved tusks

ASIATIC ELEPHANT
Elephas maximus
Domesticated for forest operations and ceremonial use, this species has smaller ears and a more arched back than African species. Females usually lack tusks, as do some males.

AFRICAN SAVANNA ELEPHANT
Loxodonta africana
The largest living land animal, this elephant has a large head and ears. Both sexes have well-developed, curved tusks.

semicircular toenails

6½–8¼ ft
2–2.5 m

muscular trunk projections

AFRICAN FOREST ELEPHANT
Loxodonta cyclotis
With five toenails on the forelimbs and four on the hind limbs, this small elephant with straight tusks from the forests of Africa resembles the Asiatic elephant.

ARMADILLOS

Recognizable by their armored shell— unique among mammals—armadillos occur in various shapes, sizes, and colors. All are native to the Americas.

These animals, sole members of the order Cingulata, range across a variety of habitats, feeding primarily on insects and other invertebrates. Despite having short legs, armadillos can run quickly and burrow with their strong claws to avoid predators. Their main defense consists of a bony carapace, covered with horny plates, across their upperparts. In most species, there are rigid shields over the shoulders and hips, with a varying number of bands separated by flexible skin covering the back and flanks. This allows some species to roll into a

ball to protect the vulnerable, furry underparts. Armadillos have few natural predators. Despite the threats from human hunters and habitat loss, the range of some armadillo species is expanding.

Although their armor is heavy, armadillos are effective swimmers. By inflating the stomach and intestines with air, they can increase their buoyancy and cross small bodies of water. They can also remain underwater for several minutes, another mechanism for avoiding predation.

HABITS

Most species of armadillo are nocturnal, though they occasionally emerge during the day as well. These animals are largely solitary, associating with one another only during the mating season. Males sometimes display aggression toward rivals.

PHYLUM	CHORDATA
CLASS	MAMMALIA
ORDER	CINGULATA
FAMILIES	1
SPECIES	21

Contrary to popular belief, not all armadillos can roll into a ball in defense— only species of the genus *Tolypeutes* can.

ARMADILLOS

The only surviving family in their order, the Dasypodidae are found in the Americas. Covered with bony plates across their upperparts, they use their sharp claws to dig for invertebrates and to excavate burrows. Of the 20 or so species, some are able to roll into a ball when threatened, in order to protect their vulnerable, soft, furry underparts.

tail long in proportion to body

9½–22½ in
24–57 cm

LONG-NOSED ARMADILLO
Dasypus sp.
The six species of this genus inhabit shaded, rocky areas. Unlike other armadillos, they have sparse, yellowish fur, mainly on their bellies.

long, pointed snout used for foraging

well-developed claws

9–16 in
22–40 cm

ANDEAN HAIRY ARMADILLO
Chaetophractus nationi
Unusual in having copious hair between its scales, this inhabits high-altitude South American grasslands. It is hunted for its meat and shell.

9–16 in
22–40 cm

LARGER HAIRY ARMADILLO
Chaetophractus villosus
From arid habitats in southern South America, this species has long, coarse hairs between the 18 or so plates of protective armor across its back.

10–13 in
26–33 cm

PICHI
Zaedyus pichiy
A small, dark armadillo with thick dorsal plates, this species wedges itself into its burrow when threatened, presenting its jagged scales for defense.

16–19½ in
40–49 cm

SIX-BANDED ARMADILLO
Euphractus sexcinctus
More active by day than most armadillos, this brownish yellow species forages in grassland and forest, eating plant and animal matter.

3½–4½ in
9–11.5 cm

PINK FAIRY ARMADILLO
Chlamyphorus truncatus
This tiny armadillo from central Argentina is mainly subterranean, "swimming" through loose sand; its head is shielded to reduce abrasion.

30–35 in
75–90 cm

12–28 in
30–70 cm

rounded, pink ears

jointed bands on blackish upperparts

NORTHERN NAKED-TAILED ARMADILLO
Cabassous centralis
The protective plates of this Central and South American armadillo do not extend to its tail. It relies mainly on dense cover to avoid predators.

GIANT ARMADILLO
Priodontes maximus
The largest armadillo, this species has tough armor and uses its long, curved third front claw to dig for food and in defense.

SIX-BANDED ARMADILLO
Euphractus sexcinctus

Despite its name, this species of armadillo may have anything from six to eight armored bands around its middle. The upper part of the animal's body is covered by a sturdy carapace, which is made up of plates of bone, topped with a thin covering of horny material, and is embedded in the armadillo's skin. The bands act as joints in the carapace, giving the animal flexibility, although it is unable to curl completely into a ball like some other species. The six-banded armadillo is diurnal, spending the daylight hours trundling around its home range in search of food. It has a varied diet, ranging from roots and shoots to invertebrates and carrion.

SIZE 16–19½ in (40–49 cm)
HABITAT Forest
DISTRIBUTION South America, mainly south of Amazon Basin
DIET Omnivorous

< SHIELDED EYE
The armadillo's eyes are protected by the headshield at the expense of a slightly restricted field of vision. Because it has poor eyesight anyway, this makes little difference.

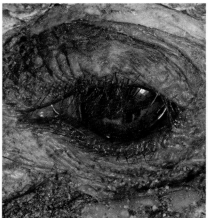

< NOSE
Armadillos have an acute sense of smell. They sniff and snuffle their way to most meals, and can easily detect food buried in the soil.

head is narrow and pointed and protected by a shield made of fused bony plates —————

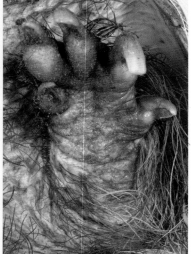

< STRONG LEGS
The armadillo has short but powerful legs. When digging, soil loosened by the front feet is kicked out of the hole by the hind feet.

∧ HAIRY SKIN
The bony carapace is interrupted by six to eight bands, which are separated by flexible skin. Long, bristly hairs sprout from between the bands—this is one of the hairy armadillos.

TAIL >
Glands at the base of the armadillo's tail release scent through small holes in the armor plates. The scent marks the armadillo's territory.

∧ CLAWS
Long, strong claws allow the armadillo to dig rapidly into hard ground. In a matter of minutes it is able to excavate a trench deep enough to bury itself in.

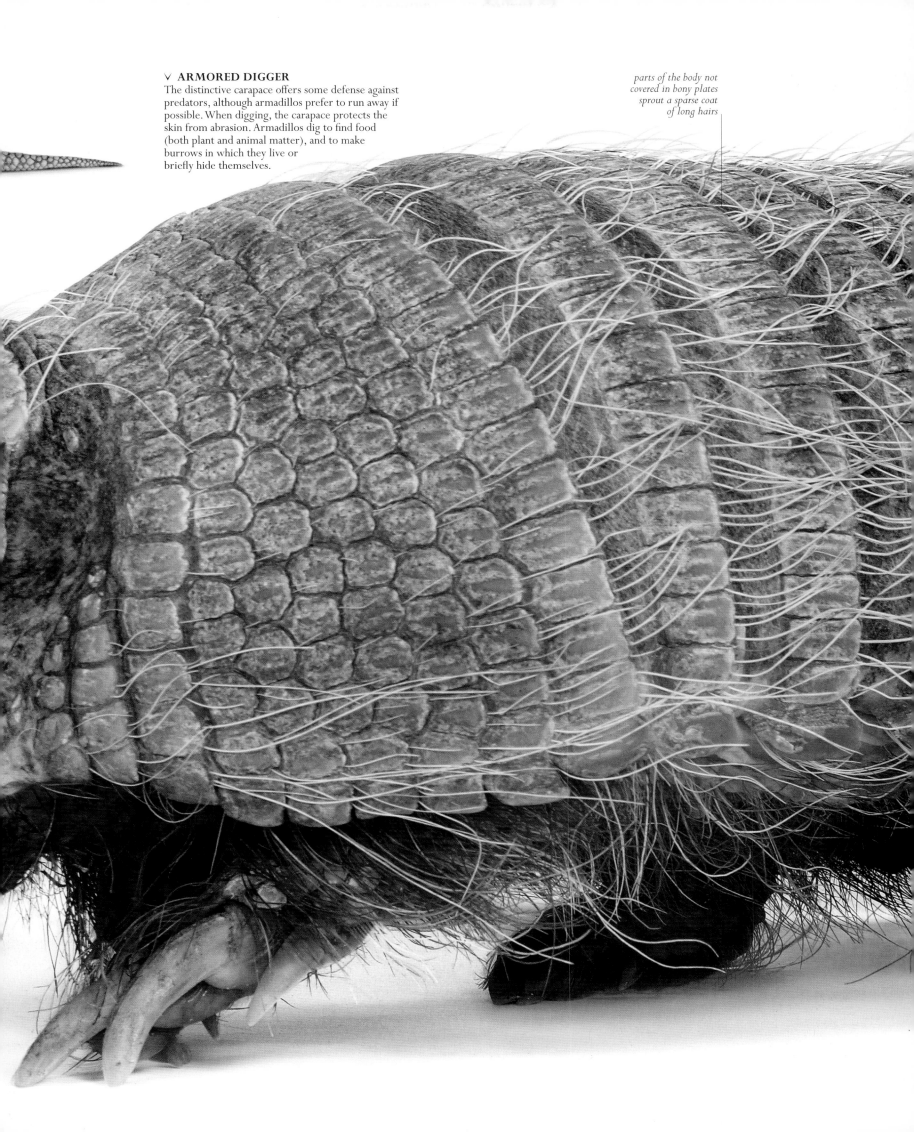

∨ ARMORED DIGGER

The distinctive carapace offers some defense against predators, although armadillos prefer to run away if possible. When digging, the carapace protects the skin from abrasion. Armadillos dig to find food (both plant and animal matter), and to make burrows in which they live or briefly hide themselves.

parts of the body not covered in bony plates sprout a sparse coat of long hairs

SLOTHS AND ANTEATERS

Although very different in form and habits, sloths and anteaters have one characteristic in common: they share a lack of normal mammalian dentition.

With just one exception, members of this order, the Pilosa, are largely arboreal creatures: only the giant anteater is terrestrial. They are all native to Central and South America.

The two anteater families feed on ants, termites, and other insects. Anteaters have no teeth at all, relying on their long sticky tongue to capture insects, which are then crushed in the mouth before being swallowed.

The slow-moving sloths are herbivorous. They have no incisors or canines, but instead have a number of cylindrical, rootless teeth, used to grind up their food. A sloth may take about a month to completely digest a meal, as the fibrous leaf matter passes slowly through several stomach compartments and is broken down by bacteria.

ARBOREAL LIFE

The long, prehensile tails of the arboreal anteaters allow them to lead a more active lifestyle than the sloths. On the ground, all the arboreal species move only with difficulty, although some are excellent swimmers. When walking, their mobility is restricted by their enlarged, curved claws, which the sloths use like hooks to hang from branches. Anteaters have enlarged claws only on their front feet. These claws are used to tear open insect nests in search of food, and are formidable defense weapons.

PHYLUM	CHORDATA
CLASS	MAMMALIA
ORDER	PILOSA
FAMILIES	4
SPECIES	10

To defend themselves members of the genus *Tamandua* stand up on their hind legs and lash out with their powerful front limbs.

THREE-TOED SLOTHS

Generally smaller and slower moving than two-toed sloths, the arboreal and mostly nocturnal Bradypodidae have three toes on each foot, bearing long, curved claws that help them hang from branches. Their long, shaggy fur has a green tinge, due to algae growing on it.

18–20 in
45–50 cm

MANED SLOTH
Bradypus torquatus
This small Brazilian sloth has long, dark fur—especially around its head and neck—which often harbors algae, ticks, and moths.

BROWN-THROATED SLOTH
Bradypus variegatus
An inhabitant of Central and South American forests, this sloth is the most widespread in its family. The female attracts her mate with a shrill scream.

16½–32 in
42–80 cm

18–30 in
45–76 cm

shaggy, coarse coat

PALE-THROATED SLOTH
Bradypus tridactylus
Found in the rain forests of South America, this species has an unusual appearance, with almost no tail or external ears.

TWO-TOED SLOTHS

Unlike three-toed sloths, the Megalonychidae have only two toes on their forefeet, a more prominent snout, and no tail. Similarly arboreal and mainly nocturnal, they usually descend from trees head first.

SOUTHERN TWO-TOED SLOTH
Choloepus didactylus
This large herbivore from South America swims well, even crossing rivers. Its main predators are large raptors, such as harpy eagles.

two toes on front feet

three toes on hind feet

wavy, variegated hair adds camouflage to body

21–29 in
53–74 cm

SILKY ANTEATERS

The Cyclopedidae, though well-represented in the fossil record, has only one extant species. Large, curved front claws and a prehensile tail help the silky anteater live in trees, where it nests in holes. It feeds on ants and termites.

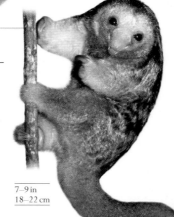

SILKY ANTEATER
Cyclopes didactylus
Arboreal, nocturnal, and slow moving, this is the smallest anteater. When threatened, its sharp claws provide an effective defense against predators.

7–9 in
18–22 cm

GIANT ANTEATER AND RELATIVES

Native to Central and South America, the Myrmecophagidae have elongated snouts and long tongues. They use their powerful claws to rip open termite mounds and ant hills. Lacking teeth, they catch insects with their sticky saliva and spine-covered tongues.

21–35 in
53–88 cm

COLLARED ANTEATER
Tamandua tetradactyla
This solitary species has a prehensile tail. Only the southern populations in South America have black markings.

huge,
bushy tail

stiff, strawlike
hair

long, tubular
snout

3¼–4 ft
1–1.2 m

GIANT ANTEATER
Myrmecophaga tridactyla
The largest anteater, with a tail almost as long as its body, this species uses its long, sticky tongue to catch up to 30,000 insects a day.

RABBITS, HARES, AND PIKAS

Two herbivorous families make up the order known as Lagomorpha. Their apparent similarities to rodents result from adaptation to a common lifestyle.

These species occupy a variety of habitats, from tropical forest to Arctic tundra. All are terrestrial herbivores or browsers. They are gnawing animals and share the diet of many rodents. Like those of rodents, the teeth of rabbits, hares, and pikas continue to grow throughout life—they are worn down by chewing. However, there are fundamental differences between the groups: rabbits and their relatives have four incisors in the upper jaw, while rodents have only two.

Relying on food that is relatively slow to break down, these animals have a modified digestive system. They produce two types of feces: moist pellets, which are eaten to gain further nutrients from the food, and dry pellets passed as waste.

ESCAPING PREDATORS

Pikas, the most rodentlike species in this order, take refuge from predators in burrows and crevices after raising the alarm with a whistle. Rabbits and hares, in contrast, have long ears to detect danger and powerful limbs to flee from predators. These animals have large eyes positioned high on each side of the head to provide almost 360-degree vision. When a predator is observed, hares drum their hind legs on the ground in warning.

PHYLUM	CHORDATA
CLASS	MAMMALIA
ORDER	LAGOMORPHA
FAMILIES	3
SPECIES	92

To survive the winter, the pika collects various plants, creates a hay pile of dried food, and stores it in its den.

RABBITS AND HARES

Native across much of the world, the Leporidae have long, movable ears and enlarged hind legs to detect and escape predators. Large eyes reflect their mainly nocturnal habits. Rabbits often inhabit permanent burrow systems, whereas the more solitary hares make only transitory shelters.

EUROPEAN RABBIT
Oryctolagus cuniculus
Native to the Iberian peninsula, this species has been introduced for meat and fur worldwide, with devastating effects on local habitats and wildlife.

13½–18 in
34–45 cm

furry feet

6–12 in
15–30 cm

LOP-EARED RABBIT
Oryctolagus cuniculus
One of the oldest domesticated forms, this type has long, lop-eared (floppy) ears. There is variation in both color and size.

5–7 in
13–18 cm

DWARF RABBIT
Oryctolagus cuniculus
One of the smallest breeds, the dwarf rabbit occurs in many colors and patterns. Its rounded face and small ears make it a popular pet.

10–15 in
25–38 cm

CREAM ANGORA RABBIT
Oryctolagus cuniculus
Valued for its long, soft hair, which is spun into yarn, this breed originates from Anatolia (in modern-day Turkey).

brown fur, sometimes turns reddish

long, black-tipped ears

20–28 in
50–70 cm

9–22 in
22–55 cm

COTTONTAIL
Sylvilagus sp.
Distributed across the Americas, most of the 17 species in this genus have stubby white tails and nest above ground, not in burrows.

MARSH RABBIT
Sylvilagus palustris
Inhabiting wetlands of North America, this species is a strong swimmer. It walks instead of hopping, unlike most of its relatives.

16½–17½ in
42–44 cm

EUROPEAN HARE
Lepus europaeus
A grazer in summer and a browser of bark and buds in winter, this hare is shy and solitary, except during its springtime "boxing" courtship. Females fight off males until ready to mate.

powerful hind legs

14–18 in
36–46 cm

SNOWSHOE HARE
Lepus americanus
Adapted to harsh North American winters, this species has a white winter coat for camouflage, and large hind feet to allow movement on soft snow.

22–28 in
55–70 cm

22–26 in
56–66 cm

WHITE-TAILED JACKRABBIT
Lepus townsendii
This species ranges extensively across western North America. The northern population turns white in winter, while the southern animals develop whitish patches only on their flanks.

ARCTIC HARE
Lepus arcticus
Adapted to polar and mountainous areas, this hare survives with thick fur that turns white in winter, and by digging holes in the snow for shelter.

18½–25 in
47–63 cm

BLACK-TAILED JACKRABBIT
Lepus californicus
Widespread on western North American prairies and farmland, this species is prone to huge local fluctuations. It has a black tail and black ear-tips.

18–23½ in
45–60 cm

ANTELOPE JACKRABBIT
Lepus alleni
Very long ears and insulated, reflective fur help this large hare to keep cool in its desert grassland habitats in Mexico.

18–26 in
46–65 cm

MOUNTAIN HARE
Lepus timidus
From polar regions and mountains in Europe and Asia, this hare has a white winter coat and a tail that stays white all year round.

20½–23½ in
52–60 cm

CAPE HARE
Lepus capensis
Common in open habitats of Africa and the Middle East, this species is very similar to the closely related European hare.

long legs

PIKAS

The Ochotonidae are small herbivores that live in rocky mountain sides and open steppes in North America and Asia. Alert to predators, they produce high-pitched alarm calls as they run for cover in crevices and burrows.

6½–8½ in
16–21 cm

AMERICAN PIKA
Ochotona princeps
This North American montane scree dweller dries food piles in the sun, and stores them in its den for the winter.

RODENTS

Ranging from tiny mice to animals the size of a pig, rodents live in almost every habitat type. They make up nearly half of all mammal species.

Members of the order Rodentia are distinguished by having a pair of prominent upper and lower incisor teeth (often orange or yellow in color), which continue to grow throughout life. Gnawing—a characteristic behavior of all rodents—wears the teeth away at the same rate as they grow. Rodents have no canines, just a long space between the incisors and the three or four cheek teeth in each jaw. The scientific classification of rodents is also based on other features of their teeth and jaws that are not visible externally.

SPECIES DIVERSITY

To suit their various lifestyles, rodents often have special adaptations such as webbed toes, big ears, or long hind feet for a bounding gait. Some rodents burrow, others live in trees or in the water, but none live in the sea. Many species inhabit desert regions, where they may never drink, deriving all the water they need from food.

THE IMPACT OF RODENTS

Several rodent species carry fatal diseases that have killed millions of people. Others consume or contaminate huge quantities of stored food meant for humans. The house mouse has become the world's most widespread wild mammal through its close association with humans. This species inhabits every continent, including Antarctica, even surviving in mines and cold stores. Some rodents cause damage to crops or trees, or burrow in inconvenient places. Beavers can transform entire habitats, affecting hundreds of other animal and plant species.

On the positive side, rodents are a vital food source for predatory animals and, in some countries, for humans as well. Many small rodents, such as hamsters, are specially bred as pets.

PHYLUM	CHORDATA
CLASS	MAMMALIA
ORDER	RODENTIA
FAMILIES	33
SPECIES	2,277

DEBATE
RODENT ORGANIZATION

The huge diversity of species in the order Rodentia makes the task of classifying them complicated. With 33 families of modern rodents to organize, experts in zoological classification have divided them into two subgroups, according to differences in their skulls, teeth, and jaws. The Sciurognathi (squirrel-jaws) are globally widespread, and include all squirrels, beavers, and mouselike rodents. The Hystricognathi (porcupine-jaws) includes the cavies, porcupines, chinchillas, and the capybara, and these are mostly restricted to the Southern Hemisphere and the tropics.

MOUNTAIN BEAVER

Only one species of the once widespread family Aplodontiidae has survived into modern times. The mountain beaver, or sewellel, lives in burrows in the damp forests and plantations of western North America.

flattened head

12–16 in
30–40 cm

MOUNTAIN BEAVER
Aplodontia rufa
This most primitive type of living rodent inhabits forested coastal mountains of western Canada and the USA.

SQUIRRELS AND CHIPMUNKS

Occurring almost everywhere, except in the polar regions, Australia, and the Sahara Desert, members of the family Sciuridae range from tropical rain forests to Arctic tundra, and from the tops of trees to underground tunnels. This family includes typical bushy-tailed tree squirrels, many species of burrow-digging ground squirrels, chipmunks, and marmots. They feed mainly on nuts and seeds.

AMERICAN RED SQUIRREL
Tamiasciurus hudsonicus
The chattering calls and sharp barks of the red squirrel are common sounds of coniferous woodlands in Canada and the northern USA.

6½–8 in
17–20 cm

9–12 in
23–30 cm

GRAY SQUIRREL
Sciurus carolinensis
Common in the eastern USA, this species was introduced to parts of Europe, where it is slowly displacing the native red squirrel.

7–9½ in
18–24 cm

EURASIAN RED SQUIRREL
Sciurus vulgaris
When this squirrel molts into its summer coat, the long hairs on its ears are lost.

SOUTHERN FLYING SQUIRREL
Glaucomys volans
A strictly nocturnal native of the eastern USA, this squirrel lives in tree holes and attics, often in groups during winter.

gliding membrane

5–6 in
13–15 cm

10–18 in
25–46 cm

PALE GIANT SQUIRREL
Ratufa affinis
Of the four species of giant squirrel, this species inhabits the forests of the Malay Peninsula, Borneo, and Sumatra.

10–18 in
25–45 cm

GRIZZLED GIANT SQUIRREL
Ratufa macroura
One of the world's largest tree squirrels, this species occurs in southern India and Sri Lanka. It eats fruits, flowers, and insects.

5–11 in
13–28 cm

PREVOST'S SQUIRREL
Callosciurus prevostii
There are 15 species of these tricolored squirrels. Prevost's is found in Malaysia, Borneo, Sumatra, and nearby islands, such as Sulawesi.

»

» SQUIRRELS AND CHIPMUNKS

long, bushy tail

9½ in
24 cm

6½–10½ in
17–27 cm

GAMBIAN SUN SQUIRREL
Heliosciurus gambianus
A common species in savanna woodlands across Africa, from Senegal to Zimbabwe, this squirrel feeds mainly on the seeds of Acacia trees.

CAPE GROUND SQUIRREL
Xerus inauris
This coarsely haired squirrel escapes the extreme temperatures of the semideserts of South Africa by sheltering in burrows.

6–8 in
15–20 cm

COLUMBIAN GROUND SQUIRREL
Urocitellus columbianus
A rather large, bushy-tailed ground squirrel, this species forms colonies in meadows and forest edges from Idaho, USA, north into western Canada.

GOLDEN-MANTLED GROUND SQUIRREL
Callospermophilus lateralis
Appearing like a larger version of the chipmunk, this species is a common sight in forests and mountains of the western USA.

10–12 in
25–30 cm

short hairy tail

4¾–6 in
12–15 cm

4¾–6 in
12–15 cm

HOPI CHIPMUNK
Tamias rufus
Chipmunks from different parts of North America form separate species. This species occurs in Utah, Colorado, and adjacent parts of Arizona.

EASTERN CHIPMUNK
Tamias striatus
This striped, ground-living species is a common sight in forest campgrounds of the eastern USA, where it has become quite tame.

BEAVERS

There are only two species of beavers in the family Castoridae. Both are exploited for their fur. One is widespread across North America, and the other in patches throughout most of Europe. Both build dams out of stones, mud, and trees, creating habitats for other species.

glossy, brown fur

BEAVER
Castor sp.
European and North American beavers are different species, but both live a similar, semiaquatic life in rivers and lakes.

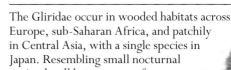

2½–4 ft
0.8–1.2 m

flat, scaly tail

DORMICE

The Gliridae occur in wooded habitats across Europe, sub-Saharan Africa, and patchily in Central Asia, with a single species in Japan. Resembling small nocturnal squirrels, all but one are soft-furred, arboreal species. Many are threatened or are declining in numbers.

2¾–6 in
7–15 cm

AFRICAN DORMOUSE
Graphiurus sp.
There are 14 species of African dormice. They are all similar in appearance and inhabit forested areas of sub-Saharan Africa.

bushy tail

GOPHERS

Members of the family Geomyidae, found in North America, live alone in shallow burrows that allow them to access roots and leaves. They carry their food in cheek pouches and store it in underground chambers.

5–9½ in
13–24 cm

BOTTA'S POCKET GOPHER
Thomomys bottae
A small burrowing creature of soft soil and grassy areas, this species digs up mounds of earth, which can damage farm machinery.

14–20 in
35–50 cm

gingery brown fur

MARMOT
Marmota sp.
Several species of marmots inhabit mountain grasslands and rocky areas of North America, while some occur in Eurasia. Marmots hibernate for up to nine months.

POCKET MICE AND KANGAROO RATS

Mostly common, with a few rare species, the Heteromyidae are found in varied habitats from Canada to Central America, with kangaroo rats living mainly in deserts. Pocket mice typically run on four feet, but kangaroo rats hop on two large hind feet.

4 in
10 cm

longer hairs of tail tuft

2¾–3½ in
7–9 cm

DESERT POCKET MOUSE
Chaetodipus penicillatus
This species is one of many small nocturnal rodents that inhabit the open sandy deserts of the southwest USA and northern Mexico.

MERRIAM'S KANGAROO RAT
Dipodomys merriami
Bounding away with its tail outstretched, this nocturnal inhabitant of the North American deserts looks like a miniature kangaroo.

10½–12½ in
27–32 cm

BLACK-TAILED PRAIRIE DOG
Cynomys ludovicianus
Active during daytime, this ground squirrel lives in large groups in "towns" consisting of extensive communal burrows.

JUMPING MICE

The family Dipodidae is made up of jerboas, birch mice, and jumping mice. Their powerful hind feet and long tail enable them to bound around like miniature kangaroos.

MEADOW JUMPING MOUSE
Zapus hudsonius
Found in the cool grasslands of northern North America, this species also has isolated populations in the mountains of Arizona and New Mexico.

2¾–4¼ in
7–11 cm

5½–6½ in
14–16 cm

HARRIS'S ANTELOPE SQUIRREL
Ammospermophilus harrisii
An agile inhabitant of the Sonora Desert and northern Mexico, this rodent is active in the heat of the day, but hibernates during winter.

LESSER EGYPTIAN JERBOA
Jaculus jaculus
This desert species is found across North Africa, from Senegal to Egypt, south to Somalia, and east to Iran.

3½–6½ in
9–16 cm

short ears

EDIBLE DORMOUSE
Glis glis
Known in Germany as "the seven-sleeper", this dormouse hibernates for over six months. It is eaten in Mediterranean countries, such as Slovenia.

4¾–6½ in
12–17 cm

VOLES, LEMMINGS, AND MUSKRATS

About 680 chubby, short-tailed rodents, including hamsters, voles, lemmings, and muskrats, make up the family Cricetidae. They are found worldwide, from western Europe to Siberia and the Pacific coast.

russet fur on back

3½–4¼ in
9–11 cm

BANK VOLE
Myodes glareolus
Active mostly at dawn and night, this typical vole inhabits scrub, woodlands, and gardens across most of western Europe and east into Russia.

3½–4¾ in
9–12 cm

COMMON VOLE
Microtus arvalis
A common burrowing inhabitant of grasslands across most of northern Europe, this vole also occurs farther east and in Russia.

4¾–9 in
12–23 cm

EURASIAN WATER VOLE
Arvicola amphibius
In Britain and parts of Europe, this vole occurs near waterside habitats, but in Russia and Iran, it lives away from water, digging extensive burrow systems.

COMMON DORMOUSE
Muscardinus avellanarius
An excellent climber and jumper, this dormouse feeds on flowers, fruits, and insects at night, in shrubby woodland across Europe.

2¼–3½ in
6–9 cm

densely furred tail

2¾–6½ in
7–16 cm

NORWAY LEMMING
Lemmus lemmus
The main small mammal of the European tundra, the lemming's fluctuating population affects the breeding success of many Arctic predators.

3¼–4¾ in
8–12 cm

STEPPE VOLE
Lagurus lagurus
The only species in its genus, this vole lives in the dry, grassy plains that stretch from Ukraine to western Mongolia.

9–13 in
23–33 cm

MUSKRAT
Ondatra zibethicus
Originally an inhabitant of rivers, ponds, and streams in N. America, the muskrat has been introduced to Europe and is now widespread.

»

≫ VOLES, LEMMINGS, AND MUSKRATS

2¾–4¾ in
7–12 cm

relatively long tail

STRIPED DWARF HAMSTER
Cricetulus barabensis

By raiding crops to gather seeds and grain, this rodent can become a serious pest in farming areas. Farmers destroy its burrows while plowing.

2–4 in
5–10 cm

8–13½ in
20–34 cm

ROBOROVSKI'S DESERT HAMSTER
Phodopus roborovskii

From the dry grasslands of central Asia, this tiny animal is popular as a pet. It breeds rapidly, up to 15 litters a year.

COMMON HAMSTER
Cricetus cricetus

This burrowing species lives alone and hibernates underground through the winter. Its burrows may contain up to 145 lb (65 kg) of stored food.

golden orange fur

6½–7 in
17–18 cm

6½–7 in
17–18 cm

LONG-HAIRED GOLDEN HAMSTER
Mesocricetus auratus

The golden hamster has been selectively bred to create many exotic forms, such as this variety, that would not survive in the wild.

GOLDEN HAMSTER
Mesocricetus auratus

Originally from Syria, where it is now an endangered species, this hamster has become a favorite household pet in Europe and North America.

3½–4¼ in
9–11 cm

WHITE-FOOTED MOUSE
Peromyscus leucopus

A very common and adaptable creature, this mouse inhabits almost every type of land habitat in central and the eastern USA.

4¾–8 in
12–20 cm

HISPID COTTON RAT
Sigmodon hispidus

This short-lived, surface-dwelling herbivore lives in grassland habitats in the southern USA and Mexico.

MICE, RATS, AND RELATIVES

One fifth of all mammal species belong to this huge family. The Muridae are distributed almost worldwide, including the polar regions. Some species carry serious diseases, and others are significant agricultural pests. However, several are used in medical research and are commonly kept as pets.

2¾–4¾ in
7–12 cm

NORTHEAST AFRICAN SPINY MOUSE
Acomys cahirinus

Like other spiny mice, this rodent has stiffened hairs on its body for protection and very thin skin, which permits easy cooling in its hot, dry habitat.

2¾–4¾ in
7–12 cm

ARABIAN SPINY MOUSE
Acomys dimidiatus

This mouse was previously considered the same species as the northeast African one, except that it occurs east of the Red Sea.

4–7 in
10–18 cm

3½–7 in
9–18 cm

SHAW'S JIRD
Meriones shawi

Common in the deserts of North Africa and the Middle East, this rodent does not hibernate, but survives the winter by storing up to 22 lb (10 kg) of food in its burrows.

MONGOLIAN JIRD
Meriones unguiculatus

In the wild, this gerbil lives in the dry steppes of central Asia, forming large social groups. Many are now kept as pets.

2–4¾ in
5–12 cm

PALLID GERBIL
Gerbillus perpallidus

Like most small desert rodents, this gerbil has a pale body. It is widespread in North Africa and the Middle East, where its coloration provides good camouflage.

4–5 in
10–13 cm

FAT-TAILED JIRD
Pachyuromys duprasi

Like many small, desert-living mammals, this species stores fat in its tail. The tail has no fur, allowing it to radiate excess body heat.

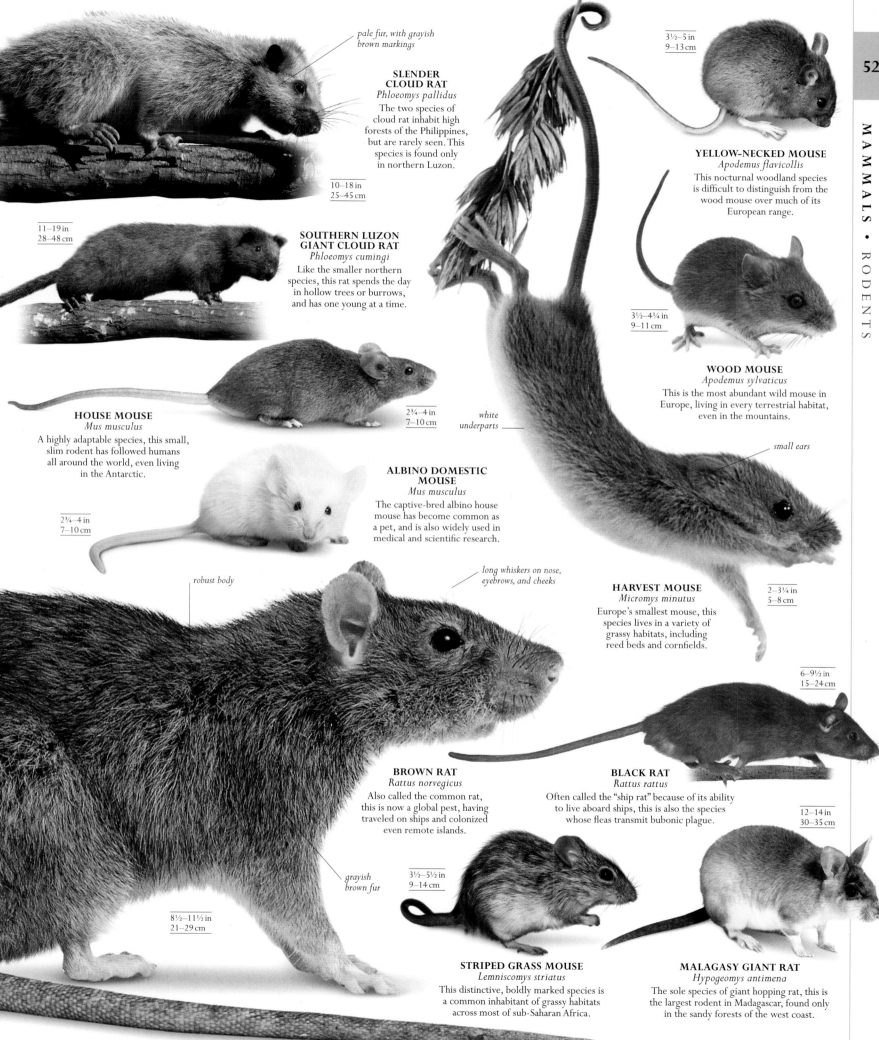

pale fur, with grayish brown markings

SLENDER CLOUD RAT
Phloeomys pallidus

The two species of cloud rat inhabit high forests of the Philippines, but are rarely seen. This species is found only in northern Luzon.

10–18 in
25–45 cm

YELLOW-NECKED MOUSE
Apodemus flavicollis

This nocturnal woodland species is difficult to distinguish from the wood mouse over much of its European range.

3½–5 in
9–13 cm

SOUTHERN LUZON GIANT CLOUD RAT
Phloeomys cumingi

Like the smaller northern species, this rat spends the day in hollow trees or burrows, and has one young at a time.

11–19 in
28–48 cm

WOOD MOUSE
Apodemus sylvaticus

This is the most abundant wild mouse in Europe, living in every terrestrial habitat, even in the mountains.

3½–4¼ in
9–11 cm

small ears

HOUSE MOUSE
Mus musculus

A highly adaptable species, this small, slim rodent has followed humans all around the world, even living in the Antarctic.

2¾–4 in
7–10 cm

ALBINO DOMESTIC MOUSE
Mus musculus

The captive-bred albino house mouse has become common as a pet, and is also widely used in medical and scientific research.

2¾–4 in
7–10 cm

white underparts

HARVEST MOUSE
Micromys minutus

Europe's smallest mouse, this species lives in a variety of grassy habitats, including reed beds and cornfields.

2–3¼ in
5–8 cm

robust body

long whiskers on nose, eyebrows, and cheeks

BROWN RAT
Rattus norvegicus

Also called the common rat, this is now a global pest, having traveled on ships and colonized even remote islands.

grayish brown fur

8½–11½ in
21–29 cm

BLACK RAT
Rattus rattus

Often called the "ship rat" because of its ability to live aboard ships, this is also the species whose fleas transmit bubonic plague.

6–9½ in
15–24 cm

STRIPED GRASS MOUSE
Lemniscomys striatus

This distinctive, boldly marked species is a common inhabitant of grassy habitats across most of sub-Saharan Africa.

3½–5½ in
9–14 cm

MALAGASY GIANT RAT
Hypogeomys antimena

The sole species of giant hopping rat, this is the largest rodent in Madagascar, found only in the sandy forests of the west coast.

12–14 in
30–35 cm

BAMBOO RATS AND RELATIVES

The family Spalacidae includes blind mole rats, bamboo rats, root rats, and zokors. Big, protruding incisors characterize the blind mole rats. Adapted to an underground life, they have no external eyes and ears. The bamboo rats of eastern Asia have visible eyes.

6–10 in
15–26 cm

LESSER BAMBOO RAT
Cannomys badius
Found from Nepal to Vietnam, this single species of its genus digs deep burrows in forests, grassy areas, and sometimes gardens.

6½–14 in
17–35 cm

GREATER MOLE RAT
Spalax microphthalmus
This blind species has sensory bristles running from its snout to its eye sockets. It is native to the steppes of Ukraine and southeastern Russia.

SPRINGHARES

Resembling rabbits in size and behavior, the Pedetidae differ in producing only one young at a time. They breed all year round, but living in dry, open habitats makes them vulnerable to predators.

14–17 in
35–43 cm

EAST AFRICAN SPRINGHARE
Pedetes surdaster
Springhares escape their noctural predators with kangaroo-like leaps. Less common than *Pedetes capensis*, this species lives on the Serengeti plain in Africa.

SOUTH AFRICAN SPRINGHARE
Pedetes capensis
This species leaves its burrow at night to nibble at the grass and herbs found in the dry regions of southern Africa.

14–17 in
35–43 cm

long tufted tail

AFRICAN MOLE-RATS

Living entirely underground, the Bathyergidae have only rudimentary eyes. They burrow in sand and soft soils to feed on roots, using their large, protruding front teeth as shovels. Their lips close behind the teeth to save the mouth filling with dirt.

3¼–4 in
8–10 cm

long, protruding incisor teeth

NAKED MOLE-RAT
Heterocephalus glaber
This highly social animal lives in a colony, where each individual performs different, specialized jobs to help the colony as a whole.

long, rounded tail

3½–10½ in
9–27 cm

COMMON MOLE-RAT
Cryptomys hottentotus
Found from Tanzania to South Africa, this common species lives in soft soils and farmland, where it feeds mainly on roots.

NAMAQUA DUNE MOLE-RAT
Bathyergus janetta
Native to Namibia and southwestern South Africa, this species uses its forefeet rather than its teeth to dig burrows.

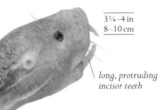

7–13 in
17½–33 cm

NEW WORLD PORCUPINES

Tree-dwelling animals in American forests, the Erethizontidae have short spines, which are generally less than 4 in (10 cm) long. Most have a prehensile tail to grip branches.

26–32 in
65–80 cm

NORTH AMERICAN PORCUPINE
Erethizon dorsata
This forest dweller is found throughout North America, from Alaska to Mexico. Its spines are hidden among its shaggy fur.

12–23½ in
30–60 cm

BRAZILIAN PORCUPINE
Coendou prehensilis
This nocturnal species inhabits forested areas of South America and Trinidad. It sleeps by day and forages for leaves and shoots at dusk.

OLD WORLD PORCUPINES

Found across most of Africa and southern Asia, the 14 species of the family Hystricidae family live in burrows. They are covered with long, stiff quills that protect them against most predators. When attacked, they often rattle their quills as a reminder of how sharp and impenetrable these are.

CRESTED PORCUPINE
Hystrix cristata
Widespread across the northern half of Africa, except the Sahara, the crested porcupine is a familiar nocturnal rodent.

23½–39 in
60–100 cm

spines are modified hair

30–39 in
75–100 cm

CAPE PORCUPINE
Hystrix africaeaustralis
Found in savanna across most of southern Africa, this species forages at night, alone or in groups, sniffing out roots and berries.

VISCACHAS AND CHINCHILLAS

All seven members of the family Chinchillidae from South America have a prominent tail and large hind feet. They normally live in social groups, inhabiting burrows or rocky outcrops. Most species are now quite rare because of their exploitation for fur and their status as pests.

large ears help regulate body temperature

long whiskers aid spatial awareness

CHINCHILLA
Chinchilla sp.
Widely prized for their fur, chinchillas have a fine, thick coat that insulates them against the cold in their Andean home.

9–15 in
22–38 cm

bushy tail balances the body

12–18 in
30–45 cm

MOUNTAIN VISCACHA
Lagidium viscacia
This agile rodent has dense fur to protect it from the cold at night. It lives on steep, rocky, mountain slopes.

DASSIE RAT

This single species in the family Petromuridae is found only in southern Africa. Its peculiar, flat skull and soft ribs are adaptations to living in crevices and under stones, and distinguish it from all other rodents.

DASSIE RAT
Petromus typicus
Living on dry, rocky hillsides, this rodent emerges from crevices at dawn and dusk to forage for seeds and shoots.

5½–8 in
14–20 cm

CANE RAT

The two species of the family Thryonomyidae have pale brown, coarse, flattened hair that blends with their surroundings in dry grass and cane thickets. Cane rats produce two small families of well-developed young per year.

14–23½ in
35–60 cm

CANE RAT
Thryonomys sp.
There are two species of African cane rats: one lives in savanna grasslands and the other in reedbeds and marshes.

PACARANA

The single species of the family Dinomyidae is a shy, bulky, slow-moving, and almost defenseless animal. It lives alone or in pairs in montane forests, where it is preyed upon by jaguars and humans.

PACARANA
Dinomys branickii
Threatened by loss of its South American forest habitat, and hunted for food, the pacarana is now an endangered species.

28–32 in
70–80 cm

GUINEA PIGS AND PATAGONIAN HARES

Including some of South America's most widespread and abundant rodents, the Caviidae occur from mountain meadows to tropical floodplains, and breed all year round. All but the mara are short-legged and dumpy animals.

BRAZILIAN GUINEA PIG
Cavia aperea
Guinea pigs mostly live in lowland habitats, but this species is also found in the Andes, from Peru to Chile.

8–16 in
20–40 cm

ROSETTE GUINEA PIG
Cavia porcellus
Several coat variations occur in pet guinea pigs. In this one, the fur forms substantial spiky whorls over the body.

8–16 in
20–40 cm

long ears

27–30 in
69–75 cm

LONG HAIRED GUINEA PIG
Cavia porcellus
Often kept as a pet, in captivity this long-haired breed needs grooming to prevent its fur from becoming tangled.

8–16 in
20–40 cm

DOMESTIC GUINEA PIG
Cavia porcellus
First domesticated more than 500 years ago for food, guinea pigs have since become popular pets worldwide.

8–16 in
20–40 cm

MARA
Dolichotis sp.
Extensive shared burrow systems and communal breeding habits are among the unusual features of these long-legged rodents.

CRESTED PORCUPINE
Hystrix cristata

With its long, barbed quills raised in threat, the crested porcupine is an intimidating prospect. A predator that has had the painful experience of being spiked is unlikely ever to attempt another attack. Lions, hyenas, and even people have been known to die from infected quill injuries. Despite this impressive means of defense, the porcupine is a peaceable and rather nervous animal, easily startled, and more likely to flee danger than stand its ground. Porcupines live alone or in family groups, sharing extensive burrow systems. Crested porcupines occur throughout much of northern Africa. The species was also once widespread in southern Europe—the population found in Italy may be a relic of former times or the result of a more recent introduction, perhaps by the Romans.

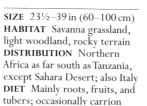

SIZE 23½–39 in (60–100 cm)
HABITAT Savanna grassland, light woodland, rocky terrain
DISTRIBUTION Northern Africa as far south as Tanzania, except Sahara Desert; also Italy
DIET Mainly roots, fruits, and tubers; occasionally carrion

∨ EYE
A porcupine's eyesight is poor, but then there is often little to see in the darkness of an African night. It uses its sharp senses of hearing and smell to find its way.

< SPIKY HAIRDO
Raising the quills creates an illusion of size. A cornered porcupine will stand tall and try to bluff its way out of trouble. If this fails, it will turn tail and charge backward, shoving its spiky rear end in the face of its attacker.

EAR >
The small ears are largely hidden in the coarse hair. The porcupine has good hearing, which it uses to avoid danger, shuffling off into the night if it hears another animal approaching.

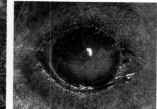

∨ MOUTH AND TEETH
Porcupines have the specialized gnawing teeth typical of rodents, enabling them to chew tough roots and tubers. The muscles that control the jaws are immensely powerful.

QUILLS-A-QUIVER >
Porcupine quills are greatly enlarged hairs. They can be raised by larger versions of the tiny muscles that create goosebumps in our own skin.

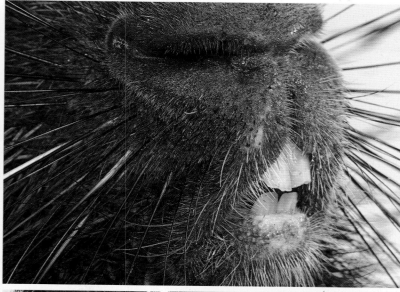

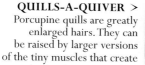

like its quills, the porcupine's thin coat of coarse hair can also be raised in alarm

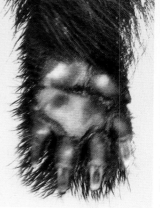

RATTLING TAIL ∧
The quills of the tail are swollen and hollow. The porcupine shakes them when alarmed and the soft rattling sound warns its enemies that this is a well-armed opponent.

< ∧ FEET AND CLAWS
Porcupines walk on the flat soles of their feet, with a slightly clumsy, shambling gait. The soles are hairless and padded. They have short toes and strong claws, well suited to digging.

CAPYBARA

The subfamily Hydrochoerinae contains four species, including the largest rodent, the capybara. Capybaras have one family per year; the offspring are born at the end of the wet season, when the grass is most nutritious. They may live up to six years.

coarse fur
dries quickly

small,
rounded ears

CAPYBARA
Hydrochoeris
hydrochaeris
As big as a pig, this is
the world's largest rodent.
It lives a semiaquatic
life in the swamps of
South America.

3¼–4¼ ft
1–1.3 m

SPINY RATS

Mainly herbivores, some of these rodents also eat insects. Most have bristly fur, but not all. The Echimyidae are widely distributed in South America, but most are poorly known and some are extinct.

SPINY RAT
Proechimys sp.
These rodents have a spiny
protective coat. This feature has evolved
in parallel to the spiny mice of Africa.

6½–12 in
16–30 cm

HUTIAS

The seven extant species of the Capromyidae live in forests on different islands of the Caribbean. Six species are threatened with extinction; only the Cuban hutia is relatively secure.

12–17 in
30–43 cm

**DESMAREST'S
CUBAN HUTIA**
Capromys pilorides
This hutia is common in Cuba. Other surviving
hutia species face extinction because of a loss
of their habitat and hunting.

PACAS

This family includes two species of nocturnal rodent from Central and South America. The Cuniculidae resemble small pigs as they root about on the forest floor for fruits, seeds, and roots.

23½–32 in
60–80 cm

PACA
Cuniculus paca
Mainly a forest
inhabitant, this
rodent lives in
Central and South
America, from
Mexico to Paraguay.

DEGUS, ROCK RATS, AND RELATIVES

The molar teeth of these small, silky furred rats form a figure-of-eight shape when worn, giving them their Latin name. The Octodontidae are widespread in the southern part of South America.

DEGU
Octodon degus
This species is found on the western
slopes of the Andes in Chile. Its tail
breaks off easily if caught by a predator.

4¾–7½ in
12–19 cm

AGOUTIS

Active during the day, these long-legged, running rodents of the family Dasyproctidae are very shy. They breed all year round but have only two young at a time, which can run within an hour of birth.

RED-RUMPED AGOUTI
Dasyprocta leporina
Found in forested areas of
northeastern South America
and the Lesser Antilles, this
species can be identified by
its light orange rump.

16–24 in/41–62 cm

COYPU

The only member of the family Myocastoridae, the coypu was originally from the swamps of South America. It was introduced to Europe for the fur trade, and individuals that escaped established wild populations.

long,
rounded tail

14–26 in
36–65 cm

prominent
incisor teeth

COYPU
Myocastor coypus
This animal has
distinctively shaggy fur
and huge orange front
teeth. It also has webbed
hind feet for swimming
and a thick, scaly tail.

16–24 in
41–62 cm

**CENTRAL AMERICAN
AGOUTI**
Dasyprocta punctata
With a range that extends from
Mexico south to Argentina, this
agouti mainly feeds on fruits,
but may also eat crabs. Pairs are
thought to stay together for life.

16½–24 in
42–62 cm

AZARA'S AGOUTI
Dasyprocta azarae
An inhabitant of the forests
of southern Brazil,
Paraguay, and northern
Argentina, this agouti barks
when alarmed. It eats a
variety of seeds and fruits.

TREE SHREWS

These small mammals look and behave like squirrels. They are active in the daytime and spend much of their time foraging for food on the ground.

Tree shrews are native to the tropical rain forests of Southeast Asia. They are unrelated to the true shrews and have their own order, Scandentia. Sharp claws on all their fingers and toes enable tree shrews to climb trees rapidly but, despite their name, most of them are only partly arboreal. Their mixed diet includes insects, worms, fruits, and sometimes small mammals, reptiles, and birds.

Some tree shrews are solitary, while others live in pairs or groups. They are rapid breeders, raising their young in nests in tree crevices or on branches. The female pays her brood little attention, making only the occasional short visit to suckle them.

PHYLUM	CHORDATA
CLASS	MAMMALIA
ORDER	SCANDENTIA
FAMILIES	2
SPECIES	20

PEN-TAILED TREE SHREW

The family Ptilocercidae has only one member, the Southeast Asian pen-tailed tree shrew. The species is named for its long, feathery tail, which resembles a quill pen. This aids balance when climbing.

4–5½ in
10–14 cm

PEN-TAILED TREE SHREW
Ptilocercus lowii
This species has a rather spindly tail with a brushy tip, unlike most other tree shrews, which have a thick, bushy tail.

TREE SHREWS

Members of the family Tupaiidae have long snouts, which are used to locate insects and other invertebrates as well as fruits and leaves. Tree shrews also have sharp claws on all their fingers and toes, helping them climb trees rapidly.

GREATER TREE SHREW
Tupaia tana
Tree shrews are nocturnal inhabitants of Southeast Asian forests. This species is found in Borneo and Sumatra and on nearby islands.

long snout

6–9 in
15–23 cm

elongated claws help hold onto tree branches

COLUGOS

The two species that form the order Dermoptera are gliding, rather than flying, mammals. They are found in the rain forests of Southeast Asia.

The colugos are distinguished by a furry membrane that stretches from the neck to the tips of the fingers and tail. When the limbs are splayed this membrane spreads out, allowing the lemur to maneuver in the air as it glides from tree to tree. The distance covered in a glide may be more than 330 ft (100 m). Colugos live in the canopy, hanging upside down from branches or sheltering in cracks or hollow trees by day and emerging at night to feed on fruits and leaves. They are almost unable to move on the ground.

The teeth of colugos are unlike those of any other mammal. In the lower jaw, the teeth are arranged like combs and are believed to be used for grooming as well as for feeding.

PHYLUM	CHORDATA
CLASS	MAMMALIA
ORDER	DERMOPTERA
FAMILIES	1
SPECIES	2

FLYING LEMURS

The only two members of the Cynocephalidae have a furry membrane stretching from the neck to the tips of fingers and tail, which helps them glide from tree to tree. They also have peculiar comblike lower teeth, used to strain food such as fruits and flowers.

COLUGO
Cynocephalus variegatus
This lemur lives alone or in small groups, inhabiting tree holes or resting high on treetops in the tropical forests of Southeast Asia and the Indonesian islands.

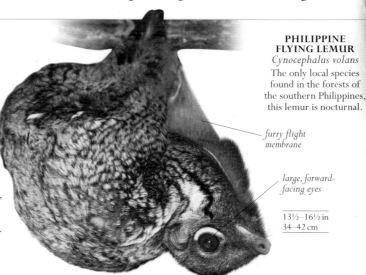

furry flight membrane

large, forward-facing eyes

13½–16½ in
34–42 cm

PHILIPPINE FLYING LEMUR
Cynocephalus volans
The only local species found in the forests of the southern Philippines, this lemur is nocturnal.

13½–16½ in
34–42 cm

PRIMATES

The order to which humans belong, primates have a large brain relative to their size and forward-facing eyes which give them 3-D vision.

With the exception of a few species, including humans, primates are restricted to tropical and subtropical regions of the Americas, Africa, and Asia. They range in size from a mouse lemur, weighing 1 oz (30 g), to a gorilla weighing up to 440 lb (200 kg).

Primates rely more on vision than on smell. Many are arboreal and have characteristics such as stereoscopic vision, allowing good judgement of distance for jumping between trees; opposable thumbs and prehensile tails for grasping branches; long legs for leaping; and long arms for swinging. Some primates have specialized diets, but many other species are omnivorous.

MAJOR GROUPS

Primates are divided into two suborders. The Strepsirrhini includes the mostly nocturnal lemurs, lorises, galagos, and their relatives. These species have a better developed sense of smell than other primates. The Haplorrhini includes the New and Old World monkeys and the apes, many of which are diurnal, and so more reliant on vision than the Strepsirrhini.

SOCIAL PATTERNS

Most primates are highly social, living in small family groups, single-male harems, or large mixed-sex troops. Many species are characterized by high levels of male competition for females. Sexual selection favouring the largest or most dominant males has led to sexual dimorphism—differences between males and females in body size or features such as canine teeth. The different sexes may also be differently colored, a form of dimorphism known as sexual dichromatism.

Most New World monkeys are monogamous, with both parents sharing responsibility for raising their young. Old World monkeys tend to live in groups dominated by related females, and the males take little or no part in parental duties. Primates are generally slow to mature fully and slow to reproduce, but they are relatively long-lived. The larger apes have a potential lifespan of up to 45 years in the wild, and longer in captivity.

PHYLUM	CHORDATA
CLASS	MAMMALIA
ORDER	PRIMATES
FAMILIES	13
SPECIES	About 250

DEBATE

PRIMATE HOTSPOTS

Today scientists list more species of primates than a decade ago—most the result of subspecies (geographical variants) being recognized as new species. In the Amazon basin populations of monkeys separated by rivers and mountain ranges can differ in subtle ways—such as in the structure of their chromosomes. Some scientists believe that these monkeys diverged when their forest habitats became isolated due to geological activity many thousands of years ago. Such forests may contain many species isolated in this way and are a particular focus of conservation groups aiming to protect biodiversity.

BUSHBABIES AND GALAGOS

Native to sub-Saharan Africa, the members of the family Galagidae inhabit a wide range of woodland and forest habitats, including scrub and wooded savanna. They have longer hind legs than forelegs, which help them take large leaps between trees in forests. These primates often wash their hands and feet with urine, which may help improve their grip and leave scent trails. All are nocturnal.

large, moveable ears

thick fur

huge eyes

SILVERY GREATER GALAGO
Otolemur monteiri
This species has the same range as the brown greater galago but prefers denser vegetation. Despite its name, melanistic (black) forms are common.

11–18½ in
28–47 cm

BROWN GREATER GALAGO
Otolemur crassicaudatus
One of the largest galagos, this primate lives in a variety of forests in southern Africa. This omnivore uses its comblike teeth to scrape gum from trees.

12–14½ in
30–37 cm

LORISES AND POTTOS

These small, nocturnal omnivores have short tails, and their forelimbs and hind limbs are of equal length. The Lorisidae have opposable thumbs for grasping branches and they move much more slowly and sedately through the trees than the galagos—climbing rather than leaping.

dense fur

RED SLENDER LORIS
Loris tardigradus
Native to Sri Lanka, this slender primate moves carefully through the forest canopy on its long limbs.

opposable thumb

2¾–10 in
7–26 cm

dark rings around huge eyes

vicelike grip

9–12 in
22–31 cm

10–15 in
26–38 cm

6–10 in
15–25 cm

12–16 in
30–40 cm

SLOW LORIS
Nycticebus coucang
As its name suggests, this loris moves slowly and deliberately through the trees. It inhabits tropical forest in Southeast Asia.

PYGMY SLOW LORIS
Nycticebus pygmaeus
This species inhabits thick tropical rainforest and bamboo forest in Laos, Cambodia, Vietnam, and southern China.

POTTO
Perodicticus potto
A shy species, this primate lives in thick rain forest across equatorial Africa. A bony shield at the nape of its neck protects it against predators.

GOLDEN ANGWANTIBO
Arctocebus aureus
Also called the golden potto, this species inhabits the understory of moist lowland forest in equatorial West and central Africa.

SENEGAL BUSHBABY
Galago senegalensis
Widespread across central Africa and parts of East Africa, the lesser, or Senegal, bushbaby inhabits arid savanna woodland.

large ears

TARSIERS

The small Tarsiidae are named for their greatly elongated anklebone or tarsus. These arboreal species have long limb bones, elongated fingers, and a long, thin tail. Their round heads have very large eyes, which help them see at night while hunting for insects.

4¾–8 in
12–20 cm

5½–6½ in
14–17 cm

silky fur

PHILIPPINE TARSIER
Tarsius syrichta
Endemic to a variety of rain forest and scrub in the Philippines, this tarsier possesses the largest eyes relative to its body size among all mammals.

MOHOLI BUSHBABY
Galago moholi
This small, shy galago lives in small groups in southern Africa. Lithe and agile, it leaps through woodland, where it eats insects and tree gum.

3¼–6½ in
8.5–16.5 cm

3¼–6½ in
8.5–16 cm

DEMIDOFF'S BUSHBABY
Galago demidoff
Also called the dwarf bushbaby, this small primate uses its long hind legs to jump through rain forest canopies of West and central Africa.

4¼–5 in
10.5–12.5 cm

long tail

HORSFIELD'S TARSIER
Tarsius bancanus
Adapted for clinging and leaping between trees, this species, also called the western tarsier, lives in the tropical rain forests of Sumatra and Borneo.

long, slender tail

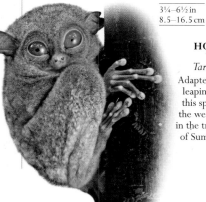

LEMURS

Found in forests across Madagascar, the family Lemuridae is made up of the most typical lemurs: mainly arboreal and quadrupedal. Most are cathemeral, meaning they are active during the day and night. In several species, males and females are differently colored.

15–16 in
38–40 cm

thick
woolly fur

16–16½ in
40–42 cm

GREATER BAMBOO LEMUR
Prolemur simus
One of the rarest lemurs, this species lives in southeastern Madagascar, where it feeds almost exclusively on giant bamboo.

15½–18 in
39–46 cm

RING-TAILED LEMUR
Lemur catta
Living in groups of up to 25 individuals, this lemur often spends time on the ground. Its diet consists of fruits, vegetation, sap, and bark.

20–23½ in
51–60 cm

BLACK AND WHITE RUFFED LEMUR
Varecia variegata
The largest lemur, this species eats a high proportion of fruit. Unusual for lemurs, it makes a leafy nest for its young.

BANDRO
Hapalemur alaotrensis
Also called the Alaotran bamboo lemur, this critically endangered species lives only in papyrus marsh and reedbeds around Lake Alaotra, Madagascar's largest lake.

14–16½ in
35–42 cm

RED-BELLIED LEMUR
Eulemur rubriventer
A monogamous species, this lemur lives in small groups that consist of a mated pair and their dependent young.

RED-COLLARED LEMUR
Eulemur collaris
This lemur has a scent gland on its wrist, with which it perfumes its long, furry tail for use in communication.

15–20 in
38–50 cm

tail is as
long as body

15½–16½ in
39–42 cm

WHITE-HEADED LEMUR
Eulemur albifrons
Only the male of this species has distinctive white fur around its black face; the female has a gray face.

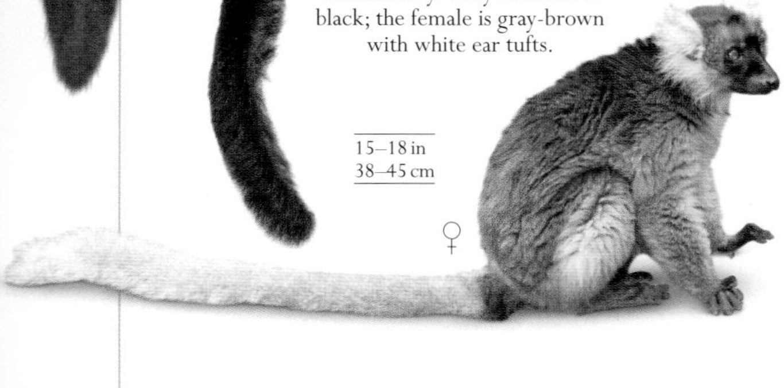

12½–14½ in
32–37 cm

red cheek
patch in male

MONGOOSE LEMUR
Eulemur mongoz
Primarily nocturnal during the dry season, the mongoose lemur becomes more diurnal at the start of the wet season.

♂

grasping hands

BLACK LEMUR
Eulemur macaco
This is a sexually dichromatic species: the colors of males and females vary. Only the male is black; the female is gray-brown with white ear tufts.

15–18 in
38–45 cm

♀

DWARF LEMURS AND MOUSE LEMURS

Among the smallest of all primates, the members of the family Cheirogaleidae have short limbs and large eyes. All are nocturnal, arboreal inhabitants of forests in Madagascar, and become torpid to survive the dry season.

PALE FORK-MARKED LEMUR
Phaner pallescens
Adapted to eat tree gum, this lemur has a long tongue and large premolars for chiseling bark.

9–12 in
22–30 cm

6½–10 in
17–26 cm

short limbs

GREATER DWARF LEMUR
Cheirogaleus major
This solitary lemur feeds mainly on fruits and nectar. It stores fat in its tail during the wet season.

4¾–6 in
12–15 cm

GRAY MOUSE LEMUR
Microcebus murinus
An omnivore, this lemur's diet includes insects, flowers, and fruits. The female carries the infant in her mouth for the first few weeks.

EASTERN RUFOUS MOUSE LEMUR
Microcebus rufus
Inhabiting a range of forest habitats, this omnivorous species consumes a variety of fruits, insects, and gum.

4–8 in
10–20 cm

SPORTIVE LEMURS

The family Lepilemuridae consists of the sportive lemurs of Madagascar. These medium-sized lemurs have prominent muzzles and large eyes, and are strictly arboreal and nocturnal. Their low-energy, leafy diet means that they are among the least active of all primates.

9–10 in
22–26 cm

10 in
26 cm

WHITE-FOOTED SPORTIVE LEMUR
Lepilemur leucopus
Grayish above and white below, this lemur spends lengths of time perched vertically on tree trunks, in between bouts of foraging for food.

BLACK-STRIPED SPORTIVE LEMUR
Lepilemur dorsalis
This species lives in moist forest in northwestern Madagascar and nearby islands. It has a blunt muzzle and small ears.

SIFAKAS AND RELATIVES

The largest lemurs—the indri and sifakas—and the smaller avahis or woolly lemurs form the Indriidae of Madagascar. All members have long, powerful limbs for leaping between trees, and, with the exception of the indri, a long tail.

face mostly bare, with white fur along muzzle

16–20 in
40–50 cm

16½–20 in
42–50 cm

long hind limbs

VERREAUX'S SIFAKA
Propithecus verreauxi
Found in southwestern Madagascar, this sifaka is capable of leaping through cactuslike vegetation on its long legs without being injured.

16½–20½ in
42–52 cm

MILNE-EDWARD'S SIFAKA
Propithecus edwardi
Occurring in small family groups in southeastern Madagascar, this sifaka has a large opposable thumb for clinging to tree trunks.

COQUEREL'S SIFAKA
Propithecus coquereli
Like other sifakas, this primate crosses open ground by skipping on its hind legs, using its forearms for balance.

23½–28 in
60–72 cm

INDRI
Indri indri
The largest member of the lemur group, this is the only species to have a short, vestigial tail.

AYE-AYE

There is only one species in the Daubentoniidae—the aye-aye of Madagascar. This nocturnal primate has large, bare ears; a shaggy coat; very long fingers; and continually growing incisor teeth.

12–16 in
30–40 cm

AYE-AYE
Daubentonia madagascariensis
The aye-aye uses its elongated middle finger for locating and extracting insect larvae from dead wood.

HOWLER, SPIDER, AND WOOLLY MONKEYS

The howler, spider, and woolly monkeys of the Atelidae are the largest New World monkeys. All have a prehensile tail that is used as a fifth limb when moving through the trees. Spider monkeys have longer limbs than the other members of the family.

GUATEMALAN HOWLER MONKEY
Alouatta pigra
Found in the Yucatán Peninsula of Mexico, Belize, and Guatemala, this species forms groups of up to 11 individuals.

20–28 in
50–71 cm

19–25 in
48–63 cm

BLACK HOWLER
Alouatta caraya
This species lives in central South American rain forests. Only the males are black; females are a buff-golden color.

20–26 in
50–65 cm

19–27 in
48–68 cm

MANTLED HOWLER MONKEY
Alouatta palliata
Named after the "mantle" of long guard hair on its flanks, this howler monkey lives in Central America and northern South America.

large larynx for howling

VENEZUELAN RED HOWLER MONKEY
Alouatta seniculus
This howler monkey has an enlarged hyoid bone in the neck that allows it to make calls that can be heard several miles away.

16–22 in
40–55 cm

COLOMBIAN SPIDER MONKEY
Ateles fusciceps rufiventris
Like other spider monkeys, this subspecies does not have thumbs on its hands. It is found in Colombia and Panama.

12–25 in
31–63 cm

GEOFFROY'S SPIDER MONKEY
Ateles geoffroyi
Active during the day, this fruit-eating species lives in relatively large groups of up to 35 individuals in forests throughout Central America.

18–31 in
46–78 cm

SOUTHERN MURIQUI
Brachyteles arachnoides
Also known as the woolly spider monkey, this species, found in Brazilian forests, is critically endangered due to the loss of its habitat.

hairless tail tip

20–26 in
50–65 cm

GRAY WOOLLY MONKEY
Lagothrix cana
A powerfully built species, this monkey lives in large troops in the primary forests of Brazil, Bolivia, and Peru.

16–27 in
40–69 cm

BROWN WOOLLY MONKEY
Lagothrix lagotricha
One of the largest New World monkeys, this species lives in lowland primary forest in the upper Amazon basin.

NIGHT MONKEYS

Also called owl monkeys, the members of the Aotidae are the only nocturnal monkeys in the New World. They are small, with large eyes set in flat, round faces with dense woolly fur. They have a well-developed sense of smell.

9½–16½ in
24–42 cm

BLACK-HEADED NIGHT MONKEY
Aotus nigriceps
This monogamous species inhabits the central and upper Amazon basin in primary and secondary forests of Brazil, Bolivia, and Peru.

9½–19 in
24–48 cm

NORTHERN NIGHT MONKEY
Aotus trivirgatus
Also called the three-striped monkey, this species is most active on moonlit nights. It lives in forests in Venezuela and northern Brazil.

TITI, SAKI, AND UAKARI MONKEYS

This family contains a range of small- to medium-sized monkeys. The Pitheciidae are diurnal, arboreal, and social. All share a common dental makeup, including large, splayed canine teeth that help them deal with tough seeds and fruits.

black hair tipped with white

12–28 in
30–70 cm

WHITE-FACED SAKI
Pithecia pithecia
Male white-faced or Guianan sakis are black with pale fur around the face, whereas females are gray-brown.

15–19 in
38–48 cm

BLACK-BEARDED SAKI
Chiropotes satanas
This species is known from South Amazonia. The males possess a beard and large swellings on their foreheads.

15–16½ in
38–42 cm

MONK SAKI
Pithecia monachus
A shy primate, this species lives in the high canopy of forests in northwest Brazil, Peru, Colombia, and Ecuador.

14½–19 in
37–48 cm

RIO TAPAJÓS SAKI
Pithecia irrorata
Also called the bald-faced saki, this species lives in western Brazil, northern Bolivia, and eastern Peru. Its diet is predominantly seeds.

hairless, red face

12–16½ in
31–42 cm

BLACK-FRONTED TITI
Callicebus nigrifrons
A fruit-eater, this titi is an inhabitant of Atlantic coastal forest around São Paulo in southeastern Brazil.

9–14 in
23–36 cm

COLLARED TITI MONKEY
Callicebus torquatus
Also known as the yellow-handed titi, this species prefers unflooded forests on sandy soils in Brazil. It feeds mostly on fruits and seeds.

long, shaggy coat

10½–17 in
27–43 cm

RED-BELLIED TITI
Callicebus moloch
A strongly monogamous species, this titi maintains family territories. It lives in the low canopy of forests in central Brazil.

12–20 in
30–50 cm

BLACK-HEADED UAKARI
Cacajao melanocephalus
A highly social species, this uakari lives in groups of 30 or more individuals in the upper Amazon basin.

14–22½ in
36–57 cm

RED BALD-HEADED UAKARI
Cacajao calvus rubicundus
Several subspecies of the bald-headed uakari inhabit seasonally flooded forests of the Amazon basin. The red faces of these monkeys are thought to signal their health.

MARMOSETS, SQUIRREL MONKEYS, AND RELATIVES

The Cebidae consists of relatively small social monkeys, which inhabit a variety of forests throughout tropical and subtropical Central and South America. They are all diurnal and arboreal and have forward-facing eyes and short muzzles. All except the capuchins have long, nonprehensile tails. Marmosets and tamarins have claws instead of nails, and lack the third molar.

8½–12 in
21–31 cm

GOELDI'S MONKEY
Callimico goeldii
Inhabiting dense forest undergrowth, such as that of bamboo forest in the upper Amazon, this monkey makes forays into the tree canopy for fruits.

8–9 in
20–23 cm

SILVERY MARMOSET
Callithrix argentata
A specialized gum-eater (gumivore), this marmoset has huge ears, narrow jaws, and short canine teeth for cutting into tree bark.

COMMON MARMOSET
Callithrix jacchus
The female of this species usually mates with two males, both of whom assist with parental care of the usually twin offspring.

4¾–6 in
12–15 cm

long, curved claws

8 in
20 cm

GEOFFROY'S MARMOSET
Callithrix geoffroyi
Also called the white-headed marmoset, this species uses scent-marks to deter others from using the holes it makes in tree bark to extract gum.

9–11 in
23–28 cm

BLACK TUFTED-EAR MARMOSET
Callithrix penicillata
A monogamous species, this primate is diurnal and lives high in the rain forest canopy, where it feeds on tree sap.

4¾–6 in
12–15 cm

PYGMY MARMOSET
Callithrix pygmaea
The smallest monkey in the world, this species eats tree gum in seasonally flooded forests of the upper Amazon basin.

fine, yellow hair on back

9–10 in
23–26 cm

white mustache

EMPEROR TAMARIN
Saguinus imperator
Distinguished by its long, white mustache, this tamarin lives in the tropical forests in Peru, Brazil, and Bolivia.

9–12 in
23–30 cm

WHITE-LIPPED TAMARIN
Saguinus labiatus
The dominant female of this species releases chemical signals, called pheromones, that suppress reproduction in other female members of the group.

8½–11 in
21–28 cm

BARE-FACED TAMARIN
Saguinus bicolor
Also called the pied tamarin, this tree-dwelling primate lives in lowland forest in the central Amazon region near Manaus, Brazil.

COTTON-TOP TAMARIN
Saguinus oedipus
Inhabiting a very restricted range in northwest Colombia and Panama, this species feeds mainly on insects and fruits.

8–10 in
20–25 cm

GOLDEN-HANDED TAMARIN
Saguinus midas
Found in northeast South America, this tamarin has brightly colored hands and feet, with claws on all digits except the big toe, which has a nail.

SADDLE-BACK TAMARIN
Saguinus fuscicollis
This tamarin lives in secondary forest and forest edges in the upper Amazon basin, foraging for insects, fruits, nectar, and tree sap and gum.

8–10½ in
20–27 cm

banded tail

8½–11 in
21–28 cm

GOLDEN LION TAMARIN
Leontopithecus rosalia
Among the most endangered monkeys in the world, this species is found only in the Atlantic coastal forest of southeast Brazil.

8–13½ in
20–34 cm

GOLDEN-HEADED LION TAMARIN
Leontopithecus chrysomelas
Restricted to the Atlantic forest of southern Bahia in northeast Brazil, this primate raises its mane to look bigger when in danger.

naked face

8–10 in
20–25 cm

WEEPER CAPUCHIN
Cebus olivaceus
Native to north-central South America, this capuchin often uses its prehensile tail to support its body while feeding with its hands.

WHITE-HEADED CAPUCHIN
Cebus capucinus
The only capuchin found in Central America, the range of this primate extends from Honduras to the coasts of Colombia and Ecuador.

12–22½ in
31–57 cm

10½–14½ in
27–37 cm

14½–18 in
37–46 cm

prehensile tail

LARGE-HEADED CAPUCHIN
Cebus apella macrocephalus
Capable of using tools to open hard fruits, this primate lives in the forests of western South America.

13–22½ in
33–57 cm

bright yellow limbs

10½–12½ in
27–32 cm

long tail with a black tip

BLACK-CAPPED SQUIRREL MONKEY
Saimiri boliviensis
Males of this species become fatter around the neck and shoulders during the breeding season, when they compete for mates.

COMMON SQUIRREL MONKEY
Saimiri sciureus
A gregarious species, this monkey lives in large groups in a wide variety of forests in northern South America.

nails on digits

OLD WORLD MONKEYS

Widely distributed across Africa and Asia, the Cercopithecidae have closely spaced, downward-facing nostrils (catarrhine) and flattened nails. Most are diurnal and arboreal; however, baboons are mainly terrestrial. Guenons, baboons, and macaques are omnivorous, with strong jaws, cheek pouches, and a simple stomach. Colobus and leaf monkeys are leaf-eaters (folivorous) with complex stomachs, and they lack cheek pouches.

forward-facing eyes give 3-D vision

18–25 in
45–64 cm

17–21 in
43–53 cm

TOQUE MACAQUE
Macaca sinica
The smallest of the macaques, this species is endemic to the wet forests on the island of Sri Lanka.

18–28 in
45–70 cm

20½–22½ in
52–57 cm

CELEBES CRESTED MACAQUE
Macaca nigra
Endemic to Indonesia's Sulawesi Island, this monkey has bare, pink swellings on its rump. These become particularly swollen in females that are ready to mate.

BARBARY MACAQUE
Macaca sylvanus
The only macaque outside Asia, the barbary macaque lives in the high cedar and oak forests of Algeria and Morocco.

lionlike mane

14½–25 in
37–63 cm

16–24 in
40–61 cm

CRAB-EATING MACAQUE
Macaca fascicularis
Besides eating crabs, as its name suggests, this omnivorous southeast Asian species also forages for insects, frogs, fruits, and seeds.

LION-TAILED MACAQUE
Macaca silenus
Endemic to the Western Ghats, mountains in southwest India, this arboreal macaque lives mainly in wet, monsoon forests.

RHESUS MACAQUE
Macaca mulatta
This macaque lives in dry, open areas from western Afghanistan through India to northern Thailand and China. Adults swim distances of up to ½ mile (0.8 km) between islands.

19½–28 in
49–70 cm

sandy coloration

STUMP-TAILED MACAQUE
Macaca arctoides
Both arboreal and terrestrial, this species is found in tropical and subtropical wet forests in southeast Asia.

18½–23½ in
47–60 cm

*forearms of similar
length to hind legs*

14–23½ in
35–60 cm

thick fur

**SOUTHERN
PIG-TAILED MACAQUE**
Macaca nemestrina
This monkey inhabits humid
areas, including rain forest and
swamps in southeast Asia. Its
diet consists mainly of fruits.

BONNET MACAQUE
Macaca radiata
Often found close to
human habitation, this
resident of southern India
is omnivorous and may
rely on humans for food.

18½–23½ in
47–60 cm

JAPANESE MACAQUE
Macaca fuscata
Inhabiting the most northerly range
of any nonhuman primate, the
Japanese macaque uses hot springs
to keep warm in winter.

*opposable
thumbs*

17½–28 in
44–70 cm

16–19 in
41–48 cm

**WHITE-THROATED
MONKEY**
Cercopithecus albogularis
Also called the Sykes'
monkey, this tree-dwelling
omnivore is distributed in
eastern and southeastern
Africa, including Zanzibar
and the Mafia Islands.

RED-TAILED MONKEY
Cercopithecus ascanius
An arboreal species of moist forest habitats in
Central Africa, the red-tailed monkey has large
cheek pouches, in which it stores fruits.

18–22 in
46–56 cm

*long,
furry tail*

16–22 in
40–55 cm

16–25 in
40–64 cm

L'HOEST'S MONKEY
Cercopithecus lhoesti
Also known as the mountain
monkey, this arboreal guenon
lives in moist primary forests in
the highlands of central Africa.

DIANA MONKEY
Cercopithecus diana
An inhabitant of the high canopy
of primary forests in West Africa,
the Diana monkey rarely
descends to the ground.

DE BRAZZA'S MONKEY
Cercopithecus neglectus
This is a semiterrestrial species of central
African swamp forests. Males are larger than
females and have a distinctive blue scrotum.

19½–26 in
49–66 cm

12½–22 in
32–56 cm

17½–25 in
44–63 cm

ANGOLAN TALAPOIN
Miopithecus talapoin
The smallest of the Old
World monkeys, the arboreal
talapoin lives in wet and
swampy forests in western
and central Africa.

BLUE MONKEY
Cercopithecus mitis
Social groups of this African native
number up to 40 individuals and
consist of a dominant alpha male,
several females, and their offspring.

MONA MONKEY
Cercopithecus mona
A tree-dwelling species, this
monkey inhabits rain forests
and mangrove forests from
Ghana to Cameroon.

TANA RIVER MANGABEY
Cercocebus galeritus
Named after the Tana River in Kenya,
where it is endemic, this terrestrial
mangabey has large incisors for
opening hard seeds.

12½–18 in
32–45 cm

GRIVET
Chlorocebus aethiops
A semiterrestrial species, the grivet lives in northeastern Africa. It is characterized by a green tinge on its upper face.

16–26 in
40–66 cm

14–26 in
35–66 cm

VERVET MONKEY
Chlorocebus pygerythrus
An inhabitant of savanna and open woodland, this monkey is distributed from Ethiopia through East Africa to South Africa.

60–88 in
23½–35 in

PATAS MONKEY
Erythrocebus patas
The long limbs and short digits of this primate make it well adapted for running. It is found from West to East Africa.

BLACK-CRESTED MANGABEY
Lophocebus aterrimus
An arboreal species that prefers rain forest, this mangabey is found in the Democratic Republic of Congo.

15–35 in
38–89 cm

speckled olive-gray fur

blue flanges on either side of the nose

25–32 in
63–81 cm

MANDRILL
Mandrillus sphinx
Found in rain forest in western central Africa, this species has distinctive facial markings that are duller in females and juveniles than in males.

24–30 in
61–77 cm

DRILL
Mandrillus leucophaeus
A large terrestrial monkey of lowland mature rain forest, the drill is found only in Cameroon, Nigeria, and the Republic of Equatorial Guinea.

baboon tail typically appears broken

olive-gray body

20–43 in
51–114 cm

20–43 in
50–114 cm

CHACMA BABOON
Papio ursinus
One of the largest baboons, this species inhabits woodland, savanna, steppe, semidesert, and montane habitats across southern Africa.

YELLOW BABOON
Papio cynocephalus
An opportunistic baboon, this omnivore's diet includes seedpods, roots, insects, and other monkeys. It lives in southern and eastern Africa.

19–34 in
48–86 cm

red-brown face

24–30 in
61–76 cm

24–30 in
61–76 cm

HAMADRYAS BABOON
Papio hamadryas
This species is found in East and central Africa, particularly Ethiopia. The male has a long silver-gray shoulder cape.

GUINEA BABOON
Papio papio
One of the smallest baboons, this species also has one of the smallest ranges, confined to western equatorial Africa.

muscular limbs help it run fast

OLIVE BABOON
Papio anubis
This species forms troops of up to 100 baboons in the savanna and grassland steppe of central sub-Saharan Africa.

20–29 in
50–74 cm

GELADA
Theropithecus gelada
A grassland grazer from the highlands of
Ethiopia, this primate has a distinctive
bare patch of skin on its chest.

**GOLDEN SNUB-NOSED
MONKEY**
Rhinopithecus roxellana
This monkey has thick fur,
which helps it survive in
high montane forests in
west and central China.

18–28 in
45–72 cm

mantle of
white hair

white "cloak"
over back

18½–31 in
47–78 cm

21½–30 in
54–76 cm

PROBOSCIS MONKEY
Nasalis larvatus
A competent swimmer, this monkey
inhabits mangrove and lowland riverine
rain forest in Borneo. It is named for
the male's large nose.

18½–27 in
47–68 cm

ANGOLA COLOBUS
Colobus angolensis
A predominantly arboreal
species, this monkey inhabits
a variety of forest habitats
in Angola, Congo, and other
neighboring countries.

GUEREZA
Colobus guereza
Also called the eastern
black-and-white colobus,
this species is widespread in
the moist tropical forests of
central and East Africa.

long tail
with a white,
bushy tip

16–31 in
41–78 cm

17–26 in
43–65 cm

NORTHERN PLAINS GREY LANGUR
Semnopithecus entellus
Also called the Hanuman langur, this gray monkey is
found in southern Asia, including India and Pakistan.

bright orange
coloration

24 in
61 cm

JAVAN LANGUR
Trachypithecus auratus
Most males and females
of this species are black
in color. However, some
individuals retain their
juvenile orange coloration
into adulthood.

TUFTED GREY LANGUR
Semnopithecus priam
Found in southeast India and Sri Lanka,
this langur lives in many different habitats.
Its diet consists mainly of tree foliage.

MANDRILL
Mandrillus sphinx

The mandrill is the largest of all monkeys, and the male is particularly spectacular. Mandrills live in troops that usually contain a dominant male, several females, a gaggle of youngsters, and various nonbreeding, lower-ranking males. Sometimes several groups merge to form troops of 200 or more. Mandrill society has a strict hierarchy. Individuals advertise their status with colorful patches of skin on the face and rump. A dominant male is fearsome-looking, and he has a temperament to match. The brightness of the skin pigments is controlled by hormones, and color is a good indicator of strength and ferocity. A rival must be very sure of himself before challenging such a magnificent animal, and serious fights tend to occur only between well-matched individuals.

SIZE 25–32 in (63–81 cm)
HABITAT Dense rain forest
DISTRIBUTION Western Central Africa, from Nigeria to Cameroon
DIET Mainly fruit

∨ **LONG, HARD GLARE**
Forward-facing eyes provide the mandrill with stereoscopic vision. They see in full color, which helps them locate ripe fruit and to distinguish the visual signals of other individuals.

< **NOSTRILS**
In a mature male, the skin around the nostrils and the center of the snout is scarlet. Females and young mandrills have a black nose.

< **TEETH**
The long canine teeth are used principally for fighting and in display. The cheek teeth are smaller, with knobbly surfaces, which are used for grinding up plant material.

grooves on either side of the nose

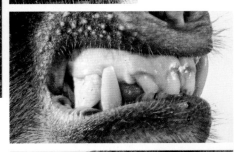

∧ **GRASP**
Mandrill thumbs are short, but fully opposable—like those of the great apes—for grasping and manipulating objects. The fingers are long and very strong, with stout nails.

∧ **HIND FEET**
The hind feet resemble the hands, in that the toes are long and able to grasp. Mandrills are good climbers, and often sleep among the branches of trees.

∧ **PLAIN RUMP**
All mandrills have a hairless rump and a short tail. The rump of the subordinate male has less color than the alpha male.

short, tufted tail

relatively short hind legs

ALL FOURS >
Mandrills spend most of their time at ground level, where they move around on all fours, and typically roam 3–6 miles (5–10 km) a day.

long, powerful arms

ALPHA MALE >
The skin color displayed by an individual changes depending on breeding condition and mood. The alpha male displays the most vivid red and blue hues on its face and rump.

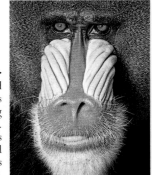
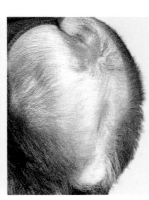

prominent brow ridges
shade the eyes from
bright sunlight

coat is made up of
long, coarse hairs

> **POWERFUL
PROFILE**
The mandrill's head is
bulky, accommodating
powerful jaw muscles
needed to crush tough
plant foods. Its ears are
relatively small, though
it has excellent hearing.
Only high-ranking males
have an orange beard.

GIBBONS

The gibbons, or lesser apes, of the family Hylobatidae are medium-sized fruit-eating primates. They have no tail and characteristically move by brachiation—swinging between trees using their very long forearms. Gibbons usually form monogamous pair bonds that are reinforced by duetting songs, which also serve to announce territorial ownership. Some species have enlarged throat sacs to amplify the sound.

AGILE GIBBON
Hylobates agilis
Although its coat coloration varies, all agile gibbons have white eyebrows, and males also have white cheeks. This species lives in Thailand, Indonesia, and Malaysia.

18–25 in
45–64 cm

black cap of female

PILEATED GIBBON
Hylobates pileatus
Female pileated gibbons are silver-grey with a black face, chest, and cap; males are black. They are found in Thailand, Cambodia, and Laos.

17½–25 in
44–64 cm

SILVERY GIBBON
Hylobates moloch
Endemic to western Java, Indonesia, males and females of the silvery gibbon are silver-gray with a dark cap.

18–25 in
45–64 cm

17½–25 in
44–64 cm

MULLER'S BORNEAN GIBBON
Hylobates muelleri
Also known as the gray gibbon, this species lives in Borneo. Monogamous pairs spend an average of 15 minutes per day singing duets.

white feet and hands

18–25 in
45–64 cm

HOOLOCK GIBBON
Bunopithecus hoolock
Male hoolock gibbons are black, but females are tan with dark brown cheeks. They are found in China, northeast India, and northwest Myanmar.

16½–23 in
42–59 cm

pronounced hair on crown of male

naked palms

NORTHERN WHITE-CHEEKED GIBBON
Nomascus leucogenys
Northern white-cheeked gibbons are creamy colored when they are born. They change color by two years of age.

18–25 in
45–64 cm

♂

♀

18–25 in
45–64 cm

LAR GIBBON
Hylobates lar
Variable in coat coloration, the white-handed or lar gibbon lives in forests in Thailand, Malaysia, Sumatra, Myanmar, and Laos.

silvery white saddle of hair

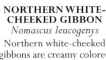

BUFF-CHEEKED GIBBON
Nomascus gabriellae
Male buff-cheeked gibbons are black with pale cheeks; females are buff with a black cap. They are found in Cambodia, Laos, and Vietnam.

28–35 in
71–90 cm

SIAMANG
Symphalangus syndactylus
The largest of the gibbons, the siamang lives on the island of Sumatra, Indonesia, and on the Malay Peninsula.

HUMANS AND APES

The family Hominidae contains the largest primates—the great apes and humans. Orangutans are arboreal, whereas chimpanzees, gorillas, and humans spend the majority of their time on the ground. Chimpanzees and gorillas move on all fours by "knuckle-walking." None of the great apes has a tail. Males are generally larger than females. and all species have relatively large braincases.

BORNEAN ORANGUTAN
Pongo pygmaeus
A large arboreal fruit eater, the Bornean orangutan lives in the canopy of primary rainforest on the island of Borneo.

31–59 in
78–150 cm

very long arms

coarse, shaggy red-brown coat

grasping hands and feet

2½–6 ft
0.8–1.8 m

EASTERN GORILLA
Gorilla beringei
The two subspecies of the largest primate, the eastern gorilla, inhabit montane cloud forest and lowland forest in the eastern Democratic Republic of Congo, Rwanda, and Uganda.

5–6 ft
1.5–1.8 m

domed forehead

heavily built body

SUMATRAN ORANGUTAN
Pongo abelii
The largest arboreal primate, the Sumatran orangutan is restricted to fragments of primary tropical forest in northern Sumatra, Indonesia.

28–33 in
70–83 cm

BONOBO
Pan paniscus
Slighter than the common chimpanzee, the bonobo or pygmy chimpanzee lives in humid tropical forest in the Democratic Republic of Congo.

4–7 ft
1.2–2.1 m

♂ ♀

WESTERN GORILLA
Gorilla gorilla
Two subspecies of western gorilla live in lowland tropical and swamp forests in western central Africa. Adult males are called silverbacks.

4–6 ft
1.3–1.8 m

strong hands and feet

25–37 in
64–94 cm

COMMON CHIMPANZEE
Pan troglodytes
Four subspecies of the common chimpanzee are distributed in dry and moist forests and savanna woodlands in equatorial Africa.

HUMAN
Homo sapiens
Characterized by a bipedal posture and a lack of body hair, humans permanently inhabit every terrestrial habitat, with the exception of Antarctica.

BATS

The only mammals capable of powered flight, bats are primarily nocturnal. Many species use echolocation to navigate and find food.

Bats are found worldwide in many different habitats, including tropical, subtropical, and temperate forests, savanna grasslands, deserts, and wetlands. The order is often divided into two suborders—Megachiroptera ("megabats") and Microchiroptera ("microbats"). Most megabats are fruit-eaters, while most microbats eat insects. However, some bats drink nectar and eat pollen, a few suck blood, and some eat vertebrates, such as fish, frogs, and bats.

Their greatly elongated arm, hand, and finger bones support an elastic wing membrane for flight. Many bats also have a tail membrane between their legs. Bats typically rest upside down, hanging from their strong toes and claws.

THE SENSE OF ECHOLOCATION

Megabats rely predominantly on vision and smell, whereas microbats use a specialized sense called echolocation to avoid hitting objects, and to detect prey in the dark. They emit pulses of sound through

their mouth or nose and form a "sound" image of their surroundings from returning echoes. Those species that emit echolocation calls through their noses usually have elaborate facial ornamentation, called noseleaves, for focusing the sound. Bats have very sensitive hearing, often highly tuned to the frequency of their returning echoes. Some species listen to prey-generated sounds, such as the rustle that an insect makes when walking on a leaf.

HABITS AND ADAPTATIONS

Bats are very social animals, living in colonies of hundreds or thousands and, exceptionally, millions of animals. They roost in trees and caves or in buildings, bridges, and mines. Temperate species either migrate to warmer climates or hibernate during winter. They may also become torpid if food becomes short in other seasons. Many interesting reproductive adaptations have developed, including the storage of sperm, delayed fertilization, and delayed implantation to ensure that the young are born at the optimum time of year.

PHYLUM	CHORDATA
CLASS	MAMMALIA
ORDER	CHIROPTERA
FAMILIES	18
SPECIES	900+

FRUIT BATS

The bats of the family Pteropodidae are distributed across the tropical and subtropical regions of the Old World. They have doglike faces with simple ears and large eyes. Most use vision and smell to locate their food, but members of the genus *Rousettus* echolocate by producing tongue clicks. These bats feed on fruits, nectar, and pollen. They have claws on both the thumb and the second finger.

2–3 in
5–7.5 cm

BLOSSOM BAT
Syconycteris australis
This specialized nectar-feeder occurs from Papua New Guinea to eastern coastal Australia. It has a pointed muzzle and a brush-tipped tongue for probing flowers.

elastic skin membrane

4¼–7 in
11–18 cm

FRANQUET'S EPAULETTED BAT
Epomops franqueti
Also called the singing fruit bat because of the high-pitched call of the males, this species occurs in West and central Africa.

2¼–3¼ in
6–8.5 cm

LONG-TONGUED FRUIT BAT
Macroglossus minimus
Occurring in Southeast Asia, this fruit bat feeds on nectar and pollen from flowers, using its long tongue.

2¾–5 in
7–13 cm

SHORT-NOSED FRUIT BAT
Cynopterus sphinx
This is the only fruit bat that makes tents from palm leaves. It is found across Southeast Asia and the Indian subcontinent.

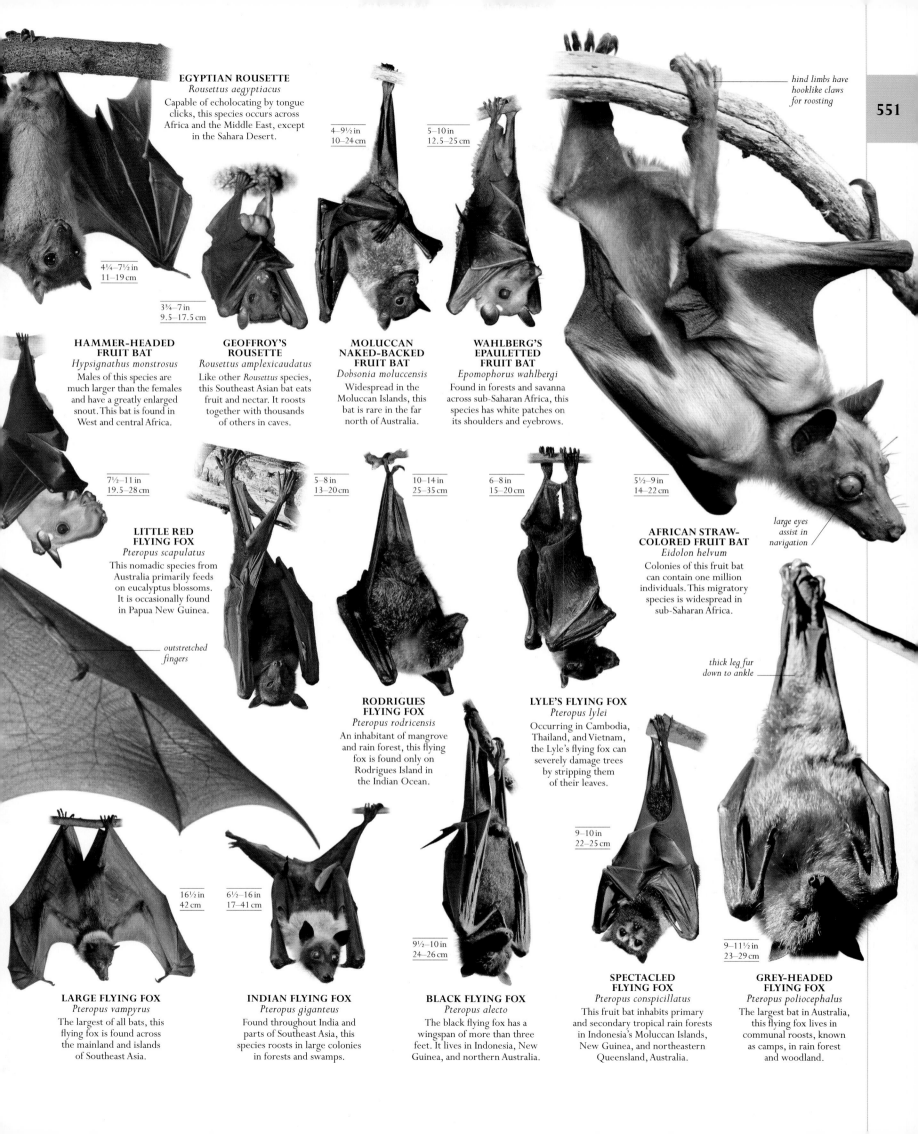

hind limbs have hooklike claws for roosting

EGYPTIAN ROUSETTE
Rousettus aegyptiacus
Capable of echolocating by tongue clicks, this species occurs across Africa and the Middle East, except in the Sahara Desert.

4¼–7½ in
11–19 cm

3¾–7 in
9.5–17.5 cm

4–9½ in
10–24 cm

5–10 in
12.5–25 cm

HAMMER-HEADED FRUIT BAT
Hypsignathus monstrosus
Males of this species are much larger than the females and have a greatly enlarged snout. This bat is found in West and central Africa.

GEOFFROY'S ROUSETTE
Rousettus amplexicaudatus
Like other *Rousettus* species, this Southeast Asian bat eats fruit and nectar. It roosts together with thousands of others in caves.

MOLUCCAN NAKED-BACKED FRUIT BAT
Dobsonia moluccensis
Widespread in the Moluccan Islands, this bat is rare in the far north of Australia.

WAHLBERG'S EPAULETTED FRUIT BAT
Epomophorus wahlbergi
Found in forests and savanna across sub-Saharan Africa, this species has white patches on its shoulders and eyebrows.

7½–11 in
19.5–28 cm

5–8 in
13–20 cm

10–14 in
25–35 cm

6–8 in
15–20 cm

5½–9 in
14–22 cm

large eyes assist in navigation

AFRICAN STRAW-COLORED FRUIT BAT
Eidolon helvum
Colonies of this fruit bat can contain one million individuals. This migratory species is widespread in sub-Saharan Africa.

LITTLE RED FLYING FOX
Pteropus scapulatus
This nomadic species from Australia primarily feeds on eucalyptus blossoms. It is occasionally found in Papua New Guinea.

outstretched fingers

thick leg fur down to ankle

RODRIGUES FLYING FOX
Pteropus rodricensis
An inhabitant of mangrove and rain forest, this flying fox is found only on Rodrigues Island in the Indian Ocean.

LYLE'S FLYING FOX
Pteropus lylei
Occurring in Cambodia, Thailand, and Vietnam, the Lyle's flying fox can severely damage trees by stripping them of their leaves.

9–10 in
22–25 cm

16½ in
42 cm

6½–16 in
17–41 cm

9½–10 in
24–26 cm

9–11½ in
23–29 cm

LARGE FLYING FOX
Pteropus vampyrus
The largest of all bats, this flying fox is found across the mainland and islands of Southeast Asia.

INDIAN FLYING FOX
Pteropus giganteus
Found throughout India and parts of Southeast Asia, this species roosts in large colonies in forests and swamps.

BLACK FLYING FOX
Pteropus alecto
The black flying fox has a wingspan of more than three feet. It lives in Indonesia, New Guinea, and northern Australia.

SPECTACLED FLYING FOX
Pteropus conspicillatus
This fruit bat inhabits primary and secondary tropical rain forests in Indonesia's Moluccan Islands, New Guinea, and northeastern Queensland, Australia.

GREY-HEADED FLYING FOX
Pteropus poliocephalus
The largest bat in Australia, this flying fox lives in communal roosts, known as camps, in rain forest and woodland.

LYLE'S FLYING FOX
Pteropus lylei

Lyle's flying fox is a medium-sized representative of the Old World fruit bat family, the Pteropodidae. Fruit bats are social animals; many hundreds may gather in roost trees during the day to rest and groom, dispersing at dusk to find ripe fruit. Although they may cause some damage to trees, many fruit bats are important pollinators and seed dispersers of tropical plants, including many commercial crops. Fruit bats are found in the tropical regions of Africa, Asia, and Australia, although this species is only found in Cambodia, Thailand, and Vietnam. They inhabit forested areas, including mangroves and fruit orchards.

SIZE 6–8 in (15–20 cm)
HABITAT Forests
DISTRIBUTION Southeast and East Asia
DIET Fruits and leaves

all bats have a thumb claw, but only Old World fruit bats have a claw on the second finger too

∨ **CANINE FACE**
Fruit bats have large eyes for navigation and large noses for sniffing out fruit, pollen, and nectar—giving them doglike faces. Species that use echolocation have larger ears and smaller eyes.

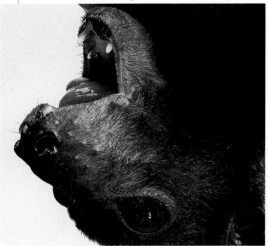

< **CLINGING CLAWS**
Bats' sharp, curved claws are perfect for clinging to tree branches. When roosting, tendons lock in place to ensure that the claws stay bent without the muscles needing to contract.

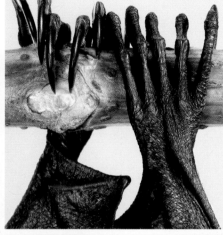

∧ **HAND WING**
A bat's wing is formed of elongated forearm and finger bones supporting a thin elastic membrane attached to the sides of the body. It has a large surface area, giving lift in flight.

∧ **TAIL OR NO TAIL?**
Pteropus bats lack tails, but some have a partial membrane supported by cartilaginous spurs called calcars, which protrude from the ankle.

WALKING UPSIDE DOWN >
Fruit bats use their large clawed thumb to help them move along the branches of their roost tree. The claw can also be used to manipulate fruit when feeding.

∧ **WELL WRAPPED**
When resting, most fruit bats hang upside down with their leathery wings folded around their bodies. Bats roosting in the open risk overheating, so they flap their wings and cover themselves in saliva to help keep cool in hot weather.

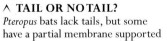

reddish brown,
foxy coloring

< HANGING AROUND
Few bats are able to move
well on the ground, let
alone take off from a flat
surface. By hanging upside
down they are able to
take flight quickly. During
the day, they hang upside
down to sleep, nestled
together for warmth.

large eyes are adapted
for excellent vision,
especially at night

upright, foxlike ears
detect sounds beyond the
range of human hearing

HORSESHOE BATS

Species of the family Rhinolophidae are distributed throughout southern Europe, Africa, Asia, and Australasia. Their noseleaf is characteristically horseshoe-shaped. They have the most sophisticated echolocation of all bats, involving highly specialized sound transmission and reception.

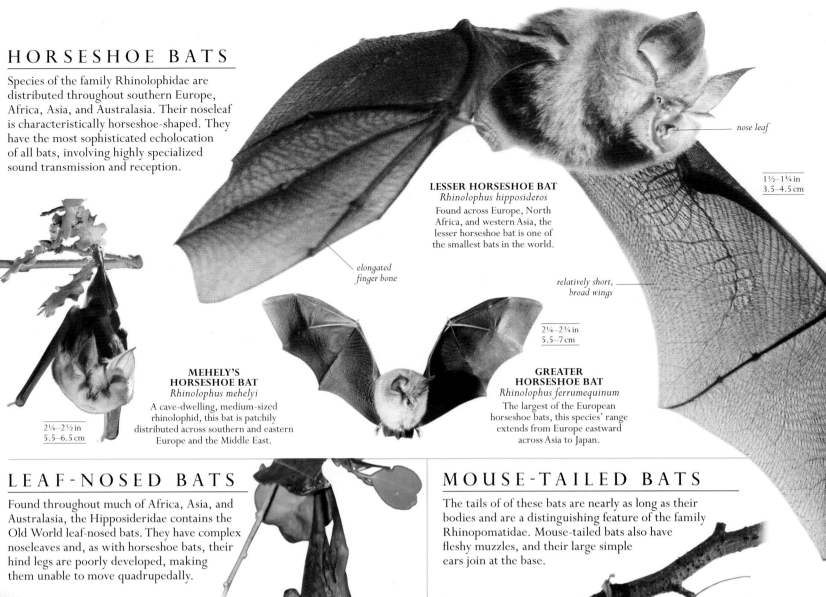

nose leaf

LESSER HORSESHOE BAT
Rhinolophus hipposideros
Found across Europe, North Africa, and western Asia, the lesser horseshoe bat is one of the smallest bats in the world.

1½–1¾ in
3.5–4.5 cm

elongated finger bone

relatively short, broad wings

2¼–2¾ in
5.5–7 cm

MEHELY'S HORSESHOE BAT
Rhinolophus mehelyi
A cave-dwelling, medium-sized rhinolophid, this bat is patchily distributed across southern and eastern Europe and the Middle East.

2¼–2½ in
5.5–6.5 cm

GREATER HORSESHOE BAT
Rhinolophus ferrumequinum
The largest of the European horseshoe bats, this species' range extends from Europe eastward across Asia to Japan.

LEAF-NOSED BATS

Found throughout much of Africa, Asia, and Australasia, the Hipposideridae contains the Old World leaf-nosed bats. They have complex noseleaves and, as with horseshoe bats, their hind legs are poorly developed, making them unable to move quadrupedally.

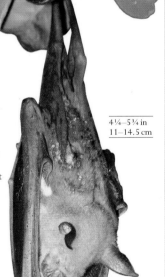

3–3½ in
8–9 cm

COMMERSON'S LEAF-NOSED BAT
Hipposideros commersoni
This forest-dwelling bat roosts in hollow trees across the island of Madagascar. It is one of the largest leaf-nosed bats, weighing up to 6¼ oz (180 g).

4¼–5¾ in
11–14.5 cm

SUNDEVALL'S LEAF-NOSED BAT
Hipposideros caffer
A savanna species, this leaf-nosed bat roosts in caves and buildings throughout Africa except in the Sahara and central forested regions.

MOUSE-TAILED BATS

The tails of of these bats are nearly as long as their bodies and are a distinguishing feature of the family Rhinopomatidae. Mouse-tailed bats also have fleshy muzzles, and their large simple ears join at the base.

2–3½ in
5–9 cm

first finger (thumb)

MOUSE-TAILED BAT
Rhinopoma sp.
Four species of fast-flying insectivorous mouse-tailed bat live in arid and semiarid regions of North Africa, the Middle East, and India.

HOG-NOSED BAT

The sole member of the family Craseonycteridae, the tiny hog-nosed bat has long broad wings that allow it to hover. It lacks a tail and calcars (cartilaginous extensions of the ankle).

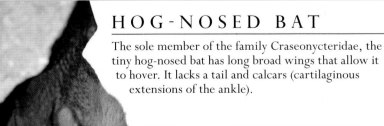

1–1¼ in
3–3.5 cm

KITTI'S HOG-NOSED BAT
Craseonycteris thonglongyai
One of the world's smallest mammals, and often called the bumblebee bat, Kitti's hog-nosed bat lives in riverside caves in Thailand and Myanmar.

SLIT-FACED BATS

The slit-faced bats of the family Nycteridae have a furrow from their nostrils to a pit between their eyes. The cartilage at the end of their tail ends in a "Y" shape.

2–2½ in
5–6.5 cm

MALAYAN SLIT-FACED BAT
Nycteris tragata
The facial furrow of this slit-faced bat helps direct its echolocation calls. It is found in tropical forest in Myanmar, Malaysia, Sumatra, and Borneo.

SHEATH-TAILED BATS

The Emballonuridae are more commonly known as sheath-tailed bats because just the tip of their tail protrudes through the tail membrane, giving it the appearance of being sheathed. Many have glandular odor-storing sacs on their wings.

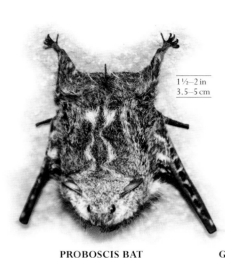

2¼–4 in
6–10 cm

1½–2 in
4–5 cm

1½–2 in
3.5–5 cm

3–3¼ in
7.5–8 cm

HILDEGARDE'S TOMB BAT
Taphozous hildegardeae
Dependent on caves for roosting, Hildegarde's tomb bat feeds on insects in the coastal forests of Kenya and Tanzania.

LESSER SHEATH-TAILED BAT
Emballonura monticola
The short tail of this bat appears to retract into a sheath when it stretches its legs. It lives in Indonesia, Malaysia, Myanmar, and Thailand.

PROBOSCIS BAT
Rhynchonycteris naso
These bats spend the day roosting in groups on the underside of branches. They live in tropical forest in Central and South America.

GREATER SAC-WINGED BAT
Saccopteryx bilineata
Male greater sac-winged bats attract females by a pungent secretion in their wing sacs. The species is found in Central and South America.

AMERICAN LEAF-NOSED BATS

The Phyllostomidae contains species that are distributed from southwestern USA to northern Argentina. Most have large ears and a noseleaf, shaped like a spearhead, to enhance echolocation.

3¼ in / 8.5 cm

2¼–2½ in
6–6.5 cm

2–2½ in
5–6.5 cm

2¾–4 in
7–10 cm

PALE SPEAR-NOSED BAT
Phyllostomus discolor
This species emits echolocation calls through its nose. It is found in Central and northern South America.

COMMON TENT-MAKING BAT
Uroderma bilobatum
Found in lowland forests from Mexico to central South America, this bat creates shelters by biting palm and banana leaves.

GERVAIS' FRUIT-EATING BAT
Artibeus cinereus
Preferring to roost in palm trees, Gervais' fruit-eating bat lives in South America, including Venezuela, Brazil, and the Guianas.

SEBA'S SHORT-TAILED BAT
Carollia perspicillata
Inhabiting moist evergreen and dry deciduous forests across much of Central and South America, this species is a generalist fruit eater.

2–2½ in
5–6.5 cm

2–2½ in
5–6.5 cm

2¼–3 in
6–7.5 cm

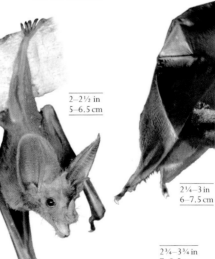

GEOFFROY'S TAILLESS BAT
Anoura geoffroyi
A specialized nectar feeder, this species has a long muzzle, elongate cheek teeth, and a brush-tipped tongue. It lives in Central and South America.

SILKY SHORT-TAILED BAT
Carollia brevicauda
This leaf-nosed bat is widespread in Central America and the Amazon region. It helps restore degraded forests by dispersing the seeds of fruit trees.

2½–3¼ in
6.5–9 cm

2–4 in
5–10 cm

2¾–3¾ in
7–9.5 cm

FRINGE-LIPPED BAT
Trachops cirrhosus
The fringe-lipped bat catches frogs by listening for their calls. It lives in tropical forest in Central and northern South America.

WHITE-LINED BROAD-NOSED BAT
Platyrrhinus lineatus
Named for the white lines on its face and back, this broad-nosed bat roosts in damp forests of central South America.

CALIFORNIAN LEAF-NOSED BAT
Macrotus californicus
Primarily hunting for moths by sight rather than echolocation, the Californian leaf-nosed bat lives in northern Mexico and the southeastern USA.

COMMON VAMPIRE BAT
Desmodus rotundus
Famous for feeding on other mammals' blood, the common vampire bat occurs in a variety of habitats in Mexico and Central and South America.

MORMOOPIDAE

The family Mormoopidae contains the naked-backed and mustache bats. Several species have wings that join across their backs and a fringe of stiff hairs around the muzzle.

1½–2¼ in
4–5.5 cm

DAVY'S NAKED-BACKED BAT
Pteronotus davyi
Unlike most other bats, this bat's wing membranes meet across its back. It is found from Mexico to South America.

NOCTILIONIDAE

The two species of bulldog bats from the family Noctilionidae have long legs and large feet and claws. They have full lips and cheek pouches for storing food while flying.

4–5 in / 10–13 cm

GREATER BULLDOG BAT
Noctilio leporinus
Specialized for snatching fish from the water's surface with its claws, the greater bulldog or fishing bat lives in tropical Central and South America.

NATALIDAE

The funnel-eared bats of the family Natalidae are small and slender with large ears. Adult males have a sensory structure, called a natalid organ, on their foreheads.

1½–1¾ in
4–4.5 cm

MEXICAN FUNNEL-EARED BAT
Natalus stramineus
An insectivorous, normally cave-roosting species, the Mexican funnel-eared bat lives in Central and South America, including islands of the Lesser Antilles.

MEGADERMATIDAE

The family Megadermatidae contains a handful of fairly large microchiropterans—echolocating microbats. Carnivorous and insectivorous, they have large ears and eyes and a broad tail membrane but little or no tail.

4–5 in
10–13 cm

GHOST BAT
Macroderma gigas
Endemic to northern Australia, the ghost bat is one of the largest microbats. It preys on vertebrates, such as frogs and lizards.

MYSTACINIDAE

The one extant member of the Mystacinidae has extra talons on its thumb and toe claws. Its tough, leathery wings can be folded against the body when moving on the ground.

2¼–3¼ in
6–8 cm

NEW ZEALAND SHORT-TAILED BAT
Mystacina tuberculata
An agile mover on the ground, the New Zealand short-tailed bat sniffs out prey in leaf litter on the forest floor.

THYROPTERIDAE

The family Thyropteridae contains three species of disk-winged bats. Suction cups on the wrists and ankles allow them to attach to smooth-surfaced tropical leaves when roosting.

1–2¼ in
2.5–5.5 cm

SPIX'S DISK-WINGED BAT
Thyroptera tricolor
An insectivorous inhabitant of lowland forests from southern Mexico to southeast Brazil, Spix's disk-winged bat roosts inside furled leaves.

MOLOSSIDAE

Free-tailed bats of the family Molossidae have a distinctive tail that extends beyond the edge of the tail membrane. They are robust, stocky bats with long narrow wings for fast flight. Their wing and tail membranes are particularly leathery.

BROAD-EARED FREE-TAILED BAT
Nyctinomops laticaudatus
Occupying a range of different habitats, this free-tailed bat is found in tropical and subtropical Central and South America.

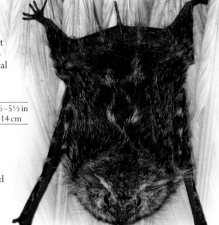

3¼–3½ in
8–9 cm

3¾ in
9.5 cm

MEXICAN FREE-TAILED BAT
Tadarida brasiliensis
Colonies of the Mexican free-tailed bat roosting in caves and beneath bridges in Texas and Mexico can number in the millions.

3½–5½ in
9–14 cm

EUROPEAN FREE-TAILED BAT
Tadarida teniotis
The only free-tailed bat to live in Europe, this bat's range extends from the Mediterranean to southern and southeast Asia.

2–4½ in
5–11.5 cm

MILLER'S MASTIFF BAT
Molossus pretiosus
An insectivorous bat of lowland dry forests, open savanna, and cactus scrub, Miller's mastiff bat is distributed from Mexico to Brazil.

VESPERTILIONIDAE

Numbering more than 300 species, the Vespertilionidae is the largest family of bats. Known as evening or common bats, they are distributed worldwide excluding the polar regions, and are mostly insectivorous. They typically have plain noses and small eyes.

2–3¼ in / 5–8 cm

SCHREIBERS'S BAT
Miniopterus schreibersii
Possessing long finger bones and broad wings, Schreiber's bat is patchily distributed in southwestern Europe and North and West Africa.

ears joined at base

1½–2¼ in
4–6 cm

4–5 in
10–13 cm

GREY LONG-EARED BAT
Plecotus austriacus
The grey long-eared bat's ears are almost the same length as its body. It lives in southern and central Europe and North Africa.

BIG BROWN BAT
Eptesicus fuscus
Often found roosting in buildings, the insectivorous big brown bat is distributed from southern Canada to northern Brazil and some Caribbean islands.

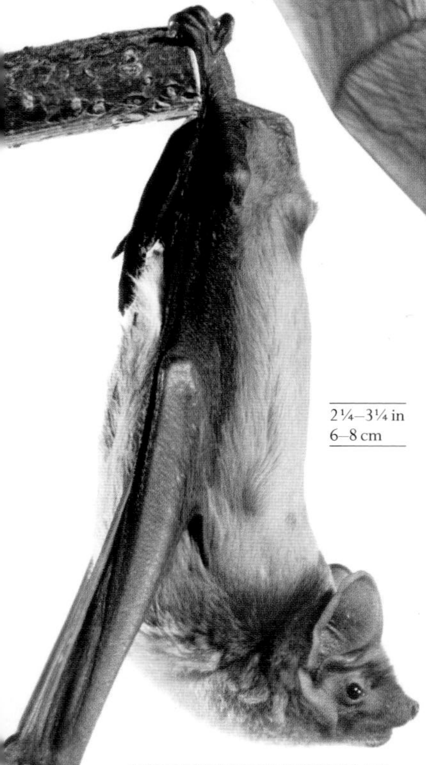

2¼–3¼ in
6–8 cm

COMMON NOCTULE BAT
Nyctalus noctula
A fast and powerful flier with narrow, blackish brown wings, the noctule is found across northeast Europe and parts of Asia.

2–2½ in
5–6.5 cm

PARTI-COLOURED BAT
Vespertilio murinus
This pale-bellied, dark-backed bat is native to mountain, steppe, and forest habitats from eastern and central Europe to Asia.

1½–2 in
4–5 cm

NATTERER'S BAT
Myotis nattereri
From northwest Africa, across Europe to southwest Asia, Natterer's bat catches insects in its fringed tail membrane during a slow, hovering flight.

3¼–3¾ in
8–9.5 cm

FRINGED MYOTIS
Myotis thysanodes
Named after a fringe of hairs along the edge of its tail membrane, this myotis bat lives in western North America.

3–3¾ in
7.5–9.5 cm

1½–2 in / 3.5–4.5 cm

1¾–2¼ in / 4.5–5.5 cm

1½–2¼ in
4–6 cm

EASTERN PIPISTRELLE
Perimyotis subflavus
The eastern pipistrelle hibernates underground during winter. Its range covers eastern North America, from southern Canada to northern Honduras.

COMMON PIPISTRELLE
Pipistrellus pipistrellus
The most widespread pipistrelle species, the common pipistrelle's range extends from western Europe to the Far East and North Africa.

NATHUSIUS' PIPISTRELLE
Pipstrellus nathusii
A long-distance migrant capable of journeys of over 1,200 miles (1,900 km) in spring and fall, this species mainly occurs in eastern and central Europe.

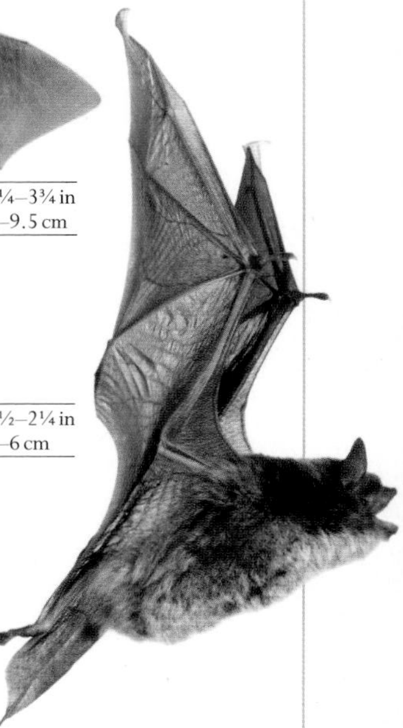

DAUBENTON'S BAT
Myotis daubentonii
This Eurasian bat has relatively large feet, which help it catch flying insects emerging from the surface of bodies of water.

HEDGEHOGS AND MOONRATS

There is only one extant family of rather primitive mammals in the order Erinaceomorpha. Native to Eurasia and Africa, it includes the hedgehogs.

With long snouts, hedgehogs resemble—and were once classified with—the much smaller shrews and moles. Large eyes and excellent hearing reflect their crepuscular (active at dawn and dusk) or nocturnal habits. Their ears are often prominent, especially in desert species where they also serve to cool the animal. Generally with five toes on each foot, many species have sharp claws for digging burrows and shelters.

The 16 species of hedgehogs are covered in stout spines, made up of modified hairs. These provide an effective defense against predators,

especially combined with their habit of rolling into a ball when threatened. In contrast, the eight moonrat species (also known as gymnures) are covered in normal fur and have longer, naked tails.

A VARIED DIET

All members of the order are described as omnivores, on the basis that their diet contains at least a small proportion of fruit and fungi, while hedgehogs in particular relish carrion and birds' eggs. However, the bulk of their diet is made up of living animals, from earthworms, mollusks, and other terrestrial invertebrates, to small reptiles, amphibians, and mammals. Food is detected by sight and their very keen sense of smell; the numerous, sharply pointed teeth are well suited to such a generalized diet.

PHYLUM	CHORDATA
CLASS	MAMMALIA
ORDER	ERINACEOMORPHA
FAMILIES	1
SPECIES	24

Immunity to snake venom allows a hedgehog to take advantage of any snake it comes across as a potential food source.

HEDGEHOGS AND MOONRATS

With long, sensitive snouts and short, hairy tails, members of the family Erinaceidae eat almost anything, from invertebrates and fruit to birds' eggs and carrion. The hedgehogs, from Eurasia and Africa, are covered in sharp protective spines, while the southeast Asian moonrats have normal hair and look more like rats or opossums. Moonrats have especially well-developed scent glands that exude a strong garliclike odor for marking territories.

LONG-EARED HEDGEHOG
Hemiechinus auritus
Long ears help radiate heat and keep this nocturnal hedgehog cool in the deserts of North Africa and central Asia.

pale spines with darker bands

large ears

6–8 in / 15–20 cm

SOUTHERN AFRICAN HEDGEHOG
Atelerix frontalis
Inhabiting grassland, scrub, and gardens in southern Africa, this species has a white band across its forehead, contrasting with its dark face.

7–10 in
18–25 cm

5½–11 in
14–28 cm

5½–11 in
14–28 cm

8–12 in
20–30 cm

NORTH AFRICAN HEDGEHOG
Atelerix algirus
Found in varied habitats in the Mediterranean region, this long-legged hedgehog with a pale face and underparts has a spineless "parting" on its crown.

DESERT HEDGEHOG
Paraechinus aethiopicus
This small African and Middle Eastern hedgehog is immune to the venom of snakes and scorpions, which form a large part of its diet.

12–16 in
30–40 cm

EUROPEAN HEDGEHOG
Erinaceus europaeus
Found throughout western Europe, this species inhabits woodland, farmland, and gardens. In cooler areas, it hibernates in a nest of leaves and grass.

MOONRAT
Echinosorex gymnura
Resembling a large rat, the nocturnal, white-coated moonrat lives in swamps and other wet habitats in Malaysia.

MOLES AND RELATIVES

The three families that make up the order Soricomorpha are insectivores, and have an elongated snout with sharp teeth, a long tail, and velvety fur.

About 90 percent of the species in the order Soricomorpha fall within just one family, the shrews (Soricidae). Members of this order are among the most primitive of the placental mammals, with small brains relative to their body size. There is also a considerable size range, from the tiny Etruscan shrew that weighs just $\frac{1}{16}$ oz (2 g) to the Cuban solenodon at $2\frac{1}{4}$ lb (1 kg).

Well adapted to their insectivorous way of life, the long, mobile, cartilaginous snout of shrews, moles, and solenodons contains numerous simple, sharp-pointed teeth, which they use to catch,

kill, and dismember earthworms, other invertebrates, and even small vertebrates. In addition, some species produce venomous saliva, which runs along a groove in one of their lower incisor teeth, and helps subdue larger prey before it is killed and consumed.

SIMILAR BUT DIFFERENT

While most species in the order have short, velvety fur, the solenodons have a coarser, shaggy coat. Most also have a long tail, especially those species that forage in trees and bushes, where it provides balance. Many of the moles have only a short tail, one of several characteristics typical to mammals that have a tunneling way of life. Other features include small eyes and front limbs modified for digging.

PHYLUM	CHORDATA
CLASS	MAMMALIA
ORDER	SORICOMORPHA
FAMILIES	3
SPECIES	428

A mother shrew is followed by her young. They form a chain, each holding on to the rear of the sibling in front of it.

SOLENODONS

Showing features of the earliest mammals, which evolved during the age of the dinosaurs 225–65 million years ago, the Solenodontidae resemble very large shrews. They have elongated, flexible, cartilaginous snouts; long, naked, scaly tails; small eyes; and coarse, dark fur. Unlike most mammals, they also have venomous saliva, which is used to subdue prey ranging from invertebrates to small reptiles.

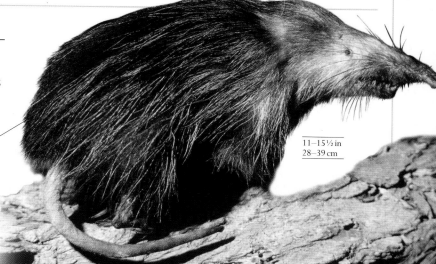

shaggy, brown coat

11–13 in/28–33 cm

HISPANIOLAN SOLENODON
Solenodon paradoxus
One of two extant species in its family, this solenodon is known only on Hispaniola, an island to the east of Cuba in the Caribbean.

CUBAN SOLENODON
Solenodon cubanus
With longer, finer fur than its Hispaniolian relative, this secretive, nocturnal burrower was erroneously believed to be extinct during the 20th century.

11–15½ in 28–39 cm

MOLES AND DESMANS

Small, dark insectivores with cylindrical bodies, short, dense fur, and very sensitive, hairless, tubular snouts, moles of the family Talpidae are well adapted for a life of burrowing—the forelegs have powerful claws, and the hands are turned permanently outward in the form of a shovel. In contrast, the aquatic desmans have webbed paws fringed with stiff hairs and long, flattened tails to aid in swimming.

4¼–6½ in 11–16 cm

4¼–6½ in 11–17 cm

EASTERN MOLE
Scalopus aquaticus
A burrowing mole from North America, typically in moist sandy soils, this species has ears and eyes covered by skin and fur respectively.

EUROPEAN MOLE
Talpa europaea
Secretive due to its burrowing lifestyle, this mole creates extensive networks of permanent tunnels, often marked by distinctive surface molehills.

6–8 in 15–20 cm

STAR-NOSED MOLE
Condylura cristata
A semiaquatic North American mole, this species has 11 pairs of pink fleshy appendages around its snout, which detect prey by touch.

4¼–6½ in 11–16 cm

4–5½ in 10–14 cm

7–8½ in 18–21 cm

dense, waterproof coat

SMALL JAPANESE MOLE
Mogera imaizumii
This small mole from Japan is found in soft, deep soils, and is distinguished from close relatives by dental features.

PYRENEAN DESMAN
Galemys pyrenaicus
Feeding in Pyrenean mountain streams, this desman rarely digs burrows; it shelters in rock crevices or the holes of water voles.

RUSSIAN DESMAN
Desmana moschata
The largest member of its family, this desman's webbed hind feet and long, flattened tail are adaptations for swimming in search of food.

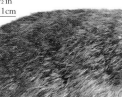

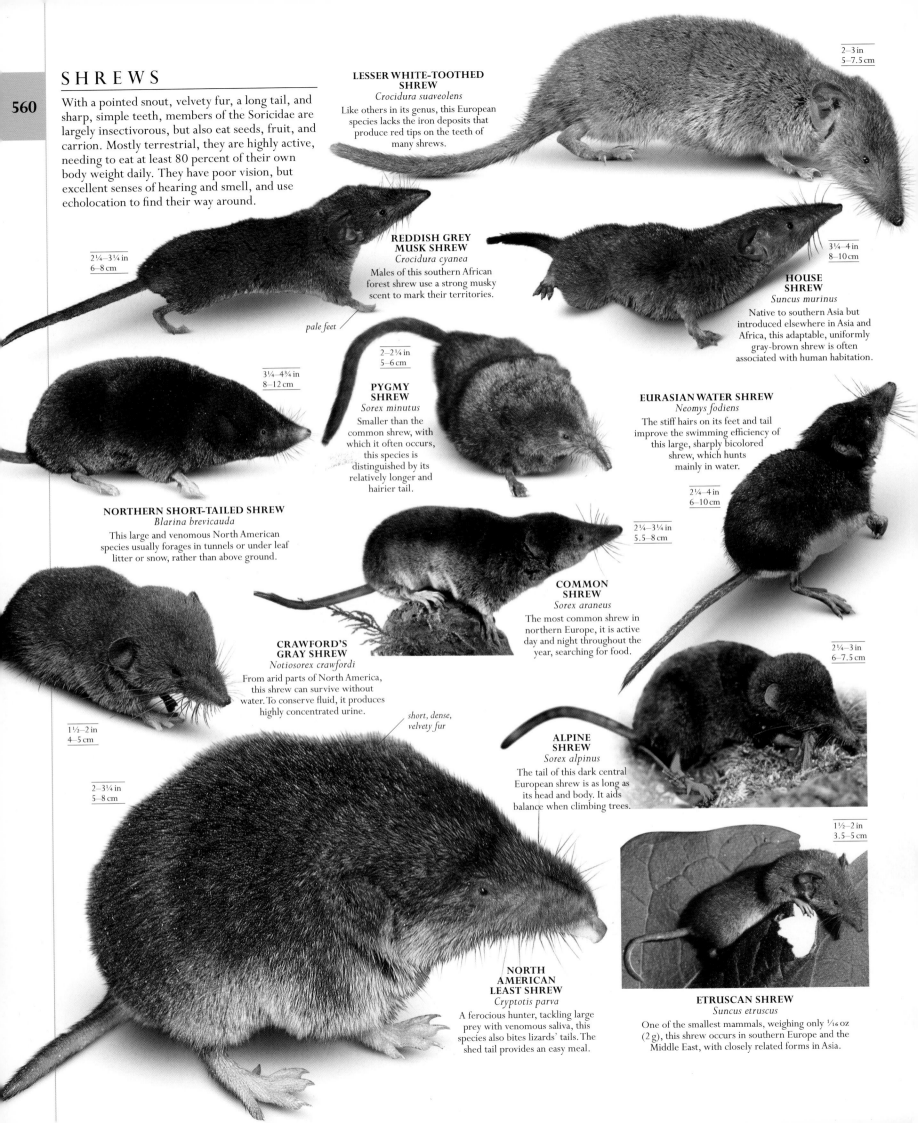

SHREWS

With a pointed snout, velvety fur, a long tail, and sharp, simple teeth, members of the Soricidae are largely insectivorous, but also eat seeds, fruit, and carrion. Mostly terrestrial, they are highly active, needing to eat at least 80 percent of their own body weight daily. They have poor vision, but excellent senses of hearing and smell, and use echolocation to find their way around.

LESSER WHITE-TOOTHED SHREW
Crocidura suaveolens
Like others in its genus, this European species lacks the iron deposits that produce red tips on the teeth of many shrews.

2–3 in
5–7.5 cm

REDDISH GREY MUSK SHREW
Crocidura cyanea
Males of this southern African forest shrew use a strong musky scent to mark their territories.

2¼–3¼ in
6–8 cm

HOUSE SHREW
Suncus murinus
Native to southern Asia but introduced elsewhere in Asia and Africa, this adaptable, uniformly gray-brown shrew is often associated with human habitation.

3¼–4 in
8–10 cm

pale feet

PYGMY SHREW
Sorex minutus
Smaller than the common shrew, with which it often occurs, this species is distinguished by its relatively longer and hairier tail.

2–2¼ in
5–6 cm

NORTHERN SHORT-TAILED SHREW
Blarina brevicauda
This large and venomous North American species usually forages in tunnels or under leaf litter or snow, rather than above ground.

3¼–4¾ in
8–12 cm

EURASIAN WATER SHREW
Neomys fodiens
The stiff hairs on its feet and tail improve the swimming efficiency of this large, sharply bicolored shrew, which hunts mainly in water.

2¼–4 in
6–10 cm

CRAWFORD'S GRAY SHREW
Notiosorex crawfordi
From arid parts of North America, this shrew can survive without water. To conserve fluid, it produces highly concentrated urine.

COMMON SHREW
Sorex araneus
The most common shrew in northern Europe, it is active day and night throughout the year, searching for food.

2¼–3¼ in
5.5–8 cm

1½–2 in
4–5 cm

short, dense, velvety fur

2–3¼ in
5–8 cm

ALPINE SHREW
Sorex alpinus
The tail of this dark central European shrew is as long as its head and body. It aids balance when climbing trees.

2¼–3 in
6–7.5 cm

NORTH AMERICAN LEAST SHREW
Cryptotis parva
A ferocious hunter, tackling large prey with venomous saliva, this species also bites lizards' tails. The shed tail provides an easy meal.

ETRUSCAN SHREW
Suncus etruscus
One of the smallest mammals, weighing only ⅟₁₆ oz (2 g), this shrew occurs in southern Europe and the Middle East, with closely related forms in Asia.

1½–2 in
3.5–5 cm

PANGOLINS

With bodies covered in large keratin scales, pangolins are also known as scaly anteaters, reflecting their appearance and diet.

Mainly nocturnal and with small eyes, pangolins use their excellent sense of smell to find food. Although somewhat similar in appearance to the American armadillos and sharing a similar diet, pangolins belong to a different order, Pholidota, and are most closely related to the carnivores.

The horny scales, which cover all exposed parts of the body, can make up one fifth of a pangolin's body weight; despite this, they are excellent swimmers. There are both terrestrial and arboreal species: ground-dwellers live in deep burrows, while tree-dwellers live in holes in trees.

Pangolins have powerful claws that are used to excavate insect nests. Food is captured on the sticky tongue, which extends up to 16 in (40 cm) from the toothless mouth. Because of the size of their front claws, pangolins walk on their wrists, the forepaws curled over to protect the claws.

SAFE FROM ATTACK

The scaly coat provides protection against predators, and is enhanced by the tendency of pangolins to roll into a ball when threatened or sleeping. Further defense is provided by their ability to emit a foul-smelling chemical from their anal glands. Nevertheless, pangolins are heavily exploited for their meat, scales, and perceived value in traditional Chinese medicine.

PHYLUM	CHORDATA
CLASS	MAMMALIA
ORDER	PHOLIDOTA
FAMILIES	1
SPECIES	8

The long, sticky tongue of this African ground-dwelling pangolin is used for drinking as well as capturing insects.

MANIDAE

The eight pangolin species of the family Manidae live in tropical Africa and Asia. They are unique among mammals in having their skin protected by large horny scales. When threatened, they can curl up into a ball, the sharp plate edges providing additional defense. Large, powerful front claws are used to excavate ant hills and termite mounds in search of food, captured by sticky saliva on their long, probing tongues.

30 scales on tail

20–26 in
50–65 cm

long fleshy nose

SUNDA PANGOLIN
Manis javanica
Semiarboreal, this Asian pangolin moves clumsily on the ground, unless threatened, then it runs on its hind feet, using its tail for balance.

TREE PANGOLIN
Manis tricuspis
This arboreal pangolin from equatorial Africa has pale fur and distinctive three-pointed scales, although the points can become worn with age.

14–18 in
35–46 cm

LONG-TAILED PANGOLIN
Manis tetradactyla
This small pangolin from West Africa lives high in the forest canopy. It has a prehensile tail two-thirds of its total body length.

INDIAN PANGOLIN
Manis crassicaudata
Its overlapping scales and ability to emit a noxious defensive fluid can protect this pangolin even from the attentions of tigers.

18–30 in
45–75 cm

broad, rounded scales

12–16 in
30–40 cm

GROUND PANGOLIN
Manis temminckii
The only pangolin in southern and eastern Africa, this species is very secretive. It is hunted for its scales, which are used to make love charms.

16–28 in / 40–70 cm

CARNIVORES

Predominantly meat-eaters, carnivores have bodies that are adapted for hunting and their teeth are specialized for grasping and killing prey.

The first carnivores evolved about 50 million years ago. They were small, tree-dwelling, rather catlike mammals, but their descendants, which include some of the largest predators on Earth, display a range of forms and lifestyles. The order Carnivora ranges in size from the least weasel, 5½ in (14 cm) long, to the southern elephant seal, which measures up to 23 ft (7 m) from nose to tail. The order includes the world's fastest land animal, the cheetah, and the famously idle giant panda. Carnivores are naturally present on every continent except Australia, where they have been introduced by man. They are not restricted to dry land—more than 30 species of seal and sea lion are more at home in the oceans.

CHARACTERISTICS

With such diversity, it can be difficult to ascertain what the animals of this order have in common. The most important shared characteristic of the group is their teeth. All members of Carnivora have four long canine teeth and a distinctive set of cheek teeth known as carnassials, which are modified for cutting meat. The sharp edges of the carnassials work like a pair of scissor blades when the animal opens and shut its jaws.

Most carnivores eat at least some meat, but few are exclusively carnivorous. Several species, such as the foxes and raccoons, are omnivores, eating a wide range of plant and animal foods. One species, the giant panda, is completely herbivorous, with a diet consisting mainly of bamboo.

SOLITARY OR SOCIAL

Carnivores may lead solitary lives, like most weasels and bears, or be highly social, like wolves, lions, and meerkats. Animals in the latter category live in highly organized cooperative groups, sharing responsibility for hunting, rearing young, and protecting their territory. Seals and sea lions are generally colonial during the breeding season, when they are obliged to return to dry land to mate and give birth; some species gather in hundreds or even thousands on favored beaches.

PHYLUM	CHORDATA
CLASS	MAMMALIA
ORDER	CARNIVORA
FAMILIES	15
SPECIES	286

DEBATE
PINNIPED OR CARNIVORE?

Taken at face value, it seems unlikely that the aquatic seals, sea lions, and walruses, known as pinnipeds from the Latin for "web-footed," could belong to the same group as weasels and wildcats. But the structure of their skulls and teeth and the information coded in their DNA tells a different story. Pinnipeds have limbs modified for swimming, but—unlike whales—they are not wholly aquatic and must return to land in order to breed. Fossil and molecular evidence suggests that seals, sea lions, and the walrus share a common bearlike ancestor, which diverged from the other carnivores about 23 million years ago.

DOGS, FOXES, AND RELATIVES

The Canidae are medium-sized, long-legged mammals, most of which have a bushy tail and erect ears. They are swift, intelligent predators, although most also eat plant foods. The highly social gray wolf is the ancestor of the domestic dog, a species domesticated by man over 10,000 years ago, and now hugely varied in size and form.

15–20 in
38–50 cm

large ears

BLANFORD'S FOX
Vulpes cana
A strictly nocturnal species, this fox inhabits steppe country of the Arabian Peninsula and the Middle East. It feeds upon invertebrates and fruits.

15½–22½ in
39–57 cm

BENGAL FOX
Vulpes bengalensis
This agile, omnivorous fox inhabits open country in Nepal and India. Pairs remain together from year to year and rear several litters together.

short, pointed face

18–23½ in
45–60 cm

yellowish white underside

CORSAC FOX
Vulpes corsac
A social, pack-dwelling animal of the Asian steppes, the Corsac fox is an opportunistic hunter of small animals. It also eats vegetable matter.

20–30 in
50–75 cm

ARCTIC FOX
Vulpes lagopus
This sturdy fox lives in the world's northernmost regions. Its variable color forms include snowy white.

KIT FOX
Vulpes macrotis
Found in the southwestern USA, the kit fox is an efficient digger. Families live in burrows with up to 20 entrances.

14½–20 in
37–50 cm

SWIFT FOX
Vulpes velox
This close relative of the kit fox occurs in the central USA and has been reintroduced in Canada, where it became extinct in 1938.

14½–21½ in
37–54 cm

FENNEC FOX
Vulpes zerda
The distinctive ears of this tiny, nocturnal North African fox serve a dual purpose: providing keen hearing and shedding excess body heat.

pointed ears, black on top

13–16 in
33–41 cm

14–22 in
35–55 cm

RÜPPELL'S FOX
Vulpes reuppellii
Ranging from North Africa to Pakistan, this small, social fox survives desert conditions by eating a wide range of plants and animals.

RED FOX
Vulpes vulpes
This highly adaptable hunter and scavenger is the world's most widespread carnivore, occurring throughout most of the Northern Hemisphere.

20–35 in
50–90 cm

18–23½ in
46–60 cm

black lower limb

BAT-EARED FOX
Otocyon megalotis
Inhabiting open grassland and scrubby savanna in southern and eastern Africa, this social fox feeds chiefly on termites and beetles.

large, bushy tail, with a white tip

PAMPAS FOX
Lycalopex gymnocercus
A native of temperate grassland in South America, this solitary fox hunts a variety of small animals and occasionally sheep.

20–29 in
50–74 cm

21½–26 in
54–66 cm

brown-black body fur

19½–28 in
49–70 cm

GRAY FOX
Urocyon cinereoargenteus
This fox is a relatively common inhabitant of forested areas throughout the Americas, although it avoids areas where coyotes and bobcats live.

18–36 in
45–92 cm

22½–30 in
57–77 cm

CULPEO
Lycalopex culpaeus
A large, solitary fox from the highlands of South America, the culpeo hunts larger prey than most fox species.

CRAB-EATING FOX
Cerdocyon thous
An adaptable omnivore from the grasslands and temperate and tropical forests of South America, this species feeds on fruits, carrion, and small animals.

RACCOON DOG
Nyctereutes procyonoides
This unusual dog inhabits wetlands in eastern Asia. It climbs well and is the only species of dog to hibernate.

»

ginger ears

26–32 in
65–80 cm

18–35 in
45–90 cm

26–41 in
65–105 cm

BLACK-BACKED JACKAL
Canis mesomelas
The largest of the African jackals, this adaptable,
omnivorous dog lives in family groups.
It is active both day and night.

SIDE-STRIPED JACKAL
Canis adustus
A widespread African jackal, this species is a
nocturnal scavenger and hunter. It is often
persecuted by farmers.

GOLDEN JACKAL
Canis aureus
Sometimes described as the archetypal
dog, this swift-footed, social, and
opportunistic hunter and forager is
widespread across Africa and Asia.

33–39 in
84–101 cm

COYOTE
Canis latrans
Native to North and Central
America, this widespread and
common species travels in
packs and occasionally mates
with the gray wolf.

ETHIOPIAN WOLF
Canis simensis
Currently listed as the
world's rarest dog, wild
packs of this threatened wolf
survive only in the remote
highlands of Ethiopia.

28–38 in
70–97 cm

RED WOLF
Canis lupus rufus
This highly endangered
subspecies from the southeastern
USA survives on special
reserves, thanks to captive
breeding in the 1980s.

3½–4½ ft
1.1–1.4 m

3¼–4 ft
1–1.2 m

ARCTIC WOLF
Canis lupus arctcos
Distinguished by a pale coat, this
subspecies of the gray wolf lives in parts
of Canada, Alaska, and Greenland.

*thick fur
traps heat*

3–5½ ft
0.9–1.6 m

sharp teeth

GRAY WOLF
Canis lupus lupus
The ancestor of the domestic
dog, this adaptable wolf has a
vast distribution, covering most
of the Northern Hemisphere.

*large feet
and claws*

1–4 ft
0.9–1.2 m

DINGO
Canis lupus dingo
This large subspecies of the gray wolf was introduced
to Australia about 4,000 years ago, where it rapidly
became the continent's top predator.

GOLDEN RETRIEVER
Canis lupus familiaris
Bred in Scotland to retrieve game, this dog is loyal and intelligent, and makes an excellent pet. It loves being in water.

2¾–3¼ ft
85–100 cm

25–31 in
63–79 cm

plumelike tail

dense coat

DALMATIAN
Canis lupus familiaris
Initially bred as guard dogs and hunting companions, dalmatians subsequently became coach dogs, escorting horsedrawn carriages. Today they are kept as pets.

23½ in
60 cm

BASSET HOUND
Canis lupus familiaris
This dog was bred as a tracking animal. Its short legs allow it to move easily through dense vegetation.

31–38 in
79–96 cm

MALAMUTE
Canis lupus familiaris
Bred in Alaska as a sled dog, the malamute resembles the ancestral wolf, and may be among the earliest domestic breeds.

large, erect, mobile ears

MANED WOLF
Chrysocyon brachyurus
An opportunistic omnivore, the maned wolf inhabits South American savanna, where its long legs help it see over tall grasses.

16 in
40 cm

rough outer fur with downy undercoat

dark muzzle

SMOOTH FOX TERRIER
Canis lupus familiaris
An enthusiastic digger and small enough to enter fox burrows, this lively dog was bred to keep farms free of vermin.

26–34 in
67–85 cm

8–12 in
20–30 cm

CHIHUAHUA
Canis lupus familiaris
The smallest of domestic dog breeds, but bold-natured nonetheless, the chihuahua originated in Mexico. Its coat can be of almost any color.

ROUGH COLLIE
Canis lupus familiaris
Originally bred as a herding dog in the Scottish Highlands, this dog has an athletic build disguised by a thick coat.

3–4¼ ft
1–1.3 m

long legs have earned the nickname "fox on stilts"

2½–4½ ft
0.8–1.4 m

22½–30 in
57–75 cm

3–4½ ft
0.9–1.4 m

DHOLE
Cuon alpinus
Known throughout its Asian range as a fierce predator, this wild dog lives and hunts in packs, targeting large mammals, such as deer and goats.

AFRICAN WILD DOG
Lycaon pictus
This endangered dog lives in highly organized packs, hunting and rearing young cooperatively and supporting sick and injured relatives.

BUSH DOG
Speothos venaticus
This distinctive, short-legged species is a predator of the Amazon region. It hunts mainly rodents, working alone or in a pack.

BEARS

The Ursidae are large, stocky-yet-agile animals native to Europe, Asia, and the Americas. Most bears are omnivorous, with plant material making up most of their diet. However, the polar bear is a specialized meat-eater and the giant panda, sometimes placed in a separate family, is an herbivore. Bears live alone except for mothers caring for their young.

BROWN BEAR
Ursus arctos
This bear's vast range, including North America, northern Europe, and Asia, demonstrates its dietary flexibility. Seasonal foods include berries and spawning salmon.

4–6 ft
1.2–1.8 m

5–9¼ ft
1.5–2.8 m

five claws grow to 4 in (10 cm)

white face with black eyes and ears

GIANT PANDA
Ailuropoda melanoleuca
This endangered bear lives in central Chinese forests. Despite being a carnivore, it eats mostly bamboo; this lack of adequate nutrition results in a low-energy lifestyle.

EARED SEALS

Members of this family are distinguished from true seals by their small external ears and by their limbs, which can be used for moving on land—they are able to walk on their flippers, albeit in an ungainly manner. The Otariidae are excellent swimmers, but their dives are short and shallow compared with those of some true seals. They occur in most oceans except the North Atlantic.

STELLER'S SEA LION
Eumetopias jubatus
Also known as the northern sea lion, this North Pacific species is the largest of the eared seals. It eats mainly fish, but may take smaller seals.

thick neck

6½–11 ft
2–3.3 m

AUSTRALIAN SEA LION
Neophoca cinerea
This relatively rare species has breeding colonies restricted to Western and South Australia. Small groups stay together even outside the breeding season.

4¼–8¼ ft
1.3–2.5 m

5¼–8¼ ft
1.6–2.5 m

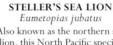

dark brown or black pup

black flippers

NEW ZEALAND SEA LION
Phocarctos hookeri
Found only in the waters around New Zealand, this rare species breeds on just a few offshore islands.

JUVENILE

ADULT

doglike muzzle with whiskers

6–8½ ft
1.8–2.6 m

4½–7¼ ft
1.4–2.2 m

CALIFORNIA SEA LION
Zalophus californianus
Widely recognized as the "performing seal," this species is agile on land and in water, often leaping partly clear of the surface.

streamlined body tapers from shoulder to tail

5–8¼ ft
1.5–2.5 m

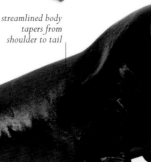

SOUTHERN SEA LION
Otaria flavescens
Thickset with a blunt face, this resident of South America and the Falkland Islands in the South Atlantic occasionally hunts cooperatively and may enter rivers in search of fish.

NORTHERN FUR SEAL
Callorhinus ursinus
Except when breeding, this species lives far offshore in the North Pacific. Males can be over five times as heavy as females.

all-white
body

6–9¼ ft
1.8–2.8 m

relatively
long neck

partially furred
paw-pads provide
extra grip on ice

4–6¼ ft
1.2–1.9 m

3½–6¼ ft
1.1–1.9 m

AMERICAN BLACK BEAR
Ursus americanus

Numbering about 900,000 individuals
across varied habitats in North America,
this is the world's most common bear.
Some have brown or blond fur.

ASIATIC BLACK BEAR
Ursus thibetanus

A widespread forest-dweller,
this bear is variable in appearance,
habitat, and behavior. In the tropics,
only pregnant females hibernate.

POLAR BEAR
Ursus maritimus

One of the largest predators on land, the polar
bear is equally at home in water. It spends most
of its life on Arctic sea ice.

4½–6¼ ft
1.4–1.9 m

3¼–5 ft
1–1.5 m

3¼–6 ft
1–1.8 m

SUN BEAR
Helarctos malayanus

This shy Southeast Asian
forest bear eats insects,
honey, fruits, and plant
shoots. It is active by day,
becoming nocturnal
if disturbed.

SPECTACLED BEAR
Tremarctos ornatus

A vulnerable species of
Andean cloud forests, this
bear is an excellent climber.
Its varied diet includes
fruits, shoots, and meat.

SLOTH BEAR
Melursus ursinus

This shaggy Indian bear can
live in a range of habitats. It
uses its huge claws to open
termite mounds, sucking
up the panicking insects.

BROWN FUR SEAL
Arctocephalus pusillus

This species has two distinct
populations—one South African,
the other Australian. Both have
suffered from excessive hunting.

4–5¼ ft
1.2–1.6 m

GALAPAGOS
FUR SEAL
Arctocephalus galapagoensis

The smallest of the eared seals,
this species is also the least
variable, with males only
slightly larger than females.

pointed muzzle

AUSTRALIAN
FUR SEAL
Arctocephalus forsteri

This seal breeds on the
rocky coasts of New
Zealand and Australia.
Now protected by law, it is
no longer hunted and its
numbers are increasing.

*velvety fur traps
warm air next
to the skin*

GUADALUPE FUR SEAL
Arctocephalus townsendi

Distinguished by its long, tapering nose, this
fur seal breeds on rocky beaches and caves
accessible only from the sea.

4¼–8¼ ft
1.3–2.5 m

4½–7¾ ft
1.4–2.4 m

4½–6½ ft
1.4–2 m

*massive
neck with a
coarse mane*

4¼–6½ ft
1.3–2 m

SOUTH AMERICAN
FUR SEAL
Arctocephalus australis

This voracious predator of
fish, squid, and crustaceans
breeds on the rocky beaches
of South America and the
Falkland Islands.

4½–6¼ ft
1.4–1.9 m

ANTARCTIC FUR SEAL
Arctocephalus gazella

This species breeds on the
scattered islands of the Southern
Ocean. It is recovering from
excessive hunting in the past.

fore flippers

POLAR BEAR
Ursus maritimus

unlike other bears, the polar bear has a slightly convex "Roman" nose

This magnificent animal is as at home at sea as on land. It is the world's largest terrestrial predator, and its huge bulk can make it seem lumbering. In the water, however, it is transformed into a creature of effortless grace. Polar bears are true nomads and spend much of the year far from land, roaming the frozen expanse of the Arctic Ocean. In summer the melting ice forces them to retreat to land, where they sometimes come into contact with humans. Young polar bears are born in the middle of winter in a hibernation den dug by their mother. She scarcely wakes from hibernation when they arrive, but nurses them in her sleep for three months, breaking down her own body reserves to create rich, fatty milk. Come the spring, the cubs have increased their birth weight dramatically, while the mother is close to starvation. She spends the next two years teaching them to swim, hunt seals, defend themselves, and build their own snow dens. Climate change is threatening the habitat and feeding habits of polar bears, which could result in the bears' extinction.

SIZE 6–9¼ ft (1.8–2.8 m)
HABITAT Arctic ice fields
DISTRIBUTION Arctic Ocean; polar regions of Russia, Alaska, Canada, Norway, and Greenland
DIET Mainly seals

FURRY EARS >
The small ears are completely covered with fur to prevent frostbite. Polar bears have good hearing but rely mainly on their sense of smell to locate their prey.

< DARK EYE
The dark eyes and nose are the most conspicuous parts of the bear's body. Polar bears have good eyesight, comparable to that of humans.

^ CRUSHING BLOW
The front paws are the principle means of dispatching prey. With their excellent sense of smell, polar bears can detect young seals located in dens below the ice, and then crush them with a mighty blow from above.

^ PADDLE PAWS
The huge, furry-soled paws make effective paddles when swimming, propelling the bear at around 4 miles (6 km) per hour for hours at a time.

^ STUBBY TAIL
The polar bear has little use for a long tail, so the appendage is reduced to a stub—all but hidden in the dense fur.

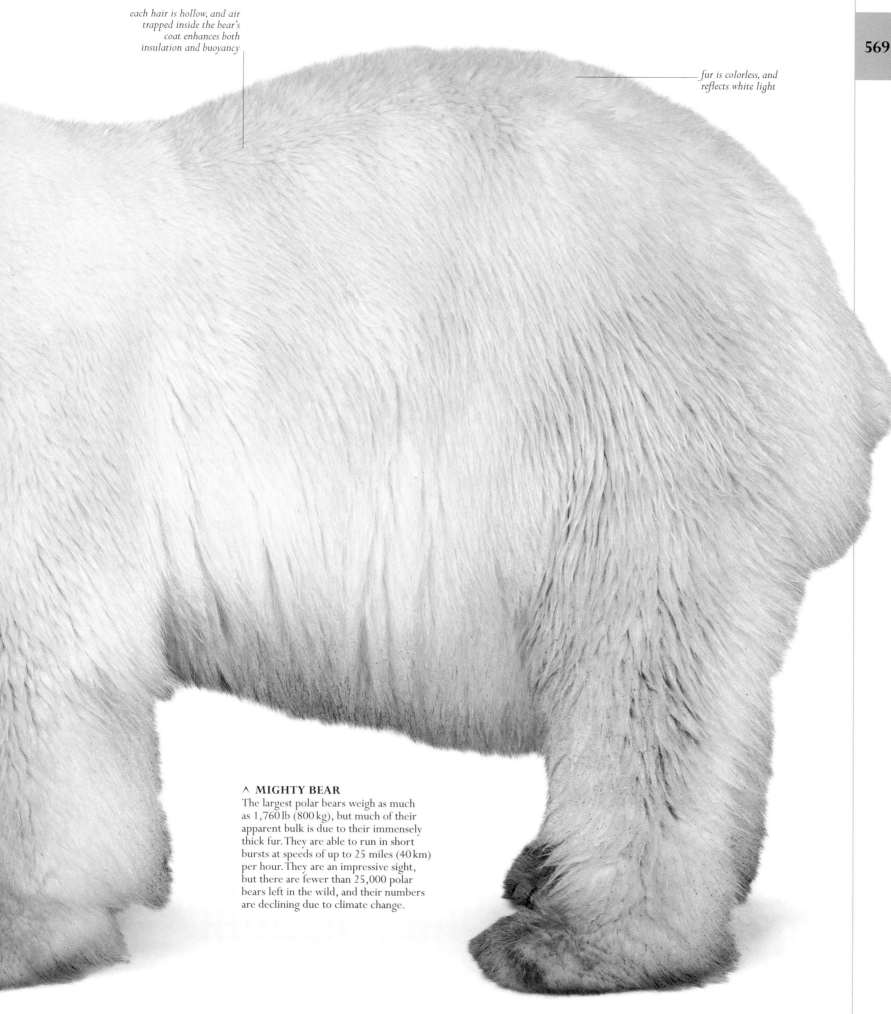

each hair is hollow, and air trapped inside the bear's coat enhances both insulation and buoyancy

fur is colorless, and reflects white light

∧ MIGHTY BEAR

The largest polar bears weigh as much as 1,760 lb (800 kg), but much of their apparent bulk is due to their immensely thick fur. They are able to run in short bursts at speeds of up to 25 miles (40 km) per hour. They are an impressive sight, but there are fewer than 25,000 polar bears left in the wild, and their numbers are declining due to climate change.

WALRUS

The walrus is the only member of the family Odobenidae. This enormous Arctic seal has a huge, blubbery body, long tusks in both sexes, and a mustache of sensitive whiskers for detecting food. Thousands of walruses often haul out of the water in large groups on beaches and ice floes.

long, ivory tusks

7½–12 ft
2.3–3.6 m

thick, creased skin

paddlelike front flippers

blunt body tapering at tail

WALRUS
Odobenus rosmarus
The walrus frequents shallow waters in the Arctic. The male walrus is twice the size of the female, which it attracts with underwater bellows.

EARLESS SEALS

Also called true seals, the Phocidae are more adapted to life in water than their cousins, walruses and sea lions. In place of ear flaps, they have tiny holes on the sides of their heads. Their flippers are useless on land, but propel them with speed and agility in water. Mostly inhabiting cool temperate and polar waters, seals prey on fish and invertebrates, while leopard seals hunt penguins.

5½–8¼ ft
1.7–2.5 m

ROSS SEAL
Ommatophoca rossii
This rare, fast-swimming seal spends most of its life hunting squid beneath Antarctic pack ice. Males are smaller than females.

short flippers

small head

8¼–11 ft
2.5–3.3 m

few, short whiskers

WEDDELL SEAL
Leptonychotes weddellii
The most southerly dwelling wild mammal, this expert diver specializes in long, deep dives under Antarctic ice shelves. Females are larger than males.

8¼–11 ft
2.5–3.4 m

LEOPARD SEAL
Hydrurga leptonyx
This fearsome predator mostly ambushes smaller seals, fish, and penguins at the edge of Antarctic pack ice, although it also eats krill.

5½–11 ft
1.7–3.3 m

GRAY SEAL
Halichoerus grypus
The gray seal lives in the North Atlantic, with large numbers breeding on a small island off Nova Scotia. Males are up to three times as heavy as females.

6½–8½ ft
2–2.6 m

BEARDED SEAL
Erignathus barbatus
This large, Arctic seal feeds on bottom-dwelling fish and invertebrates, which it locates partly by touch, using its long, stiff whiskers.

6½–7¾ ft
2–2.4 m

5½–6¼ ft
1.7–1.9 m

short flippers

HAWAIIAN MONK SEAL
Monachus schauinslandi
Along with its Mediterranean relative, this is one of only two surviving, highly endangered and protected species of monk seals. Fewer than 1,400 of this species remain.

HARP SEAL
Pagophilus groenlandicus
This small seal migrates south in winter and north in summer in noisy groups, following the Arctic pack ice, on which it hauls out and rests.

6½–7¾ ft
2–2.4 m

CRABEATER SEAL
Lobodon carcinophaga
Despite its name, this agile Antarctic seal eats
mainly krill, which it filters from the water,
using specially modified teeth.

6½–8¾ ft
2–2.7 m

inflatable muzzle of male

**HOODED
SEAL**
Cystophora cristata
This solitary, Arctic seal has
an unusual, inflatable nose that
droops over its mouth. The
pups become independent
at just five days old.

6½–16 ft
2–5 m

6½–23 ft
2–7 m

**SOUTHERN
ELEPHANT SEAL**
Mirounga leonina
The huge male of this
Southern Ocean species has
a trunklike nose and is the
largest carnivore, weighing
up to 11,000 lb (5,000 kg).

♀

NORTHERN ELEPHANT SEAL
Mirounga angustirostris
This is a large seal of the North Pacific. The male has a
long, trunklike nose. Like its southern relative, it was
almost hunted to extinction, but is now recovering.

♂

*large eyes set
back on head*

4–6½ ft
1.2–2 m

4½–5½ ft
1.4–1.7 cm

SPOTTED SEAL
Phoca largha
This small seal is mainly found on
ice floes off the northern coasts of
Siberia and Yukon, Canada. Adults
form stable pairs in order to breed.

*spots, rings, and
blotches on body*

**COMMON
SEAL**
Phoca vitulina
Also known as the
harbor seal, this dainty seal
is widespread along temperate
coasts. It hauls out onto sandy
beaches and sheltered reefs.

3¼–5½ ft
1–1.7 m

RINGED SEAL
Pusa hispida
This small seal mainly inhabits
Arctic ice shelves. Pups are born
in lairs beneath the ice to protect
them from predators.

3½–4½ ft
1.1–1.4 m

BAIKAL SEAL
Pusa sibirica
This small, freshwater
seal is found in Lake
Baikal, Siberia. In winter,
it uses its teeth and claws
to maintain breathing
holes in the ice.

5 ft
1.5 m

CASPIAN SEAL
Pusa caspica
About half a million of this small species live in the
Caspian Sea. Unlike other seals, the male is thought
to take only one mate, not fighting off other males.

SKUNKS AND RELATIVES

This small group of cat-sized mammals from the Americas is named for its most distinctive behavior—that of squirting foul-smelling musk at an aggressor. The family name, Mephitidae, comes from the Latin word for "bad smell."

8–12½ in
20–32 cm

PALAWAN STINK BADGER
Mydaus marchei
This bumbling cousin of the American skunks is found only on the Philippine islands of Palawan and Calamian. It eats mainly invertebrates.

12½–19½ in
32–49 cm

HUMBOLDT'S HOG-NOSED SKUNK
Conepatus humboldtii
A native of southern Chile and Argentina, this small skunk detects underground invertebrate prey by scent and digs it up.

10–14 in
25–35 cm

EASTERN SPOTTED SKUNK
Spilogale putorius
A relatively small, weasel-like skunk from the eastern USA, this species is more agile than other skunks and climbs well.

9–13 in
23–33 cm

HOODED SKUNK
Mephitis macroura
This widespread Central American skunk occupies a range of habitats, feeding on fruits, eggs, and small animals.

white stripe warns predators of foul secretions

long hairs on tail and back rise when animal is alarmed

STRIPED SKUNK
Mephitis mephitis
This nocturnal omnivore ranges from Canada to Mexico. Its northern populations hibernate in winter.

9–16 in
23–40 cm

RACCOONS AND RELATIVES

Raccoons, kinkajous, and olingos of the family Procyonidae are nimble New World carnivores. Most species are omnivorous and eat mainly plant material (especially fruits) and also insects, snails, small birds, and mammals. The raccoon is the largest of the group.

SOUTH AMERICAN COATI
Nasua nasua
This species lives in loose female-dominated colonies. An agile climber, it mainly eats fruits, supplemented out of season with animal prey.

17–27 in
43–68 cm

16–26 in
41–67 cm

WHITE-NOSED COATI
Nasua narica
This social Central American omnivore spends the day foraging at ground level. It climbs well, and often sleeps in trees.

tail has faint dark rings

RED PANDA

This tree-dwelling, herbivorous mammal is usually placed in its own family, the Ailuridae, though some zoologists classify it with the raccoons, others place it in a group with the giant panda.

LESSER PANDA
Ailurus fulgens
Also known as the red panda, this raccoonlike mammal inhabits temperate forest in the Himalayas. It eats fruit and some animal matter.

20–29 in
50–73 cm

WEASELS AND RELATIVES

Members of the family Mustelidae occur throughout Eurasia, Africa, and the Americas. They are characterized by a sinuous body and short legs, although the badgers and wolverines are more robust. Most are active predators and good swimmers.

8–14 in
20–36 cm

EUROPEAN MINK
Mustela lutreola
A semiaquatic predator, this species was once widespread in central and western Europe, but is now much less common than the introduced American mink.

12–17 in
30–43 cm

AMERICAN MINK
Neovison vison
A fierce and stealthy predator, this mink is also a skilled swimmer. It has been introduced worldwide by fur farmers.

tail has conspicuous black tip

KINKAJOU
Potos flavus
A nocturnal tree-dweller from Central and South America, the kinkajou uses its long tongue to pluck fruits and collect honey from bee nests.

OLINGO
Bassaricyon gabbii
Olingos are shy, nocturnal fruit-eaters. This is one of the three species found in the forests of Central America and northern South America.

14–19½ in
35–49 cm

16–30 in
41–76 cm

prehensile tail

long fur is grizzled

RACCOON
Procyon lotor
This adaptable opportunist favors wooded and scrubby habitats across North America, but often thrives in towns by scavenging human refuse.

black "mask" covers small eyes

17½–24 in
44–62 cm

12–14½ in
30–37 cm

RINGTAIL
Bassariscus astutus
An agile omnivore from Central America, this species forages by night for fruits and small animal prey, pausing often to scent-mark territory.

8–18 in
20–46 cm

EUROPEAN POLECAT
Mustela putorius
This lively nocturnal species, found in forests and meadows of central and western Europe, is the ancestor of the domestic ferret.

9–10 in
23–26 cm

slender, elongated neck

4¼–10 in
11–26 cm

LEAST WEASEL
Mustela nivalis
The smallest carnivore, yet a fierce and highly successful predator, this weasel specializes in hunting mice.

pointed muzzle

6½–12½ in
17–32 cm

STOAT
Mustela erminea
This lithe and ferocious little predator is found throughout much of the Northern Hemisphere. Its most northerly populations turn white in winter.

BLACK-FOOTED FERRET
Mustela nigripes
Extinct in the wild in the late 20th century, this slender burrowing ferret has been reintroduced to reserves in the midwestern USA.

14–20 in
35–50 cm

LONG-TAILED WEASEL
Mustela frenata
A widespread species of the Americas, this weasel hunts mice and voles. Northern individuals turn white in winter.

»

22–28 in
55–70 cm

HOG BADGER
Arctonyx collaris
Native to Southeast Asia, this
badger uses its elongated snout
to root around for edible
morsels on forest floors.

four white stripes
from head to tail

AMERICAN BADGER
Taxidea taxus
This burrowing animal inhabits grassland and
woodland throughout central North America.
It eats a huge variety of plants and animals.

16½–28 in
42–72 cm

white stripe
from nose,
down the back
to the rump

22–35 in
56–90 cm

EURASIAN BADGER
Meles meles
This stocky badger inhabits
wooded terrain across much
of Europe and Asia. It lives
communally in extensive
burrow systems called setts.

29–38 in
74–96 cm

HONEY BADGER
Mellivora capensis
An exceptionally feisty animal, this species lives in western
and southern Asia, and Africa. It raids beehives for honey,
but also eats termites, scorpions, and porcupines.

18½–22 in
47–55 cm

GREATER GRISON
Galictis vittata
This adaptable omnivorous
weasel with badgerlike
markings is found
in tropical forest and
grassland in Central and
South America.

26–41 in
65–105 cm

SABLE
Martes zibellina
This fierce predator from the
forests of Siberia, China, and
Japan is hunted in turn by humans
for its luxuriantly soft, silky fur.

14–22 in
35–56 cm

WOLVERINE
Gulo gulo
This large mustelid is widespread
in North America and Eurasia. Its
voracious feeding habits have earned
it its other name, "glutton."

dense, dark
brown coat

flattened
tapering tail

18–23 in
45–58 cm

16–21½ in
40–54 cm

18–26 in
45–65 cm

FISHER
Martes pennanti
Found in dense North
American forests, this large
marten rarely eats fish, and is
one of the few predators
to tackle porcupines.

**EUROPEAN
PINE MARTEN**
Martes martes
Present in forests throughout
much of Europe, this lively hunter
is seldom seen, as it is nocturnal
and wary of humans.

BEECH MARTEN
Martes foina
Widespread in Eurasia,
this marten emerges from
a rock crevice or hollow
log at dusk in search of
small mammals, birds,
and seasonal fruits.

AFRICAN ZORILLA
Ictonyx striatus
A striped polecat, this African
species hunts a range of prey
by night. It rests by day in
hollow logs or burrows.

11–15 in
28–38 cm

*dorsal stripes meet
at the tail*

9½–13 in
24–33 cm

**AFRICAN
STRIPED WEASEL**
Poecilogale albinucha
Found in central and southern
Africa, this species lives in self-dug
burrows. It emerges at night to
hunt for smaller animals, especially
rodents, which it tracks by scent.

*long front
claws to dig up
buried insects*

GIANT OTTER
Pteronura brasiliensis
This South American predator needs 2¼ lb (3 kg)
of fish a day. Fewer than 5,000 individuals of
this endangered species remain.

3¼–4¼ ft
1–1.3 m

AFRICAN CLAWLESS OTTER
Aonyx capensis
Found alongside water in forests and wetland
throughout much of sub-Saharan Africa, this
large otter eats mainly crabs, frogs, and fish.

29–35 in
73–88 cm

14–18½ in
36–47 cm

*grayish white
flecks on face
and throat*

**ASIAN SMALL-CLAWED
OTTER**
Aonyx cinerea
The world's smallest otter,
this species lives in wetlands
in India and Southeast Asia. Its
survival is threatened by habitat
loss and pollution.

*short,
blunt claws*

20–35 in
50–90 cm

23–29 in
58–73 cm

2½–4 ft
75–120 cm

EURASIAN RIVER OTTER
Lutra lutra
Eurasian otters thrive in both river and
coastal habitats, as long as they have access
to freshwater for drinking and washing.

NORTH AMERICAN RIVER OTTER
Lontra canadensis
Widespread in North America, this species inhabits
well-vegetated rivers and lake shores. It eats mainly
fish and crayfish, but also hunts small land animals.

SEA OTTER
Enhydra lutris
The sea otter hunts fish and shellfish
in the cool North Pacific, relying on its
incredibly dense fur to keep warm.

CATS

Members of the cat family, or Felidae, are among the most specialized of meat-eaters, and many species consume no vegetable material at all. As a group, cats are supremely athletic, with supple, muscular bodies that are well adapted to running, climbing, leaping, and swimming. Their short jaws contain sharp teeth adapted for stabbing (canines) and slicing (carnassials). They have retractile claws.

LEOPARD
Panthera pardus
This most adaptable of the big cats occurs widely throughout Africa and southern Asia. It often hides its prey in trees, away from other hunters.

3–6¼ ft
0.9–1.9 m

CLOUDED LEOPARD
Neofelis nebulosa
A large nocturnal southeast Asian forest cat with cloud-shaped markings, this species is declining due to hunting and habitat loss.

26–42 in/67–107 cm

BLACK LEOPARD
Panthera pardus
Melanistic black coloration is not uncommon among leopards. Also known as black panthers, they are found mainly in dense, moist forests in southeast Asia.

3–6¼ ft
92–190 cm

TIGER
Panthera tigris
The world's largest cat hunts using stealth and raw power to take prey as large as oxen. Fewer than 5,000 remain in the wild in Asia.

4½–9½ ft
1.4–2.8 m

JAGUAR
Panthera onca
The America's only big cat, the jaguar is an excellent climber and swimmer. Its broad prey base includes deer, turtles, and fish.

4–5½ ft
1.2–1.7 m

thick mane

LION
Panthera leo
Africa's top predator lives in family groups known as prides. Females work together to bring down prey including zebra and antelope.

♂

♀

5¼–8¼ ft
1.6–2.5 m

2¾–4 ft
86–125 cm

SNOW LEOPARD
Uncia uncia
At home in the remote high mountains of central
Asia, the snow leopard lives alone and hunts wild
sheep and goats, deer, and marmots.

32–43 in
80–110 cm

EURASIAN LYNX
Lynx lynx
This large lynx is big
enough to tackle small
deer. One kill will feed
an individual for about
a week.

27–32 in
68–82 cm

IBERIAN LYNX
Lynx pardinus
Now being bred in
captivity, this lynx is
probably the world's
most endangered cat,
with fewer than 150
remaining wild
in Spain.

large tufted ears

24–42 in
61–106 cm

CARACAL
Caracal caracal
Nocturnal hunters of
medium-sized prey, such
as hyraxes and small
antelope, caracals occupy
dry scrublands in Africa
and southwestern Asia.

short tail

26–41 in
65–105 cm

BOBCAT
Lynx rufus
This adaptable stalk-and-pounce predator is named
for its short "bob" tail. It occurs throughout North
America and hunts mainly rabbits.

small head with
high-set eyes

21–26 in
53–67 cm

32–42 in
80–106 cm

BAY CAT
Catopuma badia
Fewer than eight of this little-known forest cat,
found only on the island of Borneo, have been
caught since 1928.

CANADIAN LYNX
Lynx canadensis
The Canadian lynx inhabits
dense forests and tundra. Its
numbers fluctuate with the
availability of its preferred
prey, the snowshoe hare.

ASIAN
GOLDEN CAT
Catopuma temminckii
This large golden
brown, occasionally
spotty, cat inhabits
forested parts of
Southeast Asia. Pairs
cooperate to hunt
and rear young.

29–41 in
73–105 cm

CHEETAH
Acinonyx jubatus
The fastest animal on four
legs, reaching 64 mph
(104 kph), the cheetah uses
speed to catch antelope on
the African savanna.

4–5 ft / 1.2–1.5 m

long tail
aids balance

furry ears can swivel independently to scan surroundings for sounds of prey or danger

TIGER
Panthera tigris

The largest and most striking of the big cats, the tiger is a powerful predator with almost supernatural grace and agility. Its natural range extends from the tropical jungles of Indonesia to the snowy expanses of Siberia, where the largest individuals are found. A full-grown male may weigh anything up to 660 lb (300 kg), but despite this bulk it can leap up to 33 ft (10 m) in a single bound. Adult tigers live alone, except for females with cubs—mothers watch over their cubs for two years or more, while teaching them vital survival skills.

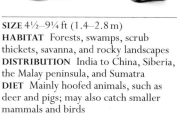

SIZE 4½–9¼ ft (1.4–2.8 m)
HABITAT Forests, swamps, scrub thickets, savanna, and rocky landscapes
DISTRIBUTION India to China, Siberia, the Malay peninsula, and Sumatra
DIET Mainly hoofed animals, such as deer and pigs; may also catch smaller mammals and birds

sense of smell is surprisingly poor, although tigers do use scent marks to define their territory

long whiskers allow the tiger to feel its way through dense undergrowth in almost total darkness

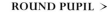

ROUND PUPIL >
Unlike small cats, whose pupils contract to vertical slits, tiger pupils are always round. They expand to provide excellent night vision and contract to pinpricks in bright light.

WHITE EAR FLASH ∨
The prominent white spot on the back of each ear is thought to aid communication. Cubs following their mother will notice ear movements that may signal danger.

∧ STAB AND SLICE
Four long canine teeth deliver the killing bite to a tiger's prey. Blade-edged cheek teeth called carnassials slice through the meat with ease.

FORELIMBS >
The tiger has long legs and large feet, which allow it to run fast, leap large distances, and knock prey as large as oxen to the ground with a single, deadly swipe.

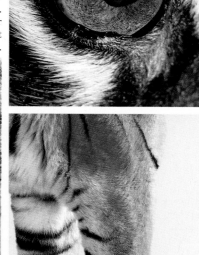

retracted claw

nonslip pad

∧ PADDED PAW
Tigers have five toes—four on the sole of the foot and the fifth forming a dew claw. The claws are withdrawn fully into the paw when not in use.

< KILLER IN STRIPES
A fiery orange coat marked with bold black stripes provides superb camouflage as the tiger moves through sun-dappled vegetation. The white tigers seen in zoos are usually bred in captivity and are extremely rare in the wild. Indeed, tigers have been hunted almost to extinction; globally, there are fewer than 8,000 left in the wild.

< TAIL END
The long tail is typically held curved just above the ground. The tiger uses it to aid balance when chasing prey or climbing.

SIAMESE CAT
Felis catus
This elegant and sociable breed originated in Thailand. Siamese kittens are born cream and develop their dark extremities as they grow.

14–20 in
35–50 cm

TABBY CAT
Felis catus
The tabby is not a breed but a coat pattern found in many breeds. It resembles the pattern seen in the ancestral wild cat.

14–20 in
35–50 cm

SPHYNX CAT
Felis catus
Hairless except for a peachy fuzz, sphynx cats were developed in Canada. They are sociable and like to be cuddled because they feel the cold.

14–20 in
35–50 cm

14–20 in
35–50 cm

CORNISH REX
Felis catus
This is an unusual breed in which the guard hairs of the fur are missing, leaving only downy underfur.

16–26 in
40–66 cm

EUROPEAN WILD CAT
Felis silvestris silvestris
This elusive but ferocious predator is declining due to persecution, habitat loss, and interbreeding with feral domestic cats.

PERSIAN CAT
Felis catus
A long-established and popular breed, this domestic cat is characterized by long hair and a short muzzle.

14–20 in
35–50 cm

14–20 in
35–50 cm

MANX CAT
Felis catus
Short-tailed cats appeared naturally on the Isle of Man over 300 years ago. The trait spread rapidly in the small island population.

yellowish gray to reddish brown fur

16–20 in
40–50 cm

INDIAN DESERT CAT
Felis silvestris ornata
Also known as the Asian steppe wildcat, this subspecies is distinguished from its European relatives by its smaller size and spotted golden coat.

JUNGLE CAT
Felis chaus
A large and relatively common wild cat occuring from Egypt to Indonesia, the jungle cat prefers grassland and swampy habitats.

24–34 in
61–85 cm

SAND CAT
Felis margarita
A small desert specialist found in North Africa, Arabia, and Kazakhstan, the sand cat hunts gerbils and other nocturnal rodents.

10–12 in
23–31 cm

BLACK-FOOTED CAT
Felis nigripes
An opportunistic solitary hunter, the black-footed cat is threatened by persecution and habitat loss in its native southern Africa.

14–20½ in
36–52 cm

PALLAS'S CAT
Felis manul
This short-legged cat lives in the stony deserts of central Asia. Its coat provides useful camouflage when stalking pikas, gerbils, and grouse.

18–26 in
46–65 cm

SERVAL
Leptailurus serval
Found in grassland habitats throughout much of Africa, this distinctive-looking cat is an agile predator of small mammals.

23–36 in
59–92 cm

MARBLED CAT
Pardofelis marmorata
A rare forest species adapted to life in the trees, the southeast Asian marbled cat climbs well and preys mainly on birds.

18–24 in
45–62 cm

RUSTY-SPOTTED CAT
Prionailurus rubiginosus
This lively cat from India and Sri Lanka stalks most of its prey on the ground, but is also an excellent climber.

14–19 in
35–48 cm

pointed ears

LEOPARD CAT
Prionailurus bengalensis
This variable species has a huge range. Russian leopard cats are up to three times heavier than those in the forests of Indonesia.

18–30 in
45–75 cm

retractile claws

FISHING CAT
Prionailurus viverrinus
A large cat with a patchy distribution in southern and southeast Asia, the endangered fishing cat also eats waterfowl and land animals.

22½–45 in
57–115 cm

FLAT-HEADED CAT
Prionailurus planiceps
This unusual water-loving cat from southeast Asia hunts mainly fish and crustaceans, which it finds by dipping its head or groping with its paws.

18–20½ in
45–52 cm

>>

sandy
colored
coat

round head
with erect ears

large canine
teeth for
killing prey

19½–33 in
49–83 cm

2¾–5 ft
0.9–1.6 m

long hind legs for
exceptional sprinting
and leaping power

JAGUARUNDI
Puma yagouaroundi
One of the largest and
most widespread
South American cats, the
jaguarundi hunts small
mammals by day in
diverse habitats.

PUMA
Puma concolor
Also known as cougar, panther, and
mountain lion, the puma occupies
rugged terrain across a vast range
from Canada to Argentina.

HYENAS AND AARDWOLF

The small family of Hyaenidae contains three species of scavenging
hyena and the aardwolf, which specializes in eating insects. Hyenas
are characterized by a stocky, doglike body with short hind legs and
powerful bone-crushing jaws. They are highly intelligent and live
in family groups, called clans.

AARDWOLF
Proteles cristata
This dainty, weak-jawed cousin
of the hyenas eats only insects.
It lives in East and southern
Africa in the dry grassland
favored by termites.

22–31in
55–80 cm

powerful neck
and forequarters

SPOTTED HYENA
Crocuta crocuta
This efficient scavenger
is also a proficient hunter
of ungulates. It lives in
unforested areas of
sub-Saharan Africa.

3–5½ ft
1–1.7 m

3¼–4 ft
1–1.2 m

3½–4½ ft
1.1–1.4 m

STRIPED HYENA
Hyaena hyaena
The small striped hyena is found in
open country from North Africa to
India. It eats a mixed diet of scavenged
meat, small prey animals, and fruit.

BROWN HYENA
Hyaena brunnea
This social southern African scavenger
survives desert conditions by foraging
at night and supplementing its diet
with watery fruits.

ANDEAN MOUNTAIN CAT
Leopardus jacobitus
This extremely rare species is restricted to remote uplands, where it is seldom seen by humans. It eats chinchilla-like rodents called viscachas.

23–25 in
58–64 cm

GEOFFROY'S CAT
Leopardus geoffroyi
An adaptable hunter of small mammals, fish, and birds, Geoffroy's cat stalks grasslands, forests, and wetlands from Bolivia to southern Argentina.

17–35 in
43–88 cm

spotted coat varies from gray to golden

COLOCOLO
Leopardus colocolo
Highly adaptable, the mainly nocturnal colocolo is found in a wide variety of habitats in South America, from forest to grassland and swamp.

16½–31 in
42–79 cm

15½–22 in
39–55 cm

ONCILLA
Leopardus tigrinus
This spotted forest dweller is widespread from Costa Rica to Argentina, where it hunts rodents, opossums, and birds. It is solitary and nocturnal.

lithe body adapted for climbing

large eyes adapted to nocturnal lifestyle

17–31 in
43–79 cm

MARGAY
Leopardus wiedii
Seldom seen because of its rarity and preference for dense forest cover, the margay is found from Mexico to northern South America.

22–39 in
55–100 cm

OCELOT
Leopardus pardalis
Ocelots occupy forested parts of Central and South America. They hunt nocturnally, targeting rodents and other prey on land and in water.

strong claws are hidden within fleshy sheaths

MALAGASY CARNIVORES

The large island of Madagascar has been separate from other land masses for about 100 million years, allowing its resident mammals to evolve independently. Its native carnivores are now classified in their own family, the Eupleridae. They are very varied in appearance, having diversified to fill the niches occupied elsewhere by predators such as cats, weasels, and mongooses.

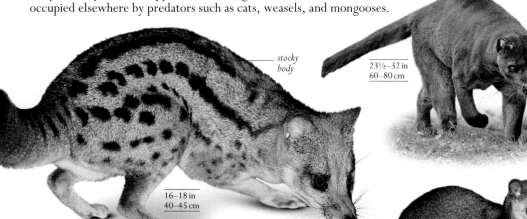

stocky body

FOSSA
Cryptoprocta ferox
The catlike fossa is the largest Madagascan carnivore. It hunts mainly lemurs, but it will eat almost any small animal it can catch.

23½–32 in
60–80 cm

12–15 in
30–38 cm

MALAGASY FANALOKA
Fossa fossana
A small civetlike animal, the fanaloka inhabits Madagascar's wet forests. It hunts invertebrate prey on land and in water.

16–18 in
40–45 cm

foxlike pointed muzzle

18–20 in
45–50 cm

FALANOUC
Eupleres goudotii
A ground-dwelling, nocturnal forest animal, the falanouc is an adept digger. It uses its large feet to unearth invertebrate prey.

MALAGASY RING-TAILED VONTSIRA
Galidia elegans
The Madagascan equivalent of mongooses, vontsiras are active forest animals. They feed opportunistically on most plant and animal foods.

∨ **DEMOLITION EXPERT**
This tough scavenger can reduce a
carcass to nothing in minutes. It will
attempt to bolt great chunks of meat
before other scavengers arrive. Then,
if time allows, they dismember the
remains and stash them nearby. Often,
all that is left of a herbivore, such as an
antelope, is the green stomach contents.

*crest of long hairs along
the back lies flat when
the animal is relaxed*

*shorter hind legs add
to the front-heavy,
skulking appearance*

SIZE 3¼–4 ft (1–1.2 m)
HABITAT Open country, savanna
grassland, and scrub to semidesert
DISTRIBUTION North and East
Africa, Middle East to East India
DIET Mainly carrion

muscular shoulders and neck allow the hyena to carry or drag its own bodyweight in carrion

large ears follow sound from any direction

STRIPED HYENA
Hyaena hyaena

Maligned in folklore as skulking, cowardly scavengers, most hyenas are also adept hunters in their own right. The striped hyena is less bold and social than its spotted cousin, and, in Africa at least, individuals tend to live a solitary life except when breeding. Elsewhere, such as in Israel and India, striped hyenas more often live in family groups, which are dominated by a single adult female. Carrion forms the bulk of the striped hyena's diet, but it is also partial to fruit, especially melons, which provide valuable water. Striped hyenas are unpopular with farmers, who fear attacks on livestock and damage to crops, but these animals increasingly need conservation, with perhaps fewer than 10,000 remaining in the wild.

∨ THROAT
The black throat patch seems to have a social function. When two hyenas meet, they sniff one another and often paw at each other's throat—their equivalent of shaking hands.

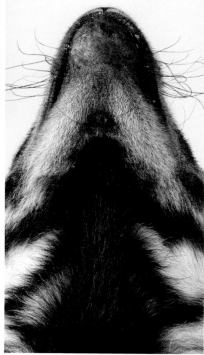

∧ NOSE FOR TROUBLE
Scent is an important sense to hyenas. Individuals often pause in their daily routine to daub scent from a gland under the tail on rocks and tussocks around their home range.

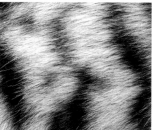

∧ MOUTH AND TEETH
The hyena's stout jaws are operated by immensely powerful muscles. The large cheek teeth (molars and premolars) are easily capable of crushing bones.

< FUR
The striped hyena is smaller than the spotted or brown hyenas, and is distinguished by black stripes on its legs and flanks. The pattern aids camouflage in its dusty habitat.

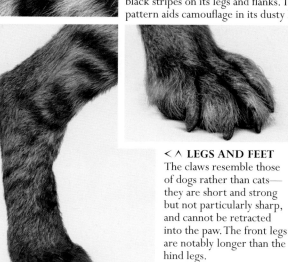

< ∧ LEGS AND FEET
The claws resemble those of dogs rather than cats—they are short and strong but not particularly sharp, and cannot be retracted into the paw. The front legs are notably longer than the hind legs.

MONGOOSES

Mongooses are small, slender-bodied, mostly ground-dwelling carnivores in the family Herpestidae. They live in warm temperate to tropical parts of Africa and Eurasia. Some species live in complex social groups. Mongooses may be the nearest surviving relatives of the first carnivores.

wedge-shaped head

9½–18 in
24–46 cm

18½–27 in
47–69 cm

YELLOW MONGOOSE
Cynictis penicillata
Inhabiting dry savanna in southern Africa, this mongoose lives in groups led by a dominant male, but forages independently.

WHITE-TAILED MONGOOSE
Ichneumia albicauda
Found in dry habitats in Africa and the southern part of the Arabian Peninsula, this large insect-eater also feeds on vertebrates and ripe berries.

sharp, nonretractile claws

6½–9 in
16–23 cm

COMMON DWARF MONGOOSE
Helogale parvula
This small, energetic carnivore forages in packs in African grassland, woodland, and brush country. It eats large invertebrates, such as crickets and scorpions.

12–14½ in
30–37 cm

COMMON CUSIMANSE
Crossarchus obscurus
This forest-dwelling animal from West Africa lives and hunts in nomadic troops. It is said to make an excellent pet.

12–16 in
30–40 cm

BANDED MONGOOSE
Mungos mungo
Found in troops in sub-Saharan woodland, this mongoose lives in dens, often excavated in termite mounds. Group members are led, but not dominated, by older females.

SLENDER MONGOOSE
Galerella sanguinea
A widespread species in Africa, the slender mongoose generally lives alone. It is active by day and busiest just before dusk.

12½–13½ in
32–34 cm

MEERKAT
Suricata suricatta
This species lives in groups in semidesert habitats. All members help babysit, maintain the burrow, and take turns standing sentry while others forage.

pointed nose

9½–14 in
24–35 cm

22–24 in
56–61 cm

EGYPTIAN MONGOOSE
Herpestes ichneumon
Not restricted to Egypt, this grizzled gray mongoose occurs in open grassy habitats from Spain to South Africa.

coat marked with broken bands

18–21 in
45–53 cm

INDIAN GREY MONGOOSE
Herpestes edwardsi
Frequenting forests and plantations, this species often hunts close to human habitation, where it makes itself useful by killing mice and rats.

13–19 in
33–48 cm

INDIAN BROWN MONGOOSE
Herpestes fuscus
This uncommon species occupies the jungles of southern India and Sri Lanka. Like other mongooses, it can kill snakes, but prefers easier prey.

15½–18½ in
39–47 cm

RUDDY MONGOOSE
Herpestes smithii
This little-known Indian forest-dweller hunts birds, reptiles, and smaller mammals. Its tail is sometimes longer than its body.

slender tail

AFRICAN PALM CIVET

The secretive and nocturnal African palm civet is the sole member of the Nandiniidae, which is thought to have split from civet and catlike ancestors between 36 and 54 million years ago.

14½–25 in
37–63 cm

AFRICAN PALM CIVET
Nandinia binotata
This is a very common but shy tree-dwelling species from central Africa. Although an omnivore, it eats mostly fruits.

CIVETS, GENETS, AND LINSANGS

With most species possessing boldly patterned coats, these attractive animals resemble long-tailed cats, but are less specialized meat-eaters. When threatened, these shy and nocturnal animals of the family Viverridae squirt a jet of foul-smelling fluid from stink glands near the base of the tail.

24–38 in
61–97 cm

BINTURONG
Arctictis binturong
A prehensile tail helps this Southeast Asian species move methodically through the forest canopy in search of fruits and small animals.

26–33 in
67–84 cm

AFRICAN CIVET
Civettictis civetta
An opportunistic animal, this large, ground-dwelling omnivore lives alone and marks its territory with a strong musky scent.

16½–28 in
42–70 cm

ASIAN PALM CIVET
Paradoxurus hermaphroditus
With a natural range extending from Pakistan to Indonesia, this fruit-loving civet is considered a pest in palm and banana plantations.

20–34 in
51–87 cm

MASKED PALM CIVET
Paguma larvata
This agile, solitary, tree-dwelling native of Indochina eats fruits, insects, and small vertebrates.

long tail aids balance when climbing trees

rows of black spots

19½–27 in
49–68 cm

SMALL INDIAN CIVET
Viverricula indica
This small ground-dwelling civet inhabits forests, grasslands, and bamboo thickets from Pakistan to China and Indonesia.

18–20½ in
46–52 cm

SMALL-SPOTTED GENET
Genetta genetta
Also called the common genet, this widespread predator of small mammals and birds occupies scrub and forests of Africa and southern Europe.

large eyes to see in the dark

17–23 in
43–58 cm

CAPE GENET
Genetta tigrina
This genet is found in eastern South Africa and Lesotho. It eats mostly invertebrates, but will tackle prey as large as geese.

soft, velvety fur

21½–30 in
54–77 cm

BANDED LINSANG
Prionodon linsang
Also called the tiger civet, this shy species lives in tree holes in S.E. Asian jungle, hunting rats, squirrels, lizards, and birds.

13–18 in
33–45 cm

long, thick tail

ORIENTAL CIVET
Viverra tangalunga
Restricted to the tropical forests of Malaysia, Indonesia, and the Philippines, this nocturnal forager mostly hunts for prey on the ground.

ODD-TOED UNGULATES

All members of the Perissodactyla are browsers and grazers of plants. They appear to have little in common, but are linked through a series of extinct intermediate forms.

Modern odd-toed ungulate families range from graceful horses to piglike tapirs and massive rhinoceroses. Unlike the Artiodactyla, they have a relatively simple stomach, and rely on bacteria in the cecum, a pocketlike extension of the large intestine, and colon to digest plant cellulose.

In prehistoric times, odd-toed ungulates were among the most important herbivorous mammals, sometimes becoming the dominant herbivores in grassland and forest ecosystems. For a variety of reasons, including competition with even-toed ungulates, most of these species are now known only from the fossil record.

WEIGHT-BEARING TOES

Odd-toed ungulates rest their weight primarily on the third toe of each foot. Indeed, horses have lost the rest of their toes, and the single remaining digit is protected with a well-developed horny hoof. The other two families have retained more toes—three on all four feet in the case of rhinoceroses, and three on the hind legs and four on the front in the case of tapirs.

HORSE POWER

Once distributed across the world, apart from Antarctica and Australasia, extant species in this order are mainly native to Africa and Asia. Only some of the tapir species are found in the Americas, for although the horse family evolved in that region, it had died out by the end of the Pleistocene around 10,000 years ago. The Spanish conquistadors reintroduced the modern domestic horse to the Americas in the 15th century.

Horses have a long history of domestication, especially as transport and pack animals and to provide power for agriculture and forestry. The first equid to be domesticated appears to have been the wild ass, some 7,000 years ago, with horses following about a millennium later. Today there are over 250 horse breeds worldwide.

PHYLUM	CHORDATA
CLASS	MAMMALIA
ORDER	PERISSODACTYLA
FAMILIES	3
SPECIES	17

The black rhinoceros is a browser and has a prehensile upper lip, which it uses to grasp twigs and leaves.

RHINOCEROSES

A huge, barrel-shaped body and a large head bearing one or two horns make the five species of the family Rhinocerotidae unmistakable. While their bulk, horns, and protective skin ensure that they have few natural enemies, these mostly solitary animals are threatened by hunting and the destruction of their habitat. These herbivores can ferment food in their hindgut, which enables them to eat woody as well as leafy matter.

rough skin with few hairs

long front horn

4–5 ft
1.2–1.5 m

SUMATRAN RHINOCEROS
Dicerorhinus sumatrensis
This critically endangered species from Southeast Asian forests is the smallest rhinoceros. The smaller of its two horns is usually a mere stub.

4½–5½ ft
1.4–1.7 m

prehensile upper lip

BLACK RHINOCEROS
Diceros bicornis
Smaller but more aggressive than the white rhinoceros, this critically endangered animal of sub-Saharan Africa has a prehensile upper lip to draw twigs and leaves into its mouth.

TAPIRS

Inhabiting tropical forests, the Tapiridae are found in Southeast Asia, and South and Central America. These large herbivores have a flexible proboscis, a long nose that can grasp foliage overhead. Other distinctive features include oval, white-tipped ears and a protruding rump with a short tail. Young tapirs have a striped and spotted coat. Splayed hooves, with four toes on the front feet and three on the rear, help tapirs walk on soft ground.

dual coloration adds camouflage to body

MALAYAN TAPIR
Tapirus indicus
This striking, two-toned tapir is the largest and only Asian species. Males and females create overlapping trails around their rain forest territory in Southeast Asia.

white, saddle-shaped marking

3–3½ ft
90–105 cm

long, flexible snout

2½–3½ ft
77–108 cm

SOUTH AMERICAN TAPIR
Tapirus terrestris
Despite its size, this shy tapir is unobtrusive. It moves easily through jungle undergrowth, and is a favored prey of crocodiles.

2½–4 ft
0.8–1.2 m

BAIRD'S TAPIR
Tapirus bairdii
The largest land mammal in South America, it favors waterside habitats in dense jungles and swamps for swimming and wallowing.

30–39 in
75–100 cm

MOUNTAIN TAPIR
Tapirus pinchaque
The smallest of all tapirs, this species inhabits montane cloud forest in the northern Andes. It has a thick, woolly coat and a white lower lip.

characteristic hump in front of shoulder

5–6¼ ft
1.5–1.9 m

elongated head

square mouth

WHITE RHINOCEROS
Ceratotherium simum
Found in African savanna, this social species is also the heaviest rhinoceros. "White" is a corruption of "wide"—for the animal's square mouth, adapted for grazing.

three toes

5½–6½ ft
1.7–2 m

INDIAN RHINOCEROS
Rhinoceros unicornis
Living in grasslands, forests, and wetlands of India and Nepal, this solitary, single-horned rhinoceros has thick folds of skin around its neck.

4½–5½ ft
1.4–1.7 m

JAVAN RHINOCEROS
Rhinoceros sondaicus
Once widespread in Southeast Asia, this solitary, nocturnal browser is one of the world's rarest animals. Its small horn does not exceed 8 in (20 cm) in length.

»

WHITE RHINOCEROS
Ceratotherium simum

Despite its unnerving appearance, this giant of the African plains is a mild-tempered vegetarian. The huge horns are used almost entirely in self-defense or for protecting the young. Adult rhinoceroses usually live alone, although they sometimes form loose groups to share feeding areas. Males are territorial and mark their territory with pungent urine and droppings. They will compete for the right to mate with a certain female, but most disputes are solved with a bout of showing off, after which the weaker male backs down. This species has suffered a drastic decline in numbers and range as a result of hunting and habitat loss.

SIZE 5–6¼ ft (1.5–1.9 m)
HABITAT Savanna grassland
DISTRIBUTION Central and South Africa
DIET Grass

∨ NEARSIGHTED
Rhinoceroses are somewhat nearsighted. Having eyes located on the sides of the head gives them a wide field of vision, but limits their ability to see straight ahead.

∨ HAIRY EARS
The ears are the hairiest part of a rhinoceros's body. They provide excellent hearing, and swivel to detect sounds from any direction.

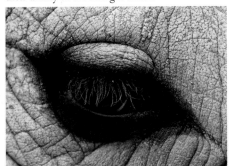

WIDE MOUTH >
The white rhinoceros is the largest animal to survive on an exclusive diet of grass. Its wide, straight mouth is shaped for maximum efficiency when cropping short grass.

< HORN OF HAIR
Rhinoceros horn is made of a protein called keratin. The same substance is found in hair and nails—in fact, the rhinoceros's horn is little more than a mass of compacted hair.

< RHINO HIDE
Despite the species' name, the tough, wrinkled skin of the white rhinoceros is gray, not white. Up to ¾ in (2 cm) thick, the skin is made up of layers of collagen, arranged in a crisscross pattern.

the horn of a mature white rhino may grow up to 5 ft (1.5 m) long— and attract poachers

< UNDERSIDE OF FOOT
The unusual shape of a rhinoceros's foot makes the animal relatively easy to track. Some experienced trackers can even recognize individual animals from their footprint.

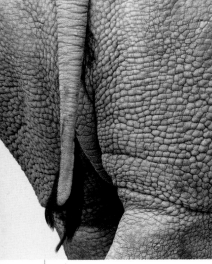

∧ TELLING TAIL
The short tail hangs down when the rhinoceros is relaxed and curls up like that of a pig at times of excitement, such as when mating.

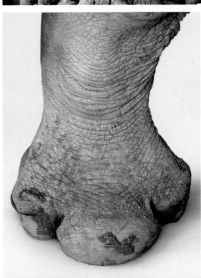

< THREE TOES
Rhinoceroses have three toes on each foot. Most of the weight is borne on the middle toe, with the smaller side toes aiding balance and grip.

> GENTLE GIANT
White rhinoceroses suffer from an undeserved reputation for bad temper. In reality they are peace-loving, even timid, animals. Under normal circumstances they charge only when provoked or confused. The black rhinoceros is much more aggressive.

large nostrils and an
excellent sense of smell
make up for weak eyesight

HORSES AND RELATIVES

Although the Equidae are a large family in the fossil record, just seven species of horse, donkey, and zebra remain. They live in herds, inhabiting open grassland and desert. They have all-round vision and their movable, sensitive ears alert them to predators. These swift animals have slender legs with a single hoofed toe, and their coat is short except for longer hair on the mane and tail.

stiff, striped mane

narrow, intermediate stripes

3¼–4½ ft
1–1.4 m

MOUNTAIN ZEBRA
Equus zebra
Found in southwest Africa, this species lives in dry, rocky, mountainous habitats. It has broad stripes on the hindquarters that contrast with its narrow stripes elsewhere.

4¼–4½ ft
1.3–1.4 m

CHAPMAN'S ZEBRA
Equus burchelli antiquorum
This zebra from southern Africa is a subspecies of the plains zebra. It has characteristic shadow stripes on its body that alternate with the black ones.

4–4½ ft
1.2–1.4 m

GRANT'S ZEBRA
Equus burchellii boehmi
The smallest of the six subspecies of the plains zebra, this mammal has broad, defined stripes. It is a native of the savanna in East Africa.

5–5¼ ft
1.5–1.6 m

GREVY'S ZEBRA
Equus grevyi
The largest wild species in its family, this big-eared zebra from East Africa has narrow, variable stripes on its body and a white belly.

4–4¼ ft
1.2–1.3 m

KHUR
Equus hemionus khur
A fast runner from arid Asian grasslands, this subspecies is now found in the wild only in a sanctuary in Gujarat, India.

dorsal stripe

4–4¼ ft
1.2–1.3 m

KULAN
Equus hemionus kulan
A little larger than most donkeys, this subspecies of the Asiatic wild ass has a black dorsal stripe, edged with white, and a short, black mane.

4–5 ft
1.2–1.5 m

PERSIAN ONAGER
Equus hemionus onager
Extinct in much of its former Asian range and now found only in parts of Iran, this subspecies has a brown dorsal stripe on its back.

4½–5 ft
1.4–1.5 m

KIANG
Equus kiang
The largest wild ass, the kiang inhabits the Tibetan plateau. Its chestnut-colored coat is the darkest of all Asian wild asses, and is woolly in winter.

gray-brown coat

striped legs

4¼–5 ft
1.3–1.5 m

SOMALI WILD ASS
Equus asinus somalicus
The ancestor of the donkey, this wild ass from northeast Africa has a short, gray coat, and striped, zebralike legs.

35–51 in
0.9–1.3 m

DONKEY
Equus asinus asinus
The domesticated form of the African wild ass, this subspecies has a worldwide distribution as a transport and pack animal.

PRZEWALSKI'S HORSE
Equus caballus przewalskii
The last true wild horse, this species often bears faint leg stripes. It once survived only in captivity, but is now being reintroduced into the wild in Mongolia.

pale muzzle

4¼–4½ ft
1.3–1.4 m

pale, yellowish brown flanks

brown legs, often with faint stripes

single toe encased in a hoof

3½–4¼ ft
1.1–1.3 m

EXMOOR PONY
Equus caballus caballus
This rare, primitive, and hardy pony survives semiwild in Exmoor, England. It is always dun, bay, or brown with black points.

SHIRE HORSE
Equus caballus caballus
A large, powerful horse, bred in England from Dutch stock, this breed is still used in agriculture and forestry for pulling heavy loads.

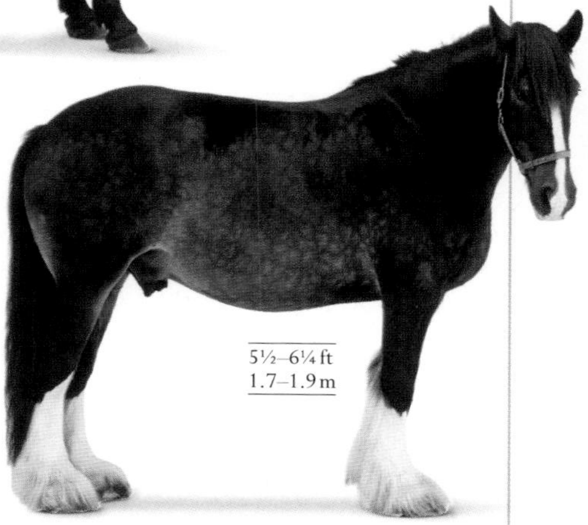

5½–6¼ ft
1.7–1.9 m

distinctive "dished" face

ARABIAN HORSE
Equus caballus caballus
A swift desert horse, the Arabian has had a huge influence on the development of the modern racehorse.

5¼ ft
1.5–1.6 m

PAINT HORSE
Equus caballus caballus
This American horse combines a dark coat, ranging from chestnut to black, with variable white areas.

5–5¼ ft
1.5–1.6 m

silky mane

3½–5 ft
1.1–1.5 m

MULE
Equus asinus x *E. caballus*
Built like a horse but with the paternal donkey's head and long ears, this generally sterile hybrid is a strong pack animal.

HINNY
Equus caballus x *E. asinus*
The offspring of a male horse and a female donkey, the hinny has a donkeylike body and the head, ears, and mane of a horse.

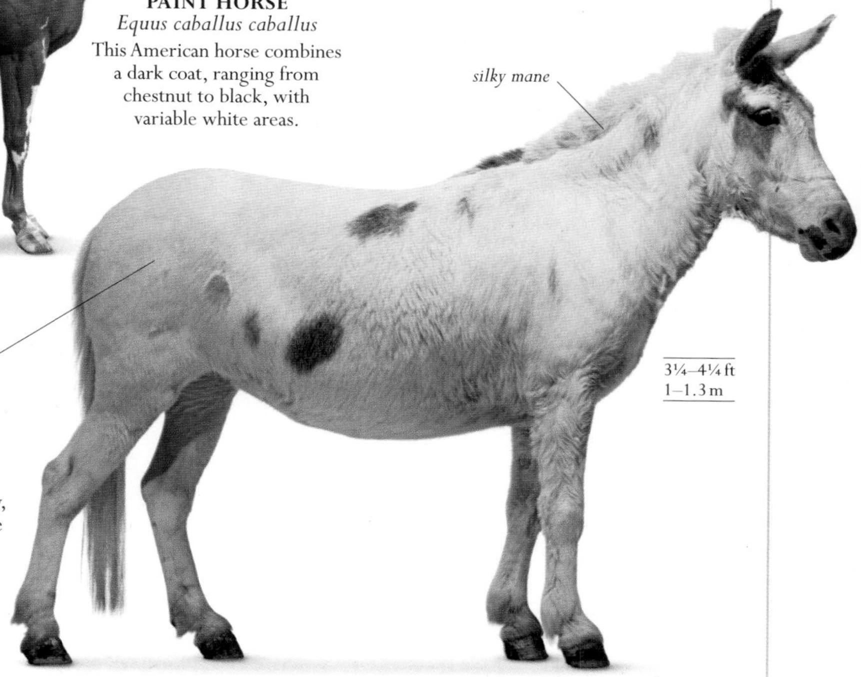

variable coat color

3¼–4¼ ft
1–1.3 m

EVEN-TOED UNGULATES

One of two orders of hoofed mammals, the Artiodactyla stand on two or four toes. Most are herbivorous, and have a multichambered, fermenting stomach.

Even-toed ungulates are often described as cloven hoofed. This is because the tip of each toe has its own hoof—a hard, rubbery, sole surrounded by a thick nail. The hoof wears down through use, but grows continually. Only the camelids lack hoofs—although they walk on two toes, the hoof is reduced to a small nail.

CHEWING THE CUD
With long legs and their feet protected with hoofs, even-toed ungulates often range extensively across grassland and forested habitats in search of food. Most families are grazers of grass and herbs or browsers of shoots and leaves. Their digestive system is well adapted to a diet of tough vegetable matter: the stomach has three or four chambers and contains bacteria that digest the cellulose in plant cell walls, releasing the nutrients inside. To help with this, partly digested food (cud) is regurgitated for further chewing, a process

known as rumination. Artiodactyls also have large, broad cheek teeth to grind up their tough food, and long intestines. Pigs and peccaries are somewhat different: with a more omnivorous diet, they have two-chambered stomachs and more general-purpose teeth. In some species, the canine teeth are enlarged into tusks for defense, fighting, and rooting for food.

DOMESTICATION
Introduced into Australasia, members of this order are found on every continent except Antarctica. They vary greatly in shape and in size, with shoulder height ranging from 8 in (20 cm) in the mouse-deer to almost 13 ft (4 m) in the giraffe. Many species are hunted in the wild as food by humans, but others have been domesticated and are economically important—cattle, llamas, sheep, and pigs are used for meat, leather, wool, dairy products, and transport. Domestication has led to the development of animal breeds with a distinctive appearance that matches their specific purpose.

PHYLUM	CHORDATA
CLASS	MAMMALIA
ORDER	ARTIODACTYLA
FAMILIES	10
SPECIES	240

DEBATE
ONE SPECIES OR SIX?

Taxonomists, scientists who classify living things, are often described as being either "splitters" or "lumpers." In the heyday of zoological exploration of Africa, there was a tendency to split taxa into too many species, many of which were later lumped back together. Giraffes are a good example, with some dozens of described "species" ending up under the single name, *Giraffa camelopardalis*. More recently, however, genetic evidence has begun to suggest that six subspecies of giraffe, recognizable by their different coat patterns, may indeed be separate species, some of them highly endangered.

PECCARIES

Found in the Americas, the Tayassuidae share many features with pigs, such as small eyes and a snout that ends in a cartilaginous disc. But they have a more complex, chambered stomach, and short, straight tusks.

white neck collar

3–3½ ft
90–110 cm

CHACOAN PECCARY
Catagonus wagneri
Found in the Chaco region of central South America, this large peccary was first described from fossils, and was discovered to be a living species only in 1975.

17½–22½ in
44–57 cm

12–20 in
30–50 cm

long snout

WHITE-LIPPED PECCARY
Tayassu pecari
Living in large herds in Central and South America, this species is considered dangerous: when threatened, it charges rather than flees.

COLLARED PECCARY
Pecari tajacu
Widespread in the tropical and subtropical Americas, this diurnal pig is extremely social It is considered destructive in agricultural areas because it often raids crops.

PIGS

Unique in their order, the Old World Suidae have four toes, although they walk only on the middle two. They have a simple stomach, unlike the chambered stomach of most of their relatives. Largely omnivorous, they dig with their snout and tusks for food. They have bristly coats and a short tail that ends in a long tassel.

shaggy mane

26–32 in
65–80 cm

25–34 in
64–85 cm

BURU BABIRUSA
Babyrousa babyrussa
Males of this species have remarkable tusks: the upper canines grow upward, pierce the snout, and curve backward. It is native to some Indonesian islands.

2½–3½ in
75–110 cm

GIANT FOREST HOG
Hylochoerus meinertzhageni
Unlike most pigs, this large, nocturnal wild species from Africa has a thick covering of black and ginger fur.

WARTHOG
Phacochoerus africanus
This pig has a warty face and two pairs of tusks. It often feeds on its front knees when grazing, and runs with its tail erect.

23½–34 in
60–85 cm

rounded back

BUSHPIG
Potamochoerus larvatus
Found in African forests and reedbeds, this pig has brown fur and a pale mane, which stands upright when the animal is alarmed.

8–10 in
20–25 cm

PYGMY HOG
Sus salvanius
Once spread across India and Nepal, this tiny, dark brown pig with a sharply tapered snout is now critically endangered.

12–26 in
30–65 cm

23½–30 in
60–75 cm

RED RIVER HOG
Potamochoerus porcus
This richly colored pig from central Africa has a pair of lumps on its snout, and distinctive white facial markings.

VISAYAN WARTY PIG
Sus cebifrons
Three pairs of fleshy warts on the face protect this boar against the tusks of rival males while fighting.

28–32 in
71–81 cm

BEARDED PIG
Sus barbatus
Found in the forests of southeast Asia, this pig often moves in herds. It has a whitish "beard" and distinct tail tassels.

28–32 in
70–80 cm

narrow mane of longer hair

23½–32 in
60–80 cm

23½–43 in
60–110 cm

WILD BOAR
Sus scrofa
Widespread in Eurasia, the bristly wild boar is the main ancestor of domestic pigs. The piglets have stripes along the body for camouflage in dense thickets.

long snout ends in a large disk of cartilage

PIÉTRAIN PIG
Sus scrofa domesticus
A Belgian domestic breed that produces high-quality, lean meat, this pig has dark spots set within paler blotches on its body.

MIDDLE WHITE PIG
Sus scrofa domesticus
Reared in England for pork, this pig is an unpigmented domestic breed, characterized by its round form and short, upturned nose.

MUSK DEER

Found mainly in Asian montane forests, the solitary, nocturnal members of the family Moschidae are named for the musk gland of adult males. These small, stocky deer have enlarged upper canines and longer hind limbs that help them climb rough terrain.

brittle sandy brown coat

elongated, tusklike canines used for fighting

20–21 in
51–53 cm

ALPINE MUSK DEER
Moschus chrysogaster
Inhabiting Chinese highland forests, this small deer has long, harelike ears. The male's musk sac produces musk, which is used to mark territory.

CHEVROTAINS AND MOUSE DEER

Found in African and Asian tropical forests, the Tragulidae are somewhat similar to small deer, but lack horns or antlers. In males, the enlarged upper canines project from either side of the lower jaw. Like pigs, these deer have four toes on each foot and rather short legs. The stomach is chambered, to ferment and digest tough plant food.

10–12 in
25–31 cm

INDIAN SPOTTED CHEVROTAIN
Moschiola meminna
Nocturnal and secretive, this tiny chevrotain from India and Sri Lanka has spots and white stripes on its body.

12–16 in
30–40 cm

distinctive markings

WATER CHEVROTAIN
Hyemoschus aquaticus
Large and boldly marked with stripes and spots, this mouse deer of West and central African forests is a good swimmer and diver.

12–14 in
30–35 cm

8–10 in
20–25 cm

GREATER MOUSE DEER
Tragulus napu
Although quite small, this is the largest Asian mouse deer. Its pointed head has dark stripes running from the large eyes to the black nose.

JAVA MOUSE DEER
Tragulus javanicus
This is the world's smallest hoofed mammal. Its relationship with other mouse deer in Southeast Asian forests is poorly understood.

DEER

Distributed almost globally, the Cervidae are largely absent from Africa and introduced in Australia. They occupy forested and open habitats, but many prefer the transitional zones in between. The size and form of the antlers of each species is unique. Unlike permanently horned animals, such as antelopes, male deer of all but one species shed and regrow antlers each year; females usually lack horns or have only short stubs.

3¼–5¼ ft
1–1.6 m

33–42 in
83–110 cm

RUSA DEER
Rusa timorensis
This Indonesian forest deer was introduced into arid Australian bushlands, where it thrives. It has large ears and antlers for its size.

SAMBAR
Rusa unicolor
A large, dark brown deer with a prominent mane, the sambar browses in woodlands throughout southern Asia, up to the Himalayan foothills.

28–30 in
70–76 cm

VISAYAN SPOTTED DEER
Rusa alfredi
Endemic to the Philippines, this nocturnal deer has short legs, a distinctive crouching posture, and a dense pattern of pale spots.

17–18 in
43–45 cm

short tail

REEVES' MUNTJAC
Muntiacus reevesi
Native to eastern Asia, this species was introduced in western Europe. Despite its small size, it damages woodlands by grazing and browsing.

16–26 in
40–65 cm

long, slender legs

RED MUNTJAC
Muntiacus muntjak
This South Asian deer uses its short, one-branched antlers and elongated upper canines to defend itself against predators and its territory against rivals.

PÈRE DAVID'S DEER
Elaphurus davidianus
Known only from captive populations, this presumed native of China was described in 1865 by a French missionary, Father Armand David.

3½–4 ft
1.1–1.2 m

AXIS DEER
Axis axis
Introduced in Australia and North America, this common Indian forest deer has lyre-shaped antlers. It is a favorite prey of tigers.

23½–39 in
60–100 cm

HOG DEER
Axis porcinus
Found in Asian forests, this deer is named after its hoglike habit of running head down, going under rather than leaping over obstacles.

24–28 in
61–70 cm

palmate antlers

tail white below

30–39 in
75–100 cm

long, pointed, branched antlers

4–4½ ft
1.2–1.4 m

BARASINGHA
Rucervus duvaucelii
This wetland deer from India was introduced in the USA for sport; the multipointed, branched antlers of males are highly prized.

20–28 in
50–70 cm

TUFTED DEER
Elaphodus cephalophus
This small deer lives in Asian montane forests. Males of this species have small antlers and short tusks, with a tuft of black hair on the forehead.

FALLOW DEER
Dama dama
Often domesticated as a source of venison, this species is distinguished by its spotted coat and flattened, palmate antlers.

3¼–4½ ft
1–1.4 m

RED DEER
Cervus elaphus elaphus
From Europe, Turkey, and North Africa, this deer varies across its range in bulk, antler size, and the prominence of its mane.

3½–4½ ft
1.1–1.4 m

WAPITI
Cervus elaphus canadensis
Although this North American and Asian subspecies resembles the red deer, genetic analysis suggests it may be a species in its own right.

shaggy fur on neck

reddish brown coat

20–37 in
50–95 cm

SIKA DEER
Cervus nippon
Distinguished by its stout, upright antlers, the sika breeds with red deer, especially when introduced outside its native east Asia.

3¼–3½ ft
1–1.1 m

MULE DEER
Odocoileus hemionus
This species has a black-tipped tail and forked antlers, which distinguish it from the white-tailed deer that shares its range in western North America.

32–39 in
80–100 cm

grayish brown coat

WHITE-TAILED DEER
Odocoileus virginianus
Found from Canada to Peru, and introduced in Europe and New Zealand, this deer flashes the white underside of its tail when alarmed.

EUROPEAN ROE DEER
Capreolus capreolus
A small European scrub and forest
deer, its fur is reddish brown in
summer, becoming darker, sometimes
almost black, in winter.

26–30 in
65–75 cm

antlers vary
greatly in
shape

thick neck

paddle used for
scraping away snow

hairy
nose pad

REINDEER
Rangifer tarandus
The footpads of this Arctic deer
shrink in winter, exposing the
hoof rim and enabling it to dig
through snow to reach food.

32–59 in
80–150 cm

MARSH DEER
Blastocerus dichotomus
Adapted to wetland habitats, the marsh
deer is the largest South American deer.
It swims well and has membranes between
its toes to walk on soft ground.

3¼–4 ft
1–1.2 m

MOOSE
Alces americanus
The world's largest deer, inhabiting the
forests of North America, the male's
palmate antlers contrast with the more
typical branched ones of the female.

6–7 ft
1.8–2.1 m

**SOUTH AMERICAN
BROWN BROCKET**
Mazama gouazoubira
A solitary deer of
scrubland and forest
thickets in Central and
South America, it feeds
mainly on fruit
and cacti during
the dry season.

22–28 in
55–70 cm

**SOUTH AMERICAN
RED BROCKET**
Mazama americana
From the jungles of South America, this small,
solitary deer prefers fruit rather than leaves.
Males have short, unbranched antlers.

14–30 in
35–75 cm

SOUTHERN PUDU
Pudu puda
One of the world's smallest deer, this
stocky animal lives in the temperate
rain forests of Argentina and Chile.

14–18 in
35–45 cm

18–22 in
45–55 cm

PAMPAS DEER
Ozotoceros bezoarticus
A slender deer of South
American grasslands
and wetlands, the
pampas deer stands
on its hind legs
to browse from
tree branches.

28–30 in
70–75 cm

**CHINESE WATER
DEER**
Hydropotes inermis
Unlike all other deer,
neither sex grows antlers.
Their long upper canines
protrude as tusks, up to
3¼ in (8 cm) long in males.

PRONGHORN

Widely represented in the North
American fossil record, the
Antilocapridae included species with
bizarrely shaped or multiple horns.
Today only the pronghorn survives.
With a similar body shape and cloven
hooves, the pronghorn resembles
antelopes (Bovidae), but unlike them
it lacks lateral toes and sheds its
horns outside the breeding season.

white neck
band

32–41 in
81–104 cm

PRONGHORN
Antilocapra americana
The fastest mammal in the
New World, pronghorns
form large herds in open
grassland and are the
ecological counterpart
of Old World antelopes.

BOVIDS

A large and varied family, the Bovidae are found on all continents except Antarctica. Despite their diversity, they share common features: males (and in some species, females) have permanent unbranched horns, often twisted or fluted, and a complex, four-chambered, ruminant's stomach.

COMMON ELAND
Taurotragus oryx
With spiral horns, the eland is the largest antelope, found in open grassland from Ethiopia to South Africa. Males sometimes develop white stripes down their flanks.

3½–5 ft
1–1.5m

prominent dewlap

loosely spiralled horns

large ears

NILGAI
Boselaphus tragocamelus
The largest Asian antelope, with a robust body sloping down from the shoulder, male nilgai develop a blue-gray coat; females are yellow-brown in color.

4–5 ft
1.2–1.5m

22–26 in
55–65 cm

FOUR-HORNED ANTELOPE
Tetracerus quadricornis
This solitary Asian forest species usually develops two pairs of horns; one between the ears and the other on the forehead.

3–3½ ft
90–110 cm

NYALA
Tragelaphus angasii
This southern African forest antelope has spiral horns and a dark brown coat, with vertical white stripes on the sides.

white chest patch

24–39 in
60–100 cm

BUSHBUCK
Tragelaphus scriptus
Widespread in sub-Saharan forests, this species has varied stripes and spots, especially on the face, ears, and tail.

2½–4 ft
75–125 cm

SITATUNGA
Tragelaphus spekii
Inhabiting Central African swamps, this excellent swimmer will often take refuge in pools of water when threatened by a predator.

chestnut coat with white stripes

3½–4¼ ft
1.1–1.3 m

BONGO
Tragelaphus eurycerus
Well camouflaged in the dense forests of west and central Africa, female bongos are usually more brightly colored than males. Both sexes have spiral horns.

GREATER KUDU
Tragelaphus strepsiceros
Male greater kudu have some of the most magnificent horns of any antelope, with two and a half twists when fully grown.

LESSER KUDU
Tragelaphus imberbis
A nocturnal antelope of arid scrubland in northeast Africa, both males and females have about ten white stripes on their coat.

3¼–5 ft
1–1.5m

3–3½ ft
90–110 cm

AFRICAN BUFFALO
Syncerus caffer
Unpredictable and dangerous, this buffalo cannot be domesticated. Grassland forms have more curved horns than the smaller forest-dwellers.

3¼–5½ ft
1–1.7 m

5–6¼ ft
1.5–1.9 m

ASIAN WATER BUFFALO
Bubalus bubalis
Predominantly domesticated, especially for their power and milk, a few wild water buffalo remain in southern Asia. Coloring and horns are variable.

32–35 in
80–90 cm

LOWLAND ANOA
Bubalus depressicornis
The smallest of all wild cattle, native to the rain forests of Sulawesi, the horns are straight and upright compared with other buffalo.

distinctive hump over the shoulders

short, curved horns

shorter hair on rear

large head

AMERICAN BISON
Bison bison
Once roaming North America in huge herds, the few remaining wild bison are dwarfed by captive populations raised for meat and hides.

6–6½ ft
1.8–2 m

shaggy, dark brown coat

EUROPEAN BISON
Bison bonasus
With shorter hair but longer horns than the American bison, this species is now restricted to eastern European and Russian primary forests.

6–7¼ ft
1.8–2.2 m

BANTENG
Bos javanicus
Native to southeast Asia and domesticated locally as draft animals, this bovid is brown, with white lower legs, muzzle, rump, and eye spots.

5¼–5½ ft
1.6–1.7 m

YAK
Bos grunniens
Inhabiting montane parts of Central Asia, insulated with long, shaggy fur, wild yaks are usually brown, but domesticated ones are more varied and many have white markings.

6½–7¼ ft
2–2.2 m

GAUR
Bos frontalis
The largest species of wild cattle, this muscular Asian forest bovine is dark brown, but paler on the muzzle and lower legs.

5½–7¼ ft
1.7–2.2 m

long horns from which this breed gets its name

4½–5 ft
1.4–1.5m

ANKOLE
Bos taurus

Originating in Africa, the ankole's huge horns, up to 3½ ft (1.8 m) long and very thick, help keep it cool in hot conditions.

4–5 ft
1.2–1.5m

TEXAS LONGHORN
Bos taurus

With a wide variety of colors, and impressive spreading horns, this breed is hardy and well suited to an extensive ranching system.

4–5 ft
1.2–1.5m

HEREFORD
Bos taurus

Originating in England, the Hereford is a beef breed that has deep, muscular forequarters and a docile temperament.

4–4½ ft
1.2–1.4 m

BRAHMAN
Bos taurus indicus

Originating in Asia, the brahman is now farmed throughout the tropics. Also known as the zebu, it has a characteristic hump on the back.

14–16½ in
35–42 cm

MAXWELL'S DUIKER
Philantomba maxwellii

With a gray-brown coat, this small West African rain forest species has few distinctive features, apart from its pale facial markings.

12½–16 in
32–41 cm

BLUE DUIKER
Philantomba monticola

This small forest antelope from Africa has simple conical horns. It eats eggs, rodents, and ants, in addition to a browser's usual diet.

16–20 in
40–50 cm

ZEBRA DUIKER
Cephalophus zebra

This is the only duiker without a predominantly plain coat. The stripes provide camouflage in its West African forest-edge habitat.

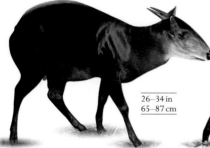

26–34 in
65–87 cm

YELLOW-BACKED DUIKER
Cephalophus silvicultor

A large Central African duiker, its dark gray coat has a conspicuous white or yellow patch on the back.

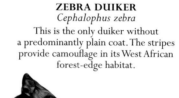

18–23 in
45–58 cm

BLACK-FRONTED DUIKER
Cephalophus nigrifrons

A dark forehead and eye glands, contrasting with paler brows, give this forest dweller from Central Africa a distinctive facial appearance.

4–4¼ ft
1.2–1.3 m

JERSEY
Bos taurus

Famed for its rich, creamy milk, this breed was developed on the island of Jersey from stock imported from France.

21½–22 in
55–56 cm

OGILBY'S DUIKER
Cephalophus ogilbyi

From the rain forests of West Africa, this duiker has well-developed hindquarters, with a red-brown rump.

18–28 in
45–70 cm

COMMON DUIKER
Sylvicapra grimmia

A widespread sub-Saharan antelope with tiny horns, this species occupies a wide range of habitats, often scavenging on fallen fruit.

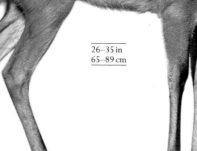

26–35 in
65–89 cm

26–41 in
65–105 cm

BOHOR REEDBUCK
Redunca redunca

Ranging across the rolling grasslands of Central Africa, the slender female contrasts with the thick-necked, horned male.

SOUTHERN REEDBUCK
Redunca arundinum

A robust antelope in grasslands in southern Central Africa, this species has conspicuous black foreleg markings. Only the males have horns.

≫

3¼–4¼ ft
1–1.3 m

3¼–4¼ ft
1–1.3 m

LECHWE
Kobus leche
A highly gregarious antelope that prefers southern Central African marshlands, its long legs enable the lechwe to run well through shallow water.

32–39 in
80–100 cm

34–43 in
85–110 cm

UGANDA KOB
Kobus kob thomasi
A sociable antelope from East Africa, the Uganda kob has a distinct white throat patch. Males have ridged, lyre-shaped horns.

30–33 in
77–83 cm

COMMON WATERBUCK
Kobus ellipsiprymnus
ellipsiprymnus
Despite its name, this African waterbuck is a savanna and woodland antelope, although it will seek refuge from predators in water.

DEFASSA WATERBUCK
Kobus ellipsiprymnus
defassa
From West and Central Africa, this subspecies of waterbuck has a solid white rump, rather than a white crescent surrounding the tail.

PUKU
Kobus vardonii
Very similar to the kob, but from southern Central Africa, the puku is slightly smaller and more stocky.

ROAN ANTELOPE
Hippotragus equinus
The ridged horns of this sub-Saharan savanna antelope grow in a smooth curve. The animal has a distinctive black and white face.

4–5 ft
1.2–1.5 m

3–3¾ ft
95–115 cm

long, thin,
curved horns

ADDAX
Addax nasomaculatus
This endangered antelope from the Sahara has long horns with two or three spiral twists, and a pale sandy or whitish coat.

SCIMITAR-HORNED ORYX
Oryx dammah
Formerly ranging across the Sahara, this whitish oryx was almost hunted to extinction in the 20th century, but has been reintroduced to some areas.

4–4½ ft
1.2–1.4 m

35–55 in
90–140 cm

SABLE ANTELOPE
Hippotragus niger
A powerful antelope with a dark coat and striking white face markings, its magnificent horns can reach more than 3 ft (1 m) in length.

ARABIAN ORYX
Oryx leucoryx
White-coated, with long, straight horns, this antelope has recently been reintroduced to parts of its former Middle Eastern range.

rusty colored
chest and neck

3¼–4¼ ft
1–1.3 m

relatively short,
sturdy legs

BEISA ORYX
Oryx beisa
Gray-brown, marked with black on the face, tail, flanks, and front legs, this East African desert oryx has very long, barely curved horns.

3¼–4 ft
1–1.25 m

4–4½ ft
1.2–1.4 m

GEMSBOK
Oryx gazella
The largest oryx, the gemsbok frequents arid habitats in southern Africa, but has also been successfully established in North America.

HARTEBEEST
Alcelaphus buselaphus
A large, long-faced antelope that lives on open grassland in eastern Africa. There are several subspecies, differing in color and horn shape.

3½–5 ft
1.1–1.5 m

3½–5 ft
1.1–1.5 m

RED HARTEBEEST
Alcelaphus caama
Chestnut brown, with a darker face and tail, this antelope is sometimes treated as a subspecies of the hartebeest.

KLIPSPRINGER
Oreotragus oreotragus
Meaning "rock jumper" in Afrikaans, the klipspringer inhabits rocky outcrops. It can leap more than ten times its own height.

17–23 in
43–58 cm

LICHTENSTEIN'S HARTEBEEST
Alcelaphus lichtensteinii
Distinguished by its strongly curved horns that turn inward at their tips, this hartebeest is found in savanna and floodplain-grassland in Central Africa.

3½–5 ft
1.1–1.5 m

20–26 in
50–66 cm

heavy, recurved horns

ORIBI
Ourebia ourebi
A graceful, long-necked antelope from sub-Saharan Africa, the oribi has a distinctive white brow line and a large, dark facial gland.

sloping back

black face

32–39 in
80–100 cm

long, black throat hair

BONTEBOK
Damaliscus pygargus dorcas
With a striking white face patch, this South African antelope was hunted almost to extinction. It is now found only in protected areas.

dark gray coat

TOPI
Damaliscus korrigum
Inhabiting eastern and Central Africa, male topi often stand on termite mounds to defend their territory and look for predators.

4–5 ft
1.2–1.5 m

BLUE WILDEBEEST
Connochaetes taurinus taurinus
A highly gregarious, African antelope, also known as a brindled gnu, this species is found in savanna regions of southern Africa.

3½–4¼ ft
1.1–1.3 m

28–34 in
70–87 cm

SPRINGBOK
Antidorcas marsupialis
With its numbers much reduced
by hunting, this agile antelope,
with lyre-shaped horns, browses
the arid lands of southern Africa.

21–26 in
53–67 cm

THOMSON'S GAZELLE
Eudorcas thomsonii
The most numerous gazelle
on the plains of East Africa,
this species has a broad black
stripe on its sides.

23½–34 in
60–85 cm

BLACKBUCK
Antilope cervicapra
With the ability to reach speeds
of 50 mph (80 kph), the
blackbuck inhabits grassland
and open woodland in India
and Pakistan.

*strongly
ridged horns*

22–32 in
56–80 cm

**GOITERED
GAZELLE**
Gazella subgutturosa
Named for the bulging larynx of the
rutting male, this Central Asian gazelle is
unique in that only the male has horns.

21–26 in
53–65 cm

DORCAS GAZELLE
Gazella dorcas
Inhabiting deserts from
North Africa to the Middle
East, this small gazelle can
survive without drinking,
taking moisture from its food.

23½–28 in
60–70 cm

MOUNTAIN GAZELLE
Gazella gazella
From the mountains and plains
of the Middle East, this gazelle
has several isolated subspecies,
some extremely rare and
threatened by poaching.

13½–15 in
34–38 cm

14–18 in
35–45 cm

GÜNTHER'S DIK-DIK
Madoqua guentheri
A dik-dik of the East African
semideserts, this species has a
long, elastic muzzle that can
be inflated in order to
regulate its temperature.

KIRK'S DIK-DIK
Madoqua kirkii
A tiny antelope, named for
its sharp alarm note, Kirk's
dik-dik has an elongated,
mobile snout. It forms
territorial pairs.

23½–32 in
60–80 cm

STEPPE SAIGA
Saiga tatarica
Now highly endangered and
found only in the central Asian
steppes, the saiga's enlarged,
flexible nose warms cold winter
air and filters out summer dust.

32–41 in
80–105 cm

GERENUK
Litocranius walleri
Long-necked and standing
on its hind legs, the East
African gerenuk browses
leaves growing out of the
reach of other antelopes.

SUNI
Neotragus moschatus
A tiny, reddish antelope
from southeast Africa, the
suni is nocturnal and spends
much of its time hidden in
dense scrub.

12–16½ in
30–43 cm

*lower leg
lacks muscle*

35–43 in
90–110 cm

DAMA GAZELLE
Nanger dama
This rare Saharan gazelle is
strikingly bicolored. The
amount of white varies
between subspecies, although
all have a white throat patch.

30–36 in
76–91 cm

GRANT'S GAZELLE
Nanger granti
Frequenting the East African
plains, this gazelle can survive
without drinking water, so it
does not follow the migrations
of many of its relatives.

24–35 in
60–90 cm

SOEMMERRING'S GAZELLE
Nanger soemmerringii
Similar to, but much rarer
than Grant's gazelle, this East
African species is distinguished
by its stronger face pattern and
larger white rump patch.

18–24 in
45–60 cm

STEENBOK
Raphicerus campestris
A small East African
bush antelope, the steenbok
has particularly large,
white-lined ears, with
black margins and
internal markings.

18–24 in
45–60 cm

**SHARPE'S
GRYSBOK**
Raphicerus sharpei
Shy and solitary, with
stubby horns, this
nocturnal antelope from
eastern Africa takes refuge
from predators in the
burrows of aardvarks.

29–36 in
73–92 cm

IMPALA
Aepyceros melampus
An African plains antelope,
with lyre-shaped horns in the
male, the impala is an
important source of food for
the big cats of the region.

**ALPINE
CHAMOIS**
Rupicapra rupicapra
Living high among
mountain rocks in the
upland blocks of southern
Europe and Asia Minor,
this species occurs in
isolated populations.
Each has a subtly
different appearance.

28–34 in
70–85 cm

22½–31 in
57–78 cm

HIMALAYAN GORAL
Naemorhedus goral
A rough-haired, goatlike browser
with curved horns, the goral
lives in small herds in the forests
of the Himalayas.

white coat has
dense underfur
for warmth

32–37 in
80–95 cm

MOUNTAIN GOAT
Oreamnos americanus
A sure-footed climber from the
northern Rocky Mountains, this
goat has a dense, woolly white
coat to protect it from low
temperatures and high winds.

**SUMATRAN
SEROW**
Capricornis sumatraensis
A goat-antelope, with
coarse fur and a distinct
mane, the Sumatran serow
lives on forested mountian
slopes where it feeds on
grass and leaves.

30–36 in
76–92 cm

BARBARY SHEEP
Ammotragus lervia
Also known as the aoudad, this native of arid mountain regions in North Africa stands motionless when threatened, making it difficult to see.

30–44 in
75–112 cm

curved horns

long hair on throat and front legs

MUSKOX
Ovibos moschatus
An inhabitant of Arctic tundra, the muskox has a shaggy coat, with dense insulating underwool, which provides protection against the elements.

4–4½ ft
1.2–1.4 m

TAKIN
Budorcas taxicolor
Living in small herds in the mountain forests of China and Bhutan, the takin has shaggy fur and a broad, arched muzzle.

3¼–4¼ ft
1–1.3 m

HIMALAYAN TAHR
Hemitragus jemlahicus
Living on rocky Himalayan mountain slopes, the tahr has hooves with rubbery soles, which give additional grip on precipitous or unstable ground.

24–35 in
60–90 cm

BHARAL
Pseudois nayaur
From rocky deserts to mountain slopes on the Tibetan Plateau, this species stays close to cliffs, its refuge from predators.

30–35 in
75–90 cm

WALIA IBEX
Capra walie
Due to the the lack of seasonal changes in the Ethiopian mountains, this rare species of ibex, unlike other ibexes, breeds all year round.

26–43 in
65–110 cm

ALPINE IBEX
Capra ibex
Living above the tree line in the Alps, ibex have recurved horns up to 3¼ ft (1 m) long that are especially impressive on males.

20–41 in
50–105 cm

24–35 in
60–90 cm

NUBIAN IBEX
Capra nubiana
Closely related to, and perhaps a subspecies of, the Alpine ibex, this species inhabits desert mountains of the Middle East.

MARKHOR
Capra falconeri
The largest wild goat, from the mountains of central Asia, is endangered by hunting for its impressive corkscrew horns and meat.

26–45 in
65–115 cm

ANGORA GOAT
Capra hircus
This breed of goat, originating in Turkey, has a fleece that is much valued as the source of mohair, a durable silky fiber.

36–44 in
92–112 cm

BAGOT GOAT
Capra hircus
One of more than 300 goat breeds, the Bagot was bred in England from stock brought home in the 13th century by the returning Crusaders.

28–39 in
70–100 cm

GOLDEN GUERNSEY GOAT
Capra hircus
A small goat, often with long hair, this rare breed is kept for milk and showing, and originated on Guernsey in the Channel Islands.

28–35 in
70–90 cm

MOUFLON
Ovis aries orientalis
With a reddish coat and pale saddle, this native of Asia Minor has been established on several Mediterranean islands since Neolithic times.

35–39 in
90–100 cm

MANX LOAGHTAN SHEEP
Ovis aries
Originating on the Isle of Man, this primitive, hardy breed, kept for its meat, has brown wool and usually four horns.

26–32 in
65–80 cm

COTSWOLD SHEEP
Ovis aries
Originating in England, this white-faced breed is hardy and dual-purpose, kept for its long wool and its meat.

26–39 in
65–100 cm

26–32 in
65–80 cm

JACOB SHEEP
Ovis aries
This ancient, hardy breed, with a patterned coat, is said to have originated in Palestine. It has as many as three pairs of horns.

ARGALI
Ovis ammon
First described by Marco Polo (1254–1324), this is a large Asian mountain sheep with long corkscrew horns in mature males.

35–49 in
90–125 cm

FAT-TAILED SHEEP
Ovis aries
This breed, found mainly in Africa and Asia, tolerates dry conditions, relying on fat stored in its swollen tail and hindquarters.

26–43 in
65–110 cm

massive, curving horns

DALL SHEEP
Ovis dalli
Creamy white or brown, with curved yellowish horns, dall sheep live in the subarctic mountains of Canada and Alaska.

32–35 in
80–90 cm

short legs

BIGHORN SHEEP
Ovis canadensis
Found in mountains and deserts in North America, males use their impressive horns to establish a dominance hierarchy. Only high-ranking males secure access to females.

30–41 in
75–105 cm

SNOW SHEEP
Ovis nivicola
With pale wool and dark legs, this Siberian sheep is extremely agile and can move rapidly over rough, mountainous terrain.

37–43 in
95–112 cm

GIRAFFE AND OKAPI

Diverse in the fossil record, the Giraffidae is represented now by just two species from sub-Saharan Africa. Although very different in shape and habitat, they share some common features, including a long, dark tongue, ossicones (horns) covered in skin, and lobed canine teeth. Otherwise they resemble the bovids, with cloven hooves, a four-chambered stomach, and the incisors (upper front teeth) replaced by a horny pad.

5–6½ ft
1.5–2 m

OKAPI
Okapia johnstoni
Restricted to the rain forests of Central Africa, the okapi's long neck and flexible blue tongue show clear similarities to the giraffe.

short horns covered in skin

large ears

short upright mane

short body with sloping back

8¼–11 ft
2.5–3.3 m

ROTHSCHILD'S GIRAFFE
Giraffa camelopardalis rothschildi
Unlike most giraffe races, this subspecies has white "socks"—its blotches do not extend onto the lower leg.

8¼–12 ft
2.5–3.6 m

MASAI GIRAFFE
Giraffa camelopardalis tippelskirchi
With a neck up to 7¾ ft (2.4 m) long on top of the shoulder height, this is the largest subspecies of the world's tallest mammal.

8¼–11 ft
2.5–3.3 m

irregular blotches

8¼–11 ft
2.5–3.3 m

RETICULATED GIRAFFE
Giraffa camelopardalis reticulata
Ranging from northern Kenya to Ethiopia, this race of giraffe has large polygonal blotches, often with pale centers, on a pale background.

THORNICROFT'S GIRAFFE
Giraffa camelopardalis thornicrofti
With star-shaped or lobed blotches that extend onto the lower leg, this race is found in eastern Zambia.

CAMELS AND RELATIVES

Unique within their order, members of the family Camelidae have just two toes, but no hooves. Each toe has a small nail at its tip and a soft footpad, which can shift slightly to help maintain grip in mountainous terrain and prevent sinking in soft sand. They also have distinctive teeth, oval red blood cells, a three-chambered stomach, and a leg musculature that entails lying down resting on their knees.

DROMEDARY
Camelus dromedarius
Used for transport in arid regions, this Arabian species is supremely adapted to desert life. Only the feral Australian populations show wild characteristics.

5½–6½ ft
1.7–2 m

30–34 in
75–85 cm

BACTRIAN CAMEL
Camelus bactrianus
Widely domesticated, as a wild animal the two-humped camel is restricted to critically small populations in the Asian deserts.

6–7½ ft
1.8–2.3 m

LLAMA
Lama glama glama
Derived from the guanaco, the llama is a valuable pack and meat animal. Originally from the Andes, it is now more widely distributed in Europe and North America.

5½–6 ft
1.7–1.8 m

long, woolly coat

GUANACO
Lama glama guanicoe
Native to the arid mountains of South America, the guanaco has high levels of oxygen-carrying hemoglobin in the blood, enabling it to thrive at extreme altitudes.

3½–4 ft
1.1–1.2 m

nail on tip of toe

VICUÑA
Vicugna vicugna
The smaller of the two wild Andean camelids, the vicuña produces a fine wool, which led to its domestication and development of the alpaca.

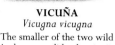 30–35 in
75–90 cm

ALPACA
Vicugna pacos
Kept in herds that graze high in the Andes, and now elsewhere in the world, this species is an important source of wool.

HIPPOPOTAMUSES

Among the Artiodactyla, family Hippopotamidae is unique in that its members walk on four toes on each foot. They also have huge, barrel-shaped bodies, short but stout legs, and large heads. Their wide mouths, with tusklike canine teeth, are used in feeding, fighting, and defense. Reflecting an amphibious lifestyle, the nostrils and eyes are on the top of the snout and the skin is smooth, lacking sweat glands.

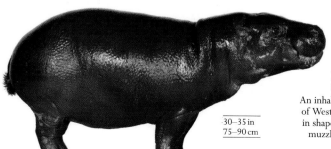

PYGMY HIPPOPOTAMUS
Choeropsis liberiensis
An inhabitant of the forested swamps of West Africa, this species is similar in shape to its larger relative, but its muzzle is proportionally smaller.

30–35 in
75–90 cm

4¼–5½ ft
1.3–1.7 m

HIPPOPOTAMUS
Hippopotamus amphibius
A solitary nocturnal grazer and social daytime wallower, this species is now found mainly in eastern and southern Africa.

BACTRIAN CAMEL
Camelus bactrianus

The Bactrian camel is an exceptionally hardy animal, built to survive in the harsh desert landscapes of southern Asia, where temperatures range between 104°F (40°C) in summer and -20°F (-29°C) in winter. Camels are adapted to travel long distances over difficult terrain in search of food—such as grasses, leaves, and shrubs—which is scarce. When water is available, the Bactrian camel is able to gulp more than 26 gallons (100 liters) in 10 minutes. It can also survive by drinking salty water if necessary. Almost all of the world's Bactrian camels are domesticated; fewer than 1,000 animals remain in the wild, in remote and inhospitable parts of China and Mongolia. Recent studies suggest that these wild camels are genetically distinct from the domestic variety, making it all the more urgent that they be conserved.

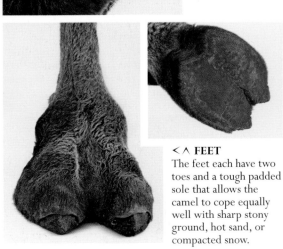

closable nostrils keep sand out

small, furry ears

thick fur keeps the camel warm, and also protects it from sunburn

shaggy mane

SIZE 6–7½ ft (1.8–2.3 m)
HABITAT Stony desert, steppe, and rocky plains
DISTRIBUTION Asia
DIET Herbivorous

∨ EYELASHES
Two rows of extra-thick eyelashes protect the eyes from strong sunlight, and windblown sand and grit, saving precious water from being blinked away as tears.

∨ KNEE PADS
The thick pads of skin on the knees are used for kneeling. Camels rest on their knees, with their legs folded beneath them.

∧ TEETH
Camels swallow their food whole, then regurgitate and rechew it to aid digestion. Starving individuals have been known to eat rope and leather, which are hard to break down.

∧ MOUTH
A groove connecting to the upper lip channels any valuable moisture that trickles from the camel's nostrils back into its mouth—nothing is wasted.

< ∧ FEET
The feet each have two toes and a tough padded sole that allows the camel to cope equally well with sharp stony ground, hot sand, or compacted snow.

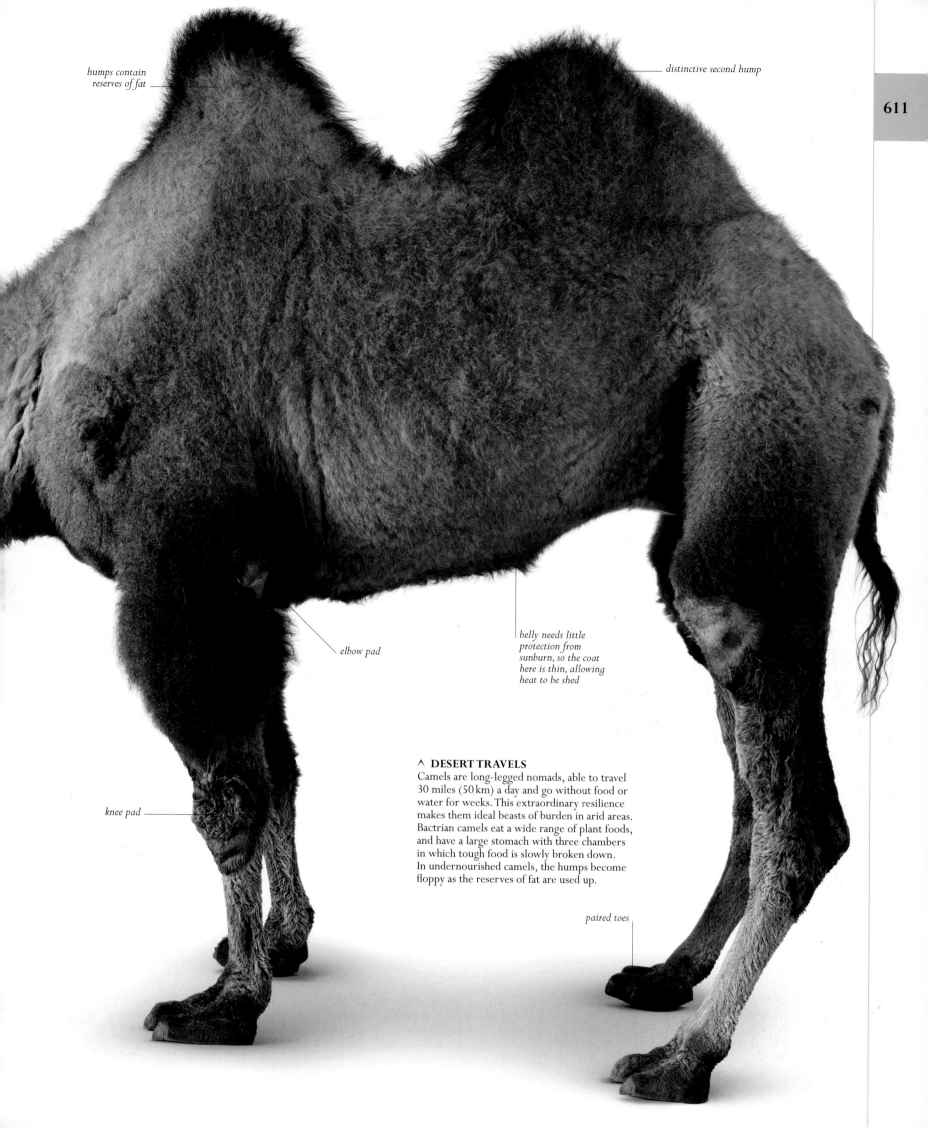

humps contain
reserves of fat

distinctive second hump

belly needs little
protection from
sunburn, so the coat
here is thin, allowing
heat to be shed

elbow pad

knee pad

paired toes

∧ DESERT TRAVELS

Camels are long-legged nomads, able to travel
30 miles (50 km) a day and go without food or
water for weeks. This extraordinary resilience
makes them ideal beasts of burden in arid areas.
Bactrian camels eat a wide range of plant foods,
and have a large stomach with three chambers
in which tough food is slowly broken down.
In undernourished camels, the humps become
floppy as the reserves of fat are used up.

WHALES, PORPOISES, AND DOLPHINS

Collectively known as cetaceans, this group of mammals is wholly aquatic and all but four species in the order are found in coastal and marine waters.

Supremely adapted to life in water, the cetacean body is tapered and streamlined, with the forelimbs modified into flippers. There are no visible hind limbs, but the tail has horizontal flukes for added propulsion. Many species also have a dorsal fin. The skin is almost hairless and the body is insulated by a layer of blubber, which is especially thick in species living in cold water.

BREATHING AND COMMUNICATING

Capable of diving to great depths and for long periods because of their ability to store oxygen in their muscle tissues, cetaceans must come to the surface to breathe. Breathing is through nostrils, known as blowholes, which are located on the top of the head. When stale air is expelled, it may create a spout of condensation—the spout's size, angle, and shape can be used to distinguish some species, even when their body is almost fully submerged.

Most species produce sound. Some use a series of clicks for echolocation. The clicks bounce off nearby objects, alerting them to any obstacles on their path. Others communicate with each other using vocalizations that range from whistles and groans to the complex songs of many great whales. While their hearing is good, their ears are reduced to simple openings behind the eyes. External ears are both undesirable—they would affect streamlining—and unnecessary, since water is an effective medium for sound transmission.

HUNTER OR FILTER-FEEDER

Cetaceans are divided into two main groups based on their feeding habits. Toothed species, predators of fishes, large invertebrates, seabirds, seals, and sometimes smaller cetaceans, catch prey with their sharp teeth and usually swallow it whole, without chewing. In contrast, filter-feeding baleen whales have fibrous baleen plates suspended from their upper jaw that act like sieves. Water, containing invertebrates and small fishes, is taken in and then forced out through the plates by the tongue, leaving behind the organisms.

PHYLUM	CHORDATA
CLASS	MAMMALIA
ORDER	CETACEA
FAMILIES	12
SPECIES	About 90

DEBATE
LAND-LIVING ANCESTORS

The taxonomic position of cetaceans has long been debated; their extreme adaptation to a wholly aquatic lifestyle masks anatomical features that are shared with other orders. Today it is generally accepted that Cetacea is most closely related to Artiodactyla—specifically the hippopotamus family—so they are grouped together as Cetartiodactyla, at order or superorder level. Evidence comes mainly from genetic and molecular studies, but there is increasing anatomical support, including strong similarities between the ankle bones of artiodactyls and some fossil ancestors of whales.

RIGHT WHALES

Found in cool temperate and polar waters, the Balaenidae were considered the "right" ones to hunt, as they are easily approached, often close to shore, and have a thick layer of oily blubber—an adaptation to their cool water habitats. Right whales lack dorsal fins and throat grooves and their strongly curved jaws support the longest baleen plates of any whales.

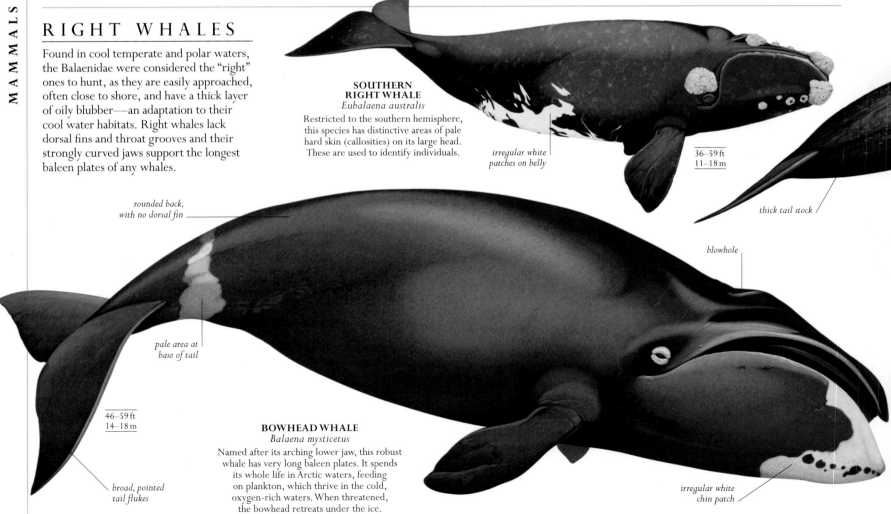

SOUTHERN RIGHT WHALE
Eubalaena australis
Restricted to the southern hemisphere, this species has distinctive areas of pale hard skin (callosities) on its large head. These are used to identify individuals.

irregular white patches on belly

36–59 ft
11–18 m

thick tail stock

blowhole

rounded back, with no dorsal fin

pale area at base of tail

46–59 ft
14–18 m

broad, pointed tail flukes

BOWHEAD WHALE
Balaena mysticetus
Named after its arching lower jaw, this robust whale has very long baleen plates. It spends its whole life in Arctic waters, feeding on plankton, which thrive in the cold, oxygen-rich waters. When threatened, the bowhead retreats under the ice.

irregular white chin patch

PYGMY RIGHT WHALE

Consisting of one species, the family Neobalaenidae is confined to waters in the southern hemisphere. Unlike most true right whales, the pygmy right whale has a small, prominent dorsal fin, but no callosities on the head. Its baleen plates are ivory in colour.

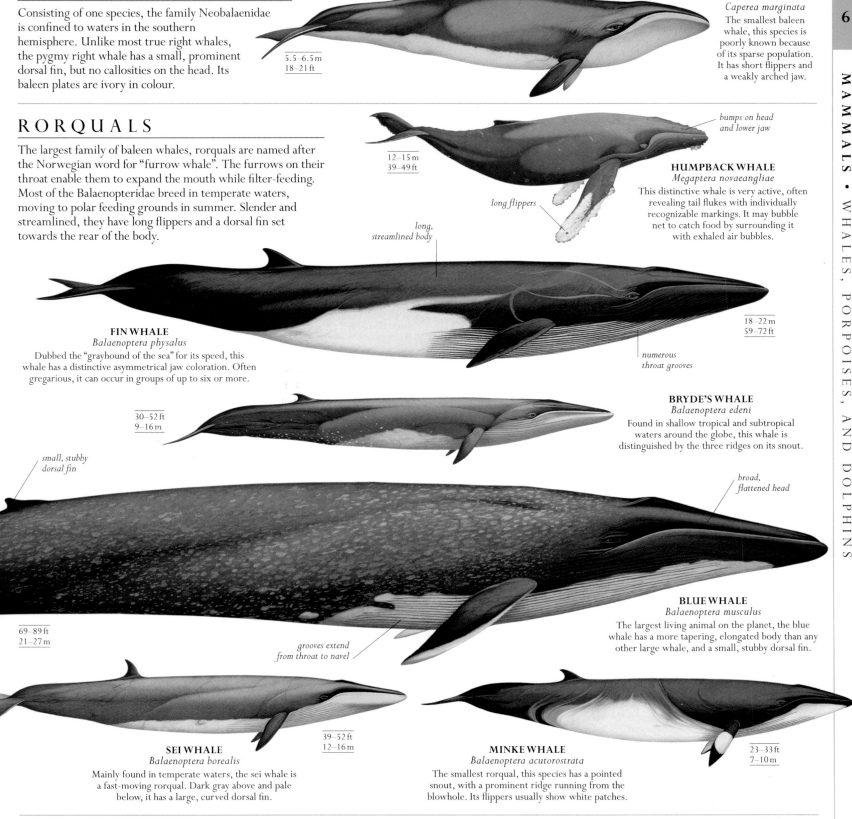

5.5–6.5 m
18–21 ft

PYGMY RIGHT WHALE
Caperea marginata
The smallest baleen whale, this species is poorly known because of its sparse population. It has short flippers and a weakly arched jaw.

RORQUALS

The largest family of baleen whales, rorquals are named after the Norwegian word for "furrow whale". The furrows on their throat enable them to expand the mouth while filter-feeding. Most of the Balaenopteridae breed in temperate waters, moving to polar feeding grounds in summer. Slender and streamlined, they have long flippers and a dorsal fin set towards the rear of the body.

bumps on head
and lower jaw

12–15 m
39–49 ft

long flippers

HUMPBACK WHALE
Megaptera novaeangliae
This distinctive whale is very active, often revealing tail flukes with individually recognizable markings. It may bubble net to catch food by surrounding it with exhaled air bubbles.

long, streamlined body

FIN WHALE
Balaenoptera physalus
Dubbed the "grayhound of the sea" for its speed, this whale has a distinctive asymmetrical jaw coloration. Often gregarious, it can occur in groups of up to six or more.

18–22 m
59–72 ft

numerous throat grooves

30–52 ft
9–16 m

BRYDE'S WHALE
Balaenoptera edeni
Found in shallow tropical and subtropical waters around the globe, this whale is distinguished by the three ridges on its snout.

small, stubby dorsal fin

broad, flattened head

69–89 ft
21–27 m

grooves extend from throat to navel

BLUE WHALE
Balaenoptera musculus
The largest living animal on the planet, the blue whale has a more tapering, elongated body than any other large whale, and a small, stubby dorsal fin.

39–52 ft
12–16 m

SEI WHALE
Balaenoptera borealis
Mainly found in temperate waters, the sei whale is a fast-moving rorqual. Dark gray above and pale below, it has a large, curved dorsal fin.

23–33 ft
7–10 m

MINKE WHALE
Balaenoptera acutorostrata
The smallest rorqual, this species has a pointed snout, with a prominent ridge running from the blowhole. Its flippers usually show white patches.

GRAY WHALE

The sole species in the family Eschrichtiidae is now restricted to the eastern North Pacific, having been hunted to extinction in the Atlantic. It undertakes vast annual migrations— the longest of any mammal—from the Bering Sea to breed in subtropical waters, especially off Baja California, Mexico.

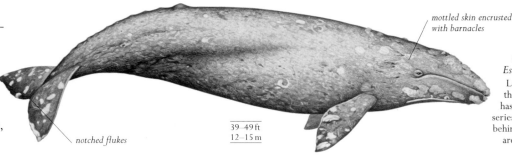

mottled skin encrusted with barnacles

notched flukes

39–49 ft
12–15 m

GRAY WHALE
Eschrichtius robustus
Lacking a dorsal fin, the gray whale's back has a low hump, with a series of smaller "knuckles" behind. The throat grooves are poorly developed.

BEAKED WHALES

The Ziphiidae are found in the open ocean, typically in small groups that congregate around underwater canyons. They feed near the sea floor, diving for an hour or more. Their facial beaks usually have one or two pairs of teeth, which are for display only—food is simply sucked in. Because of their remote habitat and ability to dive deep, few of the 20 species in the family are familiar; some have never been sighted alive.

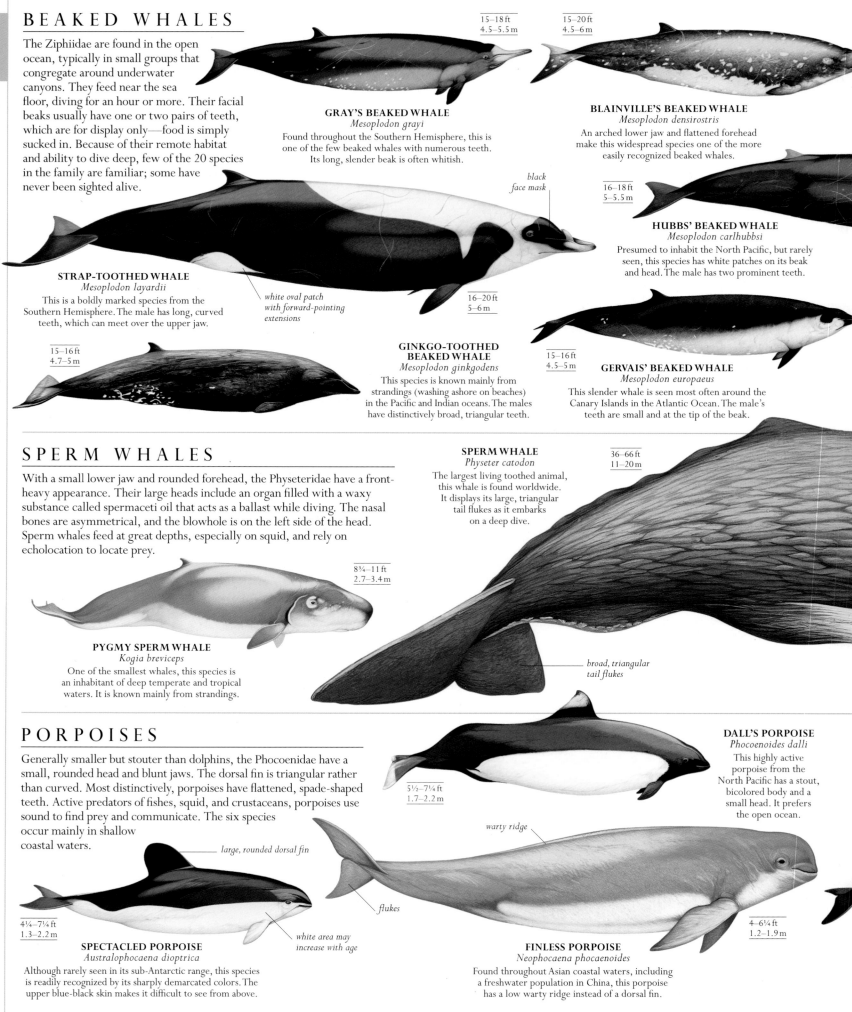

15–18 ft
4.5–5.5 m

15–20 ft
4.5–6 m

GRAY'S BEAKED WHALE
Mesoplodon grayi
Found throughout the Southern Hemisphere, this is one of the few beaked whales with numerous teeth. Its long, slender beak is often whitish.

BLAINVILLE'S BEAKED WHALE
Mesoplodon densirostris
An arched lower jaw and flattened forehead make this widespread species one of the more easily recognized beaked whales.

black face mask

16–18 ft
5–5.5 m

HUBBS' BEAKED WHALE
Mesoplodon carlhubbsi
Presumed to inhabit the North Pacific, but rarely seen, this species has white patches on its beak and head. The male has two prominent teeth.

STRAP-TOOTHED WHALE
Mesoplodon layardii
This is a boldly marked species from the Southern Hemisphere. The male has long, curved teeth, which can meet over the upper jaw.

white oval patch with forward-pointing extensions

16–20 ft
5–6 m

15–16 ft
4.7–5 m

GINKGO-TOOTHED BEAKED WHALE
Mesoplodon ginkgodens
This species is known mainly from strandings (washing ashore on beaches) in the Pacific and Indian oceans. The males have distinctively broad, triangular teeth.

15–16 ft
4.5–5 m

GERVAIS' BEAKED WHALE
Mesoplodon europaeus
This slender whale is seen most often around the Canary Islands in the Atlantic Ocean. The male's teeth are small and at the tip of the beak.

SPERM WHALES

With a small lower jaw and rounded forehead, the Physeteridae have a front-heavy appearance. Their large heads include an organ filled with a waxy substance called spermaceti oil that acts as a ballast while diving. The nasal bones are asymmetrical, and the blowhole is on the left side of the head. Sperm whales feed at great depths, especially on squid, and rely on echolocation to locate prey.

SPERM WHALE
Physeter catodon
The largest living toothed animal, this whale is found worldwide. It displays its large, triangular tail flukes as it embarks on a deep dive.

36–66 ft
11–20 m

8¾–11 ft
2.7–3.4 m

PYGMY SPERM WHALE
Kogia breviceps
One of the smallest whales, this species is an inhabitant of deep temperate and tropical waters. It is known mainly from strandings.

broad, triangular tail flukes

PORPOISES

Generally smaller but stouter than dolphins, the Phocoenidae have a small, rounded head and blunt jaws. The dorsal fin is triangular rather than curved. Most distinctively, porpoises have flattened, spade-shaped teeth. Active predators of fishes, squid, and crustaceans, porpoises use sound to find prey and communicate. The six species occur mainly in shallow coastal waters.

DALL'S PORPOISE
Phocoenoides dalli
This highly active porpoise from the North Pacific has a stout, bicolored body and a small head. It prefers the open ocean.

5½–7¼ ft
1.7–2.2 m

warty ridge

large, rounded dorsal fin

4¼–7¼ ft
1.3–2.2 m

flukes

SPECTACLED PORPOISE
Australophocaena dioptrica
Although rarely seen in its sub-Antarctic range, this species is readily recognized by its sharply demarcated colors. The upper blue-black skin makes it difficult to see from above.

white area may increase with age

FINLESS PORPOISE
Neophocaena phocaenoides
Found throughout Asian coastal waters, including a freshwater population in China, this porpoise has a low warty ridge instead of a dorsal fin.

4–6¼ ft
1.2–1.9 m

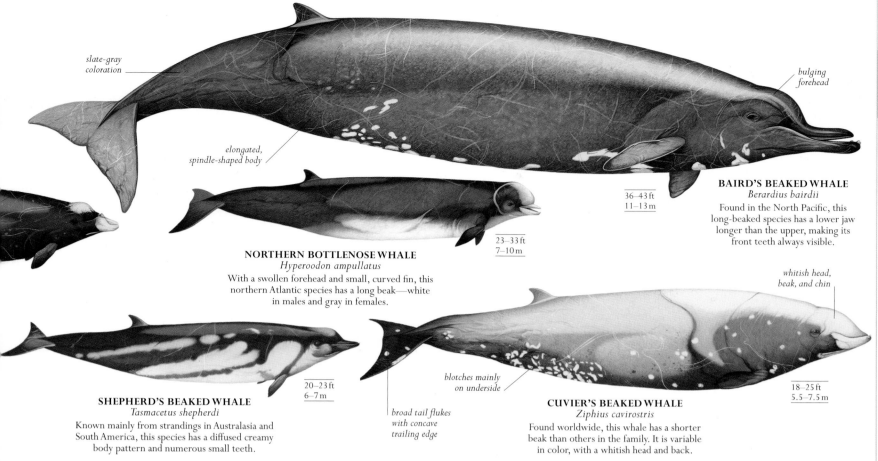

slate-gray coloration

elongated, spindle-shaped body

bulging forehead

36–43 ft
11–13 m

BAIRD'S BEAKED WHALE
Berardius bairdii
Found in the North Pacific, this long-beaked species has a lower jaw longer than the upper, making its front teeth always visible.

23–33 ft
7–10 m

NORTHERN BOTTLENOSE WHALE
Hyperoodon ampullatus
With a swollen forehead and small, curved fin, this northern Atlantic species has a long beak—white in males and gray in females.

whitish head, beak, and chin

20–23 ft
6–7 m

SHEPHERD'S BEAKED WHALE
Tasmacetus shepherdi
Known mainly from strandings in Australasia and South America, this species has a diffused creamy body pattern and numerous small teeth.

blotches mainly on underside

broad tail flukes with concave trailing edge

18–25 ft
5.5–7.5 m

CUVIER'S BEAKED WHALE
Ziphius cavirostris
Found worldwide, this whale has a shorter beak than others in the family. It is variable in color, with a whitish head and back.

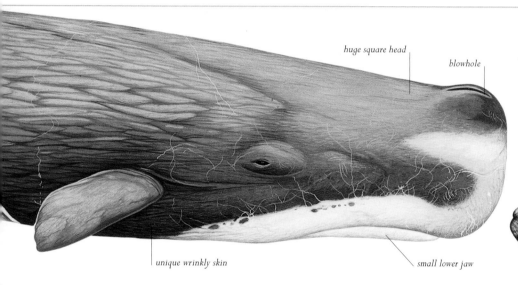

huge square head

blowhole

unique wrinkly skin

small lower jaw

NARWHAL AND BELUGA

The Monodontidae form a small family of two medium-sized Arctic whales that are rather unusual in appearance. They are highly gregarious species and are found in bays, estuaries, and fjords, and along the edges of pack ice, sometimes in herds of several hundred. Neither species has a true dorsal fin; both have a swollen, rounded forehead, which can change shape when they produce sounds, of which they have a wide range.

tusk protrudes from upper jaw of males

12–16 ft
3.8–5 m

NARWHAL
Monodon monoceros
Mottled gray and brown in color, the narwhal has only two teeth, one of which grows into a twisted tusk up to 9¾ ft (3 m) long in adult males.

4½–6¼ ft
1.4–1.9 m

HARBOR PORPOISE
Phocoena phocoena
Widespread across the Northern Hemisphere, the harbor porpoise is one of the most familiar cetaceans. Often found in estuaries, it sometimes ventures upriver.

body scarred by polar bear

ADULT

JUVENILE

9¾–18 ft
3–5.5 m

4½–6½ ft
1.4–2 m

BURMEISTER'S PORPOISE
Phocoena spinipinnis
This dark porpoise, with a distinctive backward-pointing dorsal fin, is one of the most abundant cetaceans around the coasts of South America.

4–5 ft
1.2–1.5 m

VAQUITA
Phocoena sinus
Unique to the Sea of Cortéz (Gulf of California), this is the smallest and rarest porpoise, often inhabiting lagoons so shallow that its back protrudes above the surface.

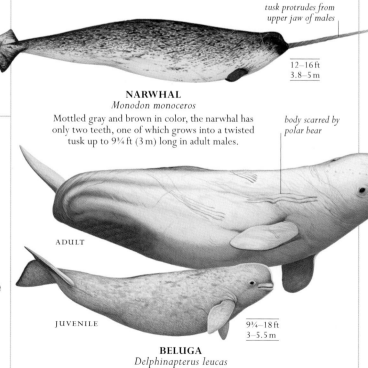

BELUGA
Delphinapterus leucas
The beluga has a circumpolar distribution in Arctic and subarctic waters, spending the winter around and under pack ice. It is the only whale entirely white as an adult.

OCEANIC DOLPHINS

Found worldwide, often in shallow seas over continental shelves, the Delphinidae typically have a curved dorsal fin, protruding beak, and swollen forehead. Small- to medium-sized, their colors and patterns vary widely. Most mainly eat fish and travel in herds or "pods." Larger species are collectively known as blackfish.

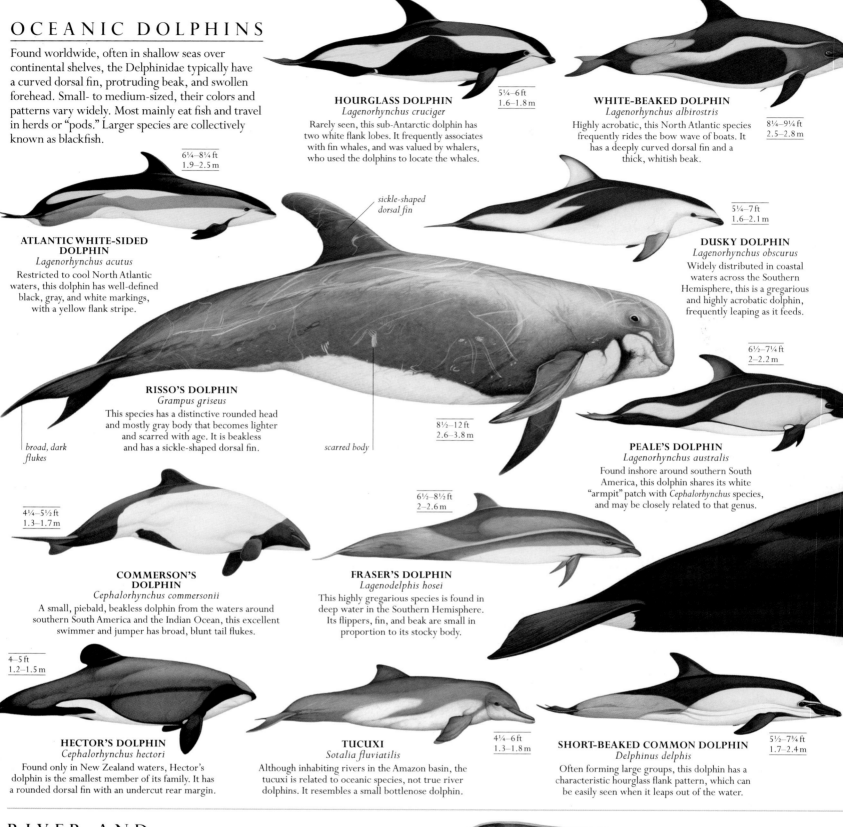

HOURGLASS DOLPHIN
Lagenorhynchus cruciger
5¼–6 ft
1.6–1.8 m
Rarely seen, this sub-Antarctic dolphin has two white flank lobes. It frequently associates with fin whales, and was valued by whalers, who used the dolphins to locate the whales.

WHITE-BEAKED DOLPHIN
Lagenorhynchus albirostris
8¼–9¼ ft
2.5–2.8 m
Highly acrobatic, this North Atlantic species frequently rides the bow wave of boats. It has a deeply curved dorsal fin and a thick, whitish beak.

ATLANTIC WHITE-SIDED DOLPHIN
Lagenorhynchus acutus
6¼–8¼ ft
1.9–2.5 m
Restricted to cool North Atlantic waters, this dolphin has well-defined black, gray, and white markings, with a yellow flank stripe.

sickle-shaped dorsal fin

DUSKY DOLPHIN
Lagenorhynchus obscurus
5¼–7 ft
1.6–2.1 m
Widely distributed in coastal waters across the Southern Hemisphere, this is a gregarious and highly acrobatic dolphin, frequently leaping as it feeds.

RISSO'S DOLPHIN
Grampus griseus
This species has a distinctive rounded head and mostly gray body that becomes lighter and scarred with age. It is beakless and has a sickle-shaped dorsal fin.

broad, dark flukes

scarred body

8½–12 ft
2.6–3.8 m

PEALE'S DOLPHIN
Lagenorhynchus australis
6½–7¼ ft
2–2.2 m
Found inshore around southern South America, this dolphin shares its white "armpit" patch with *Cephalorhynchus* species, and may be closely related to that genus.

COMMERSON'S DOLPHIN
Cephalorhynchus commersonii
4¼–5½ ft
1.3–1.7 m
A small, piebald, beakless dolphin from the waters around southern South America and the Indian Ocean, this excellent swimmer and jumper has broad, blunt tail flukes.

FRASER'S DOLPHIN
Lagenodelphis hosei
6½–8½ ft
2–2.6 m
This highly gregarious species is found in deep water in the Southern Hemisphere. Its flippers, fin, and beak are small in proportion to its stocky body.

HECTOR'S DOLPHIN
Cephalorhynchus hectori
4–5 ft
1.2–1.5 m
Found only in New Zealand waters, Hector's dolphin is the smallest member of its family. It has a rounded dorsal fin with an undercut rear margin.

TUCUXI
Sotalia fluviatilis
4¼–6 ft
1.3–1.8 m
Although inhabiting rivers in the Amazon basin, the tucuxi is related to oceanic species, not true river dolphins. It resembles a small bottlenose dolphin.

SHORT-BEAKED COMMON DOLPHIN
Delphinus delphis
5½–7¾ ft
1.7–2.4 m
Often forming large groups, this dolphin has a characteristic hourglass flank pattern, which can be easily seen when it leaps out of the water.

RIVER AND ESTUARINE DOLPHINS

The family Iniidae has three species, including the probably extinct Chinese river dolphin, which can be recognized by their small eyes, bulging foreheads, and long beaks. The Iniidae share their long, slender beak with the Indian river dolphins, although their teeth are not visible when the mouth is closed.

AMAZON RIVER DOLPHIN
Inia geoffrensis
6–8¼ ft
1.8–2.5 m
The largest river dolphin, this species may be identified by its lack of a dorsal fin and often pinkish color. Its range overlaps with that of the tucuxi.

slender, slightly down-curved bill

ROUGH-TOOTHED DOLPHIN
Steno bredanensis

Occurring in most warm waters, this species has a conical head that extends into a long, slender beak, and a broad-based, pointed dorsal fin.

7–8½ ft
2.1–2.6 m

BOTTLENOSE DOLPHIN
Tursiops truncatus

This widespread dolphin frequently interacts with humans. Offshore populations are larger and darker than those inshore, and have shorter fins and beaks.

6¼–13 ft
1.9–3.9 m

short, pronounced beak

long, slender flippers

ATLANTIC SPOTTED DOLPHIN
Stenella frontalis

From the tropical and subtropical Atlantic, this dolphin has dark spots below and pale spots above. These increase in density as it matures.

5½–7½ ft
1.7–2.3 m

STRIPED DOLPHIN
Stenella coeruleoalba

Found in temperate and tropical parts of all oceans, this acrobatic species has a distinctive pattern of blue stripes and wedges.

6–8¼ ft
1.8–2.5 m

SOUTHERN RIGHT WHALE DOLPHIN
Lissodelphis peronii

Within its range, the only dolphin to lack a dorsal fin, this distinctive black and white species occurs throughout the cool oceans of the Southern Hemisphere.

6–9½ ft
1.8–2.9 m

FALSE KILLER WHALE
Pseudorca crassidens

This uniformly dark species is widespread in shallow temperate and tropical waters. It feeds on cetaceans and large fishes, often from fishing nets.

14–20 ft
4.3–6 m

MELON-HEADED WHALE
Peponocephala electra

An offshore tropical species, this whale has a rounded head and a tall, pointed dorsal fin. It is uniformly gray, with a darker face mask.

7–8¾ ft
2.1–2.7 m

tall dorsal fin

PYGMY KILLER WHALE
Feresa attenuata

Lacking a beak, this small, dark, robust species can be very aggressive. It kills and eats other dolphins within its circumtropical range.

7–8½ ft
2.1–2.6 m

KILLER WHALE
Orcinus orca

Boldly marked, with a tall dorsal fin, this species is at the top of marine food chain. It feeds on fishes, seals, sharks, and other cetaceans.

18–33 ft
5.5–10 m

white underparts

large, broad flippers

LONG-FINNED PILOT WHALE
Globicephala melas

Prone to mass stranding, this social species is widespread in temperate waters. Its bulbous forehead can be seen as it lifts its head out of the water to look around.

11–23 ft
3.5–7 m

FRANCISCANA
Pontoporia blainvillei

Although inhabiting estuaries and coastal waters of eastern South America, this species is in the river dolphin family. It has, proportionally, the longest beak of any cetacean.

4¼–5½ ft
1.3–1.7 m

INDIAN RIVER DOLPHINS

The sole species in the family Platanistidae has two virtually identical subspecies in the Indus and Ganges rivers. Their long teeth are visible even when the mouth is closed. Their tiny eyes lack lenses and they are effectively blind.

proportionally broad flukes

INDUS RIVER DOLPHIN
Platanista minor

From gray to pale blue to brown, this dolphin has a distinctive long beak, large flippers, and triangular dorsal hump. It navigates and hunts using echolocation.

robust, uniformly colored body

sharp teeth

5–8¼ ft
1.5–2.5 m

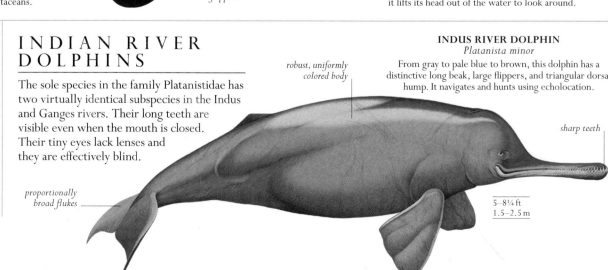

GLOSSARY

ABDOMEN
The hind part of the body. The abdomen lies below the ribcage in mammals, and behind the thorax in arthropods.

ACCESSORY FRUIT
A fruit that is formed from the flower's ovary and another structure, such as the swollen base of the flower. Examples of accessory fruits are apples and figs.

ADIPOSE FIN
A small fin in fishes behind the dorsal fin. It is mostly fatty tissue covered with skin.

AGARIC
Mushroomlike fruitbody of a fungus, consisting of stem and cap.

ALKALOID
Bitter, sometimes poisonous, chemical produced by certain plants and fungi.

ALLUVIAL (DEPOSIT)
Concentrations of material that have been separated by weathering from the host rock, then deposited in rivers or streams.

AMPHIBOLE
Group of common rock-forming minerals, often with complex composition. Most are ferromagnesian silicates.

ANAL FIN
The unpaired fin of a fish on the lower side of its body, behind the anus.

ANGIOSPERM
A seed-producing plant that encloses its seeds in a fruit and produces flowers. *See also* gymnosperm.

ANNUAL
A plant that completes its life cycle, from germination to death, in a single season of growth.

ANTENNA (PL. ANTENNAE)
A sensory feeler on the head of arthropods and some other invertebrates, such as mollusks. Antennae are always present in pairs, and can be sensitive to touch, sound, heat, and taste. Their size and shape varies widely according to the way in which they are used.

ANTHER
In flowering plants, the podlike structure on the stamen that produces pollen.

ANTLER
A bony growth on the head of deer. Unlike horns, antlers often branch, and in most cases they are grown and shed every year in a cycle linked with the breeding season.

ARBOREAL
Living fully or partly (known as semiarboreal) in trees.

ASCUS (PL. ASCI)
Microscopic saclike structure of sac fungi (Ascomycota) that produces spores.

ASEXUAL REPRODUCTION
A form of reproduction that involves just one parent, producing offspring that are genetically identical to each other (clones). It is most common in microbes, plants, and invertebrates.

BALEEN
A fibrous substance used by some whales for filtering food from water. Baleen grows in the form of plates with frayed edges, which hang from a whale's upper jaw. The baleen plates trap food, which the whale then swallows.

BASAL ANGIOSPERM
Order of flowering plant with certain primitive features that diverged away from the main evolutionary line of flowering plants before others. It includes water lilies.

BASIDIUM (PL. BASIDIA)
Microscopic clublike structure of mushrooms and relatives (Basidiomycota) that produces spores.

BEAK
A set of narrow protruding jaws, usually without teeth. Beaks have evolved separately in many groups of vertebrates, including birds, tortoises, and some whales. The beak of a bird is also called a bill.

BERRY
A fleshy many-seeded fruit of a plant that develops from a single ovary. Many fruits popularly called berries are not true berries but compound fruits—an example of which is raspberry.

BIENNIAL
A plant that completes its life cycle, from germination to death, in two years. It typically stores up food in the first season that can be used for reproduction in the second season.

BIOLUMINESCENCE
The production of light by living organisms.

BIPEDAL
An animal that moves on two legs.

BRACKET
Shelflike fruitbody of a fungus.

BRACT
Modified leaf, often brightly colored, beneath a flower or flower cluster.

BROMELIAD
Flowering plant of the family Bromeliaceae. Almost exclusively tropical American, most live as rain forest epiphytes: perched in the rain forest branches without getting nourishment from the trees. Many form rosettes of leaves that collect rainwater, forming treetop pools that are important nurseries of insect larvae and frog tadpoles.

BROWSER
An animal that feeds on the leaves of trees and shrubs rather than on grasses.

BULB
The underground shoot of a plant, consisting of modified leaves and used for food storage during periods of dormancy or for asexual reproduction.

CALYX
Outer cuplike whorl of a flower, made up of sepals.

CAMOUFLAGE
Colors or patterns that enable an animal to merge with its background. It is often used for protection against predators and for concealment when approaching prey.

CANINE (TOOTH)
In mammals, a tooth with a single sharp point that is shaped for piercing and gripping prey. Canine teeth are located toward the front of the jaws, and are highly developed in carnivores.

CARAPACE
A hard shield found on the back of the bodies of animals, invertebrates, and crustaceans. In turtles and tortoises the carapace is the upper part of the shell.

CARBOHYDRATE
Foodstuff, such as sugar or starch, that provides energy.

CARNASSIAL (TOOTH)
In mammalian carnivores, a bladelike upper premolar and lower molar tooth that have evolved for slicing through flesh.

CARNIVORE
Any animal that eats meat. The word carnivore can also be used in a more restricted sense to mean mammals of the order Carnivora.

CARPEL
Female reproductive part of a flower, divided into ovary, style, and stigma. Also known as the pistil.

CARTILAGE
A rubbery substance forming part of vertebrate skeletons. In most vertebrates, it lines the joints, but in cartilaginous fish it forms the whole skeleton.

CATKIN
Hanging cluster of flowers, usually with simple flowers all of the same sex.

CECUM
Pouch of the digestive tract, often used in the digestion of plant food.

CELL
The smallest unit of an organism that can exist on its own.

CELLULOSE
A complex carbohydrate found in plants. It is used by plants as building material, and has a resilient chemical structure that animals find hard to digest. Plant-eating animals break it down in their stomach with the aid of microorganisms.

CHITIN
A tough carbohydrate making up cell walls of fungi and certain animal exoskeletons.

CHLOROPHYLL
Green pigment found in chloroplasts used to trap light energy in photosynthesis.

CHLOROPLAST
Granule inside the cell of a photosynthesizing eukaryote used for photosynthesis. *See* photosynthesis.

CHROMOSOME
Microscopic filament inside a cell that carries genetic information (DNA).

CLIMBER
A plant that grows up a vertical surface, such as a rock or a tree, using it for support. Climbers do not gain nourishment from other plants, and may weaken them by blocking out light.

CLOACA
An opening toward the rear of the body that is shared by several body systems. In some vertebrates—such as bony fish and amphibians—the gut, kidneys, and reproductive systems all use this opening, mainly for waste.

CLONES
Genetically identical individuals. Two or more identical organisms that share exactly the same genes.

CLOVEN-HOOFED
Having hooves that look as though they are split in half. Most cloven-hoofed mammals, such as deer and antelope, actually have two hooves, arranged on either side of a line that divides the foot in two.

COCOON
A case made of open, woven silk. Many insects spin a cocoon before they begin pupation, and many spiders spin one to hold their eggs.

COLONY
A group of animals belonging to the same species that spend their lives together, often dividing up the tasks involved in survival. In some colonial species, particularly aquatic invertebrates, the colony members are permanently fastened together. In others, such as ants, bees, and wasps, members forage for food independently but live in the same nest.

COMMENSAL
Living in close relationship with an another species without either helping or damaging it.

COMPOUND
A substance made up of two of more elements chemically reacted together.

COMPOUND EYE
An eye that is divided into separate compartments, each with its own set of lenses. The number of compartments they can contain varies from a few dozen to thousands. Compound eyes are a common feature of arthropods.

COMPOUND LEAF
A leaf with a blade divided into smaller leaflets. *See also* simple leaf.

CONE
The reproductive structure of a plant consisting of a cluster of scales or bracts used for producing spores, ovules, or pollen. Cones are found on various trees, particularly pine trees.

CORM
The underground storage organ of a plant, formed from a swollen stem base.

COROLLA
Inner whorl of a flower made up of petals.

CRYPTIC (COLORATION)
Coloration and markings that make an animal hard to see against its background.

CRYSTAL
A solid that has a definite internal atomic structure, producing a characteristic external shape and certain physical and optical properties.

CYTOPLASM
The jellylike interior of a cell. In the cells of eukaryotes it is restricted to the region around the nucleus.

DECIDUOUS
A plant that seasonally loses its leaves. For instance, many temperate plants lose their leaves in winter.

DELAYED IMPLANTATION
In mammals, a delay between the fertilization of an egg and the subsequent development of an embryo. This delay allows birth to occur when conditions, such as food availability, are favorable for raising young.

DNA (DEOXYRIBONUCLEIC ACID)
The chemical substance found in the cells of all living organisms that determines their inherited characteristics.

DICOTYLEDON (DICOT)
Group of flowering plants with two cotyledons (seed leaves).

DIOECIOUS
A plant that has male and female parts on different individuals.

DOMESTICATED
An animal that lives fully or partly under human control. Some such animals look identical to their counterparts in the wild, but many have been bred to produce artificial varieties not found in nature.

DORSAL FIN
The unpaired fin of a fish. The dorsal fin is located on the fish's back.

ECHOLOCATION
A method of sensing nearby objects by using pulses of high-frequency sound. Echoes bounce back from obstacles and other animals, allowing the sender to build up a picture of its surroundings. It is used by some bats, cave-dwelling birds, and marine mammals.

ECLIPSE PLUMAGE
In some birds, particularly waterfowl, an unobtrusive plumage adopted by the males once the breeding season is over.

ECOSYSTEM
A collection of species living in the same habitat, together with their physical surroundings.

ECTOPARASITE
An organism that lives parasitically on the surface of another organism's body. Some animal ectoparasites spend all their lives on their hosts, but many—including fleas and ticks—develop elsewhere and climb onto the host in order to feed.

ECTOTHERMIC
Having a body temperature that is dictated principally by the temperature of the surroundings. Ectothermic is also known as cold-blooded.

ELEMENT
Chemical substance that cannot be broken down into a simpler chemical form.

EMBRYO
A young animal or plant at a rudimentary stage of development.

ENDEMIC
A species native to a particular geographic area, such as an island, forest, mountain, state, or country, and which is found nowhere else.

ENDOPARASITE
An organism that lives parasitically inside the body of another organism, either feeding directly on its tissues or stealing some of its food. Endoparasites frequently have complex life cycles involving more than one host.

ENDOSKELETON
An internal skeleton, typically made of bone. Unlike an exoskeleton, this kind of skeleton can grow with the rest of the body.

ENDOTHERMIC
Able to maintain a constant, warm body temperature, regardless of external conditions. Also known as warm-blooded.

ENZYME
A group of substances produced by all living things that promotes a chemical process, such as photosynthesis or digestion.

EPIPHYTE
A plant or plantlike organism (such as alga or lichen) that lives on the body of another plant without getting any nourishment from it.

EUDICOTYLEDON (EUDICOT)
Any one of the higher dicotyledon flowering plant orders that together contain the majority of flowering plants.

EUKARYOTE
An organism whose cells have a nucleus. Protists, fungi, plants, and animals are eukaryotes.

EVAPORITE (DEPOSIT)
Sedimentary rock or mineral resulting from the evaporation of water from mineral-bearing fluids, usually seawater.

EVERGREEN
A plant that does not lose its leaves seasonally. An example is conifer.

EXTERNAL FERTILIZATION
In reproduction, a form of fertilization that takes place outside the female's body, usually in water, for example among coral.

EXOSKELETON
An external skeleton that supports and protects an animal's body. The most complex exoskeletons, formed by arthropods, consist of rigid plates that meet at flexible joints. This kind of skeleton cannot grow, and has to be shed and replaced at periodic intervals.

FERAL
An animal that comes from domesticated stock but which has subsequently taken up life in the wild. Examples include city pigeons, cats, and horses.

FERTILIZATION
The union of an egg cell and sperm, which creates a cell capable of developing into a new organism.

FLAGELLUM (PL. FLAGELLA)
A whiplike structure of a cell used for propulsion. It is the main structure of locomotion in flagellate protoctists.

FLOWER
Reproductive structure of the largest group of seed plants, typically consisting of sepals, petals, stamens, and carpels.

FOOD CHAIN
A series of organisms, each of which is eaten by the next.

FOSSIL
Any record of past life preserved in Earth's crust. They include bones, shells, footprints, excrement, and ground borings.

FRUIT
A fleshy structure of a plant that develops from the ovary of the flower and contains one or more seeds. Fruits can be simple, like berries, or compound, where the fruits of separate flowers are merged. See also accessory fruit.

FRUITBODY
Fleshy spore-producing structure of a fungus, typically in the shape of a mushroom (agaric) or bracket.

GALL
A tumorlike growth in a plant that is induced by another organism (such as a fungus or an insect). By triggering the formation of galls, animals provide themselves with a safe hiding place and a convenient source of food.

GAMETE
A sex cell. In animals, this is either a sperm cell or an unfertilized egg cell.

GENE
The basic unit of heredity in all living things, typically a segment of DNA that provides the coded instructions for a particular protein.

GERMINATION
Developmental stage where a seed or a spore begins to grow.

GESTATION (PERIOD)
Length of time from fertilization to birth in a live-bearing animal.

GILL
In fish, amphibians, crustaceans, and mollusks, an organ used for extracting oxygen from water. Unlike lungs, gills are outgrowths of the body. In fungi, gills are the bladelike spore-producing structures under the cap of agaric fungi.

GRAZER
An animal that feeds on grass or algae.

GYMNOSPERM
A seed plant that doesn't enclose its seeds in a fruit. Many gymnosperms carry their seeds on cones. See also angiosperm.

HERB (HERBACEOUS PLANT)
A nonwoody plant, usually much shorter than shrubs or trees.

HERBIVORE
An animal that feeds on plants or algae.

HERMAPHRODITE
An organism that has both male and female sex organs.

HIBERNATION
A period of dormancy in winter. During hibernation, an animal's body processes drop to a low level to conserve energy.

HORMONE
A chemical signal produced by one part of the body that changes the behavior of another part of the body.

HORN
In mammals, a pointed growth on the head. True horns are hollow, and are often curved.

HOST
An organism on or in which a parasite, or symbiont, feeds.

HYDROTHERMAL VEIN
Sheet-shaped mass of material altered or deposited by water heated by igneous activity in rock.

HYPHA (PL. HYPHAE)
Microscopic threadlike structure making up the body of a fungus. Many such hyphae make up a mass that is known as a mycelium.

IGNEOUS ROCK
Rock formed from erupted volcanic lava or solidified magma.

INCISOR (TOOTH)
In mammals, a flat tooth at the front of the jaw that is shaped for slicing or gnawing.

INCUBATION
In birds, the period when a parent sits on the eggs and warms them, allowing them to develop. Incubation periods range from under 14 days to several months.

INFLORESCENCE
A flower cluster or a single flower.

INORGANIC
Chemical substance that is not based on the element carbon.

INSECTIVORE
An animal that feeds on insects.

INTERNAL FERTILIZATION
In reproduction, a form of fertilization that takes place inside the female's body. Internal fertilization is a characteristic of many land animals, particularly including insects and vertebrates.

KEEL
In birds, an enlargement of the breastbone that is anchored to the muscles that are used in flight.

KERATIN
A tough structural protein found in hair, claws, and horns.

LARVA (PL. LARVAE)
An immature but independent animal that looks completely different from an adult of the species. A larva develops the adult shape by metamorphosis; in many insects, the change takes place in a resting stage that is called a pupa.

LAVA
Molten rock that has erupted from a volcano, and then hardens.

LEGUME
Plants of the pea family of flowering plants, the Fabaceae. They are important because they have root nodules that contain nitrogen-fixing bacteria.

LEK
A communal display area used by male animals (particularly birds) during courtship. The same location is often revisited for many years.

LICHEN
Mutualistic association between a fungus and a photosynthetic algae. From this relationship, the fungus obtains sugars, and the alga obtains minerals.

LIFE CYCLE
The developmental sequence of an organism from gametes (sex cells) to death.

LOCOMOTION
Movement from place to place.

LUSTER
The shine or look of a mineral due to the reflection of light off its surface.

MAGMA
Rock in a molten state and lying below Earth's surface.

MAGNOLIIDAE
Group of flowering plants consisting of orders with certain primitive characteristics, such as undifferentiated tepals instead of sepals and petals.

MANDIBLE
The paired jaws of an arthropod, or, in vertebrates, the bone that makes up all or part of the lower jaw.

METABOLISM
The complete array of chemical processes that takes place inside an animal's body. Some of these processes release energy by breaking down food, while others use energy, for example, by making the body's muscles contract.

METAMORPHIC ROCK
Rock that has been changed by heat, pressure, or both to form new rock consisting of new minerals.

METAMORPHOSIS
A change in body shape undergone by many animals—particularly invertebrates—as they grow from being a juvenile to an adult. In insects, metamorphosis can be complete or incomplete. Complete metamorphosis involves a total shape change during a resting stage, which is called a pupa. Incomplete metamorphosis involves a series of less drastic changes, and these occur each time the young animal molts.

MIGRATION
A journey undertaken to a different region, following a well-defined route. Most migratory animals move with the seasons to take advantage of good breeding conditions in one place, and a suitable wintering climate in another.

MIMICRY
A form of camouflage in which an animal resembles another animal or an inanimate object, such as a twig or a leaf. Mimicry is very common in insects, with many harmless species imitating ones that have dangerous bites or stings.

MINERAL
An inorganic, naturally occurring material with a constant chemical composition and regular internal atomic structure.

MITOCHONDRION (PL. MITOCHONDRIA)
Granule inside the cell of a eukaryote used for respiration. Mitochondria use up oxygen to release energy.

MOLAR (TOOTH)
In mammals, a tooth at the rear of the jaw. Molar teeth often have a flattened or ridged surface, and deep roots. They are normally used for chewing.

MONOCOTYLEDON (MONOCOT)
Group of flowering plants with a single seed leaf (cotyledon).

MONOECIOUS
A plant that has separate male and female parts on the same individual.

MONOGAMOUS
Mating with a single partner, either during the course of one breeding season or throughout life. Monogamous partnerships are common in animals that care for the development of their young.

MOLT
The shedding of fur, feathers, or skin so that it can be replaced. Mammals and birds molt to keep their fur and feathers in good condition, to adjust their insulation, or so they can be ready to breed. Arthropods, such as insects, molt their exoskeleton in order to grow.

MUTUALISM
A relationship between two different species within an ecological community where both species benefit. For instance, a flowering plant and a pollinating insect share a mutualistic relationship.

MYCELIUM
Mass of threadlike hyphae making up the body of a fungus.

MYCORRHIZA (PL. MYCORRHIZAE)
The mutualistic association between a fungus and the roots of a plant. The fungus obtains sugars, while the plant increases its mineral uptake by absorbing minerals via the fungus' extensive mycelium.

NICHE
An organism's place and role in its habitat. Although two species may share the same habitat, they never share the same niche.

NITROGEN FIXATION
A chemical process whereby atmospheric nitrogen is converted into more complex nitrogen-containing substances, such as protein. Nitrogen fixation is undertaken by certain types of microorganisms.

NODE
A swelling marking the point where two sections of a stem join. They often sprout leaves at the node.

NUCLEUS (PL. NUCLEI)
The structure inside the cell of a eukaryote that contains chromosomes.

NUT
Dry, hard-shelled fruit of some plants, usually a single seed.

NYMPH
An immature insect that looks similar to its parents but which does not have functioning wings or reproductive organs. A nymph develops its adult form by metamorphosis, changing slightly each time it molts.

OMNIVORE
An animal that eats both plants and other animals as its primary food source.

OPERCULUM
A cover or lid. In some gastropod mollusks, an operculum is used to seal the shell when the animal has withdrawn inside. In bony fish, an operculum on each side of the body protects the chamber containing the gills.

OPPOSABLE
Able to be pressed together from opposite directions. For example, many primates have opposable thumbs, which can be pressed against the other fingers so that objects can be grasped.

ORGAN
The structure of an organism that carries out a particular function. Examples are the heart, skin, or a leaf.

ORGANELLE
A specialized structure that forms part of a plant or animal cell.

ORGANIC
Chemical substance based on carbon.

OVIPAROUS
Reproducing by laying eggs.

OVIPOSITOR
An egg-laying tube extending out of the body of some female animals, especially in insects.

OVULE
Structure that contains the egg of a seed plant. These are encased in an ovary in flowering plants, but naked in gymnosperms. After fertilization the ovule becomes the seed.

PARASITE
An organism that lives on or in another host organism, gaining advantage (such as nourishment) while causing the host harm. Most parasites are much smaller than their host and have complex life cycles involving the production of huge numbers of young. Parasites often weaken their host but generally do not kill them.

PARTHENOGENESIS
A form of reproduction in which an egg cell develops into a young animal without having to be fertilized, producing offspring that are genetically identical to the parent. In animals that have separate sexes, young produced by parthenogenesis are always female. Parthenogenesis is common in invertebrates.

PECTORAL FIN
One of the two paired fins positioned toward the front of the fish's body, often just behind its head. Pectoral fins are usually highly mobile and are normally used for maneuvering.

PELVIC FIN
The rear paired fins in fish, which are normally positioned close to the underside, sometimes near the head but more often toward the tail. Pelvic fins are generally used as stabilizers.

PERENNIAL
A plant that normally lives for more than one season.

PERIANTH
The outer two whorls of a flower (the calyx, made up of sepals, and the corolla, made up of petals); especially where the two are undifferentiated.

PETAL
One of the parts of the corolla of a flower. It is often brightly colored to attract pollinating animals.

PETIOLE
The stalk of a leaf.

PHEROMONE
A chemical produced by one animal that has an effect on other members of its species. Pheromones are often volatile substances

that spread through the air, triggering a response from animals some distance away.

PHOTOSYNTHESIS
A process whereby organisms use light energy to make food and oxygen; photosynthesis occurs in plants, algae, and many microorganisms.

PLACENTA
An organ developed by an embryo animal that allows it to absorb nutrients and oxygen from its mother's bloodstream before it is born.

PLANKTON
Floating organisms, many of them microscopic, that drift in open waters, particularly near the surface of the sea. Planktonic organisms can often move, but most are too small to make any headway against strong currents. Planktonic animals are known as zooplankton. Planktonic algae are called phytoplankton.

PLASTRON
The lower part of the shell structure of tortoises and turtles.

POLLEN
Tiny grains produced by seed plants that contain male gametes for fertilizing the female egg of either a flowering plant or a coniferous plant.

POLYGAMOUS
A reproductive system in which individuals mate with more than one partner during the course of a single breeding season.

PREDATOR
An animal that catches and kills other animals, referred to as its prey. Some predators attack and catch their prey by lying in wait, but most actively pursue and attack other animals.

PREHENSILE
Able to curl around objects and grip them.

PREMOLAR (TOOTH)
In mammals, a tooth positioned midway along the jaw between the canine and the molar teeth. In carnivores, specialized premolar teeth act like shears, slicing through flesh.

PROBOSCIS
An animal's nose, or set of mouthparts with a noselike shape. In insects that feed on fluids, the proboscis is often long and slender, and can usually be stowed away when not in use.

PROKARYOTE
An organism whose cells do not have a nucleus. Examples of prokaryotes are archaea and bacteria.

PRONOTUM
The part of an insect's cuticle that covers the first segment of its thorax, often hardening into a shell.

PROTEIN
Substance found in food, such as meat, fish, cheese, and beans, that is used for growth and for carrying out a variety of biological functions.

PSEUDOPOD
Projection of a cell that can change shape. It is used for creeping forward or for catching food. It occurs in protists, such as amoebas.

PUPA (PL. PUPAE)
In insects, a stage during which the larval body is broken down and rebuilt as an adult. During the pupal stage the insect does not feed and usually cannot move, although some pupae may wriggle if they are touched. The pupa is protected by a hard case, which itself is sometimes wrapped in silk.

QUADRUPEDAL
An animal that walks on four legs.

RAPTOR
A bird of prey.

RHIZOME
A creeping or underground stem that can send out new shoots.

ROCK
Material made up of one or more minerals.

ROOT NODULE
Spherical swelling on the root of a legume that contains nitrogen-fixing bacteria.

RUMINANT
A hoofed mammal that has a specialized digestive system with several stomach chambers. One of these—the rumen—contains large numbers of microorganisms which help break down plant food. To speed this process, a ruminant usually regurgitates its food and rechews it, a process called chewing the cud.

SALLY
A bird's short flight from a perch to catch an invertebrate, often in mid-air.

SCUTE
A shieldlike plate or scale that forms a bony covering on some animals.

SEDIMENTARY ROCK
Rock formed by the consolidation and hardening of rock fragments, organic remains, or other material.

SEED
Developmental stage of a seed plant, consisting of an encapsulated embryo.

SEPAL
One of the parts of the calyx of a flower, usually small and leaflike and enveloping the unopened flower bud.

SEXUAL DIMORPHISM
Showing physical differences between males and females. In animals that have separate sexes, males and females always differ, but in highly dimorphic species, such as elephant seals, the two sexes look very different and are often unequal in size.

SHOOT
The aerial part of plant. It is a new growth usually growing upward.

SHRUB
Woody perennial plant with multiple stems.

SIMPLE LEAF
A leaf with an undivided blade.

SPERMATOPHORE
A packet of sperm that is transferred either directly from male to female, or indirectly—for example, by being left on the ground. Spermatophores are produced by a range of animals, including salamanders and some arthropods.

SPIRACLE
In rays and some insects, an opening behind the eye that lets water flow into the gills. In insects and myriapods, the spiracle is an opening on the body wall that lets air into the tracheal system.

SPORE
A single cell containing half the quantity of genetic material of typical body cells. Unlike gametes, spores can divide and grow without being fertilized. Spores are produced by fungi and plants.

SPOROPHYTE
The spore-producing stage of a plant. It is the dominant (visible) stage of ferns and seed plants.

STAMEN
Male reproductive part of a flower. It has an anther borne on a long filament.

STEREOSCOPIC (VISION)
Vision in which the two eyes face forward, giving overlapping fields of view and allowing the animal to judge distance. Also called binocular vision.

STOMA (PL. STOMATA)
Tiny adjustable pore on the surface of a plant that allows the exchange of gases for photosynthesis and respiration.

SWIM BLADDER
A gas-filled bladder that most bony fish use to regulate their buoyancy. By adjusting the gas pressure inside the bladder, a fish can become neutrally buoyant, meaning that it neither rises nor sinks.

SYMBIOSIS
Any relationship between two different species within an ecological community. Examples of symbiotic relationships are predator–prey, parasite–host, and mutualism.

TEPAL
The outer part of a flower undifferentiated into sepals and petals. Tepals collectively make up the perianth.

TERRESTRIAL
Living always or mainly on the ground.

TERRITORY
An area defended by an animal, or group of animals, against other members of the same species. Territories often include resources, such as food supply, that help the male attract a mate.

THORAX
The middle region of an arthropod's body. The thorax contains powerful muscles and, if the animal has any, bears legs and wings.

The thorax is the chest in vertebrates with four limbs.

TORPOR
A sleeplike state in which body processes slow to a fraction of their normal rate. Animals usually become torpid to survive difficult conditions, such as extreme cold or lack of food.

TRACHEAL SYSTEM
A system of minute tubes that arthropods (for example, insects) use to carry oxygen into their bodies. Air enters the tubes through openings called spiracles, and then flows through the tracheae to reach individual cells.

TREE
Woody perennial plant with a single stem, known as a trunk.

UNGULATE
A hoofed mammal.

UTERUS
In female mammals, the part of the body that contains and usually nourishes developing young. In placental mammals, the young are connected to the wall of the uterus via a placenta.

VECTOR
An organism that transmits a disease-causing parasite from one host to another.

VEIL
Thin skin- or weblike tissue that protects the fruitbody of a fungus.

VESTIGIAL
Relating to an organ that is atrophied or non-functional.

VIVIPAROUS
Reproducing by giving birth to live young.

VOLVA
A saclike remnant of a veil at the base of a fungus's fruitbody stem.

WOODY PLANT
A plant that has wood—a type of strengthening tissue found in plants consisting of thick-walled water-transporting vessels.

ZYGODACTYL FEET
A specialized arrangement of the feet in which the toes are arranged in pairs, with the second and third toes facing forward and the first and fourth toes facing backward. This adaptation helps birds climb and perch on tree-trunks and other vertical surfaces. Several groups of birds have zygodactyl feet, including parrots, cuckoos and turacos, owls, toucans, and woodpeckers and their relatives.

INDEX

Page numbers in **bold** type refer to feature pages or introductions to animal groups.

ACKNOWLEDGMENTS

Consultants at the Smithsonian Institution:

Dr. Don E. Wilson, Senior Scientist/Chair of the Department of Vertebrate Zoology;
Dr. George Zug, Emeritus Research Zoologist, Department of Vertebrate Zoology, Division of Amphibians and Reptiles;
Dr. Jeffrey T. Williams: Collections Manager, Department of Vertebrate Zoology

Dr. Hans-Dieter Sues, Curator of Vertebrate Paleontology/Senior Research Geologist, Department of Paleobiology

Paul Pohwat, Mineral Collection Manager, Department of Mineral Sciences;
Leslie Hale, Rock and Ore Collections Manager, Department of Mineral Sciences;
Dr. Jeffrey E. Post, Geologist/Curator, National Gem and Mineral Collection, Department of Mineral Sciences

Dr. Carla Dove, Program Manager, Feather Identification Lab, Division of Birds, Department of Vertebrate Zoology

Dr. Warren Wagner, Research Botanist/ Curator, Chair of Botany, and Staff of the Department of Botany

Gary Hevel, Museum Specialist/Public Information Officer, Department of Entomology;
Dana M. De Roche, Department of Entomology

Department of Invertebrate Zoology:
Dr. Rafael Lemaitre: Research Zoologist/ Curator of Crustacea;
Dr. M. G. (Jerry) Harasewych, Research Zoologist;
Dr. Michael Vecchione, Adjunct Scientist, National Systemics Laboratory, National Marine Fisheries Service, NOAA;
Dr. Chris Meyer, Research Zoologist;
Dr. Jon Norenburg, Research Zoologist;
Dr. Allen Collins, Zoologist, National Systemics Laboratory, National Marine Fisheries Service, NOAA;
Dr. David L. Pawson, Senior Research Scientist;
Dr. Klaus Rutzler, Research Zoologist;
Dr. Stephen Cairns, Research Scientist / Chair

Additional consultants:

Dr. Diana Lipscomb, Chair and Professor Biological Sciences, George Washington University

Dr. James D. Lawrey, Department of Environmental Science and Policy, George Mason University

Dr. Robert Lücking, Research Collections Manager/Adjunct Curator, Department of Botany, The Field Museum

Dr. Thorsten Lumbsch, Associate Curator and Chair, Department of Botany, The Field Museum

Dr. Ashleigh Smythe, Visiting Assistant Professor of Biology, Hamilton College

Dr. Matthew D. Kane, Program Director, Ecosystem Science, Division of Environmental Biology, National Science Foundation

Dr. William B. Whitman, Department of Microbiology, University of Georgia

Andrew M. Minnis: Systematic Mycology and Microbiology Laboratory, USDA

Dorling Kindersley would like to thank the following people for their assistance with this book:
David Burnie, Kim Dennis-Bryan, Sarah Larter, and Alison Sturgeon for structural development; Hannah Bowen, Sudeshna Dasgupta, Jemima Dunne, Angeles Gavira Guerrero, Cathy Meeus, Andrea Mills, Manas Ranjan Debata, Paula Regan, Alison Sturgeon, Andy Szudek, and Miezan van Zyl for additional editing; Helen Abramson, Niamh Connaughton, Manisha Majithia, and Claire Rugg for editorial assistance; Sudakshina Basu, Steve Crozier, Clare Joyce, Edward Kinsey, Amit Malhotra, Neha Sharma, and Nitu Singh for additional design; Amy Orsborne for jacket design; Richard Gilbert, Ann Kay, Anna Kruger, Constance Novis, Nikky Twyman, and Fiona Wild for proofreading; Sue Butterworth for the index; Claire Cordier, Laura Evans, Rose Horridge, and Emma Shepherd from the DK picture library; Peter Cross, Julia Harris-Voss, Sarah Hopper, Liz Moore, Rebecca Sodergren, Jo Walton, Debra Weatherley, and Suzanne Williams for picture research; Mohammad Usman for production; Stephen Harris for reviewing the plants chapter; and Derek Harvey, for his tremendous knowledge and unstinting enthusiasm for this book.

The publisher would also like to thank the following companies for their generosity in allowing Dorling Kindersley access to their collections for photography:
Anglo Aquarium Plant Co LTD, Strayfield Road, Enfield, Middlesex EN2 9JE, http://anglo-aquarium.co.uk; **Cactusland**, Southfield Nurseries, Bourne Road, Morton, Bourne, Lincolnshire PE10 0RH, www.cactusland.co.uk; **Burnham Nurseries Orchids**, Burnham Nurseries Ltd, Forches Cross, Newton Abbot, Devon TQ12 6PZ, www.orchids.uk.com; **Triffid Nurseries**, Great Hallows, Church Lane, Stoke Ash, Suffolk IP23 7ET, www.triffidnurseries.co.uk; **Amazing Animals**, Heythrop, Green Lane, Chipping Norton, Oxfordshire OX7 5TU, www.amazinganimals.co.uk; **Birdland Park and Gardens**, Rissington Rd, Bourton-on-the-Water, Gloucestershire GL54 2BN, www.birdland.co.uk; **Virginia Cheeseman F.R.E.S.**, 21 Willow Close, Flackwell Heath, High Wycombe, Buckinghamshire HP10 9LH, www.virginiacheeseman.co.uk; **Cotswold Falconry Centre**, Batsford Park, Batsford, Moreton in Marsh, Gloucestershire GL56 9AB, www.cotswold-falconry.co.uk; **Cotswold Wildlife Park**, Burford, Oxfordshire OX18 4JP, www.cotswoldwildlifepark.co.uk; **Emerald Exotics**, 37A Corn Street, Witney, Oxfordshire OX28 6BW, www.emerald-exotics.co.uk; **Shaun Foggett**, www.crocodilesoftheworld.co.uk.

Picture credits
Alamy Images: The Africa Image Library 545; Amazon Images 539; Arco Images GmbH / Huetter C 587; Art Directors & TRIP 143; Barrett & MacKay / All Canada Photos 29, 31; blickwinkel 144, 162, 303, 321, 322, 557, 601; Penny Boyd 586; Tom Brakefield 29; Brandon Cole Marine Photography 515; BSIP SA 93; James Caldwell 265; Rosemary Calvert 20; CuboImages srl 147; Andrew Darrington 287; Danita Delimont 151; Garry DeLong 105; Paul Dymond 454; Emilio Ereza 344; David Fleetham 320; Florapix 148; Florida Images 148; Martin Fowler 156; Les Gibbon 301; Rupert Hansen 29; Chris Hellier 143; Imagebroker / Florian Kopp 566; Indiapicture / P S Lehri 597; Interphoto 29; Barbara Jordan 29; T. Kitchin & V. Hurst 23; S & D & K Maslowski / FLPA 28; Carver Mostardi 555; Tsuneo Nakamura / Volvox Inc 571; The Natural History Museum, London 256; Nic Hamilton Photographic 29; Pictorial Press Ltd; 28; Matt Smith 176; Stefan Sollfors 264; Natural Visions 149; Joe Vogan 564; Wildlife GmbH 28, 130, 151. **Maria Elisabeth Albinsson:** CSIRO 95, 100. **Algaebase.org:** Robert Anderson 104; Ignacio Bárbara 104; Mirella Coppola di Canzano (c) University of Trieste 104; Razy Hoffman 103; E.M. Tronchin and O. De Clerck 104. **Ardea:** Ian Beames 535; John Cancalosi 390; John Clegg 257, 270; Steve Downer 312, 554; Jean-Paul Ferrero 272, 505, 587; Kenneth W Fink 551, 598; Francois Gohier 598; Joanna Van Gruisen 596; Steve Hopkin 257, 261, 299; Tom & Pat Leeson 575; Ken Lucas 34, 271, 300, 313; Ken Lucas 581; Thimas Marent 582; John Mason 287; Pat Morris 35, 513, 555, 560; Pat Morris 501, 581; Gavin Parsons 268; David Spears (Last Refuge) 263; David Spears / Last Refuge 269; Peter Steyn 513; Andy Teare 575; Duncan Usher 265; M Watson 596, 608. **Australian National Botanic Gardens:** © M.Fagg 179. **Nick Baker, ecologyasia:** 554. **Jón Baldur Hlíðberg (www.fauna.is):** 323, 334, 335, 336, 337, 340, 503. **Michael J Barritt:** 505. **Dr. Philippe Béarez / Muséum national d'histoire naturelle, Paris:** 332. **Bioimages:** Malcolm Storey 242. **Biosphoto:** Jany Sauvanet 529. **Ashley M. Bradford:** 289. **David Bygott:** 35, 515. **Ramon Campos:** 504. **David Cappaert:** 280. **CDC:** Janice Haney Carr 33, 93; Janice Carr 93; Dr Richard Facklam 93; Courtesy of Larry Stauffer, Oregon State Public Health Laboratory 93. **Tyler Christensen:** 298. **Josep Clotas:** 324. **Patrick Coin:** Patrick Coin 261. **Niall Corbett:** 528. **Corbis:** 13, 22; Theo Allofs 19, 122, 404; Steve Austin 122; Hinrich Baesemann 420; E. & P. Bauer 467; Tom Bean 14; Annie Griffiths Belt 428; Biodisc 33, 236; Biodisc / Visuals Unlimited 282; Jonathan Blair 38; Michael Blajenov 597; Tom Brakefield 21, 24; Frank Burek 19; Janice Carr 90; W. Cody 19, 107, 118; Brandon D. Cole 317; Richard Cummins 20; Tim Davis 31; Renee DeMartin 24; Dennis Kunkel Microscopy, Inc. 33, 93; Dennis Kunkel Microscopy, Inc / Visuals Unlimited 33, 100; DLILLC 24, 31, 412, 438; Pat Doyle 26; Wim van Egmond 98; Ron Erwin 24; Eurasia Press / Steven Vidler 407; Neil Farrin / JAI 27; Andre Fatras 26; Natalie Fobes 209; Patricia Fogden 350; Christopher Talbot Frank 16; Stephen Frink 346; Jack Goldfarb / Design Pics 19; C. Goldsmith / BSIP 22; Mike Grandmaison 118; Franck Guiziou / Hemis 19; Don Hammond / Design Pics 19; Martin Harvey 19; Martin Harvey / Gallo Images 31; Helmut Heintges 24; Imagebroker 608; Pierre Jacques / Hemis 19; Peter Johnson 18, 452; Don Johnston / All Canada Photos 262; Mike Jones 404; Wolfgang Kaehler 18, 26, 236; Karen Kasmauski 27; Steven Kazlowski / Science Faction 16; Layne Kennedy 38; Antonio Lacerda / EPA 15; Frans Lanting 14, 18, 19, 23, 31, 247, 248, 372, 421, 456; Frederic Larson / San Francisco Chronicle 20; Lester Lefkowitz 18; Charles & Josette Lenars 21; Library of Congress - digital ve / Science Faction 28; Bob Marsh / Papilio 210; Chris Mattison 370; Joe McDonald 350, 523; Momatiuk / Eastcott 31; moodboard 16, 319, 371; Sally A. Morgan 25; Werner H. Mueller 19; David Muench 110; David A. Northcott 385; Owaki - Kulla 15, 19; William Perlman 32; Photolibrary 30; Patrick Pleau / EPA 19; Louie Psihoyos / Science Faction 16; Ivan Quintero / EPA 20; Radius Images 107, 112; Lew Robertson 19; Derek Rogers 570; Jeffrey Rotman 328; Kevin Schafer 27, 421; David Scharf / Science Faction 28; Scubazoo 370; Dr. Peter Siver 89, 91; Paul Souders 13, 19, 24; Keren Su 18; Glyn Thomas / moodboard 19; Steve & Ann Toon / Robert Harding World Imagery 411; Craig Tuttle 319, 405; James Urback / Superstock 571; Jeff Vanuga 22; Visuals Unlimited 14, 33, 92, 98, 100, 105; Kennan Ward 13; M Watson 370; Michele Westmorland 18; Stuart Westmorland 501; Norbert Wu / Science Faction 316, 319; Norbert Wu 321, 328; Yu Xiangquan / Xinhua Press 21; Robert Yin 272; Robert Yinn 318; Frank Young / Papilio 209; Frank Young 237. **Alan Couch:** 505. **David Cowles:** David Cowles at http://rosario.wallawalla.edu/inverts 256. **Whitney Cranshaw :** 286. **Alan Cressler:** Alan Cressler 260. **CSIRO:** 332. **Michael J Cuomo:** www.phsource.us 256. **Ignacio De la Riva:** 361. **Frances Dipper :** 250, 251. **Dive Gallery / Jeffrey Jeffords (www.divegallery.com):** David B Fleetham 314. **Dorling Kindersley:** Demetrio Carrasco / Courtesy of Huascaran National Park 145; Frank Greenaway / Natural History Museum, London 280. **Jane K. Dolven:** 98. **Dreamstime.com:** 600; Amskad 299; John Anderson 344; Steve Byland 524; Bonita Chessier 538; Musat Christian 544; Clickit 599; Colette6 597; Ambrogio Corralloni 522; Cosmln 298; Davthy 537; Dbmz 297; Destinyvispro 607; Docbombay 524; Edurivero 598; Stefan Ekernas 501; Stefan Ekernas 545; Michael Flippo 601; Joao Estevao Freitas 261; Geddy 270; Eric Geveart 548; Daniel Gilbey 604; Maksum Gorpenyuk 597; Jeff Grabert 583; Iorboaz 596; Eric Isselee 35, 521, 523, 525, 532, 600, 601, 603; Isselee 526; Jontimmer 598; Jemini Joseph 597; Juliakedo 604; Valery Kraynov 33, 172; Adam Larsen 503; Aleksander Lorenz 172; Sonya Lunsford 597; Wayne Lynch / All Canada Photos 520; Stephen Meese 595; Milosluz 261; Jason Mintzer 522; Mlane 180; Nina Morozova 133, 148; Derrick Neill 522; Duncan Noakes 605; outdoorsman 577; Pancaketom 557; Natalia Pavlova 567; Susan Pettitt 589; Xiaobin Qiu 566; Rajahs 570; Laurent Renault 538; Dmitry Rukhlenko 600; Steven Russell Smith Photos 557; Ryszard 299; Benjamin Schalkwijk 544; Olga Sharan 595; Paul Shneider 597; Sloth92 543, 544; Smellme 537, 606; Nico Smit 582, 603; Nickolay Stanev 603; Vladimirdavydov 290; Oleg Vusovich 571; Leigh Warner 581; Worldfoto 551; Judy Worley 588; Zaznoba 543. **Andy Murch / Elasmodiver.com:** 324, 337; Andy Murch 328, 329. **eye of science:** 236. **Shane Farrell:** 262, 280. **Fauna & Flora Internatioanl: . David Fenwick (www.aphotoflora.com):** 339. **Carol Fenwick (www.carolscornwall.com):** 105. **Hernan Fernandez:** 504. **Flickr.com:** 294; Bar Aviad 562; Ana Cotta 539; Barbara J. Coxe 169; Pat Gaines 572; Sonnia Hill 152; Barry Hodges 131; Emilio Esteban Infantes 131; Marj Kibby 136; Ron Kube, Calgary, Alberta, Canada 573; John Leverton 589; John Merriman 180; Marcio Motta MSc. Biologist of Maracaja Institute for Mammalian Conservation 583; Jerry R. Oldenettel 171; Jennifer Richmond 169. **Florida Museum of Natural History:** Dr Arthur Anker 512. **Dr Peter M Forster:** 329. **FLPA:** 30; Nicholas and Sherry Lu Aldridge 105; Ingo Arndt / Minden Pictures 209, 243, 316; Fred Bavendam / Minden Pictures 270, 273, 309, 314, 315; Fred Bavendam 309, 324; Stephen Belcher / Minden Pictures 273; blickwinkel 146; Neil Bowman 551; Jim Brandenburg 563; Jonathan Carlile / Imagebroker 269; Christiana Carvalho 507; B. Borrell Casals 264; Nigel Cattlin 258, 263; Robin Chittenden 308; Arthur Christiansen 525; Hugh Clark 557; D.Jones 270, 308; Flip De Nooyer / FN / Minden 201; Tui De Roy / Minden Pictures 400, 406, 598; Tui De Roy / Minden Pictures 570; Dembinsky Photo Ass 560; Reinhard Dirscher 309; Jasper Doest / Minden Pictures 374; Richard Du Toit / Minden Pictures 35, 501, 512; Michael Durham / Minden Pictures 517, 557; Gerry Ellis 507, 558; Gerry Ellis / Minden Pictures